FLIGHT THEORY AND AERODYNAMICS

FLIGHT THEORY AND AERODYNAMICS

A Practical Guide for Operational Safety

FIFTH EDITION

Brian A. Johnson
Guilford Technical Community College and Embry-Riddle Aeronautical University
Greensboro, North Carolina, United States

Philip R. Fittante
Guilford Technical Community College
Greensboro, North Carolina, United States

WILEY

Published by John Wiley & Sons, Inc., Hoboken, New Jersey.
Published simultaneously in Canada.

Library of Congress Cataloging-in-Publication Data Applied for:

Hardback ISBN: 9781394282296

Cover Design: Wiley
Cover Image: © Devrimb/Getty Images

Set in 9/11pt and TimesLTStd by Straive, Chennai, India

MWEP420978171225

The authors would like to thank their contacts at Wiley for their continuous support throughout this revision, as well as the support of colleagues and families. Without the support and patience of our wives, Jane Fittante and Kim Johnson, this latest revision would not have been possible.

Finally, the authors would like to acknowledge the previous work of Charles E. Dole and James E. Lewis, the original authors for the first two editions of this textbook, as well as the late Joseph R. Badick, co-author of the third and fourth editions of this textbook who was taken from this life way too early. It is with respect and admiration that we acknowledge the contribution of all three previous authors to improving aviation safety through education and practical application.

Contents

About the Authors

Philip R. Fittante is a retired US Air Force Lieutenant Colonel. He is a graduate of Euro-NATO Joint Jet Pilot Training and the US Air Force Test Pilot School. He served as a T-38 Instructor Pilot, B-1B Evaluator Pilot, and Chief Developmental Test Pilot for both the B-1B and B-2 bombers. As a US Navy civilian Flight Test Engineer and Test Pilot, he conducted developmental testing of the P-8A and MQ-4C UAS. A member of the Society of Experimental Test Pilots since 1998, he is a current flight instructor and aircraft owner with an airline transport pilot certificate. He holds an MS in aeronautical science from Embry-Riddle Aeronautical University, an MS in computer science from Midwestern State University and an MA in National Security and Strategic Studies from the Naval War College. An author of two novels, he currently serves as an Adjunct Professor for the Career Pilot/Aviation Management degree program at Guilford Technical Community College in Greensboro, NC.

Brian A. Johnson is a former regional airline and corporate pilot who holds an Airline Transport Pilot (ATP) certificate with a multi-engine land rating, in addition to commercial pilot single-engine land/single-engine sea ratings and a small Unmanned Aircraft System (sUAS) certificate. He is an active instrument and multiengine gold seal flight instructor and serves as a regional FAASTeam representative. He holds a master's degree in aeronautical science from Embry-Riddle Aeronautical University. He most recently served as program director and chief ground instructor for Guilford Technical Community College's Career Pilot/Aviation Management degree program in Greensboro, NC, and as of 2025, is in his 14th year as an adjunct faculty member for the Aeronautics Department serving Embry-Riddle Aeronautical University's Worldwide campus. He currently serves as a full-time company pilot based out of Greensboro, NC.

Preface

The fifth edition of *Flight Theory and Aerodynamics* was revised to further refine the textbook's goal of serving as an introductory guide to the principles of flight, focused on entry-level collegiate aeronautical and engineering programs, which require a practical application of physics and aerodynamics to the "hands-on" world of flight. Each chapter balances the guiding aerodynamic principles in use today using formulas and examples as they apply to not only the design of aircraft but also the maneuvering and operational safety of modern-day aircraft. This textbook was enhanced from the previous versions to not only serve as a foundation textbook for those with a limited aviation background interested in the basic principles of flight, but also as a resource for more advanced engineers and aviators who may want a review of guiding aerodynamic principles after years of industry experience.

The fifth edition of the textbook includes multiple figure and content updates to all chapters and includes the addition of a new chapter based on Unmanned Aerial Vehicles (UAVs) due to the exponential increase of UAVs within the aviation industry, and the trend of unmanned aircraft to takeover roles once performed by manned aircraft. The step-by-step "math" examples introduced in the fourth edition have been reinforced in most chapters to make the formula introduction even easier for students to follow and apply. Additional *Application* examples have been added to further prove the correlation of the introduced aerodynamic principles to real-world flight application. In many chapters, additional end-of-chapter questions have been added, along with an entirely new section, *Correlation Through Scenario-Based Learning*, which requires the application of chapter material to the operation of a Diamond DA50.

Enhancements to the fifth edition:

- Chapter 1 adds an expanded introduction to the history of the key pioneers of flight who contributed significantly to the subject of aerodynamics. The chapter also includes an introduction to geometry when evaluating vector resolution.
- Chapters 2 offers updated figures and the expanded use of example problems, while Chapter 3 adds additional sections on secondary control systems and the classification of airfoils.
- Chapter 4 includes a more detailed review of angle of attack and coefficient of lift, an accident review of Air France 447, updates to the lift equation examples, added information concerning slats, and a section on lift as applied to flight of hot air balloons.
- Chapter 5 expands on the impact of aspect ratio on drag, laminar flow airfoils, lift to drag ratio for gliding flight, and winglet design, in addition to multiple figure updates.
- Chapter 6 includes multiple minor updates on T-38 performance data, and an added section on afterburners and vectored thrust.
- Chapter 7 expands on the explanation of horsepower as related to propeller and rotor-driven aircraft and includes multiple figure updates.
- Chapters 8 and 9 build on the takeoff and landing aerodynamic principles introduced within previous editions, with notable updates on declared distance introduction, rejected takeoff aerodynamics, accident brief on Southwest Airlines Flight 1248, and an assessment on runway braking action.

- Chapter 10 adds a detailed discussion on uncoordinated flight as related to stalls and spin entry, updates to the spin recovery section, expansion of wind shear to include vertical shear, and a breakdown of low-level turbulence.
- Chapter 11 includes multiple figure updates as well as minor updates to each section, then ends with a detailed introduction to energy management and the airplane as an energy management system.
- Chapters 12 and 13 offer multiple updates on aircraft stability and respective application to modern aircraft. Updates include a summary of stability derivatives, a section on mean aerodynamic chord, an introduction to aircraft accidents related to Boeing's MCAS, additions to the static margin and neutral point discussion, accident brief on a B-1A static margin test, and expansion of the section on single-engine flight due to engine failure and the impact on lateral stability for multi-engine aircraft.
- Chapter 14 includes a new section on flight control augmentation systems, low boom flight development, and multiple figure updates.
- Chapter 15 includes minor updates to content and figures.
- Chapter 16 adds a new chapter to the latest textbook edition, Unmanned Aerial Vehicle Flight Theory. This chapter summarizes the categorization of UAVs, the aerodynamics of UAV fuselage and power plant design, and the future of UAV design and aerodynamics.

Phillip R. Fittante, Lt Col (Ret)
Adjunct Faculty, Guilford Technical Community College,
Greensboro, NC, USA

Brian A. Johnson
Adjunct Faculty, Embry-Riddle Aeronautical University,
Daytona Beach, FL, USA

About the Companion Website

This book is accompanied by a companion website:

www.wiley.com/go/flighttheoryandaerodynamics5e

This website includes:

1. End-of-Chapter Solutions
2. Summary of DA50 End of Chapter Questions
3. Test Questions
4. Presentation Slides

1 Introduction to the Flight Environment

CHAPTER OBJECTIVES

After completing this chapter, you should be able to:

- Define basic units of measurement used in the introduction to aerodynamics in flight and convert from one unit of measurement to another.
- Identify the four forces on an airplane in constant altitude, unaccelerated flight.
- Calculate the mass of an aircraft.
- Define vector addition and apply it to an aircraft in a climb.
- Describe Newton's laws of motion and recognize how they apply to an introduction to aerodynamics.
- Define the purpose of linear motion in relation to constant acceleration, and then calculate aircraft acceleration, takeoff distance, and takeoff time.
- Describe the difference between energy and work and calculate the potential and kinetic energy of an aircraft in flight.
- Calculate the equivalent horsepower of an aircraft from a known thrust and speed.
- Define friction as it applies to an aircraft.

A basic understanding of the physical laws of nature that affect aircraft in flight and on the ground is a prerequisite for the study of aerodynamics. Modern aircraft have become more sophisticated, and more automated, using advanced materials in their construction requiring pilots to renew their understanding of the natural forces encountered during flight. Understanding how pilots control and counteract these forces better prepares pilots and engineers for the art of flying for harnessing the fundamental physical laws that guide them. Although at times this textbook will provide a quantitative approach to various principles and operating practices with formulas and examples using equations, it is more important that the reader understand WHY a principle of flight theory is discussed and how that subject matter intertwines with other materials presented; thus, a qualitative approach is used throughout this textbook.

Perhaps your goal is to be a pilot, who will "slip the surly bonds of earth," as John Gillespie Magee wrote in his classic poem "High Flight." Or you may wish to build or maintain aircraft as a skilled technician.

Flight Theory and Aerodynamics: A Practical Guide for Operational Safety, Fifth Edition.
Brian A. Johnson and Philip R. Fittante.

Companion website: www.wiley.com/go/flighttheoryandaerodynamics5e

Or possibly you wish to serve in another vital role in the aviation industry, such as manager, dispatcher, meteorologist, engineer, teacher, or unmanned aerial vehicle (UAV) operator. Whichever area you might be considering, this textbook will build on what you already know and will help prepare you for a successful aviation career.

INTRODUCTION

This chapter begins with a review of the basic principles of physics and concludes with a summary of mechanical energy, power, and linear motion. A working knowledge of these areas, and how they relate to basic aerodynamics, is vital as we move past the rudimentary "four forces of flight" and introduce thrust and power-producing aircraft, lift and drag curves, stability and control, maneuvering performance, slow-speed flight, and other topics.

You may already have been introduced to the four basic forces acting on an aircraft in flight: lift, weight, thrust, and drag. Now, we must understand how these forces change as an aircraft accelerates down the runway, or descends on final approach to a runway and gently touches down even when traveling twice the speed of a car on the highway. Once an aircraft has safely made it into the air, what effect does weight have on its ability to climb, and should the aircraft climb up to the flight levels or stay lower and take "advantage" of the denser air closer to the ground?

By developing an understanding of the aerodynamics of flight, and of the ways in which design, weight, load factors, and gravity affect an aircraft during flight maneuvers from stalls to high-speed flight, the pilot learns how to control the balance between these forces. This textbook will help clarify these concepts among others, leaving you with a better understanding of the flight environment.

HISTORY OF AERODYNAMICS

The history of aerodynamics started well before the Wright brothers' first flight on December 17, 1903. In fact, a full accounting of the history of aerodynamics would warrant an entire textbook, so we offer a brief history of the primary pioneers and groups that had a hand in introducing aerodynamic principles covered in this textbook. Throughout the subsequent chapters, we will introduce additional historical tie-ins to relevant material when applicable.

During the Italian Renaissance the painter Leonardo da Vinci sketched many designs of various types of flying machines, including a design of a wing-flapping machine called the ornithopter. Many of his designs were based around the flight of birds but were only theoretical in nature as the flapping wings of a heavier-than-air device driven by only human muscle would never fly. Leonardo da Vinci even sketched an early design of what later would resemble a helicopter. Contracted by the Milanese court to research military technology, da Vinci eventually became intrigued by aerial reconnaissance and flying machines. Because his work was not published until centuries later in 1892 in a notebook entitled *Codice sul volo degli uccelli (Codex on the Flight of Birds)*, his sketches were not part of the design basis for nineteenth-century aviation pioneers.

Sir George Cayley is widely credited with designing the modern airplane, an aircraft with vertical and horizontal control surfaces on the rear of the aircraft, a fuselage with a place for the pilot, and a propulsion system that he termed "assisters." He is also considered, by many, to be the world's first aeronautical engineer. Like many aviation pioneers to follow, Cayley used the concept of windmills and kites to develop gliders. He built a glider that included an empennage, a tail with a rudder, and an elevator, and

Figure 1.1. Flying replica of Sir George Cayley's 1853 glider.
Source: U.S. Department of Transportation (2023a)/Public Domain.

he recognized the importance of the center of gravity (CG). In the mid-nineteenth century, Cayley even designed a glider large enough to carry a person and is credited with the earliest-known manned heavier-than-air flight (Figure 1.1). In Chapter 3, we will further introduce the structure of an airplane using the same design principles as Cayley's early gliders, and in Chapter 12, we will expand upon the importance of CG in modern aircraft.

The first notable attempt at providing a propulsion system beyond using gravity or human-generated power was by two Englishmen, John Stringfellow and William Henson. They designed a 20-foot model that used a steam engine to propel two six-bladed pusher propellers. The model failed to achieve sustained flight under its own power, but it proved that with modifications (lighter weight steam engine) it may be possible to sustain heavier-than-air flight. Stringfellow continued with his research and eventually launched a ten-foot model of an unmanned aircraft that hung on a wire. With improvements in the power-to-weight ratio of the steam engine, pioneers were now getting closer to heavier-than-air, sustained, manned flight.

Otto Lilienthal, considered by Wilbur Wright to be the most influential aviation pioneer in the nineteenth century, completed as many as 2 000 glider flights before dying in a glider crash in 1896. Many of those who followed him used his notes on aerodynamic data to develop their own wings or gliders. Experimenting with cambered wings, a rear elevator, and ornithopter wingtips, Lilienthal used hilltops to prove man could fly heavier-than-air aircraft without an engine (Figure 1.2). The development of a new type of engine, along with an improved power-to-weight ratio, gave way to the most famous aviation pioneers in history, the Wright brothers.

The two bicycle mechanics from Dayton, OH, Wilbur and Orville Wright, first developed a biplane kite in 1899, based predominately on the aeronautical writings of aviation pioneers such as Otto Lilienthal and engineer and glider designer Octave Chanute. The glider included wings that could be warped

Figure 1.2. Otto Lilienthal in flight.
Source: U.S. Department of Transportation (2023a)/Public Domain.

in flight using cords to bank the glider, a fixed horizontal stabilizer, and wings that could be adjusted forward and backward to adjust for the CG. By moving to the Outer Banks of North Carolina seasonally to take advantage of consistently strong winds off the ocean, and the tall sand dunes for experimenting with gliders, the Wright brothers found that the stronger the headwind, the less power would be required from the engine to develop forward thrust. In Chapters 2–4, we will introduce a more detailed review of how air moves around a wing and why this methodology used by the Wright brothers was successful. The lift formula we will introduce in Chapter 4 is directly related to the equations used by Lilienthal, Chanute, and the Wright brothers. By 1902, the Wright brothers had solved the problem of control and could successively control the glider around the three axes of the airplane (introduced in Chapter 12). By December 1903, their mechanic in Ohio, Charles Taylor, had built a four-cylinder, 12-horsepower water-cooled gasoline engine that finally solved the final piece of the puzzle, and with two counter-rotating pusher propellers, the Wright brothers achieved heavier-than-air, sustained manned flight with the Wright Flyer (Figure 1.3).

Other aviation pioneers deserve recognition for their contribution to aerodynamics. In the United States, Glenn Curtiss arguably may have contributed more to the development of modern aircraft than the Wright brothers, using his expertise in engine development to improve propulsion systems including both pusher and tractor propellers. From his use of ailerons instead of wing-warping to the development of early seaplanes (hydroplanes), Curtiss' influence on military and commercial aviation was instrumental to the development of a robust aviation industry in the US post-World War I. Glenn Curtiss has even been referred to as the "Father of Naval Aviation." In Europe in the early twentieth century and leading up to World War I, many aviation pioneers led the way with aerodynamic innovation, including the development of the first monocoque shell, surpassing the advancements made in the United States and placing the forefront of aircraft development in Europe. In 1906, Alberto Santos-Dumont completed the first airplane flight in Europe and was believed by many for a short time to be the "first to fly," eventually moving on to setting records in dirigibles. Gabriel and Charles Voisin, influenced by the Wright brothers, developed a pusher biplane with a forward elevator and went on to design aircraft for the French military in preparation for World War I. In Great Britain, the Short brothers built upon the Wright brothers' design and progressed to developing seaplanes with folding wings. Igo Etrich, an Austrian, designed the *Taube* monoplane with a bird-like wing (washed-out wingtips) and tail design, which eventually became a German production airplane leading up

Figure 1.3. Orville Wright at the controls of the first flight.
Source: U.S. Department of Transportation (2023a)/Public Domain.

to World War I. In Russia, Igor Sikorsky built the S-2 and taught himself to fly, even experimenting with the first helicopter design before continuing with fixed-wing aircraft. In 1913, Sikorsky designed and flew the first four-engine airplane, *The Grand*, and in 1914, he designed the largest seaplane in the world. And most notably, returning to rotorcraft flight, but now in the United States, in 1939, Sikorsky flew the VS-300 helicopter with a single main rotor and tail rotor, which marked the birth of the entire helicopter industry. Chapter 15 will explore more about the aerodynamics surrounding rotorcraft design.

Finally, in the United States, no other group had a larger influence on modern aerodynamics, aircraft, and engine development than the National Advisory Committee for Aeronautics (NACA). Lagging Europe in airplane technology in every area, Congress founded the NACA in March 1915 as an independent government agency reporting directly to the President. NACA conducted many flight tests of aircraft models in wind tunnels (Figure 1.4) as well as full-scale aircraft, even winning the Collier Trophy in 1929 for the NACA "low drag" cowling which involved streamlining the front of the aircraft to increase speed. As we will expand upon in Chapter 3, the NACA was instrumental in airfoil design, especially leading up to and during World War II and many modern airfoil designs can still be traced back to research completed at this time. After World War II, and many modern airfoil designs can still be traced back to research completed at this time. After World War II, the NACA was instrumental to early supersonic aircraft design, including modifying fuselage and tail design as well as "sweeping" the wings. In Chapter 14, we will introduce Dr. Richard Whitcomb, a NACA engineer, who in 1951 designed a "trimmed" midsection of a fuselage that allowed an aircraft to finally reach supersonic speeds. The NACA even experimented with Vertical Takeoff and Landing (VTOL) airplanes, which will be expanded on in Chapter 16. And in 1958, at the beginning of the space race when the United States once again began to lag in aviation development, this time to the Soviet Union, the NACA mission and technology were turned over to the National Aeronautics and Space Agency (NASA).

Figure 1.4. Supersonic wind tunnel experiment.
Source: NASA/Public Domain.

BASIC QUANTITIES

An introduction to aerodynamics must begin with a review of physics, and, in particular, the branch of physics that will be presented here is called *mechanics*. We will examine the fundamental physical laws governing the forces acting on an aircraft in flight, and what effect these natural laws and forces have on the performance characteristics of aircraft. To control an aircraft, whether it is an airplane, helicopter, glider, or balloon, the pilot must understand the principles involved and learn to use or counteract these natural forces.

We will start with the concepts of work, energy, power, and friction, and then build upon them as we move forward in future chapters.

Because the metric system of measurement has not yet been widely accepted in the United States, the English system of measurement is used in this book. The fundamental units are

Force	Pounds (lb)
Distance	Feet (ft)
Time	Seconds (s)

From the fundamental units, other quantities can be derived:

Velocity (distance/time)	ft/s or fps
Area (distance squared)	square ft or ft^2
Pressure (force/unit area)	lb/ft^2 or psf
Acceleration (rate of change in velocity)	ft/s/s or fps^2

Aircraft measure airspeed in knots (nautical miles per hour) or in Mach number (the ratio of true airspeed to the speed of sound). Rates of climb and descent are measured in feet per minute, so quantities other than those above are used in some cases. Some useful conversion factors are listed below:

Multiply	by	to get
knots (kts.)	1.69	feet per second (fps)
fps	0.5925	kts.
miles per hour (mph)	1.47	fps
fps	0.6818	mph
mph	0.8690	kts.
kts.	1.15	mph
nautical miles (nm)	6076	feet (ft)
nm	1.15	statute miles (sm)
sm	0.869	nm
kts.	101.3	feet per minute (fpm)

EXAMPLE

Convert 110 kts. to fps: 110 kts. × 1.69 = 185.9 fps
Convert 50 kts. to fpm: 50 kts. × 101.3 = 5 065 fpm
Convert 450 fps to kts.: = 450 fps × 0.5925 = 267 kts.
Convert 25 sm to nm: 25 sm × 0.869 = 21.7 nm

Application 1.1

An airplane flight manual (AFM) states a given aircraft should be rotated at 65 kts. Indicated airspeed (IAS), yet the pilot misinterprets the airspeed indicator and rotates at 65 mph (IAS).

Does the aircraft rotate at a faster or slower airspeed than the manufacturer recommends? What are the implications?

FORCES

A force is a push or a pull tending to change the state of motion of a body. Figure 1.5 shows the four separate components of aerodynamic forces during straight-and-level, unaccelerated flight. The component that is 90° to the flight path and acts toward the top of the airplane is called *lift*. The component that is parallel to the flight path and acts toward the rear of the airplane is called *drag*; while the opposing forward force is

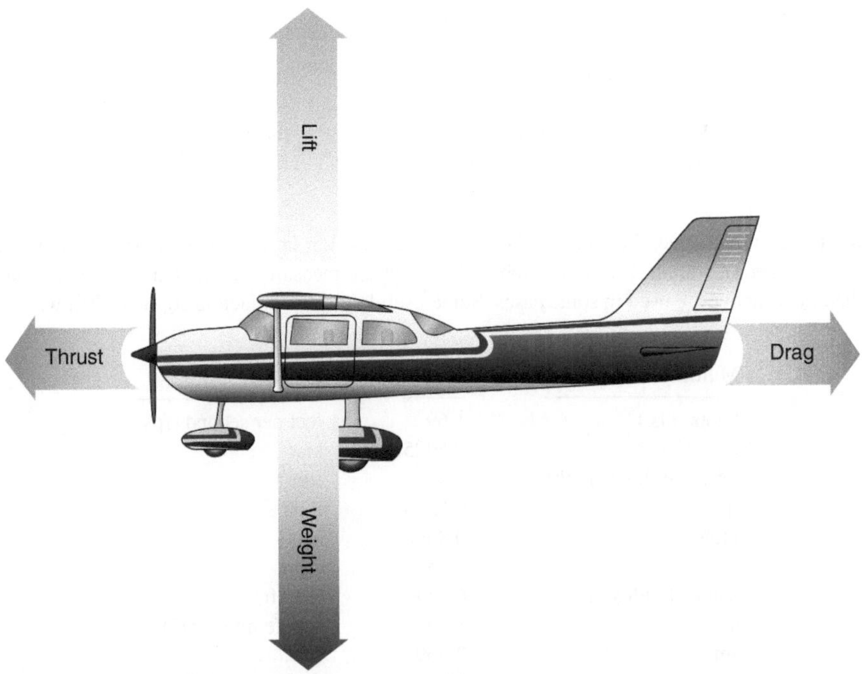

Figure 1.5. Four forces on an airplane in unaccelerated, level flight.
Source: Pilot's Handbook of Aeronautical Knowledge (2008a)/U.S. Department of Transportation Federal Aviation Administration/Public domain.

thrust and is usually created by the engine. *Weight* opposes lift and as we will see is a function of the mass of the aircraft and gravity.

The sum of the opposing forces is always zero in steady flight, but this does not mean the four forces are equal. In future chapters of this textbook, we will further demonstrate the following statement regarding forces acting on an airplane in steady flight: The sum of all upward components of forces equals the sum of all downward components of forces, and the sum of all forward components of forces equals the sum of all backward components of forces. Figure 1.6 depicts the resultant force between lift and drag of an airplane in flight.

MASS

Mass is a measure of the amount of material contained in a body, usually measured in kilograms; we will use slugs as the unit in this textbook. *Weight*, on the other hand, is a force caused by the gravitational attraction of the earth ($g = 32.2\ \text{ft/s}^2$), moon, sun, or other heavenly bodies. Weight will vary depending on where the body is located in space (specifically, how far from the source of gravitational attraction), but mass will not vary with position.

$$\text{Weight}(W) = \text{Mass}(m) \times \text{Acceleration of gravity}(g)$$

$$W = mg \tag{1.1}$$

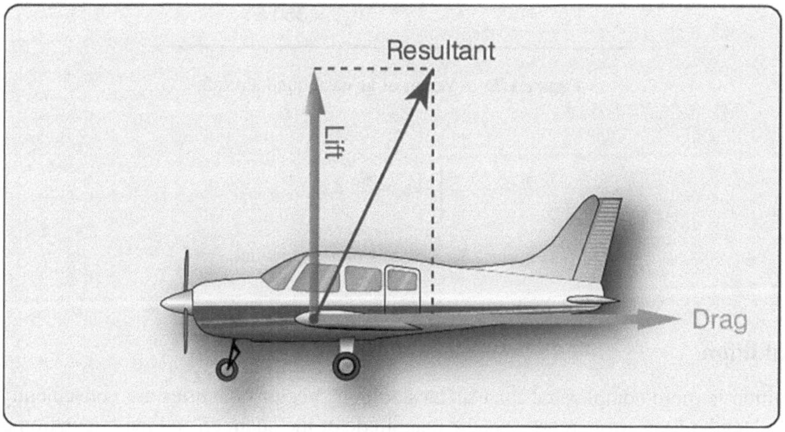

Figure 1.6. Resolved forces of lift and drag in steady flight.
Source: Aviation Maintenance Technician Handbook – Airframe (2023a)/U.S. Department of Transportation Federal Aviation Administration/Public domain.

Rearranging gives

$$m = \frac{W}{g}\,\frac{\text{lb}}{\text{ft/s}^2} = \frac{\text{lb}\cdot\text{s}^2}{\text{ft}}$$

This mass unit is called the *slug*.

EXAMPLE

Calculate the mass of an aircraft that weighs 2 576 lb.

$$m = \frac{W}{g} \rightarrow m = \frac{2\,576\ \text{lb}}{32.2\ \text{ft/s}^2}$$

$$m = 80.0\ \text{slugs}$$

SCALAR AND VECTOR QUANTITIES

A quantity that has size or magnitude only is called a *scalar* quantity. The quantities of mass, time, and temperature are examples of scalar quantities. A quantity that has both magnitude and direction is called a *vector* quantity. Forces, accelerations, and velocities are examples of vector quantities. Speed is a scalar, but if we consider the direction of the speed, then it is a vector quantity called *velocity*. If we say an aircraft traveled 100 nm, the distance is a scalar, but if we say an aircraft traveled 100 nm on a heading of 360°, the distance is a vector quantity.

Scalar Addition

Scalar quantities can be added (or subtracted) by simple arithmetic. For example, if you have 5 gal of gas in your car's tank and you stop at a gas station and top off your tank with 9 gal more, your tank now holds 14 gal.

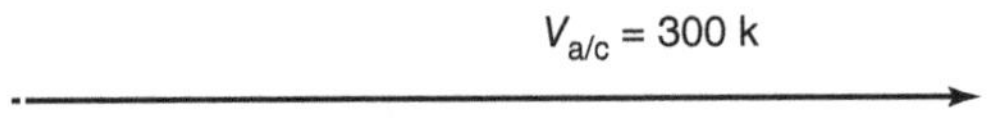

Figure 1.7. Vector of an eastbound aircraft.

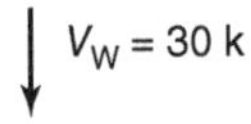

Figure 1.8. Vector of a north wind.

Vector Addition

Vector addition is more complicated than scalar addition. Vector quantities are conveniently shown by arrows. The length of the arrow represents the magnitude of the quantity, and the orientation of the arrow represents the directional property of the quantity. For example, if we consider the top of this page as representing north and we want to show the velocity of an aircraft flying east at an airspeed of 300 kts., the velocity vector is as shown in Figure 1.7 and in this example would be at a 90° angle to the ground track of the aircraft. If there is a 30-kts. wind from the north, the wind vector is shown in Figure 1.8.

To find the aircraft's flight path, ground speed, and drift angle, we add these two vectors as follows. Place the tail of the wind vector at the head of the arrow of the aircraft vector and draw a straight line from the tail of the aircraft vector to the head of the arrow of the wind vector. This *resultant* vector represents the path of the aircraft over the ground. The length of the resultant vector represents the ground speed, and the angle between the aircraft vector and the resultant vector is the drift angle (Figure 1.9).

The ground speed is the hypotenuse of the right triangle and is found by use of the Pythagorean theorem $c^2 = a^2 + b^2$ or $V_r^2 = V_{a/c}^2 + V_w^2$:

$$\text{Groundspeed} = V_r = \sqrt{(300)^2 + (30)^2} = 302 \text{ kts.}$$

Note that the Pythagorean theorem is only applicable if there is a direct crosswind resulting in a 90° angle. The drift angle is the angle whose tangent is $V_w/V_{a/c} = 30/300 = 0.1$, which is 5.7° to the right (south) of the aircraft heading.

Vector Resolution

It is often desirable to replace a given vector with two or more other vectors. This is called *vector resolution*. The resulting vectors are called component vectors of the original vector and, if added vectorially, they will produce the original vector. A review of trigonometry may be needed at this point. Figure 1.10 depicts

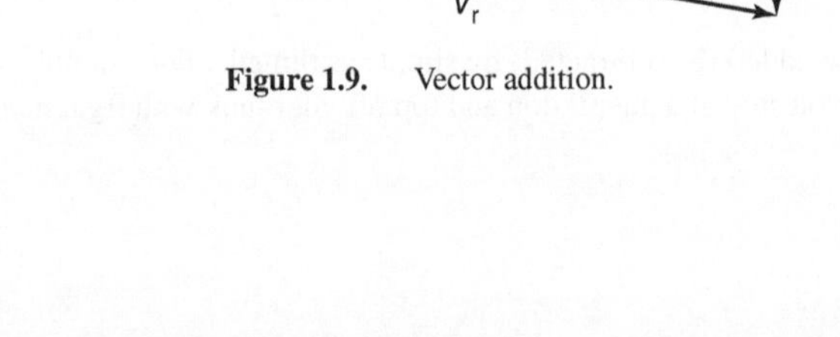

Figure 1.9. Vector addition.

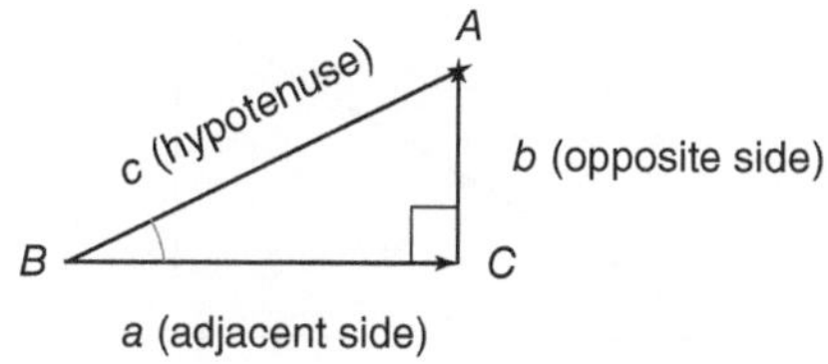

Figure 1.10. Right triangle.

a right triangle, the lowercase letters indicate the three sides of a triangle, and the uppercase letters indicate the angle for each corner. In this example, Side a is opposite Angle B, Side b is adjacent to Angle B, and Side c is always across from the 90° angle and is referred to as the hypotenuse. The three trigonometric functions are ratios that compare two sides of a triangle as follows:

$$\text{Sine (sin)} = \frac{\text{Opposite side}}{\text{Hypotenuse side}} \quad \text{Cosine (cos)} = \frac{\text{Adjacent side}}{\text{Hypotenuse side}} \quad \text{Tangent (tan)} = \frac{\text{Opposite side}}{\text{Adjacent side}}$$

Using a calculator, if the angle is known, then the sin, cos, or tan keys can be used to solve the problem. How we apply these trigonometric functions to vector resolution can be found in the example for Figure 1.11.

Figure 1.11 uses vector resolution to determine ground speed and rate of climb. For example, if an aircraft is in a steady climb, at an airspeed of 200 kts., and the flight path makes a 30° angle with the horizontal, the ground speed and rate of climb can be found by vector resolution. The flight path and velocity are shown by vector $V_{a/c}$ in Figure 1.11.

In Figure 1.12, to resolve the vector $V_{a/c}$ into a component V_h parallel to the horizontal, which will represent the ground speed, and a vertical component, V_v, which will represent the rate of climb, we simply draw a straight line vertically upward from the horizontal to the tip of the arrow $V_{a/c}$. This vertical line represents the rate of climb and the horizontal line represents the ground speed of the aircraft. If the airspeed $V_{a/c}$ is 200 kts. and the climb angle is 30°, we can use the trigonometric functions discussed earlier in this section to resolve the problems:

$$\text{Groundspeed} \rightarrow \cos 30° = \frac{V_h}{V_{a/c}} \rightarrow V_h = V_{a/c} \cos 30° \rightarrow V_h = 200(0.866) \rightarrow V_h = 173.2 \text{ kts.}$$

$$\text{Rate of climb} \rightarrow \sin 30° = \frac{V_v}{V_{a/c}} \rightarrow V_v = V_{a/c} \sin 30° \rightarrow V_v = 200(0.500) \rightarrow V_v = 100 \text{ kts. or } 10130 \text{ fpm}$$

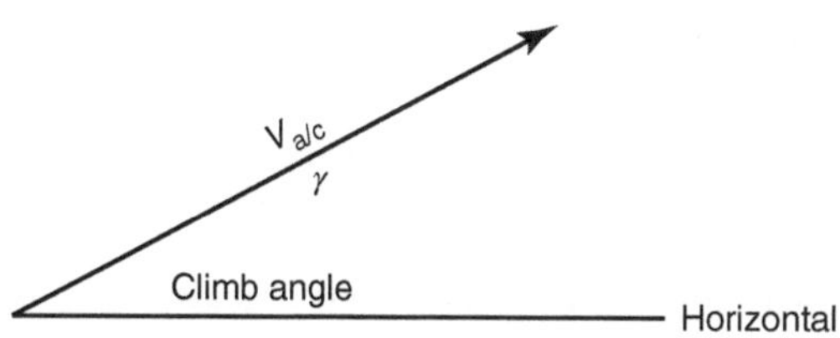

Figure 1.11. Vector of an aircraft in a climb.

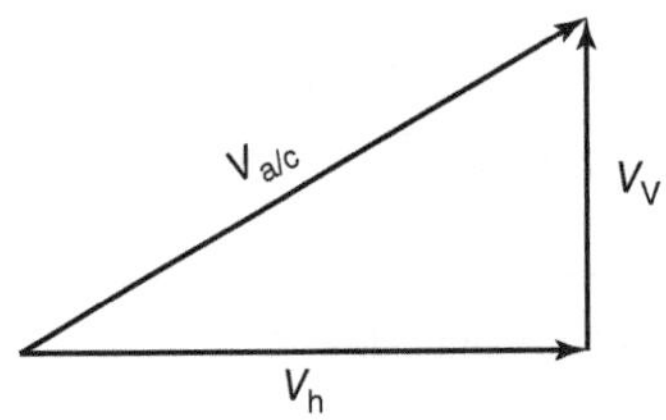

Figure 1.12. Vectors of groundspeed and rate of climb.

MOMENTS

If a mechanic tightens a nut by applying a force to a wrench, a twisting action, called a *moment*, is created about the center of the bolt. This particular type of moment is called *torque* (pronounced "tork"). Moments, M, are measured by multiplying the amount of the applied force, F, by the *moment arm*, L:

$$\text{Moment} = \text{Force} \times \text{Arm} \quad \text{or} \quad M = FL \tag{1.2}$$

The moment arm is the perpendicular distance from the line of action of the applied force to the center of rotation. Moments are measured as foot–pounds (ft-lb) or as inch–pounds (in.-lb). If a mechanic uses a 10 in.-long wrench and applies 25 lb of force, the torque on the nut is 250 in.-lb.

The aircraft moments that are of particular interest to pilots include pitching moments, yawing moments, and rolling moments. If you have ever completed a weight and balance computation for an aircraft, you have calculated a moment, where weight was the *force* and the *arm* was the inches from datum. Pitching moments, for example, occur when an aircraft's elevator is moved. Air loads on the elevator, multiplied by the distance to the aircraft's CG, create pitching moments, which cause the nose to pitch up or down. As you can see from Eq. 1.2, if a force remains the same but the arm is increased, the moment increases.

Several forces may act on an aircraft at the same time, and each will produce its own moment about the aircraft's CG. Some of these moments may oppose others in direction. Therefore, it is necessary to classify each moment, not only by its magnitude but also by its direction of rotation. One such classification could be by *clockwise* or *counterclockwise* rotation. In the case of pitching moments, a *nose-up* or *nose-down* classification seems appropriate.

Mathematically, it is desirable that moments be classified as positive (+) or negative (−). For example, if a clockwise moment is considered to be a + moment, then a counterclockwise moment must be considered to be a − moment. By definition, aircraft nose-up pitching moments are considered to be + moments.

EQUILIBRIUM CONDITIONS

Equilibrium occurs when there is a state in which opposing forces or actions are balanced so that one is not stronger or greater than the other. A body must meet two requirements to be in a state of equilibrium:

1. There must be no unbalanced forces acting on the body. This is written as the mathematical formula $\Sigma F = 0$, where Σ (cap sigma) is the Greek symbol for "sum of." Figure 1.5 illustrates the situation where this condition is satisfied (lift = weight, thrust = drag, etc.)
2. There must be no unbalanced moments acting on the body. Mathematically, $\Sigma M = 0$.

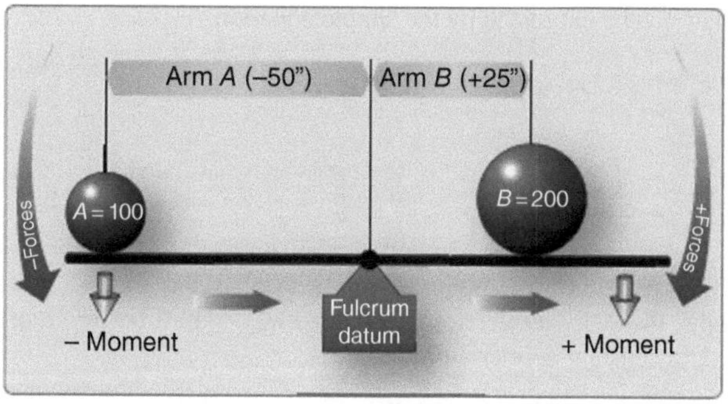

Figure 1.13. Balance Lever.
Source: Weight and Balance Handbook FAA-H-8083-1B (2016d)/U.S. Department of Transportation Federal Aviation Administration/Public domain.

Moments at the fulcrum in Figure 1.13 are 5 000 ft-lb clockwise and 5 000 ft-lb counterclockwise. The weight (force) of A is 100 lb and is located 50 in. (″) to the left of the datum (fulcrum), thus $100\ \text{lb} \times -50'' = -5\,000\ \text{lb - in.}$ The weight of B is 200 lb and is located 25 in. to the right of the datum, thus $200\ \text{lb} \times 25'' = 5\,000\ \text{lb - in.}$ So, $\Sigma M = 0$.

NEWTON'S LAWS OF MOTION

Sir Isaac Newton summarized three generalizations about force and motion. These are known as the *laws of motion.*

Newton's First Law

In simple language, the first law states that *a body at rest will remain at rest and a body in motion will remain in motion, in a straight line, unless acted upon by an unbalanced force.* The first law implies that bodies have a property called *inertia.* Inertia may be defined as the property of a body that results in maintaining its velocity unchanged unless it interacts with an unbalanced force. For example, an aircraft parked on the ramp would not even need chocks unless an unbalanced force (such as wind, or gravity if parked on a slope) acted on it. The measure of inertia is what is technically known as *mass.*

Newton's Second Law

The second law states that *if a body is acted on by an unbalanced force, the body will accelerate in the direction of the force and the acceleration will be directly proportional to the force and inversely proportional to the mass of the body. Acceleration* is the change in the velocity of a body in a unit of time. Consider an aircraft accelerating down the runway or decelerating after touchdown. For our discussion, the primary forces acting on an aircraft accelerating or decelerating down a runway are thrust, drag, and friction. Future chapters will discuss the net force on an aircraft during the takeoff and landing regimes.

The amount of acceleration a is directly proportional to the unbalanced force, F, and is inversely proportional to the mass, m, of the body. For a constant mass, force equals mass times acceleration.

Newton's second law can be expressed by the simple equation:

$$F = m\,a \tag{1.3}$$

Then, solving for a,

$$a = \frac{F}{m}$$

EXAMPLE

An airplane that weighs 14 400 lb accelerates down a runway with a net force of 4 000 lb, what is the acceleration (a) assuming constant acceleration?

$$m = \frac{W}{g} = \frac{14\,400\ \text{lb}}{32.2\ \text{ft/s}^2} = 447.2\ \text{slugs}$$

$$a = \frac{F}{m} \rightarrow a = \frac{4\,000\ \text{lb}}{447.2\ \text{slugs}} \rightarrow a = 8.9\ \text{ft/s}^2$$

Newton's Third Law

The third law states that *for every action force, there is an equal and opposite reaction force*. Note that for this law to have any meaning, there must be an interaction between the force and the body. For example, the gases produced by burning fuel in a rocket engine are accelerated through the rocket nozzle. The equal and opposite force acts on the interior walls of the combustion chamber, and the rocket is accelerated in the opposite direction. As a propeller aircraft pushes air backward from the propeller, the aircraft is pushed forward.

ENERGY AND WORK

Energy is the ability to do work. In physics, work has a meaning different from the popular definition. You can push against a solid wall until you are exhausted but, unless the wall moves, you are not doing any work. Work requires that a force must move an object (displacement) in the direction of the force. Another way of saying this is that *only the component of the force in the direction of movement does any work:*

$$\text{Work} = \text{Force} \times \text{Distance}$$

There are many kinds of energy: solar, chemical, heat, nuclear, and others. The type of energy that is of interest to us in aviation is *mechanical energy*. For our discussion, the units of work will be measured in ft-lb.

There are two kinds of mechanical energy: The first is called *potential energy of position*, or more simply *potential energy* (PE). No movement is involved in calculating PE so this type of energy can be referred to as stored energy of motion. A good example of this kind of energy is water stored behind a dam. If released, the water would be able to do work, such as running a generator. As a fighter aircraft zooms to a zenith point, it builds PE; once it starts to accelerate downward, it converts PE to KE. PE equals the weight, W, of an object multiplied by the height, h, of the object above some base plane:

$$\text{PE} = Wh\ \text{(ft-lb)} \tag{1.4}$$

Note: Some Federal Aviation Administration (FAA) references may list PE = mgh *as related to Eq.* 1.1.

The second kind of mechanical energy is called *kinetic energy* (KE). As the name implies, KE requires the movement of an object (energy of motion). It is a function of the mass, m, of the object and its velocity, V:

$$\mathrm{KE} = \frac{1}{2}mV^2 \ \text{(ft-lb)} \tag{1.5}$$

The total mechanical energy (TE) of an object, is the sum of its PE and KE:

$$\mathrm{TE} = \mathrm{PE} + \mathrm{KE} \ \text{(ft-lb)} \tag{1.6}$$

The law of conservation of energy states that the total energy (of a closed system) remains constant. Both PE and KE can change in value, but the total energy must remain the same. For example, when a ball is thrown upward if the height of the thrower is the reference plane, its energy is all kinetic when it leaves the thrower's hand. As it rises, PE is continually increasing, but KE is always decreasing by the same amount, so the sum remains constant. At the top of its travel, PE is at its maximum (the same amount as the KE it had when it left the thrower's hand) and KE is zero. *Energy cannot be created or destroyed but it can change in form.*

EXAMPLE

An aircraft that weighs 15 000 lb is flying at 10 000 ft altitude at an airspeed of 210 kts. Calculate the potential energy, kinetic energy, and the total energy.

$$\mathrm{PE} = Wh \rightarrow \mathrm{PE} = 15\,000 \ \text{lb} \times 10\,000 \ \text{ft} \rightarrow \mathrm{PE} = 1.5 \times 10^8 \ \text{ft-lb}$$

$$\mathrm{KE} = \frac{1}{2}mV^2 \rightarrow \mathrm{KE} = \frac{1}{2}(465.8 \ \text{slugs})(354.9 \ \text{ft/s})^2 \rightarrow \mathrm{KE} = 2.9 \times 10^7 \ \text{ft-lb}$$

$$\text{Total energy: } \mathrm{TE} = \mathrm{PE} + \mathrm{KE} \rightarrow \mathrm{TE} = 1.79 \times 10^8 \ \text{ft-lb}$$

Application 1.2

Consider a general aviation airplane that weighs 3 000 lb with a designated approach speed over the runway threshold of 65 kts., and calculate the KE. Now, consider if that same airplane approaches the runway with an extra 10 kts. of speed due to poor planning, and calculate the new KE.

Why does only a 10 kts. change in approach speed results in such a wide margin of KE? What are the consequences of this "extra" energy?

POWER

In our discussion of work and energy, we have not mentioned time. *Power* is defined as "the rate of doing work" or work/time. We know:

$$\text{Work} = \text{Force} \times \text{Distance}$$

and

$$\text{Speed} = \text{Distance/Time}$$

$$\text{Power} = \frac{\text{Work}}{\text{Time}} \rightarrow \text{Power} = \frac{\text{Force} \times \text{Distance}}{\text{Time}} \rightarrow \text{Power} = \text{Force} \times \text{Speed} \quad (\text{ft-lb/s})$$

One unit of power = ft-lb/s, and James Watt defined the term *horsepower* (HP) as 550 ft-lb/s, which represents the power to move 550 lbs, 1 ft in one second, or 33 000 lbs, 1 ft in one minute. To convert fps to kts., divide 550 by 1.69 which equals 325. So, if the speed is measured in knots, V_k, and the force (lbs) is the *thrust*, T, of a jet engine, then:

$$\text{HP} = \frac{\text{Thrust} \times V_k}{325} \rightarrow \text{HP} = \frac{TV_k}{325} \tag{1.7}$$

Equation 1.7 is very useful in comparing thrust-producing aircraft (turbojets) with power-producing aircraft (propeller aircraft and rotorcraft); a more expansive comparison will be introduced in Chapters 7 and 8.

EXAMPLE

An aircraft's turbojet engine produces 8 000 lb of thrust at 180 kts., what is the equivalent horsepower that engine is producing?

$$\text{HP} = \frac{TV_k}{325} \rightarrow \text{HP} = \frac{(8\,000\ \text{lb})\,(180\ \text{kts.})}{325} \rightarrow \text{HP} = 4\,431\ \text{HP}$$

Application 1.3

Consider the example calculation provided to solve for horsepower (HP).

Would the horsepower remain the same if the thrust remained 8 000 lb but the aircraft slowed to a speed of 160 kts.? Why or why not? How can the equation be altered to solve for thrust (T) if an aircraft is maintaining a constant speed with a known HP?

FRICTION

If two surfaces are in contact with each other, then a force develops between them when an attempt is made to move them relative to each other. This force is called *friction*. Generally, we think of friction as something to be avoided because it wastes energy and causes parts to wear. In our discussion on drag, we will discuss the parasite drag on an airplane in flight and the thrust or power to overcome that force. Friction is not always our enemy; however, without it, there would be no traction between an aircraft's tires and the runway. Once an aircraft lands, the lift is reduced and a portion of the weight contributes to frictional force. Depending on the aircraft type, aerodynamic braking, thrust reversers, and spoilers will be used to assist the brakes and shorten the landing, or rejected takeoff distance.

At the microscopic level, as in the surface of a wing, friction causes resistance and slows down the velocity of the air as it passes over it. The layer of air that is impacted by the friction of the wing, or any other surface of the aircraft, is referred to as the boundary layer.

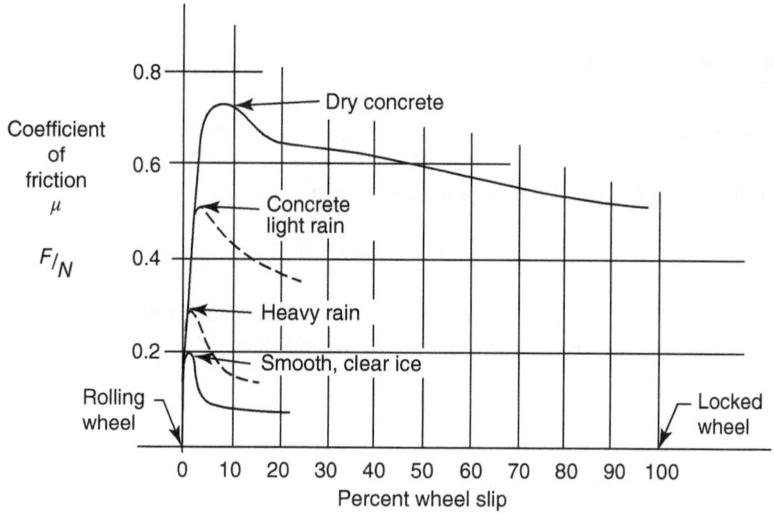

Figure 1.14. Coefficients of friction for airplane tires on a runway.

Several factors are involved in determining friction effects on aircraft during takeoff and landing operations. Among these are runway surfacing material, condition of the runway, tire material and tread, and the amount of brake slippage. All of these variables determine a *coefficient of friction* μ (mu). The actual braking force, F_b, is the product of this coefficient μ (Greek symbol mu) and the normal force, N, between the tires and the runway (Eq. 1.8):

$$F_b = \mu N \text{ (lb)} \tag{1.8}$$

Figure 1.14 shows typical values of the coefficient of friction for various conditions. Note the value of μ for dry concrete is ~0.7 with ~10% wheel slip, while the μ on smooth, clear ice is ~0.2. This means that an airplane wheel rolling on smooth, clear ice will experience much lower friction (increased stopping distance) than a wheel rolling on dry concrete.

EXAMPLE

Calculate the braking force on dry concrete when the normal force (N) is 2 000 lb.

$$F_b = \mu N \rightarrow F_b = (0.7)(2\,000 \text{ lb}) \rightarrow F_b = 1\,400 \text{ lb}$$

INTRODUCTION TO LINEAR MOTION

Newton's laws of motion express relationships among force, mass, and acceleration, but they stop short of discussing velocity, time, and distance. These are covered here. In the interest of simplicity, we assume here that acceleration is constant. Then,

$$\text{Acceleration } a = \frac{\text{Change in velocity}}{\text{Change in time}} = \frac{\Delta V}{\Delta t} = \frac{V_f - V_i}{t_f - t_i} \quad \text{or} \quad \frac{V_f - V_i}{t}$$

where

Δ (cap delta) means "change in"

V_f = final velocity at time t_f

V_i = initial velocity at time t_i.

If we start the time at $t_i = 0$ and rearrange the above, then

$$V_f = V_i + at \tag{1.9}$$

If we start the time at $t_i = 0$ and $V_i = 0$ (brakes locked before takeoff roll) and rearrange the above where V_f can be any velocity given, for example, liftoff velocity, then

$$t = \frac{V_f}{a} \tag{1.10}$$

The distance s traveled in a certain time is

$$s = V_i t + \frac{1}{2}at^2 \tag{1.11}$$

Solving Eqs. 1.9 and 1.11 simultaneously and eliminating t, we can derive a third equation:

$$s = \frac{V_f^2 - V_i^2}{2a} \tag{1.12}$$

A more expansive derivation of Eqs. 1.9–1.12 will be shown in Chapter 8: *Takeoff Performance* and Chapter 9: *Landing Performance* as these equations will be used in calculating takeoff and landing performance.

EXAMPLE

An aircraft that weighs 15 000 lb begins from a brakes-locked position on the runway and then accelerates down the runway with a net force of 5 000 lb until liftoff at a velocity of 110 kts. Calculate the average acceleration down the runway, the average time it takes to reach liftoff speed, and the total takeoff distance on the runway.

First, to calculate the acceleration, we need to find the force (F) and the mass of the aircraft during the takeoff roll, Eq. 1.3: $F = ma$

$$\text{Finding the mass: } m = \frac{W}{g} \rightarrow m = \frac{15\,000\ \text{lb}}{32.2\ \text{ft/s}^2} \rightarrow m = 465.8\ \text{slugs}$$

$$\text{Finding the average acceleration: } a = \frac{F}{m} \rightarrow a = \frac{5\,000\ \text{lb}}{465.8\ \text{slugs}} \rightarrow a = 10.7\ \text{ft/s}^2$$

Average time to liftoff:

$$V_f = V_i + at \rightarrow \text{where } V_i = 0$$

so,

$$110\ \text{kts.} \times 1.69 = 185.9\ \text{ft/s}$$

$$t = \frac{V_f}{a} \rightarrow t = \frac{185.9\ \text{ft/s}}{10.7\ \text{ft/s}} \rightarrow t = 17.4\ \text{seconds}$$

$$\text{Total takeoff distance: } s = \frac{V_f^2 - V_i^2}{2a} \rightarrow s = \frac{(185.9)^2 - (0)}{2(10.7\ \text{ft/s}^2)} \rightarrow s = 1\,614.9\ \text{ft}$$

INTRODUCTION TO ROTATIONAL MOTION

Without derivation, some of the relationships among tangential (tip) velocity, V_t; radius of rotation, r; revolutions per minute, rpm; centripetal forces, CF; weight of rotating parts, W; and acceleration of gravity, g; are shown below. A more detailed discussion regarding rotorcraft, electric vertical take-off and landing (eVTOLs) aircraft, and unmanned aerial vehicles (UAVs) can be found in Chapter 15 of this textbook.

$$V_t = \frac{r\ (\text{rpm})}{9.55}\ (\text{fps}) \tag{1.13}$$

$$\text{CF} = \frac{WV_t^2}{gr}\ (\text{lb}) \tag{1.14}$$

$$\text{CF} = \frac{Wr\ (\text{rpm})^2}{2\,930} \tag{1.15}$$

EXAMPLE

A UH60 Black Hawk has a rotor diameter of 53.6 ft (radius = 26.8 ft) and operates in a hover at 260 rpm. Find the tip speed V_t of the rotor.

$$V_t = \frac{r\ (\text{rpm})}{9.55} \rightarrow V_t = \frac{26.8\ \text{ft}\,(260\ \text{rpm})}{9.55} \rightarrow V_t = 730\ \text{fps}$$

SYMBOLS

a	acceleration (ft/s^2)
E	Energy (ft-lb)
KE	Kinetic energy (ft-lb)
PE	Potential energy (ft-lb)
TE	Total energy (ft-lb)
F	Force (lb)
F_b	Braking force (lb)
g	Acceleration of gravity (ft/s^2)
h	Height (ft)

HP	Horsepower
L	Moment arm (ft or in.)
m	Mass (slugs, lb-s^2/ft)
M	Moment (ft-lb or in.-lb)
N	Normal force (lb)
r	Radius (ft)
rpm	Revolutions per minute
s	Distance (ft)
T	Thrust (lb)
t	Time (second)
V	Velocity (ft/s) or (kts.)
V_f	Final velocity (ft/s)
V_k	Velocity (kts.)
V_i	Initial velocity (ft/s)
V_t	Tangential (tip) speed (ft/s)
W	Weight (lb)
μ (mu)	Coefficient of friction (dimensionless)

EQUATIONS

1.1 $W = mg$

1.2 $M = FL$

1.3 $F = ma$

1.4 $\text{PE} = Wh$

1.5 $\text{KE} = \frac{1}{2}mV^2$

1.6 $\text{TE} = \text{PE} + \text{KE}$

1.7 $\text{HP} = \frac{TV_k}{325}$

1.8 $F_b = \mu N$

1.9 $V_f = V_i + at$

1.10 $t = \frac{V_f}{a}$

1.11 $s = V_i t + \frac{1}{2}at^2$

1.12 $s = \frac{V_f^2 - V_i^2}{2a}$

1.13 $V_t = \frac{r\,(\text{rpm})}{9.55}$

1.14 $\text{CF} = \frac{WV_t^2}{gr}$

1.15 $\text{CF} = \frac{Wr\,(\text{rpm})^2}{2\,930}$

KEY TERMS

Acceleration
Area
Arm

Coefficient of friction
Centripetal force
Component vector
Energy
Equilibrium
Force
Friction
Kinetic energy
Laws of motion
Linear
Mass
Mechanical energy
Motion
Potential energy
Power
Pressure
Resultant vector
Rotational motion
Scalar quantity
Torque
Velocity
Vector quantity
Vector resolution
Work

PROBLEMS

Note: Answers to the problems in the first two sections are given at the end of the book.

Application of Chapter Theory

1. Based on the reading, who was considered the first aeronautical engineer?
   a. Sir George Cayley
   b. Orville Wright
   c. Leonardo da Vinci
   d. Otto Lilienthal

2. Which Wright brother was at the controls of the first manned, heavier-than-air flight on December 17, 1903?
   a. Wilbur Wright
   b. Orville Wright

3. Which group in the early twentieth century had a profound impact on the development of aerodynamics, including the development of a "low drag" cowling and extensive research on airfoil design?
   a. National Aeronautics and Space Administration
   b. Aerial Experiment Association
   c. Federal Aviation Agency
   d. National Advisory Committee for Aeronautics

4. Newton's third law of motion states:
   a. A body at rest will remain at rest and a body in motion will remain in motion, in a straight line, unless acted upon by an unbalanced force.
   b. For every action force there is an equal and opposite reaction force.
   c. If a body is acted on by an unbalanced force, the body will accelerate in the direction of the force, and the acceleration will be directly proportional to the force and inversely proportional to the mass of the body.

5. An aircraft parked on an airport ramp would be an example of Newton's _______ law of motion.
   a. first
   b. second
   c. fourth
   d. third

6. An airplane in level flight increases thrust, resulting in an acceleration until once again thrust equals:
   a. aerodynamic force
   b. lift
   c. weight
   d. drag

7. Theoretically, an airplane that weighs 2 300 lb in straight-and-level, unaccelerated flight, must produce at least _________ in total lift to maintain a constant altitude, assuming no additional forces are involved.
   a. 2 000 lb.
   b. 2 300 lb.
   c. 1 150 lb.
   d. >2 300 lb

Calculations Using Chapter Principles

8. Convert 65 kts. to fps.
9. Convert 200 fps to kts.
10. Convert 35 kts. to fpm.
11. Convert 52 nm to sm.
12. An airplane weighs 16 000 lb. The local gravitational acceleration g is 32.2 fps^2. What is the mass of the airplane?
13. The airplane in Problem 12 accelerates down the runway with a net forward force (thrust less drag) of 6 000 lb. Find the acceleration of the airplane.
14. The airplane in Problem 13 starts from a brakes-locked position on the runway. The airplane takes off at an airspeed of 200 fps. Find the time for the aircraft to reach takeoff speed.
15. Under no-wind conditions, what takeoff roll is required for the aircraft in Problem 14?
16. Upon reaching a velocity of 100 fps, the pilot of the airplane in Problem 14 decides to abort the takeoff, applies brakes, and stops the airplane at 1 000 ft. Find the airplane's deceleration.
17. An airplane is towing a glider to altitude. The tow rope is 20° below the horizontal and has a tension force of 300 lb exerted on it by the airplane. Find the horizontal drag of the glider and the amount of lift that the rope is providing to the glider. Sin 20° = 0.342; cos 20° = 0.940.

18. A jet airplane is climbing at a constant airspeed in no-wind conditions. The plane is directly over a point on the ground that is 4 statute miles from the takeoff point and the altimeter reads 15 840 ft. Find the airplane's climb angle and the distance that it has flown through the air.

19. Find the distance s and the force F on the seesaw fulcrum shown in the figure. Assume that the system is in equilibrium.

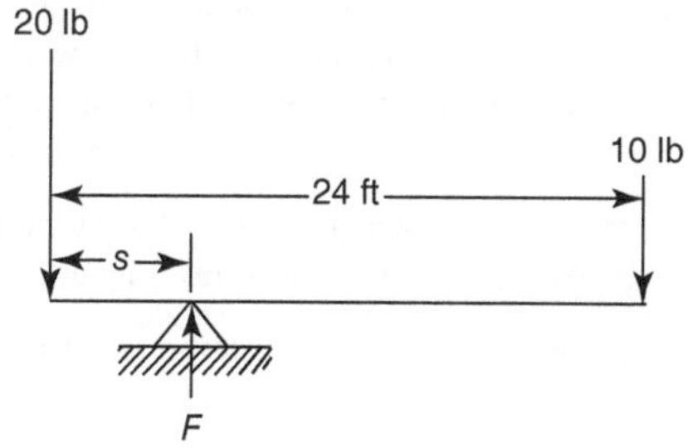

20. A helicopter has a rotor diameter of 30 ft and operates in a hover at 286.5 rpm. Find the tip speed V_t of the rotor.

21. An airplane weighs 16 000 lb and is flying at 5 000 ft altitude and at an airspeed of 200 fps. Find (a) the potential energy, (b) the kinetic energy, and (c) the total energy. Assuming no extra drag on the airplane, if the pilot drove until the airspeed was 400 fps, what would the altitude be?

22. An aircraft's turbojet engine produces 10 000 lb of thrust at 162.5 kts. true airspeed. What is the equivalent power that it is producing?

23. An aircraft weighs 24 000 lb and has 75% of its weight on the main (braking) wheels. If the coefficient of friction is 0.7, find the braking force F_b on the airplane.

Correlation Through Scenario-Based Learning

**For this section, this textbook will focus on one make/model of aircraft and provide scenarios in each chapter that build on the principles from that chapter, so there is continuity with one specific airplane. When the information is not provided, additional research outside this textbook is warranted.*

The airplane that will be used throughout this textbook for the end of the chapter *Correlation Through Scenario-Based Learning* sections will be a *Diamond DA50RG*. This aircraft is powered by a 300 HP diesel engine, has retractable gear, provides Full Authority Digital Engine Control (FADEC), twin-turbochargers, flaps, FIKI-approved, and meets the FAA qualifications for a technically advanced aircraft (TAA), high-performance, and complex airplane. We recommend you reference the latest revision of the DA50RG airplane flight manual as found on the Diamond Aircraft Technical Publication's website.

1. During takeoff the rotation speed will be 70 kts., what is rotation speed when converted to MPH?

2. Using the maximum takeoff weight of 4 407 lb, calculate the mass of the DA50RG at the beginning of the takeoff roll.

3. The DA50 AFM uses both kg and lb to describe mass (weight), how many kg are in 1 lb?

4. Using your answer from Problem #2, and an assumed average net force during the takeoff roll of 700 lb, calculate the average acceleration of the DA50RG.

5. Using only the calculations and discussions provided in this chapter, calculate the takeoff distance if the aircraft is started from a brakes-locked position and rotates at 70 kts., with the average acceleration as found in Problem #4.

6. Using your average acceleration from Problem #3 and the rotation speed from Problem #4, calculate the time in seconds from brakes release to rotation.

7. Assuming all 300 HP are available for takeoff, and a velocity of 70 kts. at rotation, calculate the thrust being produced during rotation. During climb-out and after a power reduction to 90% load and an assumed availability of 272 HP due to twin turbochargers, at 90 kts., what is the thrust being produced during the climb?

8. Assuming the aircraft still weighs 4 407 lb when at 1 000 ft above ground level, calculate the potential energy, kinetic energy, and total energy if the aircraft is traveling at 110 kts.

9. Calculate the tip velocity in ft/s of the propeller during the takeoff roll, assuming 2 300 rpm and a propeller radius of 3.4 ft.

2 Atmosphere, Altitude, and Airspeed Measurement

CHAPTER OBJECTIVES

After completing this chapter, you should be able to:

- Identify the important properties of the atmosphere that influence the aerodynamics of flight.
- Define standard pressure and temperature, and calculate pressure and temperature ratios when a standard atmosphere is not encountered.
- Summarize the relationship between pressure altitude and density altitude.
- Analyze the standard atmosphere table and recognize the change in atmospheric properties with a change in altitude.
- Define and compare the definitions for various types of altitude used in aerodynamics and illustrate why each type is important.
- Explain the relationship between the continuity equation and Bernoulli's equation, and show how they apply to an aircraft in flight.
- Define and compare the definitions for various types of airspeed used in aerodynamics and illustrate why each type is important.
- Determine the true airspeed of an aircraft in flight.

PROPERTIES OF THE ATMOSPHERE

The aerodynamic forces and moments acting on an aircraft in flight are due, in great part, to the properties of the air mass in which the aircraft is flying. By volume, the atmosphere is composed of approximately 78% nitrogen, 21% oxygen, and 1% other gases. The most important properties of air that affect aerodynamic behavior are its static pressure, temperature, density, and viscosity.

It is important to remember at this point that air is a fluid and, like other gases, takes on the shape of its container. Just as a liquid can fill a container, air has the capacity to expand and fill the container as well, though the density will differ significantly. Throughout this textbook, we will expand on this introduction

Flight Theory and Aerodynamics: A Practical Guide for Operational Safety, Fifth Edition.
Brian A. Johnson and Philip R. Fittante.

Companion website: www.wiley.com/go/flighttheoryandaerodynamics5e

to the fluid properties of air, especially as it relates to an airfoil and ultimately the impact on aircraft performance calculations.

Static Pressure

The *static pressure* of the air, P, is simply the weight per unit area of the air above the level under consideration. Air has mass and, as we have discussed, thus has weight, which means it exerts a force. For instance, the weight of a column of air with a cross-sectional area of 1 ft^2 and extending upward from sea level through the atmosphere is 2 116 lb. The sea level static pressure is, therefore, 2 116 pounds per square foot (psf) or 14.7 pounds per square inch (psi). Another commonly used measure of static pressure is *inches of mercury*. On a standard sea level day, the air's static pressure will support a column of mercury (Hg) that is 29.92″ high. Weather reports express pressure in *millibars*; standard atmospheric pressure is 1 013.2 mb. In addition to these rather confusing systems, there are metric measurements in use throughout most of the world. For the discussion of performance problems in this textbook, we will primarily use the measurement of static pressure in inches of mercury as the standard used unless stated otherwise.

Static pressure is reduced as altitude is increased because there is less air weight above. At 18 000 ft altitude, the static pressure is about half that at sea level, the higher you go the less air there is above. The accepted standard pressure lapse rate is approximately 1″ Hg decrease in pressure for every 1 000 ft gain in altitude from sea level (Figure 2.1). This change in atmospheric pressure with altitude is an important concept during the evaluation of aircraft performance as well as the operation of aircraft flight instruments. Figure 2.2 indicates the approximate decrease in pressure and temperature of a standard atmosphere with an increase in altitude above the earth's surface.

In aerodynamics, it is convenient to use pressure ratios, rather than actual pressures; thus, the units of measurement are canceled out. When at sea level on a standard day, the pressure ratio can be determined using the equation:

$$\text{Pressure ratio } \delta \text{ (delta)} = \frac{P}{P_0} \tag{2.1}$$

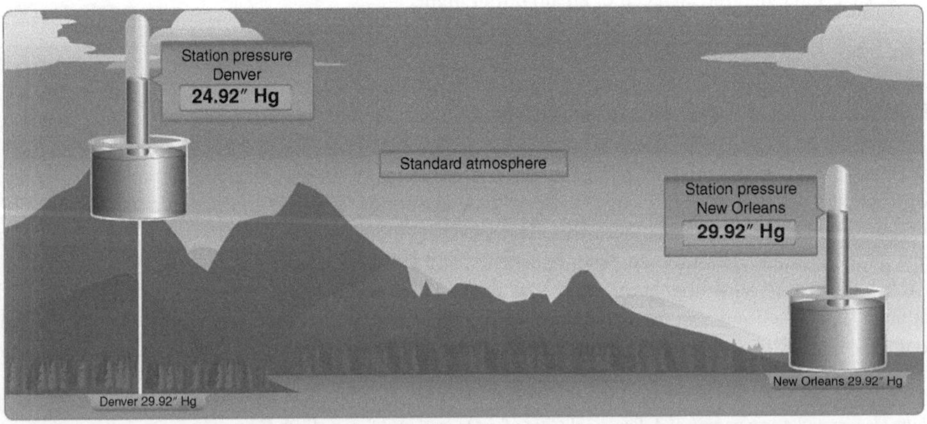

Figure 2.1. Station pressure converted to sea level pressure.
Source: Pilot's Handbook of Aeronautical Knowledge (2023c)/U.S. Department of Transportation Federal Aviation Administration/Public domain.

Standard atmosphere			
Altitude (ft)	Pressure (Hg)	Temperature	
		(°C)	(°F)
0	29.92	15.0	59.0
1 000	28.86	13.0	55.4
2 000	27.82	11.0	51.9
3 000	26.82	9.1	48.3
4 000	25.84	7.1	44.7
5 000	24.89	5.1	41.2
6 000	23.98	3.1	37.6
7 000	23.09	1.1	34.0
8 000	22.22	−0.9	30.5
9 000	21.38	−2.8	26.9
10 000	20.57	−4.8	23.3
11 000	19.79	−6.8	19.8
12 000	19.02	−8.8	16.2
13 000	18.29	−10.8	12.6
14 000	17.57	−12.7	9.1
15 000	16.88	−14.7	5.5
16 000	16.21	−16.7	1.9
17 000	15.56	−18.7	−1.6
18 000	14.94	−20.7	−5.2
19 000	14.33	−22.6	−8.8
20 000	13.74	−24.6	−12.3

Figure 2.2. Properties of a standard atmosphere.
Source: U.S. Department of Transportation Federal Aviation Administration (2016b)/U.S. Department of Transportation/Public Domain.

where P_0 is the sea level standard static pressure (2 116 psf or 29.92″ Hg). Thus, a pressure ratio of 0.5 means that the ambient pressure is one-half of the standard sea level value. At 18 000 ft, on a standard day, the pressure ratio is 0.4992, which translates into ~50% of the pressure at sea level.

EXAMPLE

An airplane is operating from an airport at sea level with a barometric pressure of 29.67″. Calculate the pressure ratio (δ):

$$\delta = \frac{P}{P_0} \rightarrow \delta = \frac{29.67^{\varepsilon}}{29.92^{\varepsilon}} \rightarrow \delta = 0.992$$

Temperature

The commonly used measures of temperature are the Fahrenheit, °F, and Celsius, °C scales. Aviation weather reports for pilots, as well as performance calculation tables, will usually report the temperature in °C. In a standard atmosphere, the sea level surface temperature is 15 °C or 59 °F.

Because neither of these scales has absolute zero as a base, neither can be used in calculations; the absolute temperature must be used instead (Figure 2.3). Absolute zero is −460 °F, or −273 °C. To convert from the Fahrenheit system to the absolute system, called Rankine, R, add 460 to the °F. To convert from

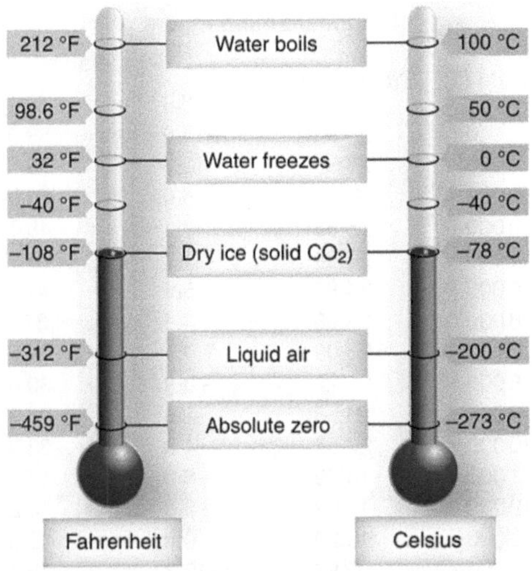

Figure 2.3. Fahrenheit and Celsius temperature scale comparison.
Source: Balloon Flying Handbook (2008c)/U.S. Department of Transportation Federal Aviation Administration/Public domain.

the Celsius system to the absolute system, called Kelvin, K, add 273 to the °C. The symbol for absolute temperature is T and the symbol for sea level standard temperature is T_0:

$$T_0 = 519\,°\text{R} \;\; (59\,°\text{F} + 460)$$

$$T_0 = 288\,°\text{K} \;\; (15\,°\text{C} + 273)$$

By using temperature ratios, instead of actual temperatures, the units cancel. The temperature ratio is the Greek letter theta, θ:

$$\theta = \frac{T}{T_0} \tag{2.2}$$

At sea level, on a standard day, $\theta_{0°} = 1.0$. The temperature in a standard atmosphere decreases with altitude until the tropopause is reached (36 089 ft on a standard day). The rate of change of temperature with altitude is known as the lapse rate. The standard lapse rate is approximately a 2 °C decrease in temperature for every 1 000 ft increase in altitude from sea level (Figure 2.2) up to the tropopause. Temperature then remains constant until an altitude of about 82 023 ft.

As an example, using an outside air temperature (OAT) with Fahrenheit, if the temperature at a pressure altitude of 5 000 ft is −10 °F, then θ:

$$\theta = \frac{(-10\,°\text{F} + 460)}{519\,°\text{R}} = 0.867$$

As an example, using an OAT with Celsius, if the temperature at a pressure altitude of 5 000 ft is −10 °C, then θ:

$$\theta = \frac{(-10\,°\text{C} + 273)}{288\,°\text{K}} = 0.913$$

Density

Density is the most important property of air in the study of aerodynamics, and is directly impacted by pressure, temperature, and humidity changes. Because air can be compressed and expanded, the lower the pressure the less dense the air: density is directly proportional to pressure. Increasing the temperature of the air (giving the particles greater kinetic energy) also decreases the density of the air, so in this case density and temperature have an inverse relationship.

Less dense, thinner air has a lower air density and is said to be a *higher density altitude* (decreasing aircraft performance); more dense, thicker air is said to be a *lower density altitude* (improved aircraft performance). Decreasing the density of the air results in a higher density altitude, while increasing the density of the air results in a lower density altitude.

Atmospheric pressure decreases as altitude increases. Temperature also decreases with increasing altitude, with two exceptions: first, when a temperature inversion layer exists, and second, in the stratosphere, where the temperature remains constant and may even rise with increasing altitude. The discussion above would indicate that greater altitude (less dense air) and colder temperature (more dense air) would result in a conflict in regard to density. But usually, the effect of a decrease in pressure with altitude overcomes any improvement in performance the colder, dense air may have, and a lower density altitude is the rule of thumb the higher in altitude an aircraft climbs.

The effect of moisture content on performance will be largely ignored in this textbook because most textbooks treat the effect of humidity as being negligible for practical purposes, but it is important to understand that water vapor is lighter than air, so moist air is lighter than dry air. As the amount of water vapor increases, the density of the air decreases, resulting in a higher density altitude (decrease in aircraft performance).

Application 2.1

Consider an aircraft departing from an airport located at sea level in Florida on a hot, humid day. Now consider the same aircraft with the same weight, temperature, and pressure conditions on a much drier day from the same airport, the only change is the amount of water vapor in the atmosphere.

Using an online density altitude calculator such as the one provided by the National Oceanic and Atmospheric Administration (NOAA), calculate the impact on density altitude by adjusting the dewpoint values. Could an increase in water vapor affect an aircraft's climb rate?

Density is the mass of the air per unit of volume. The symbol for density is ρ (rho):

$$\rho = \frac{\text{Mass}}{\text{Unit volume}} \quad (\text{slugs/ft}^3)$$

The standard sea level density is $\rho_0 = 0.002\,377$ slugs/ft^3. Density decreases with an increase in altitude. At 22 000 ft, the density is 0.001 183 slugs/ft^3 (about one-half of sea level density).

It is desirable in aerodynamics to use density ratios instead of the actual values of density. The symbol for the density ratio is σ (sigma):

$$\sigma = \frac{\rho}{\rho_0} \tag{2.3}$$

The universal gas law shows that density is directly proportional to pressure and inversely proportional to absolute temperature:

$$\rho = \frac{P}{RT} \tag{2.4}$$

Forming a ratio gives

$$\frac{\rho}{\rho_0} = \frac{P/RT}{P_0/RT_0} = \frac{P/P_0}{T/T_0}$$

R is the gas constant and cancels, so the density ratio, or sigma, is a function of pressure and temperature:

$$\sigma = \frac{\delta}{\theta} \tag{2.5}$$

EXAMPLE

An airplane is operating from a sea-level airfield that has a barometric pressure of 29.67″ Hg and a runway temperature of 65 °F. Find the pressure ratio, temperature ratio, and density ratio:

Pressure ratio = $\delta = 0.9920$ (See previous example)
Temperature ratio = Using Eq. 2.2, $\theta = \frac{(65\,^\circ\mathrm{F}+460)}{519\,^\circ\mathrm{R}} = 1.012$
Density ratio = Using Eq. 2.5,

$$\sigma = \frac{\delta}{\theta} \rightarrow \sigma = \frac{0.9920}{1.012} \rightarrow \sigma = 0.980$$

Viscosity

Viscosity can be simply defined as the internal friction of a fluid caused by the molecular attraction that makes it resist its tendency to flow. The viscosity of the air is important when discussing airflow in the region very close to the surface of an aircraft. This region is called the *boundary layer*. We discuss viscosity in more detail in Chapter 4 when we take up the subject of boundary layer theory.

ICAO STANDARD ATMOSPHERE

To provide a basis for comparing aircraft performance in different parts of the world and under varying atmospheric conditions, the performance data must be reduced to a set of standard conditions. These are defined by the International Civil Aviation Organization (ICAO) and are compiled in a standard atmosphere table. An abbreviated table is shown here as Table 2.1. Columns in the table show standard day density, density ratio, pressure, pressure ratio, temperature, temperature ratio, and speed of sound at various altitudes.

ALTITUDE MEASUREMENT

When a pilot uses the term altitude, the reference is usually to altitude above sea level as read on the altimeter, but it is important that the distinction is made to understand what *types* of altitude exist. When

Table 2.1. Standard atmosphere table.

Altitude (ft)	Density ratio, σ	$\sqrt{\sigma}$	Pressure ratio, δ	Temperature (°F)	Temperature ratio, θ	Speed of sound (kts.)	Kinematic viscosity, ν (ft^2/s)
0	1.0000	1.0000	1.0000	59.00	1.0000	661.7	0.000 158
1 000	0.9711	0.9854	0.9644	55.43	0.9931	659.5	0.000 161
2 000	0.9428	0.9710	0.9298	51.87	0.9862	657.2	0.000 165
3 000	0.9151	0.9566	0.8962	48.30	0.9794	654.9	0.000 169
4 000	0.8881	0.9424	0.8637	44.74	0.9725	652.6	0.000 174
5 000	0.8617	0.9283	0.8320	41.17	0.9656	650.3	0.000 178
6 000	0.8359	0.9143	0.8014	37.60	0.9587	647.9	0.000 182
7 000	0.8106	0.9004	0.7716	34.04	0.9519	645.6	0.000 187
8 000	0.7860	0.8866	0.7428	30.47	0.9450	643.3	0.000 192
9 000	0.7620	0.8729	0.7148	26.90	0.9381	640.9	0.000 197
10 000	0.7385	0.8593	0.6877	23.34	0.9312	638.6	0.000 202
15 000	0.6292	0.7932	0.5643	5.51	0.8969	626.7	0.000 229
20 000	0.5328	0.7299	0.4595	−12.32	0.8625	614.6	0.000 262
25 000	0.4481	0.6694	0.3711	−30.15	0.8281	602.2	0.000 302
30 000	0.3741	0.6117	0.2970	−47.98	0.7937	589.5	0.000 349
35 000	0.3099	0.5567	0.2353	−65.82	0.7594	576.6	0.000 405
36 089[a]	0.2971	0.5450	0.2234	−69.70	0.7519	573.8	0.000 419
40 000	0.2462	0.4962	0.1851	−69.70	0.7519	573.8	0.000 506
45 000	0.1936	0.4400	0.1455	−69.70	0.7519	573.8	0.000 643
50 000	0.1522	0.3002	0.1145	−69.70	0.7519	573.8	0.000 818

[a]The tropopause.

meteorologists refer to the height of the cloud layer above an airfield, they are usually referring to the altitude above the field elevation. When air traffic control refers to an altitude at Flight Level (FL)180 and above, they are referring to pressure altitude. A flight crew in an aircraft approaching the runway during a low visibility Category (CAT) III instrument approach will need to understand the value of a radar altimeter. Understanding what "altitudes" are important at different periods of flight, and the effect of temperature, pressure, and moisture on those altitudes, is imperative for safe flight.

Indicated Altitude

Indicated altitude is the *altitude* that is read directly from the altimeter and is uncorrected for any errors. In the United States, below FL180 the altimeter is set to the current altimeter setting of the field you are departing from or arriving to or is given by air traffic control for the current area you are flying in. In the United States, when flying at or above 18 000 ft, altitude is measured in FLs (e.g. FL180 for 18 000 ft). At FL180, the indicated altitude will be equal to pressure altitude as the altimeter setting is set to 29.92″, standard pressure, or QNE. The altitude at which the crew changes to 29.92 is called the transition altitude (TA). When the crew descends for landing, the altitude at which they return the altimeter setting to local barometric pressure corrected to sea level (QNH) is called the transition level (TL). (Remember it this way: 29.92 is selected at the TA, and the "A" stands for aloft, as in climbing or cruise. When returning to land, the TL is set on the descent, and "L" stands for low, or landing.)

When QNE is lower than 29.92, the lowest usable FL is no longer FL180. The lowest usable FL is obtained from the aeronautical publications. For instance, in the United States, if the pressure in the area of operations is between 29.91 and 29.42″, the lowest usable enroute altitude is FL185. It should also be noted that the TA and TL outside the United States will not always be 18 000. ICAO members set their own values.

Incidentally, QFE is the reference pressure set in the altimeter if the pilot wishes to know the elevation above the airfield. When the aircraft is on the airfield, the altimeter reads zero. QFE is seldom used as it would be of limited value when away from the immediate vicinity of the airfield.

Calibrated Altitude

Calibrated altitude is indicated altitude corrected for instrument and installation errors.

True Altitude

True altitude is the actual altitude above mean sea level (MSL) and is referenced as MSL. On most aeronautical charts, MSL altitudes are published for man-made objects such as towers and buildings, as well as for terrain, since this is the altitude closest to the altitude read off the altimeter. An important note is that true altitude will only be the same as the indicated altitude when flying in standard conditions, which is very rare. When flying in conditions colder than standard, the altimeter will indicate a higher altitude than you are flying, so the true altitude will be lower than the indicated altitude. The same dangerous situation can develop when you are flying from a high-pressure area to a low-pressure area and the altimeter is not corrected for the local altimeter setting. Your altimeter will interpret the lower pressure as a higher altitude and your true altitude will again be lower than your indicated altitude. From the variations in true altitude versus indicated altitude, the saying was developed "high to low, or hot to cold, look out below." Of course, this assumes that the altimeter is never reset to local pressure for an entire flight covering a long distance with varying temperatures and pressures.

Absolute Altitude

Absolute altitude is the vertical altitude above ground level (AGL), and can be measured with devices such as a radar altimeter. Of course, your absolute altitude is more critical the closer to the ground you are flying; so even when not equipped with a radar altimeter, a pilot should be aware of their AGL altitude. When conducting an instrument approach in inclement weather, knowledge of your AGL altitude is vital to the safe completion of the approach or execution of a missed approach. Your height above airport (HAA), height above touchdown zone (HAT), and decision height (DH) are all AGL altitudes and should be briefed before the approach.

Pressure Altitude

Regarding aircraft performance, two types of altitude are of most interest to a pilot: pressure altitude and density altitude.

Pressure altitude is the corrected altitude in the standard atmosphere corresponding to a certain static pressure, and is used to compute true airspeed (TAS), density altitude, and takeoff and landing data.

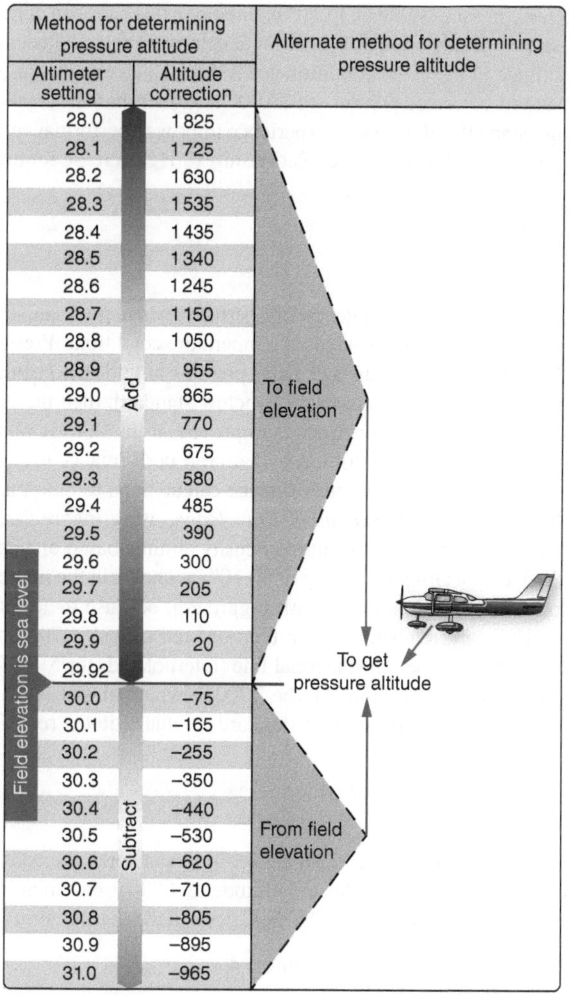

Figure 2.4. Field elevation versus pressure altitude.
Source: Pilot's Handbook of Aeronautical Knowledge (2008a)/U.S. Department of Transportation Federal Aviation Administration/Public domain.

Pressure altitude is the vertical distance above a standard datum plane where atmospheric pressure is 29.92″ of Hg. In the United States, at FL180 and above, the altimeter is always set to 29.92″ unless abnormally low pressure exists in the area. This is one of two ways to quickly find pressure altitude, set the altimeter to 29.92″ and the indicated altitude is equivalent to pressure altitude.

Figure 2.4 provides a second convenient method to determine pressure altitude by applying a correction factor to the indicated altitude. For example, if a given airport elevation is 3 500 ft and the automated weather observation states a pressure value of 30.10″ (remember, this has been corrected for sea level pressure at this location), use the altitude correction column to determine that 165 needs to be subtracted from 3 500 ft. The pressure altitude in the given condition is 3 335 ft, which makes sense as it is lower than the indicated altitude. The airport corrected pressure of 30.10″ is higher than the standard pressure of 29.92″, thus the air density is higher and the aircraft will experience improved performance. The corrections shown in Figure 2.4 may be approximated by using the relationship 1″ Hg decrease for every 1 000 ft of increase in altitude.

Density Altitude

Density altitude is the altitude used to calculate aircraft performance in most situations. Density altitude is found by correcting pressure altitude for nonstandard temperature conditions. Pressure altitude and density altitude are the same when conditions are standard. Once pressure altitude has been determined, the density altitude is calculated using OAT. If the temperature is below standard, then the density altitude is lower than the pressure altitude and aircraft performance is improved. If the OAT is warmer than standard, the density altitude is higher than the pressure altitude, and aircraft performance is degraded.

Figure 2.5 is a commonly used chart for performance calculations to determine density altitude. As provided by the *Pilot's Handbook of Aeronautical Knowledge*, using Figure 2.5, we will calculate the pressure altitude first, and then calculate the resulting density altitude based on temperature. With a given airport elevation of 5 883 ft and an altimeter setting of 30.10″, as shown in the figure, the resulting pressure altitude is calculated to be 5 718 ft. Continuing with Figure 2.5, because the density altitude is pressure altitude corrected for nonstandard temperature, we then use the provided 70 °C temperature and move from the X-axis vertically until we reach the diagonal line (interpolated) for 5 718 ft. Moving horizontally to the Y-axis, the density altitude is determined to be 7 700 ft, which makes sense as it is higher than the pressure altitude because the temperature is above standard for that altitude, resulting in lower air density (higher density altitude).

EXAMPLE

Using Table 2.1 and several of the equations from earlier in this chapter, the density altitude can also be determined using ratios. An aircraft at an indicated altitude of 1 800 ft has an altimeter setting of 29.70″ Hg (sea level) and an outside air temperature of 75 °F. Calculate the density altitude:

Pressure altitude (PA) = 2 005 ft is found using Figure 2.4.
Referencing Table for PA 2 000 ft, $\delta = 0.9298$
Using Eq. 2.2, $\theta = 1.031$
Using Eq. 2.5, solve for the density ratio

$$\sigma = \frac{\delta}{\theta} \rightarrow \sigma = \frac{0.9298}{1.031} \rightarrow \sigma = 0.902$$

Using Table 2.1, it can be interpolated that with a density ratio of 0.902, the resultant density altitude is 3 500 ft. This makes sense as the sea level pressure is lower than standard and the temperature is above standard for that altitude, which results in lower air density (higher density altitude).

As discussed, density altitude influences aircraft performance; the higher the density altitude, the lower the aircraft performance. So, in the previous problem, even though the aircraft is at an indicated altitude of 1 800 ft, the density altitude for performance calculations is 3 500 ft. Low air density equals a higher density altitude; high air density equals a lower density altitude. Therefore, aircraft performance charts are provided for various density altitudes.

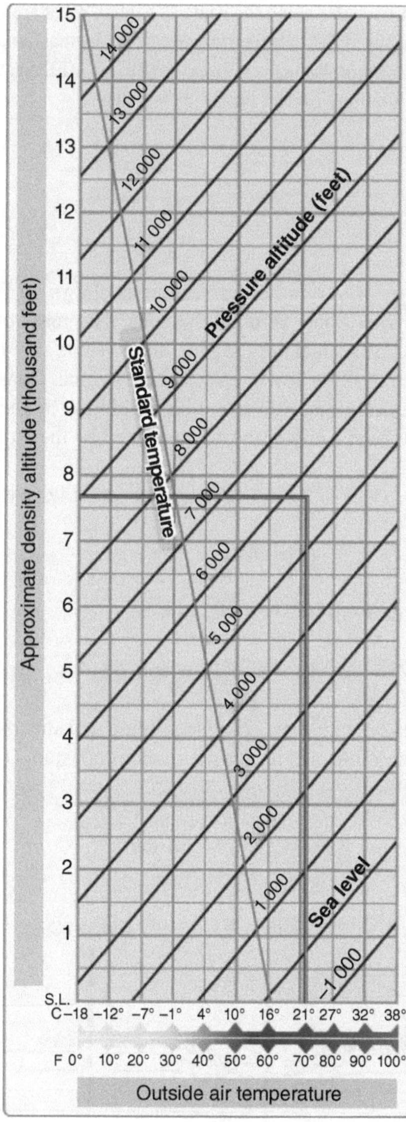

Altimeter setting ("Hg)	Pressure altitude conversion factor
28.0	1 824
28.1	1 727
28.2	1 630
28.3	1 533
28.4	1 436
28.5	1 340
28.6	1 244
28.7	1 148
28.8	1 053
28.9	957
29.0	863
29.1	768
29.2	673
29.3	579
29.4	485
29.5	392
29.6	298
29.7	205
29.8	112
29.9	20
29.92	0
30.0	−73
30.1	−165
30.2	−257
30.3	−348
30.4	−440
30.5	−531
30.6	−622
30.7	−712
30.8	−803

Figure 2.5. Pressure altitude conversion and density altitude chart.
Source: U.S. Department of Transportation Federal Aviation Administration (2016b)/Public Domain.

Application 2.2

A non-pressurized, single-engine aircraft departs for a cross-country flight at 6 500 ft. A portion of the route crosses rising terrain with peaks at 5 000 ft. The pilot is only concerned with maintaining an indicated altitude of 6 500 ft as filed on the visual flight rules (VFR) flight plan.

What other altitudes must the pilot consider along the route of flight? What are the implications of failing to adjust the altimeter setting if the barometric pressure is decreasing along the route? Consider the FAA definition of Maximum Elevation Figure (MEF) on VFR Sectionals and if nonstandard pressure and temperature may impact obstacle clearance.

CONTINUITY EQUATION

Consider the flow of air through a pipe of varying cross section as shown in Figure 2.6. There is no flow through the sides of the pipe: air flows only through the ends. The mass of air entering the pipe, in a given unit of time, equals the mass of air leaving the pipe, in the same unit of time. The mass flow through the pipe, must remain constant. The mass flow at each station is equal. Constant mass flow is called *steady-state flow*. The mass airflow is equal to the volume of air multiplied by the density of the air. The volume of air, at any station, is equal to the velocity of the air multiplied by the cross-sectional area of that station.

The mass airflow symbol Q is the product of the density, the area, and the velocity:

$$Q = \rho AV \tag{2.6}$$

The continuity equation states that the mass airflow is a constant:

$$\rho_1 A_1 V_1 = \rho_2 A_2 V_2 = \rho_3 A_3 V_3 = \text{constant} \tag{2.7}$$

The continuity equation is valid for steady-state flow, both in subsonic and supersonic flow. For subsonic flow, the air is considered to be incompressible, and its density remains constant. The density symbols can then be eliminated; thus, for subsonic flow,

$$A_1 V_1 = A_2 V_2 = A_3 V_3 = \text{constant} \tag{2.8}$$

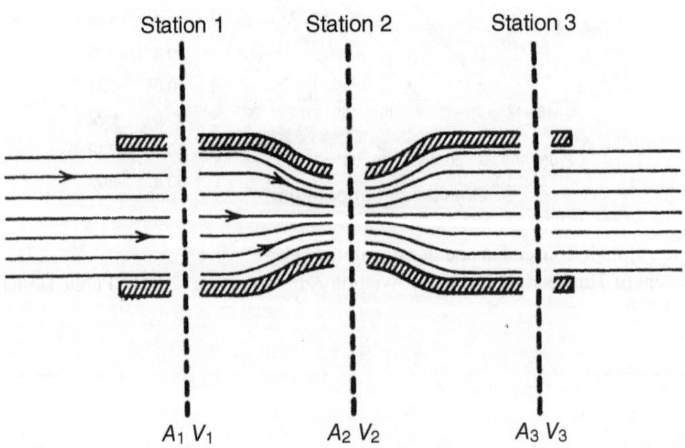

Figure 2.6. Flow of air through a pipe.

Velocity is inversely proportional to the cross-sectional area: as the cross-sectional area decreases, velocity increases.

BERNOULLI'S EQUATION

The continuity equation explains the relationship between velocity and cross-sectional area. It does not explain the differences in static pressure of the air passing through a pipe of varying cross sections. Bernoulli, using the principle of conservation of energy, developed a concept that explains the behavior of pressures in flowing gases.

Consider the flow of air through a venturi tube as shown in Figure 2.7. In Image A, as the mass of air experiences a constriction in the tube and as the velocity of the mass increases, the pressure decreases. A comparative image of a wing experiencing Bernoulli's principle during flight is in Image B. Note the decreased density toward the rear of the airfoil, this will be a discussion area in Chapters 3 and 4.

The energy of an airstream is in two forms: it has *potential energy*, which is its static pressure, and *kinetic energy*, which is its dynamic pressure. The total pressure of the airstream is the sum of the static pressure and the dynamic pressure. The total pressure remains constant, according to the law of conservation of energy. Thus, an increase in one form of pressure must result in an equal decrease in the other.

Static pressure is an easily understood concept (see the discussion earlier in this chapter). Dynamic pressure, q, is similar to kinetic energy in mechanics and is expressed by

$$q = \frac{1}{2}\rho V^2 \text{ (psf)} \tag{2.9}$$

where V is measured in feet per second. Pilots are much more familiar with velocity measured in knots instead of in feet per second, so a new equation for dynamic pressure, q, is used in this book. Its derivation

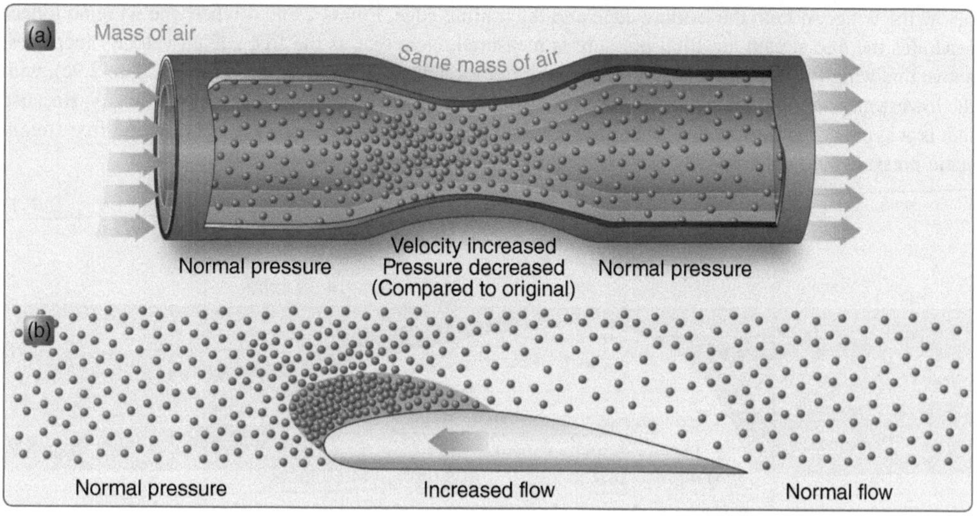

Figure 2.7. Pressure change in a venturi tube.
Source: U.S. Department of Transportation Federal Aviation Administration (2018)/U.S. Department of Transportation / Public Domain.

is shown here:

$$\text{Density ratio, } \sigma = \frac{\rho}{\rho_0} \rightarrow \sigma = \frac{\rho}{0.002\,377} \rightarrow \rho = \sigma(0.002\,377)$$

$$V_{\text{fps}} = (1.69)V_{\text{k}}$$

Accepted constant for velocity in knots and the altitude density ratio = 0.00339

$$0.00339 = \frac{1}{295}$$

Substituting in Eq. 2.9 yields

$$q = \frac{\sigma V_{\text{k}}^2}{295} \text{ (psf)} \tag{2.10}$$

Bernoulli's equation can now be expressed as
Total pressure (head pressure), H = Static pressure, P + Dynamic pressure, q:

Total pressure (head pressure), H = Static pressure, P + Dynamic pressure, q:

$$H = P + \frac{\sigma V_{\text{k}}^2}{295} \text{ (psf)} \tag{2.11}$$

To visualize how lift is developed on a cambered airfoil, draw a line down the middle of a venturi tube. Discard the upper half of the figure and superimpose an airfoil on the constricted section of the tube (Figure 2.8). Note that the static pressure over the airfoil is less than that ahead of it and behind it, so, as dynamic pressure increases, static pressure decreases. Figure 2.9 shows a more detailed change in both free stream speed and free stream static pressure around a symmetrical (top and bottom half are the same) airfoil. As the leading edge of the wing encounters the free stream air, Point 2, it is forced above and below the wing. At both the leading edge and the trailing edge, Points 2 and 4 where the wing no longer separates the free stream air, total pressure is measured. Note that as the free stream velocity increases, above this wing for this image (Figure 2.9b), the free stream static pressure decreases (Figure 2.9c), with the lowest measurement of static pressure occurring at the same location of maximum velocity. Because this is a symmetrical airfoil, the same image could be used for the freestream velocity and free stream static pressure below the wing.

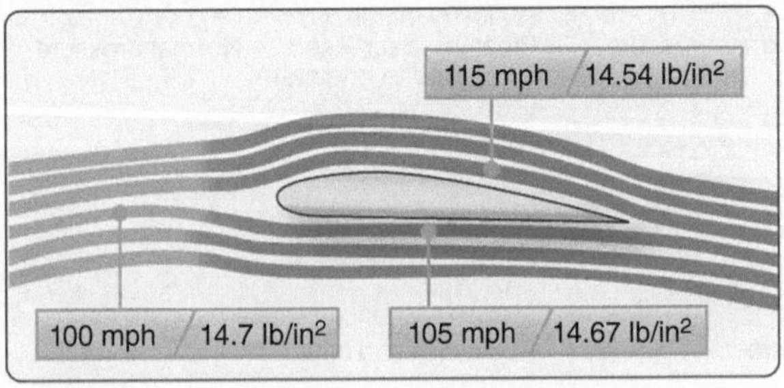

Figure 2.8. Velocities and pressures on an airfoil superimposed on a wing.
Source: Aviation Maintenance Technician Handbook – Airframe (2023a)/U.S. Department of Transportation Federal Aviation Administration/Public domain.

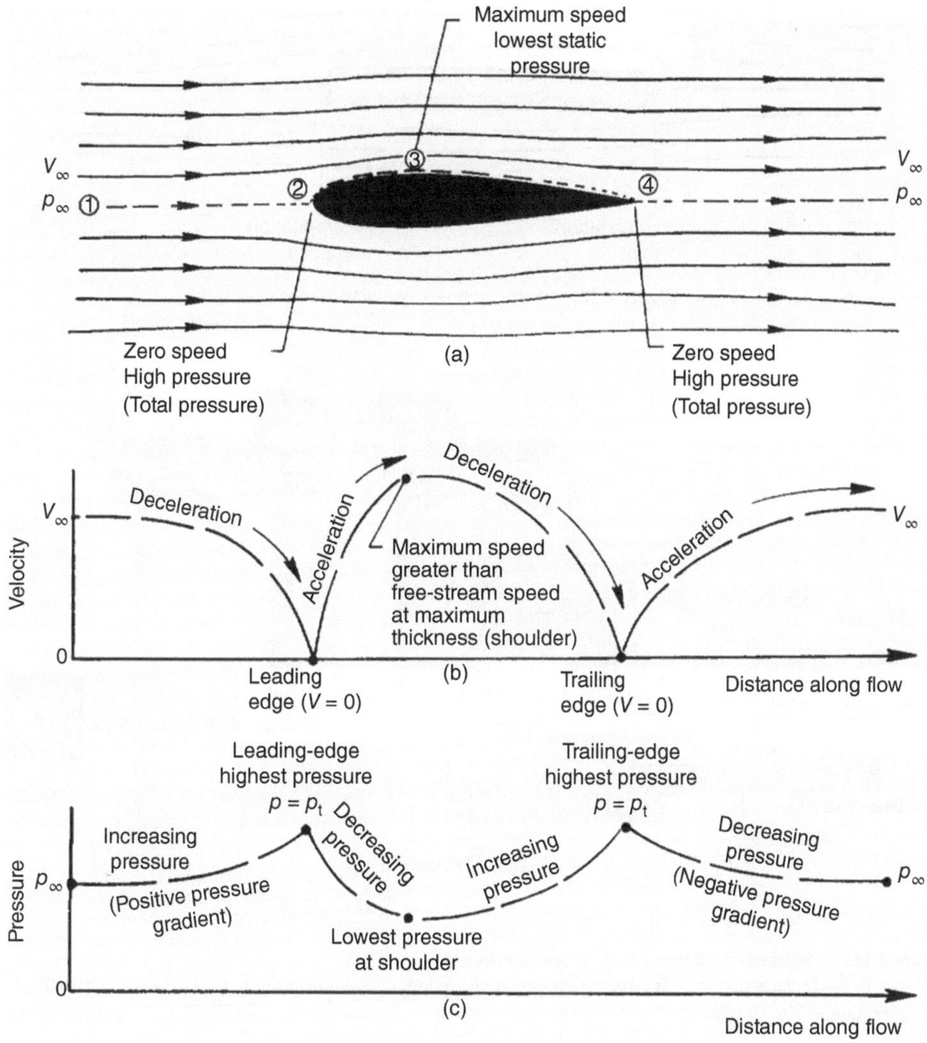

Figure 2.9. Airfoil with ideal fluid flow. (a) Airflow about a symmetric airfoil, (b) Free stream velocity, (c) Free stream pressure.
Source: NASA/Public Domain.

AIRSPEED MEASUREMENT

If a symmetrically shaped object is placed in a moving airstream (Figure 2.10), the flow pattern will be as shown. Some airflow will pass over the object and some will flow beneath it, but at the point at the nose of the object, the flow will be stopped completely. This point is called the *stagnation point*. Because the air velocity at this point is zero, the dynamic pressure is also zero. The stagnation pressure is, therefore, all static pressure and must be equal to the total pressure, H, of the airstream.

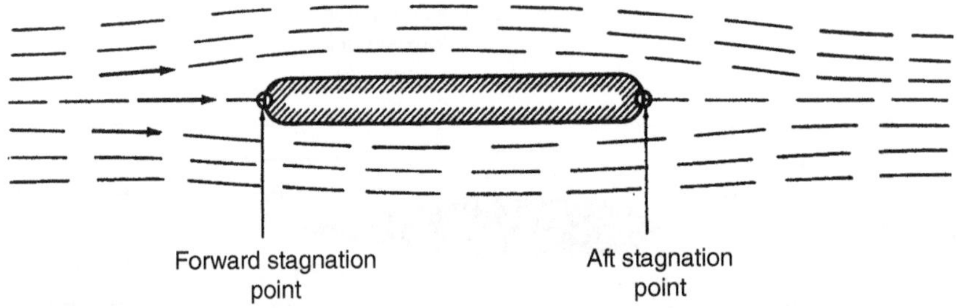

Figure 2.10. Flow around a symmetrical object.

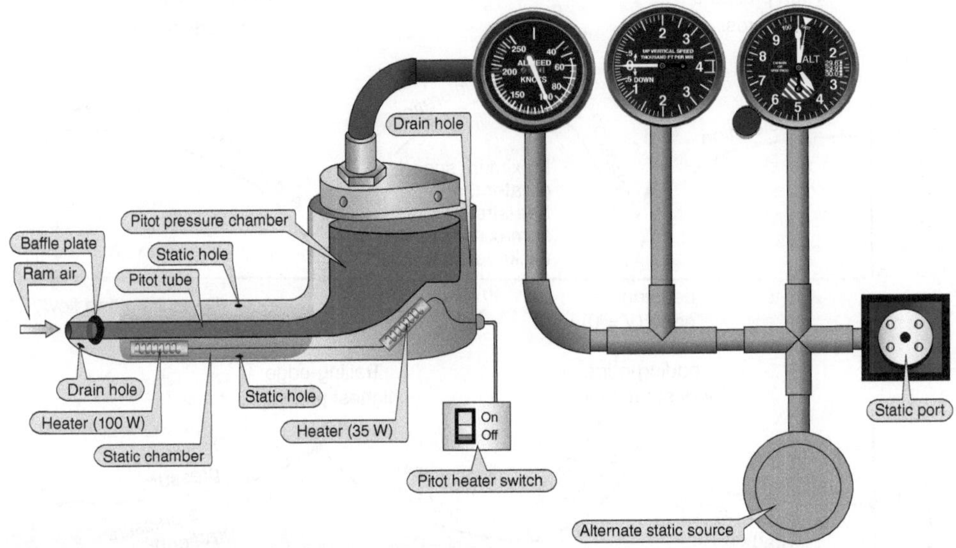

Figure 2.11. Schematic of a pitot–static airspeed indicator.
Source: U.S. Department of Transportation Federal Aviation Administration (2012c)/U.S. Department of Transportation/Public Domain.

In Figure 2.11, the free stream values of velocity and pressure are used to measure the indicated airspeed (IAS) of an aircraft. The pitot tube is shown as the *total pressure port* and must be pointed into the relative wind for accurate readings. Static pressure, sometimes referred to as ambient pressure, is the pressure around the aircraft whether it is moving or at rest, and is the same "local" pressure your body experiences. When in motion, the air entering the pitot tube comes to a complete stop and thus the static pressure in the tube is equal to the total free stream pressure, H. This pressure is ducted into a diaphragm inside the airspeed indicator.

The static pressure port(s) can be made as a part of the point tube or it can be at a distance from the pitot tube on the side of the aircraft. It should be located at a point where the local air velocity is exactly equal to the airplane velocity. The static port is made so that none of the velocity enters the port. The port measures only static pressure, and none of the dynamic pressure. The static pressure is ducted into the chamber surrounding the diaphragm within the inside of the airspeed indicator.

Now we have static pressure inside the diaphragm that is equal to total pressure (H), and then static pressure is measured from the static port outside the diaphragm. The difference between the pressure inside the diaphragm and outside the diaphragm is the differential pressure that deflects the flexible diaphragm that is geared to the airspeed pointer. The static pressures cancel each other, thus the dynamic pressure is left and indicated on the airspeed indicator. The airspeed indicator a measures differential pressure. The airspeed indicator is calibrated to read airspeed.

Figure 2.12 shows a modern pitot–static system associated with an air data computer (ADC). Though the concept with this solid-state device is the same as in traditional pitot–static systems, the signal sent to modern "glass" instruments is digital and often has the ability to generate trend vectors.

Application 2.3

Figure 2.13 indicates the impact of a blockage of the pitot tube while an aircraft is flying at altitude. The resultant effect is that the airspeed indicator acts like an altimeter: as the aircraft climbs in altitude, the airspeed increases, and when the aircraft descends, the airspeed decreases.

In a traditional airspeed indicator with a diaphragm, what is happening in reference to our discussion on q, P, and H in the situation above? What is the impact of the blockage on the other pitot–static instruments? Why?

Indicated Airspeed

Indicated airspeed (IAS) is the direct reading of the airspeed indicator and is uncorrected for errors related to installation or nonstandard atmospheric density. If there are any errors in the instrument, they may be shown on an instrument error card located near the instrument and/or in the aircraft flight manual (AFM).

Calibrated Airspeed

Calibrated airspeed (CAS) is obtained when the necessary corrections have been made to the IAS for installation errors and instrument errors. These position errors are especially prevalent at lower airspeeds, and the IAS may be "indicating" slower than the CAS. Most of the time in non-pressurized aircraft operating at lower altitudes, the CAS can be assumed to be within several knots of the IAS.

Equivalent Airspeed

Equivalent airspeed (EAS) results when the CAS has been corrected for compressibility effects. Figure 2.14 shows a compressibility correction chart. In general, if flying above 10 000 ft and 200 kts., the compressibility correction should be made. Unlike the instrument and position error charts, which vary with different aircraft, this chart is good for any aircraft.

EAS is not a significant factor in airspeed computations when aircraft fly at relatively low speeds and altitudes, but at higher speeds and altitudes, the compressibility correction must be considered. For example, if an aircraft is flying at a pressure altitude of 20 000 ft at a CAS of 400 kts., Figure 2.14 indicates a compressibility correction of −17.5 kts. The EAS for this example is 382.5 kts.

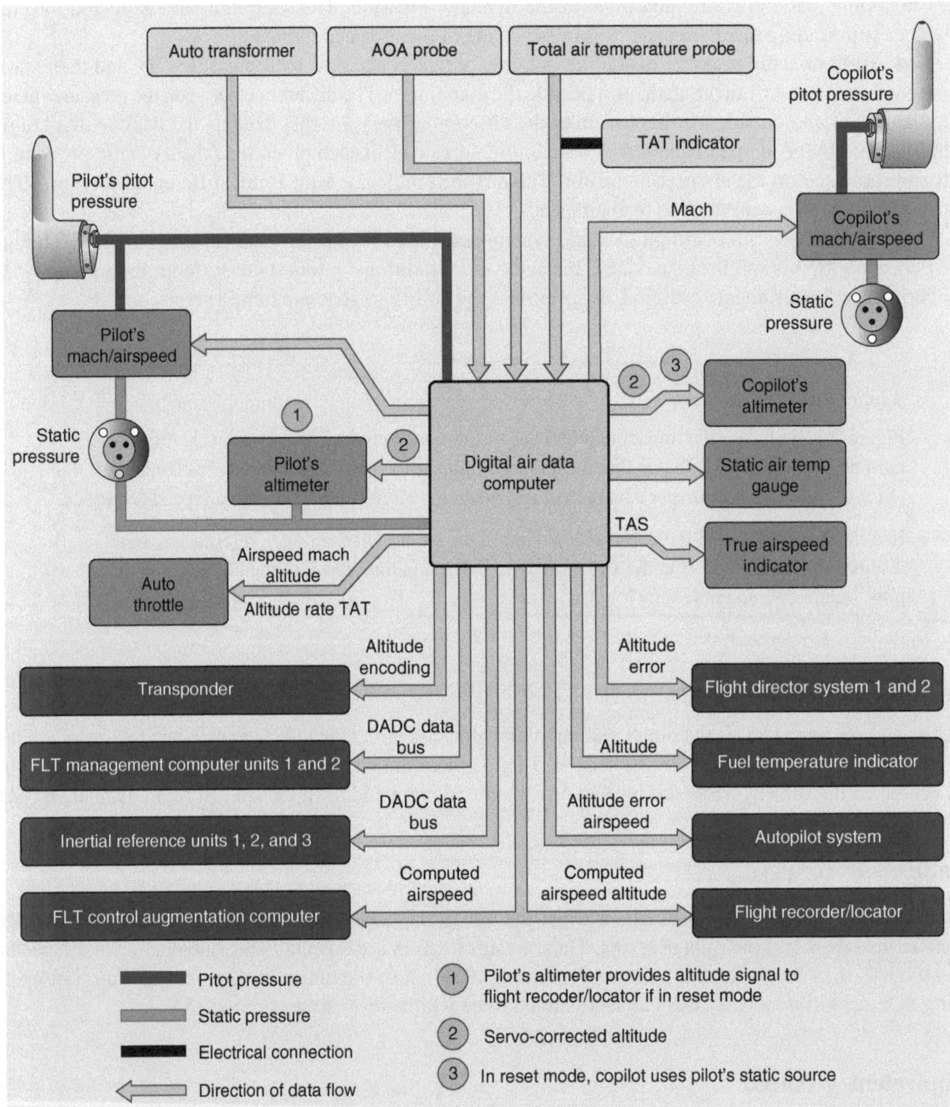

Figure 2.12. Air data computer and pitot–static sensing.
Source: U.S. Department of Transportation Federal Aviation Administration (2012a)/U.S. Department of Transportation/Public Domain.

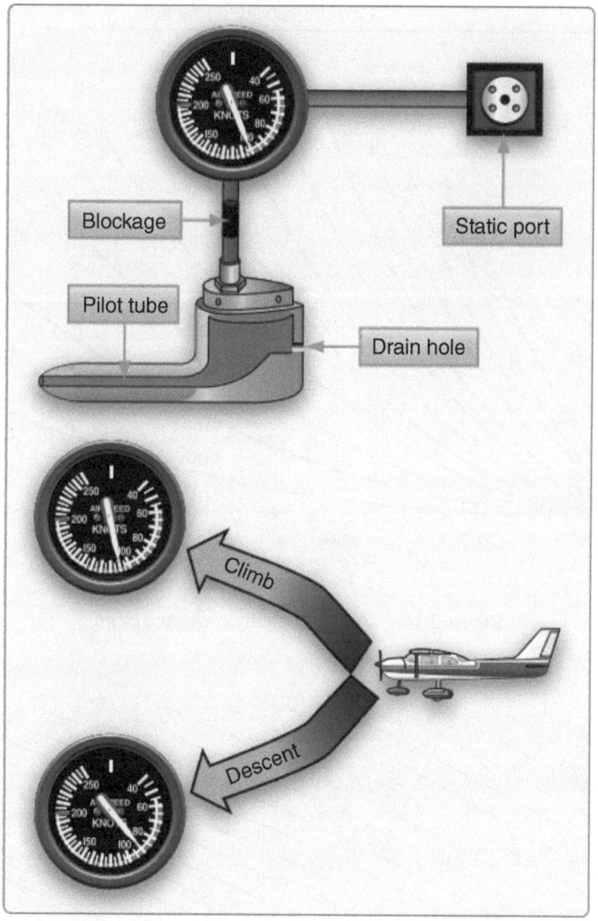

Figure 2.13. Blocked pitot tube and drain hole.
Source: U.S. Department of Transportation Federal Aviation Administration (2016b)/U.S. Department of Transportation/Public Domain.

True Airspeed

True airspeed (TAS) is obtained when EAS has been corrected for density ratio, the airspeed indicator measures dynamic pressure and is calibrated for sea level standard day density. As altitude increases, the density ratio decreases and a correction must be made. The correction factor is $\sqrt{\sigma}$:

$$\text{TAS} = \frac{\text{EAS}}{\sqrt{\sigma}} \tag{2.12}$$

Values of $\sqrt{\sigma}$ can be found in the ICAO Standard Altitude Chart (Table 2.1); values of $1/\sqrt{\sigma}$ can be found in Figure 2.15.

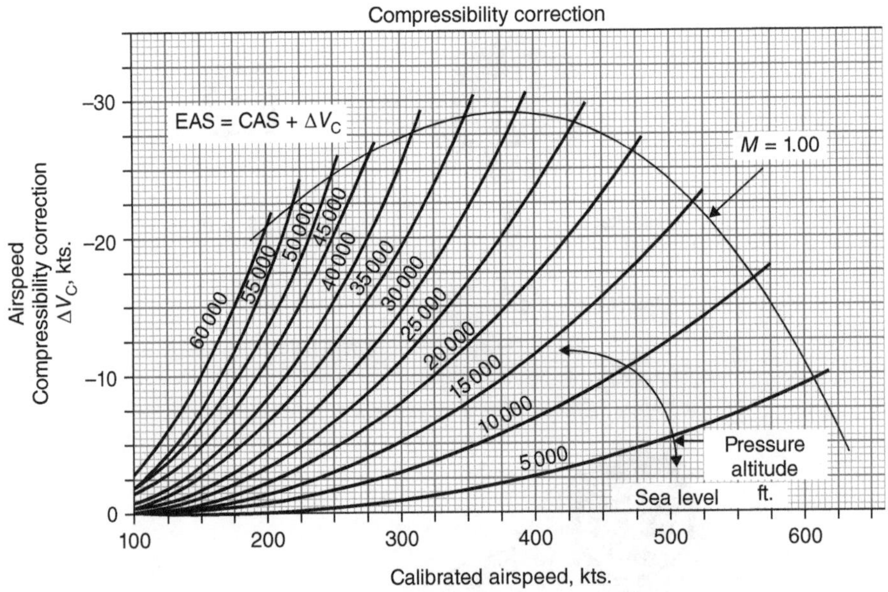

Figure 2.14. Compressibility correction chart.

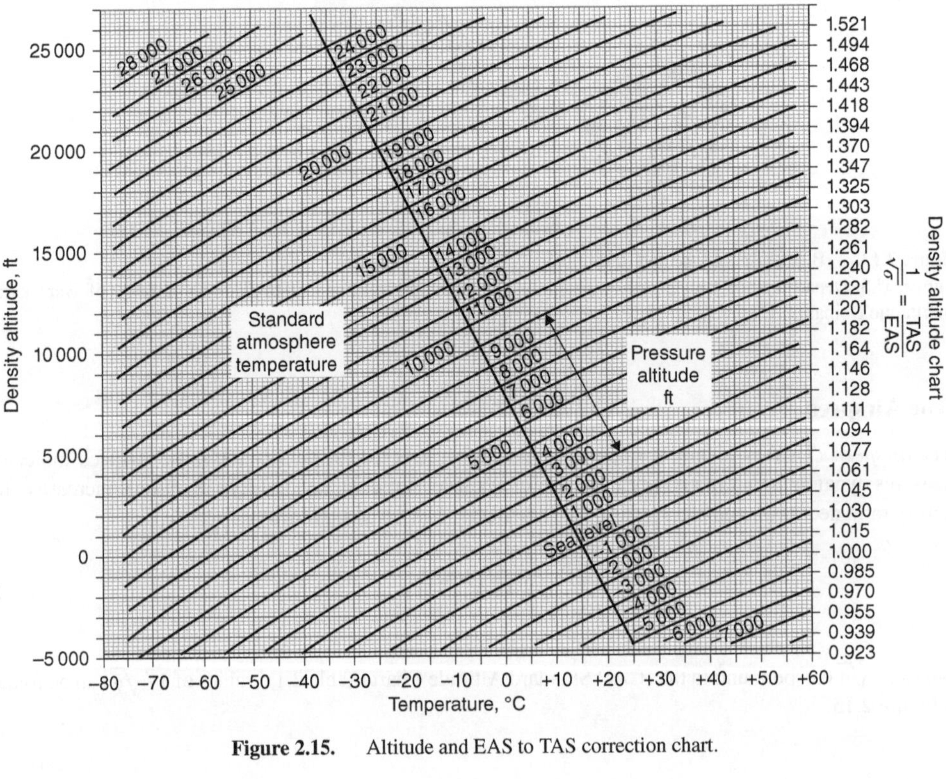

Figure 2.15. Altitude and EAS to TAS correction chart.

Due to the decrease in air density with an increase in altitude, for any given TAS, CAS will decrease as altitude increases. As higher altitudes are attained, the aircraft must fly faster to obtain the same pressure differential. For a given CAS, as an aircraft increases in altitude, the TAS will increase. The higher the aircraft travels in altitude, the greater the difference between CAS and TAS.

EXAMPLE

Using our example from the CAS discussion, EAS was determined to be 382.5 kts. If the outside air temperature at a pressure altitude of 20 000 ft is −30 °C, we can utilize Eq. 2.12 to find the TAS. From Table 2.1, the pressure ratio (δ) is 0.4595.

$$\theta = \frac{T}{T_0} \rightarrow \theta = \frac{(-30\,^\circ\text{C} + 273\,^\circ\text{K})}{288\,^\circ\text{K}} \rightarrow \theta = 0.844$$

$$\sigma = \frac{\delta}{\theta} \rightarrow \sigma = \frac{0.4595}{0.844} \rightarrow \sigma = 0.544$$

$$\text{Eq. 2.12, TAS} = \frac{382.5\text{ kts.}}{\sqrt{0.544}} \rightarrow \text{TAS} = 518.6\text{ kts.}$$

Figure 2.16 depicts the change in TAS between sea level and 15 000 ft when the IAS remains constant. For the respective IAS, the CAS calibration has been applied. As the aircraft climbs, the TAS increases as the air density decreases. The aircraft must travel at 130 kts. TAS to register 100 kts. on the airspeed indicator.

Understanding the relationship between the speeds above, and the calculation of each one, can be facilitated by remembering "ICE-T." IAS is read off the airspeed indicator, CAS is IAS corrected for installation/position errors, EAS is CAS corrected for compressibility, and finally TAS is EAS corrected for temperature and pressure.

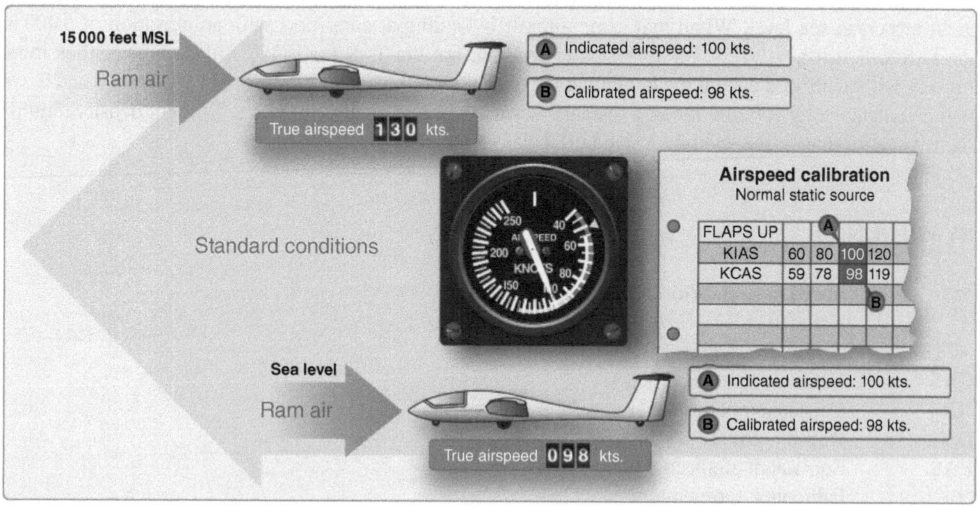

Figure 2.16. IAS, CAS, and TAS comparison.
Source: U.S. Department of Transportation Federal Aviation Administration (2013)/Public Domain.

Mach

The Mach number is found by comparing TAS to the speed of sound for a given set of conditions at a specific altitude. The speed of sound is an important factor in the study of high-speed flight and is discussed in depth in Chapter 14. Because the aircraft's speed in relation to the speed of sound is so important in high-speed flight, airspeeds are usually measured as Mach number (named after the Austrian physicist Ernst Mach). Mach number is the aircraft's TAS divided by the speed of sound (in the same atmospheric conditions):

$$M = \frac{V}{a} \tag{2.13}$$

where

M = Mach number

V = true airspeed (kts.)

a = local speed of sound (kts.)

EXAMPLE

Using the TAS from the previous example (518.6 kts.), when $a = 607.3$ kts., calculate the Mach number for the aircraft.

$$M = \frac{V}{a} \rightarrow M = \frac{518.6 \text{ kts.}}{607.3 \text{ kts.}} \rightarrow M = 0.85$$

Groundspeed

Groundspeed (GS) is the actual speed of the aircraft over the ground, either calculated manually or more commonly nowadays read off the GPS (i.e., global positioning satellite) navigational unit. The GS increases with a tailwind and decreases with a headwind, and is calculated as TAS adjusted for the wind. GS equals TAS in a no-wind situation. Consider an airplane that has an approach speed of 120 kts. IAS when landing at an airport at sea level. When that same aircraft is landing at an airport with an elevation of 5 000 ft. the IAS will still be 120 kts. on approach to landing, but the TAS (and GS) will be higher, thus more runway will be utilized during the landing at the airport with the higher elevation. In Chapters 8 and 9, we will calculate the takeoff and landing distance when landing at an airport with a different density altitude (density ratio) than at sea level on a standard day.

SYMBOLS

a	Speed of sound (local for a given condition)
A	Area (ft^2)
AGL	Above ground level
CAS	Calibrated airspeed (kts.)
°C	Celsius temperature (°)
DA	Density altitude
EAS	Equivalent airspeed
°F	Fahrenheit temperature
GS	Groundspeed
H	Total pressure (head) (psf)

IAS	Indicated airspeed
°K	Kelvin temperature
MSL	Mean sea level
M	Mach number (ratio)
P	Static pressure
PA	Pressure altitude
P_0	Sea level standard pressure
q	Dynamic pressure
R	Universal gas constant
°R	Rankine temperature
TAS	True airspeed
T	Absolute temperature
T_0	Sea level standard temperature
V	Velocity (fps)
V_k	Velocity (kts.)
δ (delta)	Pressure ratio
θ (theta)	Temperature ratio
ρ (rho)	Density
σ (sigma)	Density ratio

EQUATIONS

2.1 $\delta = \frac{P}{P_0}$

2.2 $\theta = \frac{T}{T_0}$

2.3 $\sigma = \frac{\rho}{\rho_0}$

2.4 $\rho = \frac{P}{RT}$

2.5 $\sigma = \frac{\delta}{\theta}$

2.6 $Q = \rho AV$

2.7 $\rho_1 A_1 V_1 = \rho_2 A_2 V_2$

2.8 $A_1 V_1 = A_2 V_2$

2.9 $q = \frac{1}{2}\rho V^2$ (psf)

2.10 $q = \frac{\sigma V_k^2}{295}$ (psf)

2.11 $H = P + \frac{\sigma V_k^2}{295}$

2.12 $\text{TAS} = \frac{\text{EAS}}{\sqrt{\sigma}}$

2.13 $M = \frac{V}{a}$

KEY TERMS

Aerodynamic forces
Absolute altitude

Air data computer
Altimeter
Altitude
Airspeed indicator
Bernoulli's equation
Calibrated airspeed
Compressibility
Continuity equation
Density
Density altitude (high/low)
Equivalent airspeed
Groundspeed
International Civil Aviation Organization (ICAO)
Indicated airspeed
Indicated altitude
Mean sea level (MSL)
Millibars
Pressure altitude
Stagnation point
Standard atmosphere
Static pressure
True airspeed
True altitude
Temperature (Fahrenheit/Celsius)
Venturi
Viscosity

PROBLEMS

Note: Answers to the problems in the first two sections are given at the end of the book.

Application of Chapter Theory

1. An increase in static air pressure
   a. affects air density by decreasing the density.
   b. does not affect air density.
   c. affects air density by increasing the density.
2. A decrease in temperature
   a. affects air density by decreasing the density.
   b. does not affect air density.
   c. affects air density by increasing the density.
3. Pressure ratio is
   a. ambient pressure divided by sea level standard pressure measured in the same units.
   b. ambient pressure in millibars divided by 29.92.
   c. ambient pressure in pounds per square inch divided by 2 116.
   d. sea level standard pressure in inches of mercury divided by 29.92.

4. Density ratio, σ (sigma) is
   a. equal to pressure ratio divided by temperature ratio.
   b. measured in slugs per cubic foot.
   c. equal to the ambient density divided by sea level standard pressure.
   d. None of the above.

5. According to Table 2.1, the value for the density ratio on a standard day at sea level is:
   a. 0.8617
   b. 1.0
   c. 1.5
   d. 0.002377

6. Bernoulli's equation for subsonic flow states that
   a. if the velocity of an airstream within a tube is increased, the static pressure of the air increases.
   b. if the area of a tube decreases, the static pressure of the air increases.
   c. if the velocity of an airstream within a tube increases, the static pressure of the air decreases, but the sum of the static pressure and the velocity remains constant.
   d. None of the above.

7. Dynamic pressure of an airstream is
   a. directly proportional to the square of velocity.
   b. directly proportional to the air density.
   c. Neither (a) nor (b).
   d. Both (a) and (b).

8. In this book, we use the formula for dynamic pressure, $q = \sigma V_k^2/295$, rather than the more conventional formula, $q = \frac{1}{2}\rho V^2$, because
   a. V in our formula is measured in kts.
   b. density ratio is easier to handle (mathematically) than the actual density (slugs per cubic foot).
   c. Both (a) and (b).
   d. Neither (a) nor (b).

9. True altitude equals indicated altitude:
   a. below FL180.
   b. during standard conditions.
   c. when the altimeter is set to 29.92″.
   d. only when the temperature is standard regardless of pressure.

10. Which of the following determines the vertical altitude above the ground:
    a. pressure altitude.
    b. indicated altitude.
    c. true altitude.
    d. absolute altitude.

11. The corrections that must be made to indicated airspeed (IAS) to obtain calibrated airspeed (CAS) are
    a. position error and compressibility error.
    b. instrument error and position error.
    c. instrument error and density error.
    d. position error and density error.

12. The correction that must be made to CAS to obtain equivalent airspeed (EAS) is called compressibility error, which
- **a.** is always a negative value.
- **b.** can be ignored at high altitudes.
- **c.** can be ignored at high airspeed.
- **d.** can be either a positive or a negative value.

13. The correction from EAS to true airspeed (TAS) is dependent on
- **a.** temperature ratio alone.
- **b.** density ratio alone.
- **c.** pressure ratio alone.
- **d.** None of the above.

14. Groundspeed equals ______ when zero wind at altitude.
- **a.** equivalent airspeed.
- **b.** indicated airspeed.
- **c.** true airspeed.
- **d.** calibrated airspeed.

Calculations Using Chapter Principles

15. An airplane operates from an airfield that has a barometric pressure of 27.82″ Hg and a runway temperature of 100 °F. Calculate (or find in Table 2.1) the following:
- **a.** Pressure ratio.
- **b.** Pressure altitude.
- **c.** Temperature ratio.
- **d.** Density ratio.
- **e.** Density altitude.

16. Fill in the values below for the stations in the drawing.

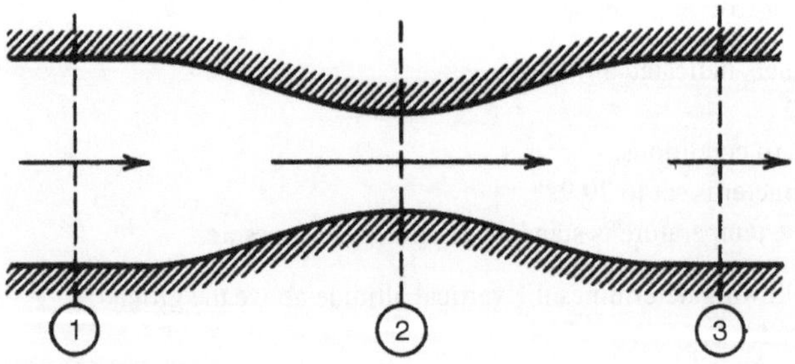

At station 1	At station 2	At station 3
$A_1 = 10$ ft^2	$A_2 = 5$ ft^2	$V_3 = 80$ kts.
$V_1 = 100$ kts.	$V_2 = ?$	$A_3 = ?$
$P_1 = 2\,030$ psf	$P_2 = ?$	$P_3 = ?$
$\sigma = 0.968$		
$H = ?$	$q_2 = ?$	$q_3 = ?$

17. Using Table 2.1, calculate the dynamic pressure, q at 5 000 ft density altitude and 200 kts. TAS.

18. The airspeed indicator of an airplane reads 355 kts. There are no instrument or position errors. If the airplane is flying at a pressure altitude of 25 000 ft, find the equivalent airspeed (EAS).

19. Find the true airspeed (TAS) of the airplane in Problem 18 if the outside air temperature is −40 °C.

20. Using the TAS from the previous example, when $a = 594.5$ kts., calculate the Mach number for the aircraft.

Correlation Through Scenario-Based Learning

Aircraft – *Diamond DA50RG*

1. Using Eqs. 2.1, 2.2, and 2.5, as well as Table 2.1, calculate the pressure ratio, temperature ratio, and density ratio for an aircraft departing from KEYW when the ATIS states a barometric pressure of 29.85″ and outside air temperature (OAT) of 32 °C.

2. Using Figure 2.5, calculate the pressure altitude and density altitude using the same information from the previous problem, how do your answers compare to your answers from the previous problem when using the chapter equations and Table 2.1?

3. Research a reputable website that includes a calculator for pressure and density altitude, as well as relative humidity. Using the same barometric pressure, temperature, and airport elevation, how does the density altitude compare to your calculations from the first two problems? Using 0% and 100% humidity values within the online calculator, how does the density altitude change based on water vapor content? Why?

4. Using a DA50RG aircraft flight manual (AFM) found online (recommend the Diamond Aircraft website for the most up-to-date manual), and the equations and examples provided in this chapter, calculate the following airspeeds given:

Aircraft in cruise flight at 10 000 ft (pressure altitude)

Flaps up

OAT −5 °C

IAS 175 kts.

20 kts. direct headwind

a. CAS (use CAS conversion in Section 5 of the AFM)
b. EAS (use Figure 2.14)
c. TAS
d. GS
e. Mach

3 Structures, Airfoils, and Aerodynamic Forces

CHAPTER OBJECTIVES

After completing this chapter, you should be able to:

- Identify and provide examples of primary and secondary flight controls and their purpose on an aircraft.
- Describe the importance of an airfoil and summarize airfoil terminology.
- Identify the variables in the geometry of an airfoil and how airfoils are classified.
- Characterize the development of forces on airfoils and the relationship to Bernoulli's principle.
- Determine the relationship between aerodynamic force and the center of pressure.
- Compare and contrast the development of pitching moments on a symmetrical airfoil and a cambered airfoil.
- Define the aerodynamic center and identify the relationship of the aerodynamic center to the angle of attack and pitching moments.
- Review an aircraft accident summary and correlate it to the importance of primary and secondary flight controls for the safety of flight.

AIRCRAFT STRUCTURES

Up to this point, you may have considered the wing to be the only aerodynamically important structure of an aircraft. But in fact, the entire structure of the airplane plays a role in the efficiency of an aircraft in flight, and identifying how, and to what extent, each part of an airplane structure plays a role is an important first step. We will begin with a review of the more prominent structures discussed in aerodynamics, because their direct role in lift and drag provides the foundation for more complicated discussions in the future.

Flight Theory and Aerodynamics: A Practical Guide for Operational Safety, Fifth Edition.
Brian A. Johnson and Philip R. Fittante.

Companion website: www.wiley.com/go/flighttheoryandaerodynamics5e

Flight control systems are of two types: primary and secondary. Primary control systems on fixed-wing aircraft include the ailerons, elevator (stabilator), and rudder, and, depending on the aircraft type and aircraft speed, give the pilot a "feel" of how the aircraft is performing. Secondary systems such as trim systems, flaps, spoilers, and leading-edge devices are used in relieving control pressures for the pilots, assisting primary control surfaces in high-speed flight, or improving the performance characteristics of the aircraft in general.

Airplanes are flown in various configurations of gear and flaps, but for this textbook, we will commonly refer to *clean* and *dirty* as the two reference configurations. In a clean configuration, the gear is retracted (when applicable), and the flaps and other high-lift devices are retracted. In the dirty configuration, the gear is considered down and locked, and the high-lift devices are partially or fully deployed.

Figure 3.1 shows examples of airfoils found in many transport category aircraft still in use today. Figure 3.2 details the flight control system of a helicopter, incorporating anti-torque pedals, a cyclic, and a collective. An expanded discussion of these primary flight controls can be found in Chapter 15.

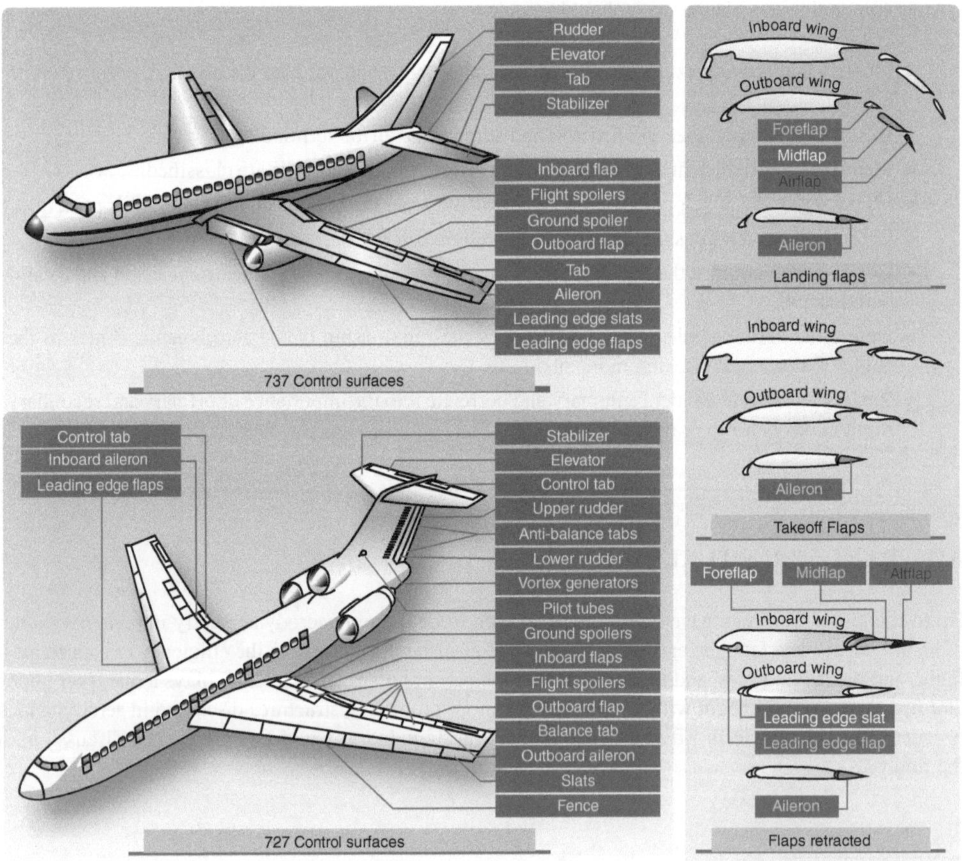

Figure 3.1. Modern transport category control surfaces.
Source: *Pilot's Handbook of Aeronautical Knowledge* (2008a)/U.S. Department of Transportation Federal Aviation Administration/Public domain.

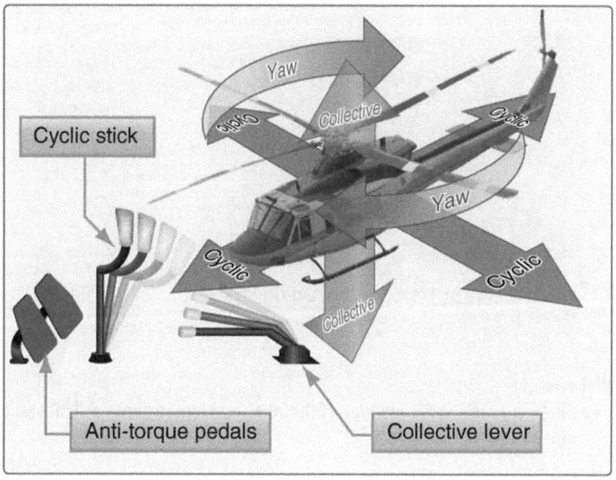

Figure 3.2. Helicopter flight controls.
Source: Pilot's Handbook of Aeronautical Knowledge (2016b)/U.S. Department of Transportation Federal Aviation Administration/Public domain.

Primary Flight Controls

Ailerons

The ailerons are usually located on the trailing edge of each wing closer to the outboard area by the tip. Ailerons control *roll* about the longitudinal axis as they move opposite of each other when the pilot banks left or right with the yoke or stick, with the "up" aileron on the downward moving wing. Some aircraft have ailerons that are connected by mechanical means to the aircraft yoke through cables, bell cranks, pulleys, and/or push–pull tubes. Some jet aircraft have additional ailerons located in the midwing area for high-speed maneuvering, to reduce roll rate. Many advanced aircraft employ digital flight control systems to control the ailerons and other control surfaces. These are known as "fly-by-wire" aircraft and will be discussed in more detail in Chapter 14.

When the yoke/stick is moved to the left, the ailerons deflect in opposite directions, the aileron on the left wing rises and the aileron on the right wing deflects downward. This action results in more lift on the right wing and a resultant roll to the left around the longitudinal axis. Later in Chapter 4, we will explain why this happens.

There are many styles of ailerons used throughout the fixed-wing aircraft industry today, including differential ailerons, frise-type ailerons, and flaperons. Differential ailerons work by deflecting the up aileron more than the down aileron, increasing drag on the downward wing and counteracting adverse yaw (Figure 3.3). Frise-type ailerons project the leading edge of the raised aileron into the wind, increasing drag on the lowered wing to once again minimize the effects of adverse yaw (Figure 3.4). Flaperons combine the control surfaces of the ailerons with the function of the trailing edge flaps, as they can be lowered like flaps but still control bank angle like traditional ailerons. Chapters 12 and 13 in this textbook will discuss these items, as well as adverse yaw, in more depth regarding aircraft stability.

Elevator/Stabilator

An elevator or stabilator controls *pitch* about the lateral axis, allowing for varying angles of attack during flight. An elevator is attached to the trailing edge of the horizontal stabilizer, which is usually fixed to the

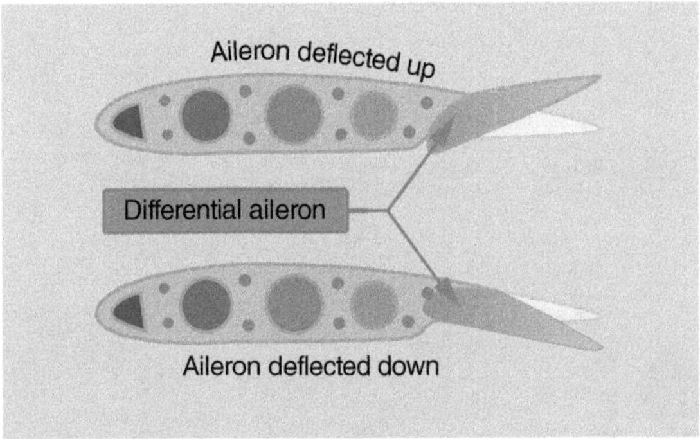

Figure 3.3. Differential ailerons.
Source: Pilot's Handbook of Aeronautical Knowledge (2008a)/U.S. Department of Transportation Federal Aviation Administration/Public domain.

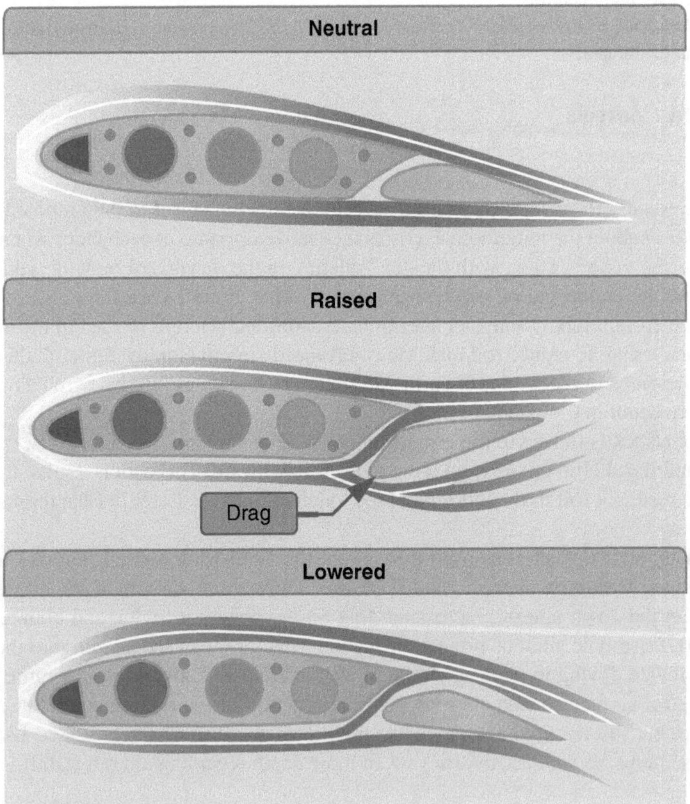

Figure 3.4. Frise-type ailerons.
Source: Pilot's Handbook of Aeronautical Knowledge (2008a)/U.S. Department of Transportation Federal Aviation Administration/Public domain.

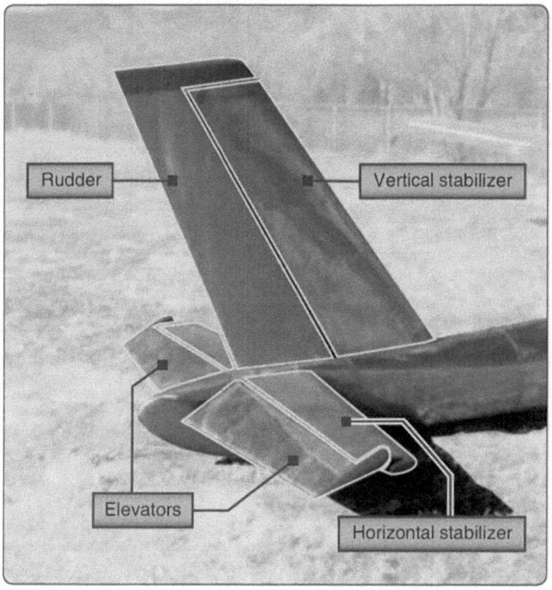

Figure 3.5. Horizontal stabilizer with elevators.
Source: *Glider Flying Handbook* (2013)/U.S. Department of Transportation Federal Aviation Administration/Public Domain.

empennage, sometimes with an angle of incidence built in. A stabilator is a one-piece horizontal stabilizer that moves as a unit around a pivot point in order to allow the pilot to control the angle of attack (AOA) by adjusting the tail-down force resulting in pitch variations of the nose of the aircraft.

The elevator is controlled by the pilot through various mechanical linkages; when the pilot pulls aft on the stick, the elevator forces the tail down, so the nose pitches up, and when the pilot pushes forward the elevator forces the tail up, so the nose goes down. As discussed in Chapter 2, the tail-down force provides a moment that moves the nose of the aircraft around the aircraft's center of gravity (CG). In the example of an up elevator, when the pilot pulls aft on the stick, a larger "camber" is created on the tail and thus a greater AF is created. Figure 3.5 shows an example of an empennage with a fixed horizontal stabilizer with elevators hinged on the trailing edge.

A stabilator essentially works like the elevator, but due to the fact the entire rear horizontal piece is movable, more force is created when the pilot moves the stick fore and aft and sensitivity is increased. This leads to greater chances of the pilot overcontrolling the aircraft; so, components like an antiservo tab and balance weight are added to reduce the sensitivity.

Some larger aircraft incorporate an adjustable horizontal stabilizer controlled by a jackscrew through a wheel in the cockpit or a motor. Although an elevator is still located on the trailing edge, the usually fixed horizontal stabilizer is adjustable, in this case allowing the pilot to move the stabilizer to reduce control pressures on the stick (Figure 3.6).

Elevons combine the pitch control of elevators and the roll control of ailerons and are usually found on the wing trailing edge of tailless aircraft such as flying wings, delta-winged aircraft, and blended-wing-body aircraft. If the pilot pushed down on the stick/yoke the elevons would deflect downward moving the nose up, and if the pilot banked the aircraft the elevons would deflect opposite of each other to perform the action of a traditional aileron. Coordinated movement of the yoke/stick to climb/descend while banking would deflect the elevons in the correct direction for pitch control while offsetting the elevons for the appropriate amount to bank the aircraft to the degree commanded by the pilot.

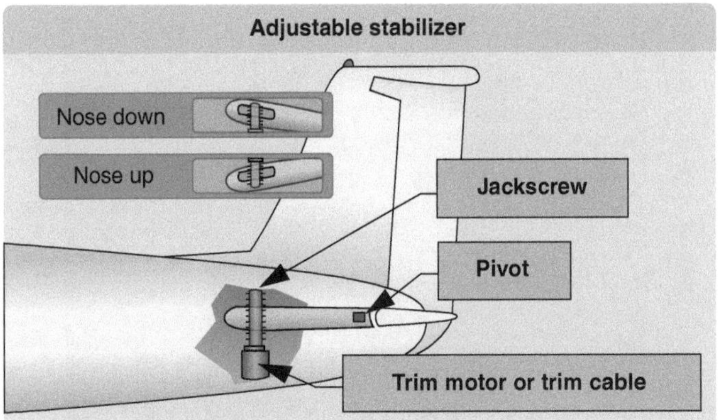

Figure 3.6. Adjustable horizontal stabilizer.
Source: Pilot's Handbook of Aeronautical Knowledge (2008a)/U.S. Department of Transportation Federal Aviation Administration/Public domain.

Rudder

The rudder controls the yaw or movement of the aircraft about its vertical axis. Similar to the smaller, movable elevator attached to the trailing edge of the horizontal stabilizer, the rudder is attached to the rear of the fixed vertical stabilizer. As with other components we discussed, the rudder is connected to the rudder pedals in the cockpit via various mechanical linkages; pressing on the left or right rudder pedal moves the rudder left or right, respectively. Most of the time, the rudder is used to maintain coordinated flight, especially when banking the aircraft. Moving the rudder creates a larger camber on the vertical stabilizer, which in turn creates a greater sideways force. This force causes the nose of the aircraft to move left and right, or yaw, around the vertical axis (Figure 3.7).

Ruddervators

Some aircraft incorporate the action of the rudder and the elevator into a V-tail structure that replaces the traditional horizontal stabilizer and elevator (Figure 3.8). When a pilot pulls back on the yoke to climb both ruddervators deflect upward so the sum forces of the right and left side move the tail down, thus the nose goes up (Figure 3.9b). When the pilot applies rudder pedal pressure the ruddervators deflect opposite to each other. If the pilot applies pressure to the right rudder pedal, the right ruddervator deflects downward (force now up and to the right) and the left ruddervator deflects upward (force now down and to the left), resulting in the tail yawing to the left, thus the nose of the airplane yaws to the right up (Figure 3.9c). If the pilot uses coordinated control inputs during a banked climb or descent then the ruddervators will defect opposite of each other but not at the same rate depending on the amount of bank or climb/descent required.

Canard

The canard design dates back to the glider designs of the Wright Flyer, and can still be found on a few modern civilian and military aircraft, and on many homebuilt aircraft. Although few were made, the Beechcraft Starship is a great example of an aircraft with a canard design (Figure 3.10). Most canard designs incorporate a horizontal stabilizer in front of the main wings, providing positive lift for the nose instead of the previously presented aft horizontal stabilizer, which provides the tail-down force. One of the advantages of

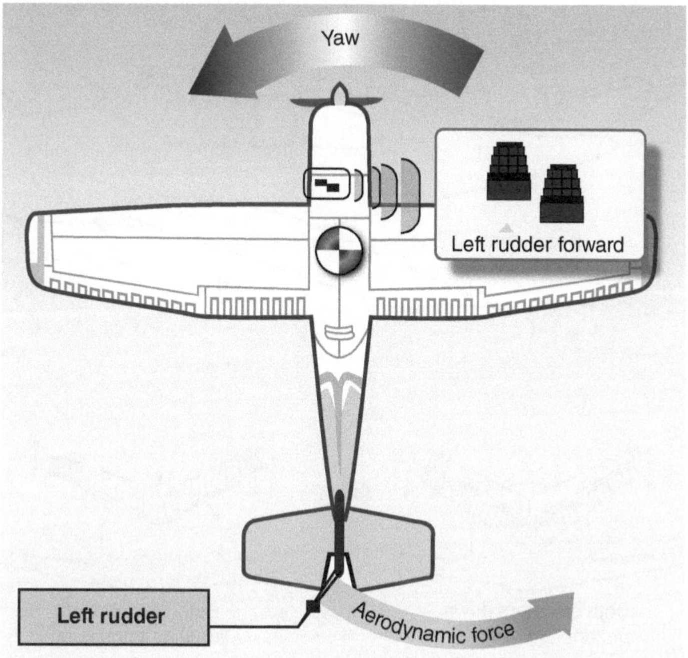

Figure 3.7. Rudder movement.
Source: Pilot's Handbook of Aeronautical Knowledge (2008a)/U.S. Department of Transportation Federal Aviation Administration/Public domain.

Figure 3.8. Ruddervators on a Beechcraft Bonanza.
Source: U.S. Department of Transportation (2023b)/Public Domain.

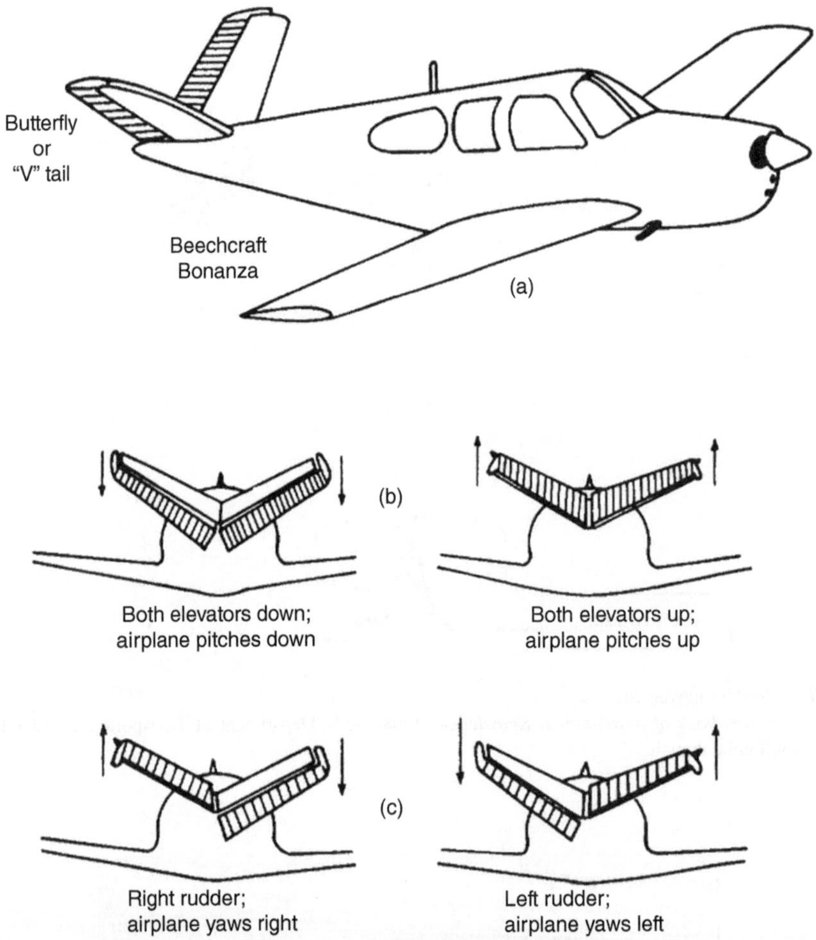

Figure 3.9. Ruddervator movement. (a) Aircraft in straight and level flight. (b) Aircraft nose up and nose down. (c) Aircraft yaw to the left and right.
Source: NASA-SP-367/Public Domain.

a canard is the additional lift provided by the canard at high angles of attack for the main wing lowering stall speed. Disadvantages include an often-required larger main wing, reduction in additional lift provided by the propeller slipstream during takeoff versus aircraft with a horizontal stabilizer (single-engine), reduced deep stall recovery, and in some aircraft decreased directional stability due to an extended fuselage ahead of the CG needed to accommodate the canard.

Secondary Flight Controls

Secondary flight control systems usually consist of wing flaps, leading edge devices, spoilers, and trim systems. These controls often support or supplement the primary controls, and their importance to understanding aerodynamic principles cannot be overstated.

Figure 3.10. Canards on a Beechcraft Starship.
Source: U.S. Department of Transportation (2023a)/Public Domain.

Flaps

The flaps are the most common high-lift devices used on aircraft, and their contribution to the amount of lift an airfoil can produce will be discussed in more detail in Chapter 4. For our discussion here, we will review their location on the aircraft, as well as the basic flap designs on the aircraft today.

Flaps for this discussion are considered to be on the trailing edge of the wing, usually inboard close to the fuselage, and are referred to as trailing edge flaps. These surfaces contribute to the camber of the wing airfoil in most cases, as well as to the area of the wing in other cases. By increasing the AOA of the wing, and in some cases the area of the wing, they allow the aircraft to fly at lower speeds, which may be needed for takeoff and landing, or for situations requiring increased maneuverability. Flaps increase the lift as well as the drag, and it is generally considered that 15° or less of flaps produce more lift than drag, and any flap setting over 15° drag exceeds lift. For this discussion, we will look at four types of flaps: plain, split, slotted, and Fowler; and as you can see, there are designs where the benefit of one design is combined with another (Figure 3.11).

The plain flap is the simplest design, and illustrates the advantages and purpose of flaps. As the flaps are deployed, the camber of the wing increases, and the lift coefficient (C_L) increases accordingly with the AOA. The lift coefficient will be discussed at length in Chapter 4. The farther the flaps are deployed, the greater the lift and the resulting drag. A split flap is deployed from underneath the wing, and results in more drag initially than the plain flap due to the disruption of the flow of air around the bottom and top of the wing.

When the flaps are slotted, at high angles of attack, high-energy air is allowed to move through the slot and energize the air on top of the deployed flap. This allows for an increase in C_L at lower speeds, allowing an aircraft to operate out of shorter landing strips or with obstacles surrounding the airport. The highly energized air also delays boundary layer separation, which lowers the stalling speed, improving performance at slow speeds. More on this topic will be discussed throughout this textbook.

Fowler flaps are commonly found on larger transport category aircraft, as they are heavier than the other flap designs and incorporate more complex systems to operate. Fowler flaps slide out and back from the

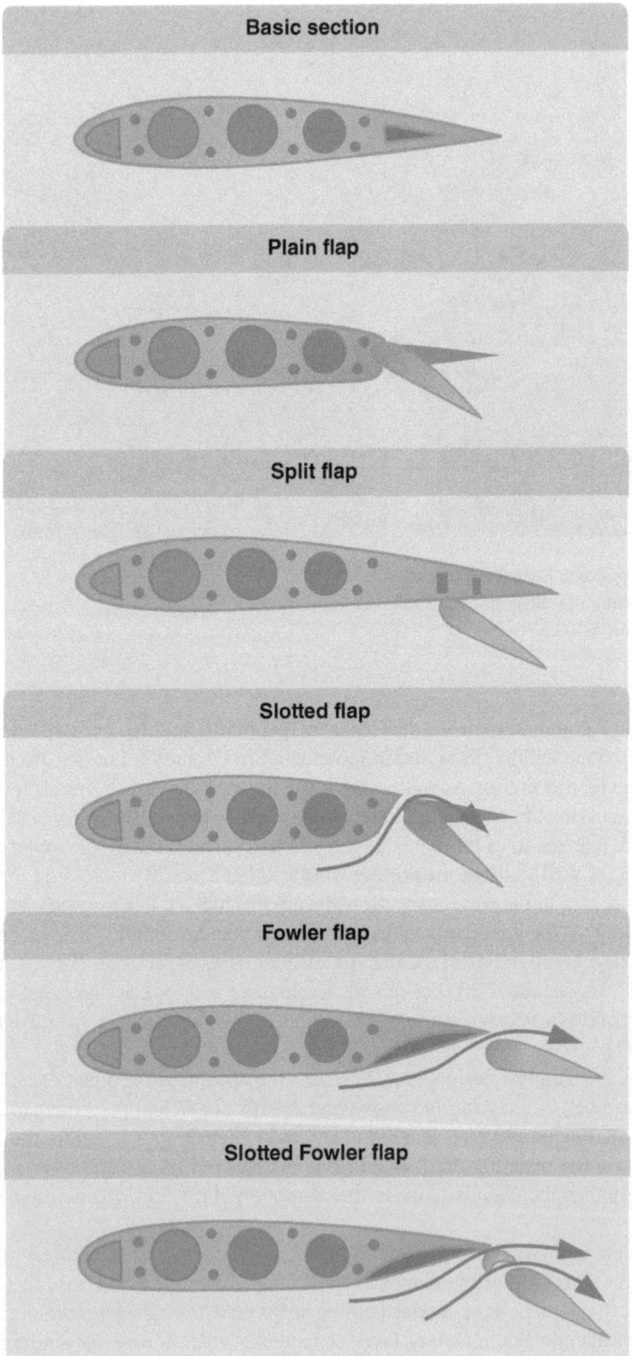

Figure 3.11. Common flap designs.
Source: Pilot's Handbook of Aeronautical Knowledge (2008a)/U.S. Department of Transportation Federal Aviation Administration/Public domain.

wing, which offers the benefit of not only increasing the camber of the wing but also of the wing area. Fowler flaps also double as slotted flaps in that they allow higher-energy air from beneath the wing to flow over the deployed flap area. Cessna high-wing, single-engine aircraft are the best example of the use of Fowler flaps on light aircraft.

Leading Edge Devices

Leading edge devices are discussed in more depth in Chapter 4; so, only a brief introduction will be presented here. As with trailing edge flaps, leading edge devices such as slots, slats, leading edge flaps (Krueger flaps), and leading-edge cuffs work to increase the C_L over a wing, allowing the aircraft to fly at slower speeds for takeoff and landing. In some cases, the camber of the wing is also increased, which allows for an increase in the AOA and increased performance. As with trailing edge flaps, the first setting results in more lift with little increase in drag, while fully deployed settings usually result in mostly drag with little lift.

Spoilers and Speed Brakes

Spoilers are used on many different types of aircraft as high-drag devices used to increase drag and reduce lift. They "spoil" the smooth air around a wing, and may be used in flight, or on the ground when landing or during a rejected takeoff. Figure 3.12 is an example of deploying ground spoilers to reduce lift over the wing and increase the effectiveness of the brakes. Some aircraft use the spoilers in flight for roll control at higher airspeeds when the action of the ailerons would be too much of a force. Sometimes spoilers are used as speed brakes to reduce lift on both wings, which allows the aircraft to descend without increasing airspeed. Gliders utilize spoilers to control their descent rate upon landing. Figure 3.13 shows various examples of speedbrakes for both transport category and military aircraft. In these examples, the speed brakes are used to control speed in a descent, whether on arrival or during the landing approach.

Figure 3.12. Ground spoilers deployed.
Source: U.S. Department of Transportation (2016b)/Public Domain.

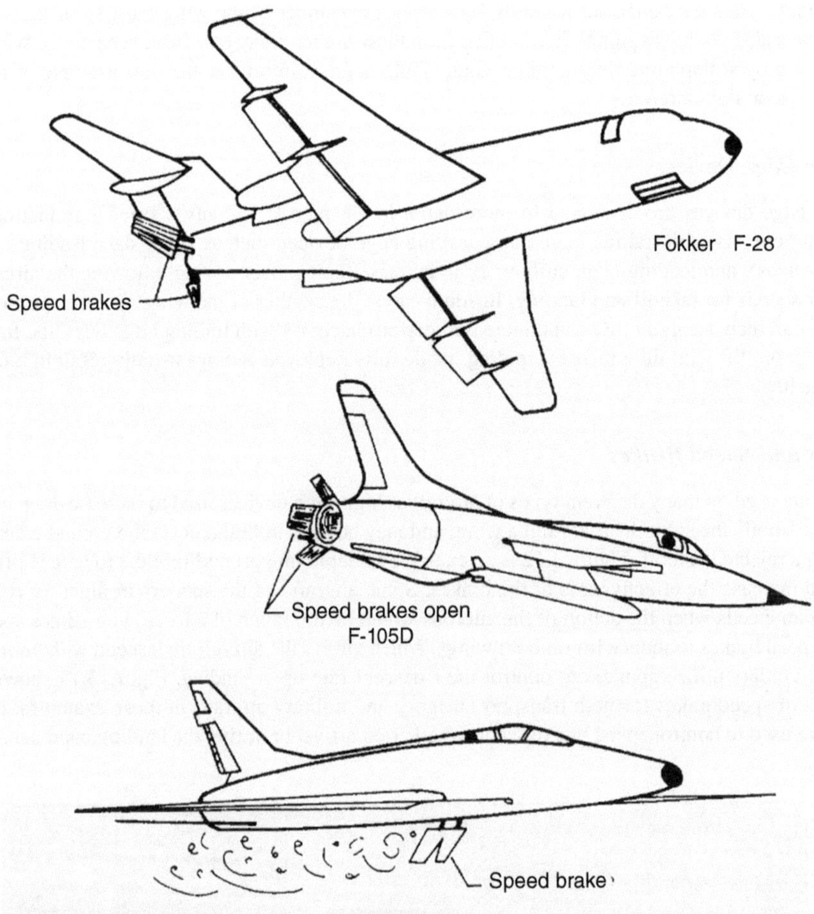

Figure 3.13. Speed brakes.
Source: NASA SP-367/Public Domain.

Trim Systems and Tabs

Trim systems are designed to alleviate the pressures on the primary flight controls as experienced by the pilot during aircraft operation. Usually located on the trailing edge of these devices, the pilot (or autopilot) operates the respective trim system in order to position the flight control where minimum pressures are exerted in the system. The two most common trim systems are trim tabs and antiservo tabs.

A trim tab is usually found on the trailing edge of the elevator or rudder, and unlike the antiservo tab, the trim tab moves in the opposite direction to that of the primary control. Figure 3.14 shows examples of trim tabs on various primary flight controls of a high-wing airplane such as a C172. An antiservo tab is found on the trailing edge of a stabilator and moves in the same direction as the primary control to which it is attached (Figure 3.15). As previously mentioned, the antiservo tab serves to reduce the sensitivity of the stabilator during the pitch to the pilot, as well as reducing control pressure as needed.

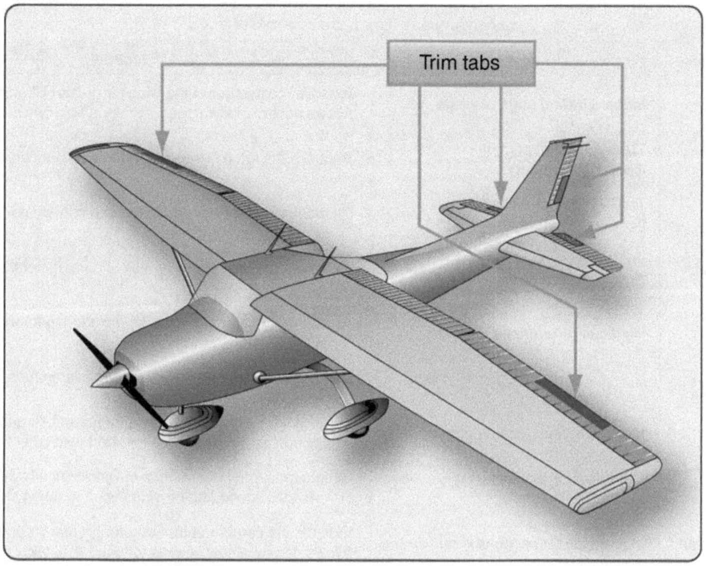

Figure 3.14. Trim tabs on primary flight controls.
Source: Aviation Maintenance Technician Handbook – Airframe (2023a)/U.S. Department of Transportation Federal Aviation Administration/Public domain.

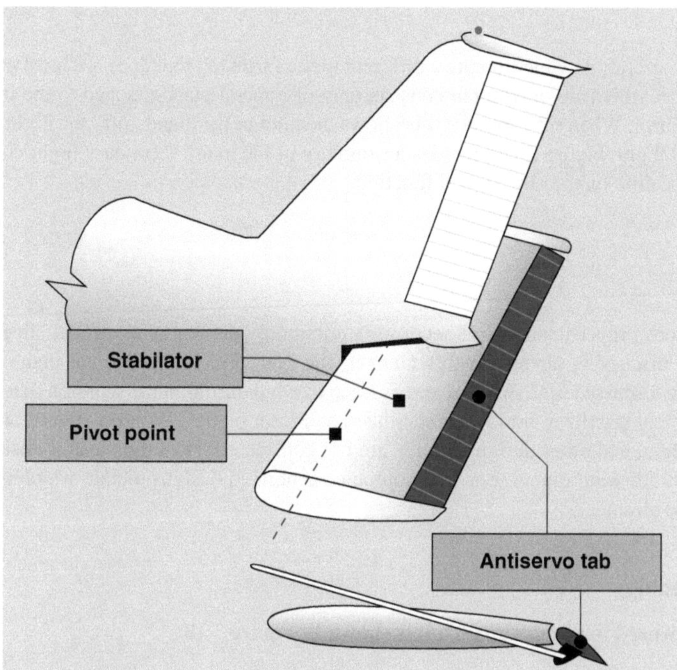

Figure 3.15. Antiservo tab.
Source: Pilot's Handbook of Aeronautical Knowledge (2008a)/U.S. Department of Transportation Federal Aviation Administration/Public domain.

Secondary/auxiliary flight control surfaces		
Name	**Location**	**Function**
Flaps	Inboard trailing edge of wings	Extends the camber of the wing for greater lift and slower flight. Allows control at low speeds for short field takeoffs and landings.
Trim tabs	Trailing edge of primary flight control surfaces	Reduces the force needed to move a primary control surface.
Balance tabs	Trailing edge of primary flight control surfaces	Reduces the force needed to move a primary control surface.
Anti-balance tabs	Trailing edge of primary flight control surfaces	Increases feel and effectiveness of primary control surface.
Servo tabs	Trailing edge of primary flight control surfaces	Assists or provides the force for moving a primary flight control.
Spoilers	Upper and/or trailing edge of wing	Decreases (spoils) lift. Can augment aileron function.
Slats	Mid to outboard leading edge of wing	Extends the camber of the wing for greater lift and slower flight. Allows control at low speeds for short field takeoffs and landings.
Slots	Outer leading edge of wing forward of ailerons	Directs air over upper surface of wing during high angle of attack. Lowers stall speed and provides control during slow flight.
Leading edge flap	Inboard leading edge of wing	Extends the camber of the wing for greater lift and slower flight. Allows control at low speeds for short field takeoffs and landings.
Note: an aircraft may possess none, one, or a combination of the above control surfaces.		

Figure 3.16. Secondary control surfaces and their location.
Source: U.S. Department of Transportation Federal Aviation Administration (2012a)/U.S. Department of Transportation/Public domain.

Some smaller aircraft may incorporate a different type of trim tab that does not involve an independent mechanical linkage, this type of trim tab consists only of a metal plate attached to the trailing edge of a primary flight control. When set correctly, this allows the pilot to fly "hands-off" the flight control while in straight and level flight. Figure 3.16 provides a summary of the many secondary flight controls available, including their location on the aircraft and function.

AIRFOILS

An *airfoil* or, more properly, an airfoil section, is commonly shown as a vertical slice of a wing (see Figure 3.17). In discussing airfoils in this chapter, the planform (or horizontal plane) of the wing is ignored as will be discussed in Chapters 4 and 5. Keep in mind during this discussion that an airfoil is also found on the vertical stabilizer, horizontal stabilizer, and rotor blades. Wingtip effects, sweepback, taper, wash/out or wash/in, and other design features are not considered. No single airfoil design is perfect for every flight situation: wind tunnel tests and computer-generated designs dictate which airfoil is best for the mission of the aircraft.

Airfoil Terminology

The terminology used to discuss an airfoil is shown in Figure 3.18:

1. *The chord line* is a straight line connecting the leading edge and the trailing edge of the airfoil.
2. *The chord* is the length of the chord line. All airfoil dimensions are measured in terms of the chord.

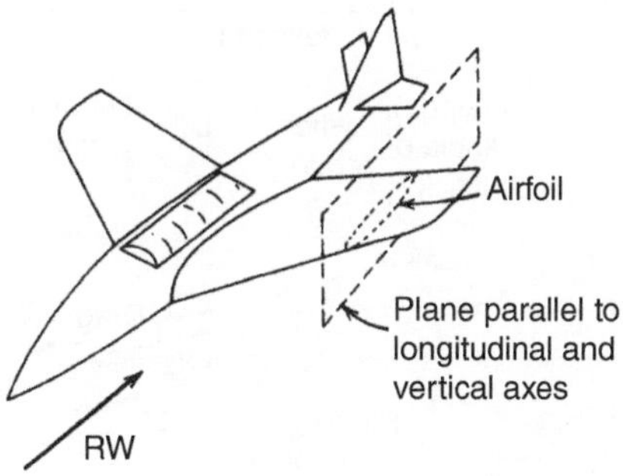

Figure 3.17. Airfoil section.

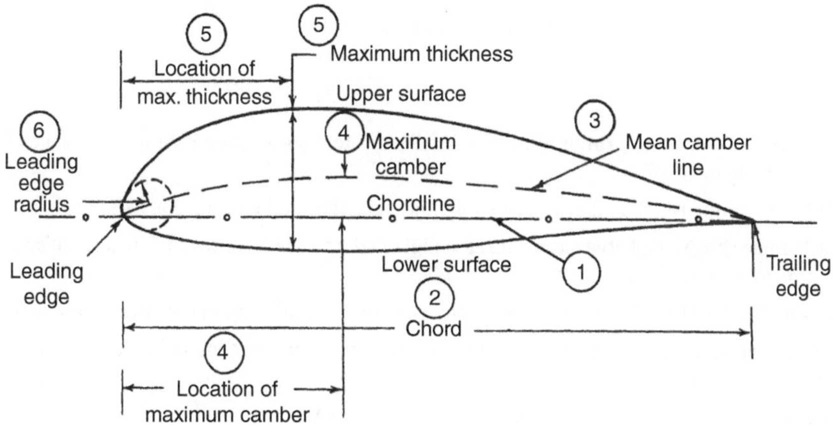

Figure 3.18. Airfoil terminology.

3. *The mean camber line* is a line drawn equidistant between the upper surface and the lower surfaces.
4. *Maximum camber* is the maximum distance between the mean camber line and the chord line. The location of the maximum camber is important in determining the aerodynamic characteristics of the airfoil.
5. *Maximum thickness* is the maximum distance between the upper and lower surfaces, and its location of maximum thickness will also be important when determining aerodynamic characteristics.
6. *The leading edge radius* is a measure of the sharpness of the leading edge. It may vary from zero for a knife-edge supersonic airfoil to about 2% (of the chord) for rather blunt leading edge airfoils.

Definitions

Flight path velocity: The speed and direction of a body passing through the air.

Relative wind (RW): The speed and direction of the air impinging on a body passing through it. It is equal and opposite in direction to the flight path vector.

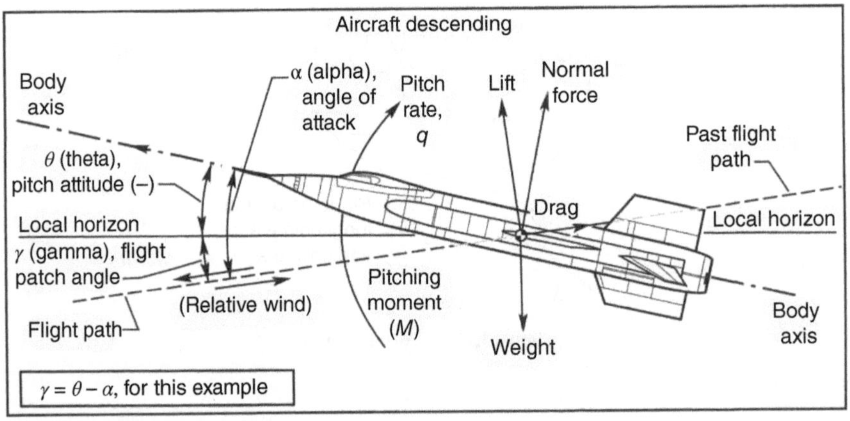

Figure 3.19. Longitudinal terms with the aircraft descending.
Source: NASA/Public Domain.

Angle of attack (AOA or α, pronounced *alpha*): The acute angle between the relative wind (RW) and the chord line of an airfoil.

Pitch angle (θ): Angle between the aircraft's longitudinal axis and the horizontal axis.

Flight path angle (γ): Angle between an aircraft's flight path vector and horizontal. It is equal and opposite in direction to the RW (Figure 3.19).

Aerodynamic force (AF): The net resulting static pressure multiplied by the planform area of an airfoil.

Aerodynamic center (AC): The point on an airfoil where the pitching moment is constant with changing AOA if the velocity is constant.

Lift: The component L of the AF that is perpendicular to the RW.

Drag: The component D of the AF that is parallel to the RW.

Center of pressure (CP): The point on the chord line where the AF acts.

Laminar flow or streamlined flow: Smooth airflow with little transfer of momentum between parallel layers.

Turbulent flow: Airflow where the streamlines break up and there is much mixing of the layers.

Geometry Variables of Airfoils

There are four main variables in the geometry of an airfoil:

1. Shape of the mean camber line
2. Thickness
3. Location of maximum thickness
4. Leading edge radius

In a cambered airfoil (nonsymmetrical), the mean camber line and chord line are different. If the mean camber line coincides with the chord line, the airfoil is said to be symmetrical. Figure 3.20 shows the

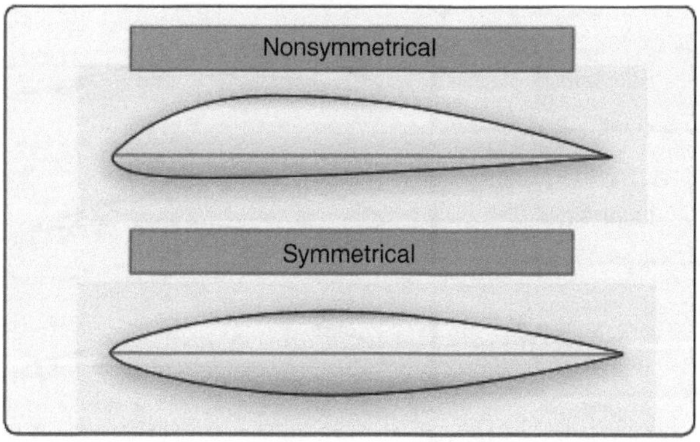

Figure 3.20. Cambered versus symmetrical airfoil.
Source: Helicopter Flying Handbook (2019)/Federal Aviation Administration/Public domain.

difference between a cambered and symmetrical airfoil, note the position of the chord line. In symmetrical airfoils, the upper and lower surfaces have the same shape and are equidistant from the chord line. Symmetrical airfoils are common in rotary-wing blades and in some aerobatic aircraft. Figure 3.21 shows examples of early airfoil design to more modern, supersonic designs.

Classification of Airfoils

Most airfoil development in the United States was done by the National Advisory Committee for Aeronautics (NACA), pronounced N-A-C-A, starting in 1929 with the use of wind tunnels. NACA was the forerunner of the National Aeronautics and Space Administration (NASA). Modern airfoil design begins with mathematical models based on flight characteristics of previously utilized designs. Computational fluid dynamics (CFD) software is then used to refine the design characteristics until high-performance wind tunnels are introduced to validate results.

From the beginning, the NACA developed families of airfoils based on the geometry variables of the airfoil discussed in the previous section, by adjusting the variables engineers were able to modify airfoil lift, drag, critical Mach number, and the desired pitching moments. Critical Mach number (supersonic flow begins somewhere on the aircraft, usually as a shock wave on top of the wing) along with airfoil design to mitigate shock wave development will be further defined and discussed in more detail in Chapter 14. The first series of airfoils investigated was the "four-digit" series. The first digit gives the amount of camber, in the percentage of chord. The second digit gives the position of maximum camber, in tenths of chord, and the last two give the maximum thickness, in percentage of chord. For example, a NACA 2415 airfoil has a maximum camber of 2% C, located at 40% C (measured from the leading edge), and has a maximum thickness of 15% C. A NACA 0015 airfoil is a symmetrical airfoil (has zero camber) and has a thickness of 15% C.

Further development led to the "five-digit" series, the "1-series," and, with the advent of higher speeds, to the so-called *laminar flow* airfoils. The NACA's 23000 series created in 1935 were very popular and are still in use today. The laminar flow airfoils are the "6-series" and "7-series" airfoils, designed to achieve higher maximum lift coefficients with lower drag. In the 6-series, the first digit indicates the series and the second gives the location of minimum pressure in tenths of chord. The third digit represents the design lift coefficient in tenths, and the last two digits (as in all NACA airfoils) show the thickness in percentage of chord. For example, NACA 64-212 is a minimum pressure of 40% C, a design lift coefficient of 0.2, and a

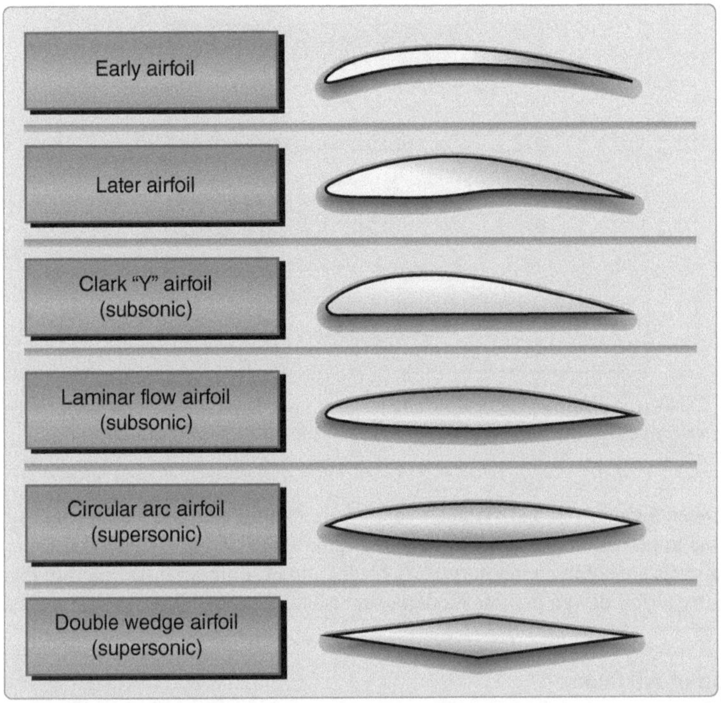

Figure 3.21. Examples of airfoil design.
Source: Aviation Maintenance Technician Handbook – Powerplant (2018)/Federal Aviation Administration/Public domain.

thickness of 12% C. One notable characteristic of the 6-series airfoils is the decrease in drag is usually only verified through a small range of angles of attack, and any deformation or surface containments can lead to a reduction in laminar flow and marked increase in drag. The laminar flow generated by the 6-series and 7-series airfoils results from moving the maximum thickness back and reducing the leading edge radius. Two things happen with this treatment:

1. The point of minimum pressure is moved backward, thus increasing the distance from the leading edge that laminar (smooth) airflow exists, which reduces drag.
2. The critical Mach number is increased, thus allowing the airspeed of the aircraft to be increased without encountering compressibility problems.

Sketches of NACA subsonic airfoil series are shown in Figure 3.22. NASA's FoilSim interactive simulation software developed by the Glenn Research Center comprises an excellent tool to explore the various shapes of airfoils and how air flows around them when airfoil characteristics are manipulated.

Figure 3.23 depicts a comparison of a cambered airfoil (NACA 2412) used on the wings of a C172 with a symmetrical airfoil (NACA 0015), typical of early rotorcraft rotor development. As you can see with the NACA 2412 cambered airfoil the mean camber line is located separately from the chord line, while the NACA 0015 airfoil illustrates that the mean camber line and chord line are located in the same position.

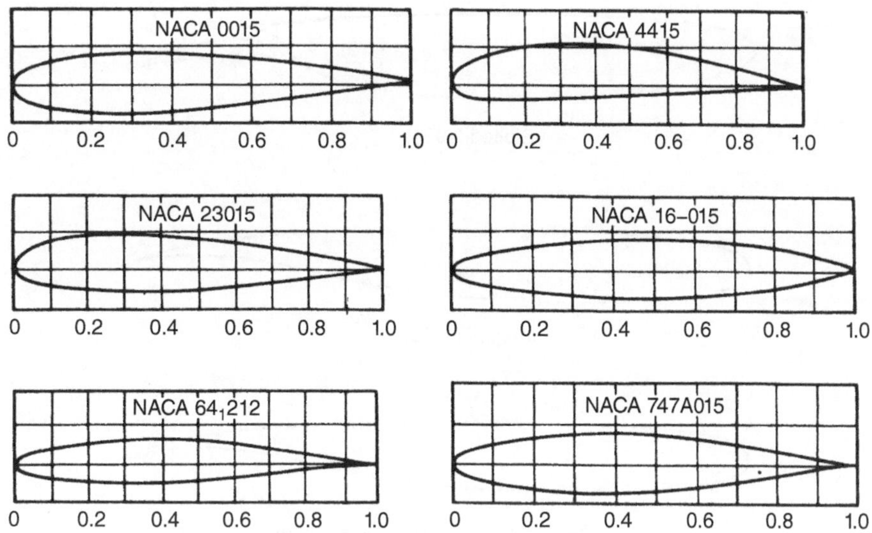

Figure 3.22. NACA airfoils (NACA data).

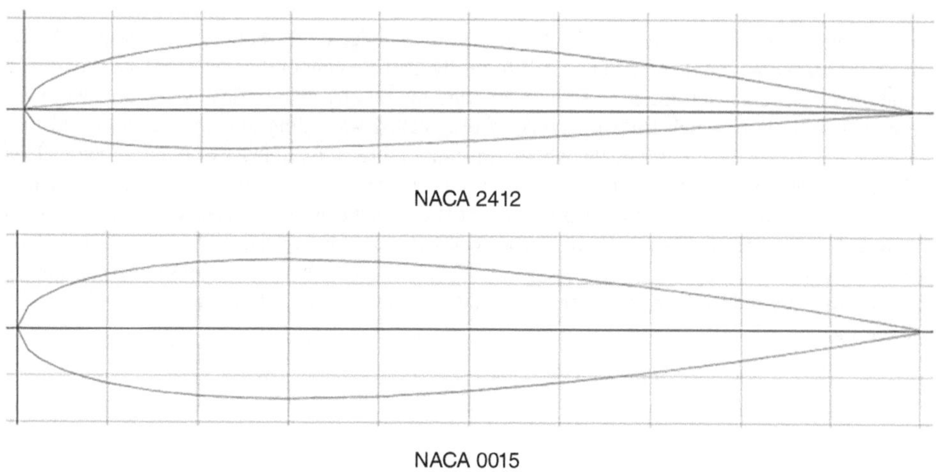

Figure 3.23. Cambered and symmetrical airfoil comparison.

Modern airfoil design using the latest CFD software still uses the same geometric variables of the airfoil to determine design characteristics, but with faster airspeeds and new structural technology such as composites used to build airfoils and high-lift devices, there are updated designs that could not have been thought of in the days of the NACA. A modern design used worldwide on corporate, military, and air transport aircraft is the *supercritical airfoil*, which is flatter on top and more rounded on the bottom than

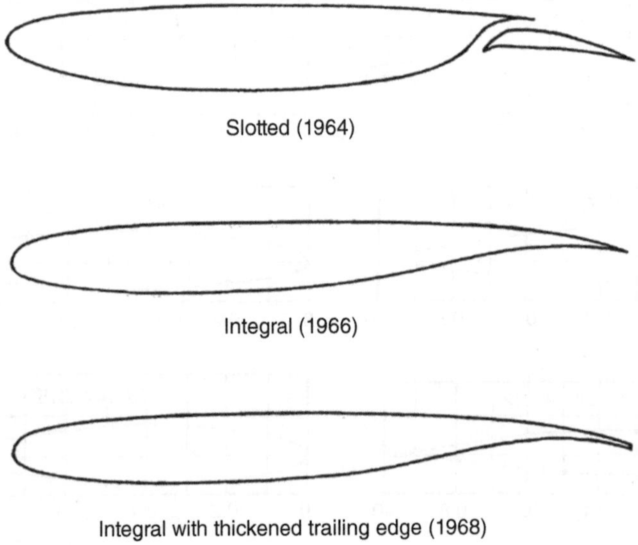

Figure 3.24. Development of the supercritical airfoil.
Source: NASA Supercritical Airfoils/Public Domain.

a conventional wing, and has a cusp located near the trailing edge on the bottom of the wing. The upper trailing edge has a downward curve to restore lift lost by the flattening of the upper surface (Figure 3.24). As modern airfoils approach the supersonic realm and the onset of the critical Mach number, a shock wave begins to develop which leads to a type of drag called wave drag that will be discussed further in Chapter 14. To delay the onset of this shock wave and mitigate the impact of the drag thus allowing the aircraft to travel faster, the supercritical airfoil reduces the strength of the shock wave compared to cambered airfoils. Airfoils that have design characteristics beneficial to slow-speed flight are not usually well-suited for high-speed subsonic flight. As we will expand upon in Chapter 14, airfoils designed for supersonic flight (at or greater than the speed of sound) are usually thinner and have sharper leading edges as opposed to more rounded leading edges as would be found on the airfoils of a Cessna 172 or Piper Cherokee 140.

Application 3.1

Symmetrical airfoils are found on many different types of aircraft, from light to heavy aircraft, and within general aviation to those designed solely for military operations. Many general aviation training aircraft may offer a cambered airfoil for the wing design but symmetrical airfoils for the horizontal and vertical stabilizers.

Identify various aircraft that incorporate symmetrical airfoils into their design. Where are these airfoils located on the aircraft, and why was the symmetrical airfoil design utilized instead of a cambered airfoil design?

DEVELOPMENT OF FORCES ON AIRFOILS

Leonardo da Vinci stated the cardinal principle of wind tunnel testing nearly 400 years before the Wright brothers achieved powered flight. Near the beginning of the sixteenth century, da Vinci said: *the action of the medium upon a body is the same whether the body moves in a quiescent medium, or whether the particles of the medium impinge with the same velocity upon the quiescent body.* This principle allows us to consider only the relative motion of the airfoil and the air surrounding it. We may use such terms as "airfoil passing through the air" and "air passing over the airfoil" interchangeably.

Pressure Disturbances on Airfoils

If an airfoil is subjected to a moving airflow, velocity and pressure changes take place that create pressure disturbances in the airflow surrounding it. These disturbances originate at the airfoil surface and propagate in all directions at the speed of sound. If the flight path velocity is subsonic, the pressure disturbances that are moving ahead of the airfoil affect the airflow approaching the airfoil (Figure 3.25). As the arrow indicates there are slight changes with the streamlines ahead of the leading edge of the airfoil.

Velocity and Static Pressure Changes About an Airfoil

The air approaching the leading edge of an airfoil is first slowed down and then speeds up again as it passes over or beneath the airfoil. Figure 3.26 compares two local velocities with the flight path velocity V_1 and with each other. As the velocity changes, so does the dynamic pressure "*q*" and, according to Bernoulli's principle, so does the static pressure "*P*." Air near the stagnation point at the leading edge of the wing has slowed down, so the static pressure in this region is higher than the ambient static pressure. Air that is passing above and below the airfoil, and thus has speeded up to a value higher than the flight path velocity (both V_2 and V_3 are greater than V_1), will produce static pressures that are lower than ambient static pressure. So as "*q*" increases, "*P*" static pressure decreases and a greater pressure differential is realized.

At a point near maximum thickness, maximum velocity and minimum static pressure will occur, note that V_2 is greater than V_3. This difference in velocity results in pressure differential, and is due to the cambered nature of the airfoil, increased AOA, or both. If the airfoil was a symmetrical airfoil at zero AOA

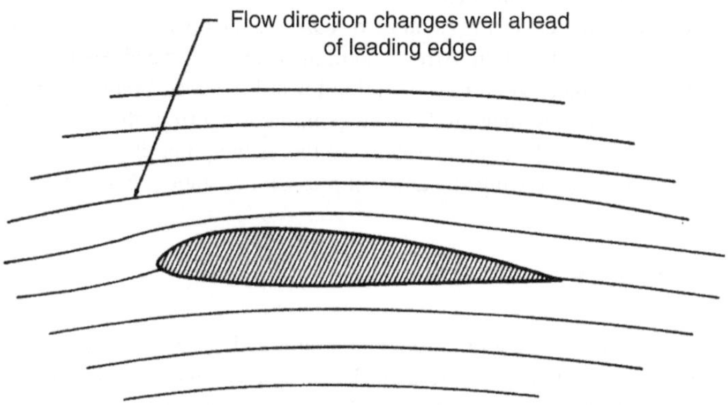

Figure 3.25. Effect of pressure disturbances on airflow around an airfoil.

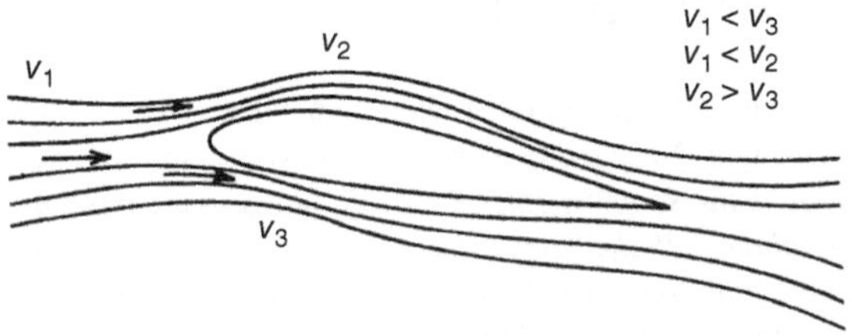

Figure 3.26. Velocity changes around an airfoil.

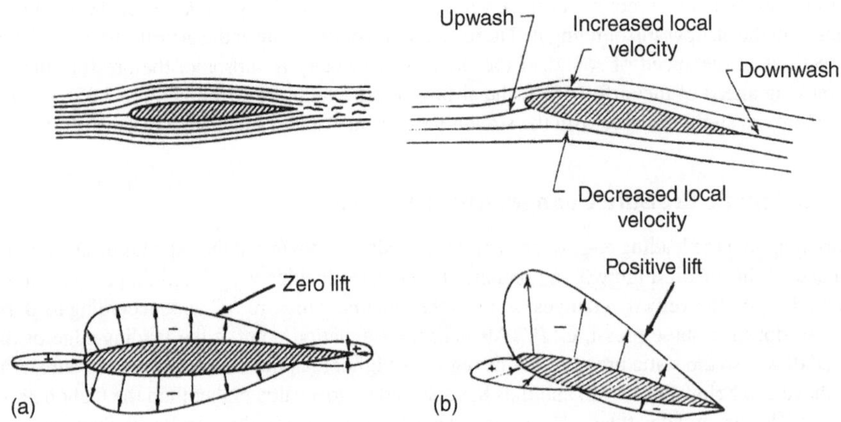

Figure 3.27. Static pressure on an airfoil (a) at zero AOA, and (b) at a positive AOA.

then the pressure would be equal on both the top and the bottom of the wing. Because air has viscosity, some of its energy will be lost to friction and a "wake" of low-velocity, turbulent air exists near the trailing edge, resulting in a small, high-pressure area.

Figure 3.27 shows a symmetrical airfoil (a) at zero AOA and the resulting pressure distribution (b) at a positive AOA and its pressure distribution. Arrows pointing away from the airfoil indicate static pressures that are below ambient static pressure; arrows pointing toward the airfoil indicate pressures higher than ambient. Note in (b) that the largest magnitude arrow pointing away from the symmetrical airfoil is near the point of the highest velocity (dynamic pressure), so this is the point of greatest pressure differential between the static pressure over the wing and ambient pressure, as static pressure is at its lowest point.

AERODYNAMIC FORCE

Aerodynamic force (AF) is the resultant of all static pressures acting on an airfoil in an airflow multiplied by the planform area that is affected by the pressure. The line of action of the AF passes through the chord line at a point called the CP. It is convenient to consider that the forces acting on an aircraft, or an airfoil, do so in some rectangular coordinate system. One such system could be defined by the longitudinal and vertical axes of an aircraft. Another could be defined by axes parallel to and perpendicular to the earth's

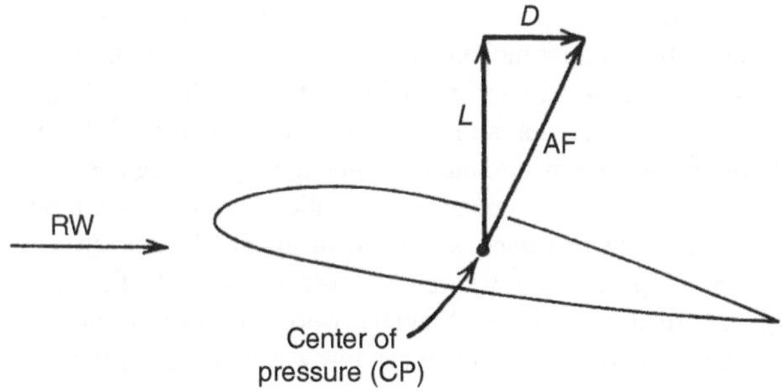

Figure 3.28. Components of aerodynamic force.

surface. A third rectangular coordinate system is defined by the RW direction and an axis perpendicular to it. This last system is chosen to define lift and drag forces. AF is resolved into two components: one parallel to the RW, called *drag* (*D*), and the other perpendicular to the RW, called *lift* (*L*). Figure 3.28 shows the resolution of AF in its components *L* and *D*.

Pressure Distribution on a Rotating Cylinder

A stationary (nonrotating) cylinder is located in a wind tunnel as shown in Figure 3.29a. The cylinder is equipped with static pressure taps. These measure the local static pressure with respect to the ambient static pressure in the test chamber. When the tunnel is started, the airflow approaches the cylinder from the left as shown by the RW vector. Arrows pointing toward the cylinder show pressures that are higher (+) than ambient static pressure; arrows pointing away from the cylinder show pressures that are less (−) than ambient static pressure. Figure 3.29a shows that the upward forces are resisted by the downward forces and no net vertical force (lift) is developed by the cylinder.

Now consider if the wind tunnel is stopped, and the cylinder begins to rotate in a motionless fluid. We begin to see the factors of viscosity and friction at work. The more viscous the fluid, the more it is resistant to flow, and because air has viscosity properties it will resist flow. Similar to a wing, the surface of the cylinder has some "roughness" to it, so as the cylinder turns some molecules adhere to the surface. The closer to the surface of the cylinder (airfoil), the greater the possibility the molecules are drawn in a

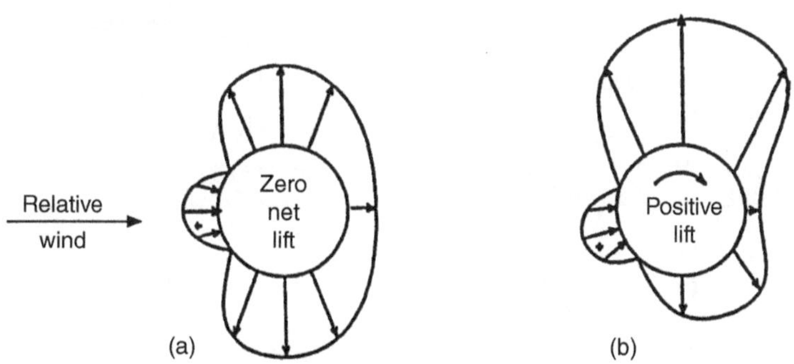

Figure 3.29. Pressure forces on (a) nonrotating cylinder and (b) rotating cylinder.

clockwise direction by viscosity, so now substituting air we see the velocity increase in the direction of rotation above the cylinder. This circular movement of the air is called *circulation*.

Finally, let us consider a rotating cylinder in a moving fluid as the cylinder continues rotating in the clockwise direction when the wind tunnel is once again started (Figure 3.29b). The air passing over the top of the cylinder will be speeded up by circulation, while the air passing over the bottom of the cylinder will be retarded. According to Bernoulli's equation, the static pressure on the top will be reduced and the static pressure on the bottom will be increased, similar to an airfoil with a positive AOA. The new pressure distribution will be shown in Figure 3.29b, where a low-pressure area produces an upward force. This is called the *Magnus effect*, named after Gustav Magnus, who discovered it in 1852. It explains why you *slice* (or *hook*) your golf ball or why a good pitcher can throw a curve. The principles of the Magnus effect are even utilized in space as a spinning satellite can maintain its orbit longer due to the Magnus force, minimizing orbital decay.

AERODYNAMIC PITCHING MOMENTS

Consider the pressure distribution of a symmetrical airfoil at zero AOA (Figure 3.30a). The large arrows show the sum of the low pressures on the top and bottom of the airfoil. They are at the CP of their respective surfaces. The CP on the top of the airfoil and the CP on the bottom are located at the same point on the chord line. The large arrows indicate that the entire pressure on the top and bottom surfaces is acting at the CP. Because these two forces are equal and opposite in direction, no net lift is generated. Note also that the lines of action of these forces coincide, so there is no unbalance of moments about any point on the airfoil. Figure 3.30b shows the pressure distribution of a symmetrical airfoil at a positive AOA. There is now an imbalance in the upper surface and lower surface lift vectors, and positive lift is being developed. However, the two lift vectors still have the same line of action, passing through the CP. There can be no moment developed about the CP. We can conclude that symmetrical airfoils do not generate pitching moments at any AOA. It is also true that the CP does not move with a change in AOA for a symmetric airfoil.

Now consider a cambered airfoil operating at an AOA where it is developing no net lift (Figure 3.31a). Upper surface lift and lower surface lift are numerically equal, but their lines of action do not coincide. A nose-down pitching moment develops from this situation. When the cambered airfoil develops positive lift (Figure 3.31b), the nose-down pitching moment still exists. By reversing the camber, it is possible to create an airfoil that has a nose-up pitching moment. Delta-wing aircraft have a reversed camber trailing edge to control the pitching moments.

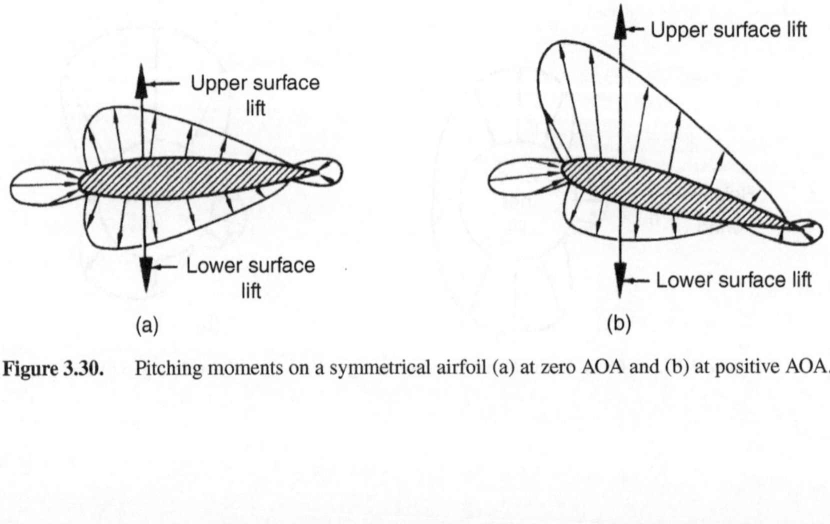

Figure 3.30. Pitching moments on a symmetrical airfoil (a) at zero AOA and (b) at positive AOA.

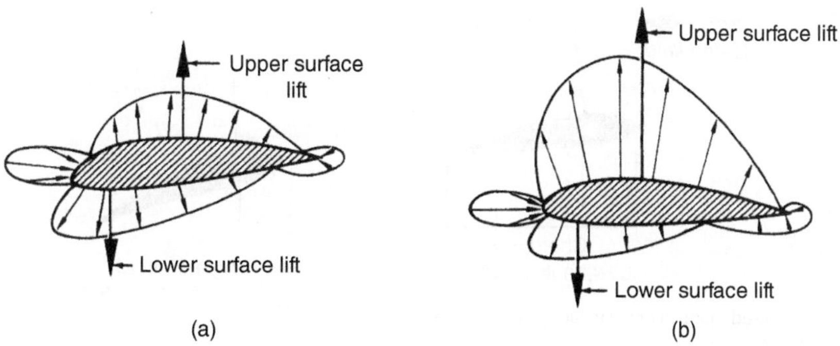

Figure 3.31. Pitching moments on a cambered airfoil: (a) zero lift and (b) developing lift.

Aerodynamic pitching moments also occur when the pilot changes the camber of an airfoil during flight, such as retracting or deploying flaps. When the trailing edge flaps are moved, the chord line changes resulting in a new AOA. When the AOA moves a pitching moment may develop, with initial flap settings, the airplane may initially "balloon" due to the immediate increase in lift, then a general nose-down pitching moment will develop. As shown in Figure 3.32, care must be taken during large power changes and significant trim settings (go-around/missed approach) as increased pitching moments may be experienced. In some aircraft, a significant nose-up pitching moment may occur with full deployment of trailing edge flaps as downwash over the horizontal stabilizer increases tail-down force (airplane nose goes up).

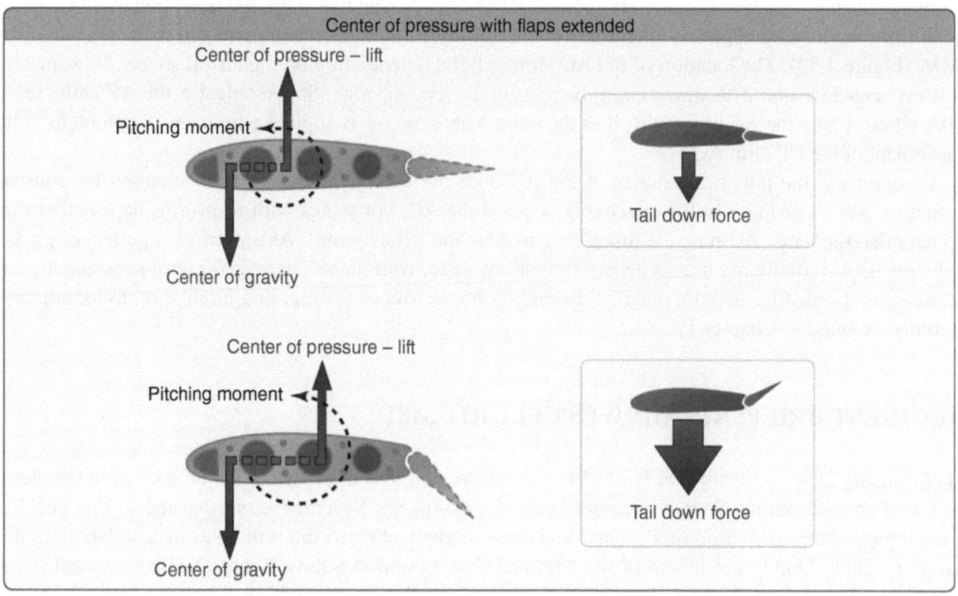

Figure 3.32. Flaps extended pitching moments.
Source: U.S. Department of Transportation Federal Aviation Administration (2016a)/U.S. Department of Transportation/Public domain.

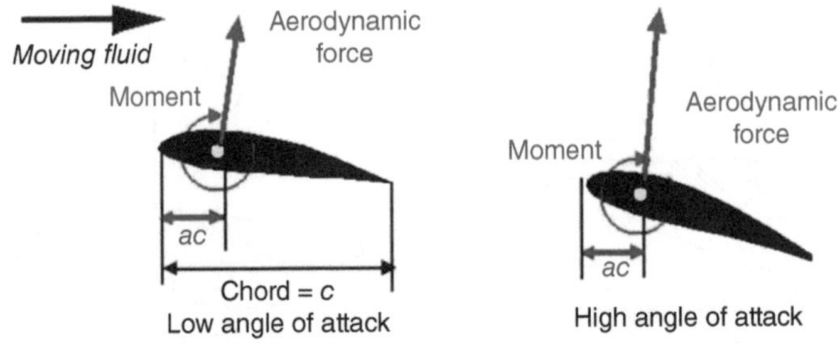

Figure 3.33. Aerodynamic center is independent of AOA.
Source: NASA/Public Domain.

AERODYNAMIC CENTER

For cambered airfoils, the CP moves along the chord line when the AOA changes. As the AOA increases, the CP moves forward and vice versa. This movement makes calculations involving stability and stress analysis very difficult from an engineering perspective. There is a point on an airfoil where the pitching moment is a constant with changing AOA if the velocity is constant. Engineers call the location where the aerodynamic moment remains constant the AC and it is a fixed point that is independent of the AOA.

The AC, unlike the CP, does not move with changes in AOA. If we consider the lift and drag forces acting at the AC, the calculations will be greatly simplified. If we know the pressure distribution around an airfoil in flight we can compute the moment at a specific point, the AF will be the same, but the value of the moment depends on where the force is applied. It has been proven that if the AF is applied at ~25% chord back from the leading edge the magnitude of the aerodynamic moment remains constant with changing AOA (Figure 3.33). The location of the AC varies slightly, depending on the airfoil shape. Subsonically, it is between 23% and 27% of the chord back from the leading edge. Supersonically, the AC shifts to the 50% chord. Using the AC of the airfoil as the point where the AF is applied eliminates the problem of the movement of the CP with AOA.

In summary, the pitching moment at the AC does not change when the AOA changes (at constant velocity), and all changes in lift effectively occur at the AC. For wings with planform shapes other than rectangular, the mean AC must be found (average for the whole wing). As an airfoil experiences greater velocity, its AC commonly moves toward the trailing edge, with the AC near 25% chord subsonically and at 50% supersonically. A more detailed discussion on the AC of a wing, and its relation to longitudinal stability, is found in Chapter 12.

ACCIDENT BRIEF: AIR MIDWEST FLIGHT 5481

On 8 January 2003, Air Midwest Flight 5481 experienced a loss of pitch control on takeoff in Charlotte, NC, and crashed, killing 2 crew members and 19 passengers. Several factors that led to this accident were topics addressed within this chapter and deserve application to the principles of aerodynamics and safety of flight. Due to the length of the National Transportation Safety Board (NTSB) accident report (NTSB/AAR-04/01), only the factors that directly relate to this chapter will be discussed, and it is recommended that the reader review the entire accident report.

The Beech 1900D has a typical horizontal stabilizer with an elevator hinged to the back. Normal travel of the elevator allows for up to 20° airplane nose up (ANU) and up to 14° airplane nose down (AND).

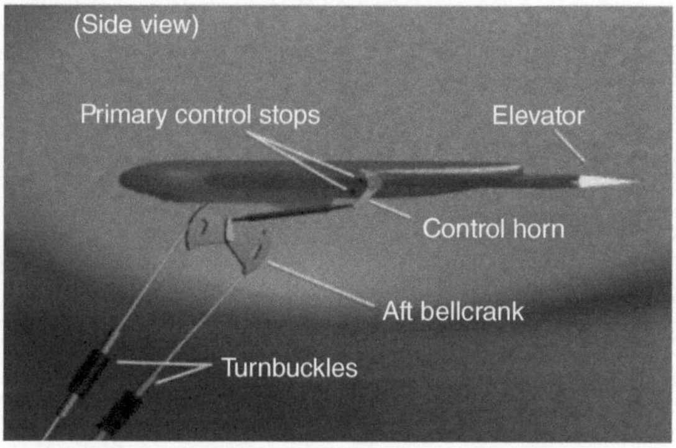

Figure 3.34. Beech 1900D pitch control system.
Source: Courtesy of National Transportation Safety Board.

As we discussed in the primary control section, the elevator allows the pilot to pitch the aircraft around the lateral axis. Figure 3.34 shows the Beech 1900D pitch control system, note the turnbuckles, aft bellcrank, and their connection to the elevator.

Due to improper maintenance on the turnbuckles within the pitch control system, the elevator was rigged improperly – (simplified for this discussion) not allowing for full elevator authority – and thus restricted downward elevator travel. The actual elevator position was restricted to 7° AND. In addition to the maintenance errors, a complicating factor was the weight and balance of the aircraft itself. Actual weight and balance calculations of the true weights (versus average weights) and position of all passengers and cargo proved the aircraft was also loaded overweight and with an AFT CG. As will be discussed later in the textbook, the position of the CG in accordance with the manufacturer's guidelines is imperative to aircraft safety. The aft CG in the accident aircraft exacerbated the maintenance error.

As the aircraft departed the runway, the nose experienced a pitch upward. Due to the aforementioned maintenance error, the pilots only had 7° AND travel with the elevator, about one-half of normal travel. This resulted in an increasing AOA and the flight crew was unable to recover. In this example, the flight crew was not at fault and could not have known about the maintenance error. Chapter 4 contains a more detailed discussion on the importance of understanding the principles of lift, and the consequences on the safety of flight if these principles are not understood.

SYMBOLS

AC	Aerodynamic center
AF	Aerodynamic force (lb)
AOA	Angle of attack (degrees)
CP	Center of pressure
D	Drag (lb)
L	Lift (lb)
RW	Relative wind
α (alpha)	Angle of attack (degrees)
γ (gamma)	Pitch angle
θ (theta)	Flight path angle

KEY TERMS

Aerodynamic center
Aerodynamic force
Aileron
Airfoil
Angle of attack
Antiservo tab
Center of pressure
Chord
Chord line
Differential aileron
Elevator
Flaps
Flight path
Lateral axis
Longitudinal axis
Magnus effect
Pitch
Relative wind
Roll
Rudder
Slots
Speed brake
Spoilers
Stabilator
Supercritical airfoil
Trim systems
Vertical axis
Yaw

PROBLEMS

Note: Answers to the problems in the first section are given at the end of the book.

Application of Chapter Theory

1. Flight control systems are of two types:
   - **a.** first and second.
   - **b.** primary and secondary.
   - **c.** trim systems and tabs.
   - **d.** balanced and imbalanced.
2. During a turn to the right, the aileron on the right wing is deflected:
   - **a.** up.
   - **b.** down.

c. in the same direction as the aileron on the left wing.
d. neither up nor down as it remains stationary.

3. During the transition from level flight to a climb attitude in a light aircraft, as the control stick is moved aft, the elevator or stabilator is positioned to produce a _______ by deflecting the elevator/stabilator _______.
   a. tail-up force/upward.
   b. tail-down force/upward.
   c. tail-down force/downward.
   d. tail-up force/downward.

4. The rudder controls movement around the:
   a. longitudinal axis.
   b. lateral axis.
   c. vertical axis.

5. Which of the following is not considered a disadvantage for an aircraft equipped with a canard:
   a. decreased longitudinal stability.
   b. elevator flutter at high speeds.
   c. reduced ability to recover from a deep stall.

6. Deploying trailing edge flaps increases lift and drag.
   a. True
   b. False

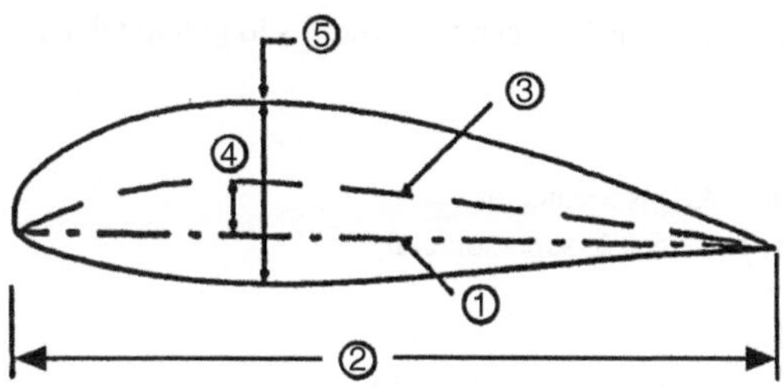

7. Number 3 on the drawing shows
   a. the chord line.
   b. the maximum camber.
   c. the thickness.
   d. the mean camber line.

8. Number 4 on the drawing shows
   a. the chord line.
   b. the thickness.
   c. the maximum camber.
   d. the mean camber line.

9. Number 5 on the drawing shows
   a. the maximum camber.
   b. the mean camber line.
   c. the upper-surface curvature.
   d. the maximum thickness.

10. Which of the following is not considered a primary variable in the geometry of an airfoil?
  a. airfoil composition.
  b. shape of the mean camber line.
  c. leading edge radius.
  d. location of the maximum thickness.

11. The Magnus effect explains why
  a. bowling ball curves.
  b. pitched baseball curves.
  c. golf ball slices.
  d. Both (b) and (c).

12. Which of the following will develop a positive lift?
  a. A symmetrical airfoil at zero AOA.
  b. A nonrotating cylinder in a wind tunnel.
  c. A cambered airfoil at zero AOA.

13. Which of the following will not produce a pitching moment?
  a. A symmetrical airfoil at a positive AOA.
  b. A cambered airfoil that is developing zero lift.
  c. A cambered airfoil that is at a positive AOA.
  d. A symmetrical airfoil at zero AOA.
  e. Both (a) and (d).

14. Pitch angle is defined as the angle between the aircraft's longitudinal axis and the horizontal axis.
  a. True
  b. False

15. The aerodynamic center (AC) is located at
  a. 50% C subsonically and 25% C supersonically.
  b. 25% C at all speeds.
  c. 50% C at all speeds.
  d. 25% C subsonically and 50% C supersonically.

16. For a cambered airfoil, the center of pressure (CP)
  a. moves to the rear of the wing at low AOA.
  b. moves backward as AOA increases.
  c. moves forward as AOA increases.
  d. Both (a) and (c).

17. Which of these statements is false? For a cambered airfoil,
  a. the AC is where all changes in lift effectively take place.
  b. there is no pitching moment at the AC.
  c. the AC is located near 25% C subsonically.
  d. the pitching moment at the AC is constant with the change of AOA (at constant airspeed).

18. For a symmetrical airfoil,
  a. the center of pressure moves forward as AOA increases.
  b. the center of pressure stays at the same place as AOA increases.
  c. there is no pitching moment about the CP.
  d. Both (b) and (c).

Correlation Through Scenario-Based Learning

Aircraft – *Diamond DA50RG*

For the following questions, use a DA50RG aircraft flight manual (AFM) found online (recommend the Diamond Aircraft website for the most up-to-date manual).

1. How are the elevator forces and rudder forces balanced during normal flight operation?
2. Referring to Section 1 of the AFM, what type of airfoil is used for the wing of the DA50? What type of airfoil is used for the horizontal and vertical stabilizers?
3. Using an online search engine and the airfoil type from the previous problem, research an airfoil tools website and create an image like what is found in Figure 3.23. One website example can be found in the bibliography, titled *Airfoil Tools*. In the textbox, enter the wing airfoil exactly as found in the AFM, but do not include */20-W4* at the end. Does the DA50 have a cambered or symmetrical wing? How can you tell? Now enter the information for the horizontal/vertical stabilizer in the text box for the same website, are these airfoils cambered or symmetrical?
4. What type of flaps are used on the DA50RG?

4 Lift

CHAPTER OBJECTIVES

After completing this chapter, you should be able to:

- Summarize the importance of angle of attack (AOA) in the production of lift and identify how AOA indicators improve the safety of flight.
- Describe boundary layer theory, contrast laminar and turbulent flow, and characterize how the Reynolds number is used to measure different flow regimes.
- Define adverse pressure gradient and illustrate how it relates to airflow separation.
- Correlate angle attack and the development of an airplane stall, identifying conditions of flight when stalls occur.
- Identify the factors involved in the development of aerodynamic force equations.
- Describe the variables within the lift equation, explain how each variable affects the production of lift, and reconstruct the lift equation to solve for each variable.
- Analyze airfoil lift characteristics as camber and thickness are varied.
- Provide examples of high coefficient of lift devices and how each device influences the production of lift.

INTRODUCTION TO LIFT

Although lift is most commonly associated with the wings of a fixed-wing aircraft, lift is also generated by propellers, helicopter rotors, rudders, sails and keels on sailboats, hydrofoils, wings on auto racing cars, and wind turbines.

While the common meaning of the word "lift" assumes that lift opposes weight, lift in the technical sense, used in this discussion, can be in any direction with respect to gravity, because it is defined with respect to the direction of flow, rather than to the direction of gravity.

When an aircraft is flying straight and level, as in cruise flight, most of the lift opposes weight, which is always directed vertically downward, regardless of the attitude of the aircraft. However, when an aircraft is climbing, descending, or maneuvering in a turn, the lift vector is tilted at an angle with respect to the

Flight Theory and Aerodynamics: A Practical Guide for Operational Safety, Fifth Edition.
Brian A. Johnson and Philip R. Fittante.

Companion website: www.wiley.com/go/flighttheoryandaerodynamics5e

vertical. The lift may also be entirely downward in some aerobatic maneuvers. In this last case, the term downforce is often used. The lift may also be largely horizontal, for instance on a sail for a sailboat.

As we have seen, lift is one of the four forces of flight, and the understanding of how this force can be controlled by the pilot is paramount to the safe, efficient operation of any aircraft. The pilot controls lift each time the stick or yoke is moved; whenever there is a change in the angle of attack (AOA), the lift must change as well. All other factors being equal, once AOA increases, the lift must also increase. Lift is proportional to the square of the aircraft's velocity, so if the AOA and other factors remain constant, an aircraft traveling 180 kts. will be producing four times as much lift as the same aircraft traveling 90 kts., but more on this later.

ANGLE OF ATTACK

The AOA is the angle between the chord line of an airfoil and the oncoming air (relative wind). Using the airfoil definitions from Chapter 3, we may also define AOA by the formula:

$$\alpha = \theta - \gamma \tag{4.1}$$

In this formula, θ (theta) represents the aircraft pitch angle and γ (gamma) represents the aircraft flight path angle. As was shown in Figure 3.19, high-performance aircraft (swept wing) typically fly at slower speeds with a positive pitch angle to maintain a desired glide path (flight path angle) on approach to landing.

EXAMPLE

An aircraft maintains a pitch angle of positive two degrees to maintain a three-degree glide path. What is the aircraft's AOA?

$$\alpha = \theta - \gamma \rightarrow \alpha = 2^\circ - (-3) \rightarrow \alpha = 5^\circ$$

Figure 4.1 shows an airfoil in level flight (note the position of the relative wind), with varying AOA. As presented in Chapter 3, note the position of the center of pressure with changing AOA.

As the AOA grows larger, the lift eventually reaches a maximum at a given angle for each airfoil, and any increase in AOA past this point causes the upper surface airflow to separate from the wing. The airfoil is said to be stalled; the AOA at which this occurs is referred to as the critical AOA.

The importance of AOA in determining aircraft performance cannot be overemphasized. Before a discussion on the importance of the stall AOA can be initiated, several other key facts of equal importance must be established:

1. An aircraft has its maximum climb angle at a certain AOA.
2. It will achieve the maximum rate of climb at another AOA.
3. It will get the maximum range at still another AOA.
4. It will get maximum endurance at a slightly higher AOA than the maximum range.

Angle of Attack Indicator

All aircraft performance depends on AOA, so modern, high-performance aircraft are fitted with AOA indicators. *These devices indicate the direction of the relative wind.* They consist of a probe, mounted on the side of the fuselage, which aligns itself with the relative wind, much like a weathervane, and cockpit

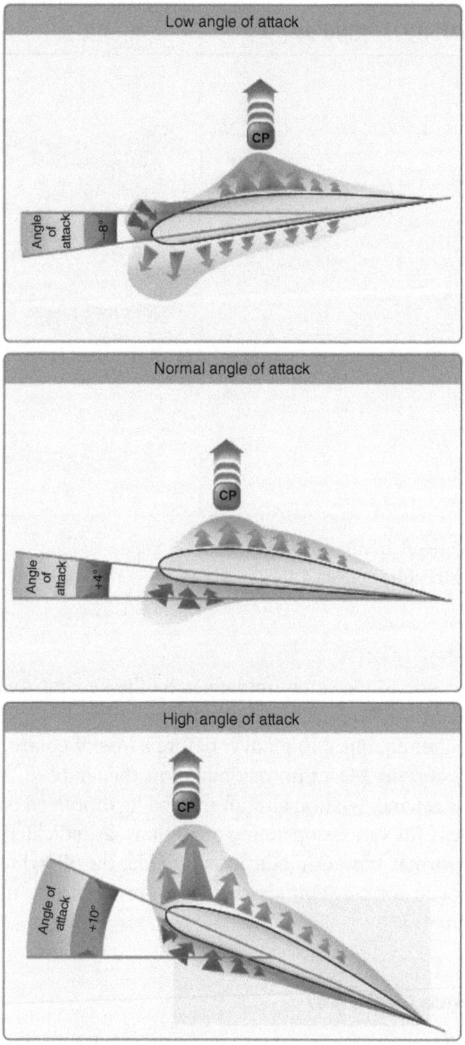

Figure 4.1. Pressure distribution on an airfoil with AOA.
Source: Pilot's Handbook of Aeronautical Knowledge (2016b)/U.S. Department of Transportation Federal Aviation Administration/Public domain.

indicators. The basic cockpit indicator merely has a pointer showing the AOA in uncalibrated units, rather than in degrees. This allows the same instrument to be used in different aircraft. The flight handbook will give the approach AOA, for instance, in units on the indicator, rather than in degrees.

In the past, only expensive high-speed fighter aircraft incorporated AOA indicators in the cockpit. In fact, some aircraft such as the F-15 and F-16 require the pilots to fly a particular AOA on the final approach as opposed to an airspeed that may vary with aircraft gross weight. In Figure 4.2, an F-16 heads-up display (HUD) shows varying AOA (α) in conjunction with HUD indicator and respective pitch attitude.

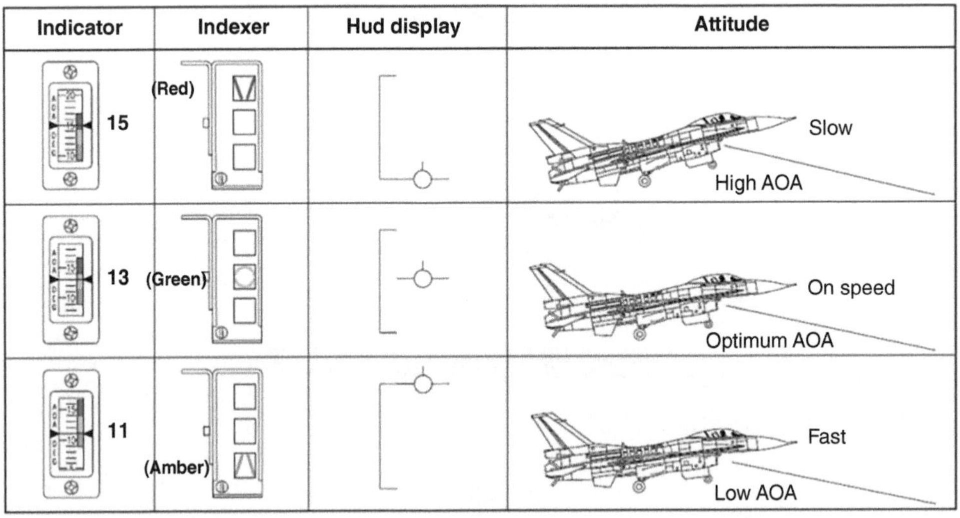

Figure 4.2. United States Air Force F-16 AOA display system.
Source: NASA TM-2014-218514/Public domain.

AOA indicators have become one of the safety initiatives by the Federal Aviation Administration (FAA) to target loss of control accidents. When the pilot has better situational awareness (SA) of the airfoil aerodynamic position, they are better equipped to recover before a loss of control incident can occur. Modern AOA indicators may display current AOA information below the airspeed "tape" within the glass panel, while AOA information for a general aviation aircraft may be incorporated on the cockpit panel itself and feature color ranges or patterns for various operating conditions, as indicated in Figure 4.3.

Regardless of the display format, the AOA indication provides the pilot direct indication of the aircraft's current AOA, reducing the need for constant airspeed monitoring during maneuvers or in high-density altitude, low-performance situations.

Accident Brief: Air France Flight 447

On 1 June 2009, Air France Flight 447 crashed into the Atlantic Ocean off the northeast coast of Brazil, killing all 228 passengers and crew. This accident is a textbook case of aircraft loss of control due to the aircrew's loss of SA concerning this Airbus A330's actual AOA.

While cruising at 35 000 ft in the vicinity of thunderstorms along the Intertropical Convergence Zone (ITCZ), the aircraft's pitot tubes iced over resulting in erroneous information to the A330's air data computers and a subsequent loss of accurate airspeed displays in the cockpit. The aircraft's autopilot disengaged and the aircraft's fly-by-wire flight control system degraded to a regime known as Alternate Control Law meaning normal envelope protections such as preventing a departure from controlled flight were no longer available.

The co-pilot in the right seat responded to the sudden loss of autopilot by pulling back on the A330's side stick controller. The aircraft immediately entered an aggressive climb and within seconds an aural stall warning sounded in the cockpit. For the next 4 minutes and 18 seconds, the Airbus A330 remained in a stalled condition until impact with the water. Though the aircraft's AOA remained close to 40° (the critical

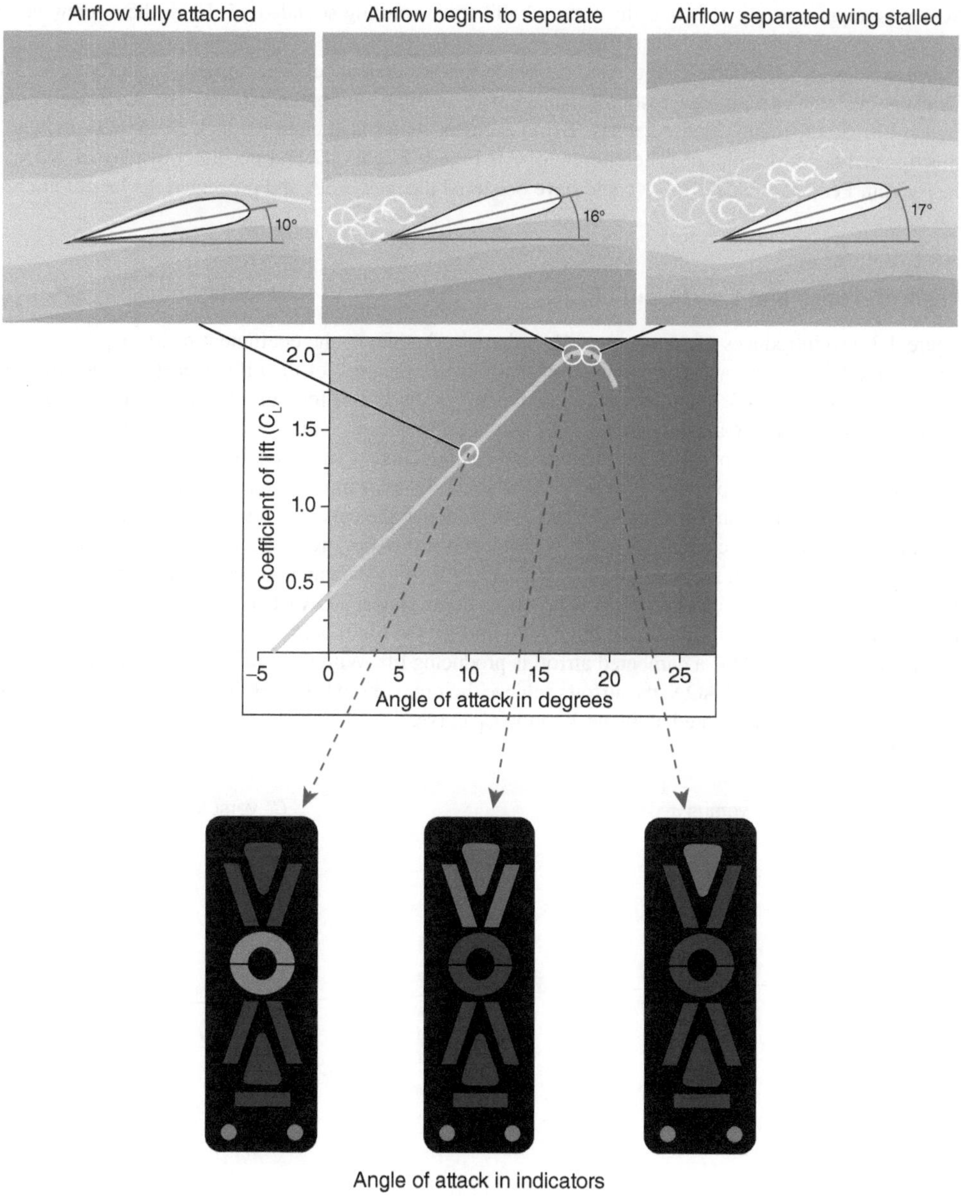

Figure 4.3. Critical angle of attack, stall, and angle of attack indications.
Source: Pilot Guide: Flight in Icing Conditions AC 91-74B (2015)/U.S. Department of Transportation Federal Aviation Administration/Public domain.

AOA was approximately 10°) and the aural "Stall" voice warning sounded 75 times, the aircrew never acknowledged that the aircraft was stalled.

The sudden loss of reliable airspeed indications in this accident uncovered numerous underlying human factors issues such as poor crew resource management, training, and organizational culture. However, the French Bureau of Enquiry and Analysis (BEA) accident report also cited the lack of an AOA indicator directly accessible to the pilot. The aircrew of Air France Flight 447 had no direct display of AOA in the cockpit. Perhaps such a display could have bolstered the crew's SA and prevented the loss of aircraft control.

Angle of Attack and Coefficient of Lift

Figure 4.3 also introduces the concept of C_L, otherwise known as the coefficient of lift. C_L is the ratio between the lift pressure and the dynamic pressure. It is a measure of the effectiveness of the airfoil to produce lift. In Figure 4.3, the values of C_L are plotted on the *Y*-axis for a given airfoil and the respective AOA are plotted on the *X*-axis, when the AOA increases so does the value of C_L. Values of C_L have been obtained experimentally for all airfoil sections and can be found in various publications and online airfoil databases. Using the Airfoils Tools website once again, we plot the C_L versus AOA curves for airfoils NACA 0015 and 2412 from Chapter 3 (Figure 4.4). Note that the curve passes through the (0, 0) point for a symmetrical airfoil, indicating that no lift is developed at 0° AOA; the curve is essentially a straight line until it approaches its stall AOA around 17°. Also note that for the cambered airfoil, the curve is shifted to the left, indicating that the cambered airfoil produces lift at a lower AOA when compared to a symmetrical airfoil, and the zero-lift point is at a *negative* AOA and not the (0, 0) point as with the symmetrical airfoil. This means that at 0° AOA, a cambered airfoil is producing lift. With both airfoils as the AOA increases, so does the C_L. At the stall AOA, the airfoils achieve their maximum lifting ability, and this point will now be termed $C_{L(max)}$ and is located at the peak of the C_L curves.

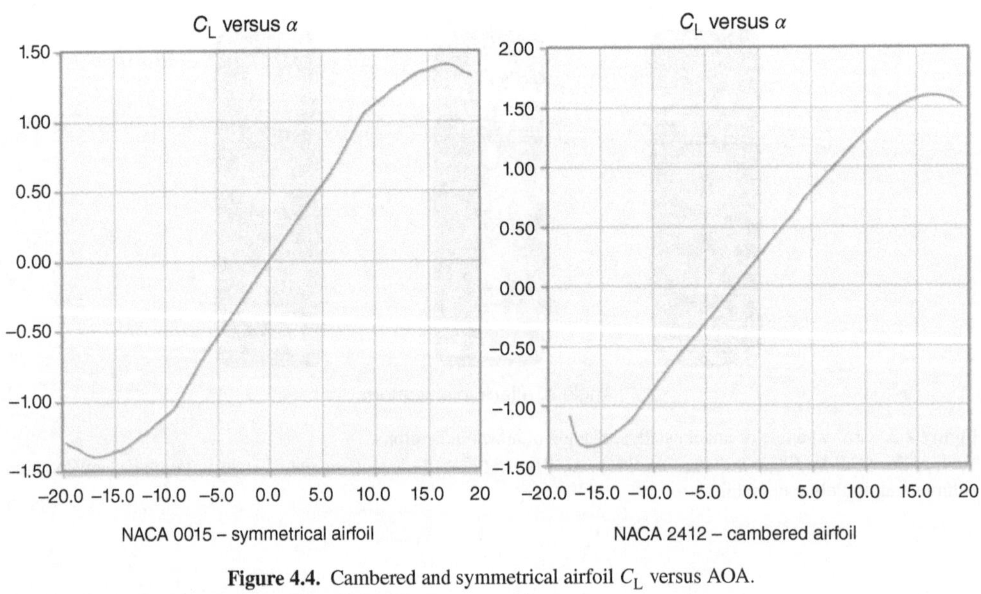

Figure 4.4. Cambered and symmetrical airfoil C_L versus AOA.

BOUNDARY LAYER THEORY

In physics and fluid mechanics, a *boundary layer* is the layer of fluid in the immediate vicinity of a bounding surface where the effects of viscosity are significant. *Viscosity* can be simply defined as the internal friction of a fluid caused by molecular attraction, and, as a reminder, gases (like air) are also considered fluids. Like a liquid, air will resist flow, and understanding the fluid properties of air is paramount to understanding boundary layer theory and principles of flight. High viscosity fluids are thicker and reduce flow while thinner fluids (less viscosity) fluids are less resistant to flow, so gas has a low viscosity and flows easily. In the Earth's atmosphere, an atmospheric boundary layer is the air layer near the ground affected by heat, moisture, or momentum transfer to or from the surface during the heating of the day. On an aircraft wing, the boundary layer is the part of the flow close to the wing, where viscous forces distort the surrounding nonviscous flow.

The viscous nature of airflow reduces the local velocities on a surface and is responsible for skin friction. The layer of air over the wing's surface that is slowed down or stopped by viscosity is the boundary layer. There are two different types of boundary layer flow: laminar and turbulent.

As we have seen, high AOA produces high coefficients of lift. Thus, lift to support an aircraft can be obtained at low airspeeds. To better understand what happens to the airflow when an aircraft is flown at or above stall AOA, we must understand the behavior of the air close to the wing (or airfoil) surface.

As a fluid moves around an object the molecules of the object "stick" to the object's surface, and then molecules moving above them collide with the slower moving molecules slowing them down, only not as much as the molecules sticking to the surface. As the distance increases away from the object's surface there are fewer and fewer collisions until eventually the free stream velocity is reached where no decrease in flow velocity is measured. The layer of air from the surface of the airfoil to the point of free-stream velocity is known as *the boundary layer*. The nature of the boundary layer determines the maximum lift coefficient and the stalling characteristics of the airfoil.

The beginning of airflow at the leading edge of a smooth airfoil surface produces a very thin layer of smooth airflow. This type of airflow is called *laminar flow* and is characterized by smooth regular streamlines. Fluid particles in this region do not intermingle. As the airflow moves back from the leading edge, the boundary layer thickens and becomes unstable. Small pressure disturbances cause the unstable airflow to tumble, and intermixing of the air particles takes place. This type of airflow is called *turbulent flow*.

Figure 4.5 illustrates the flow within the boundary layer on a flat surface and shows the increasing thickness of the boundary layer and transition from laminar to turbulent flow.

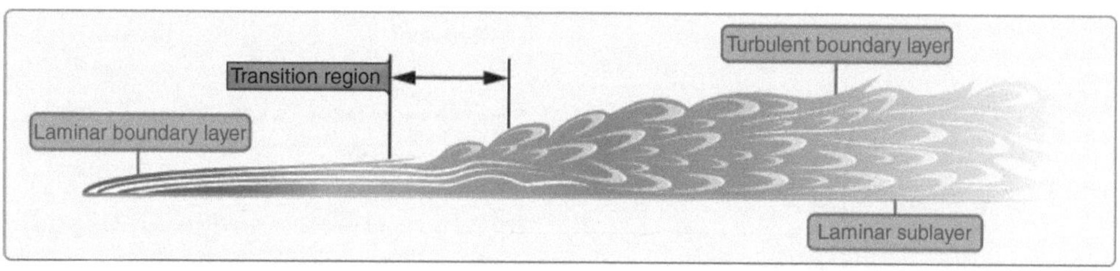

Figure 4.5. Boundary layer composition.
Source: Pilot's Handbook of Aeronautical Knowledge (2016b)/U.S. Department of Transportation Federal Aviation Administration/Public domain.

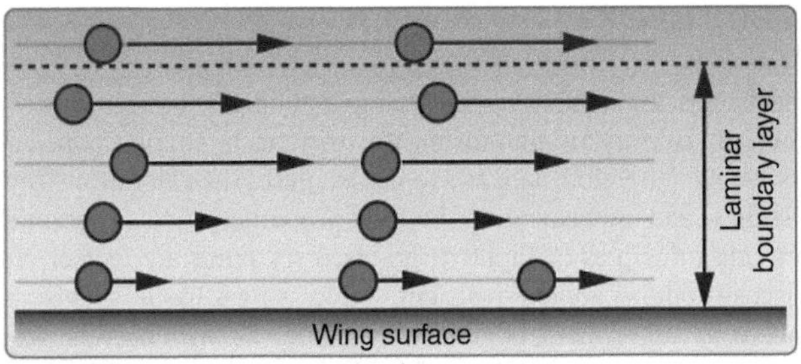

Figure 4.6. Laminar boundary layer.
Source: Glider Flying Handbook (2013)/U.S. Department of Transportation Federal Aviation Administration/Public Domain.

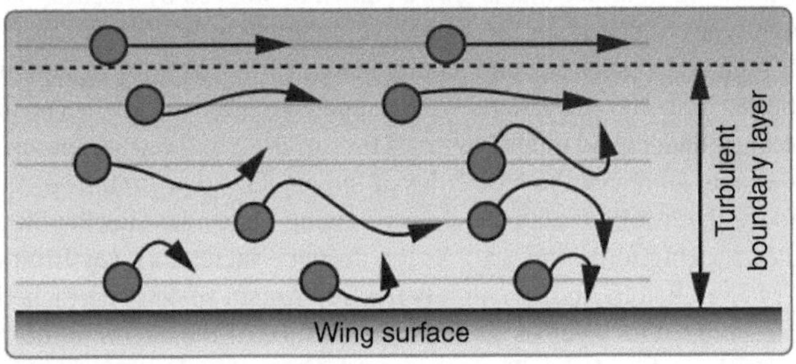

Figure 4.7. Turbulent boundary layer.
Source: Glider Flying Handbook (2013)/U.S. Department of Transportation Federal Aviation Administration/Public Domain.

Figure 4.6 demonstrates a simplification of laminar flow as each layer of air molecules smoothly moves over other layers. In Figure 4.7, the turbulent boundary layer possesses irregular "turbulent" eddies and the mixing of air levels. Velocity profiles are different in laminar and turbulent flow. Laminar profiles show a gradual decrease in the relative velocity from the outer edge of the boundary layer to the surface. Figure 4.8 depicts a comparison of fluid flow around a solid object between laminar flow and turbulent flow, in both images, the velocity of the fluid is diminished the closer the flow to the surface of the object.

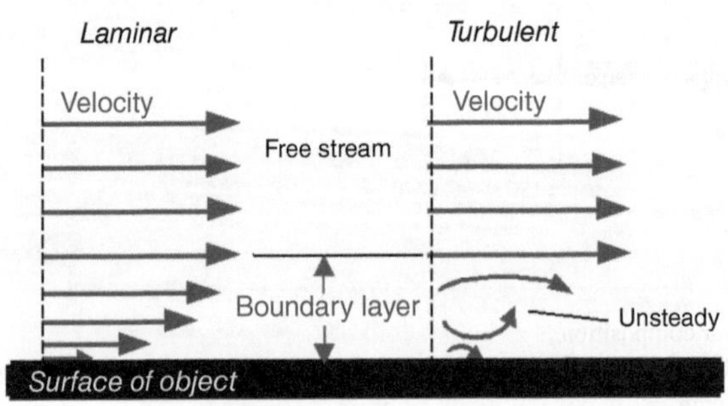

Figure 4.8. Laminar and turbulent velocity profiles.
Source: NASA Glenn Research Center/Public Domain.

Coanda Effect

The Coanda effect, named after Henri Marie Coanda (credited by many as the inventor of the first jet engine), occurs when a fluid moving over a surface adheres to the curvature of the surface. A common visual of the Coanda effect occurs when a piece of paper is held while blowing air over the top surface. The paper moves upward as the air moves over the top surface and follows the curvature of the paper downward. Adhering to Newton's third law, as the paper deflects the air downward, the air exerts an equal and opposite upward force on the paper. Water flowing off the edge of curved roof shingles into a gutter is also an example of the Coanda effect.

In aerodynamics, air moving past an airfoil is deflected downward at a sharp trailing edge, which means that the airfoil has exerted a downward force on the air moving past it. By Newton's third law, the downwardly deflected air exerts an equal and opposite (upward) force on the airfoil. This force pushes the airfoil up, creating lift. Rotorcraft experience the Coanda effect around rotating blades in certain regimes of flight, while other military aircraft use directed high-speed jet flow to obtain the same effect.

REYNOLDS NUMBER

In 1883, Osborn Reynolds carried out a series of experiments with the flow of water in pipes. By injecting dye into a stream of flowing water, he discovered that laminar and turbulent flow could be seen. He also discovered a relationship between the point where the fluid became turbulent and three properties of the fluid stream.

The Reynolds number is important in aircraft design and is defined as the ratio of momentum forces to viscous forces. Determining the Reynolds number is especially useful in wind tunnel research to verify that the airfoil being tested will show the same characteristics when applied to a full-scale aircraft. The Reynolds number quantifies the relative importance of these two types of forces for given flow conditions, and allows for the measurement of flow across an airfoil. Reynolds numbers frequently are used to characterize different flow regimes within a similar fluid, such as laminar or turbulent flow:

- Laminar flow occurs at low Reynolds numbers, where viscous forces are dominant, and is characterized by smooth, constant fluid motion.
- Turbulent flow occurs at high Reynolds numbers and is dominated by inertial forces, which tend to produce chaotic eddies, vortices, and other flow instabilities.

Fluid flow is generally chaotic, and very small changes to shape and surface roughness as is commonly seen on aircraft wings can result in very different flows. Nevertheless, Reynolds numbers are a very important guide when analyzing fluid flow and are widely used throughout the aircraft industry.

The Reynolds number is defined by the following formula:

$$R_g = \frac{Vx}{\nu} \tag{4.2}$$

where

R_g = Reynolds number
V = stream velocity (fps)
x = distance downstream (ft)
ν = kinematic viscosity (ft^2/s) (measure of a fluid's resistance to flow).

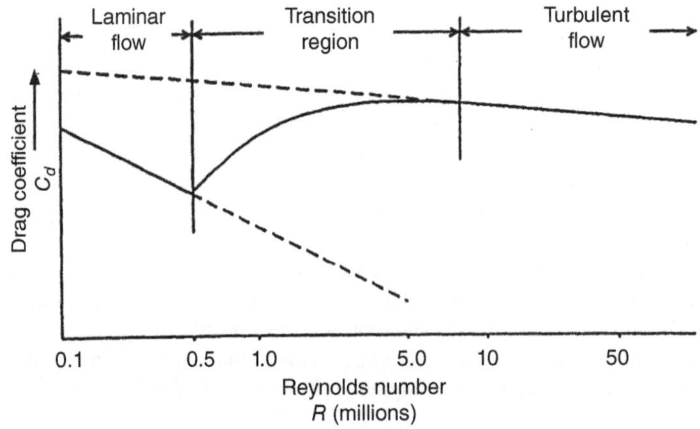

Figure 4.9. Reynolds number effect on airflow on a smooth flat plate.

EXAMPLE

An airplane has a rectangular wing with a chord length of 6 ft. If the airplane is flying at 140 kts. TAS and at a 10 000 ft density altitude, find Reynold's number at the trailing edge of the wing.

$$142 \text{ kts} = 240 \text{ fps}$$

Table 2.1: $\upsilon = 0.000\,202\ \text{ft}^2/\text{s}$ @ 10 000 ft.

$$R_g = \frac{Vx}{\nu} \rightarrow R_g = \frac{(240 \text{ fps} \times 6 \text{ ft})}{0.000\,202\ \text{ft}^2/\text{s}} \rightarrow R_g = 7.13 \times 10^6$$

Reynolds' experiments showed that at low values of R_g the flow was laminar, and at higher values the flow was turbulent. When considering airflow about an airfoil, it is not possible to define the exact R_g when the flow changes from laminar to turbulent. Both the contour of the airfoil and the smoothness of the surface influence the changeover point.

Studies on smooth, flat plates have been made, and a prediction of the type of flow is possible. Figure 4.9 shows the results of one study. Laminar airflow exists below R_g of 0.5 million, and turbulent flow exists for values greater than 10 million. The transition occurs between these two values. The consideration of the Reynolds number is very important in correlating wind tunnel data of scale models with full-sized aircraft.

ADVERSE PRESSURE GRADIENT

An *adverse pressure gradient* occurs when the static pressure increases in the direction of the flow (dynamic pressure decreases). This is important for boundary layers; increasing the fluid pressure is akin to increasing the potential energy of the fluid, leading to a reduced kinetic energy and a deceleration of the fluid. Because the fluid in the inner part of the boundary layer is slower, it is more greatly affected by the increasing pressure gradient; for a large enough pressure increase, this fluid may slow to zero velocity and even become reversed. This *flow reversal* is said to be separated from the surface and has very significant consequences in aerodynamics since flow separation significantly modifies the pressure distribution along the surface, hence the lift and drag characteristics.

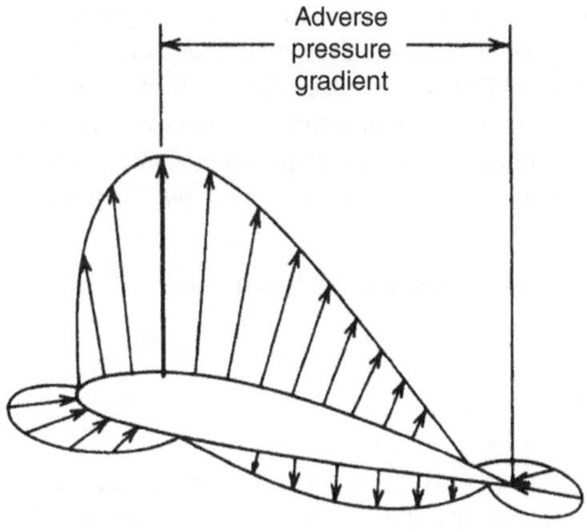

Figure 4.10. Adverse pressure gradient.

Turbulent boundary layers tend to be able to sustain an adverse pressure gradient better than an equivalent laminar boundary layer, an important factor when noting the Reynolds number of many transport category aircraft. The more efficient mixing that occurs in a turbulent boundary layer transports kinetic energy from the edge of the boundary layer to the low momentum flow at the solid surface, often preventing the separation that would occur for a laminar boundary layer under the same conditions. This physical fact has led to designs to deliberately produce turbulent boundary layers when boundary layer separation is dominant at high Reynolds numbers. Examples include the dimples on a golf ball, the fuzz on a tennis ball, or the seams on a baseball. Airplane wings are often engineered with *vortex generators* on the upper surface to produce a turbulent boundary layer.

The pressure distribution of a cambered airfoil that is producing lift is shown in Figure 4.10. The minimum pressure point is shown by the heavy arrow. Airflow approaching the leading edge of the airfoil enters the high-pressure area near the stagnation point. As the air divides to pass over the top and under the bottom surfaces of the airfoil, it moves from a high-pressure area to a lower static pressure area.

The air moving over the top of the airfoil is important to this discussion. Until the air reaches the point of minimum pressure, it is in a favorable pressure gradient. The forces of nature are acting on the air favorably and it is accelerated and sped on its way. Once the air on the top of the airfoil passes the minimum static pressure point, it enters a region of higher static pressure. This region is called an *adverse pressure gradient*. The kinetic energy of the airflow is now being transformed into higher static pressure, and the velocity is reduced.

AIRFLOW SEPARATION

All solid objects traveling through a fluid (or alternatively a stationary object exposed to a moving fluid) acquire a boundary layer of fluid around them, where viscous forces occur in the layer of fluid close to the solid surface. As we discussed in the Boundary Layer Theory section, boundary layers can be either laminar or turbulent.

Flow separation occurs when the boundary layer travels far enough against an adverse pressure gradient that the speed of the boundary layer relative to the object falls almost to zero. The fluid flow becomes detached from the surface of the object, and instead takes the forms of eddies and vortices.

In aerodynamics, flow separation can often result in increased drag, particularly pressure drag which is caused by the pressure differential between the front and rear surfaces of the object as it travels through the fluid. For this reason, *the goal is to delay flow separation and keep the local flow attached for as long as possible.*

The airflow in the boundary layer is acted on by two forces:

1. Frictional forces
2. Adverse pressure gradient forces

First, there are frictional forces between the surface of the airfoil and the air particles, and there is friction between the particles themselves. Both of these tend to reduce the relative velocity to zero. Second, the air is slowed by the adverse pressure gradient. The slowing of the air creates a stagnated region close to the surface.

Airflow from outside the boundary layer will overrun the point of stagnation and a flow reversal results: the airflow moves forward, the boundary layer separates from the surface, lift is destroyed, and drag becomes excessively high. Airflow can be seen on an airfoil operating at high AOA if lines of yarn are glued to the upper surface of a wing. The pieces of yarn will reverse direction and flap violently when a stall occurs. Later in this textbook, we will discuss how stall patterns progress across various wing planforms. Figure 4.11 shows the velocity profiles in the boundary layer during airflow separation; note the position on the airfoil where this occurs.

Consider a smooth sphere in an airflow (Figure 4.12a). If the diameter is small, the Reynolds number will be small, and laminar airflow will exist on the surface. This produces only a small amount of drag against the surface, but laminar flow allows the air to separate easily. When this happens, a wide wake forms behind the sphere. This wake produces high drag forces.

Now, assume the sphere's surface is rough (Figure 4.12b). The airflow becomes turbulent very soon as it passes around the sphere. The surface drag is now increased, but turbulent air has more energy than laminar flow and will resist separation. Thus, the width of the wake is reduced and the total drag is reduced.

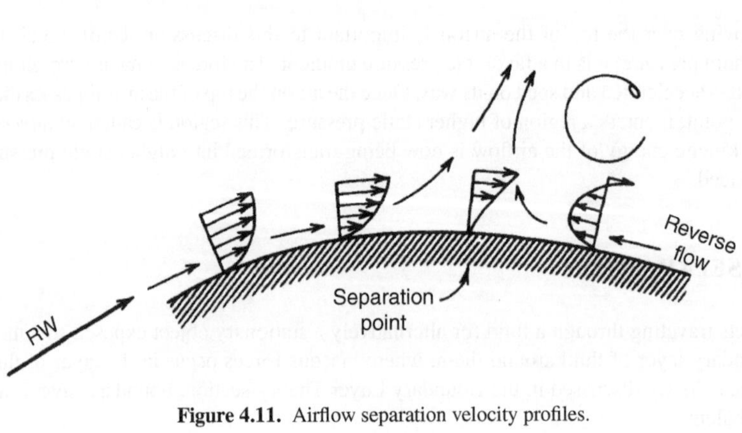

Figure 4.11. Airflow separation velocity profiles.

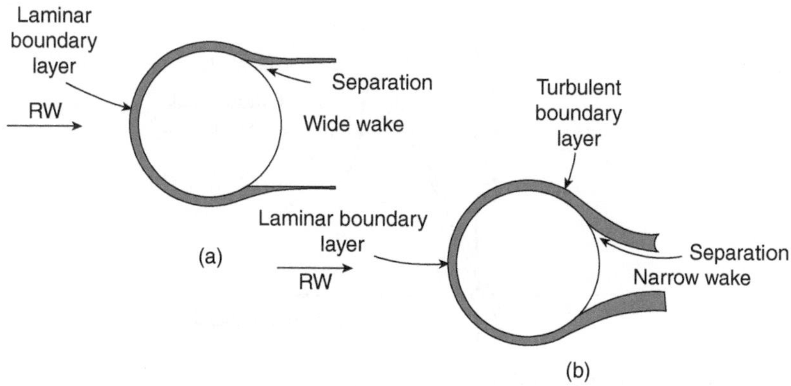

Figure 4.12. Sphere wake drag: (a) smooth sphere and (b) rough sphere.

The dimples on a golf ball and the fuzz on a tennis ball are examples of how creating turbulent flow delays separation and reduces total drag.

STALL

An airfoil stalls when the AOA increases beyond a certain point such that the lift begins to decrease. The angle at which this occurs is called the *critical* AOA. This critical angle is dependent upon the airfoil section or profile of the wing, its planform, its aspect ratio, and other factors, but is typically in the range of 14°–20° for most subsonic airfoils. Figure 4.13 shows the progression of an airfoil through increasing AOA (Image A) to a full stall and boundary layer separation when the critical AOA has been exceeded (Image C).

Flow separation begins to occur at small angles of attack while attached flow over the wing still prevails. As AOA increases, the separated regions on the top of the wing increase in size and hinder the wing's ability to create lift. At the critical AOA, separated flow is so dominant that any further increase in AOA actually produces *less* lift, and we will soon see, more drag. This is an important point: if an airfoil exceeds the

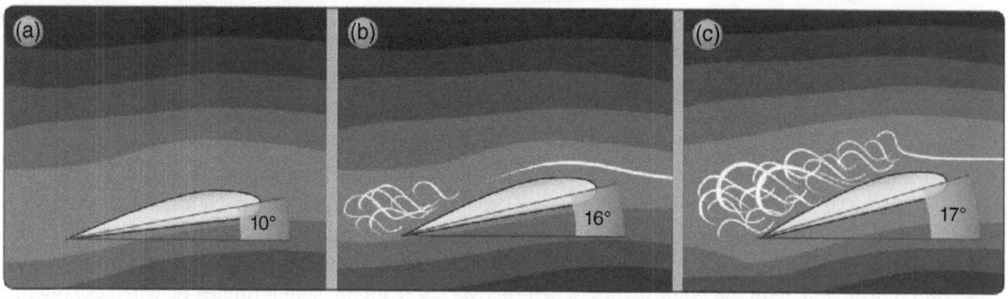

Figure 4.13. Critical angle of attack and stall. (a) Increasing AOA to 10°. (b) Increasing AOA to 16°, separated flow begins toward the trailing edge of the airfoil. (c) At 17° the airfoil has reached the critical AOA and is fully stalled. *Source:* U.S. Department of Transportation Federal Aviation Administration (2016a)/U.S. Department of Transportation/Public domain.

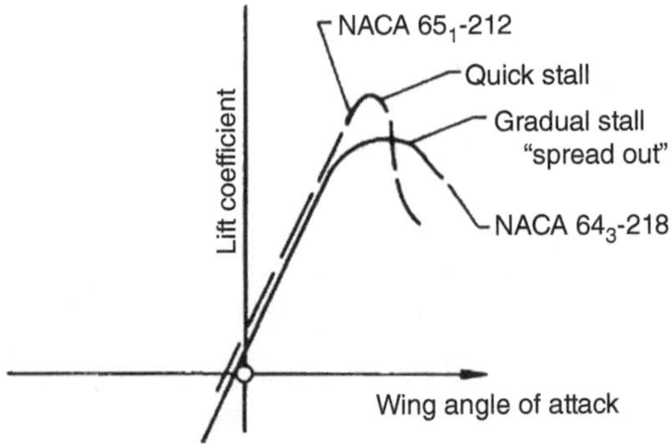

Figure 4.14. C_L versus AOA airfoil comparison.
Source: NASA-SP-367/Public Domain.

critical AOA, it may still be producing lift, just not enough lift to sustain level flight. The fact that the airfoil has stalled does not mean that the airfoil is not producing any lift. Figure 4.14 shows a C_L versus AOA comparison between NACA 65_1 - 212 and NACA 64_3 - 218. Note that NACA 64_3 - 218 has a more gradual stall indicating there will not be an abrupt stall point as with NACA 65_1 - 212.

A fixed-wing aircraft may experience buffeting or a change in attitude during a stall. Most training aircraft are designed to have a gradual stall, with characteristics that will warn the pilot and give the pilot time to react. In fact, airplane manufacturers must build into the aircraft's safety system a stall warning system that uses the aerodynamic qualities of the airplane, or a stall warning device. For example, an aircraft that does not exhibit one of the traditional characteristics of "feel" of the impending stall to the pilot may have an audible alarm or a stick shaker installed to simulate the feel of a buffet by vibrating the stick fore and aft.

Two general conditions of flight when stalls occur are introduced at this point:

1. The most common type of stall is the slow-speed, high-AOA separation. This type of stall most often occurs during the takeoff, approach, and landing phases of flight and is dangerous because of the lack of altitude for recovery. This type of stall starts at the trailing edge of a wing and progresses forward. The part of the wing that stalls first is of great interest and depends on the planform shape of the wing. Stall progression and wing planform design will be discussed in detail in Chapter 10. Note that the critical AOA for an airfoil does not change, as will be detailed in the Coefficient of Lift section later in this chapter in the textbook.
2. The second kind of stall is the accelerated stall, which occurs with a sudden rapid increase in AOA, again exceeding the critical AOA. Pilots are usually first introduced to this type of stall when conducting *steep turns* while maintaining level flight or during advanced maneuvers for the commercial pilot certificate. As the pilot pulls back on the yoke or stick during the maneuver, the aircraft may start to buffet and the stall warning device will sound or be initiated. Sharp leading edges contribute to this type of stall. The rapid increase in AOA does not allow time for the airflow to turn the corner of the leading edge, and separation takes place there. To avoid this type of stall, do not make sudden movements with the control stick when recovering from unusual attitudes or when conducting maneuvers. The onset of accelerated stalls can occur very quickly, as in a landing situation, when the pilot turns the aircraft from the base leg to the final turn on approach.

We will discuss stalls in more depth, as well as discuss various types of stalls in Chapter 10.

AERODYNAMIC FORCE EQUATIONS

The lift and drag equations depend on several factors, the most important of which are:

1. Airstream velocity V (kts.)
2. Airstream density ratio σ (dimensionless)
3. Planform area S (ft^2)
4. Profile shape of the airfoil
5. Viscosity of the air
6. Compressibility effects
7. AOA α (degrees)

The first two factors determine the dynamic pressure, q, of the airstream (see Eq. 2.10):

$$q = \frac{\sigma V^2}{295} \text{ (psf)}$$

Factors 4, 5, and 6 influence the amount of drag that an airfoil will develop at a certain AOA. It is convenient to express the lift and drag forces in terms of nondimensional coefficients that are functions of AOA. The *coefficient of lift* (C_L) was defined earlier in this chapter and the *coefficient of drag* (C_D) will be defined in Chapter 5.

LIFT EQUATION

Now that the factors influencing lift have been introduced, the lift equation can be presented with Eq. 4.3. We will use variations of this equation to examine the impact of each variable on lift, and we will see which of these variables are under the pilot's control. The unit of force for lift will be in pounds (lb).

$$L = C_L qS \rightarrow L = C_L \left(\frac{\sigma V^2}{295}\right) S \rightarrow$$

so,

$$\text{The lift equation becomes: } L = \frac{C_L \sigma V^2 S}{295} \text{ (lb)} \tag{4.3}$$

Rewriting gives

$$C_L = \frac{L}{qS} = \frac{L/S}{q} = \frac{\text{Lifting pressure}}{\text{Dynamic pressure}}$$

Coefficient of Lift and the Lift Equation

As introduced earlier in this chapter, we can find a direct comparison between C_L and AOA. Figure 4.15 shows a typical curve of C_L plotted against AOA for a symmetrical airfoil, while Figure 4.16 shows a typical curve of C_L plotted against AOA for a cambered airfoil.

An airfoil always stalls at the same AOA. For the airfoil shown in Figure 4.15, this will be at 14° AOA (note the shaded region), and for Figure 4.16, this will be at approximately 17° AOA. The stall AOA is independent of weight, attitude, and altitude if viscosity effects are ignored. When determining the AOA

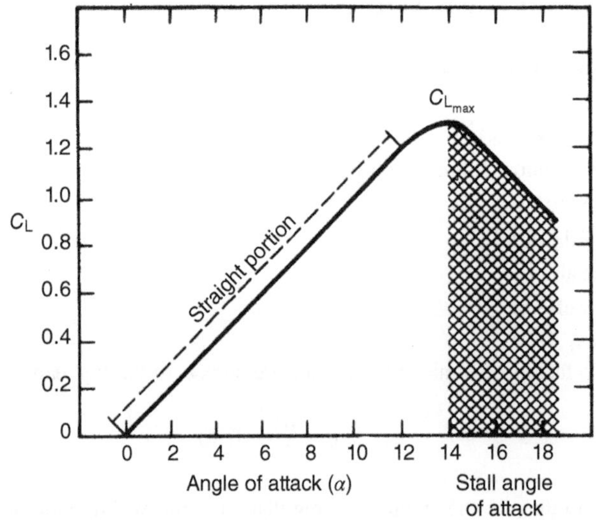

Figure 4.15. C_L versus AOA for a symmetrical airfoil.

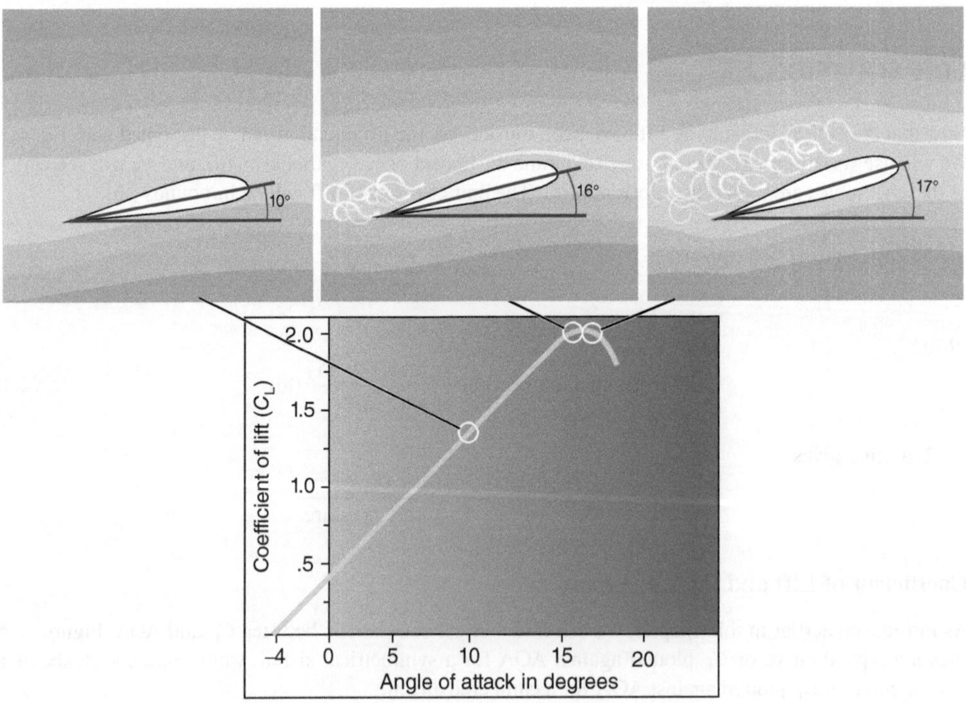

Figure 4.16. C_L versus AOA for a cambered airfoil.
Source: Airplane Flying Handbook (2004a)/U.S. Department of Transportation Federal Aviation Administration/Public domain.

for a certain given condition, if you know the C_L and have the curve of C_L plotted against the AOA for the airfoil, you can determine the resultant AOA, and vice versa.

If the aircraft is in balanced, 1*G* flight, the lift is equal to the weight. Many textbooks use "*g*" to denote *G*-loading, to differentiate from gravitational acceleration defined in Chapter 1, and to maintain continuity with previous editions of this textbook, we will continue to use "*G*" for load factor. *G*-loading refers to the load factor and is further defined in Chapter 11, at this point, we are assuming our aircraft is in level, unaccelerated flight which equals 1*G*. However, during maneuvering flight, the *G*-loading will not be equal to the weight but will be the weight multiplied by the *G* value: $L = GW$. Unless otherwise stated for our equations, we will assume the aircraft is in balanced, 1*G* flight, so the lift is equal to the weight, and 1*G* is removed from the equation, $L = W$. In Chapter 11, we will explore what happens to lift when the aircraft is not in 1*G* flight.

Equation 4.4 is the original lift equation solved for the variable C_L, remember $L = W$ for this equation:

$$L = \frac{C_L \sigma V^2 S}{295} \rightarrow C_L = \frac{295(L = W)}{\sigma V^2 S}$$

so,

$$C_L = \frac{295W}{\sigma V^2 S} \tag{4.4}$$

EXAMPLE

Find the C_L. Assume that the aircraft has an approach speed of 114 kts. at a gross weight of 15 000 lb, wing area $S = 340\ \text{ft}^2$, and is at sea level on a standard day ($\sigma = 1.0$).

Using Eq. 4.4, the aircraft develops a lift coefficient of

$$C_L = \frac{295W}{\sigma V^2 S} = \frac{(295)\ (15\,000\ \text{lb})}{(1.0)\ (114\ \text{kts.})^2\ (340\ \text{ft}^2)} = 1.0$$

Note from Figure 4.15 that a 10° AOA must be held for this flight condition.

Velocity

The speed of an aircraft is numerically equal to its velocity (ignoring the direction of the velocity vector), and in the calculations that follow we will use the terms interchangeably. Velocity has a strong effect on the lift an airfoil develops because, as mentioned earlier in this chapter, lift is proportional to the square of the velocity. If the AOA and other factors remain constant, an aircraft traveling 180 kts. will produce four times as much lift as the same aircraft traveling 90 kts. The pilot controls the velocity of the aircraft and thus controls the velocity of the air flowing over the wing. An example of this occurs when a pilot in level flight increases the true air speed (velocity) without changing heading or altitude. Referencing the original lift equation, because V increases, so then does L, so another variable must change in order to maintain level flight. Equation 4.5 is the original lift equation solved for the variable V:

$$V = \sqrt{\frac{295L}{C_L \sigma S}} \tag{4.5}$$

It is important to restate that *the critical AOA (the AOA at which the wing stalls) is independent of weight, attitude, and altitude* if viscosity effects are ignored. While the critical AOA is constant, the stall speed is not. Stall speed is affected by weight, wing loading, altitude, and other factors. Using Eq. 4.5, we can solve for the velocity at stall, introduced here as V_s. We now also know when at stall speed we are also at $C_{L(max)}$, so we can solve for the stall speed (V_s) with Eq. 4.6:

Stall speed V_s occurs at $C_{L(max)}$, so

$$V_s = \sqrt{\frac{295L}{C_{L(max)}\sigma S}} \tag{4.6}$$

EXAMPLE

Assume an aircraft with the $C_L - \alpha$ curve of Figure 4.15 and the following data: gross weight = 15 000 lb, wing area S = 340 ft², $C_{L(max)}$ = 1.3, and at sea level on a standard day (σ = 1.0). Calculate the stall speed (V_s),

$$V_s = \sqrt{\frac{(295)\ (15\,000\ \text{lb})}{(1.3)\ (1.0)\ (340\ \ \text{ft}^2)}} = 100\ \text{kts.}$$

The velocity varies as the square root of the lift (or weight under 1G flight). To calculate a new speed V_2, if all other conditions stay the same except for a new weight given (W_2), then the speed–weight relationship is calculated using Eq. 4.7:

$$V_2 = V_1\sqrt{\frac{W_2}{W_1}} \tag{4.7}$$

If the aircraft is flown at the same AOA, it will have the same lift coefficient, and not only will the stall speed vary with the square root of the weight, but all other speeds will vary by the same amount. Stall AOA does not vary with weight; stall speed does. Using Eq. 4.7 again we can now calculate a new stall speed V_{s_2} based on a new weight W_2, if we know the original stall speed V_{s_1} and original weight W_1:

$$V_{s_2} = V_{s_1}\sqrt{\frac{W_2}{W_1}}$$

EXAMPLE

If the weight of a 15 000 aircraft is increased to 20 000 lb, the AOA remains the same but the approach speed of the aircraft must be increased, as indicated by Eq. 4.7:

$$V_2 = V_1\sqrt{\frac{W_2}{W_1}} = 114\ \text{kts.}\sqrt{\frac{20\,000\ \text{lb}}{15\,000\ \text{lb}}}$$

$$= 131.6\ \text{kts.}$$

The lift equation shows that for steady unaccelerated flight, the lift is equal to the weight and is a constant. There are only two variables in the equation: C_L and V^2. The value of C_L is determined by the AOA and then the airspeed is determined. It follows that the control stick determines AOA and thus the airspeed. This is the fundamental concept in flying technique: AOA is the primary control of airspeed in steady flight. The rate of climb or descent must, therefore, be controlled by throttle position. *Pitch for airspeed, power for altitude.*

As presented in this section through the lift equation and examples, airspeed and AOA have a tightly coupled relationship in the production of lift. A new pilot quickly and naturally applies this concept when flying an aircraft, learning that their primary method to control the amount of lift produced by the aircraft is through the direct control of airspeed and AOA. If the pilot's flight instructor asks them to slow the aircraft by 20 kts. while maintaining a constant altitude, the pilot will quickly see that after a reduction in power, they will need to pull back on the stick/yoke to hold the desired altitude. The loss of lift from the reduced airspeed is offset by the increased production of lift from a higher AOA, thus maintaining the same amount of lift to offset the weight of the aircraft in level flight. This fundamental relationship between airspeed and AOA will be expanded in upcoming sections describing maneuvering flight where factors of drag, adverse yaw, centripetal force, and load factor become more pronounced. One example of a more detailed view of the relationship between pitch and power involves descending at a fixed descent angle, as with a typical Instrument Landing System (ILS) of 3°. If using a flight director with the pilot controlling power, the pitch versus power relationship is not so simplified. During the descent on an ILS approach, once the ground speed has stabilized in the theoretical situation of no wind, the approach is a constant rate of descent. If the pilot is hand flying the approach and is not on the proper airspeed and glide angle, they will pitch for glide path (airspeed) and then use power to control the altitude, and if flying on autopilot and manually controlling the power, the flight director will pitch for glide path (altitude) and the pilot will then need to use power to control airspeed. Of course, this discussion is simplified and assumes pitch and power are independent, but in most maneuvers of flight (including deviations on a glide path) the pitch and power will be used simultaneously as altitude, airspeed, and bank angle change.

Lift Formula Summary

With all other variables in the lift equation remaining the same unless otherwise noted, we can summarize the following from the above discussion:

1. If the airfoil shape and AOA (C_L) remain the same, lift increases with an increase in V (TAS).
2. If the density of the air increases, lift increases.
3. If the area of the wing (S) increases, lift increases.
4. If the AOA increases (C_L increases), then lift will increase.

Application 4.1

Rearrange Eq. 4.4 to solve for the variable of weight, maintaining the assumption the aircraft is in 1G, balanced flight.

If all other factors remain the same, what change in AOA would be needed to maintain level flight if the weight of the aircraft decreased suddenly, as when cargo is dropped out of the back of an aircraft or jumpers exit an aircraft engaged in parachute operations? What change in TAS would be required to maintain level flight if fuel was added to the aircraft during flight?

AIRFOIL LIFT CHARACTERISTICS

The lift characteristics of airfoil sections are affected by thickness and location of maximum thickness, camber and its location, leading edge radius, and other factors. Increasing the thickness, for example, results in higher local velocity airflow, which results in lower static pressure and more lift (Figure 4.17).

The mission of a proposed aircraft influences the type of airfoil selected. Both of the airfoils in Figure 4.17 have a maximum camber of 4% C, located at the 40% C position. The 4412 airfoil is 12% C

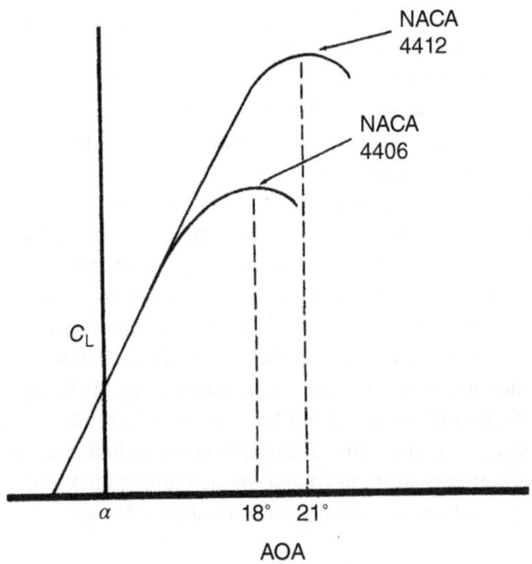

Figure 4.17. Thickness effect.

thick, and the 4 406 is 6% C thick. The thicker airfoil has a much higher value of $C_{L(max)}$, and the critical AOA is also higher. The 4 412 airfoil is superior to the 4 406 airfoil in producing lift, but it will also have more drag. Cambered airfoils produce lift at zero AOA, as can be seen in Figure 4.17.

Nearly all airfoil sections have the same slope of their $C_L - \alpha$ curves. This is about a 0.1 increase in C_L for each degree of increase in AOA. Later, when we discuss wings, rather than airfoils, we will see the effects of aspect ratio and sweepback on the slope of these curves.

Figure 4.17 shows the curves for two cambered airfoils. To better compare cambered airfoils and symmetrical airfoils, we can show them together (Figure 4.18). An increase in $C_{L(max)}$ and a decrease in stall

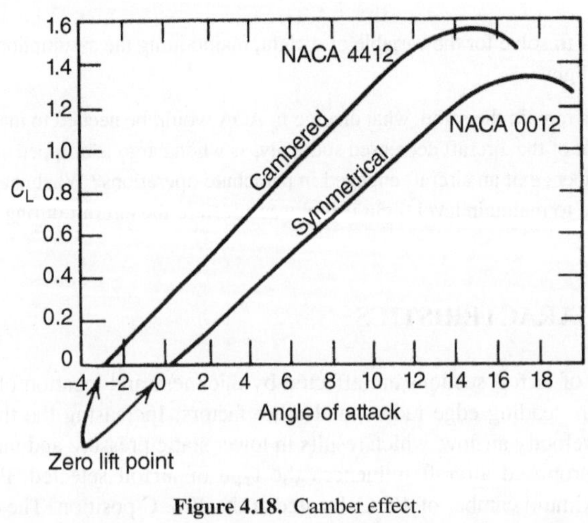

Figure 4.18. Camber effect.

AOA will be noted when the camber is increased. Also, note that the C_L for the symmetrical airfoil for any given AOA will be less than the C_L for the same AOA of the cambered airfoil.

EXAMPLE

Using the curves in Figure 4.18, calculate the stall speed at sea level standard day conditions for a 15 000 lb aircraft whose wing area is 340 ft^2. Calculate V_s for each airfoil section.

Symmetrical ($C_{L(max)} = 1.35$):

$$V_s = \sqrt{\frac{(295)(15\,000 \text{ lb})}{(1.35)(1)(340 \text{ ft}^2)}} = 98.2 \text{ kts.}$$

Cambered ($C_{L(max)} = 1.55$):

$$V_s = \sqrt{\frac{(295)(15\,000 \text{ lb})}{(1.55)(1)(340 \text{ ft}^2)}} = 91.6 \text{ kts.}$$

HIGH COEFFICIENT OF LIFT DEVICES

Landing and takeoff flaps, slots, slats, and other devices that enable an aircraft to take off and land at reduced speeds are often erroneously called high-lift devices (see Chapter 3). The lift of the aircraft during these maneuvers does not exceed its weight, so a high lift is not required. But landing and takeoff speeds can be reduced by increasing the coefficient of lift. So, the proper term is *high coefficient of lift devices.*

In our discussion of airfoil lift characteristics, we saw two ways that the $C_{L(max)}$ of an airfoil could be increased: (i) by increasing its thickness and (ii) by increasing its camber. No practical way of increasing its thickness in flight has been developed to date, so this possibility is not discussed here.

As design and material technology improves, so do the type of *camber changers* available to pilots. Figure 4.19 shows an example of the several common types of flaps and leading-edge devices used within the industry today, and their effect on C_L. Figure 4.20 shows common leading-edge devices usually found on higher-performance aircraft. Note that the designation for movable leading edge or movable slot in both figures is also referred to as a slat. Figure 4.21 shows the effects of pure camber changers on the coefficient of lift curves; as you can see the maximum C_L has increased with the addition of trailing edge flaps, and the C_L curve has shifted to the left.

Figure 4.22 shows the impact of leading-edge slats and leading-edge flaps on C_L and AOA. As leading-edge flaps are deployed the value of $C_{L(max)}$ increases as does the critical AOA. When leading-edge slats are extended, the value of both $C_{L(max)}$ and the critical AOA of attack increase even more, thus the reason these devices are found on many air transport category aircraft is to improve performance during takeoff and landing operations.

Our discussion of boundary layer theory, separation, and stalls showed that airflow separation resulted when the lower levels of the boundary layer did not have sufficient kinetic energy. The air was slowed by friction and an adverse pressure gradient. If we can find a way to increase the kinetic energy in the boundary layer, flow separation can be suppressed. Then higher angles of attack can be used and higher values of C_L can be realized. Devices that use this principle are called *energy adders.*

There are several ways of adding energy to the boundary layer. The simplest way is by stirring up the boundary layer. This replaces the slow-moving laminar airflow next to the wing skin with faster-moving

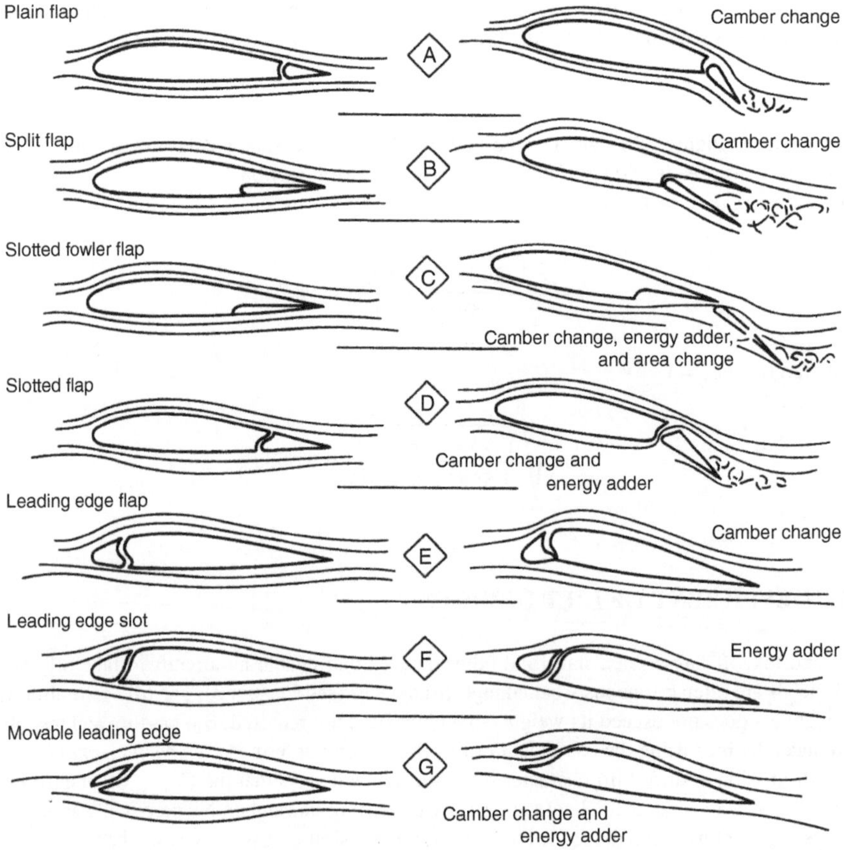

Figure 4.19. High-C_L devices.

turbulent air. One method of producing turbulence in the boundary layer is the use of *vortex generators* or micro-vortex generators (Micro-VGs). These are small airfoil-shaped vanes protruding upward from the wing or other airfoils such as the flaps on the aircraft as shown in Figure 4.23. Vortex generators mix the turbulent outer layers of the boundary layer with slow-moving laminar lower layers, thus reenergizing them, and reducing drag. The advantages of reduced landing and takeoff speeds due to the increase in lift coefficient more than compensate for the increased parasite drag of the vortex generators. During NASA's (2000) testing, Micro VGs were found to produce as much as a 10% increase in lift and a 50% decrease in drag, not only dramatically enhancing performance but also reducing noise during the landing approach.

Another method of adding energy to the boundary layer is by the use of *turbulators*, which are thin zig-zag strips added to an airfoil to energize the slower-moving laminar flow. These strips can be added to the airfoil where turbulent boundary flow is needed to mitigate airflow separation, especially at slower velocities or due to premature airflow separation due to airfoil contamination.

A third method of adding kinetic energy to the boundary layer is to route high-pressure air from the stagnation region, near the leading edge of the wing, to the low-pressure area on top of the wing. This will

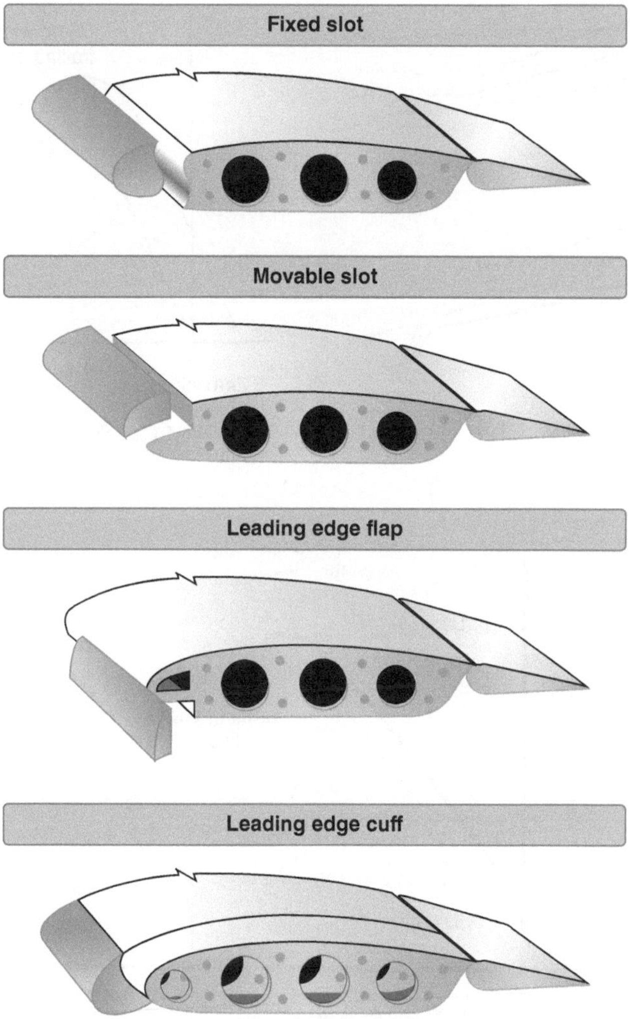

Figure 4.20. Common leading edge high-C_L devices.
Source: Pilot's Handbook of Aeronautical Knowledge (2008a)/U.S. Department of Transportation Federal Aviation Administration/Public domain.

energize the slow-moving lower layers of the boundary layer. Figure 4.24 shows this method, using a fixed slot at the leading edge of the wing. At low angles of attack, the pressure on each end of the slot is about the same, and little or no air flows through the slot (Figure 4.24a). At high angles of attack, the high-pressure stagnation point has moved down below the leading edge. The high-pressure differential causes a large airflow up through the slot and over the top of the wing (Figure 4.24b).

The effect of pure energy adders on the C_L curve is shown in Figure 4.25; note the improvement in C_L and increased critical AOA, with the use of slats. Slats incorporate a slot for higher-pressure air below

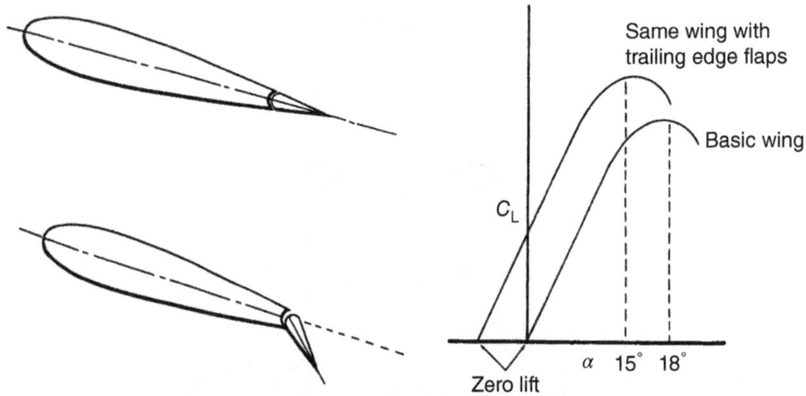

Figure 4.21. Effect of a camber changer on the $C_L - \alpha$ curve.

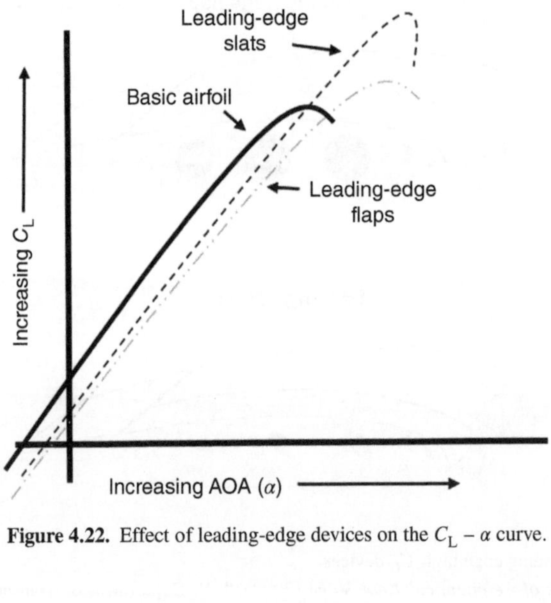

Figure 4.22. Effect of leading-edge devices on the $C_L - \alpha$ curve.

the wing to seek a lower-pressure area above the wing, energizing the boundary layer. The pure energy adders extend the curve to higher values of AOA and $C_{L(max)}$, meaning the critical AOA of the airfoil has increased. The values of C_L at lower angles of attack are not affected. Most of the more complicated high-C_L devices use a combination of camber changer and energy adder techniques. A common example of this on advanced aircraft is the use of leading-edge slats with trailing edge flaps. A slat increases wing camber while adding energy to the wing boundary layer (Figure 4.25a). The result is a higher $C_{L(max)}$ and critical AOA as shown in Figure 4.25b. Slats are typically deployed before trailing edge flap extension

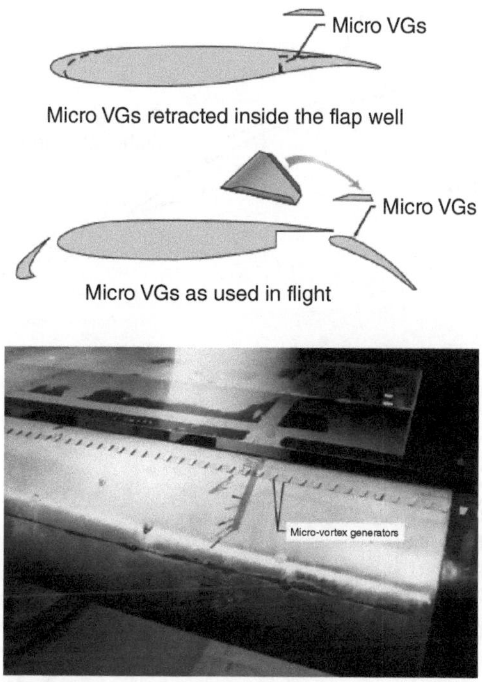

Figure 4.23. Micro-vortex generators.
Source: NASA FS-2000-06-52-LaRC/Public Domain.

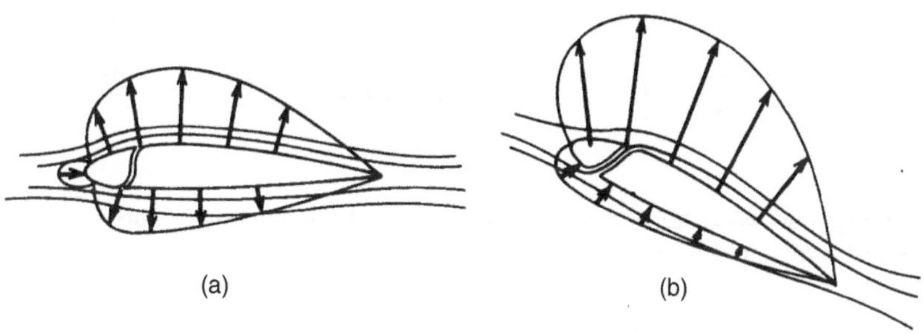

Figure 4.24. Fixed slot at (a) low AOA and (b) high AOA.

to provide a higher stall margin since trailing edge flap deployment reduces the critical AOA as shown in Figure 4.21. The combination of slats and trailing edge flaps allows an aircraft to effectively conduct slow speed operations. The location of slats on the outboard wing section and trailing edge flaps on the inboard wing section also helps ensure the wing root will stall before the wing tips. Figure 4.26 depicts an example of double-slotted flaps on the trailing edge of a Diamond DA50RG, as the flaps are deployed for takeoff and landing, the boundary layer around the flaps is energized by the slots as noted previously,

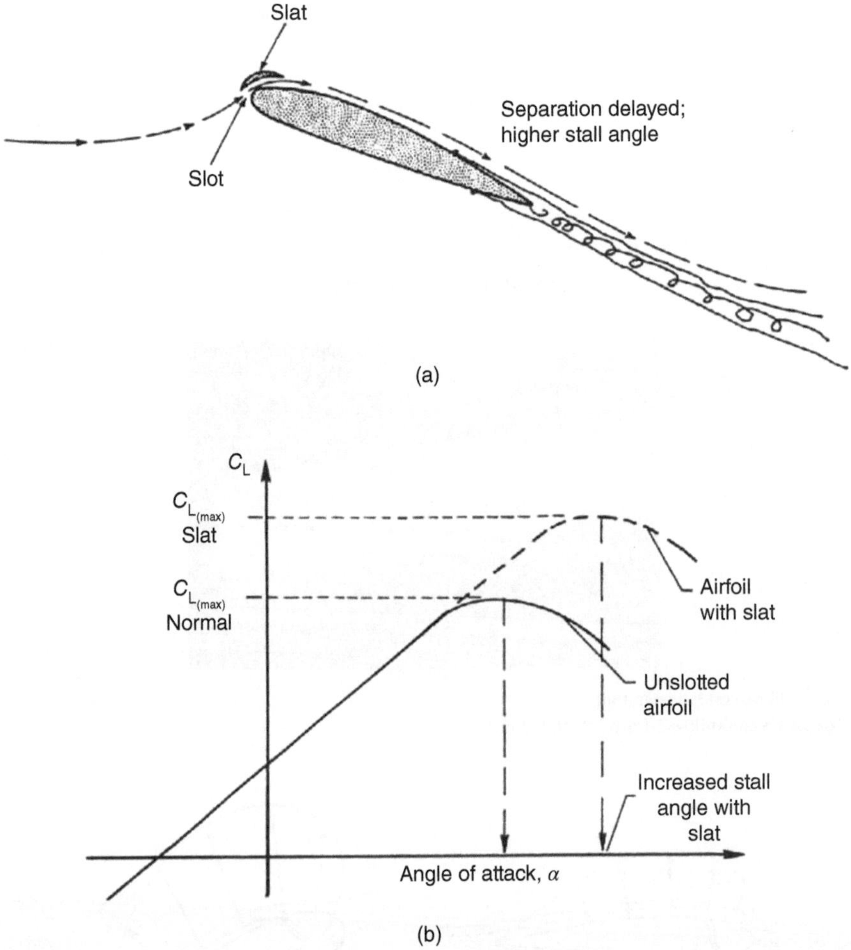

Figure 4.25. Effect of an energy adder on the $C_L - \alpha$ curve. (a) Slat. (b) Slat aerodynamic effects. *Source:* NASA-SP-367/Public Domain.

which delays separation of the boundary layer and changes the critical AOA when the aircraft is operating at slower speeds.

Application 4.2

Aircraft such as the Boeing 727 and 747 use Krueger flaps on the leading edge of the wing to increase wing camber, thus increasing C_L. The Krueger flaps are located on the inboard portion of the wings.

Why would an aircraft not employ Krueger flaps on the outboard leading edge of the wings instead? How would the $C_L - \alpha$ curve shift with the deployment of Krueger flaps?

Figure 4.26. Diamond DA50RG hinged, double-slotted flap.
Source: Courtesy of Authors.

EFFECT OF ICE AND FROST

Ice or frost on the lifting surfaces of an airplane will reduce lift and increase drag. The effects of ice and frost on the lift coefficient curves are shown in Figure 4.27. Ice is likely to accumulate near the leading edge if it is deposited during flight. This changes the contour of the airfoil and results in a considerable reduction of lift, in some cases, C_L may be reduced by 30% or more. The smoothness of the skin is also reduced, and a large increase in drag results. Figure 4.28 depicts the increase of drag on an airfoil with ice versus a clean airfoil, note the impact of drag on AOA even at very small angles of attack. In fact, airfoil drag can reach 100% and in some cases even 200%. The drag coefficient (C_D) will be further defined in the next chapter. As a result of the reduced lift and the increased drag, thrust or power required will be greater and the stall speed will increase. In addition, the weight of the ice will increase the gross weight of the aircraft, complicating the effect of the reduction in lift.

Obviously, all ice must be removed before flight, and any ice formed during flight must be removed using the aircraft's deicing equipment. If the aircraft is not equipped to handle icing conditions or moderate or severe icing is encountered, the aircraft should climb or descend as appropriate to exit those conditions as soon as possible.

The effect of frost on the wings is not so obvious. A layer of evenly distributed frost will not appreciably change the contour of the wing but it will increase the roughness. Drag will be increased, and the kinetic energy of the air will be reduced so that stall will occur at a lower AOA and at a higher airspeed. One study showed that 0.1″ of evenly distributed frost on the aircraft's wings will increase the stalling speed by 35%. This roughly doubles the required takeoff run.

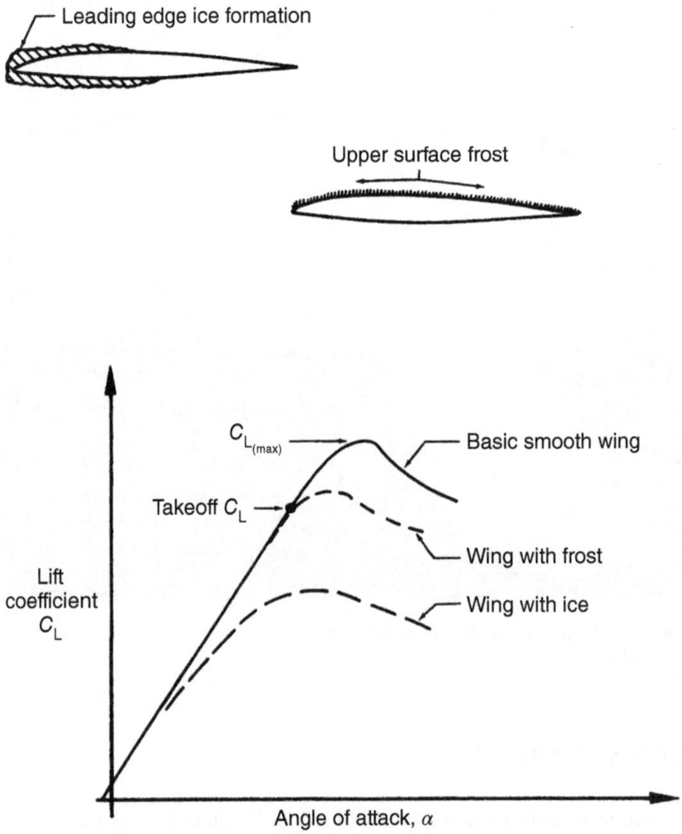

Figure 4.27. Effect of ice and frost on C_L and AOA.

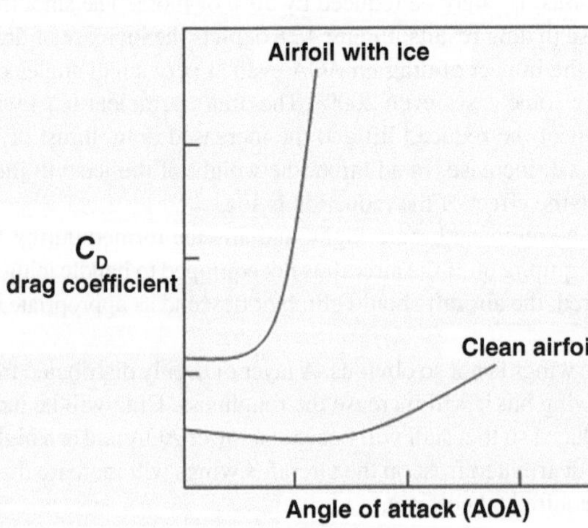

Figure 4.28. Effect of ice and frost on C_D and AOA.
Source: Pilot Guide: Flight in Icing Conditions AC 91-74B (2015)/U.S. Department of Transportation Federal Aviation Administration/Public domain.

LIFT DURING FLIGHT MANEUVERS

In level, unaccelerated flight, lift and weight are both vertical while thrust and drag are both horizontal, and all forces are balanced so that the net force on the aircraft is zero. But in maneuvering flight, which may involve some combination of climbing, descending, and turning, lift, thrust, and drag are usually not vertical. Weight always acts vertically downward, regardless of the attitude of the aircraft. When a vector is not vertical or horizontal, it can be resolved into components which may be vertical and horizontal, or if we choose, we may resolve a vector into components along the flight path and perpendicular to it. For example, during gliding flight, a component of the weight vector is directed forward and adds to thrust. In turn, increased bank angle leads to higher stall speeds and requires more lift to maintain level flight, while excessive aft stick pressure during transitions from descents to climbs results in excessive wing loading and may damage the aircraft. Understanding how these forces affect the performance of the aircraft during maneuvering flight is imperative to maintain proper safety margins.

Lift During Turns

Turning forces will be discussed in greater detail in Chapter 11, so only a limited discussion involving lift will be addressed here. As an aircraft banks, lift not only acts upward but also in the direction of the turn (Figure 4.29). Effective lift is the vertical component of lift and opposes weight (gravity). The horizontal component of lift is also known as centripetal force and is the force that causes the aircraft to deviate from a straight flight path. As an aircraft is banked and a component of lift acts as centripetal force, increased AOA is required in order to satisfy the required effective lift to maintain level flight. The greater the angle of the bank, the greater the horizontal component of lift, and the greater the rate of turn of the aircraft. If coordinated flight is not maintained, a slipping or skidding turn may develop.

Lift During Climbs and Descents

Lift during a stabilized, steady-state climb remains the same as it did when the aircraft was in steady, unaccelerated level flight as long as the airspeed remains the same. As the airspeed is the same, the AOA has not changed and the same amount of lift is being developed. In order to maintain the same airspeed, thrust must be added to overcome the component of weight that now acts in the same direction as drag. Initially during climb entry, when the pilot moves the stick aft, the AOA of the aircraft does increase momentarily, and lift does increase, but then the lift returns to the original amount after thrust is added if the original airspeed is maintained (Figure 4.30).

If a pilot pushes forward on the stick to descend, the AOA will decrease, resulting in a reduction in lift. Lift is momentarily less than weight, so the vertical forces are unbalanced, resulting in a downward acceleration (the transition from level flight to descent). In the descent, if the weight vector is resolved into components parallel to and perpendicular to the flight path, the component of weight parallel to the flight path augments engine thrust, so that less thrust is needed to maintain the same airspeed. If the thrust is reduced to maintain the same airspeed as in steady, unaccelerated flight, the AOA will remain the same and the lift will stabilize to its initial value. If in level, unaccelerated flight the pilot reduces thrust, an imbalance now exists between thrust and drag. Drag is now greater, resulting in a reduction in airspeed; thus, a decrease in lift is now also experienced if the original AOA is maintained. If the AOA is not increased with sufficient back pressure in order to reestablish the required lift to maintain level flight, the lift will be less than weight, and the aircraft will descend, eventually stabilizing in a descent at the new airspeed.

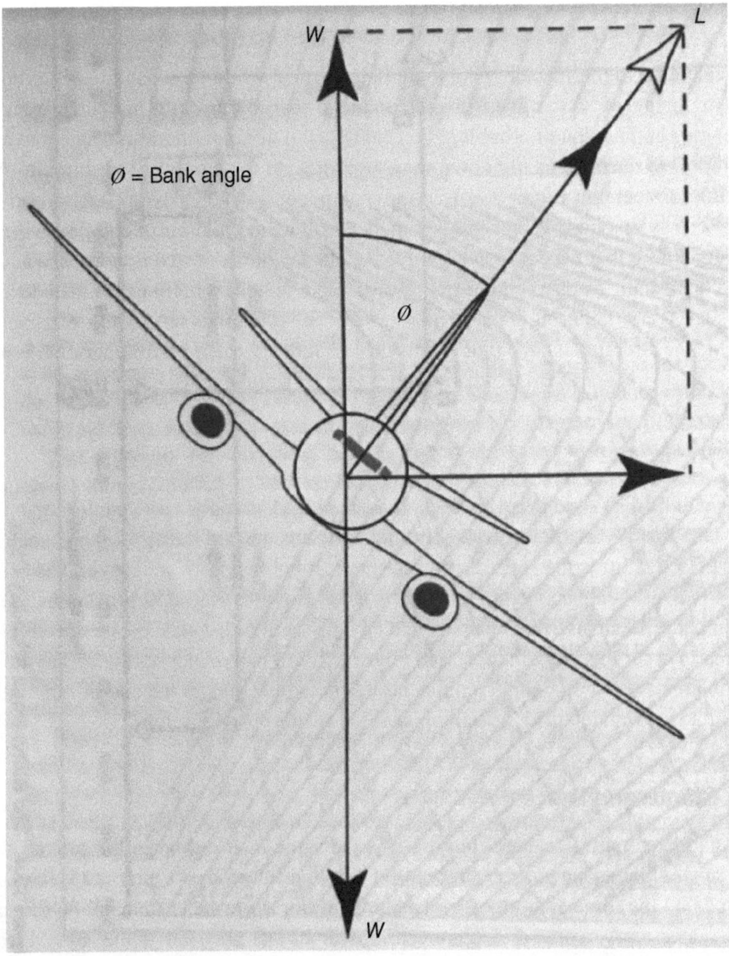

Figure 4.29. Forces in a banked turn.
Source: Swatton (2011)/Principles of Flight for Pilots./with permission of John Wiley & Sons.

Lift and Balloons

In this chapter, we have discussed lift as it applies to fixed-wing aircraft, but the same principles that allow an airplane to take flight also impact other aircraft, like balloons. As a balloon ascends or descends, it also is impacted by the medium of air that surrounds it, and as the balloon's altitude changes and it experiences various horizontal and vertical wind patterns we once again see the impact of air moving around a solid object. The three terms used to describe the different phenomena that affect a balloon in flight as a result of air moving around the balloon are false lift, false heavy, and diminished capacity.

False lift occurs due to air flowing around the top of a balloon. While on the ground a hot air balloon is in a position referred to as neutral buoyancy, there is heated air inside the balloon envelope in addition to the flow of air over the top of the balloon (Figure 4.31). While on the ground the air flowing over the

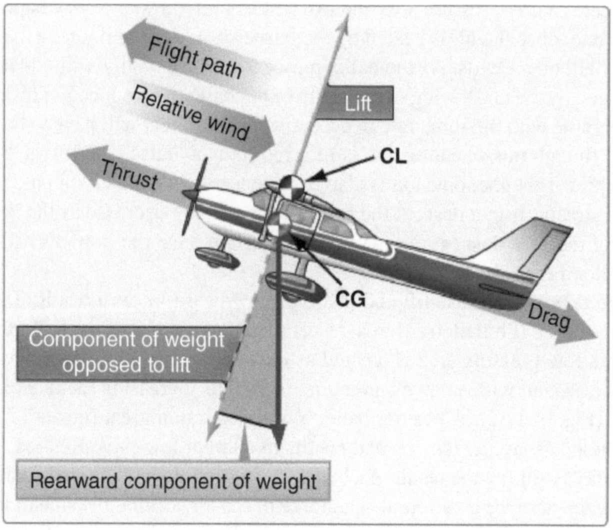

Figure 4.30. Force vectors during a stabilized climb.
Source: Pilot's Handbook of Aeronautical Knowledge (2016b)/U.S. Department of Transportation Federal Aviation Administration/Public domain.

Figure 4.31. False lift during balloon ascent.
Source: Balloon Flying Handbook (2008c)/U.S. Department of Transportation Federal Aviation Administration/Public domain.

top of the balloon creates a low-pressure area and lift, just as was introduced in Chapter 2. At equilibrium, the balloon does not rise, once the air is heated by the burner inside the balloon the balloon is no longer in equilibrium and the balloon ascends. As the balloon ascends, it eventually will equal the speed of the air mass and the lift created by the air flowing over the top of the balloon is reduced or disappears altogether. If the pilot wants to continue with the same rate of ascent, additional heat will have to be added to counteract the loss in lift. Even though this phenomenon is referred to as a "false lift" it is a "real" lift as we have discussed in this chapter. This phenomenon is also experienced when a balloon passes through an area of wind shear or the jet stream. If in a descent the rate of descent may decrease as the lift over the top of the balloon increases, but then the descent rate will increase again once the flow over the top of the balloon diminishes as the balloon exits the fast-moving air.

False heavy is the opposite of false lift and involves downward lift as a result of air flowing around a different part of the balloon. If a balloon is in a descent and the bottom of the balloon encounters an area of faster moving air, a low-pressure area is created as the air flows around the balloon and the lift vector now points toward the ground with a downward component thus increasing the rate of descent if no action is taken by the pilot (Figure 4.32). A heavier balloon will result in a great impact of false heavy because the fabric of the balloon has greater tension and results in a larger low-pressure area.

Diminished capacity is simply the result of a balloon moving into a shear environment where the speed of the horizontal movement of air is different. The force of the air striking the side of the balloon results in a side of the balloon being pushed inward forcing some of the heated air out of the mouth of the balloon, thus diminishing the capacity. As the difference in speed of the two masses increases, so does the amount of diminished capacity. If the pilot is descending and the capacity of the balloon decreases while the lift created by buoyancy decreases, then the result could be a hard landing. Another dangerous situation could develop when a pilot is descending into a slower moving air mass close to the ground, as diminished capacity increases the impact of the loss of heated air may be masked by the false lift of the lower-pressure

Figure 4.32. False heavy during balloon descent.
Source: Balloon Flying Handbook (2008c)/U.S. Department of Transportation Federal Aviation Administration/Public domain.

air on top of the balloon, but as the balloon descends out of the faster moving air and the false lift diminishes then the rate of descent may increase and result in a hard landing before the burner can be activated by the pilot to mitigate the impact in loss of lift.

SYMBOLS

C_D	Coefficient of drag (dimensionless)
C_L	Coefficient of lift
$C_{L(max)}$	Maximum value of C_L
G	Load factor (dimensionless)
R_g	Reynolds number (dimensionless)
S	Wing area (ft^2)
V_s	Stall speed velocity (kts.)
v (nu)	Kinematic viscosity (ft^2/s)

EQUATIONS

4.1 $\alpha = \theta - \gamma$

4.2 $R_g = \frac{Vx}{v}$

4.3 $L = \frac{C_L \sigma V^2 S}{295}$

4.4 $C_L = \frac{295W}{\sigma V^2 S}$

4.5 $V = \sqrt{\frac{295L}{C_L \sigma S}}$

4.6 $V_s = \sqrt{\frac{295L}{C_{L(max)} \sigma S}}$

4.7 $V_2 = V_1 \sqrt{\frac{W_2}{W_1}}$

KEY TERMS

Adverse pressure gradient
Angle of attack indicator
Boundary layer
Centripetal force
Coefficient of lift/drag
Critical angle of attack
Diminished capacity
Energy adders
False heavy
False lift
Flow reversal
High coefficient of lift devices

Laminar flow
Lift equation
Reynolds number
Stall
Turbulent flow
Vortex generators

PROBLEMS

Note: Answers to the problems in the first two sections are given at the end of the book.

Application of Chapter Theory

1. As the thickness of an airfoil is increased, the critical AOA
   a. is greater.
   b. is less.
   c. remains the same.

2. As the camber of an airfoil is increased, its C_L at any AOA
   a. is less.
   b. remains the same.
   c. is greater.

3. The Coanda effect is an example of:
   a. Newton's second law.
   b. Newton's third law.
   c. Newton's first law.
   d. Bernoulli's equation.

4. Air in the boundary layer
   a. has zero velocity at the wing surface.
   b. is turbulent near the leading edge.
   c. is more apt to stall if it is turbulent.
   d. None of the above.

5. Air in the boundary layer
   a. changes from laminar to turbulent at low Reynolds numbers.
   b. separates from the wing when its velocity is maximum.
   c. reverses flow direction when stall occurs.
   d. None of the above.

6. Two things an airfoil designer can change to increase $C_{L(max)}$ are
   a. thickness and wing area.
   b. chord length and aspect ratio.
   c. camber and wing span.
   d. thickness and camber.

7. High values of the Reynolds number will more likely
   a. occur near the leading edge of an airfoil.
   b. indicate laminar flow.

c. occur at lower airspeeds.
d. show turbulent airflow.

8. Adverse pressure gradient on an airfoil is found
a. from the point of maximum thickness to the trailing edge.
b. near the stagnation point at the leading edge.
c. from the point of minimum pressure to the trailing edge.
d. Both (a) and (c).

9. At low-velocity stall, the airflow
a. stops.
b. reverses direction.
c. speeds up.
d. moves toward the wingtips.

10. Airflow separation can be delayed by
a. making the wing surface rough.
b. using vortex generators.
c. directing high-pressure air to the top of the wing or flap through slots.
d. All of the above.

11. Show which type of high-C_L device (camber changer, energy adder, or combination) each of the below is:
a. Blowing air (boundary layer control) over a flap _____
b. Fixed slot_____
c. Slotted leading-edge flap_____
d. Plain flap_____

12. An airplane in flight encounters wing icing. The greatest danger is that
a. weight is increased.
b. C_D is increased.
c. C_L is decreased.
d. All of the above.

13. According to Figure 4.15, what is the corresponding AOA if the C_L is 0.8?
a. 4°
b. 6°
c. 8°
d. 10°

Calculations Using Chapter Principles

14. Calculate an aircraft's angle of attack if it maintains a pitch angle of positive 4° to maintain a 3° ILS glide path (descending).

15. Find the C_L on an airfoil given the following information:

Weight = 14 000 lb

Wing area, $S = 134$ ft^2

TAS = 140 kts.

Sea level, standard day

16. Find the lift on an airfoil using Figure 4.15 and Table 2.1 given the following information:

AOA = 12°

Wing area, $S = 110$ ft^2

TAS = 130 kts.

Pressure altitude = 8 000 ft

Temperature – standard day

17. An aircraft has the $C_L - \alpha$ curve shown in Figure 4.15. The following data applies:

Weight = 20 000 lb

Wing area, $S = 340$ ft^2

Density altitude = 10 000 ft

TAS = 242.4 kts.

Find the aircraft's AOA for steady flight: ________

18. Calculate the stall speed for the aircraft in Problem 17: ________

19. If the airplane in Problem 17 burns 5 000 lb of fuel during its flight, calculate its new stall speed.

20. An airplane has a rectangular wing with a chord length of 8 ft. If the airplane is flying at 200 kts. TAS and at 20 000 ft density altitude, find the Reynolds number at the trailing edge of the wing.

21. Find the stall speed (V_s) for an aircraft at sea level, the standard day with the following information:

C_L = Use figure

Wing area, $S = 110$ ft^2

Weight = 11 000 lb

1G Flight

Correlation Through Scenario-Based Learning

Aircraft – *Diamond DA50RG*

1. Using Figure 4.33, find the following:
- **a.** Cambered or symmetrical airfoil
- **b.** Approximate $C_{L(max)}$ numerical value
- **c.** Critical angle of attack

2. Calculate the Reynolds number (R_g) at the trailing edge of the wing given the following:

Density altitude = 10 000 ft.

TAS = 175 kts

Kinematic viscosity(ν) = From Table 2.1

Distance(x) = Chord of 4 ft.

3. Calculate the C_L in cruise flight based on the following:

Density altitude = 10 000 ft.

Wing area – Review DA50 Aircraft Flight Manual

Weight = 3 900 lb.

TAS = 175 kts.

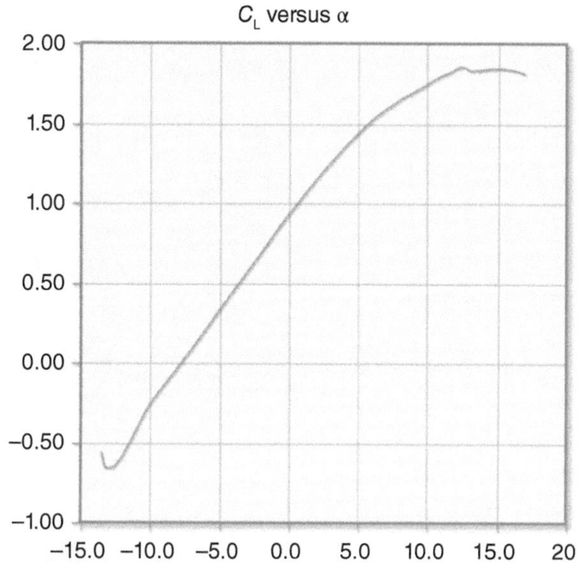

Figure 4.33. DA50RG wing airfoil C_L versus AOA curve.
Source: Created by authors using airfoiltools.com.

4. Using the C_L from Problem 3 and Figure 4.33, what is the approximate AOA for the aircraft while in cruise flight?
5. Using the given information from Problem 3 (except 175 kts.), and your answer for Problem 1b, calculate the stall speed when at 10 000 ft (assume 1*G* flight).
6. Calculate the stall speed for the aircraft at 10 000 ft under the same conditions as in Problem 5, except increase the weight to 4 407 lb. Do you think the Vs will increase, decrease, or remain the same? Why?
7. Using the same website as you used to find the DA50 airfoil in Chapter 3, review the AFM and input the airfoil type for the DA50 horizontal/vertical stabilizer. If able, using the same website, create a C_L versus AOA curve like Figure 4.33. Based on the C_L curve, is the horizontal/vertical airfoil a cambered or symmetrical airfoil? Why? How do the $C_{L(max)}$ and the critical angle of attack from the new C_L curve compare to the wing C_L curve?

5 Drag

CHAPTER OBJECTIVES

After completing this chapter, you should be able to:

- Define induced drag and explain the common terms that are used to characterize the wing planform.
- Explain aspect ratio and identify the relationship to induced drag.
- Describe the formation of wingtip vortices and correlate to wingspan.
- Define ground effect and summarize the impact of ground effect on aircraft performance.
- Define parasite drag and identify four types of parasite drag on fixed-wing aircraft.
- Explain the variables within the total drag equation and recognize how each variable influences the production of total drag.
- Explain the importance of the lift-to-drag ratio, how the lift-to-drag ratio is determined, and correlate $(L/D)_{max}$ to the safety of flight.
- Identify various methods of drag reduction on fixed-wing aircraft and relate drag reduction to aircraft performance.

Drag is the component of the aerodynamic force that is parallel to the relative wind (RW) and retards the forward motion of the aircraft. At subsonic speeds, two kinds of drag are most important: *induced drag and parasite drag*. A third kind of drag, compressibility drag, occurs at high speeds and will be discussed in Chapter 14.

Drag is generated by the interaction of a solid body with a fluid (liquid or gas), and for drag to be generated, the solid body must be in contact with the liquid or gas. Drag is generated by the difference in velocity between the solid object (wing) and the fluid (air). There must be motion between the object and the fluid, so if no motion exists, then no drag can exist. As we saw in Chapter 3, it makes no difference whether the object moves through a static fluid or whether the fluid moves past a static solid object. Because drag is a force, it is a vector quantity and possesses both a magnitude and a direction. We have seen that drag acts in a direction that is opposite to the motion of the aircraft, whereas lift acts perpendicular to the motion, and thrust acts in the direction of the motion.

Flight Theory and Aerodynamics: A Practical Guide for Operational Safety, Fifth Edition.
Brian A. Johnson and Philip R. Fittante.

Companion website: www.wiley.com/go/flighttheoryandaerodynamics5e

INDUCED DRAG

Induced drag, D_i, is the least understood type of drag, but it is the most important, especially in the critical low-speed region of flight. It is commonly but imperfectly called drag due to lift because it occurs only when lift is being developed. Induced drag occurs because the distribution of lift is not uniform across the wing as it varies from the wing root to the wingtip. As discussed in Chapter 4, there is a pressure difference between the upper and lower surfaces of the wing. Vortices are formed at the wingtips, which produce a circular flow that is very strong near the tips of the wing and decreases in intensity closer to the wing root. The magnitude of induced drag depends on the amount of lift being generated by the wing, the distribution of lift across the span, and the configuration of high-lift devices and landing gear. In general, an aircraft produces a high magnitude of induced drag when it is heavy, clean, and slow. If an aircraft is producing lift, then it is also producing induced drag.

Application 5.1

It has been stated that an aircraft produces maximum induced drag when it is heavy, clean, and slow.

In what phase(s) of flight would a fixed-wing airplane be in this position? What about a rotary-wing aircraft? Explain why induced drag is greater in each condition (i.e. heavy, slow, and clean). How would the extension of trailing edge flaps impact the critical angle of attack (AOA)?

Up to this point, we have been discussing airfoils, not wings. We must now consider the planform of the wing, or the shape of the wing as viewed from above as shown in Figure 5.1.

In describing the planform of a wing, several terms must be explained. These are shown in Figure 5.2.

1. The *wing area, S*, is the plan surface area, including the area covered by the fuselage.
2. The *wingspan, b*, is the distance from wingtip to wingtip.
3. The *average chord*, c_{av}, is the geometric average chord:

$$\text{Span} \times \text{Average chord} = \text{Wing area } (bc_{av} = S)$$

4. The ***aspect ratio*** (AR) is the span divided by the average chord, or the square of the wingspan divided by the wing area:

$$\text{AR} = b/c_{av} \quad \text{and} \quad \text{AR} = b^2/S$$

5. The *root chord*, c_r, is the chord measured at the aircraft centerline. The tip chord c_t, is the chord measured at the wingtip.
6. The *taper ratio*, λ (lambda), is the ratio of c_t to c_r:

$$\lambda = c_t/c_r$$

7. The *sweep angle*, Λ (uppercase lambda), is the angle between the line of 25% chord points and a perpendicular to the root chord.
8. The *mean aerodynamic chord* (MAC) is the chord drawn through the centroid (center of area) of the half-span area. The MAC and the c_{av} are not the same. If the actual wing were replaced by a rectangular wing having the same span and a chord equal to the MAC, the pitching moments of each would be identical.

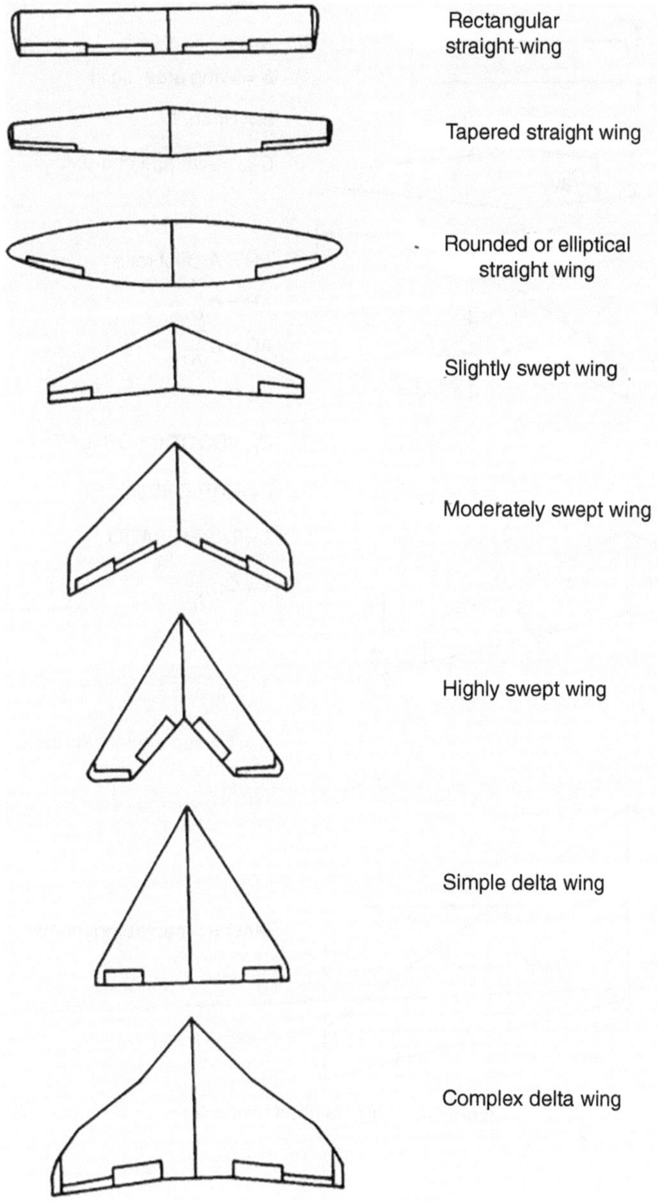

Figure 5.1. Wing planform examples.
Source: NASA-SP-367/Public Domain.

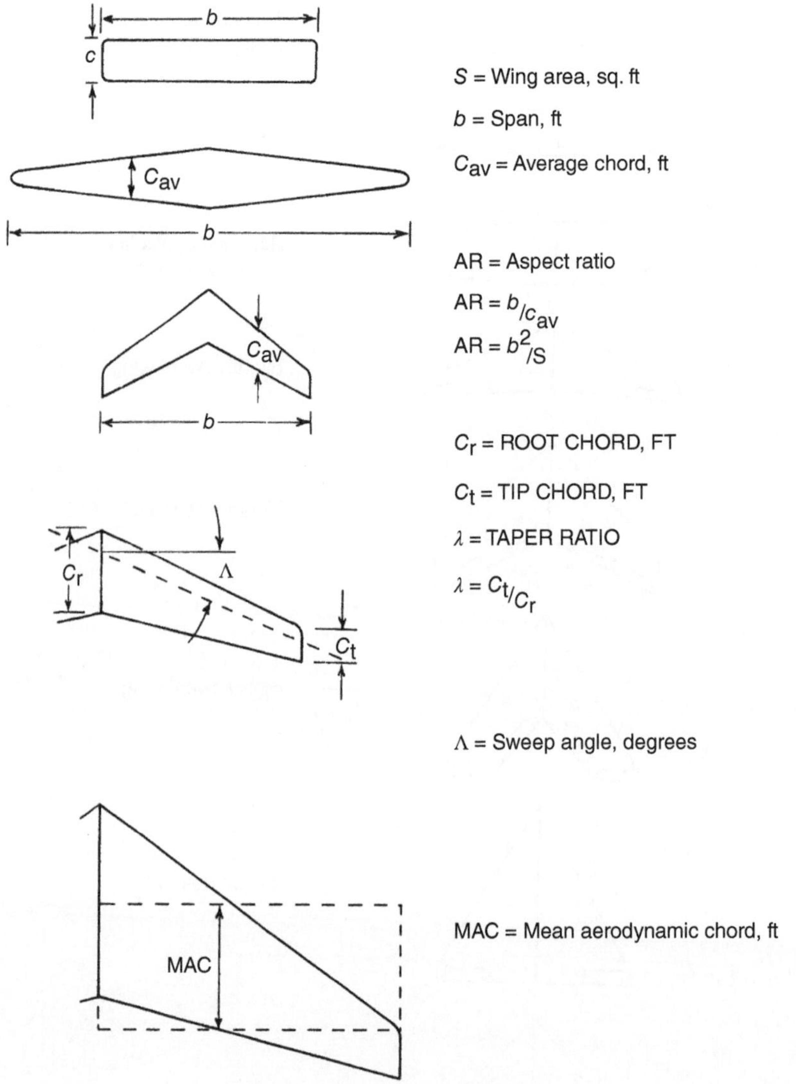

Figure 5.2. Wing planform terminology.

Aspect Ratio

As defined in Figure 5.2, the *AR* is found by dividing the wingspan by the average wing chord.

EXAMPLE

Calculate the aspect ratio given the following information:

Wingspan: 40 ft
Wing area: 176 ft^2

$$AR = b^2/S \rightarrow AR = (40\,\text{ft})^2/176\,\text{ft}^2 \rightarrow AR = 9$$

The longer the wingspan, the larger the AR. High ARs produce more lift at lower angles of attack with less induced drag than aircraft with lower ARs. A wing with a higher AR will produce smaller wingtip vortices than a wing with a lower AR, thus lower induced drag. With this improvement in efficiency also comes possible consequences due to the higher AR wing design. Longer and narrower wings may be less maneuverable, and structurally the wings may have to be stronger due to increased loading as well as bending and twisting. Longer and narrower wings may not be able to store as much fuel as more traditional shorter and wider wings. Many general aviation training aircraft have an AR lower than 10, so with a lower AR also comes more induced drag, reduced airspeed, and a reduction in takeoff and landing performance compared to aircraft with higher ARs.

The longer the wingspan, the closer the airplane is to having an infinite wing. Of course, an infinite wing is theoretical only, but we will find that the longer the wingspan the lower the impact of induced drag. Figure 5.3 shows the AR of two gliders with different wingspans, the glider on the bottom with the longer, narrower wings has a larger AR than the glider on the top, so the wingtip vortices are smaller, and the glide ratio of the bottom glider is higher. The glide ratio will be defined and expanded later in this chapter.

Application 5.2

Research the wing area and wingtip dimensions of a variable sweep-wing aircraft, examples include the General Dynamics F-111 Aardvark, the Grumman F-14 Tomcat, and the Rockwell B-1 Lancer.

What is the variable-sweep range in degrees from full forward to full-sweep? What is the change in wing area throughout the sweep regime? What is the change in wingspan throughout the sweep regime? How does the aspect ratio change from full forward to full-sweep and what impact does this have on induced drag?

Wingtip Vortices

Wingtip vortices are circular patterns of rotating air trailing behind a wing as it moves through the air and generates lift. Each wingtip vortex trails from the tip of the wing and eventually descends and dissipates. Understanding the effect of these vortices on other aircraft that may be flying in the same area leads to the core concept of wake turbulence avoidance and will be discussed in Chapter 10.

A *wingtip vortex* is formed when higher-pressure air below a wing flows around the wingtips into the low-pressure region on top of the wing that is developing lift. The vortices are strongest at the tips due to the spanwise flow underneath the airfoil moving out toward the wingtips and around them; the vortices

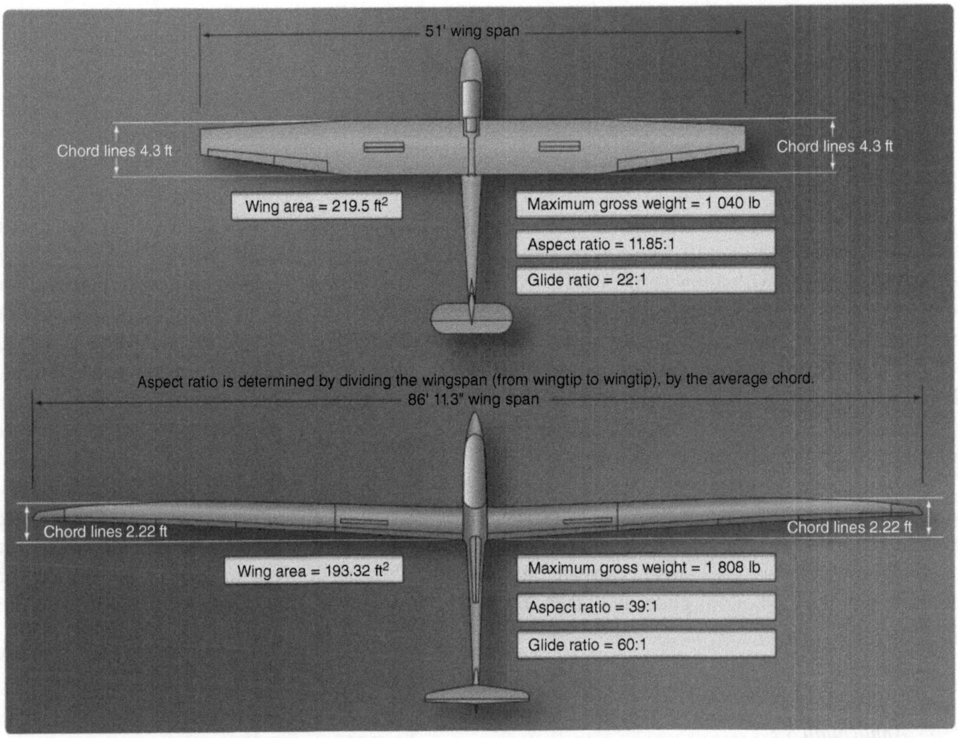

Figure 5.3. Aspect ratio.
Source: Glider Flying Handbook (2013)/U.S. Department of Transportation Federal Aviation Administration/Public Domain.

become weaker toward the centerline of the aircraft (Figure 5.4). Wingtip vortices are associated with induced drag as a consequence of downwash and are an inevitable consequence of lift generation. When a wing generates lift, the air on the top surface has a lower-pressure relative to the pressure on the bottom surface so pressure gradient force moves the air below the wing up and out around the tip to the top of the wing in a circular fashion.

If viewed from the tail of the airplane looking forward in the direction of flight, there is one wingtip vortex trailing from the left wing circulating clockwise, and another one trailing from the right wing circulating anticlockwise.

Infinite Wing

Consider an aircraft with an infinitely long wing. This *infinite wing* has no wingtips and so it has no wingtip vortices. The absence of wingtip vortices means that the upwash in front of the wing and the downwash behind the wing cancel each other, so there is no net downwash behind the wing (Figure 5.5). The RW ahead of this wing is horizontal, and the RW behind the wing is also horizontal. The RW at the aerodynamic center (AC) is the average of these two and must be horizontal. Lift is perpendicular to the RW, so it acts vertically. The vertical components of the local RW for an infinite wing are plotted in Figure 5.6. Note that the RW at the AC has no upward or downward velocity and is horizontal.

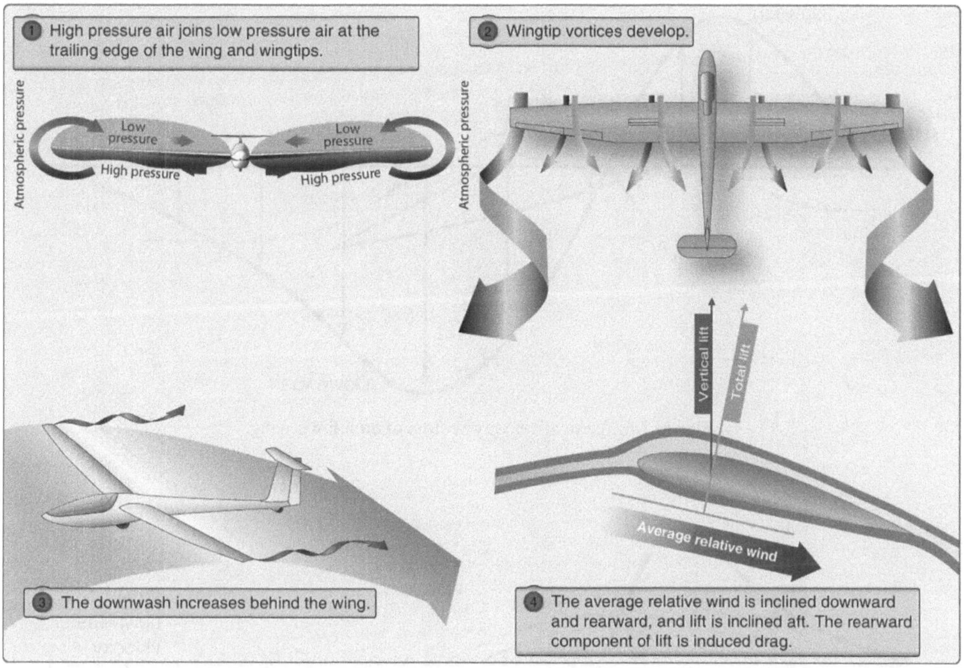

Figure 5.4. Induced drag due to wingtip vortices.
Source: Glider Flying Handbook (2013)/U.S. Department of Transportation Federal Aviation Administration/Public Domain.

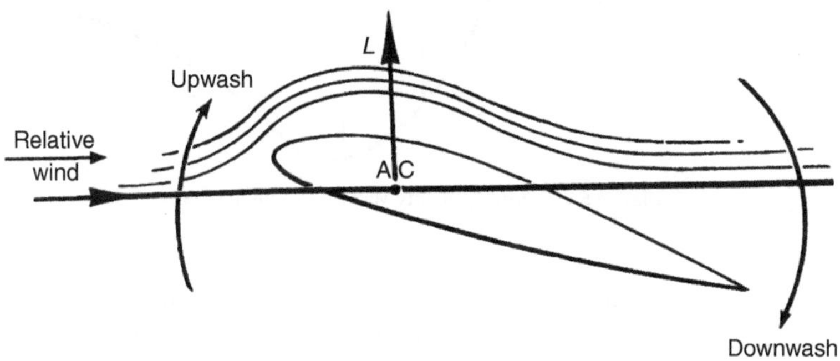

Figure 5.5. Airflow about an infinite wing.

Finite Wing

A wing with wingtips is called a *finite wing*. A finite wing that is developing lift will have tip vortices. These vortices force air down behind the wing. The vertical components of the local RW for a finite wing that is producing lift are shown in Figure 5.7.

The local RW is the vector sum of the free stream RW and the vertical velocity vectors (Figure 5.7). The RW far ahead of the aircraft is not disturbed by upwash, so for an aircraft flying horizontally, it is also

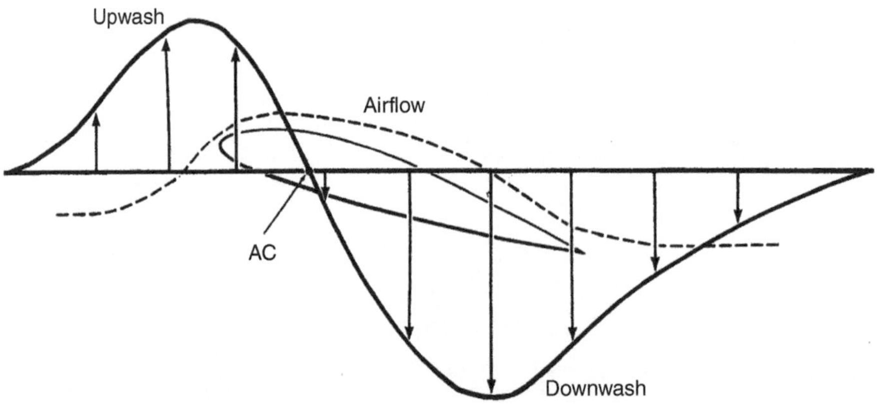

Figure 5.6. Vertical velocity vectors of an infinite wing.

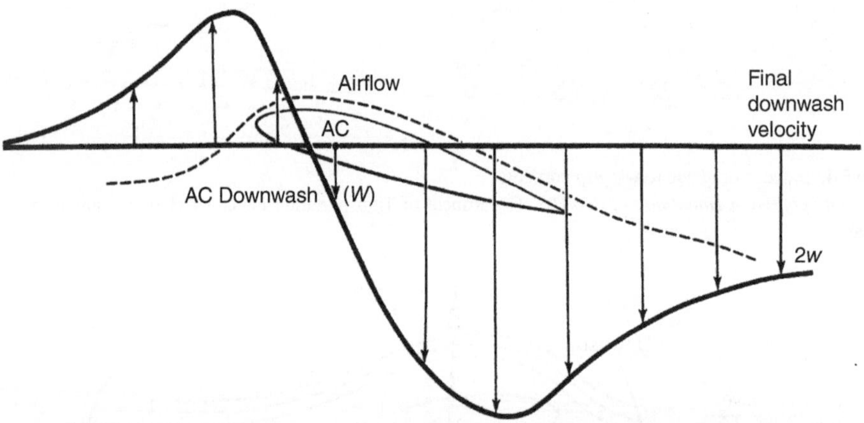

Figure 5.7. Vertical velocity vectors of a finite wing.

horizontal. The local RW behind the wing is depressed by the final downwash velocity, $2w$. The angle that the RW leaving the wing makes with the free stream RW is called the downwash angle. It can be seen in Figure 5.8.

The local RW at the AC is the average of the approaching and departing RW. The downwash angle of the local RW at the AC is, therefore, one-half of the trailing edge downwash angle. The lift vector at the AC, which is perpendicular to the local RW, is tilted back by the same angle. The angle at which the lift vector is tilted back is called the *induced angle of attack*, α_i.

The finite wing is operating at a lower AOA than the infinite wing due to the depressed RW at the AC. Thus, if the finite wing is to produce the same lift coefficient as the infinite wing, the AOA, with respect to the free stream RW, must be increased by α_i. The AOA of the infinite wing is α_0; the AOA of the finite wing is α:

$$\alpha = \alpha_0 + \alpha_i \tag{5.1}$$

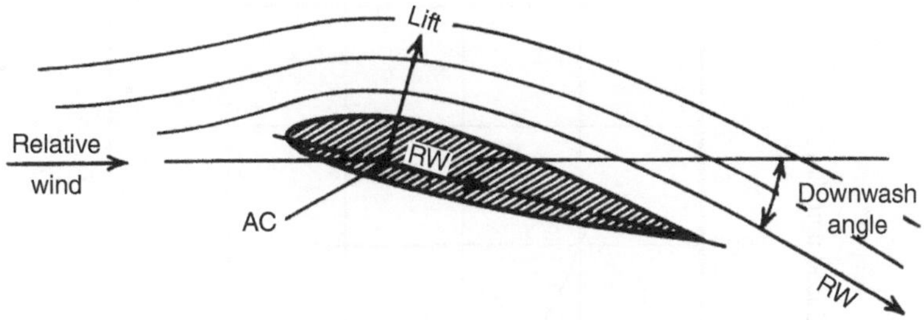

Figure 5.8. Airflow about a finite wing.

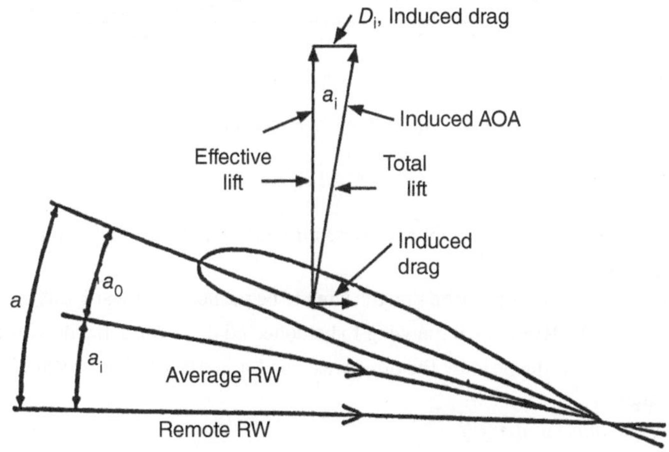

Figure 5.9. Relative wind and force vectors on a finite wing.

This is shown vectorially in Figure 5.9. The vertical vector, effective lift, is the lift vector of the infinite wing, and the tilted vector, total lift, is the lift vector of the finite wing. The component of the tilted lift vector that is parallel to the remote RW is the induced drag D_i.

Induced Drag Summary

Induced drag is influenced by the lift coefficient and AR. It increases directly as the square of C_L and inversely with AR. The most critical condition, where induced drag is greatest, is during low-speed (high-C_L) flight with a low AR (short-wingspan) aircraft. Equation 5.2 shows that *induced drag varies inversely with the square of velocity*, and Figure 5.10 shows a plot of induced drag versus airspeed. Maintaining level flight, the slower the true air speed (TAS) the greater the C_L (AOA), and thus the greater the induced drag.

$$\frac{D_{i_2}}{D_{i_1}} = \left(\frac{V_1}{V_2}\right)^2 \tag{5.2}$$

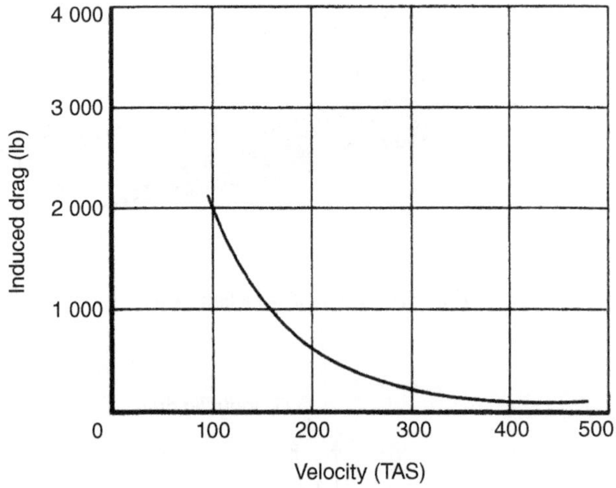

Figure 5.10. Induced drag versus velocity.

1. Wingtip vortices are formed by higher-pressure air beneath a wing moving into the lower-pressure air above the wing.
2. Wingtip vortices cause the airflow behind the wing to be pushed downward; this is called downwash.
3. Downwash causes the RW behind the wing to be deflected downward at a downwash angle.
4. The RW at the AC is influenced by the downwash, and it is deflected downward by one-half of the downwash angle.
5. Lift acts perpendicular to the RW.
6. The lift vector is tilted backward by the induced angle of attack, α_i, which is numerically equal to one-half of the downwash angle.
7. The rearward component of the tilted lift vector is induced drag.

GROUND EFFECT

When an aircraft is close to the earth's surface, a phenomenon occurs that changes the flying characteristics discussed thus far. Three effects can occur when an airplane is flown close to the runway or to a water surface; collectively they are called *ground effect* and they are noticeable when the aircraft is about one wingspan or less above the surface. The three effects are:

1. Airflow alteration around the wing
2. Pitching moments
3. Pitot–static influence

Figure 5.11 shows the change in the size of the vortices when the airplane is close to the ground versus at a higher altitude. Figure 5.12 shows that the closer the aircraft is to the surface, the greater the influence of the ground on the downwash, and thus the greater the reduction in the impact of induced drag. When

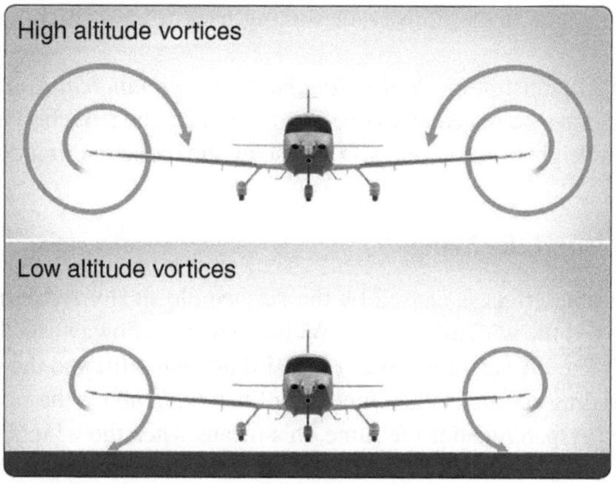

Figure 5.11. Wingtip vortex at altitude versus near the ground.
Source: Pilot's Handbook of Aeronautical Knowledge (2016b)/U.S. Department of Transportation Federal Aviation Administration/Public domain.

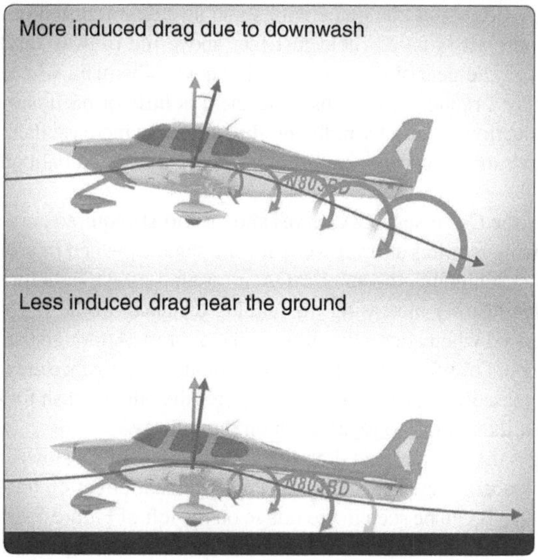

Figure 5.12. Downwash at altitude versus near the ground.
Source: Pilot's Handbook of Aeronautical Knowledge (2016b)/U.S. Department of Transportation Federal Aviation Administration/Public domain.

you have additional downwash, the RW is deflected downward to a greater degree, and as we know the lift vector is perpendicular to the RW. So, if you have more downwash, the lift vector is deflected backward to a greater degree and thus produces more induced drag.

As the ground interferes with the airflow around the aircraft, changes in pitch along with variations in throttle may be required. Many pilots have miscalculated the effects of ground effect and have found themselves floating off the end of the runway, or sinking back toward the runway with terrain quickly approaching.

The Federal Aviation Administration *Pilot's Handbook for Aeronautical Knowledge* (2016b, p. 5–11) states that when the wing is at a height equal to one-fourth its span, the reduction in induced drag is 23.5%, and when the wing is at a height equal to one-tenth its span, the reduction in induced drag is 47.6%.

Airflow Alteration Around the Wing

The first and most noticeable effect is caused by the surface physically reducing the downwash, which reduces both the upwash and the wingtip vortices. Without so much downwash, the local RW at the AC is closer to the remote RW. The lift vector is closer to the infinite wing lift and the induced drag is reduced. The reduction in induced drag coefficient is a function of height, *h*, above the runway to the wingspan, *b*. Of course, because the wingspan remains the same, this means when the wing's height above the runway decreases, the reduction in C_D increases. Pilots transitioning from an aircraft with a low AR to an aircraft with a high AR may also experience significant improvement in aircraft performance when in ground effect; so, careful monitoring of airspeed and pitch attitude is needed when entering ground effect upon landing.

Low-winged aircraft are affected more by ground effect than high-winged aircraft. This reduction of drag is clearly seen in light aircraft in their tendency to "float" during landings. To reduce this floating tendency, the aircraft approach should be made with some throttle on, instead of at idle, then the throttle should be closed as the aircraft is flared (held just feet above the runway until in a nose high attitude until at a safe speed for tricycle gear aircraft to touchdown with the main wheels first). The reduction of drag is compensated for by a reduction in thrust, and there is little or no floating action. Pilots who fail to compensate for the reduction in drag by reducing thrust greatly increase their landing distance, which may lead to additional undesirable consequences such as porpoising and ballooning, or possibly running out of runway.

Ground effect changes the C_L versus AOA curves and the thrust-required curves (Figure 5.13). According to the C_L curve, the wing requires a lower AOA for the same C_L when in ground effect than when out of ground effect. If a pilot maintains the same AOA in ground effect as before entering ground effect, the aircraft may float, significantly increasing the landing distance. In the B-2 stealth bomber, this phenomenon is quite pronounced when attempting to land the aircraft. Although the flying wing design has a relatively low AR, the amount of lift produced by the enormous wing surface area is significantly impacted by ground effect. The increase in C_L at a given AOA requires the pilot to push forward on the control stick, thus reducing AOA, to get the aircraft to touchdown on the runway.

Because induced drag is the predominant drag (versus parasite drag) at slower speeds and higher angles of attack, reducing the induced drag at this point has a significant effect on total drag and will require a similar decrease in thrust. This can be seen in the image on the left of Figure 5.13: less thrust is required to achieve a given velocity when in ground effect. The reduction of thrust required also results in a reduction in fuel flow. Though usually not a significant factor for the common training aircraft, this effect can be used in emergency flying over water or flat ground if the fuel available is critical. A word of caution to pilots of turbine engine aircraft: fuel consumption is high for these aircraft at low altitudes, so the benefits of ground effect may be nullified by poor fuel consumption.

Pitching Moments

The other effects of flying in ground effect are less well known but influence many aircraft. One of these concerns pitching moments that develop when entering or departing from ground effect. A nose up pitching moment is required to rotate an aircraft for takeoff or to flare the aircraft during landing. This requires a

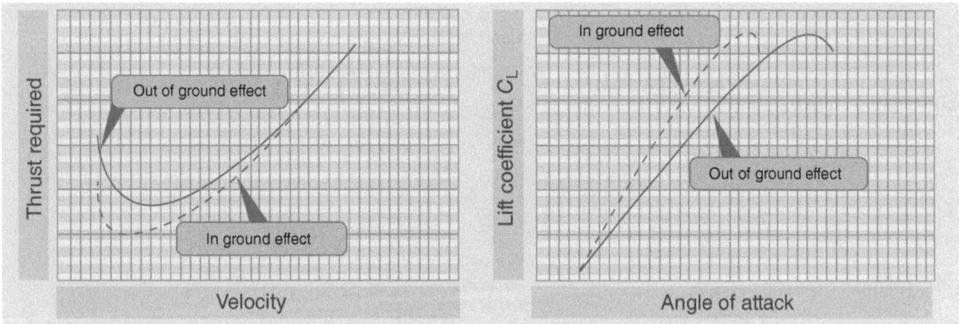

Figure 5.13. T_r and C_L curves in ground effect.
Source: Pilot's Handbook of Aeronautical Knowledge (2008a)/U.S. Department of Transportation Federal Aviation Administration/Public domain.

download on the tail. To obtain this pitching moment, the pilot applies back stick pressure so that the horizontal tail is effectively operating at a negative AOA. The downwash from the wing accentuates this negative AOA.

When the aircraft enters ground effect, on landing, the downwash is decreased, the negative AOA is reduced, the downward lift on the tail is reduced, and the aircraft experiences a nose-down pitching moment. The pilot should be aware of this pitching moment and prepare accordingly by maintaining back pressure on the stick. Hightailed aircraft will not experience this pitching moment if the tail is above the wing downwash and is not influenced by it.

During takeoff, as the aircraft leaves ground effect, the opposite pitching moment occurs. Here, the increased downwash hits the horizontal tail and causes a download on it and a nose-up pitching moment result. Unless corrective action is taken, this will cause a sharp increase in drag that, when coupled with the increase in induced drag that results when an aircraft leaves ground effect, may prevent a successful takeoff. In situations involving a takeoff with a high-density altitude, or in an overweight condition, the aircraft may become airborne within ground effect and then not be able to climb out of ground effect. As the nose pitches up when leaving ground effect, the situation is exacerbated, and the aircraft may stall.

Pitot–Static Influence

The third effect of flying in ground effect occurs in aircraft that have static airspeed ports located below the wings. There is a small, but measurable, increase in static pressure below the wings as an aircraft enters ground effect. This creates a false reading from the airspeed indicator: it will read lower than the actual airspeed and altitude on entering ground effect. This goes unnoticed during takeoffs and landings as the pilot is usually monitoring the runway, but on long flights over water, pilots trying to take advantage of ground effect may be upset to see the airspeed decrease as the aircraft approaches the water.

Ground Effect Summary

Ground effect must always be considered when landing or taking off. The consequences of a high-density altitude, overweight aircraft, CG out of limits, or poor pilot technique when combined with the effects

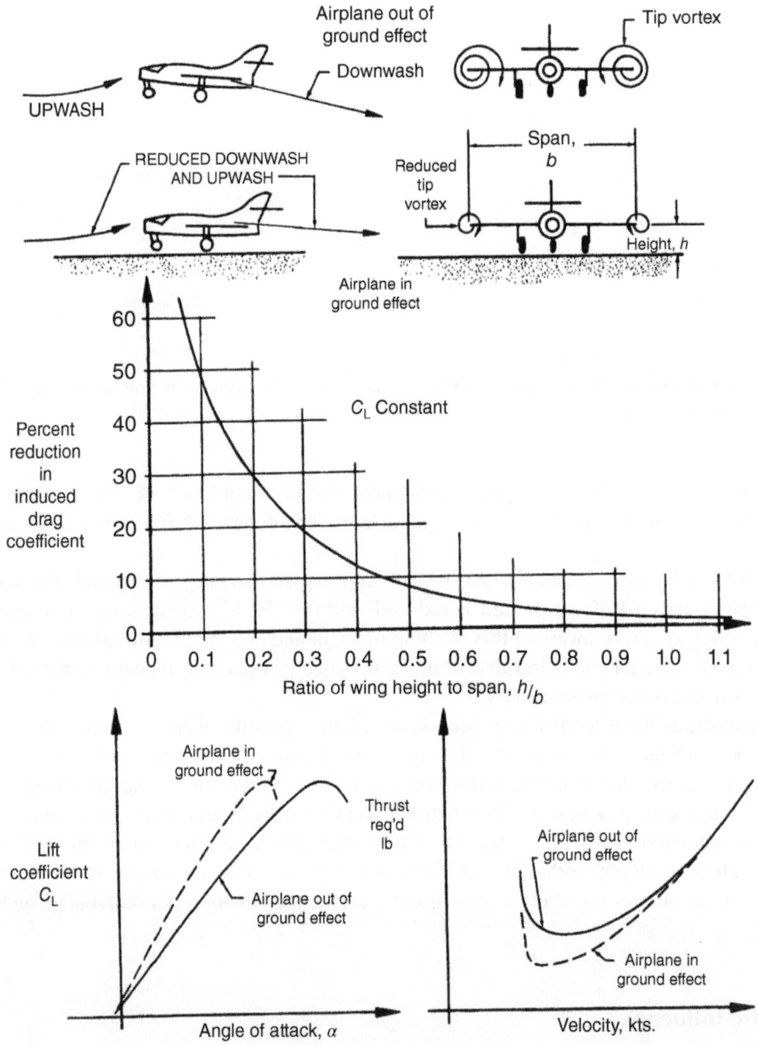

Figure 5.14. Ground effect.

of ground effect could lead to an unrecoverable situation. Figure 5.14 shows a summary of the effects of ground effect.

On entering ground effect:

1. Induced drag is decreased; reduction in thrust is required.
2. Nose-down pitching moments occur.
3. Decrease in AOA is required to maintain the same C_L.
4. The airspeed indicator reads low.

On leaving ground effect:

1. Induced drag is increased and additional thrust is required.
2. Nose-up pitching moments occur, and decrease in stability.
3. Increase in AOA is required to maintain the same C_L.
4. The airspeed indicator will read higher (correctly).

Application 5.3

Consider a low-wing versus high-wing airplane and the resulting impact of ground effect during the landing flare.

Would one type of wing design over another impact the magnitude of ground effect? If so, why? How might the "curves" in Figure 5.13 differ based on each wing type?

PARASITE DRAG

Parasite drag, D_p, is easily understood, but difficult to measure as it is the drag that is not associated with the production of lift. In flight, we have seen that induced drag results from the need to maintain lift and is greater at lower speeds when a high angle of attack is required to maintain level flight. As the speed increases, the induced drag decreases, but parasite drag will increase because the "solid body" is moving through the "fluid" with greater force and thus experiences the force of drag in a greater capacity. There are four types of parasite drag we will discuss here, and the total parasite drag is not a simple addition of each component.

Skin Friction Drag

Skin friction drag is caused by the viscous friction within the boundary layer. Regardless of how smooth the surface of the aircraft appears to be, there are small imperfections within any aircraft skin that promote aerodynamic resistance (Figure 5.15). The total area of the aircraft skin that is exposed to the airstream

Figure 5.15. Microscopic surface of a wing.
Source: Pilot's Handbook of Aeronautical Knowledge (2016b)/U.S. Department of Transportation Federal Aviation Administration/Public domain.

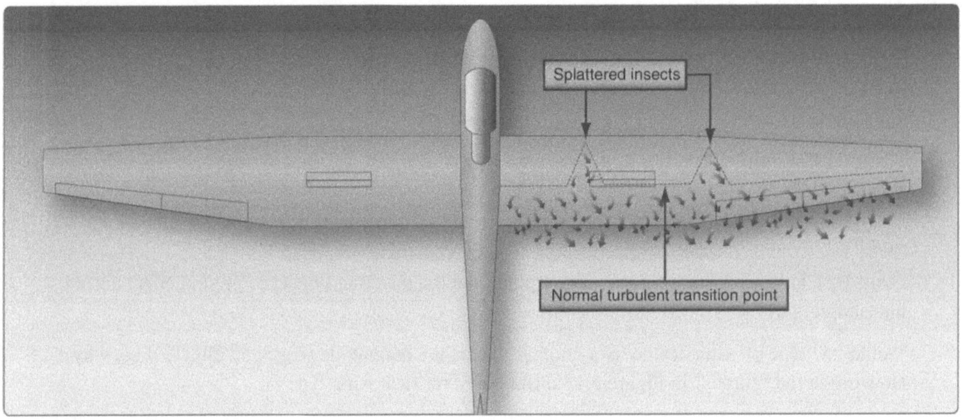

Figure 5.16. Effect of insects on skin friction drag.
Source: Glider Flying Handbook (2013)/U.S. Department of Transportation Federal Aviation Administration/Public Domain.

will be affected by this type of drag. Skin smoothness also greatly affects this drag, hence the proliferation of composite structures and streamlining technology throughout aviation. As discussed in Chapter 4, air within the boundary layer moves over the aircraft skin, air molecules are slowed down by imperfections and structural components, even remnants of insects, increasing the drag on the aircraft (Figure 5.16). These influences may also affect the boundary layer's ability to stay attached to an airfoil and result in early separation and stall. Maintaining laminar flow over the airfoil and thus preventing turbulent flow greatly reduces the magnitude of skin friction drag. The *Pilot's Handbook of Aeronautical Knowledge* (2013, p. 3–4) states that turbulent boundary layers can produce 5–10 times more skin friction drag than the equivalent laminar boundary layers. Flush-head rivets, waxed and polished surfaces, and removal of aluminum oxide help reduce skin friction drag.

Laminar Flow Airfoils

In the late 1930s, a radically new design of the airfoil was discovered, the *laminar flow airfoil.* In Figure 5.17, the $C_D - C_L$ curves are compared for two symmetrical airfoils, an NACA (National Advisory Committee for Aeronautics) "4-digit series" and an NACA "6-series" (laminar flow). The most obvious difference in these curves is the "drag bucket" that exists for the laminar flow airfoil. As long as this airfoil operates at low C_L (low AOA), the airflow remains laminar much farther back from the leading edge than on the conventional airfoil. Laminar flow has much less skin friction drag than turbulent flow, and thus less C_D. At higher C_L values, the advantage of the laminar airfoil disappears. The sharper leading edge of the laminar airfoil causes early separation and a resulting increase in the drag coefficient.

The laminar flow airfoil was used on the B-24 Liberator bombers in World War II. It was not an overwhelming success for several reasons:

1. The airplanes were operating from muddy dirt airfields. Mud thrown on the leading edges of the wings by the prop wash of other planes disrupted the laminar flow.
2. Shrapnel damage had a similar effect.

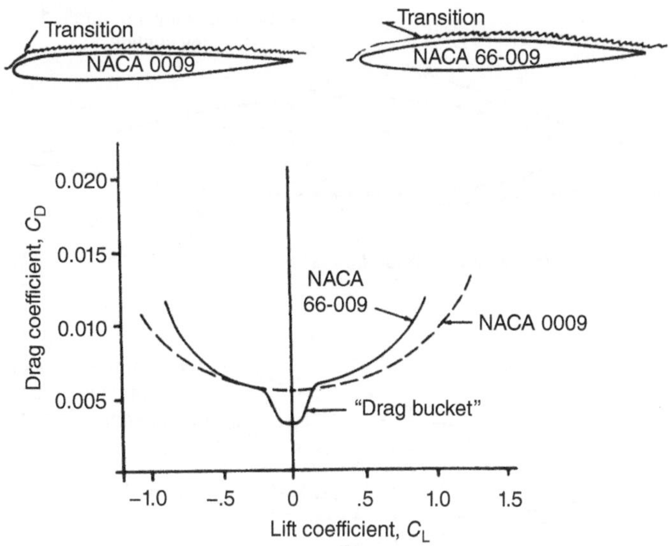

Figure 5.17. Comparison of drag characteristics of conventional and laminar flow airfoils.

3. Often, enemy gunfire caused the loss of one or two engines. This required that the airplane fly at low airspeeds and high angles of attack. Thus, flight in the "drag bucket" was impossible.

As for civilian airplanes, the Piper Aircraft Company introduced the laminar flow wing with its line of PA-28 "Cherokee" aircraft in the 1960s. In the *Owner's Handbook for the Cherokee*, the following paragraph told of the advantage of the laminar flow wing:

> The wing airfoil section is a laminar flow type, NACA 65_2-415 with a maximum thickness of about 40% aft of the leading edge. This permits the main spar carry through structure to be located under the rear seat providing unobstructed cabin floor space ahead of the rear seat.

No mention was made of flying characteristics or aircraft performance for the laminar flow wing. In the section on stalls, the owner's manual relates how stalls are "conventional." Cherokee pilots describe stalls as being docile. The aircraft does not exhibit a clean "break," but rather "mushes" and will continue to do so until back pressure is released, reducing AOA.

The major way that skin friction drag can be reduced is by maintaining the laminar airflow over the skin surfaces. Skin smoothness is one obvious method. Laminar flow airfoils represent another method of delaying boundary layer transition from laminar to turbulent flow. A third method is the use of *laminar flow control* (LFC).

LFC requires the removal of the thickening boundary layer by suction. Thin slots or holes are cut in the upper wing surfaces and suction is applied by mechanical means to remove the lower layers of the boundary layer or to add high-pressure air and laminar flow is maintained (Figure 5.18). Full-scale tests have been run and full wing laminar flow was achieved, but the results were not completely successful. The cost and complexity of cutting the tiny slots, controlling the suction amounts, the need for double skin surfaces, the added weight and complexity of the suction engines and ducting, and other problems have prevented commercial development to date.

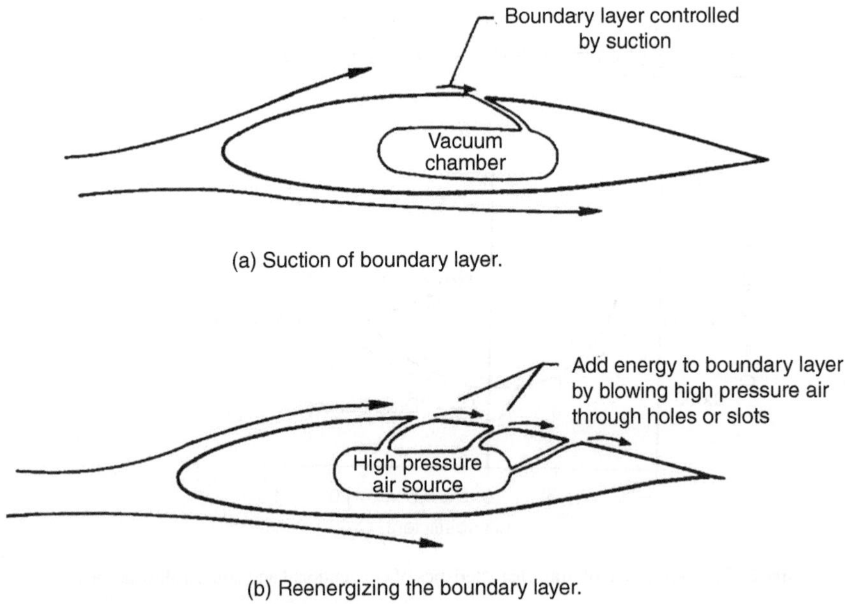

Figure 5.18. Energizing the boundary layer.
Source: NASA-SP-367/Public domain.

Table 5.1. Wing profile drag based on aircraft type.

Aircraft type	Wing profile drag / Total aircraft drag
Transport	~1/3
Business jet	~1/3
Low-speed GA	>1/3
UAV	1/3 – 1/2
Sailplane	>1/2

Source: NASA CR-2012-217559/Public domain.

NASA studies have shown that wing profile drag is the largest contributor to total aircraft drag at cruise conditions, contributing about one-third of the total drag for air transport aircraft. As the aircraft size decreases from transport down to sailplanes, the percentage of total aircraft drag due to the wing profile drag generally increases because the relative wing area increases (Table 5.1). Most business aircraft have their wing area determined by fuel volume so three approaches to the reduction of wing profile drag by increasing laminar flow have been accepted. One approach is to use a high-lift system as discussed in Chapter 4, usually leading-edge slats plus slotted Fowler flaps. Due to the slats, the laminar flow is disturbed, and combined with the complexity and weight of the high-lift systems not always the best option. A second approach is to employ LFC as discussed earlier in this section. Studies have shown that 100%-chord laminar flow can be achieved on both the upper and lower wing surfaces, but the LFC systems are complex, heavier, and more expensive than other solutions, so they are not utilized on current production aircraft.

A third approach is to utilize a natural-laminar-flow (NLF), and studies have shown that with proper airfoil shaping laminar flow can be achieved on both the upper and lower wing surfaces. One example of using a laminar flow airfoil is the business jet designed by HondaJet. HondaJet's NLF airfoil exhibits a high-drag divergence Mach number, small nose-down pitching moment, low drag for high cruise efficiency, and high $C_{L(max)}$. Fujino (2003) states the maximum thickness at 15% chord allows good wing fuel storage with laminar flow maintained to 42% chord on the upper surface and 63% chord on the bottom surface of this single-piece aluminum wing with no rivets.

Form Drag

Form drag is influenced by the shape or form of the aircraft (Figure 5.19). Antennas located around the aircraft, engine nacelles, gear, and other items comprising the total shape of the aircraft make up the objects that influence the amount of form drag present. Similar to water moving around rocks in a stream, as the moving air separates and rejoins at some point after an object, a force is created that opposes thrust and must be overcome. The overall shape of the aircraft, the number of objects exposed to the air, and how aerodynamically shaped the components are designed all influence the magnitude of the form drag. Streamlining the fuselage, engine nacelles, pods, and external stores helps reduce form drag though it will never be completely eliminated.

Interference Drag

Interference drag is caused by the interference of boundary layers from different parts of the airplane, as with the intersection of the wing and the fuselage at the wing root (Figure 5.20). If the drag of two

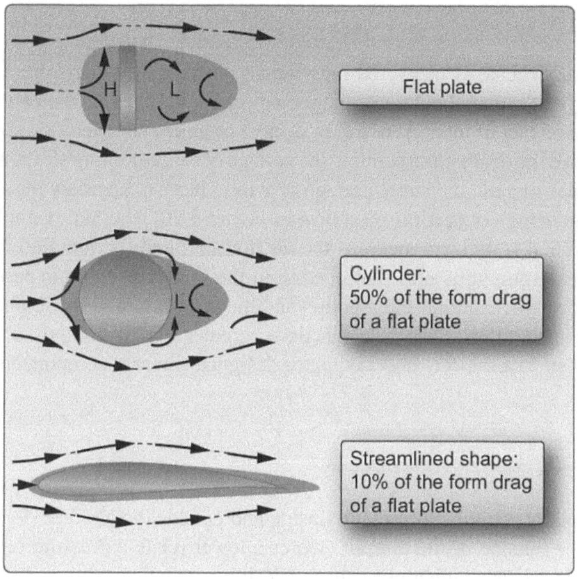

Figure 5.19. Form drag.
Source: Glider Flying Handbook (2013)/U.S. Department of Transportation Federal Aviation Administration/Public Domain.

Figure 5.20. Interference drag at the wing root.
Source: U.S. Department of Transportation (2008a)/Public Domain.

component parts of an airplane is measured individually and then the parts are assembled, the drag of the assembly will be greater than the drag of the parts because both parts when attached create their own interference drag. The impact of interference drag is most pronounced when two objects are joined at perpendicular angles, hence the desire to streamline the surface of two adjoining structures. Common areas for these undesirable angles include the strut/fuselage or wing junction, engine/wing pylon, engine/fuselage pylon, and the fuselage/wing connection, as shown in Figure 5.20. Another example can be found where the vertical and horizontal stabilizers meet, as the air flows around the junction of these two structures it is accelerated, and on some airplanes that operate in the high subsonic airspeed range this results in the formation of a shock wave around the junction and the resulting drag that comes with the supersonic flow. Smooth fairings and fillets at surface junctions, greater use of composite structures to "soften" the angle where two surfaces meet, and computer-designed aircraft components all help reduce this type of drag.

Leakage Drag

Leakage drag is caused by differential pressure inside and outside the aircraft. As we have discussed, air accelerating around the outside of the aircraft creates a lower relative pressure on the aircraft's surface. The air inside the fuselage is at a higher pressure and desires to equalize with the pressure on the outside of the aircraft. Air flowing from a higher pressure inside the fuselage through a crack or door seal will create an airstream that impinges on the airflow in the boundary layer around the aircraft and creates drag. Door and window sills are sealed with masking tape before an air race to lessen this drag.

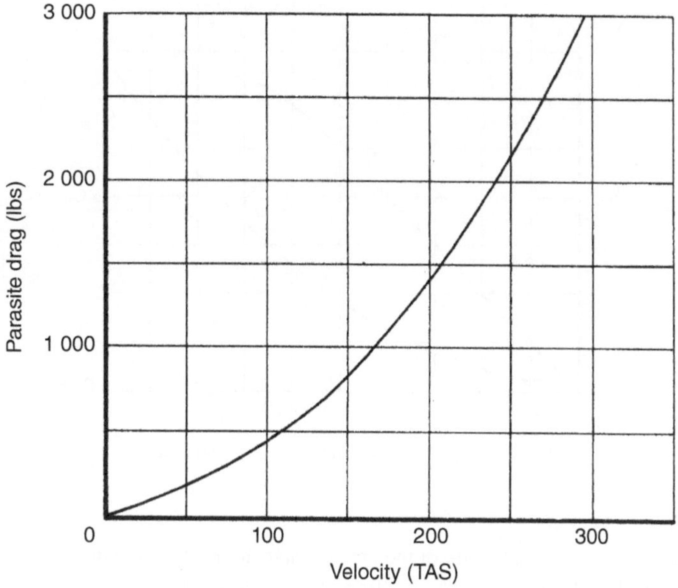

Figure 5.21. Parasite drag–airspeed curve.

Parasite Drag Summary

Parasite drag is the drag that is not associated with the production of lift. Parasite drag increases with increasing speed. As with Figure 5.10 and the plot of induced drag versus airspeed, Figure 5.21 is a plot of parasite drag versus airspeed. Parasite drag varies directly as the velocity squared, so if the airspeed doubles, then the parasite drag quadruples (Eq. 5.3).

$$\frac{D_{\mathrm{P2}}}{D_{\mathrm{P1}}} = \left(\frac{V_2}{V_1}\right)^2 \tag{5.3}$$

Figure 5.21 shows how parasite drag varies with airspeed.

DRAG EQUATION

The basic drag equation is similar to the basic lift equation (Eq. 4.3), which again incorporates Eq. 2.10 for q:

$$D = C_{\mathrm{D}} qS \rightarrow D = C_{\mathrm{D}} \left(\frac{\sigma V^2}{295}\right) S \rightarrow D = \frac{C_{\mathrm{D}} \sigma V^2 S}{295} \tag{5.4}$$

Rewriting gives

$$C_{\mathrm{D}} = \frac{D/S}{q} \rightarrow C_{\mathrm{D}} = \frac{295D}{\sigma V^2 S}$$

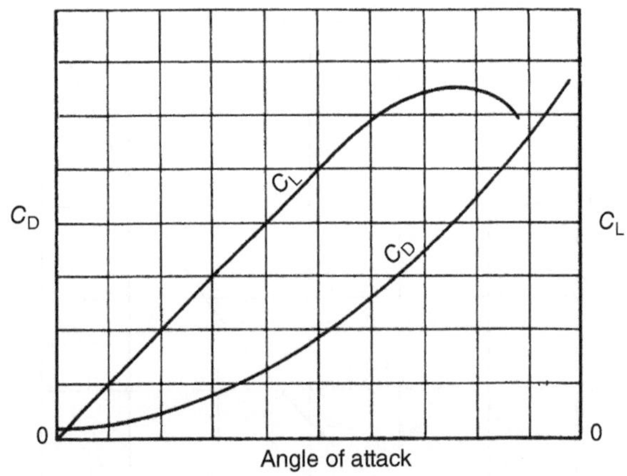

Figure 5.22. C_L versus AOA and C_D versus AOA.

The *coefficient of drag*, C_D, is the ratio of the drag pressure to the dynamic pressure. Values of C_D have been obtained experimentally for all airfoil sections and can be found in other publications. Figure 5.22 shows a C_L versus AOA curve as we have previously reviewed, but now a curve for C_D versus AOA for the same airfoil has been added.

Drag depends on multiple factors, including the density and viscosity of the air, as well as the velocity, size, and shape of the body. Determining the value of the drag coefficient is more difficult than determining the lift coefficient because there are several sources of drag, but we will introduce them here as C_{D_p} and C_{D_i}. The subscript p is added to represent parasite drag and i is added to represent induced drag. The drag coefficient includes the various forms of parasite and induced drag components and is derived from the effects of a fluid on an airfoil for this discussion. Most drag coefficients are determined experimentally for airfoils using wind tunnels and computational fluid dynamics (CFD).

Induced and Parasite Drag Calculations

Calculating the amount of drag as a result of induced drag is all about understanding the amount of lift, more specifically the C_L. Using Eq. 5.4, we can apply the subscript *i* to find the drag due to induced drag (Eq. 5.5):

$$D_i = C_{D_i} qS \tag{5.5}$$

Finding the coefficient of induced drag C_{D_i} at a given set of conditions for a specific airfoil involves measuring the amount of C_L; the AR of the wing, π (pi); and the efficiency factor (*e*), where *e* is a value of 1.0 or less given to an airfoil to determine how efficiently the airfoil produces lift (Eq. 5.6).

$$C_{D_i} = \frac{(C_L)^2}{\pi(\text{AR})e} \tag{5.6}$$

Parasite drag is hard to measure, but there is a way of simplifying this measurement. Consider the basic drag formula again (Eq. 5.4) with subscript p now added to indicate parasite drag we get:

$$D_p = C_{D_p} qS \tag{5.7}$$

At any AOA, the value of C_{D_p} will be a constant and, of course, S is a constant. A new constant, called the *equivalent parasite area*, f, is now introduced.

$$f = C_{D_p} S$$

Because C_{D_p}, is dimensionless, the dimensions of f are the same as S (ft^2). The equivalent parasite area is often called the "barn door" area of the aircraft. This is not technically correct, but it may help you visualize the area as viewed by an observer of an approaching aircraft.

The value of f for an aircraft can be determined by placing a model in a wind tunnel and adjusting the AOA until zero lift is developed, then measuring the drag. With no lift, there is no induced drag, so all the drag will be parasite drag. Remember, parasite drag is all drag that is not associated with the production of lift. If the density ratio and tunnel velocity are known, then the equivalent parasite area can be found by

$$f = \frac{D_p}{q}$$

For low angles of attack, the value of f can be considered a constant, so

$$D_p = fq \rightarrow D_p = \frac{f\sigma V^2}{295}$$

TOTAL DRAG

Total drag, D_T, is the sum of the induced drag and the parasite drag (Eq. 5.8):

$$D_T = D_i + D_p \tag{5.8}$$

As we mentioned earlier, the drag is the component of the aerodynamic force that is parallel to the free stream RW. It is made up of both the induced drag vector and the parasite drag vector (Figure 5.23).

The total drag curve for an airplane with the induced drag curve in Figure 5.10 and the parasite drag curve in Figure 5.21 is shown in Figure 5.24. The total drag curve is comprised of the total parasite drag and induced drag for each respective knot of TAS, with the unit of measurement for drag in pounds (lb).

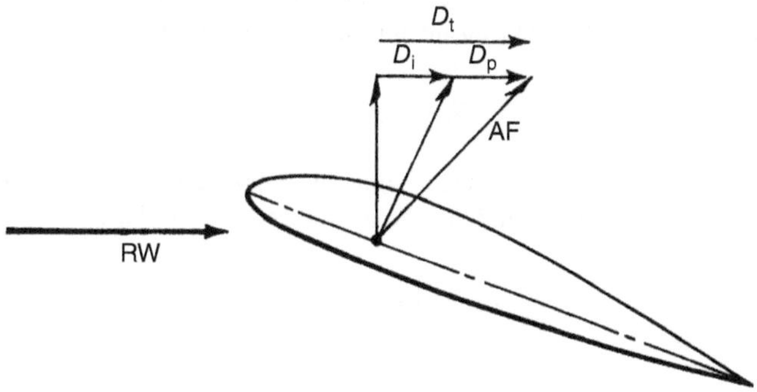

Figure 5.23. Drag vector diagram.

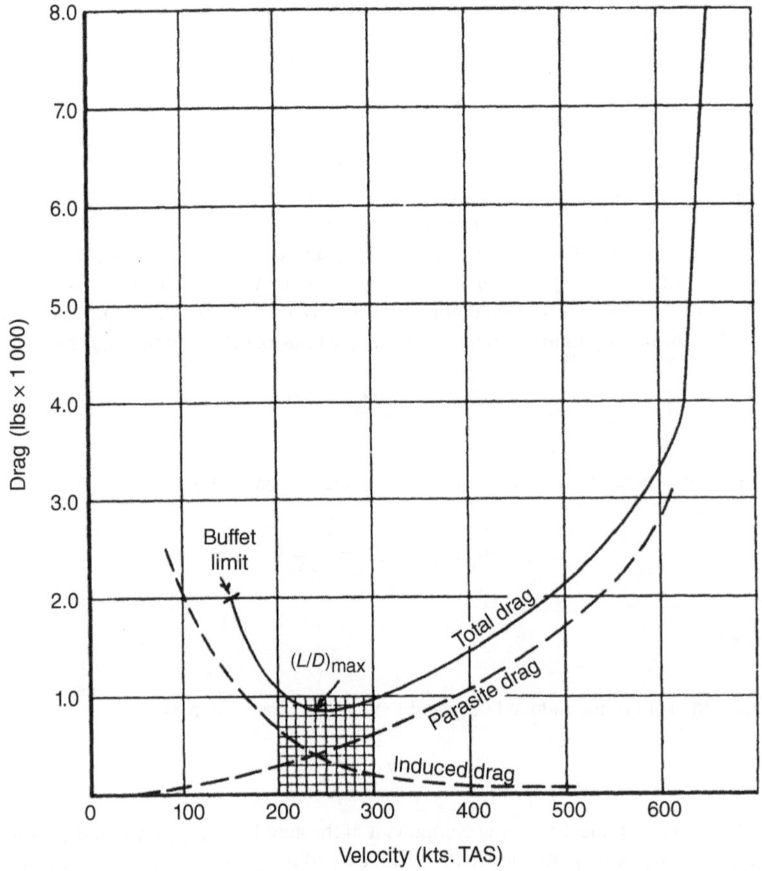

Figure 5.24. Total drag curve.

There is one location on the total drag curve where the drag minimum is located; this is known as the $(L/D)_{max}$ point, and it occurs where the induced drag and parasite drag curves intersect. Therefore, at this velocity, the induced drag is equal to the parasite drag, and can be written as:

$$D_i = D_p = \frac{1}{2} D_T \tag{5.9}$$

Equation 5.10 is an equation for calculating total drag and can be derived by combining Eqs. 5.2 and 5.3. Let condition 1 be at $(L/D)_{max}$ and condition 2 be any other point on the curve. Also, remember that $D_{i_1} = D_{p_1} = \frac{1}{2} D_{min}$.

$$D_T = \frac{1}{2} D_{min} \left[\left(\frac{V_1}{V_2} \right)^2 + \left(\frac{V_2}{V_1} \right)^2 \right] \tag{5.10}$$

The limitations of this curve must be emphasized. In the low-speed region, the high AOA required results in more "equivalent parasite area" than was calculated. The value of f is not constant. The value of the parasite and total drag will be greater than the equation shows. In the high-speed region, faster airplanes may encounter compressibility effects and resultant "wave drag." This was not considered in Eq. 5.10. The total drag curves are not accurate at airspeeds below about 1.3 V_s or above about 0.7 Mach.

LIFT-TO-DRAG RATIO

An aircraft's lift/drag ratio, *L/D*, is a measure of its efficiency. An aircraft with a high *L/D* is more efficient than one with a lower *L/D,* representing "high lift" to "low drag." An aircraft with a more efficient *L/D* ratio can carry a heavier payload, for a longer time, over a longer distance. And because this point is defined by a minimum drag-to-maximum lift ratio, this is also the point where an airplane can glide the farthest. The glide ratio can also be stated as the number of feet an aircraft travels horizontally in still air for every foot of altitude lost:

$$\text{Glide ratio} = \frac{\text{Lift}}{\text{Drag}} : 1 \tag{5.11}$$

In Figure 5.25, a given glider sink rate is used to calculate the best glide speed as well as the glide ratio. These values can be determined from the polar by locating where the curve is touched by the tangential line. In this figure, the best glide airspeed is found to be 62 kts. with a sink rate of 1.6 kts., so using Eq. 5.11 the glide ratio is found to be 38:1.

As we have seen drag also varies with speed, so the glide ratio also varies with speed. Figure 5.26 shows a glide polar graph for a glider with an airspeed of 75 kts and a given sink rate of 3.4 kts. found in the aircraft flight manual (AFM). The glide ratio at a particular airspeed can be estimated by using Eq. 5.12:

$$\text{Glide ratio} = \frac{\text{Airspeed}}{\text{Sink rate}} : 1 \tag{5.12}$$

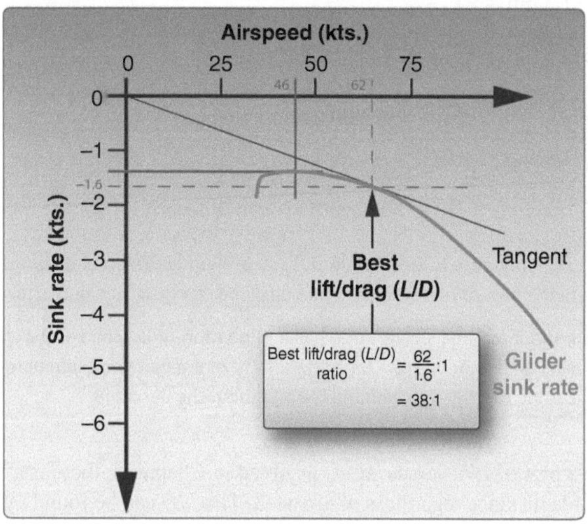

Figure 5.25. Calculating best glide airspeed and best glide ratio.
Source: Glider Flying Handbook (2013)/U.S. Department of Transportation Federal Aviation Administration/Public Domain.

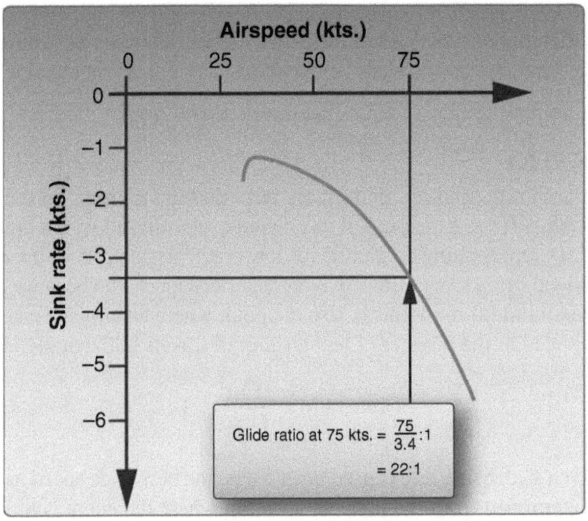

Figure 5.26. Calculating glide ratio for a given airspeed.
Source: Glider Flying Handbook (2013)/U.S. Department of Transportation Federal Aviation Administration/Public Domain.

EXAMPLE

As a rule of thumb for Cessna 172s, they glide 1.5 nm horizontally for every 1 000 ft of altitude above ground level. The glide ratio can now be calculated using this information.

1 nm = 6 076 ft, so 1.5 nm = 9 114 ft

To calculate the glide ratio, divide the horizontal distance gained by the vertical distance lost, so:

$$\text{Glide ratio} = \frac{9\,114 \text{ ft}}{1\,000 \text{ ft}} \rightarrow \text{Glide ratio} = 9.1 : 1$$

Application 5.4

Once a glide ratio is known for a specific aircraft the no-wind horizontal distance the aircraft may glide during an engine failure in a single-engine airplane can now be calculated.

If you are flying a single-engine aircraft that has a glide ratio of 8.1, and you experience an engine failure when 7 500 ft AGL, how far horizontally in nm can you glide before contacting the ground if the best angle of glide is maintained throughout the descent?

Figure 5.27 shows a plot of *L*/*D* versus AOA; as noted in Chapter 4, there is a specific AOA that will provide the greatest glide distance regardless of airspeed. The *L*/*D* can be found by dividing the values of C_L by those of C_D at the same AOA. This can be seen if we divide Eq. 4.3 by Eq. 5.4, all the factors cancel except *L*, *D*, and their coefficients:

$$\frac{L}{D} = \frac{C_L}{C_D} \tag{5.13}$$

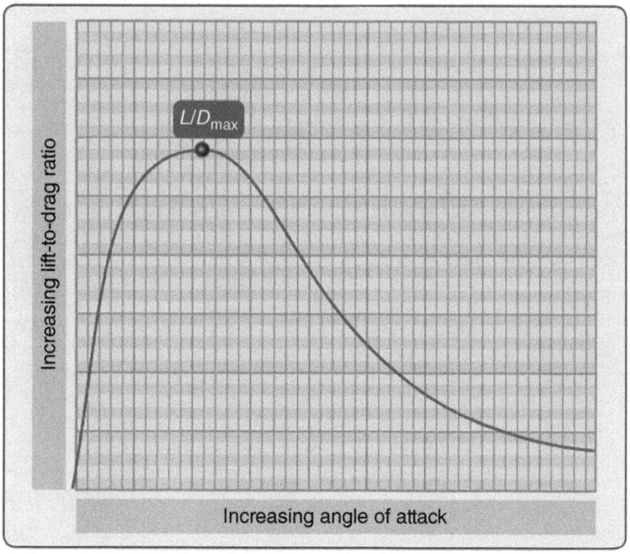

Figure 5.27. $(L/D)_{max}$.
Source: Airplane Flying Handbook (2016a)/U.S. Department of Transportation Federal Aviation Administration/Public domain.

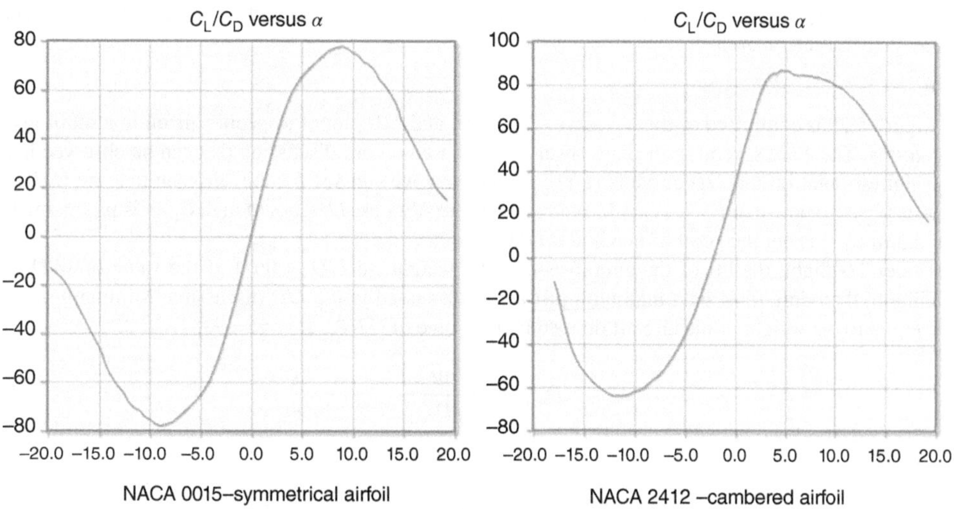

Figure 5.28. Cambered and symmetrical airfoil C_L/C_D versus AOA.

Figure 5.28 shows the C_L/C_D ratio to AOA of the same airfoils discussed in Chapter 4, note that AOA is on the X-axis. Notice that the highest ratio of lift to drag for the symmetrical airfoil of NACA 0015 is around 9°, so this would indicate the AOA that theoretically will produce the greatest lift to drag for this airfoil. For the cambered airfoil NACA 2412 the greatest ratio of C_L/C_D is found at ~5°.

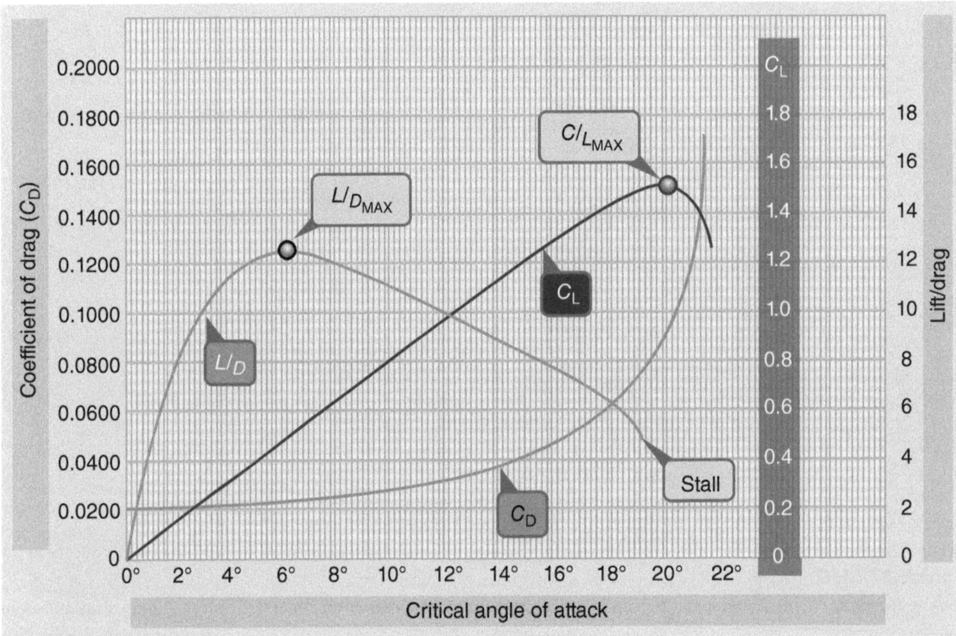

Figure 5.29. Typical lift-to-drag ratios.
Source: Pilot's Handbook of Aeronautical Knowledge (2008a)/U.S. Department of Transportation Federal Aviation Administration/Public domain.

Figure 5.29 is comprised of the C_L curve, C_D curve, and *L*/*D* plot of a specific airfoil in a set of given conditions. There are several immediate observations based on our discussion that can be observed here. The highest point on the *L*/*D* curve is $(L/D)_{max}$, and now we can see for this airplane that the $(L/D)_{max}$ AOA is 6°. Also note that $C_{L_{max}}$ is not located at the same AOA as $(L/D)_{max}$, though $C_{L_{max}}$ is producing the maximum C_L it is not the point of maximum *L*/*D*.

Under 1*G* flight, the lift of the aircraft equals the weight, so *L*/*D* = *W*/*D*. If the value of *W*/*D* is a maximum, then drag must be a minimum just as we discussed in the *L*/*D* discussion. Minimum drag is then equal to the weight of the aircraft divided by the value of $(L/D)_{max}$:

$$D_{min} = \frac{W}{(L/D)_{max}} \tag{5.14}$$

EXAMPLE

If a 12 000 lb airplane has a $(L/D)_{max}$ of 9.5, find the minimum drag.

$$D_{min} = \frac{W}{(L/D)_{max}} \rightarrow D_{min} = \frac{12000\ lb}{9.5} \rightarrow D_{min} = 1263.2\ lb$$

Now continuing with the example above, at $(L/D)_{max}$ we can use Eq. 5.9 to find the value of D_i and D_p:

$$D_i = D_p = \frac{1}{2}D_T \rightarrow D_i = D_p = 631.6\ lb$$

$(L/D)_{max}$ Summary

In summary, it is important to note that several performance characteristics are present at $(L/D)_{max}$. The best engine out glide ratio occurs here. Because lift is proportional to distance and drag is proportional to altitude, then L/D is proportional to glide ratio (distance/altitude). This means that the numerical value of L/D equals the numerical value of the glide ratio. Note that the value of $(L/D)_{max}$ and the angle of attack at which it occurs do not vary with aircraft weight or altitude. The airspeed for $(L/D)_{max}$ does vary with weight, as shown in Eq. 4.7.

DRAG REDUCTION

Most drag on modern turbofan airliners and other large jet aircraft is caused by skin friction drag and induced drag. These forms of drag represent roughly 75% of the total drag and reducing them is a major factor in decreasing fuel consumption. In an earlier discussion of induced drag, it was demonstrated that wingtip vortices cause much of the downwash and resultant backward tilting of the lift vector. Reducing the wingtip vortices has been a goal of aeronautical engineers for some time. Earlier attempts to do this are shown in Figure 5.30.

Wingtip devices are usually intended to improve the efficiency of fixed-wing aircraft though some types of older aircraft sacrificed the drag of these devices in order to increase fuel capacity. Although there are many types of drag-reducing wingtip devices in use and they function in different manners, the intended effect is always to reduce the aircraft's induced drag, thus improving thrust efficiency and thereby saving fuel. In fact, the NASA study referenced in the next section has shown using wingtip devices such as winglets can increase fuel mileage rates by 7% resulting from a 20% decrease in induced drag. Wingtip devices can also improve aircraft handling characteristics and enhance safety. Such devices increase the effective AR of a wing without materially increasing the wingspan, though incorporating these devices means accepting an increase in parasite drag, as well as higher manufacturing costs.

Winglet Design

Winglets were developed to solve the problem of the parasite drag caused by other types of wingtip vortex suppression needed to reduce induced drag. Of course, if total drag can be reduced, then higher cruise speeds and/or fuel savings can be achieved.

In 1976 NASA published a technical paper by Dr. Richard Whitcomb regarding high subsonic speeds for wingtip mounted winglets and whether the first-generation wingtip devices or wingtip extensions substantially reduced drag coefficients during lifting conditions. Dr. Whitcomb will be introduced again

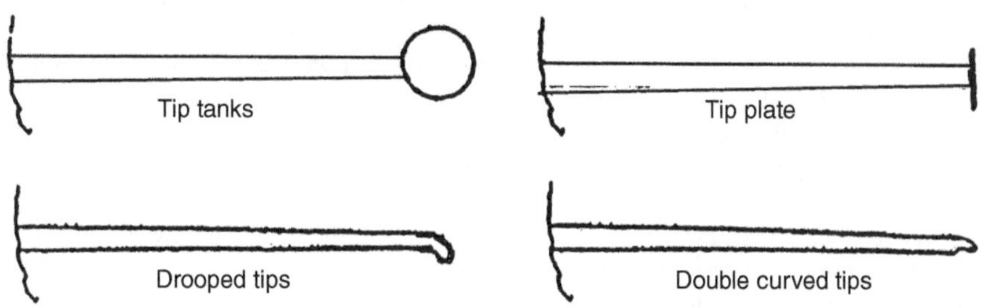

Figure 5.30. Wingtip vortex reduction methods.

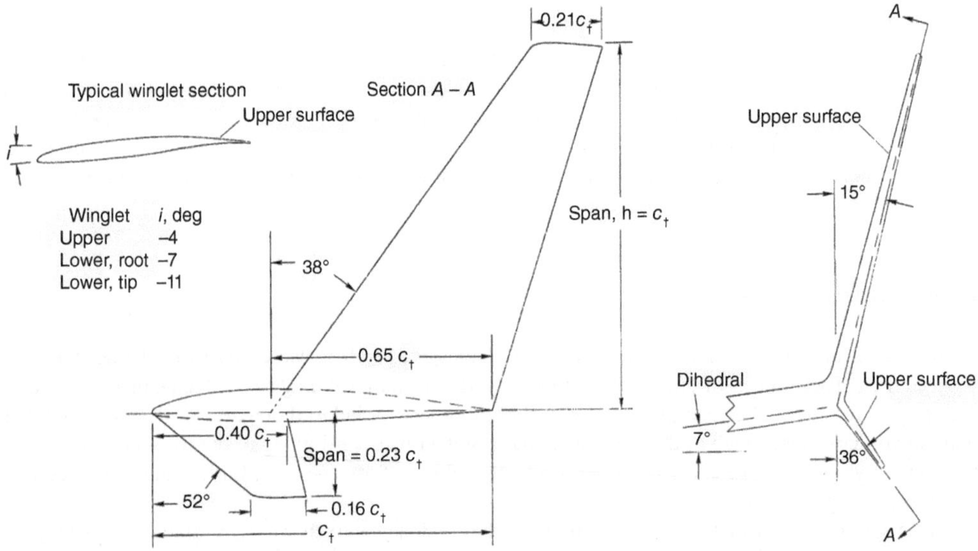

Figure 5.31. Whitcomb winglet design.
Source: NASA TN D-8260/Public domain.

in Chapter 14 for his research on the design of supersonic aircraft. The wind tunnel tests clearly showed the benefit of installing winglets on narrow-body jet transport aircraft, which resulted in the reduction of induced drag by 20% with an increase in wing lift-drag ratio of 9% for the design Mach number of 0.78. In addition, the improvement in the lift-drag ratio was double that of a wingtip extension and the resulting changes in pitching-moment coefficients were less with the winglet than with the wingtip extension. During early wind tunnel experiments, it was found that for a winglet to be fully effective the vertical surface at the tip must produce significant side forces to reduce the lift-induced inflow above the wingtip or the outflow below the wingtip.

Figure 5.31 represents the design of the theoretical winglet used in the 1976 Whitcomb study, note the position of the upper winglet in comparison to the lower winglet. The upper winglet is rearward of the surface of the wing so as not to interfere with high-speed airflow over the wing during supercritical conditions resulting in adverse interference. Experiments indicate that the winglet is most effective with the trailing edge of the winglet placed with the trailing edge of the wing. The effective sweep of the winglet should also be approximately the same as the sweep of the wing to be most effective at supercritical conditions. The optimum height of the lower winglet would be the same height as the upper winglet, but due to possible ground clearance conditions, the lower winglet is usually shorter. The placement of the lower winglet in the initial Whitcomb studies indicated the lower winglet was most effective when the leading edge was placed with the leading edge of the wing, while the trailing edge of the lower winglet was placed below the leading edge of the upper winglet.

Figure 5.32 shows the final wind-tunnel results of the 1976 NASA study of wingtip-mounted winglets at high subsonic speeds for a second-generation, wide-body jet transport wing. Note the change in C_L for the respective change from basic wing to tip extension, with the upper and lower winglets offering improved performance over a basic wing without any modification, and without the negative impacts that are a result of wing extension such as bending or twisting.

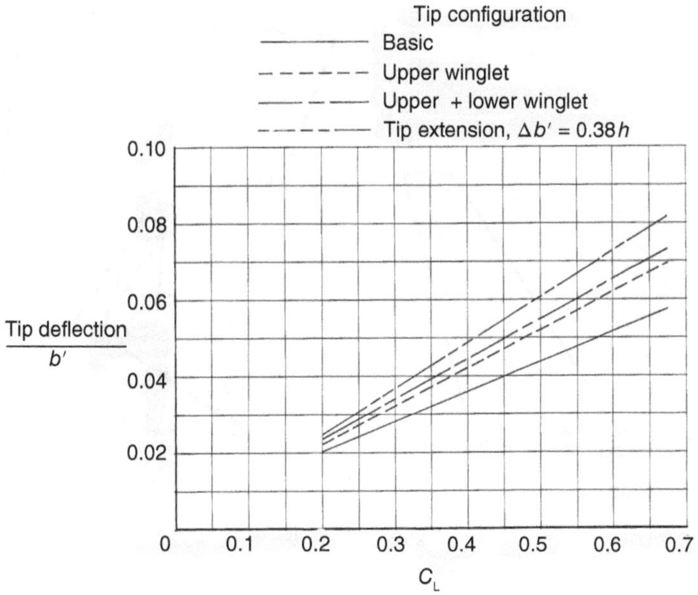

Figure 5.32. Wingtip deflection analyses with C_L.
Source: NASA TN D-8260/Public domain.

Without some type of winglet, the airflow around the tip is nearly circular. It flows outward toward the tip from beneath the wing, goes upward around the wingtip, and then back inward on the top surface. Although this rotation is reduced by placing the winglet in the path of the airflow, the winglets are usually placed some distance back from the leading edge. This is shown in the side view of Figure 5.33. The remaining circular flow creates an inward velocity, V_i, which can be vectorially added to the aircraft's relative velocity, V_{ac}, to obtain the RW shown in the top view of Figure 5.33. The aerodynamic force, drag, and lift vectors are also shown. The lift vector is tilted forward and by dropping a line from the arrow of the lift vector perpendicular to the winglet's chord line, we get a vector, $-D$. This is larger and opposite to the drag vector, showing that the winglet actually is not causing drag on the airplane, like a sailboat tacking upwind there is a component of forward thrust.

Winglet Application

During the 1980s, several companies began research on blended winglets, which combine the benefits of a winglet without the interference drag and maintenance retrofit costs of a typical winglet. As with traditional winglets, the intent was to minimize induced drag, and now with a blended design, the goal was to further reduce parasite drag (specifically interference drag) while minimizing additional manufacturing costs. With the original blended winglet design, most of the airflow at the tip was directed inward toward the fuselage with only a small amount being vectored forward. This design did reduce the lift-induced drag but not as much as future wingtip designs. An added benefit to the introduction of blended winglets was an improvement in the takeoff/landing performance of the aircraft, allowing for better obstacle clearance and the ability to derate the engine thrust. This ultimately reduces engine maintenance costs and reduces noise

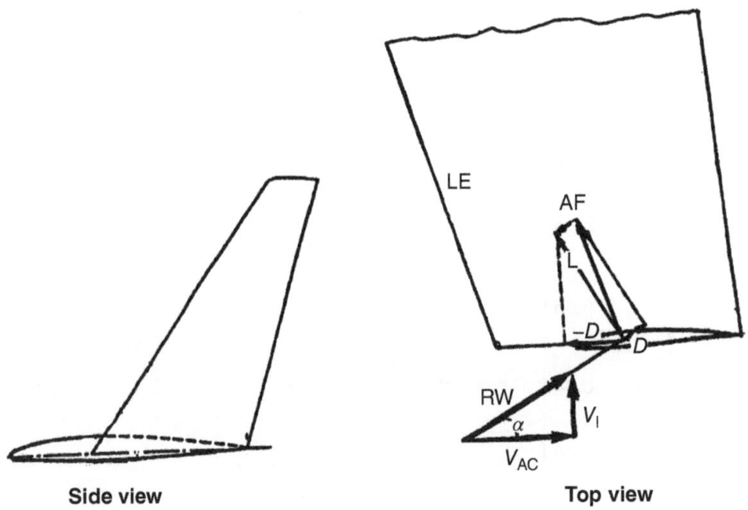

Figure 5.33. Winglets.

when operating in areas requiring noise abatement procedures. Finally, blended winglets also increase the range and payload of the aircraft, an asset to any commercial aircraft.

Many older aircraft have added structural modifications to retrofit aircraft with winglets due to drag reduction and subsequent fuel savings. Of course, a balance must be calculated between the costs of retrofit and long-term financial savings in fuel costs. Modern aircraft utilize the benefits of winglets, essentially increasing the effect of increased AR without extending the actual length of the wing, a benefit when considering terminal gate space as well as the structural strength requirements required in longer wingspans. An example of utilizing the aerodynamic benefits of a longer wing but still recognizing the confined gate space of some taxiways and commercial passenger gates is the Boeing 777X's use of a folding wingtip. From takeoff to landing, the folded wingtips are extended to utilize the benefits of a reduction in induced drag, but after landing the wingtips are retraced for taxi and parking. Other models of Boeing aircraft utilize a split scimitar winglet design which is slightly different than the Boeing 737 MAX AT™ winglets. Older generation Airbus equipment can be seen with small upper and lower wingtip devices called a wingtip fence, while newer generation Airbus aircraft incorporate a Sharklet™ design. The latest Airbus A350 and A330neo winglet designs involve a more blended winglet, curving upward and rearward resulting in a twisting, smooth extension of the wing.

Application 5.5

The Boeing 737 MAX AT™ winglet is advertised as one of the most efficient designs for a wingtip ever produced. Research the aerodynamics involved with why this claim can be made.

Based on your research, why is the MAX AT™ winglet design more effective than the blended winglet design? How does laminar flow incorporate into the design?

SYMBOLS

AR	Aspect ratio
b	Wingspan (ft)
C	Chord (ft)
C_{av}	Average chord (ft)
C_D	Coefficient of drag
C_r	Root chord (ft)
C_t	Tip chord (ft)
D_i	Induced drag (lb)
D_p	Parasite drag (lb)
D_T	Total drag (lb)
f	Equivalent parasite area (ft^2)
$(L/D)_{max}$	Maximum lift/drag ratio
MAC	Mean aerodynamic chord (ft)
$2w$	Downwash velocity (kts.)
α_0	AOA of infinite wing
α_i	Induced AOA of finite wing
λ (lambda)	Taper ratio
Λ(uppercase lambda)	Sweep angle (degrees)

EQUATIONS

5.1 $\alpha = \alpha_0 + \alpha_i$

5.2 $\frac{D_{i_2}}{D_{i_1}} = \left(\frac{V_1}{V_2}\right)^2$

5.3 $\frac{D_{p_2}}{D_{p_1}} = \left(\frac{V_2}{V_1}\right)^2$

5.4 $D = \frac{C_D \sigma V^2 S}{295}$

5.5 $D_i = C_{D_i} qS$

5.6 $C_{D_i} = \frac{(C_L)^2}{\pi ARe}$

5.7 $D_p = C_{D_p} qS$

5.8 $D_T = D_i + D_p$

5.9 $D_i = D_p = \frac{1}{2} D_T$

5.10 $D_T = \frac{1}{2} D_{min} \left[\left(\frac{V_1}{V_2}\right)^2 + \left(\frac{V_2}{V_1}\right)^2\right]$

5.11 $\text{Glide ratio} = \frac{\text{Lift}}{\text{Drag}} : 1$

5.12 $\text{Glide ratio} = \frac{\text{Airspeed}}{\text{Sink rate}} : 1$

5.13 $\frac{L}{D} = \frac{C_L}{C_D}$

5.14 $D_{min} = \frac{W}{(L/D)_{max}}$

KEY TERMS

Aspect ratio
Average chord
Drag
Drag equation
Downwash
Fairing
Fillet
Finite wing
Float
Ground effect
Induced angle of attack
Induced drag
Interference drag
Laminar flow airfoil
Laminar flow control
Leakage drag
Lift to drag ratio
Mean aerodynamic chord (MAC)
Parasite drag
Skin friction drag
Symmetrical airfoils
Sweep angle
Total drag
Vortices
Winglets

PROBLEMS

Note: Answers to the problems in the first two sections are given at the end of the book.

Application of Chapter Theory

1. The coefficient of drag can be defined as
   a. the ratio of lift forces to drag forces.
   b. a measure of the efficiency of the airplane.
   c. the ratio of drag forces to lift forces.
   d. the ratio of drag pressure to dynamic pressure.

2. Laminar flow airfoils have less drag than conventional airfoils
   a. in the high-C_L region.
   b. in the high-AOA region.
   c. in the high-speed region.
   d. in the landing phase of the flight.

3. Laminar flow airfoils have less drag than conventional airfoils because
   a. the adverse pressure gradient starts farther back on the airfoil.
   b. the airfoil is thinner.

c. more of the airflow is laminar.
d. Both (a) and (c).

4. The lift/drag ratio is
a. a measure of the aircraft's efficiency.
b. a maximum when the drag is a minimum.
c. numerically equal to the glide ratio.
d. All of the above.

5. Two aircraft with the same wing area have different wingspans, with all other airfoil characteristics being equal, the aircraft with the longer wingspan will:
a. experience a lower induced drag.
b. experience a higher induced drag.
c. wingspan does not impact the amount of induced drag if the wing area remains the same.
d. have a lower aspect ratio.

6. An airplane flying at $C_{L(max)}$ will have
a. more parasite drag than induced drag.
b. more induced drag than parasite drag.
c. equal amounts of parasite and induced drag.
d. Not enough information is given to evaluate the drag.

7. Induced drag is
a. more important to low aspect ratio airplanes than to high aspect ratio airplanes.
b. reduced when the airplane enters the ground effect.
c. reduced if the airplane has winglets.
d. All of the above.

8. Induced drag results from the lift vector being tilted to the rear, because
a. the tip vortices cause downwash behind the wing.
b. the relative wind behind the wing is pushed downward.
c. the local relative wind at the AC is depressed.
d. All of the above.

9. An airplane with a heavy load _____ it can when lightly loaded.
a. can glide farther than
b. cannot glide as far as
c. can glide the same distance as

10. If an airplane has a symmetrical wing that has an angle of incidence of 0° during takeoff, all the drag is
a. profile drag.
b. parasite drag.
c. wave drag.
d. induced drag.

11. Which of the following is not a characteristic of ground effect on an aircraft while crossing the runway threshold during landing:
a. increase in induced drag.
b. nose-down pitching moment.
c. airspeed indicator reads lower than normal.
d. decrease in AOA to maintain the same C_L.

12. A low-tailed airplane with static ports beneath the wing leaves ground effect after takeoff. It will have:

a. increased drag, nose-up pitch, lowered IAS.

b. increased drag, nose-up pitch, higher IAS.

c. decreased drag, nose-down pitch, lowered IAS.

d. decreased drag, nose-up pitch, higher IAS.

Calculations Using Chapter Principles

13. In Figure 5.29, what is the value of $C_{L_{max}}$?

a. 1.5

b. 12.5

c. 0.1500

d. 1.3

14. In Figure 5.29, what is the value of $(L/D)_{max}$?

a. 1.3

b. 0.1250

c. 12.5

d. 1.0

15. Calculate the aspect ratio given the following:

Wing span = 36 ft

Wing area = 174 ft^2

16. An aircraft has a D_i of 750 lb when at $(L/D)_{max}$, what is the value of D_p?

17. A 10 000 lb aircraft has an $(L/D)_{max}$ value of 9.8; find its minimum drag.

18. If the aircraft in Problem 17 has a wing area $S = 200$ ft^2 and the value of $C_L = 0.5$ at $(L/D)_{max}$, find the sea level airspeed where D_{min} occurs.

19. For the aircraft in Problem 17, find D_i and D_p at $(L/D)_{max}$.

20. For the aircraft in Problem 17, complete the following table and plot D_i, D_p, and D_T on a sheet of graph paper.

V_2	$(V_2/V_1)^2$	$D_p = \frac{1}{2}D_{min}(V_2/V_1)^2$	$(V_1/V_2)^2$	$D_1 = \frac{1}{2}D_{min}(V_1/V_2)^2$	$D_T = D_p + D_i$
125					
150					
172					
200					
300					
400					

Correlation Through Scenario-Based Learning

Aircraft – *Diamond DA50RG*

1. Review the DA50 AFM to find the wingspan and wing area, then calculate the aspect ratio (AR) using the example from earlier in the chapter. How does your calculated AR compare to the given AR from Section 1 of the AFM for the wing?
2. Research the AR of a DA40NG, is the value higher or lower than a DA50RG? sed on the value of the AR only, which airfoil between the two aircraft will produce comparatively less induced drag?
3. Using Figure 5.34, if a pilot experiences engine failure and pitches the aircraft for best glide, what is the best glide angle for the DA50RG if using an AOA indicator? Note that AOA is on the X-axis.
4. Referencing Section 3 in the DA50 AFM, what is the published best glide *speed* for a flaps up (fully retracted) condition? What is the best glide *speed* when the flaps are positioned in the T/O position? Why would the airspeeds be different?
5. Does the best glide *speed* change with weight? Why or why not? Does the best glide *AOA* change with weight? Why or why not?
6. Referencing Section 5 in the DA50 AFM, what is the published glide ratio? If a DA50 were in cruise flight at 12 000 ft AGL and experienced an engine failure, how many nautical miles horizontally would the aircraft theoretically be able to glide if the published glide ratio were maintained through the entire descent? How would a headwind, tailwind, or crosswind impact this distance?
7. Based on the Section 5 notes and conditions published in the AFM, if the gear was deployed, would the DA50 have a higher or lower glide ratio? What about the position of the cowl flaps?

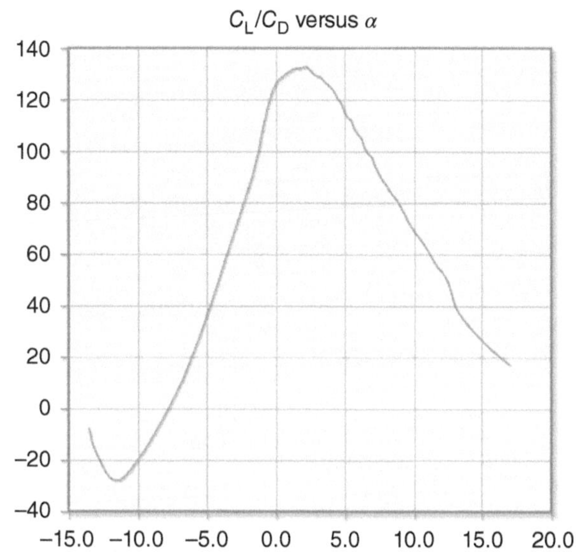

Figure 5.34. C_L/C_D versus AOA wing airfoil curve of a Diamond DA50RG.
Source: Created by authors using airfoiltools.com.

6 Jet Aircraft Performance

Chapter Objectives

After completing this chapter, you should be able to:

- Identify the type of engines that are used for thrust production.
- Explain how a thrust-required curve is produced and distinguish important performance values along the curve.
- Summarize the relationship between thrust and rpm.
- Define specific fuel consumption and calculate the variation in fuel flow with a change in altitude.
- Identify performance values that can be derived from thrust-required curves and calculate their value.
- Compare and contrast the angle of climb and the rate of climb for a thrust-producing airplane.
- Examine and illustrate the changes in a thrust-required curve as weight, configuration, and altitude are altered throughout a flight.

The performance of an aircraft depends on the relationship of the forces acting on it, where the principal forces are lift, weight, thrust, and drag. If these forces are in equilibrium, as shown in Figure 6.1, the aircraft will maintain a steady velocity and constant-altitude flight. If any of the forces acting on the aircraft change, the performance of the aircraft will also change. To better understand the relationship between these forces and performance, we will study aircraft performance curves in this chapter. Some of the items of performance that can be obtained from these curves are:

1. Maximum level flight velocity.
2. Maximum climb angle.
3. Velocity for maximum climb angle.
4. Maximum ROC.

Flight Theory and Aerodynamics: A Practical Guide for Operational Safety, Fifth Edition.
Brian A. Johnson and Philip R. Fittante.

Companion website: www.wiley.com/go/flighttheoryandaerodynamics5e

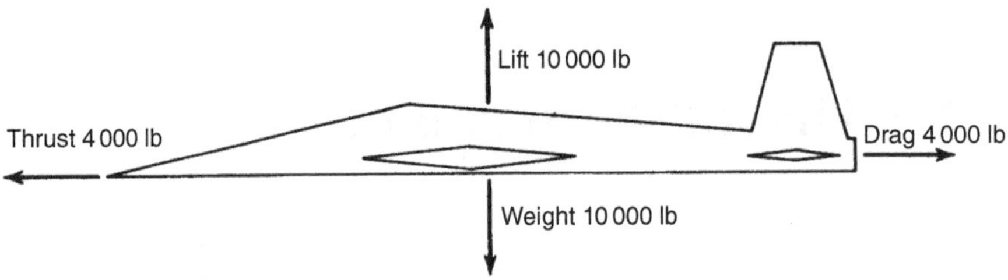

Figure 6.1. Aircraft in equilibrium flight.

5. Velocity for maximum ROC.
6. Velocity for maximum endurance.
7. Velocity for maximum range.

This chapter shows the construction and use of the performance curves for *thrust-producing aircraft*, while *power-producing aircraft* (propeller-driven aircraft) are discussed in Chapter 7.

THRUST-PRODUCING AIRCRAFT

Some aircraft produce thrust directly from the engines; turbojet, ramjet, and rocket-driven aircraft are examples of thrust producers. Of course, this thrust must be generated to overcome the drag of the airplane, or in the case of a rocket, the weight (mass × acceleration of gravity). As a mass of gas is accelerated, thrust is generated, possessing both magnitude and direction (so thrust is a vector). Depending on the mass that is accelerated and the velocity difference between the entrance and exit velocities of the jet engine, the magnitude of the thrust produced by the engine will be determined. The efficiency of this process will affect fuel consumption as well as several important flight requirements. Fuel consumption is roughly proportional to the thrust output of these aircraft, and, because range and endurance performance are functions of fuel consumption, the thrust required is of prime interest and will be the focus of our discussion.

A *turbojet engine* is a turbine engine that works by compressing subsonic air, then mixing fuel with the compressed air in the combustion chamber. After combustion, the hot, high-pressure mixture passes through a turbine to drive the compressor and then is discharged from the exit nozzle at the highest possible velocity (Figure 6.2). Thrust is produced when kinetic energy is created from the chemical potential energy of the burning gases.

A *turbofan engine* differs from a turbojet in that it contains an additional component, a bypass fan. The bypass fan adds cooler air to the engine's hot exhaust, improving thrust production and engine cooling, and reducing noise. Powered by the turbine section of the engine, the fan accelerates some of the intake flow of air around the combustion chamber and out of the nozzle, mixing with the hot exhaust gases (Figure 6.3). The amount of bypass air in relation to the hot exhaust gases is known as the bypass ratio. Modern airliners typically have high bypass ratio engines, meaning the majority of thrust produced (often 70–80%) is from the large bypass fan. The addition of the bypassed air allows the turbofan engine to create thrust more efficiently than it would with exhaust gas alone due to the greater mass of the air that is being moved. Although the form drag of the greater intake area of the turbofan engine increases parasite drag, this is

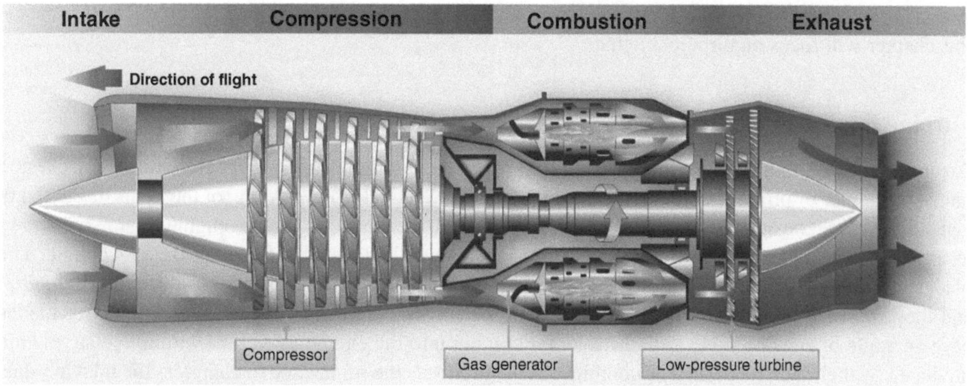

Figure 6.2. Turbojet engine.
Source: Airplane Flying Handbook (2021)/U.S. Department of Transportation Federal Aviation Administration/Public domain.

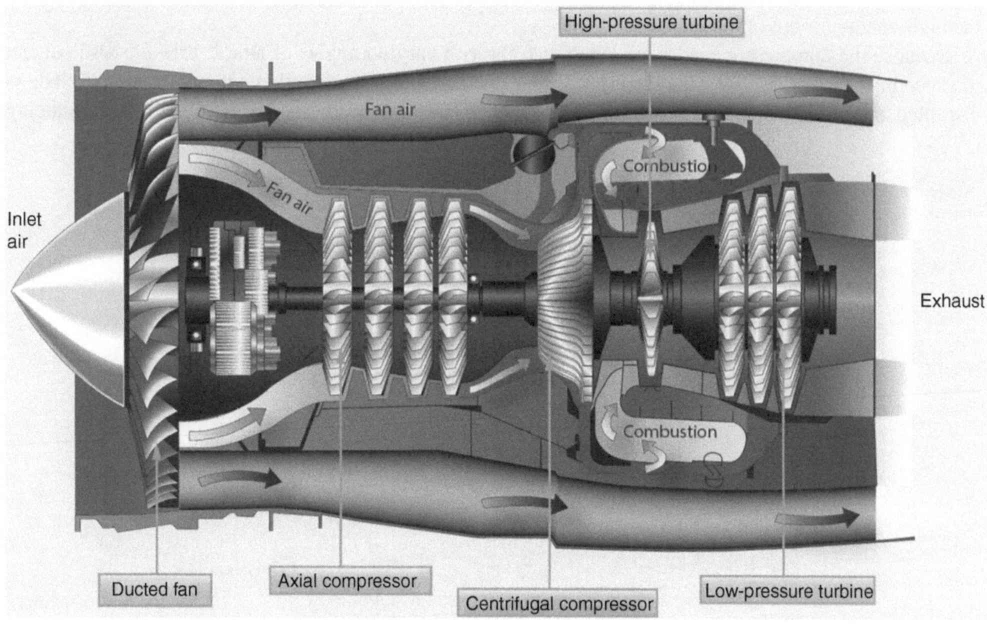

Figure 6.3. Turbofan engine.
Source: Airplane Flying Handbook (2021)/U.S. Department of Transportation Federal Aviation Administration/Public domain.

outweighed by the greatly improved engine efficiency, making the turbofan become the primary engine design on modern subsonic jet aircraft. However, for the purpose of understanding jet aircraft performance, this chapter will focus on turbojet engines.

Turbine Engine Operation

An understanding of the basic principles of turbine engine operation is important before further discussion on the principles of propulsion and jet aircraft performance can take place. Many of the items discussed in Chapters 1–5 can be applied to understanding how thrust is generated from a jet engine. Figure 6.4 shows an example of a dual-spool axial-flow compressor that would be located at the intake to a jet engine. The blades of a compressor are all unique, but they are all essentially airfoils, as we introduced in Chapter 3, and they follow the same aerodynamic principles. As the compressor is rotated, the compressor blades have an angle of attack (AOA) with the inlet air coming into the engine as well as with the relative wind produced by their own rotation. The combined AOA produces the lift needed to compress the inlet air, thus inducing a pressure change.

The term N_1 is used to represent the rotational speed of the low-pressure compressor, while N_2 represents the speed of the high-pressure compressor. Both N_1 and N_2 are shown on the engine instruments as a percentage of design rpm, and strict tolerances must be met not to damage engine components. Engine performance is commonly measured using engine pressure ratio (EPR), which is the ratio of the turbine discharge to the compressor inlet pressure. The turbine drives the two compressor sections and accelerates the high-energy air out of the exhaust.

Because the compressor blades are small airfoils with varying angles of attack, they are also subject to exceeding the critical AOA and stalling. When this happens, the smooth air flow around the blade is disrupted, and as we saw in Chapter 4, a pressure change is experienced, and the airfoil will stop producing

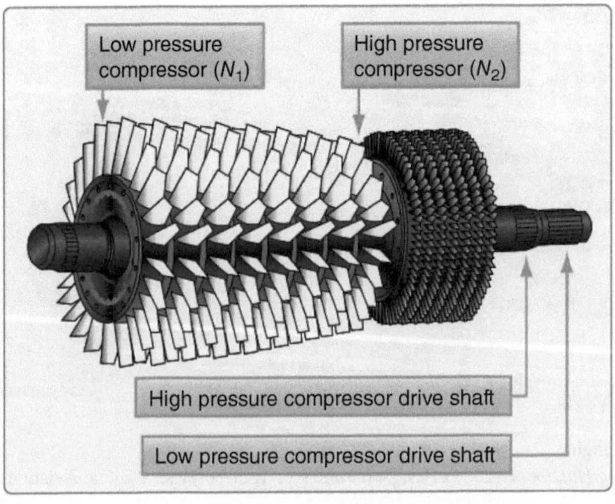

Figure 6.4. Dual-spool axial-flow compressor.
Source: Pilot's Handbook of Aeronautical Knowledge (2016b)/U.S. Department of Transportation Federal Aviation Administration/Public domain.

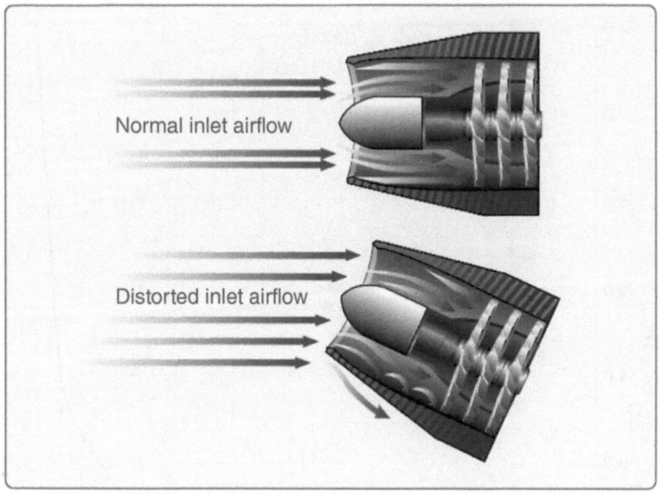

Figure 6.5. Jet engine compressor stall.
Source: Pilot's Handbook of Aeronautical Knowledge (2016b)/U.S. Department of Transportation Federal Aviation Administration/Public domain.

lift when the air flow separates. This situation is referred to as a compressor stall and can occur when the intake to the engine is located at high angles to the incoming air (Figure 6.5).

THRUST-REQUIRED CURVE

Each pound of drag requires a pound of thrust to offset it; this is a very important consideration when depicting thrust using a graph. In Chapter 5, we discussed how the drag curve for an aircraft is drawn. We can now call this same curve a *thrust-required curve*, or the minimum thrust required to maintain the aircraft in level flight. The curves used to illustrate the drag or thrust-required (T_r) curves in this chapter are for the T-38 supersonic turbojet trainer that is still in use today. Because this aircraft encounters a large amount of compressibility drag at high speeds, a sharp increase in the thrust required will be noted in this region.

Figure 6.6 shows the drag curve for the T-38. The drag curve shows the amount of drag of the aircraft at various airspeeds. Figure 6.6 shows that the aircraft has a drag of 2 000 lb at the buffet limit (stall speed), which is close to 160 kts., and at about 485 kts. the drag is also 2 000 lb. Minimum drag occurs at $(L/D)_{max}$, the bottom of the thrust-required curve, which is at 240 kts. velocity for aircraft at this weight, and has a value of 830 lb. As mentioned above, the thrust required (T_r) is equal to the drag at this speed. A thrust of 830 lb is required to offset the total drag of 830 lb in order to maintain level flight for this specific weight and flight conditions.

The curve shown in Figure 6.7 is identical to the one shown in Figure 6.6, except that the induced and parasite drag curves are omitted and the vertical scale is relabeled as "thrust required." The curves in Figures 6.6 and 6.7 are drawn for a 10 000 lb airplane in the clean configuration at sea level on a standard day.

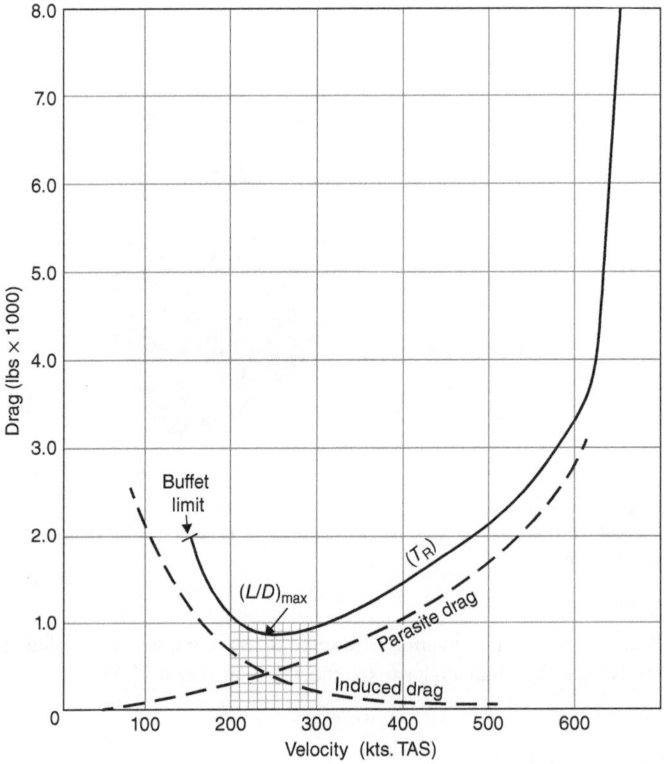

Figure 6.6. T-38 drag curve.

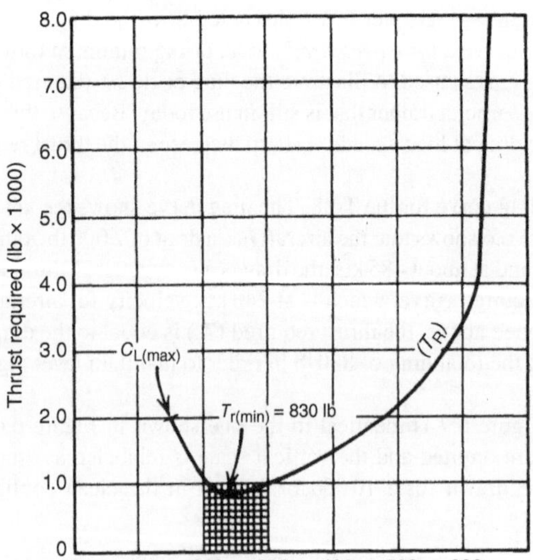

Figure 6.7. T-38 thrust required.

PRINCIPLES OF PROPULSION

The operation of all aircraft power plants follows certain common principles, based on Newton's second and third laws. You will recall that the second law can be written as $F = ma$. Simply stated, an unbalanced force F acting on a mass m will accelerate a, the mass in the direction of the force. In the case of a jet engine, the force is provided by the expansion of the burning gases in the engine. The mass is the mass of the air passing through the jet engine or through the propeller/rotor system. Thrust is directly proportional to air density; as the density increases, so will the available thrust. The acceleration is the change in the velocity of the intake air (i.e. the increase in velocity from entry to exhaust) per unit of time.

Newton's third law states that for every action force, there is an equal and opposite reaction force. This reaction force (the force of the exhaust gases on the engine) provides the thrust, T, to propel the aircraft:

$$F = T, \text{which can also be written as } F = Q(V_2 - V_1) \tag{6.1}$$

where

T = thrust (lb)

Q = mass airflow = ρAV (Eq. 2.6) (slugs/sec)

V_1 = inlet (flight) velocity (fps)

V_2 = exit velocity (fps)

Figure 6.8 shows a schematic of the process of thrust production in a jet engine. T_a represents thrust available.

Equation 6.1 means that the thrust output of an engine can be increased by either (i) increasing the mass airflow or (ii) increasing the exit velocity/flight velocity (V_2/V_1) ratio. The first alternative requires that the cross-sectional area of the engine be increased to allow more air to be processed. The second alternative requires increasing the relative velocity, $V_2 - V_1$. This imparts higher kinetic energy to the gases. The kinetic energy is wasted in the exhaust airstream, resulting in decreased efficiency of the engine. Note the section on afterburners and vectored thrust at the end of this chapter.

Propulsion efficiency, η_p, can be expressed by

$$\eta_p = \frac{2V_1}{V_2 + V_1} \tag{6.2}$$

As V_2 increases (assuming V_1 remains the same), propulsion efficiency goes down. If V_1 increases and V_2 remains the same, then the ratio gets larger. This is shown in Figure 6.9.

Early turbojet engines were used on jet fighters. To keep the drag of the engines within reason, the engines could not have a large frontal area. This limited the amount of air that could be processed, and high exhaust velocities were required to obtain the needed thrust ($V_2 - V_1$ in Eq. 6.1). As larger aircraft were developed, it was possible to use larger engines. Fanjet engines with large Q values improved the efficiency of jet engines, and now high-bypass turbofans with large bypass ratios are much more fuel efficient than turbojets and even low-bypass turbofans.

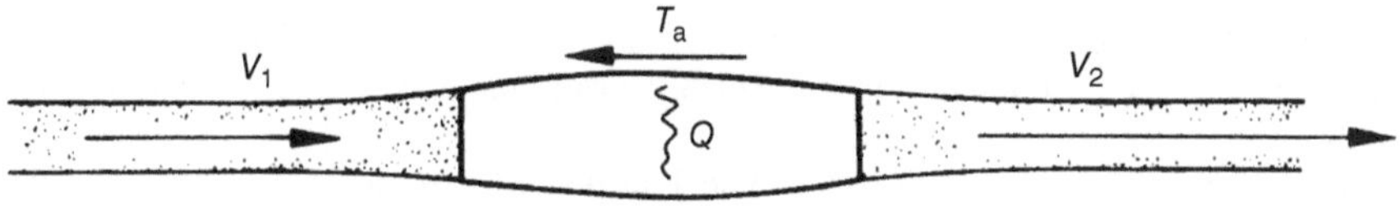

Figure 6.8. Engine thrust schematic.

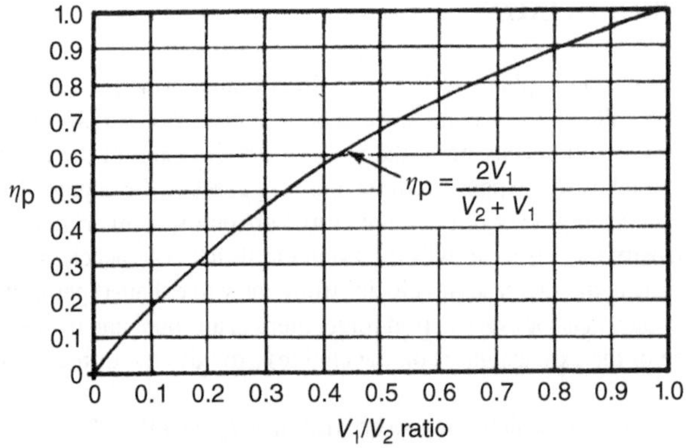

Figure 6.9. Propulsion efficiency.

THRUST-AVAILABLE TURBOJET AIRCRAFT

Turbojet aircraft engines are rated in terms of static thrust. The aircraft is restrained from moving, and the thrust is measured and converted to standard sea level conditions. It can be seen from the thrust Eq. 6.1, $T_a = Q\,(V_2 - V_1)$ or; $p\mathrm{AV}_1\,(V_2 - V_1)$, that as the aircraft gains velocity (V_1 increases), the mass flow increases, but the acceleration through the engine decreases (V_2 is nearly constant). As we can see, if density increases, then so will the thrust (Q is larger for the same intake area and V_1).

Thrust Variation and rpm

The thrust available is nearly constant with airspeed and is considered to be a constant in this discussion. Thrust varies, however, if the rpm of the engine is changed from the 100% value. Thrust is proportional to rpm, which is based on mass flow, as well as temperature, a function of fuel-to-air ratio. When rpm increases, so does mass air flow and temperature, and thus efficiency. More thrust is produced with incremental throttle movements near maximum rpm than with equivalent throttle movements near flight idle.

Figure 6.10 shows how thrust varies with rpm for a constant velocity and altitude. Note that as percent maximum rpm increases from 70%, the corresponding percent maximum thrust increases. Jet engines operate most efficiently between 85% and 100% of maximum rpm because at those speeds the compressor is working most efficiently, while flight idle is closer to 50% of maximum rpm. This change in thrust with rpm is also noted during the acceleration of the jet engine. In a propeller-driven airplane, for a given rpm, increased throttle movement equals increased thrust production as thrust is proportional to throttle lever position. In a jet airplane, throttle level positions are not proportional to thrust production throughout the full throttle range. A jet engine that is operating at a high rpm during final approach will require much less time to accelerate to full thrust than a jet engine that is operating at flight idle. Note in Figure 6.11 that after the 78% mark of engine rpm, the time to accelerate is two seconds or less. This is why early turbojet aircraft sometimes employed thrust attenuators on final approach to allow a higher engine rpm in case a go-around became necessary.

As we have discussed, the reduction of thrust is not linear with the reduction of rpm. Figure 6.12 is similar to Figure 6.10 except now we have plots for various aircraft velocities during rpm changes and thrust variation. Note the term military thrust, which indicates the maximum thrust without the use of afterburners.

Thrust is also reduced with an increase in altitude. Equation 6.1 indicates that the mass flow decreases as the density of the air decreases. This can be seen in Figure 6.13 as the thrust available decreases steadily

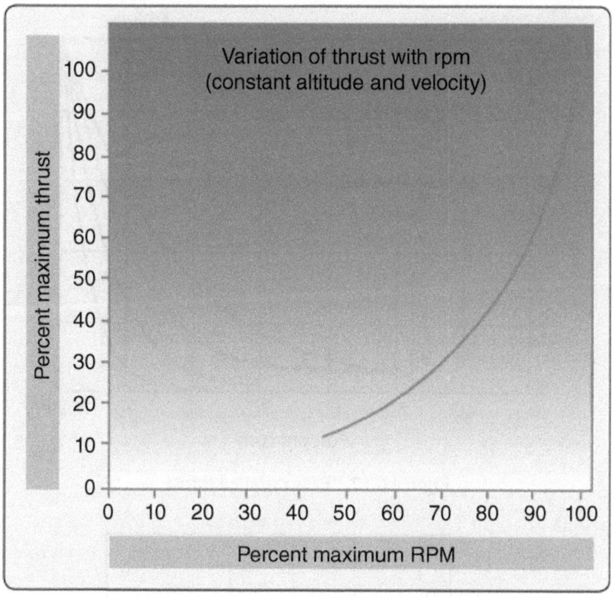

Figure 6.10. Variation of thrust with rpm.
Source: Airplane Flying Handbook (2016a)/U.S. Department of Transportation Federal Aviation Administration/Public domain.

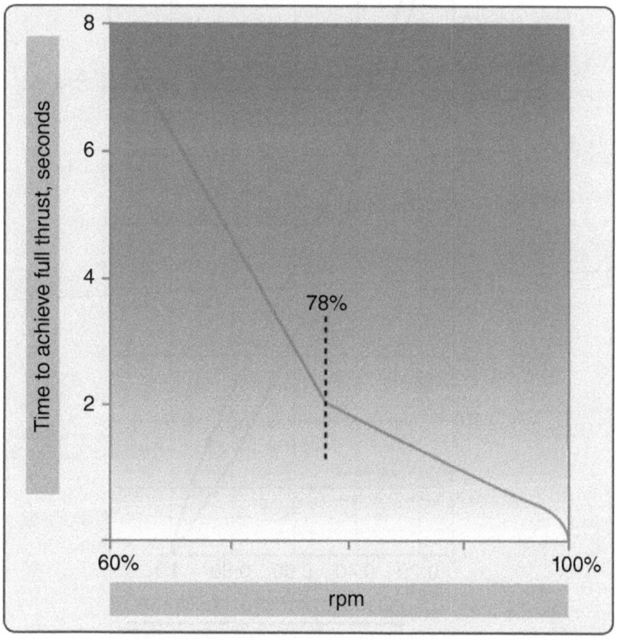

Figure 6.11. Typical jet acceleration times.
Source: Pilot's Handbook of Aeronautical Knowledge (2016b)/U.S. Department of Transportation Federal Aviation Administration/Public domain.

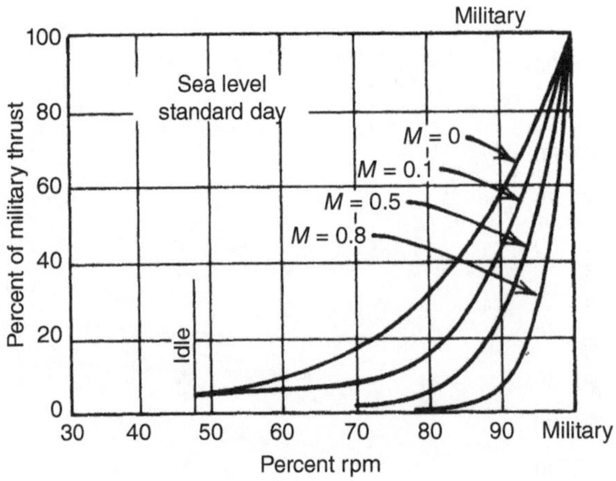

Figure 6.12. T-38 installed thrust.

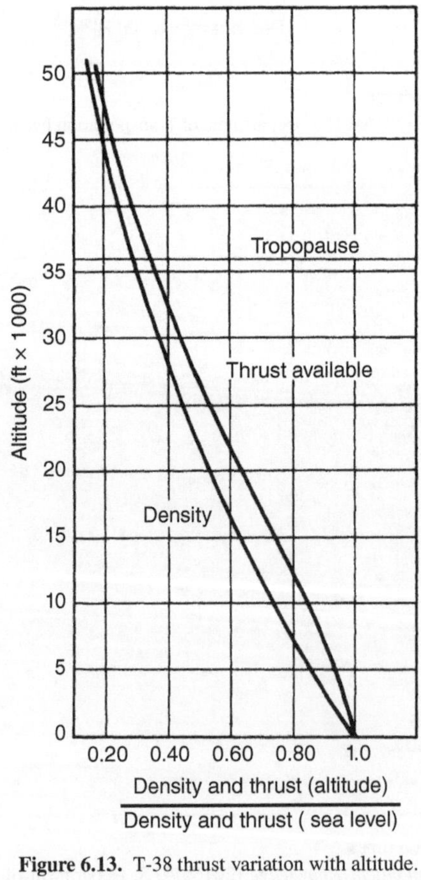

Figure 6.13. T-38 thrust variation with altitude.

with the decrease in density with altitude. If T_a is 4 200 lb at sea level at 100% rpm, then the T_a at 27 000 ft is approximately 50% of that at sea level, or:

$$T_a = (4\,200\text{ lb})\,(0.50) \rightarrow T_a = 2\,100\text{ lb}$$

SPECIFIC FUEL CONSUMPTION

An aircraft is propelled through the air by thrust, and to produce thrust, an engine burns fuel. More thrust means higher speeds but also higher fuel consumption. A pilot planning a mission must know the rate at which his engine consumes fuel at various power settings in order to know whether the aircraft will have the speed and range to complete the mission. *Specific fuel consumption* (the ratio of fuel consumption to thrust produced) is used to indicate the efficiency of an engine. Specific fuel consumption is calculated by finding the fuel flow (FF) per pound of thrust developed by a jet engine:

$$c_t = \frac{\text{Fuel flow (lb/hr)}}{\text{Thrust (lb)}} \tag{6.3}$$

Turbojet engines are designed to operate at high rpm and will produce low specific fuel consumption values (low c_t) in this region. Figure 6.14 shows the variation of c_t with rpm. The lowest values of c_t (highest

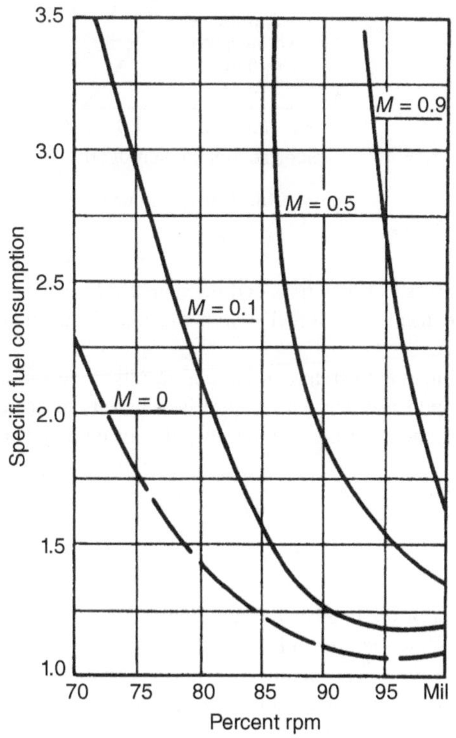

Figure 6.14. T-38 c_t–rpm.

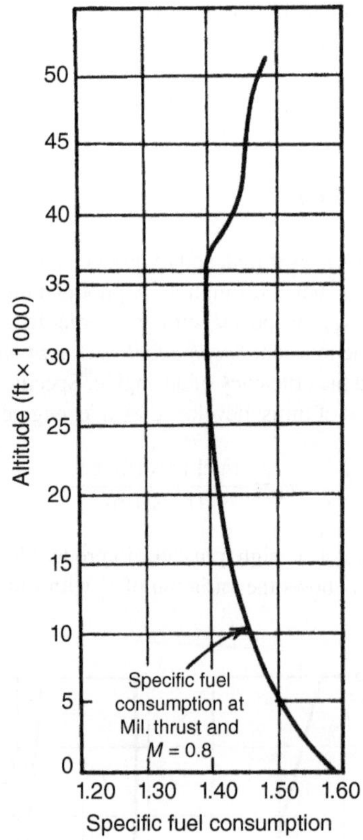

Figure 6.15. T-38 c_t–altitude.

efficiency) occur between 95% and 100% rpm, and the increase in c_t is very rapid as rpm is decreased. This does not mean you are consuming less fuel at higher levels of rpm, only that you are consuming less fuel per pound of thrust.

The beneficial effect of lower temperature at altitude results in decreased specific fuel consumption (Figure 6.15). Above the tropopause, the temperature is constant, so no reduction in c_t will occur. In fact, the compressor will not operate efficiently because the air density is reduced, causing an increase in c_t.

FUEL FLOW

The total FF equals the thrust of the engine, T, multiplied by the specific fuel consumption, c_t; hence, Eq. 6.3 has been rearranged. As we can see in Eq. 6.4, the greater the specific fuel consumption (or thrust for a given c_t), the greater the FF. For a given thrust, as the specific fuel consumption decreases (efficiency improvement), the FF decreases. Again, low values of c_t are desired.

$$FF = Tc_t \text{ (lb/hr)} \tag{6.4}$$

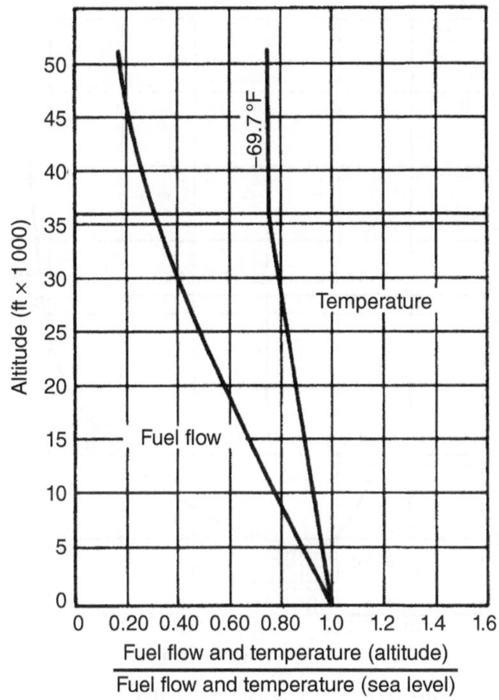

Figure 6.16. T-38 fuel flow–altitude.

Thrust varies with rpm and altitude (Figures 6.10, 6.12, and 6.13). Specific fuel consumption varies with these same factors (Figures 6.14 and 6.15). FF at altitude as a ratio of that at sea level, at a fixed rpm, is shown in Figure 6.16. As altitude increases, the FF required decreases.

Application 6.1

A Rockwell B-1 Lancer (Bone) bomber may consume as much as 360 000 lbs/hr of fuel while producing 120 000 lbs of thrust when at sea level in full afterburner.

Calculate the c_t value of the above example. Based on Fig. 6.15, what should happen to the value of c_t for the B-1 if the altitude increases? Why? What will happen to the fuel flow based on Eq. 6.4 if the thrust produced remains the same with an altitude increase, but the value of c_t changes as calculated in the previous question?

THRUST-AVAILABLE/THRUST-REQUIRED CURVES

Performance depends on the relationship of the thrust-available (T_a) and thrust-required (T_r) curves. If the thrust available is equal to the thrust required, the aircraft can fly straight and level, but cannot climb or accelerate. An excess of T_a over T_r is required in order to accelerate, climb, or even turn the aircraft

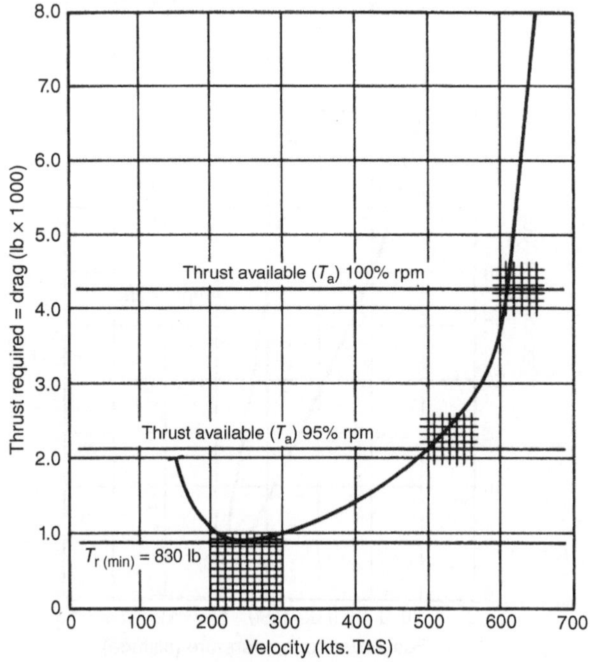

Figure 6.17. T-38 thrust available–thrust required.

if maintaining constant altitude is desired. The thrust-available and thrust-required curves are plotted for the T-38 at 10 000 lb gross weight, clean configuration at sea level, in Figure 6.17. The thrust available is plotted for two different conditions (100% rpm and 95% rpm) at sea level. The reduction of T_a with rpm was explained earlier (Figure 6.12).

Figure 6.17 shows that a 5% reduction in rpm results in a reduction of T_a from 4 200 to 2 100 lb, close to a 50% reduction in thrust. This aircraft could not maintain altitude at any speed if the thrust is reduced below 830 lb as this is the minimum thrust required for the total drag in order to maintain level flight. This is a ratio of 830/4 200 = 0.20 of the T_a at 100% rpm, sea level conditions. Figure 6.12 shows that this occurs at about 88% rpm at sea level and Mach (M) = 0.5. Absolute ceiling occurs when T_a falls to 830 lb, indicating no excess thrust is available; Figure 6.13 shows that this occurs at about 47 500 ft.

ITEMS OF AIRCRAFT PERFORMANCE

Straight and Level Flight

When an aircraft is in steady (unaccelerated) level flight, it must be in equilibrium. Lift must be exactly equal to the weight of the aircraft, and the thrust on the aircraft must be exactly equal to the drag. The thrust is equal to the drag when the thrust available is equal to the thrust required. This happens when the T_a and T_r curves intersect. There are two possibilities for such an intersection. One is in the high-speed region of flight and the other is in the low-speed area. Our focus here will be on the intersection during the high-speed region of flight; the problems in the low-speed area will be discussed in Chapter 10.

For nonafterburner flight, the thrust available for the T-38 is 4 200 lb. From Figure 6.17, it can be seen that the high-speed intersection point of T_a and T_r is at 605 kts. true air speed (TAS), and if there is a reduction in rpm to 95%, the value of V_{max} decreases to 500 kts. TAS.

Climb Performance

Every flight requires that the pilot climb the aircraft to some altitude. Mission requirements dictate the climb schedule. For instance, an intercept mission would require that altitude be gained as rapidly as possible, while for a mission of long range, the climb would be more gradual.

There are two basic types of climb: delayed climb and steady-velocity climb. In a delayed climb, the pilot builds up airspeed (kinetic energy) before starting to climb. The pilot then "zooms" the aircraft to altitude, which converts some of the kinetic energy to potential energy (height). This maneuver is used mostly by fighter aircraft pilots in intercept tactics, setting climb-to-altitude records, and other maneuvers.

Steady-velocity climb is used much more often and represents an equilibrium condition, with all the forces on the aircraft being balanced (Figure 6.18). The forces acting along the flight path are thrust, T, and drag, D. Lift acts 90° to the flight path, and the weight, W, acts toward the center of the earth. As introduced in Chapter 3, the climb angle is designated by the Greek letter gamma, γ. The weight vector is replaced by two other vectors. One is perpendicular to the flight path (W cos y) and the other is parallel to the flight path (W sin γ) and realistically acts at the aircraft's center of gravity (CG). For steady velocity, the forces along the flight path must be balanced, and as we see, thrust must now balance with the addition of the weight for a given climb angle, plus the total drag. Thus,

$$T - D - W \sin \gamma = 0 \tag{6.5a}$$

Angle of Climb

For the angle of climb, Eq. 6.5a is rewritten as

$$\sin \gamma = \frac{T - D}{W} \rightarrow \sin \gamma = \frac{T_a - T_r}{W} \tag{6.5b}$$

As an angle increases (up to 90°), the sine of the angle increases. The maximum climb angle will occur at the point where the excess thrust is at a maximum $(T_a - T_r)_{max}$. The speed at which this occurs is the

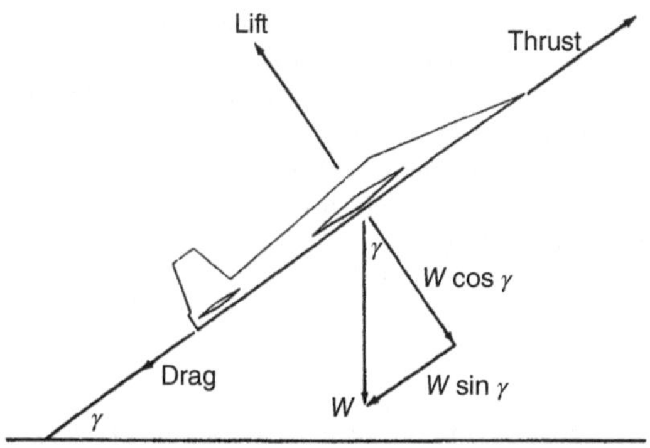

Figure 6.18. Forces acting on a climbing aircraft.

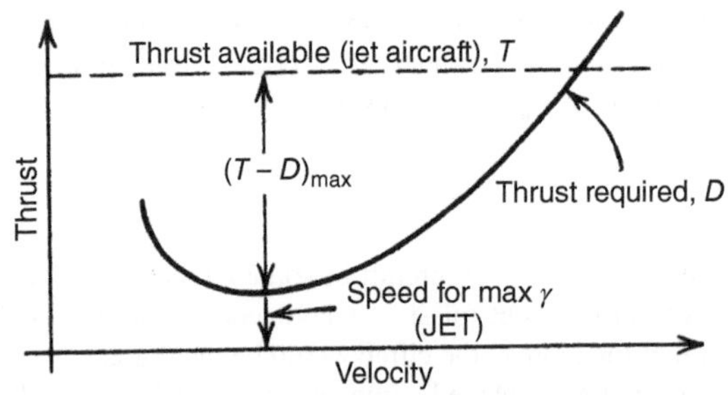

Figure 6.19. Velocity for maximum climb angle.

best angle of climb speed (V_X). The performance value for V_X will offer the greatest height in altitude over a given distance. For a turbojet, with the thrust available assumed to be constant with velocity, this will be where T_r (drag) is a minimum, which must be at $(L/D)_{max}$ (Figure 6.19). It should be noted that a turbojet aircraft such as the T-38 will typically take off and initially climb in full afterburner. In afterburner, the thrust available does increase with velocity, meaning the best angle of climb speed will not occur at $(L/D)_{max}$. Instead, it will occur at a higher velocity where there is greater excess thrust. In Chapter 7, we will see how turboprop aircraft have a reduction in power available with increasing velocity.

EXAMPLE

Using Figure 6.17, calculate the climb angle of a jet with the following conditions: $W = 10\,000$ lb, standard day, sea level, clean configuration, T_a at 100% rpm, TAS = 300 kts.

$$\sin\gamma = \frac{T_a - T_r}{W}$$

$$\sin\gamma = \frac{4\,200\text{ lb} - 1\,000\text{ lb}}{10\,000\text{ lb}}$$

$$\sin\gamma = 0.320 \rightarrow \gamma = 18.7°$$

When an aircraft is in level flight, its lift is equal to the weight of the aircraft. Figure 6.18 shows that the vertical component of lift is less than the weight if the aircraft is in a climb. This is possible because the vertical component of the thrust supports some of the weight of the aircraft. The steeper the climb, the less the lift supports the weight (i.e. the smaller the vertical component of lift becomes). The extreme case occurs when the aircraft climbs straight up ($\gamma = 90°$). Then the vertical component of lift would be zero, and the thrust would support the entire weight of the aircraft and overcome the drag. A thrust/weight ratio of at least 1 is required for this maneuver, and this performance is currently only achievable by some military aircraft. The angle of climb determines whether an aircraft can clear an obstacle, and the angle of climb is affected by headwind and tailwind, as shown in Figure 6.20. Once the airplane is airborne, the ground speed is reduced by a headwind and increased by a tailwind. With reduced ground speed, the time to reach the obstacle is increased, and more altitude is gained.

Rate of Climb

One other factor in obstacle clearance is the time required to reach the $(L/D)_{max}$ velocity. After the jet takes off, it must accelerate to this velocity and, in doing so, precious ground distance is used up during the

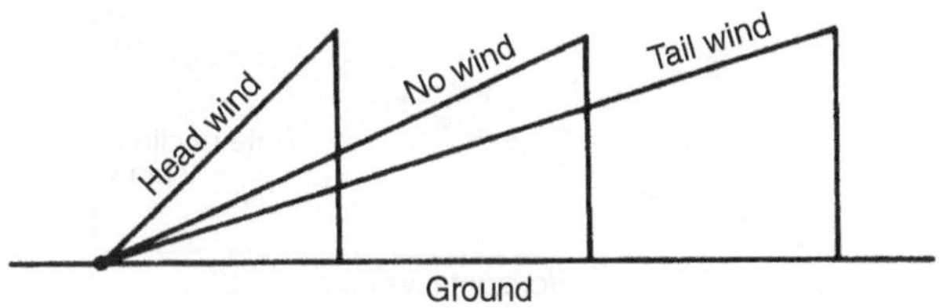

Figure 6.20. Wind effect on the climb angle to the ground.

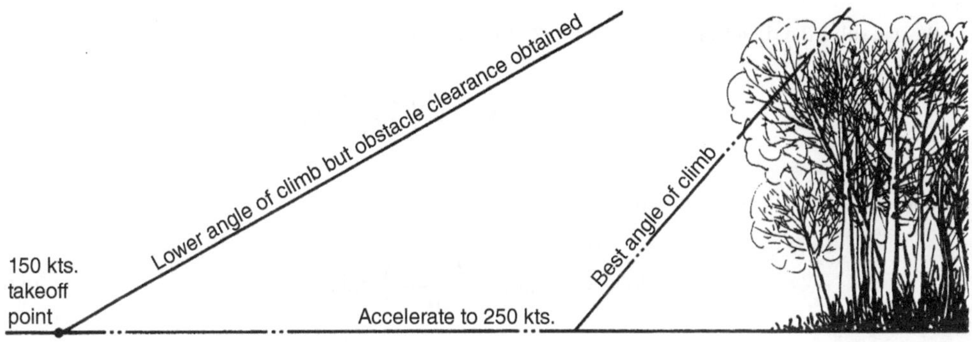

Figure 6.21. Obstacle clearance for jet takeoff.

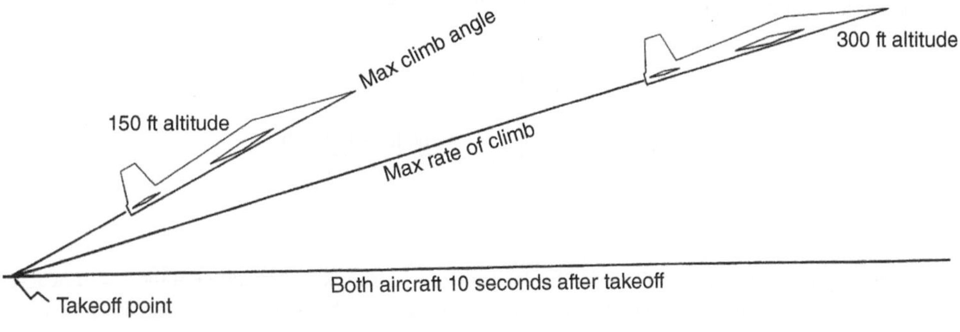

Figure 6.22. Climb angle and rate of climb.

acceleration. Once the jet reaches this velocity farther down the runway, it may not clear an obstacle in the departure path even when climbing at V_X. If the jet initiates a climb after rotation at a smaller angle of climb than V_X, but at a higher speed than at V_X, the obstacle may be cleared since the rate of climb (ROC) will be larger (Figure 6.21).

Figure 6.22 shows the difference between an aircraft climbing at V_X versus V_Y. The aircraft on the left in Figure 6.22 is flying at maximum climb angle. The airspeed of this aircraft is relatively slow compared to the aircraft on the right. The aircraft on the right is flying at a smaller climb angle, but the much higher airspeed more than compensates for the lower angle, and it is at a higher altitude. The aircraft on the right has a higher ROC (V_Y). This is shown in Figure 6.23.

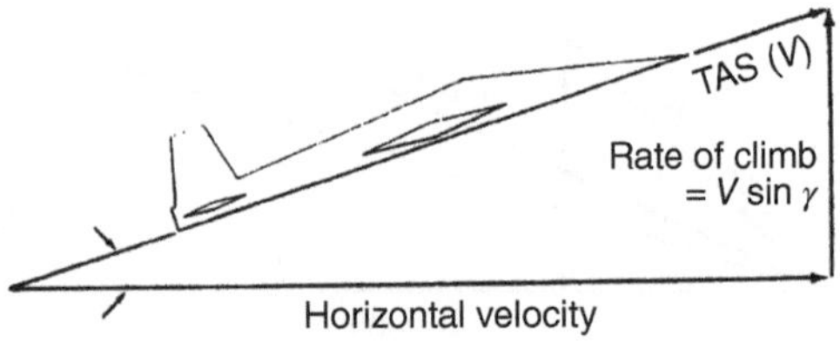

Figure 6.23. Rate of climb velocity vector.

We can solve the right triangle in Figure 6.23 for the ROC:

$$\text{ROC} = V_k \sin\,\gamma \rightarrow \text{ROC} = V_k \left(\frac{T_a - T_r}{W}\right) \text{(kts.)}$$

Converting to feet per minute (fpm),

$$\text{ROC} = 101.3\,V_k \left(\frac{T_a - T_r}{W}\right) \text{(fpm)} \tag{6.6}$$

EXAMPLE

Using Figure 6.17, calculate the rate of climb of a jet with the following conditions: $W = 10\,000$ lb, standard day, sea level, clean configuration, T_a at 100% rpm, TAS = 300 kts.

$$\text{ROC} = 101.3 V_k \frac{T_a - T_r}{W}$$

$$\text{ROC} = 101.3\,(300\text{ kts.})\frac{4\,200\text{ lb} - 1\,000\text{ lb}}{10\,000\text{ lb}}$$

$$\text{ROC} = 9\,725\text{ fpm}$$

The velocity for maximum ROC cannot be determined by a simple examination of the $T_a - T_r$ curves. Equation 6.6 shows that ROC depends on velocity and excess thrust, both of which are variables. However, by selecting various airspeeds and finding the corresponding values of T_a and T_r from Figure 6.17, then solving Eq. 6.6, a plot of ROC versus velocity can be made similar to Figure 6.24, and the velocity for maximum ROC can be found.

Application 6.2

Using Figure 6.17, calculate the angle of climb for the jet at $(L/D)_{max}$. Then, calculate the rate of climb for the jet at $(L/D)_{max}$.

Is the angle of climb higher or lower than the example provided in the book? Why? Is the ROC value found higher or lower than the ROC at 300 kts. TAS? Why?

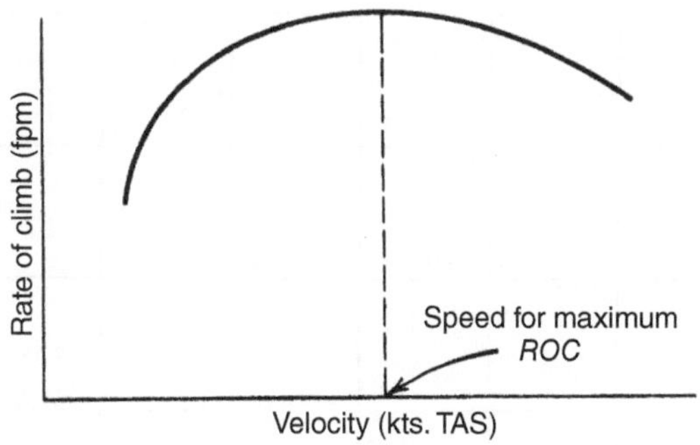

Figure 6.24. Velocity for maximum rate of climb.

Endurance

The endurance of an aircraft is the amount of *time* that it can remain airborne, and is independent of the distance covered. It is useful when holding due to weather, air traffic, and so on, or to extend the time available for alternate airport diversions and to troubleshoot an emergency. Because endurance is a function of fuel consumption, it is inversely proportional to FF: the lower the FF, the greater the endurance (Eq. 6.7).

$$\text{Endurance (in hours)} = \frac{\text{Fuel available (lb)}}{\text{Fuel flow (lb/hr)}} \tag{6.7}$$

Figure 6.25 shows the thrust-required curve for the T-38 with the FF also plotted on the vertical scale. Note that T_r is 830 lb. The minimum FF occurs at the point of minimum thrust required, which is the minimum drag point of the curve: $(L/D)_{max}$. Since FF is lowest at $(L/D)_{max}$, maximum endurance for a jet aircraft occurs at $(L/D)_{max}$. The velocity that gives maximum endurance (best endurance) is referred to as V_{BE}.

EXAMPLE

Using Figure 6.25 and Eq. (6.7), calculate the endurance of the jet aircraft with the following conditions: Fuel available = 2 100 lb, standard day, sea level, clean configuration, TAS = 500 kts.

$$\text{Endurance} = \frac{\text{Fuel available (lb)}}{\text{Fuel flow (lb/hr)}}$$

$$\text{Endurance} = \frac{2\,100\ \text{lb}}{2\,000\ \text{lb/hr}}$$

$$\text{Endurance} = 1.05\ \text{hrs}$$

Specific Range

To get maximum range, an aircraft must fly the maximum *distance* with the fuel available, so we are looking for the maximum speed per FF. The specific range (SR) can be defined by the following relationship

$$\text{Specific range} = \frac{\text{nm}}{\text{lb fuel}}\ \text{or;}\ \frac{\text{nm/hr}}{\text{lb fuel/hr}}\ \text{or;}\ \frac{\text{Speed (kts.)}}{\text{Fuel flow (lb/hr)}}$$

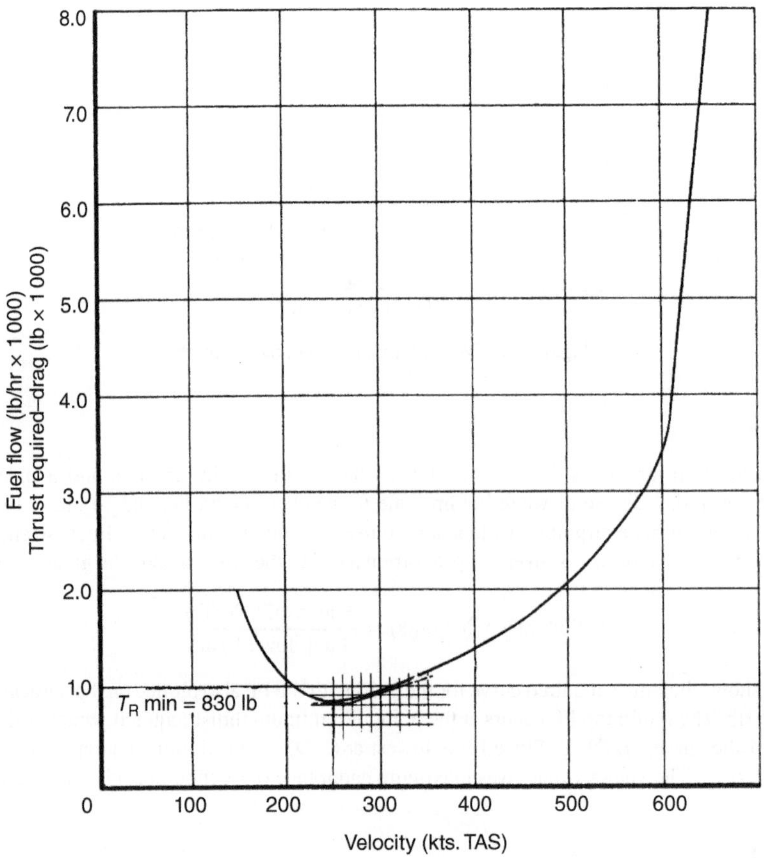

Figure 6.25. Finding maximum endurance velocity.

which represents *minimum* lb/nm as shown in Eq. 6.8:

$$(\mathrm{SR})_{max} \text{ occurs at } \left(\frac{FF}{V}\right)_{min} \tag{6.8}$$

Graphically, the maximum specific range point can be shown on a FF versus velocity curve, such as that in Figure 6.26. Select any point on the T_r curve, such as A. Draw a straight line from A to the origin and drop a vertical line from A to the velocity scale. Label the angle at the origin of the resulting right triangle, angle θ_1. The tangent of θ_1 equals the opposite side of the triangle divided by the adjacent side. The opposite side is a measure of the FF, and the adjacent side of the triangle is the aircraft's velocity:

$$\tan\theta_1 = \frac{FF}{V}$$

Hence, if FF/V is a minimum, tan θ must be a minimum, which means that θ must be a minimum. Only one point on the curve meets this requirement – the point where the line from the origin is tangent to the T_r

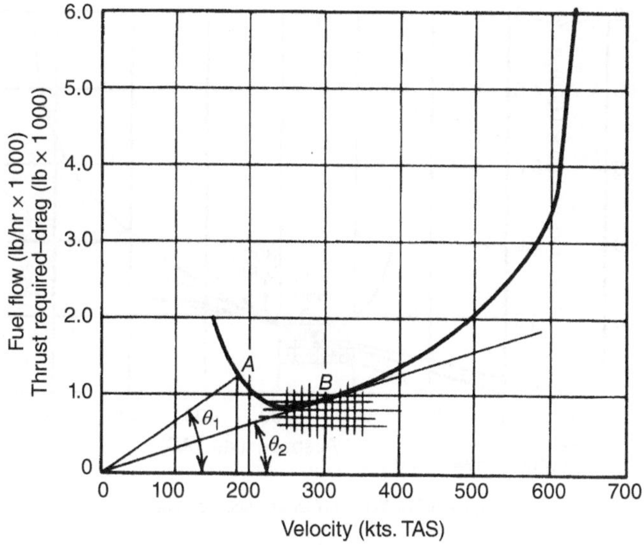

Figure 6.26. Finding the maximum SR velocity.

curve at point B. θ_2 is a minimum at B, and maximum SR will occur if the aircraft is flown at this velocity, ~300 kts. The velocity that gives maximum SR (best range) is referred to as V_{BR}. Maximum SR occurs at a velocity corresponding to the tangent point of a straight line drawn from the origin to the curve. This is not $(L/D)_{max}$, but is the point where $\sqrt{C_L/C_D}$ is a maximum.

It is also important to note at this point that the AOA to maintain maximum endurance and maximum SR is different. Referencing the lift formula from Chapter 4, if all other variables stay the same but the velocity changes (V^2), the C_L value (derived from AOA) will have to be adjusted as well to maintain the same lift. So, if the maximum endurance is found at a lower velocity than the maximum SR velocity, the AOA to maintain the maximum endurance velocity will have to be greater to generate the additional lift required due to the reduction in velocity.

Wind Effect on Specific Range

An aircraft flying into a headwind is in an unfavorable environment and thus should fly at a higher true airspeed to reduce the amount of time it will be fighting the headwind. Likewise, an aircraft flying with a tailwind is being helped on its way and thus should slow down to take advantage of the wind.

How much the airspeed should be altered is shown in Figure 6.27. This illustration shows the correct airspeed for maximum range with a 100-kts. headwind or tailwind. To find the corrected airspeed for a 100-kts. headwind, plot the headwind on the velocity scale to the right of the origin. Use this point as a new origin and draw a tangent line to the curve. This will give the new airspeed for best range. In the example, the no-wind airspeed for maximum range was 300 kts. and the corrected airspeed is 320 kts.

Correcting for a tailwind is done similarly. The tailwind is plotted on the velocity scale, but to the left of the origin. A new tangent line will show the correct airspeed of 290 kts.

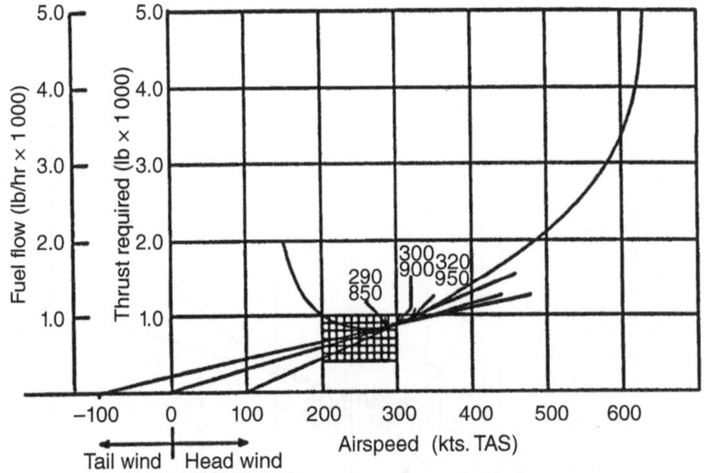

Figure 6.27. Wind effect on specific range.

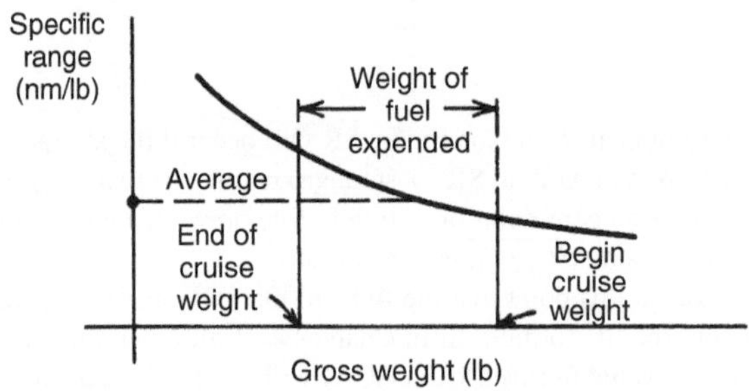

Figure 6.28. Total range calculation.

Total Range

The total range of an aircraft depends on the fuel available and the SR. SR is a variable and is affected by the gross weight of the aircraft, altitude, aircraft configuration, and pilot technique. As the weight of an aircraft decreases with fuel consumption, the SR improves. We will see in the next section how velocity must be adjusted at various weights, configurations, and altitudes in order to maintain maximum range. This variation for fuel consumption (weight reduction) is shown in Figure 6.28. The total range is measured by the area under the curve between the beginning and ending cruise weights. A way of approximating the total range is to find the average SR and multiply it by the fuel burned.

VARIATIONS IN THE THRUST-REQUIRED CURVE

The aircraft performance items that we discussed up to this point were for one weight, clean configuration, and sea level standard day conditions. Now, we will alter these conditions and examine the resulting performance, specifically as weight, configuration, and altitude are changed. To simplify the explanation, only one condition will be altered at a time.

The basic thrust-required curve for the T-38 (Figure 6.6) was drawn for a 10 000 lb airplane in the clean configuration, at sea level on a standard day. We now show how the curve changes for variations in weight, configuration, and altitude.

Weight Changes

Changing the weight of an aircraft changes the induced drag much more than the parasite drag. As weight changes, the lift must change accordingly, thus, drag (induced) due to lift will change as well. The only change in the parasite drag is caused by small changes in equivalent parasite area, f, at varying angles of attack. Only changes in induced drag are considered in Figure 6.29. Note the variation in induced drag between an airplane that weighs 12 000 lb and an airplane that weighs 8 000 lb in the lower speed regime compared to faster speeds. Both aircraft for this discussion are in level flight.

If the weight of the airplane is increased, the lift required increases, so total drag increases due to the increase in induced drag (parasite drag remains the same for our discussion). The intersection of the induced drag and the parasite drag curves still marks the point of $(L/D)_{max}$, but it is moved up and to the right. In fact, all points on the thrust-required curve are moved up and to the right. Because induced drag is greater than parasite drag in the low-speed region of flight, the curve is moved by a greater amount in the low-speed region than in the high-speed region.

The total drag curves are shown as the thrust-required curves in Figure 6.30. Both of the curves were obtained by altering the basic T_r curve for the 10 000 lb T-38 (Figure 6.6). First, the airspeed for each point on the curve was corrected using Eq. 4.7:

$$V_2 = V_1\sqrt{\frac{W_2}{W_1}}$$

Second, the drag was corrected by using the fact that the value of L/D is not changed by weight. This was discussed previously when this statement was made: "Note that the value of $(L/D)_{max}$ and the AOA at which it occurs do not vary with aircraft weight or altitude."

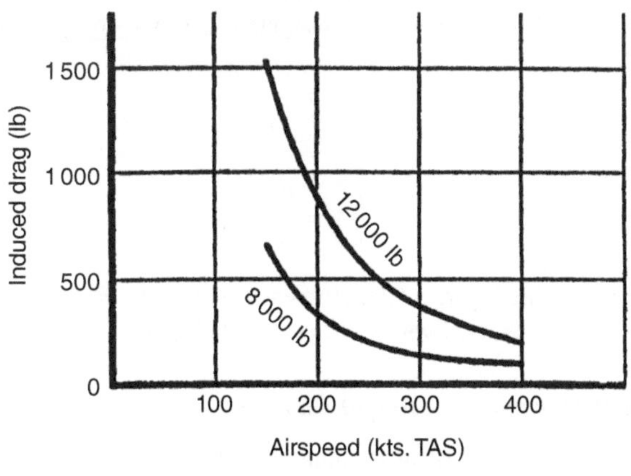

Figure 6.29. Effect of weight change on induced drag.

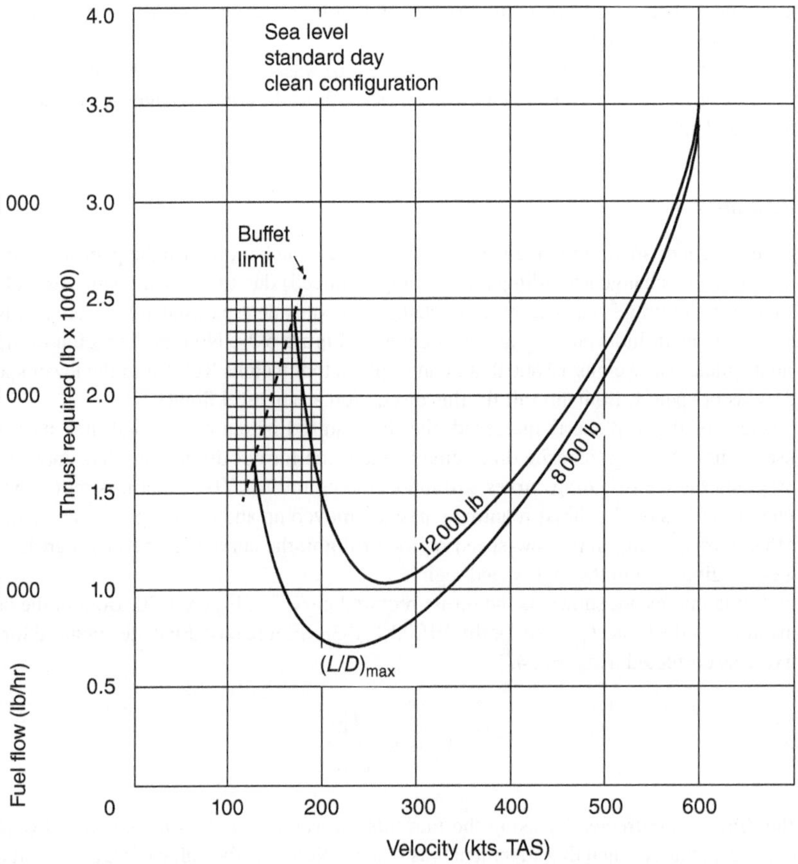

Figure 6.30. Effect of weight change on the T_r curve.

Not only does $(L/D)_{max}$ not vary, but other L/D values do not vary either. They are functions of AOA only. So, if L/D does not vary with weight, then changing the weight changes the drag proportionately, and

$$\frac{L_2}{D_2} = \frac{L_1}{D_1}$$

Under 1G flight $L = W$, so

$$\frac{D_2}{D_1} = \frac{W_2}{W_1} \tag{6.9}$$

Figure 6.30 was drawn by altering Figure 6.6 for different weights. The airspeeds were changed using Eq. 4.6 and the corresponding drag values, T_r, were changed using Eq. 6.9.

Variation of Aircraft Performance – Weight Change

Straight and Level Flight An increase in weight means an increase in total drag due to the increase in induced drag, but T_a is not changed, so V_{max} is decreased due to the change in intersection of the T_a and

T_r curves. A weight decrease will reduce the drag, and the $T_a - T_r$ intersection (V_{max}) will be at a higher airspeed.

Angle of Climb Increased weight means increased drag, while T_a is not changed. The $(T - D)/W$ term in Eq. 6.5b is reduced, and a lower angle of climb results. When weight is reduced, the opposite occurs, and the climb angle is increased as $T_a - T_r$ is increased because the amount of drag (T_r) to overcome is decreased.

Rate of Climb ROC is the speed times the sine of the climb angle. As we saw above, the $T - D$ term reduces with a weight increase, and the extra drag also slows the aircraft, so ROC is reduced with increased weight and increased with decreased weight. In the following example, compare your ROC found earlier in this chapter for the 10 000 lb aircraft at 300 kts. in Figure 6.17, to the ROC with a higher weight of 12 000 lb at the same speed. With an increase in weight, ROC decreases.

EXAMPLE

Using Figure 6.30, calculate the rate of climb of a jet with the following conditions: $W = 12\,000$ lb, standard day, sea level, clean configuration, $T_a = 4\,200$ lb, TAS = 300 kts.

$$\text{ROC} = 101.3 V_k \frac{T_a - T_r}{W}$$

$$\text{ROC} = 101.3\ (300\ \text{kts.}) \frac{4\,200 - 1\,100\ \text{lb}}{12\,000\ \text{lb}}$$

$$\text{ROC} = 7\,851\ \text{fpm}$$

Endurance More drag is developed with increased weight, so more fuel is consumed to produce the increased thrust that is required to overcome drag. Fuel flow is proportional to T_r (drag), so endurance is reduced, and vice versa.

Specific Range As fuel is burned, weight is reduced, and all points on the thrust-required curve move downward and to the left (Figure 6.31). The tangent lines indicate velocity for SR_{max}. Two important changes in the flight schedule must be made by the pilot as weight decreases if maximum range is to be obtained: the thrust and the airspeed of the aircraft must be reduced. However, a reduction in throttle will result in a reduction of thrust available, which will automatically lower the airspeed. The natural tendency of a pilot to keep the throttle setting the same as weight is reduced, allowing the airspeed to increase, but this tendency must be suppressed. As we have seen, when weight decreases, the T_r curve moves down and to the left, so the maximum speed for fuel flow also changes. For maximum range (V_{BR}), as fuel is burned, *reduce airspeed.*

EXAMPLE

From Figure 6.31, the SR at 12 000 lb is

$$\text{SR}_{12} = \frac{V}{\text{FF}} = \frac{325\ \text{kts.}}{1\,200\ \text{lb/hr}} = 0.2708\ \text{nm/lb}$$

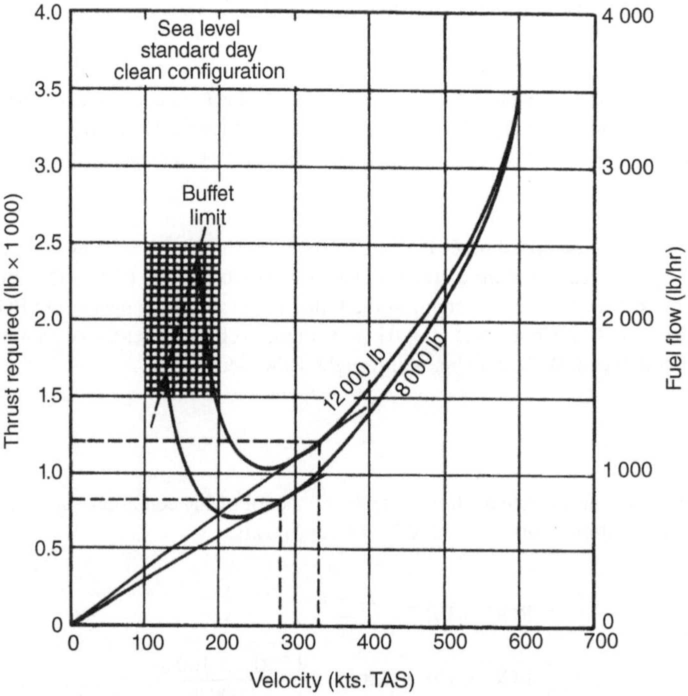

Figure 6.31. Effect of weight change on specific range.

If the pilot did not reduce thrust but allowed airspeed to increase as fuel was burned, the airspeed would be 375 kts. when the weight was 8 000 lb. Fuel flow would still be 1 200 lb/hr, and the SR would be

$$SR_8 = \frac{375 \text{ kts.}}{1\,200 \text{ lb/hr}} = 0.3125 \text{ nm/lb}$$

By reducing thrust to 825 lb and airspeed to 285 kts., the resulting SR would be

$$SR_8 = \frac{285 \text{ kts.}}{825 \text{ lb/hr}} = 0.3455 \text{ nm/lb}$$

Not reducing thrust results in a 15.4% increase in SR, but reducing thrust results in a 27.6% increase.

Application 6.3

Using Figure 6.30, calculate the angle of climb for the 12 000 lb jet at $(L/D)_{max}$. Then calculate the angle of climb for the 8 000 lb jet at $(L/D)_{max}$.

Which aircraft has the higher angle of climb? Why? Note the velocity change for $(L/D)_{max}$ between the curves using Figure 6.30, then use Eq. 4.7 to calculate the new velocity with a weight change and confirm mathematically what you resolved from the figure.

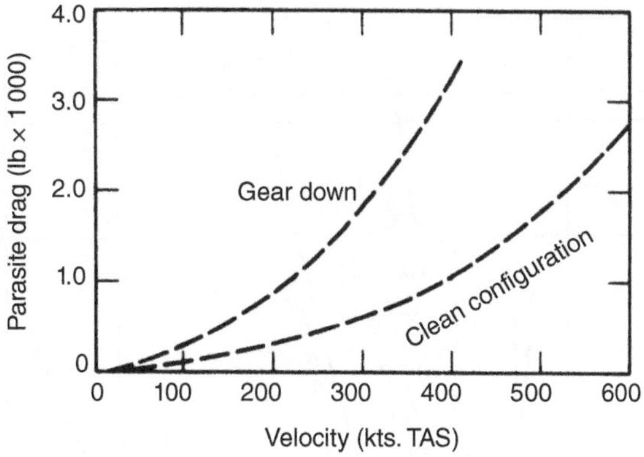

Figure 6.32. Effect of configuration on parasite drag.

Configuration Changes

When a pilot lowers the landing gear and/or flaps, the equivalent parasite area is greatly increased. This increases the parasite drag of the aircraft, as shown in Figure 6.32, but it has little effect on the induced drag because it affects the lift generated by the wings only slightly, if at all. Note the change in parasite drag as the velocity increases. The effect on the thrust-required curve is to move it upward and to the left. The amount that the curve moves is not easily calculated. The equivalent parasite area must be known for the aircraft in the dirty configuration, and the new parasite drag values must be calculated. The T_r curve can then be found by adding the new curve values to the old D_i curve values and plotting.

The thrust-required curves for several configurations of the T-38 are shown in Figure 6.33. In discussing the effect of weight change, we saw that the values of *L*/*D* remained constant at the same AOA. The value of $(L/D)_{max}$ does not change as weight varies, but it does change as the configuration changes.

EXAMPLE

For the clean configuration, the value of $(L/D)_{max}$ in Figure 6.33 is 12.05 ($L = W$, $(L/D)_{max}$, thus 10 000 lb/830 lb). In the full flaps and gear configuration, it is only 4.26 ($L = W$, $(L/D)_{max}$, thus 10 000 lb/2 350 lb). The *L*/*D* ratio is much smaller, and the curves move up and to the left as the gear is deployed, and then flaps.

Variation of Aircraft Performance – Configuration Change

Straight and Level Flight V_{max} is usually limited by the structural strength of landing gear and flaps; therefore, V_{max} is reduced by a dirty configuration. Even if structural integrity was not a factor, due to the increase in total drag due to increased parasite drag, V_{max} would be lowered.

Angle of Climb Drag increases when gear and flaps are lowered, so a smaller angle of climb results as T_a remains the same but T_r is increased. In order to maximize climb angle after takeoff or during a missed approach, retract the gear as soon as practical and retract the flaps as required.

Rate of Climb A dirty aircraft reduces climb performance. The aircraft should be cleaned up as soon as practical. As above, T_a remains the same, but T_r is increased with gear and flaps deployed. In order to

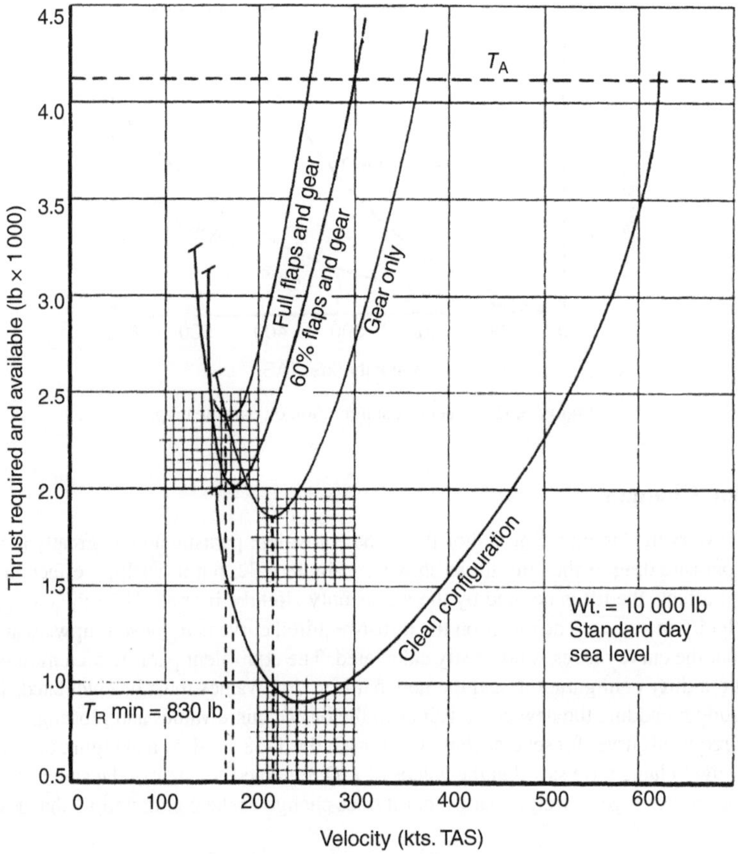

Figure 6.33. Effect of configuration on the T_r curve.

maximize ROC after takeoff or during a missed approach, retract the gear as soon as practical and retract the flaps as required. Note in the example below the reduced ROC when compared to the previous example in this chapter for an aircraft of the same weight, velocity, and flight conditions in Figure 6.17, but clean configuration instead of a gear down configuration as depicted in Figure 6.33.

EXAMPLE

Using Figure 6.33, calculate the rate of climb of a jet with the following conditions: $W = 10\,000$ lb, standard day, sea level, *gear down* only configuration, $T_a = 4\,200$ lb, TAS = 300 kts.

$$\text{ROC} = 101.3V_k \frac{T_a - T_r}{W}$$

$$\text{ROC} = 101.3\,(300\text{ kts.})\frac{4\,200 - 2\,750\text{ lb}}{10\,000\text{ lb}}$$

$$\text{ROC} = 4\,407\text{ fpm}$$

Endurance If the gear and flaps are deployed, parasite drag has increased, so total drag has increased (induced drag remains the same for this discussion). Fuel consumption will increase to compensate for the increased T_r, so endurance is reduced. When holding in an aircraft, keep the gear retracted, flaps as required, and maintain best holding airspeed to maximize endurance. If gear or flaps cannot be retracted, slow down to reduce parasite drag.

Specific Range If it is necessary to fly in the dirty configuration, slow the airplane to the new tangent point to reduce parasite drag. Note the decrease in V_{BR} from the *Clean* configuration to the *Full Flaps & Gear* configuration.

Application 6.4

Using Figure 6.33, calculate the SR at V_{BR} for the clean configuration, then the SR at V_{BR} for the *Full Flaps & Gear* configuration.

What is the percent decrease in the SR once the flaps and landing gear are deployed?

Altitude Changes

Drag on an airplane depends on the dynamic pressure (q):

$$D = C_D q S$$

At any given AOA, the value of the drag coefficient, C_D, is a constant, and the wing area, S, is also a constant. Therefore, the drag will also be a constant if the dynamic pressure is a constant. If we plotted drag versus equivalent airspeed (EAS), there would be only one drag curve for all altitudes. We could say that "drag at altitude equals drag at sea level," but we would have to qualify that statement by adding "at the same EAS," which is a function of sigma (density ratio). Because the curves are drawn for true airspeed, different curves must be drawn for each altitude. Figure 6.34 has these curves drawn for the following three altitudes: sea level, 20 000 ft, and 45 000 ft. If the drag does not change but the true airspeed at which it occurs changes, the only correction we must make is one for airspeed:

$$\frac{V_2}{V_1} = \sqrt{\frac{\sigma_1}{\sigma_2}} \rightarrow V_2 = V_1\sqrt{\frac{\sigma_1}{\sigma_2}} \tag{6.10}$$

The curves in Figure 6.34 show that the minimum thrust required is 830 lb for both altitudes. Of course, this means that the value of $(L/D)_{max}$ is not changed by altitude. The TAS for minimum drag does change as shown in Eq. 6.10. If V_1 is at sea level, where $\sigma_1 = 1.0$, and Eq. 6.10 is rewritten, then

$$V_2 = \frac{V_1}{\sqrt{\sigma_2}}$$

In addition to the T_r changes shown in Figure 6.34, we must also change the T_a in accordance with Figure 6.13, as T_a decreases with altitude. As altitude increases, σ_2 decreases and V_2 increases.

Variation of Aircraft Performance – Altitude Change

Straight and Level Flight V_{max} true airspeed is only slightly changed by altitude. The T_r curve moves to the right with an increase in altitude, and the T_a decreases. V_{max} EAS will be less at altitude as T_a decreases.

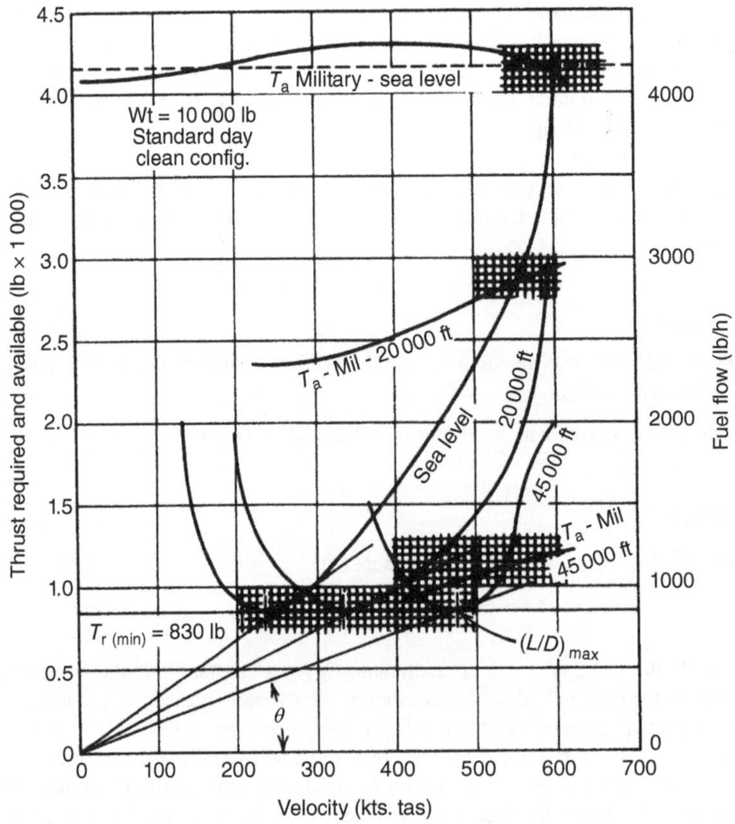

Figure 6.34. Effect of altitude on T_r and T_a, curves.

Angle of Climb With an altitude increase, thrust available decreases with an increase in altitude, and drag remains the same, so $T - D$, or $T_a - T_r$, is smaller and the angle of climb is smaller at altitude. When $T_a = T_r$, the aircraft has reached absolute altitude.

Rate of Climb With an altitude increase, T_r (drag) does not change with an increase in altitude, but T_a is decreased, so $T - D$ is reduced. True airspeed increases with altitude. However, the product of the two terms decreases. When the T_a reduces to the point where it is tangent to the T_r curve, the absolute ceiling is reached, and the ROC is zero.

Endurance With an altitude increase, fuel flow is less, so endurance is improved. A word of caution: You may burn more fuel climbing to a higher altitude than if you had remained at the lower altitude.

Specific Range

There are several advantages to flying jet aircraft at altitude:

a. Up to the tropopause, the lowering of temperature increases engine efficiency (slightly above the tropopause temperature increases as does c_t). This was discussed earlier in our study of specific fuel consumption and is illustrated in Figure 6.15.

b. A second advantage in reducing specific fuel consumption results from operating the engine at high rpm. As thrust is reduced at altitude, higher rpm is required to produce the thrust required, and lower c_t results. Remember, a turbojet engine is more efficient with a higher rpm. This is shown in Figure 6.14.

c. The third advantage can be seen in Figure 6.34. The fuel flow remains about the same for the three altitudes shown, but the velocity is substantially higher as altitude increases. The angle θ decreases as altitude increases, resulting in better SR. An 88% increase in SR is shown at 45 000 ft versus sea level.

Cruise–Climb Flight

When air traffic control allows flights that are not restricted to a set altitude, a further increase in aircraft range (for jet aircraft) can be obtained by using a cruise–climb technique. As fuel is consumed, a higher altitude will offer a higher true airspeed for approximately the same fuel flow. Cruise flight for turbojet aircraft occurs at or above the tropopause. For maximum range, an aircraft is flown at $(\sqrt{C_L}/C_D)_{max}$. With values of C_1, C_D, and TAS held constant, both lift and drag are directly proportional to the density ratio, σ.

Above the tropopause, thrust is also proportional to the density ratio. As fuel is consumed, weight is reduced and if the airplane were allowed to climb, it would remain in equilibrium because lift, drag, and thrust all vary in the same manner. Constant velocity above the tropopause means constant Mach number. Fuel flow decreases with altitude, as seen in Figure 6.16, and thus SR is increased.

The procedure is then to establish the proper maximum range airspeed at the tropopause. Maintain a constant Mach number and allow the altitude to increase as weight is reduced. The improvement in range by using cruise–climb flight over constant altitude flight is a function of the weight of fuel used during cruise, divided by the aircraft's gross weight at the beginning of cruise (Figure 6.35).

Note: In our discussion of performance in thrust-producing aircraft, the word "power" was never used. Many pilots use the terms "power" and "thrust" interchangeably. This is a gross error. *Thrust-producing aircraft do not produce power*. They do not use *power curves*. Pilots will often make statements such as "add power" or "backside of the power curve." All such statements are correct only for power-producing aircraft (discussed in Chapter 7), not for thrust-producing aircraft.

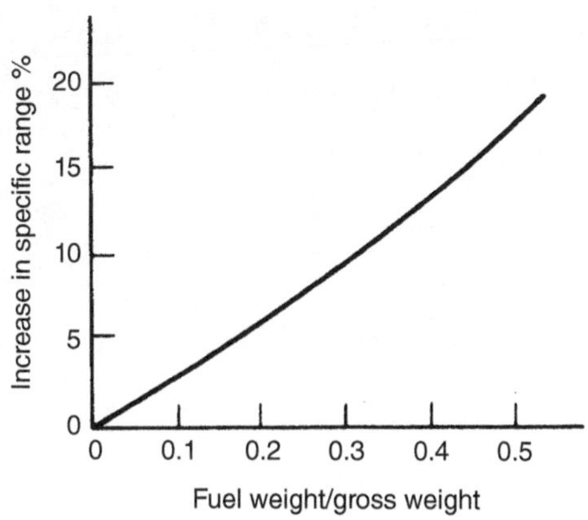

Figure 6.35. Range improvement using cruise–climb.

Afterburners and Vectored Thrust

The aerodynamics of afterburners and vectored thrust were not discussed earlier in this chapter because they are beyond the scope of this textbook, but a brief introduction is warranted. The use of afterburners, also known as thrust augmentation, pertains to military aircraft that use a variable area exhaust nozzle to expand as more fuel is added to the exhaust flow (energy and mass increase) (Figure 6.36). Operation within the afterburner is limited due to the high fuel consumption, which can reach up to twice the normal flow rate. Low-bypass turbofan engines use bypass air that flows directly into the exhaust nozzle; the additional oxygen adds to combustion, increasing the thrust during afterburner.

Thrust vectoring occurs when the aircraft's main engines direct thrust other than parallel to the longitudinal axis of the aircraft. Aircraft that are able to take off vertically and then move horizontally use vectored thrust, but more commonly, modern fighter aircraft, such as the F-22 Raptor, have vectored thrust capabilities of ±20°. By canting the exhaust thrust from the jet nozzle, the aircraft becomes much more maneuverable.

Using equations previously discussed, we can determine the horizontal and vertical force produced during vectored thrust. First, to determine the net vertical force (F_v), we simply subtract the weight (W) from the lift (L), so $F_v = L - W$. If we then take the angle of the cant of the vectored exhaust thrust, depicted as γ here, we find:

$$F_v = L - W + T(\sin \gamma) \tag{6.11}$$

Figure 6.36. Variable area exhaust nozzle.
Source: U.S. Department of Transportation Federal Aviation Administration (2023d) / Public domain.

If we then apply Eq. 1.3 and Newton's second law, we can calculate the net vertical acceleration (a_v):

$$a_v = \frac{F_v}{m} \tag{6.12}$$

To determine the net horizontal force (F_h), we now subtract the drag (D) from the thrust (T), so $F_h = T - D$. If we then again take the angle of the cant of the vectored exhaust thrust, we find:

$$F_h = T(\cos \gamma) - D \tag{6.13}$$

If we then apply Eq. 1.3 and Newton's second law, we can calculate the net horizontal acceleration (a_h):

$$a_h = \frac{F_h}{m} \tag{6.14}$$

Jet Performance Summary

For jet aircraft, we have seen that the following items of performance occur at $(L/D)_{max}$:

1. Minimum drag
2. Maximum engine-out glide range
3. Maximum climb angle
4. Maximum endurance

It is extremely important to know the airspeed or AOA for $(L/D)_{max}$ for your aircraft. The airspeed will vary with aircraft weight, but AOA will not. With respect to altitude, indicated airspeed for $(L/D)_{max}$ will not change.

Changing the weight of an aircraft changes the induced drag much more than the parasite drag. The effect of an increase in weight on the thrust-required curve is that all points are moved up and to the right. The value of $(L/D)_{max}$ does not vary with a change in weight, but changing the weight changes the drag proportionately. Changes in airplane configuration, from a clean configuration to a dirty configuration, increase the parasite drag. This increase in parasite drag moves the thrust-required curve upward and to the left. The value of $(L/D)_{max}$ does not change as weight varies, but it does change as the configuration changes. With an altitude increase, the thrust-required curves are drawn for true airspeed, so different curves must be drawn for each altitude. If the drag does not change but the true airspeed at which it occurs changes, the only correction we must make is one for airspeed. The value of $(L/D)_{max}$ is not changed by altitude.

SYMBOLS

a_h	Net horizontal acceleration
a_v	Net vertical acceleration
c_t	Specific fuel consumption (lb fuel/h per lb thrust)
FF	Fuel flow (lb/h)
F_h	Net horizontal force
F_v	Net vertical force
Q	Mass airflow (slugs/s)
ROC	Rate of climb (fpm)

SR	Specific range (nmi/lb fuel)
T_a	Thrust available (lb)
T_r	Thrust required (lb)
V_1	Intake (flight) velocity (fps)
V_2	Exit velocity (fps)
V_{BE}	Best endurance (maximum endurance)
V_{BR}	Best range (maximum specific range)
V_X	Best angle of climb
V_Y	Best rate of climb
γ (gamma)	Climb angle (°)
η_p (eta)	Propulsion efficiency (%/100)

EQUATIONS

6.1 $T = Q(V_2 - V_1)$

6.2 $\eta_p = \frac{2V_1}{V_2 + V_1}$

6.3 $c_t = \frac{\text{Fuel flow (lb/hr)}}{\text{Thrust (lb)}}$

6.4 $\text{FF} = Tc_t$ (lb/hr)

6.5a $T - D - W\sin\gamma = 0$

6.5b $\sin\gamma = \frac{T-D}{W} \rightarrow \sin\gamma = \frac{T_a - T_r}{W}$

6.6 $\text{ROC} = 101.3\,V_k \frac{T_a - T_r}{W}$ (fpm)

6.7 $\text{Endurance (hr)} = \frac{\text{Fuel available (lb)}}{\text{Fuel flow (lb/hr)}}$

6.8 $(\text{SR})_{\max} = \left(\frac{\text{FF}}{V}\right)_{\min}$

6.9 $\frac{D_2}{D_1} = \frac{W_2}{W_1}$

6.10 $\frac{V_2}{V_1} = \sqrt{\frac{\sigma_1}{\sigma_2}} \rightarrow V_2 = V_1\sqrt{\frac{\sigma_1}{\sigma_2}}$

6.11 $F_v = L - W + T(\sin\gamma)$

6.12 $a_v = \frac{F_v}{m}$

6.13 $F_h = T(\cos\gamma) - D$

6.14 $a_h = \frac{F_h}{m}$

KEY TERMS

Angle of climb
Endurance
Propulsion efficiency
Rate of climb
Specific fuel consumption
Turbojet engine

Turbofan engine
Thrust-available curve
Thrust-required curve

PROBLEMS

Note: Answers to the problems in the first two sections are given at the end of the book.

Application of Chapter Theory

1. The equation $T_a = Q(V_2 - V_1)$ shows that
   a. you can get more thrust from a jet engine by using water injection.
   b. you can get more thrust from a jet engine by decreasing the exhaust pipe area.
   c. Neither (a) nor (b).
   d. Both (a) and (b).

2. Q in $T_a = Q(V_2 - V_1)$ is the mass flow (slugs/s). The mass flow depends on
   a. the cross-sectional area of the turbojet engine inlet.
   b. the engine inlet velocity.
   c. the density of the air at the same point.
   d. All of the above.

3. The equation $\eta_p = 2V_1/(V_2 + V_1)$ shows that decreasing the exhaust pipe area will
   a. decrease the propulsive efficiency of a jet engine.
   b. increase the propulsive efficiency.
   c. increase the thrust of the engine.
   d. decrease the thrust of the engine.

4. The specific fuel consumption of a turbine engine at 35 000 ft altitude compared to that at sea level is
   a. less.
   b. more.
   c. the same.

5. The thrust available from a jet engine at 35 000 ft altitude compared to that at sea level is
   a. less.
   b. more.
   c. the same.

6. Fuel flow for a jet at 100% rpm at altitude compared to that at sea level is
   a. less.
   b. more.
   c. the same.

7. A pilot is flying an airplane at the speed for best range under no-wind conditions. A tailwind is encountered. To get the best range now, the pilot must
   a. speed up by an amount less than the wind speed.
   b. slow down by an amount equal to the wind speed.
   c. slow down by an amount more than the wind speed.
   d. slow down by an amount less than the wind speed.

8. To obtain maximum range, a jet airplane must be flown at
a. a speed less than that for $(L/D)_{max}$.
b. a speed equal to that for $(L/D)_{max}$.
c. a speed greater than that for $(L/D)_{max}$.

9. The maximum rate of climb for a jet airplane occurs at
a. a speed less than that for $(L/D)_{max}$.
b. a speed equal to that for $(L/D)_{max}$.
c. a speed greater than that for $(L/D)_{max}$.

10. The maximum climb angle for a jet aircraft occurs at
a. a speed less than that for $(L/D)_{max}$.
b. a speed equal to that for $(L/D)_{max}$.
c. a speed greater than that for $(L/D)_{max}$.

11. If the weight of a jet airplane is increased, then
a. parasite drag increases more than induced drag.
b. induced drag decreases more than parasite drag.
c. both parasite and induced drag increase by the same amount.
d. induced drag increases more than parasite drag.

12. If the weight of a jet airplane is reduced as fuel is burned, the T_r curve
a. moves down and to the right.
b. moves up and to the right.
c. moves down and to the left.
d. moves up and to the left.

13. If a jet airplane is in the gear down configuration,
a. the increase in parasite drag is more than that of the induced drag.
b. the increase in induced drag is more than that of the parasite drag.
c. both types of drag are the same.

14. If it is impossible to raise the landing gear of a jet airplane, to obtain the best range, the airspeed must be _____ from that for the clean configuration.
a. increased
b. decreased
c. not changed

15. The minimum drag for a jet airplane does not vary with altitude.
a. True
b. False

16. Figure 6.34 shows an increase in specific range with altitude because
a. T_r decreases while fuel flow decreases.
b. T_r remains the same while fuel flow decreases.
c. T_r remains the same while T_a decreases.
d. fuel flow remains about the same while airspeed increases.

17. A jet airplane is flying to obtain maximum specific range. As fuel is burned, the pilot must
a. reduce throttle but maintain the same airspeed.
b. maintain throttle setting and let the plane accelerate.
c. reduce throttle and airspeed.
d. maintain throttle and reduce airspeed.

18. A lightly loaded airplane will be able to glide farther but at a lower airspeed than when it is heavily loaded.
a. True
b. False

19. Which of the following words should never be used in the discussion of jet aircraft?
a. Power.
b. Horsepower.
c. Power curve.
d. All of the above.

Calculations Using Chapter Principles

20. From Figure 6.33, the glide ratio $(L/D)_{max}$ for the airplane in the full flaps and gear down configuration is
a. 12.05.
b. 5.4.
c. 5.0
d. 4.2.

21. A single engine turbojet aircraft is flying at 296 kts. TAS at sea level. The mass flow rate through the engine is 10 slugs/s. The exit velocity from the engine is 800 fps. Find
a. The thrust of the engine.
b. The propulsive efficiency.

22. A jet airplane has thrust available as shown in Figure 6.17 and specific fuel consumption as shown in Figure 6.14. Find the fuel flow at sea level (military rpm and $M = 0.70$).

23. Using Figure 6.16, find the fuel flow for the airplane in Problem 6.22 at the tropopause.

24. A jet airplane has the $T_a - T_r$ curves shown in Figure 6.17. The airplane data are gross weight = 10 000 lb, standard sea level day, rpm at 100% unless noted, and clean configuration. Find:
a. V_{max} at 95% rpm.
b. V for maximum climb angle.
c. Maximum climb angle.
d. Rate of climb at 400 kts.
e. V for best endurance.
f. V for maximum range.
g. V for maximum range with 100 kts. of headwind.

25. Using Figure 6.30, find the velocity for the best range for the airplane at 12 000 and 8 000 lb.

26. Using Figure 6.30, calculate the specific range for the airplane at both weights if it is flying at the best-range airspeed.

27. Using Figure 6.30, calculate the specific range of the airplane if the throttle is not retarded from the 12 000 lb best-range position, the airspeed is allowed to increase as fuel is burned, and weight is reduced to 8 000 lb. Compare your answer to that for Problem 6.26 and state your conclusions.

28. Using Figure 6.34, calculate the specific range for this airplane if it is flying at best-range airspeed at sea level and at 20 000 ft altitude.

Correlation Through Scenario-Based Learning

1. Research which aircraft still use turbojet engines. Is there one segment of the aviation industry that utilizes turbojets more than any other? Why?
2. What are some of the key differences between turbofan engines and turbojet engines?
3. What are the advantages and disadvantages of turbojet and turbofan engines?
4. Review Equations 6.1 and 6.2, how does the incorporation of large fan blades and larger nacelles relate to the variables within these equations?
5. Where is the *cold section* in a turbofan? Research the term *bypass ratio*. What does it represent, and what are the highest bypass ratios available on engines in today's commercial market? What defines low bypass ratios and high bypass ratios?
6. What are the advantages and disadvantages of turbojet and turbofan engines?
7. Research the Brayton Cycle. How does the Ideal Brayton Cycle relate to turbojet and turbofan engines?
8. What are the future innovations expected in the turbofan engine market?

7 Propeller Aircraft Performance

CHAPTER OBJECTIVES

After completing this chapter, you should be able to:

- Define horsepower (HP) and identify several types of HP.
- Explain the basic principles of a propeller and contrast the differences between a fixed-pitch propeller and a constant-speed propeller.
- Compare and contrast the principles of propulsion for a thrust-producing aircraft and a power-producing aircraft.
- Identify performance values that can be derived from power-required curves, and calculate their value.
- Summarize the relationship between power available and aircraft speed.
- Compare and contrast the angle of climb and rate of climb for a power-producing airplane.
- Examine and illustrate the changes in a power-required curve as weight, configuration, and altitude are altered throughout a flight.

All aircraft in flight must produce thrust to overcome drag. In turbojets and other thrust-producing aircraft, thrust is produced directly from the engine. In aircraft that have propellers (or rotors), the engine does not produce thrust directly. These aircraft are called *power producers* because the engine produces power, which turns the propeller. The propeller then develops an aerodynamic force as it turns through the air. This force is thrust.

Fuel consumption of power-producing aircraft is roughly proportional to the power produced, instead of the thrust produced. Range and endurance depend on fuel consumption, so the power required to fly the aircraft is of prime importance.

In Chapter 1 we defined *power* as "the rate of doing work" or work/time. We know:

$$\text{Thrust} = \text{Force} = \text{A push or pull (lb)}$$

$$\text{Work} = \text{Force} \times \text{Distance (ft-lb)}$$

Flight Theory and Aerodynamics: A Practical Guide for Operational Safety, Fifth Edition.
Brian A. Johnson and Philip R. Fittante.

Companion website: www.wiley.com/go/flighttheoryandaerodynamics5e

$$\text{Power} = \frac{\text{Work}}{\text{Time}} \rightarrow \text{Power} = \frac{\text{Force} \times \text{Distance}}{\text{Time}}\ (\text{ft-lb/s})$$

We also know: $\frac{\text{Distance}}{\text{Time}} = \text{Speed}$

$$\text{Power} = \text{Force} \times \text{Speed} \rightarrow \text{Power} = FV_{\text{fps}} \rightarrow \text{Power} = TV_{\text{fps}}\ (\text{ft-lb/s})$$

$$\text{Horsepower} = \frac{\text{Power}}{550} \rightarrow \text{Horsepower} = \frac{TV_{\text{fps}}}{550}$$

One unit of power = ft-lb/sec, and James Watt defined the term *horsepower* (HP) as 550 ft-lb/s, which represents the power to move 550 lbs, 1 ft in 1 second, or 33 000 lbs, 1 ft in 1 minute. Each pound of thrust developed at 550 ft/sec is equivalent of one HP of propulsive power. Converting ft/sec to kts. divide 550 by 1.69 which equals 325. So, if the speed is measured in knots, V_k, and the force (lbs) is the *thrust*, T, of a jet engine, then Eq. 1.7:

$$\text{HP} = \frac{\text{Thrust} \times V_k}{325} \rightarrow \text{HP} = \frac{TV_k}{325}$$

POWER AVAILABLE

Several types of horsepower (HP) are produced by a power-producing engine and propeller combination:

Indicated HP – HP developed in the combustion chambers without reference to friction losses within the engine. It is considered theoretical power because the assumption is a frictionless engine.

Brake horsepower (BHP) – Once friction is considered BHP is the actual HP delivered to the propeller for useful work. It is the HP measured at the crankshaft (piston engines) or at the turbine shaft (turbine engines). BHP is typically measured in foot-pounds of torque and is a function of HP and rpm.

Friction HP – The HP used by an engine in overcoming the friction of moving parts, driving oil pumps and fuel pumps, and expelling exhaust, could be as high as 10%–20% of indicated HP. It is the difference between BHP and indicated HP.

Shaft horsepower (SHP) – SHP is the HP measured at the propeller shaft and is less than the BHP because of gearing losses in reducing engine rpm to propeller rpm.

Thrust horsepower (THP) – THP is considered the usable HP and can be thought of as the engine and the propeller working together. If a propeller were 100% efficient the THP and the BHP would be the same. THP is less than the SHP because of propeller efficiency loss. Propeller efficiency varies with airspeed, engine speed, altitude, attitude, and temperature, and in some aircraft by propeller controls within the flight deck. *Do not confuse THP with thrust.* THP is a type of HP and must be converted to thrust units by Eq. 1.7.

Turboprop aircraft produce both power and thrust, but the amount of thrust produced directly by the engine is only about 15% of the total. They are, therefore, classified as power producers. The amount of thrust that they produce is converted into HP units and added to the SHP. The result is called *equivalent shaft horsepower* (ESHP):

$$\text{ESHP} = \text{SHP} + \frac{TV}{325\eta} \tag{7.1}$$

Propeller efficiency, η, is defined as the ratio of power output to power input:

$$\eta = \frac{\text{THP}}{\text{BHP(or SHP)}} \tag{7.2}$$

Basic Propeller Principles

An aircraft propeller consists of two or more blades, which are essentially small airfoils that produce a force to push or pull an aircraft through the air. Each blade of the propeller can be thought of as a little wing with a pitch that decreases from the hub to the tip (Figure 7.1). The blade angle usually varies from approximately 4° near the hub of the propeller to 2° or less near the outer portion.

The relative wind the propeller blade encounters is the vector sum of the aircraft's relative wind and the relative wind created by the propeller's rotational motion. The aircraft's relative wind is the same for all portions of the propeller "disc," but the relative wind created by the propeller's rotation is greater at the tip than at the hub because the rotational velocity of the tip is greater than that nearer the hub. From Figures 7.2 and 7.3, we can see that if a propeller had the same blade angle from the hub to tip, the angle of attack (AOA) at the tip would be much higher than the AOA at the hub, so that the tip might be stalled while the inboard portions of the propeller were still producing lift. The propeller's twist, with the blade angle decreasing from hub to tip, is intended to keep all portions of the blade at the same AOA thus producing a uniform amount of lift (thrust) across the entire blade.

Figure 7.3 shows the similarities of a propeller in cruise flight to the airfoil of a wing as presented in Chapters 3 and 4. The top surface is cambered and the bottom surface is flatter. The greater camber and blade angle of the propeller hub compensates for the reduced relative wind (due to reduced rotational velocity) over this section of the propeller. Conversely, the high relative wind over the tips of the propeller necessitates a reduced camber and blade angle to ensure an equal amount of lift is produced.

Although the terms propeller blade angle and propeller pitch are often used interchangeably, each has a distinct definition. The blade angle of a propeller refers to the angle between the propeller chord line and

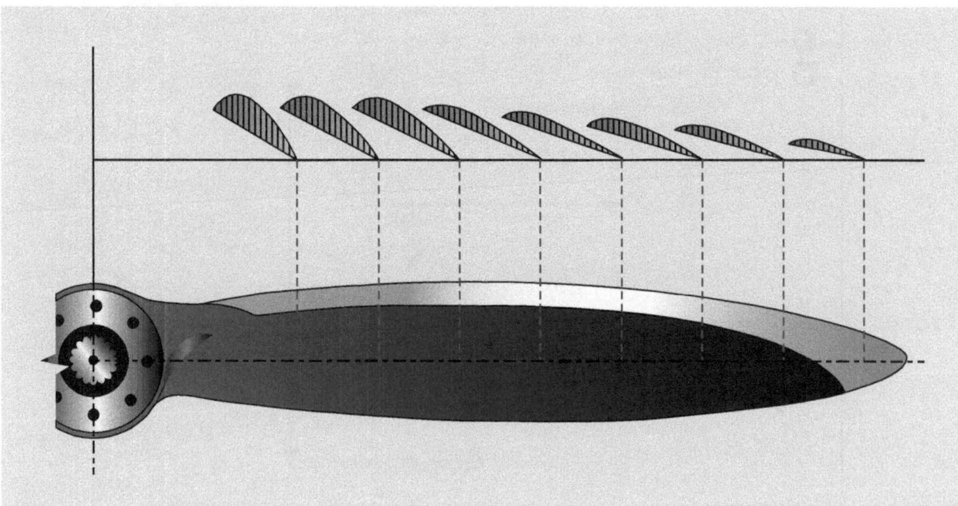

Figure 7.1. Airfoil sections of a propeller blade.
Source: Pilot's Handbook of Aeronautical Knowledge (2008a)/U.S. Department of Transportation Federal Aviation Administration/Public domain.

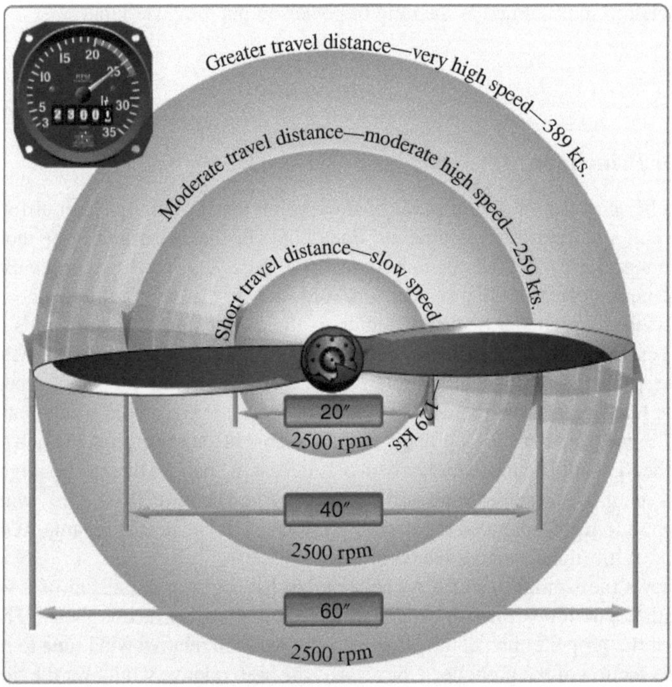

Figure 7.2. Propeller tip speed versus propeller hub.
Source: Pilot's Handbook of Aeronautical Knowledge (2016b)/U.S. Department of Transportation Federal Aviation Administration/Public domain.

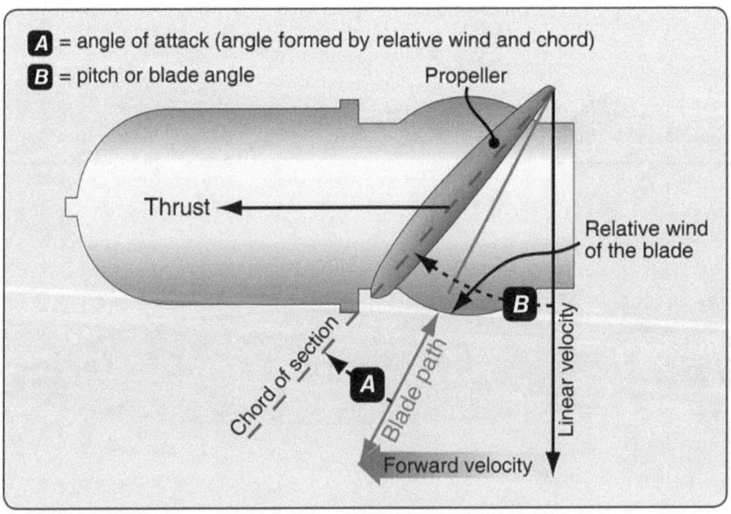

Figure 7.3. Propeller blade angle.
Source: Aviation Maintenance Technician Handbook (2023d)/U.S. Department of Transportation Federal Aviation Administration/Public domain.

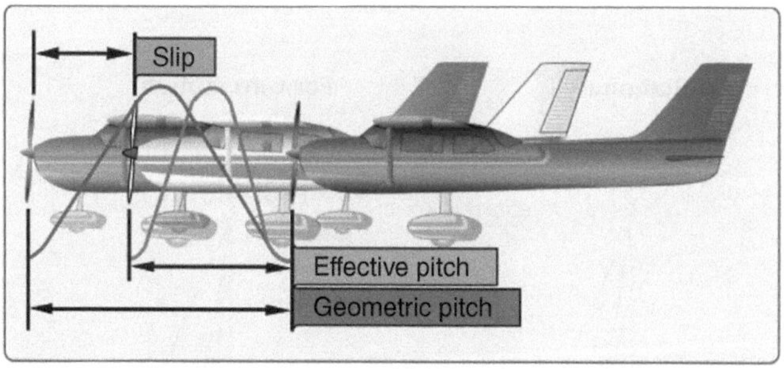

Figure 7.4. Geometric pitch versus effective pitch.
Source: Pilot's Handbook of Aeronautical Knowledge (2016b)/U.S. Department of Transportation Federal Aviation Administration/Public domain.

its plane of rotation (Figure 7.3). Propeller pitch is a measure of the distance the propeller moves forward through the air in a single revolution. As we will see in the discussion of fixed pitch versus variable pitch propellers, the propeller pitch, and thus its efficiency, is largely determined by the propeller blade angle.

Propeller efficiency, or the ratio of THP to BHP (or in some cases SHP), defines the ratio of the *useful* power output to the *actual* power input. A propeller that may be efficient during climb may not be as efficient in cruise flight. A common term for the difference between how far a propeller should theoretically advance in one revolution and how far it actually travels is slippage. This is the difference between geometric pitch and effective pitch is shown in Figure 7.4. The effective pitch is equal to the geometric pitch minus propeller slip.

Fixed-Pitch Propeller

Propellers are classified into two general groups for this discussion, *fixed-pitch* propellers and *constant-speed* propellers. A fixed-pitch propeller is usually a one-pitch-setting blade propeller, commonly of a two-blade design. Note that while the pitch is fixed, the blade angle is not constant: it decreases from hub to tip, as with any propeller. The blade achieves its maximum efficiency at only one airspeed and rotation, so the design must be a compromise between climb and cruise performance. Depending on the relative airstream the propeller encounters during different modes of flight, the propeller AOA is usually 15° or less.

There are "fixed-pitch" propeller designs where the pilot can adjust the pitch of the propeller either before takeoff or during flight, though they are considered here due to the fact the options are usually for a low pitch setting (high rpm) or high pitch setting (low rpm). *Adjustable* pitch propellers are modified before takeoff and usually set for either an efficient climb or an efficient cruise setting. *Controllable* pitch propellers are adjustable in the air to limited, preset settings. In both cases, they provide efficiency improvements at specific AOAs but reduced efficiency outside of these fixed settings.

Constant-Speed Propeller

A constant-speed governing system allows for a much more efficient operation of the propeller through varying conditions of flight. The pilot selects the desired rpm with a lever, and a governor (usually hydraulically or electrically controlled) automatically changes the blade angle to maintain a constant propeller rpm. During takeoff, the AOA of the blade angle is small, limiting the amount of mass of air the propeller can move; this allows the engine to turn at a high rpm and convert maximum fuel into mechanical energy in a given time for maximum power.

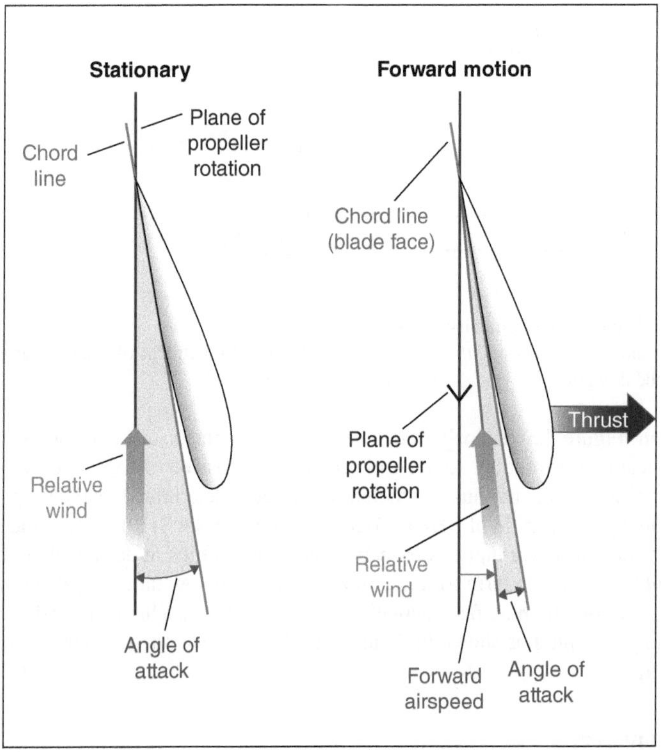

Figure 7.5. Blade angle in flight.
Source: Airplane Flying Handbook (2004a)/U.S. Department of Transportation Federal Aviation Administration/Public domain.

Figure 7.5 shows the change in the propeller AOA from a stationary position to the AOA during forward motion. As the speed of the aircraft increases after takeoff, the AOA of the propeller blade decreases as the relative wind shifts due to the forward movement. The propeller blade angle is usually changed by the pilot at this point to a higher angle, taking a bigger "bite" out of the air, yet still keeping the AOA relatively small so maximum rpm for climb can be reached without exceeding engine tolerances. Once in level flight, as forward speed increases, the relative wind shifts forward, resulting in a decrease in the blade AOA. The pilot retards the throttle and increases the propeller blade angle (reduce rpm, bigger "bite" of the air) for efficient cruise operation. Propeller efficiency is usually at a maximum during cruise flight due to less slippage.

Once the rpm is set by the pilot, the governing system will maintain the same rpm. If the nose of the aircraft were lowered and the airspeed increased beyond that for level flight, the propeller blade AOA would naturally decrease due to the forward shift in relative wind. In a fixed-pitch propeller, the rpm of the propeller would start to increase and possibly exceed tolerances set by the manufacturer. In a constant-speed propeller system, as the AOA of the propeller blade began to decrease and the rpm increase, the governing system would increase the blade angle (bigger "bite" of the air) to maintain the same rpm set by the pilot. In the opposite situation, if the nose of an aircraft were raised, placing more load on the propeller, the fixed-pitch propeller system would suffer a decrease in rpm. A constant-speed governing system would decrease the propeller blade angle allowing for a lower blade pitch, and thus increasing

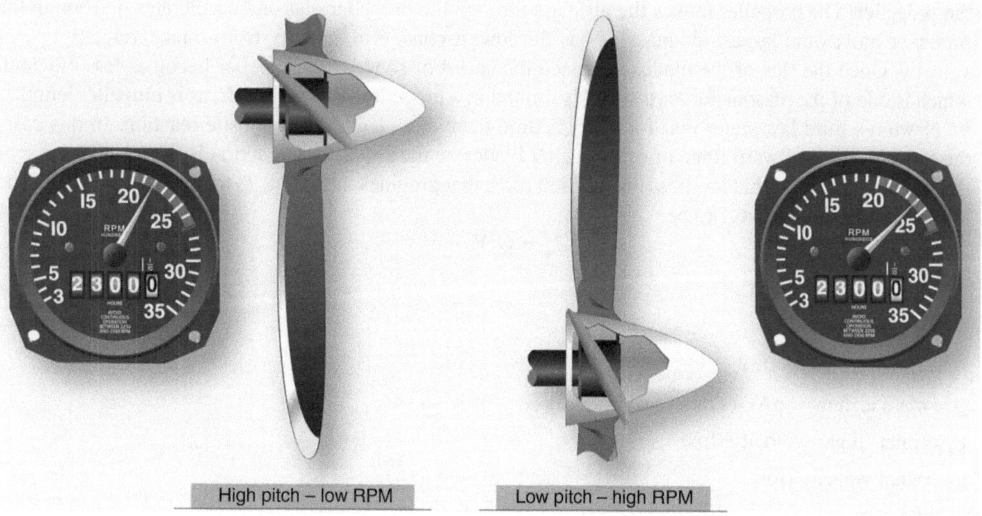

Figure 7.6. Propeller blade pitch versus rpm.
Source: U.S. Department of Transportation Federal Aviation Administration (2021)/Public Domain.

Aircraft type	Design speed (mph)	Blade angle range(°)	Pitch Low(°)	Pitch High(°)
Fixed gear	160	11½	10½	22
Retractable	180	15	11	26
Turbo retractable	225/240	20	14	34
Turbine retractable	250/300	30	10	40
Transport retractable	325	40	10/15	50/55

Figure 7.7. Various blade angle ranges.
Source: Airplane Flying Handbook (2004a)/U.S. Department of Transportation Federal Aviation Administration/Public domain.

the rpm to maintain the original rpm set by the pilot. The key relationship used by a governing system to maintain the desired propeller rpm can be summed up as low pitch = high rpm and high pitch = low rpm (Figure 7.6). The governor system automatically varies the propeller blade angle to adjust the propeller pitch in accordance with these relationships. Figure 7.7 shows examples of blade angles for different aircraft types and design speeds, typical constant-speed blade angle ranges vary from 11.5° to 40°; as the airspeed of the various aircraft listed increases, so does the blade angle range available.

PRINCIPLES OF PROPULSION

The principles of propulsion are based on Newton's second and third laws. From Chapter 1, the second law is written as

$$F = ma$$

Simply stated, an unbalanced force, F, acting on a mass, m, will result in acceleration, a, of the mass in the direction of the force. In propeller aircraft, the force is supplied by the engine shaft, which rotates the propeller. The propeller moves the air mass through the propeller disc and accelerates it. Though the propeller moves the largest air mass, it has the lowest change in velocity from intake velocity to exit velocity. Once the tips of the blades approach the speed of sound, the propeller becomes less efficient, which is one of the reasons forward speed is limited in a propeller-driven aircraft, as is propeller length.

Newton's third law states that for every action, there is an equal and opposite reaction. In this case, the action is the backward force of the propeller blades on the air, and the reaction is the forward force of the air on the propeller blades. It is this reaction force that provides the thrust, T, to propel the aircraft. The thrust was shown in Eq. 6.1 to be

$$T_a = Q(V_2 - V_1)$$

where

T_a = thrust available (lb)

Q = mass airflow = pAV (slugs/sec) (Eq. 2.6)

V_1 = inlet (flight) velocity (fps)

V_2 = final velocity (fps).

Figure 7.8 shows a schematic of the process of producing thrust from a propeller, a represents the velocity increment at the propeller disc in knots and V_2 is the final velocity of the propeller slipstream.

Propulsive efficiency of a propeller-driven system is measured in theory by the same principle, η_p is found by dividing the input power into the output power as we saw in Eq. 7.2. But because we are not using the ideal propeller and we must account for propeller disc loading and drag it is more accurate to define propeller efficiency as:

$$\eta_P = \frac{\text{Output propulsive power}}{\text{Input shaft horsepower}}$$

From Chapter 6, high exhaust velocity does produce high thrust, but it also produces low efficiency. High mass airflow also produces high thrust and does not reduce propulsion efficiency. Propeller aircraft process large quantities of air with only a small acceleration of the air compared with turbojets. Therefore, propeller aircraft are much more efficient than turbojets with respect to takeoff and landing performance, and during subsonic flight. Large diameter propellers are efficient in terms of mass airflow but suffer the consequences of high tip speeds and compressibility effects. Figure 7.9 shows an example of a partial

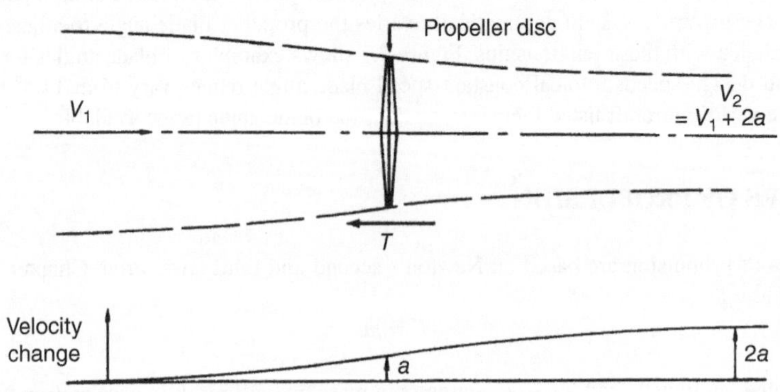

Figure 7.8. Thrust from a propeller.

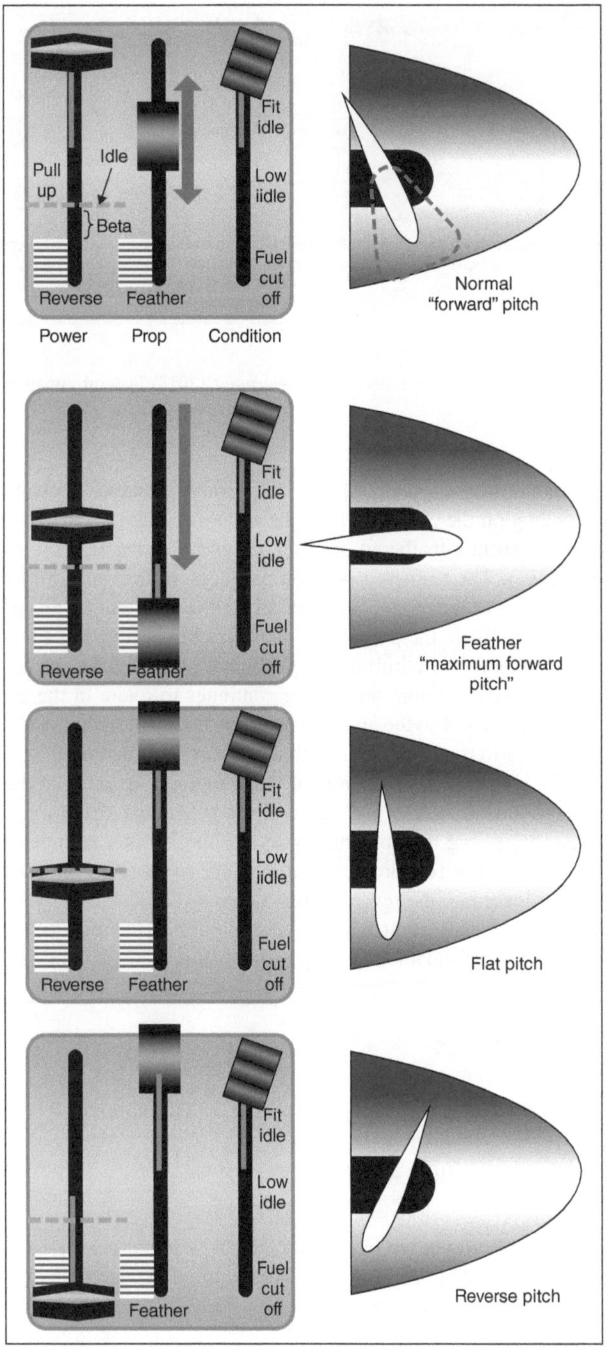

Figure 7.9. Propeller pitch angle configurations.
Source: Airplane Flying Handbook (2004a)/U.S. Department of Transportation Federal Aviation Administration/Public domain.

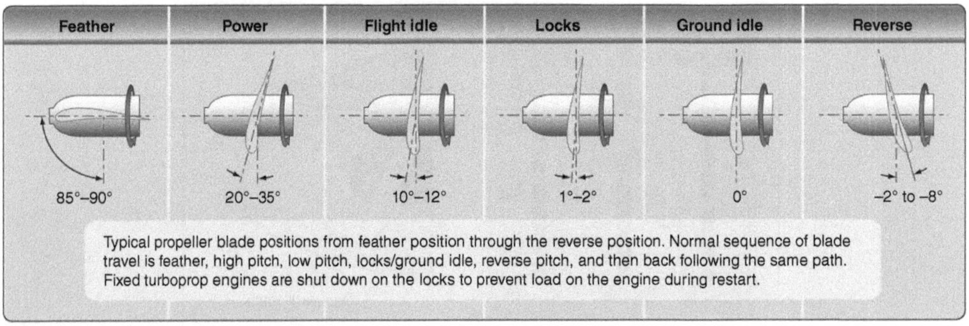

Figure 7.10. Propeller range positions.
Source: Aviation Maintenance Technician Handbook – Powerplant (2018)/Federal Aviation Administration/Public domain.

throttle quadrant in a common turboprop aircraft, and the various propeller pitch angles available to alter the mass of airflow depending on the condition of flight.

The blade positions vary from a feathered position, to normal forward pitch, to finally reverse pitch (commonly called beta range). The feathered position produces no forward thrust and is primarily used during an engine failure to secure the engine and minimize drag, and will be discussed in more detail in Chapter 13. In normal forward pitch, the propeller blades may be set between a low pitch (fine) position to provide higher rpm for takeoff and climb out and a higher pitch (coarse) position for more efficient cruise flight. In the reverse pitch position, the engine continues to rotate in the same direction, but the pitch of the propeller blades is altered so the air is pushed forward resulting in assisted braking action after landing or minor maneuvering in the ramp area. With the power lever behind the idle position, this is the beta range of operation. To accomplish this, the pilot sets the power lever aft of the idle position, which moves the propeller pitch to the "reverse thrust" range. In the beta range, the power lever directly controls the propeller blade angle, bypassing the governor system. This allows a zero-degree "flat" pitch that is useful for taxi operations to control forward speed and reduce reliance on the aircraft's brakes. Further aft movement of the power lever into the reverse thrust range results in a negative blade angle as well as increased engine torque to slow an aircraft after landing. Figure 7.10 shows an example of the normal blade pitch range for a turboprop, from a feathered position to a reverse position in sequence.

POWER-REQUIRED CURVES

Equation 1.7 allows us to convert the drag (or T_r) curve into the HP-required, P_r, curve. Equation 1.7 can be rewritten as

$$\text{HP} = \frac{TV_k}{325} \quad \text{becomes} \quad P_r = \frac{DV_k}{325}$$

Before we convert the total drag curve into a total power-required curve, let us see how the power required to overcome the induced drag, P_{ri}, and the power required to overcome the parasite drag, P_{rp}, are affected. The P_{ri} curve varies with velocity as

$$\frac{P_{ri_2}}{P_{ri_1}} = \frac{V_1}{V_2} \rightarrow P_{ri_2} = P_{ri_1}\frac{V_1}{V_2} \tag{7.3}$$

From Chapter 5, the induced drag varies inversely with the square of the velocity ratio, but *the induced power required varies inversely with the velocity ratio to the first power*. (This contrasts with the thrust-required curve that was presented in Chapter 6). To draw the P_{ri} curve, convert the induced drag values, at selected velocities, from the induced drag curve and plot them. The curve is labeled "induced power required" in Figure 7.11b. It is much flatter than the induced drag curve seen in Figure 7.11a. As speed decreases to below $(L/D)_{max}$, induced power required increases much less rapidly than induced drag. The P_{rp} curve varies with velocity:

$$\frac{P_{r_{p2}}}{P_{r_{r1}}} = \left(\frac{V_2}{V_1}\right)^3 \rightarrow P_{r_{p2}} = P_{r_{r1}}\left(\frac{V_2}{V_1}\right)^3 \tag{7.4}$$

This means that doubling airspeed requires an *eightfold* increase in power ($2^3 = 8$).

From Chapter 5, the parasite drag varies directly as the velocity ratio squared, but *the parasite power required varies directly as the velocity ratio cubed*. To draw the P_{rp} curve, convert the parasite drag values at selected velocities to the corresponding power required values and plot them. The curve is steeper than the parasite drag curve and is labeled "parasite power req'd" in Figure 7.11b.

The total power-required curve is simply the addition of the P_{ri} and the P_{rp} curves. The P_r curve is flatter in the low-speed region than the T_r curve, but it is steeper in the high-speed region. This explains why propeller aircraft perform better than turbojets in the low-speed region, but lack the high power required to fly at high speeds. The intersection of the P_{ri} and P_{rp} curves is at the velocity of the $(L/D)_{max}$ point, but this is no longer the lowest point on the curve. A tangent line from the origin to the curve locates the $(L/D)_{max}$ point.

Figure 7.11 shows both the total drag, T_r, and total power required, P_r, curves for the same aircraft. The conversion of T_r curves to P_r curves is made by using Eq. 1.7. For this discussion, we have chosen a "typical" propeller airplane whose power-required curve is shown in Figure 7.12.

Power Available Versus Velocity

In our discussion of thrust available from turbojet engines in Chapter 6, we saw that the thrust output did not vary to any great extent with the velocity of the aircraft. A similar relationship between BHP (and SHP) and aircraft velocity exists in power-producing aircraft. However, the THP does vary significantly with velocity, because the propeller efficiency varies with velocity. As discussed, propeller efficiency is also dependent on whether a fixed-pitch propeller or a constant-speed propeller is being used on the aircraft, and will alter the respective P_a curve. THP is called power available, P_a, and is related to the propeller efficiency as per Eq. 7.5:

$$P_a = (\eta_p)(\text{BHP}) \tag{7.5}$$

Figure 7.13 shows a generic P_a curve as airspeed (velocity) increases. Even though BHP is constant in this example, because the THP (T_a) decreases (propeller efficiency decreases) as velocity increases, using Eq. 1.7 we find that the P_a is not a constant during velocity changes as was T_a in Chapter 6. As airspeed increases, there is a peak in P_a, and then a decrease as airspeed continues to increase.

Figure 7.14 shows a plot of power available, and power required, note again the variation in P_a as velocity changes. The point of maximum excess power is not between the lowest point on the P_r curve and P_a. Fuel flow is also plotted on the vertical scale. Note the stall speed (V_s) location on the P_r curve, as well as the tangent line for the location of $(L/D)_{max}$.

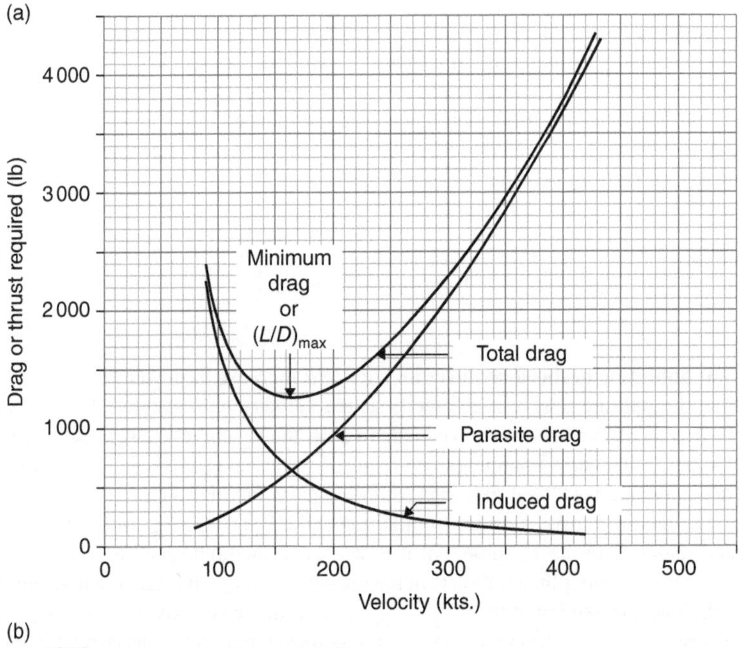

Figure 7.11. (a) Thrust-required and (b) power-required curves.

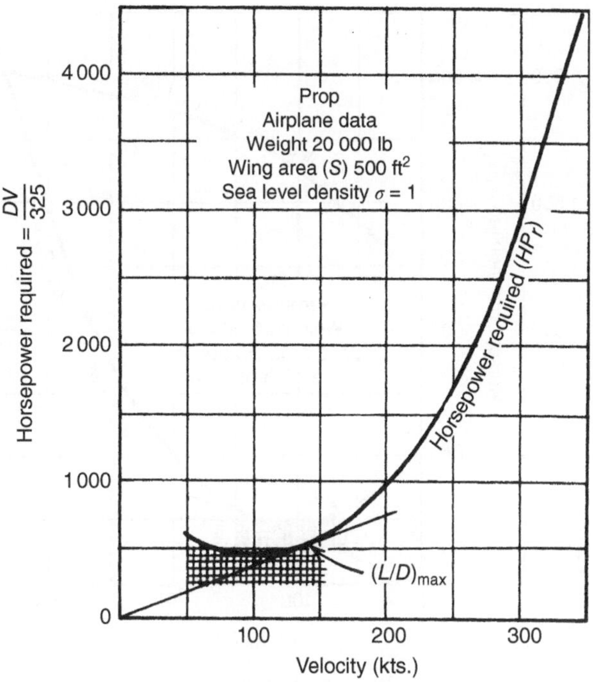

Figure 7.12. Power required.

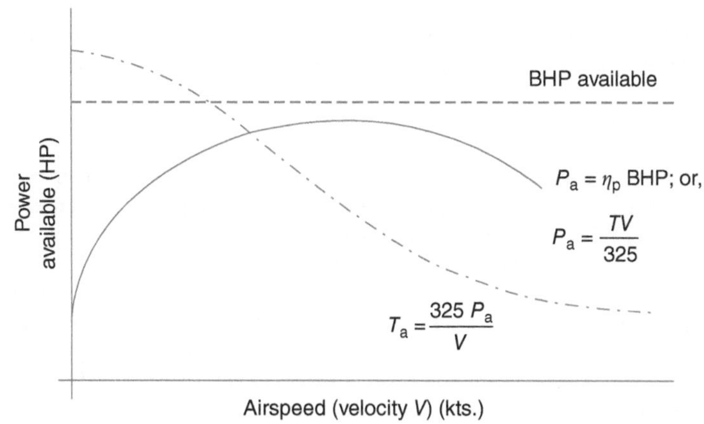

Figure 7.13. Power available curve.

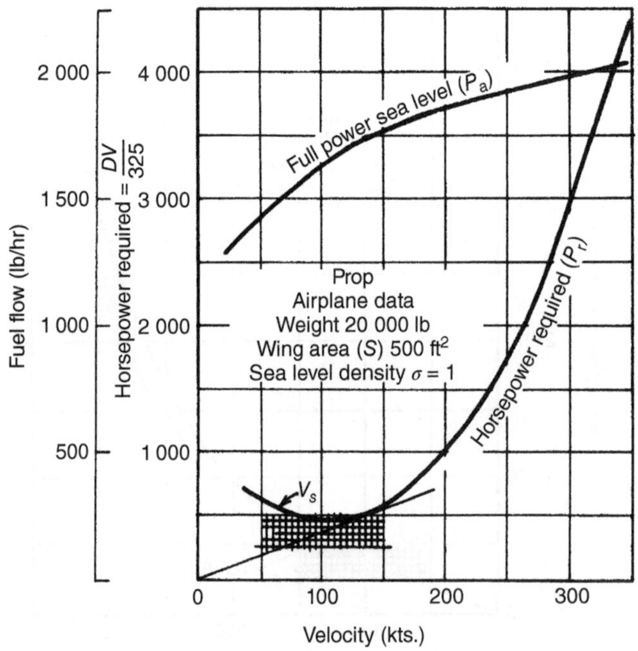

Figure 7.14. Power required and power available.

Variations with Power and Altitude

In discussing power available from power-producing engines, we must consider three types of powerplants: turboprops, turbocharged reciprocating engines, and non-turbocharged reciprocating engines.

Turboprops

Like jet engines, turboprop engines use hot exhaust gas to turn a rotating shaft in order to rotate a propeller or rotor (helicopter). Figure 7.15 is an example of a fixed-shaft turboprop engine that uses a gear-reduction mechanism between the turbine and the propeller. Figure 7.16 is an example of a free power-turbine engine such as the Pratt & Whitney PT-6 engine, where the propeller is not on the same shaft as the basic engine turbine and compressor. All mechanisms for turning the propeller at lower rpm than the turbine consume some of the engine's power output, but because optimum propeller efficiency is achieved at slower rotation speeds than the engine's operating rpm, this power "cost" is unavoidable and acceptable.

Chapter 6 discussed the effects on thrust output of turbojet engines when rpm and altitude were changed. Generally, these effects apply to any gas turbine engine (including power producers), so they will not be repeated here. In summary, we can say that as altitude increases, the power available from a turboprop engine decreases and the fuel flow is also decreased.

Turbocharged Reciprocating Engines

Reciprocating engines do not operate at high rpm settings (except for takeoff power). Turbochargers use waste gases from combustion; normally vented overboard, they are now used to power a turbine to

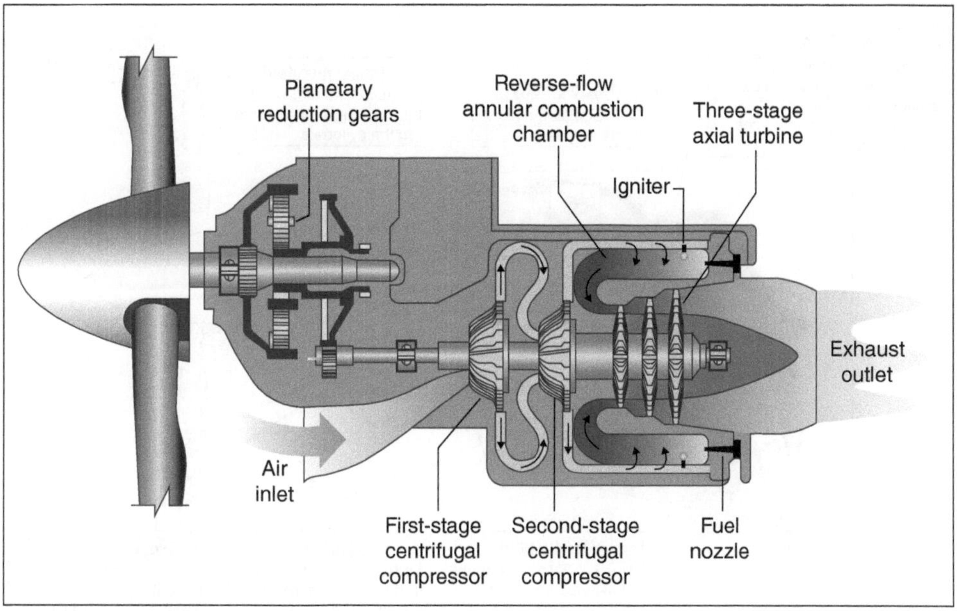

Figure 7.15. Fixed-shaft turboprop engine.
Source: Airplane Flying Handbook (2004a)/U.S. Department of Transportation Federal Aviation Administration/Public domain.

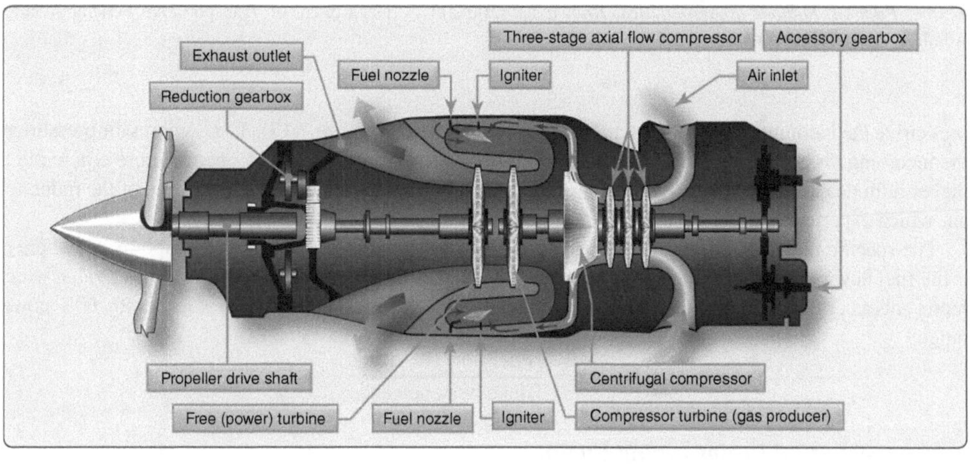

Figure 7.16. Split shaft/free turbine engine.
Source: Airplane Flying Handbook (2016a)/U.S. Department of Transportation Federal Aviation Administration/Public domain.

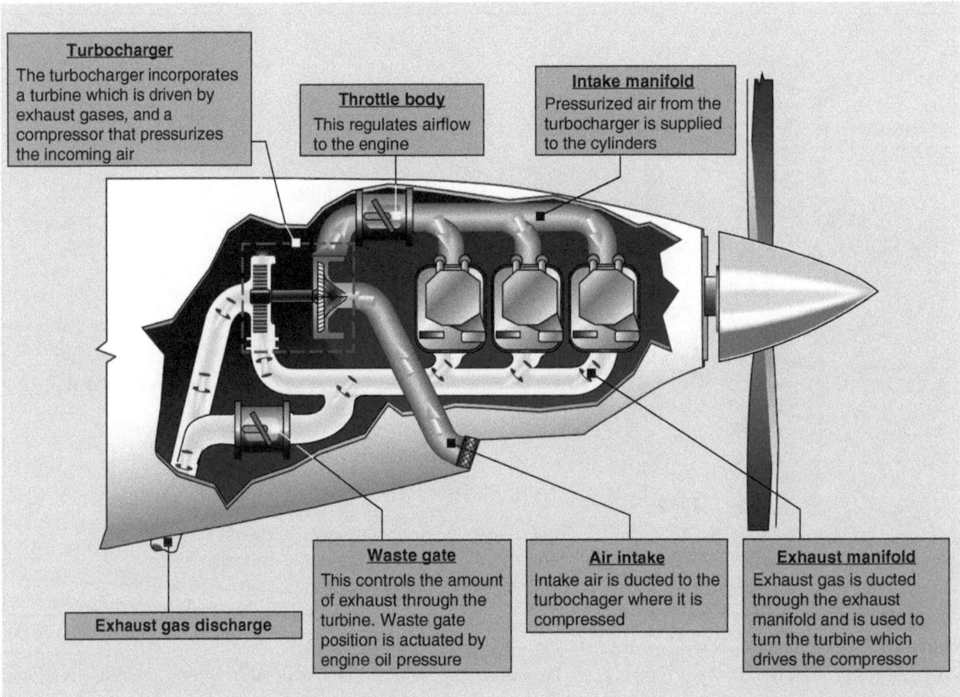

Figure 7.17. Turbocharged engine.
Source: Pilot's Handbook of Aeronautical Knowledge (2008a)/U.S. Department of Transportation Federal Aviation Administration/Public domain.

pressurize the incoming air by use of an impeller (compressor) (Figure 7.17). The compressor pressurizes the incoming air before entering the induction system, allowing the engine to operate more efficiently at higher altitudes. Many turbocharged engines use an intercooler to reduce the temperature of the induction air, which experiences a temperature rise after compression.

The specific fuel consumption of these engines is called brake-specific fuel consumption, c_b, where c_b is the fuel flow per BHP of the engine. As with c_t in Chapter 6, it is optimum to have a low c_b value which represents a combination of more efficient fuel flow for power produced. It is lowest in the 40–60% power range:

$$c_b = \frac{\text{Fuel flow (lb/h)}}{\text{Brake horsepower}} \tag{7.6}$$

Non-turbocharged Reciprocating Engines

Non-turbocharged reciprocating engines are called *normally aspirated engines*. They lose power as altitude is increased and the air density is reduced. At about 19 000 ft altitude, they can develop only about one-half of the sea level HP. This is because of the corresponding reduction in air density with altitude.

A very serious problem in reciprocating engines is called *detonation*. Detonation results from a sudden, unstable decomposition of fuel at high temperature and/or pressure. Under certain combinations of these, the mixture ahead of the advancing flame front can suddenly explode instead of smoothly burning.

This explosion will send strong pressure waves throughout the combustion chamber that are many times higher than normal combustion pressures. Heavy detonation can cause immediate severe structural damage to the engine. Detonation can be detected by a rapid rise in cylinder head pressure, a drop in manifold pressure, loss of power, and loud explosive noises (similar to the "ping" in car engines). Reducing the throttle may increase power output by stopping the detonation.

ITEMS OF AIRCRAFT PERFORMANCE

Straight and Level Flight

Equilibrium conditions on an aircraft exist when the power developed by the engine is equal to the power required by the airframe for the same set of flying conditions. When these conditions are met, the aircraft will fly at a constant altitude and airspeed. Unaccelerated maximum velocity will occur at the intersection of the full power-available curve and the power-required curve. This is shown in Figure 7.14 and for that particular aircraft is at about 340 kts. true airspeed (TAS). Of course, the assumption is made that other aircraft restrictions as to maximum velocity do not restrict flight at this airspeed.

Climb Performance

Modern propeller aircraft are not used as fighters and do not use the "zoom" techniques of jet fighters. Therefore, a steady-velocity climb is assumed in this discussion. In a steady-velocity climb, the aircraft is in equilibrium, with all the forces along the flight path in balance (Figure 7.18). The forces acting along the flight path are thrust (acting forward), drag, and $W \sin \gamma$ (acting to the rear). $W \sin \gamma$ acts at the aircraft's center of gravity, not as shown in the figure. For steady velocity to exist, these must be balanced. As with the thrust-available and thrust-required curves, we will examine the angle of climb and rate of climb (ROC), and the location of the points on a P_r curve of best angle (V_X) and best rate (V_Y) during a steady-velocity climb.

As we have now seen, there are distinct differences when analyzing power and thrust curves for a power producer, and these differences will greatly affect climb characteristics. From Eq. 6.5a we saw that, for equilibrium:

$$T - D - W \sin \gamma = 0$$

Rearranging we get:

$$\sin \gamma = \frac{T - D}{W} \quad \text{or,} \quad \sin \gamma = \frac{T_a - T_r}{W}$$

Angle of Climb

From Eq. 6.5b, the maximum climb angle is obtained when the *excess thrust is maximum*. Thrust is a force and is related to distance, not time, when discussing climb performance. In this case, the object is to maximize the climb angle over a minimal distance using the thrust produced by a propeller moving a mass of air. If a tree is off the end of a short runway, the aircraft must achieve the maximum height in the shortest distance, not necessarily the shortest time, so the pilot climbs at V_X. However, as the forward speed of the power producer increases, the efficiency of the propeller decreases and thrust is reduced (Figure 7.19).

The basic Eq. 1.7 for converting thrust or drag units to HP units can be rewritten as

$$\text{HP} = \frac{TV_k}{325} \rightarrow T = \frac{325\text{HP}}{V_k}$$

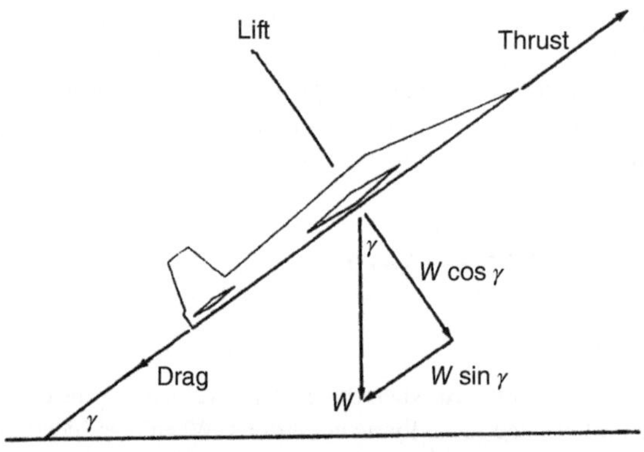

Figure 7.18. Forces on a climbing aircraft.

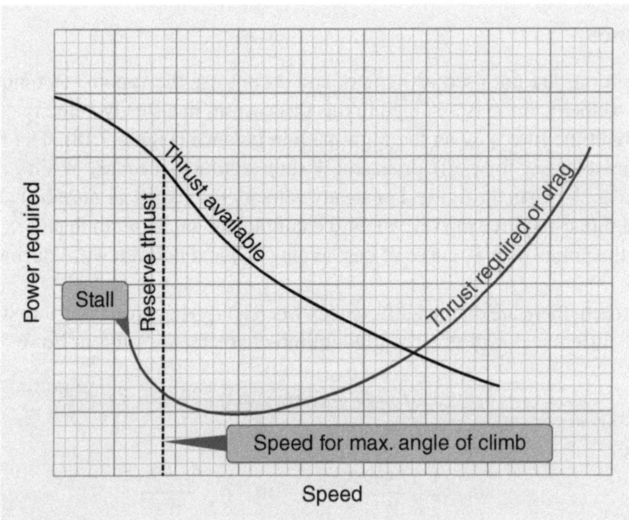

Figure 7.19. Thrust versus climb angle.
Source: Pilot's Handbook of Aeronautical Knowledge (2008a)/U.S. Department of Transportation Federal Aviation Administration/Public domain.

Adding subscripts yields

$$T_a = \frac{325P_a}{V_k} \quad \text{and} \quad T_r = \frac{325P_r}{V_k}$$

By substituting in Eq. 6.5b, we have

$$\sin\gamma = \frac{325(P_a - P_r)}{V_k W} \tag{7.7}$$

EXAMPLE

Using Figure 7.14, calculate the climb angle of the propeller aircraft with the following conditions: $W = 20\,000$ lb, standard day, sea level, clean configuration, TAS = 200 kts.

$$\sin\gamma = \frac{325(P_a - P_r)}{V_k W}$$

$$\sin\gamma = \frac{325(3\,750\text{ HP} - 1\,000\text{ HP})}{200\text{ kts.}(20\,000\text{ lb})}$$

$$\sin\gamma = 0.223$$

$$\gamma = 12.9°$$

If we select values of P_a and P_r from Figure 7.14 at several arbitrary velocities, we can calculate the corresponding climb angle using Eq. 7.7. The climb angle then can be plotted against velocity as shown in Figure 7.20. Although the maximum climb angle occurs at the stall speed for this aircraft, caution must be noted here as flying at this speed introduces other hazards to flight, so best climb angle velocity is usually published several knots above stall speed.

Contrast this to the thrust-producing aircraft, which has its maximum climb angle at $(L/D)_{max}$. Figure 7.21 shows for both a propeller-driven aircraft and a jet aircraft that the maximum angle of climb (V_X) is at the point of maximum excess thrust. On the left side of the image, the point of maximum excess thrust of a jet is at$(L/D)_{max}$, which is at the bottom of the T_r curve, while on the right side of the image, the point of maximum excess thrust for a propeller-driven aircraft is at a velocity below $(L/D)_{max}$, near stall speed in 1G flight. The advantage of the propeller aircraft over the jet in short field obstacle clearance is seen: the propeller aircraft lifts off at a lower speed (consuming less runway) and achieves the best angle of climb at a lower airspeed, and for both of these reasons, it will climb to a given altitude in a shorter distance than a jet.

The old saying "Hang it on the props" is really true. A discussion of wind effect on obstacle clearance is similar to that for thrust-producing aircraft (see Figure 6.20) so it is not repeated here.

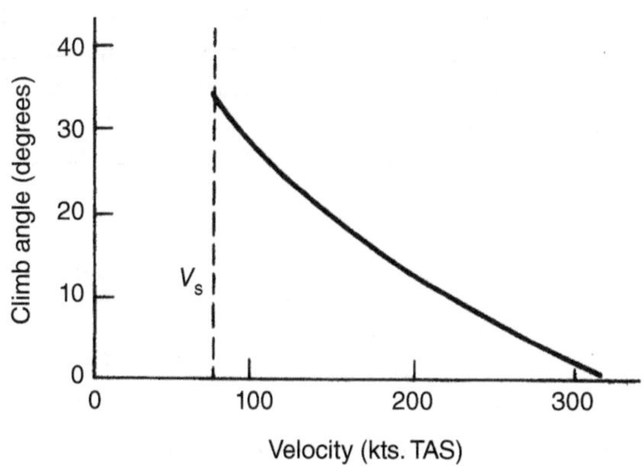

Figure 7.20. Climb angle versus velocity.

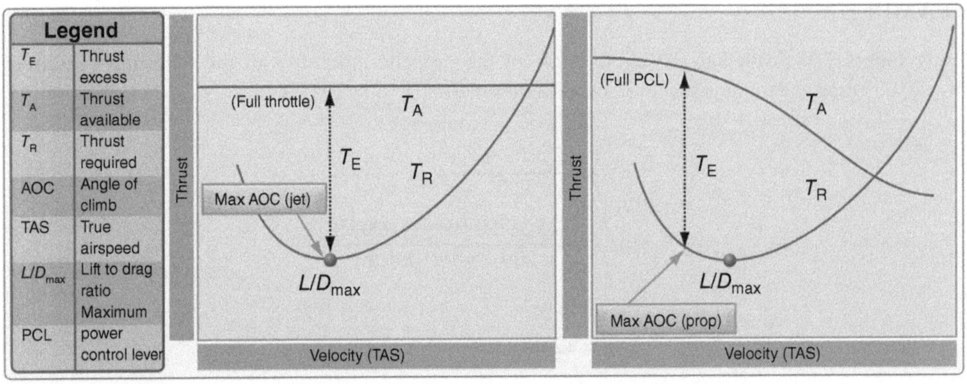

Figure 7.21. Comparison of maximum AOC between jet and propeller airplanes.
Source: Pilot's Handbook of Aeronautical Knowledge (2016b)/U.S. Department of Transportation Federal Aviation Administration/Public domain.

Application 7.1

Using Figure 7.14, calculate the angle of climb for the propeller-driven aircraft at 100 kts., then calculate the angle of climb at 75 kts.

Which airspeed provided the highest angle of climb? Why? Compared to the example angle of climb calculation at 200 kts., which of the three speeds resulted in the greatest angle of climb? Why?

Rate of Climb

The ROC is the vertical component of the flight velocity shown in Figure 7.22. Where the *angle* of climb for an aircraft for a given velocity is the altitude gained relative to the *distance traveled*, the ROC is the altitude gained relative to the *time* required to reach that altitude. A ROC is usually used during a steady velocity climb when obstacle clearance requirements do not need to be met. While the best angle of climb speed was found at the *maximum point of excess thrust*, the best ROC speed (V_Y) is found at the point of *maximum excess power*. Power is the rate at which work is done, so to obtain the greatest height for a given time, we need the best ROC, not the best *angle* of climb. Figure 7.23 shows a generic power-available

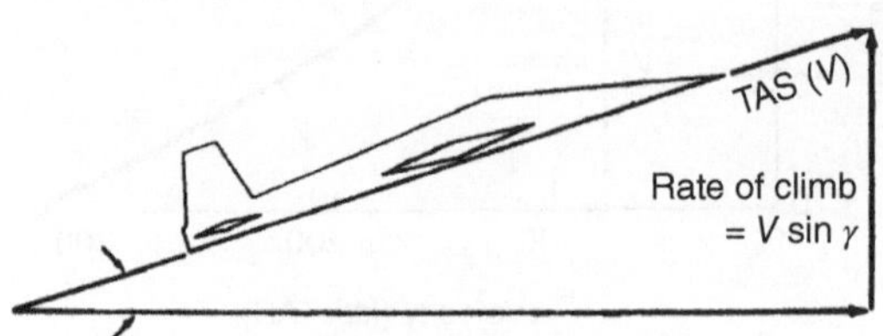

Figure 7.22. Rate of climb velocity vector.

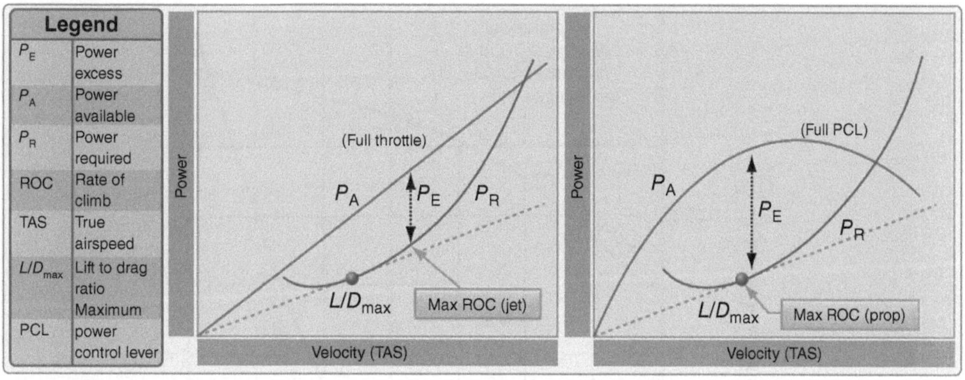

Figure 7.23. Comparison of maximum ROC between jet and propeller airplanes.
Source: Pilot's Handbook of Aeronautical Knowledge (2016b)/U.S. Department of Transportation Federal Aviation Administration/Public domain.

curve versus a power-required curve for a propeller aircraft, compared to the power-available curve and power-required curve of a jet aircraft. For this example, the conversion from T_a to P_a was made for the jet aircraft to compare the points of maximum excess power for both aircraft. Note for both aircraft the point of maximum excess power to overcome total drag. The airspeed at which this occurs is the best ROC airspeed (V_Y).

From Eq. 6.6, we can derive a new equation to calculate ROC for a power producer (remember, T_r is the thrust required to overcome drag (D)):

$$\text{ROC} = V \sin \gamma$$

$$\text{ROC} = V_k \frac{(T - D)}{W} \text{ (kts.)}$$

Converting to feet per minute (fpm):

$$\text{ROC} = 101.3 V_k \frac{(T - D)}{W} \text{ (fpm)}$$

Substituting gives

$$TV_k = 325 P_a \quad \text{and} \quad DV_k = 325 P_r$$

Therefore, when calculating ROC for a propeller aircraft, we find:

$$\text{ROC} = \frac{33\,000(P_a - P_r)}{W} \text{ (fpm)} \tag{7.8}$$

Equation 7.8 shows that the maximum ROC, ROC_{max}, occurs at the velocity where maximum excess power occurs (Figure 7.24). Note the greatest difference between P_a and P_r will produce the largest ROC, so as noted in Figure 7.24, the location of the arrow indicating the maximum ROC (V_Y) also represents the greatest difference in P_a and P_r.

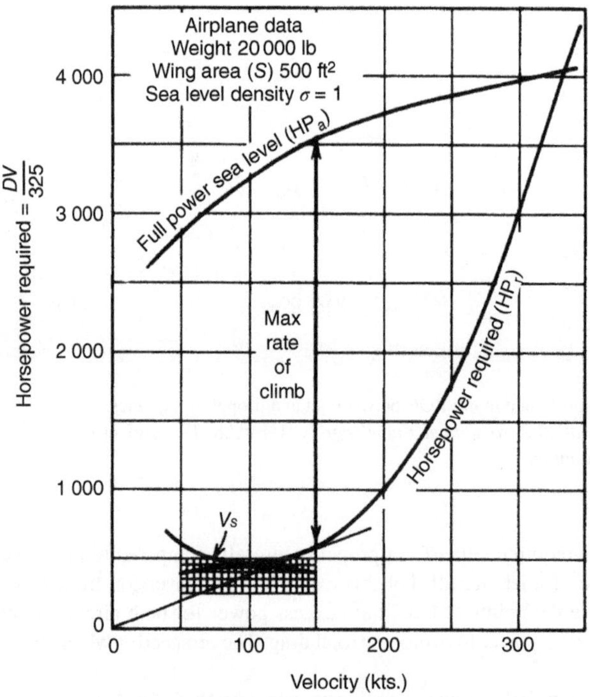

Figure 7.24. Finding the maximum rate of climb.

EXAMPLE

Using Figure 7.24, calculate the rate of climb of the propeller aircraft with the following conditions: $W = 20\,000$ lb, standard day, sea level, clean configuration, TAS = 200 kts.

$$\text{ROC} = \frac{33\,000(P_a - P_r)}{W}$$

$$\text{ROC} = \frac{33\,000(3\,750\text{ HP} - 1\,000\text{ HP})}{20\,000\text{ lb}}$$

$$\text{ROC} = 4\,538\text{ fpm}$$

Application 7.2

Using Figure 7.14, calculate the rate of climb for the propeller-driven aircraft at 100 kts., and then calculate the rate of climb at 75 kts.

Which airspeed provided the highest rate of climb? Why? Compared to your results in Application 7.1, were your results the same regarding the greatest rate of climb between the two airspeeds?

Endurance

To obtain maximum endurance (V_{BE}), as discussed in Chapter 6, minimum fuel flow is required. *Maximizing the time aloft, not the distance covered*, is the objective, and flying at V_{BE} would be useful in a holding pattern or anytime flying time is extended beyond initial planning. Returning to Eq. 6.7:

$$\text{Endurance (hours)} = \frac{\text{Fuel available (lb)}}{\text{Fuel flow (lb/hr)}}$$

Figure 7.25 is used to show the velocity to fly for both maximum endurance and maximum specific range (SR) for a propeller aircraft. Because fuel flow is roughly proportional to the power required, minimum P_r will produce minimum fuel flow, which is found at the bottom of the P_r curve. As mentioned earlier, this is not $(L/D)_{max}$ for a propeller aircraft. It is merely called the *minimum power required* point, and mathematically it is the point where $C_L^{3/2}/C_D$ is maximum.

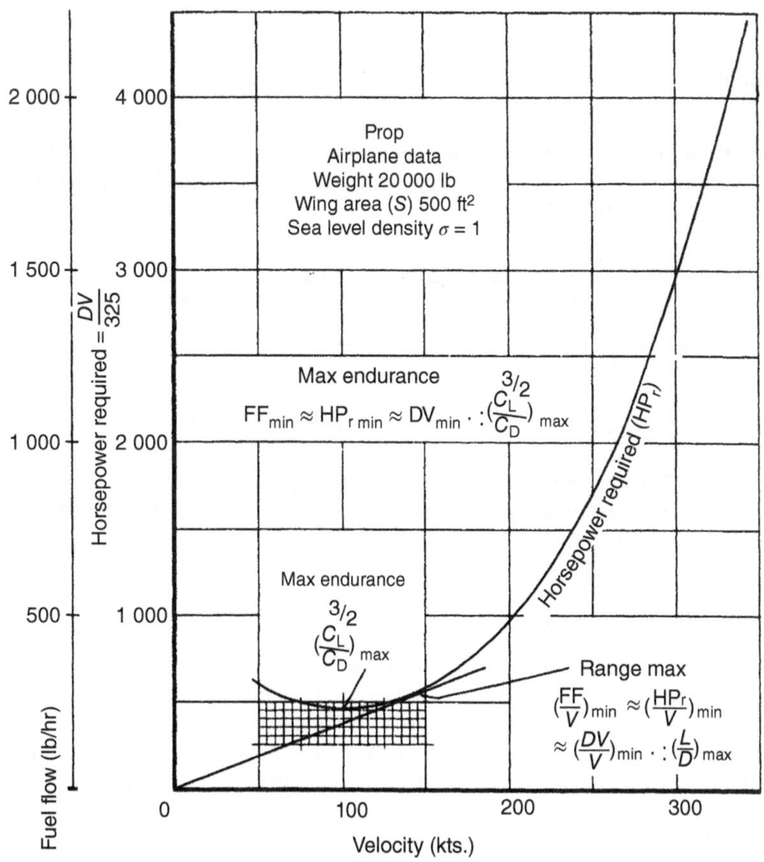

Figure 7.25. Finding the maximum endurance and range.

EXAMPLE

Using Figure 7.25 and Eq. 6.7, calculate the endurance of the propeller aircraft with the following conditions: fuel available = 4 350 lb, standard day, sea level, clean configuration, TAS = 200 kts.

$$\text{Endurance} = \frac{\text{Fuel available (lb)}}{\text{Fuel flow (lb/hr)}}$$

$$\text{Endurance} = \frac{4\,350 \text{ lb}}{500 \text{ lb/hr}}$$

$$\text{Endurance} = 8.7 \text{ hours}$$

Application 7.3

Using Figure 7.25 and the same given information in the example provided for endurance, calculate the endurance at 300 kts. instead of 200 kts.

With only 100 kts. increase in airspeed, why did the endurance of the aircraft decline so rapidly?

Specific Range

To get maximum range (V_{BR}), the aircraft must fly the maximum distance with the fuel available. The SR must be a maximum, so again we want the maximum speed per fuel flow, which occurs when the ratio of speed to power is the greatest. The SR for a propeller aircraft can be defined by the same relationship of speed and fuel flow discussed in Chapter 6:

$$\text{SR} = \frac{\text{Nautical miles flown}}{\text{Pounds of fuel used}} \quad \text{or,} \quad \text{SR} = \frac{\text{Airspeed (kts.)}}{\text{Fuel flow (lb/hr)}}$$

For maximum SR, we again use Eq. 6.8,

$$\text{SR}_{max} \left[\frac{V}{\text{FF}}\right]_{max} \text{(nm/lb)} \quad \text{or,} \quad \text{SR}_{max} \text{ occurs at } \left[\frac{\text{FF}}{V}\right]_{min} \text{(lb/nm)}$$

The point on the P_r curve in Figure 7.25 where a straight line drawn from the origin is tangent to the curve will indicate the maximum specific-range velocity. For a power-producing aircraft, this is the $(L/D)_{max}$ point. Note again that for a propeller-driven aircraft, this is not at the bottom of the P_r curve.

Wind Effect on Specific Range

As with a thrust-producing aircraft, an aircraft flying into a headwind is in an unfavorable environment. Therefore, the airspeed should be increased to reduce the time that the aircraft is exposed to the headwind. Conversely, an aircraft flying with a tailwind is being helped on its way and should slow down to take advantage of this favorable environment. The amount that the airspeed should be changed is shown in Figure 7.26. As shown in the figure, if an aircraft in a no-wind situation has a TAS of 130 kts., then encounters a 50 kts. headwind, the airspeed should be increased to 155 kts. for V_{BR}, and lowered to 115 kts. to obtain V_{BR} if a 50 kts. tailwind.

To determine how much to adjust airspeed for a headwind, plot the wind on the velocity scale to the right of the origin. Draw a new tangent line from this point to the P_r curve. The airspeed at which the tangent

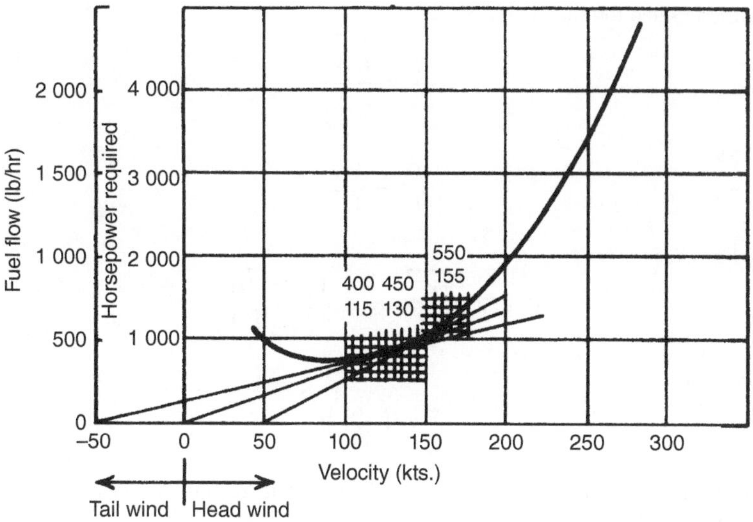

Figure 7.26. Effect of wind on range.

line touches the P_r curve is the corrected airspeed to fly. Correcting for a tailwind is done similarly. The tailwind is plotted off to the left of the origin and a new tangent line is drawn to the curve.

VARIATIONS IN THE POWER-REQUIRED CURVE

The aircraft performance items that we discussed up to this point were for one weight, clean configuration, and sea level standard day conditions. Now, we will alter these conditions and examine the resulting performance for a change in weight, altitude, and configuration. To simplify the explanation, only one condition is altered at a time.

The basic power-required curve (Figure 7.12) was drawn for a 20 000 lb airplane in a clean configuration, at sea level on a standard day. We now show how the curve changes for variations in weight, configuration, and altitude.

Weight Changes

Weight decreases during every flight as fuel is consumed. Weight increases as fuel, passengers, or cargo are added. As the mission changes, so do the operating parameters of each flight, so a pilot must be aware of how the aircraft's performance characteristics are affected by weight changes. Similar to our discussion in Chapter 6 for thrust-producing aircraft, as the weight of a power-producing aircraft is increased, the P_r to overcome the increase in drag due to the increase in lift, must also increase. As the power increases, so must the fuel flow and all related performance values.

To find out how the P_r varies with weight changes, we go back to the basic relationship between P_r and T_r (drag) that we developed in Eq. 1.7:

$$P_r = \frac{DV}{325}$$

Adding subscripts and forming a ratio gives

$$\frac{P_{r_2}}{P_{r_1}} = \frac{D_2 V_2}{D_1 V_1}$$

From Eq. 6.9,

$$\frac{D_2}{D_1} = \frac{W_2}{W_1}$$

And from Eq. 4.7,

$$V_2 = V_1 \sqrt{\frac{W_2}{W_1}} \quad \text{or,} \quad \frac{V_2}{V_1} = \sqrt{\frac{W_2}{W_1}}$$

Substituting gives

$$\frac{P_{r_2}}{P_{r_1}} = \left(\frac{W_2}{W_1}\right)^{3/2} \tag{7.9}$$

In addition to the increase in P_r, with an increase in weight, there must also be an increase in velocity. This is the same as was found in Eq. 4.7:

$$V_2 = V_1 \sqrt{\frac{W_2}{W_1}}$$

In plotting new curves for a change in weight, both Eqs. 7.9 and 4.7 must be applied. Figure 7.27 shows P_r curves for 20 000 and 30 000 lb. As weight is decreased, the curve shifts down and to the left; all previous "V speeds" for endurance, SR, and steady-velocity climbs must shift as well.

Variation of Aircraft Performance – Weight Change

Straight and Level Flight In straight and level, unaccelerated flight, all forces are balanced, so if weight is increased, then the lift must also increase, which causes the induced drag to increase. Note in Figure 7.27 that the P_r curve moves up and to the right as weight is increased, and the increase in P_r with a weight increase is much greater at lower velocities. To remain in level, flight P_r increases to balance the increase in drag, so an increase in weight means an increase in P_r. P_a is not changed; V_{max} is decreased. A reduction in weight has the opposite effect and V_{max} will be increased.

Angle of Climb Increased weight results in increased power required, while the power available is not changed. Equation 7.7 shows:

$$\sin\gamma = \frac{325(P_a - P_r)}{V_k W}$$

Both the excess power reduction and the increased weight cause a reduction in the angle of climb.

Application 7.4

Using Figure 7.27, calculate the climb angle for both the 20 000 lb curve and the 30 000 lb curve at 100 kts., and assume P_a for each curve is 3 250 HP.

How did the angle of climb change as weight increased?

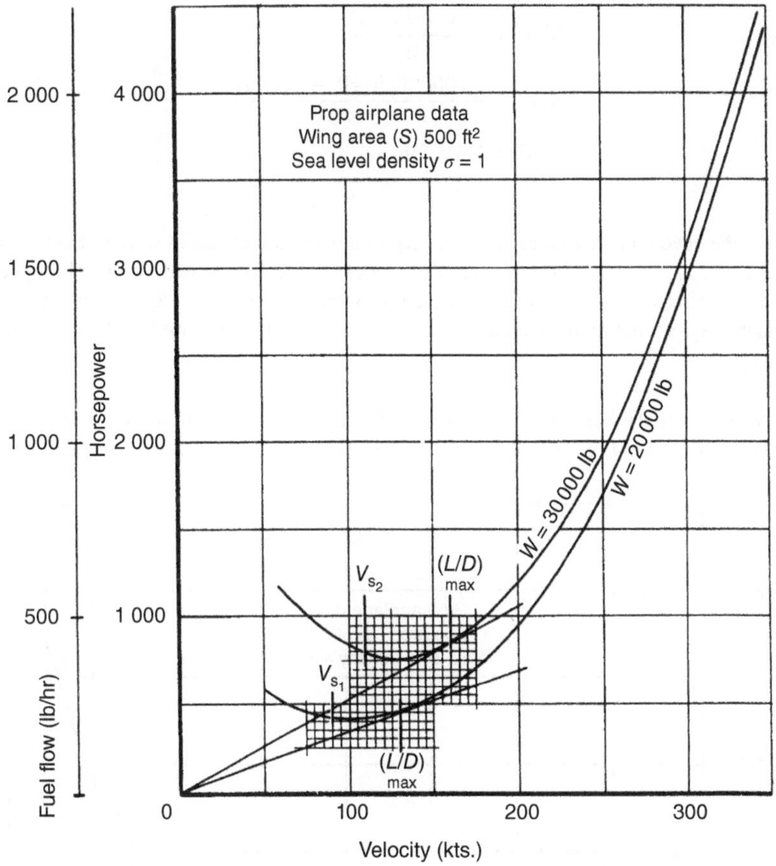

Figure 7.27. Effect of weight change on a P_r curve.

Rate of Climb As with the angle of climb, when weight increases, the drag increases due to the increase in lift required to compensate for the weight. P_r must increase accordingly. The ROC is calculated by Eq. 7.8:

$$\text{ROC} = \frac{33\,000(P_a - P_r)}{W}$$

When the value of weight in the equation above increases, P_r increases as P_a remains the same, so ROC must decrease. As weight decreases, then the ROC increases as excess power is available.

EXAMPLE

Using Figure 7.27, calculate the rate of climb of the propeller aircraft with the following conditions: $W = 30\,000$ lb, standard day, sea level, clean configuration, TAS = 200 kts. Assume the $P_a = 3\,750$ HP from Figure 7.24 at 200 kts. Compare your answer to the ROC at 20 000 lb calculated earlier in this chapter.

$$\text{ROC} = \frac{33\,000(P_a - P_r)}{W}$$

$$\text{ROC} = \frac{33\,000(3750\text{ HP} - 1200\text{ HP})}{30\,000\text{ lb}}$$

$$\text{ROC} = 2\,805\text{ fpm}$$

Endurance We have seen that more power is required with an increase in weight. Because fuel flow is proportional to P_r, any increase in power required means a decrease in endurance. The opposite is true when weight is decreased, endurance is increased. Review the fuel flow required in Figure 7.26 at the bottom of each P_r curve, noting the increase in fuel flow as weight is increased, and the change in velocity to maintain V_{BE}.

Specific Range As fuel is burned, weight is reduced, and the power-required curve moves downward and to the left. This is shown in Figure 7.28. The tangent lines intersect the P_r curves at $(L/D)_{max}$ and indicate

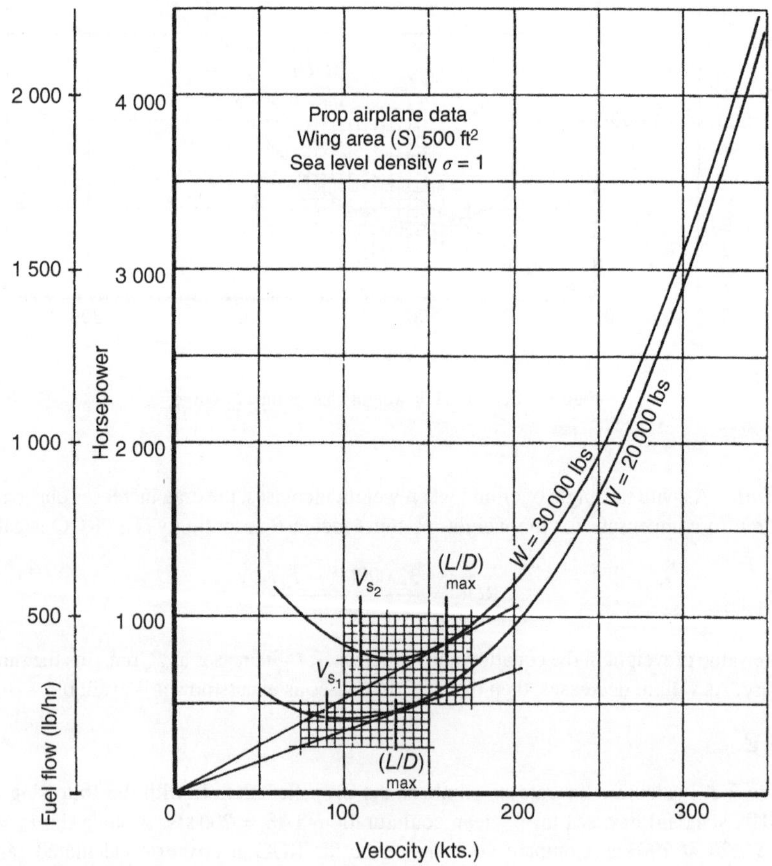

Figure 7.28. Effect of weight change on a specific range.

the maximum SR velocity and fuel flow at this point. At a gross weight (GW) of 30 000 lb, the velocity must be 160 kts. for maximum range and the fuel flow will be 425 lb/hr. V_{BR} here is 160/425 = 0.3765 nm/lb.

When the weight of the aircraft has been reduced to 20 000 lb, the aircraft must fly at 130 kts. to attain the maximum range. The fuel flow here will be 225 lb/hr and the SR = 130 kts./225 lb/hr = 0.5778 nm/lb. This is a 53.5% improvement in SR. The pilot must reduce the power output of the engine and reduce the airspeed to achieve this range. It is only necessary for the pilot to reduce the power setting because the airspeed will automatically be reduced when this is done.

The natural tendency of a pilot is to keep the power setting the same as fuel is burned, and to allow the velocity to increase as the drag decreases when fuel is consumed. This will not maintain the maximum range and will in fact reduce the range by an appreciable amount. For maximum range, as weight is reduced, *reduce airspeed*.

Configuration Change

In going from the clean configuration to the gear and flaps down (dirty) configuration, the equivalent parasite area is greatly increased. For the airplane shown in Figure 7.29, parasite drag increases by 50% (from 9.92 to 14.9 ft^2). While the configuration change increases the parasite drag, and thus the parasite power required, it has little effect on the induced drag or induced power required. As we have seen with a power producer, the faster the aircraft travels, the more power must be produced to overcome the increased parasite drag, which means more fuel must be consumed per unit of time.

There is no shortcut method for calculating the P_r curve for the dirty condition. You must calculate the parasite drag and then the total drag for the dirty condition, and then convert this to power required by Eq. 1.7. The effect of configuration change is shown on the P_r curves in Figure 7.29.

Variation of Aircraft Performance – Configuration Change

Straight and Level Flight V_{max} in the dirty configuration is usually limited by the structural strength of gear and flaps, so V_{max} is reduced. Because parasite drag increases as speed increases, and the power required to overcome the parasite drag also increases, V_{max} decreases. As the gear and flaps are retracted, parasite drag decreases for a given velocity, so now P_r also decreases to maintain level flight and V_{max} increases for a given P_a.

Angle of Climb Power required increases in the dirty configuration due to the increase in total drag (primarily parasite drag); thus, the angle of climb is reduced. However, the increase in P_r is much less than the increase in T_r for the jet aircraft (see Figures 7.29 and 6.33). Again, the superior performance of propeller aircraft at low speeds is demonstrated.

EXAMPLE

Using Figure 7.29, calculate the climb angle of the propeller aircraft for both curves with the following conditions: TAS = 150 kts., P_a = 3 500 HP.

$$\sin\gamma = \frac{325(P_a - P_r)}{V_k W}$$

$$f_1 \rightarrow \sin\gamma = \frac{325(3\,500\text{ HP} - 550\text{ HP})}{150\text{ kts.}(20\,000\text{ lb})} \rightarrow \gamma = 18.7^\circ$$

$$f_2 \rightarrow \sin\gamma = \frac{325(3\,500\text{ HP} - 750\text{ HP})}{150\text{ kts.}(20\,000\text{ lb})} \rightarrow \gamma = 17.3^\circ$$

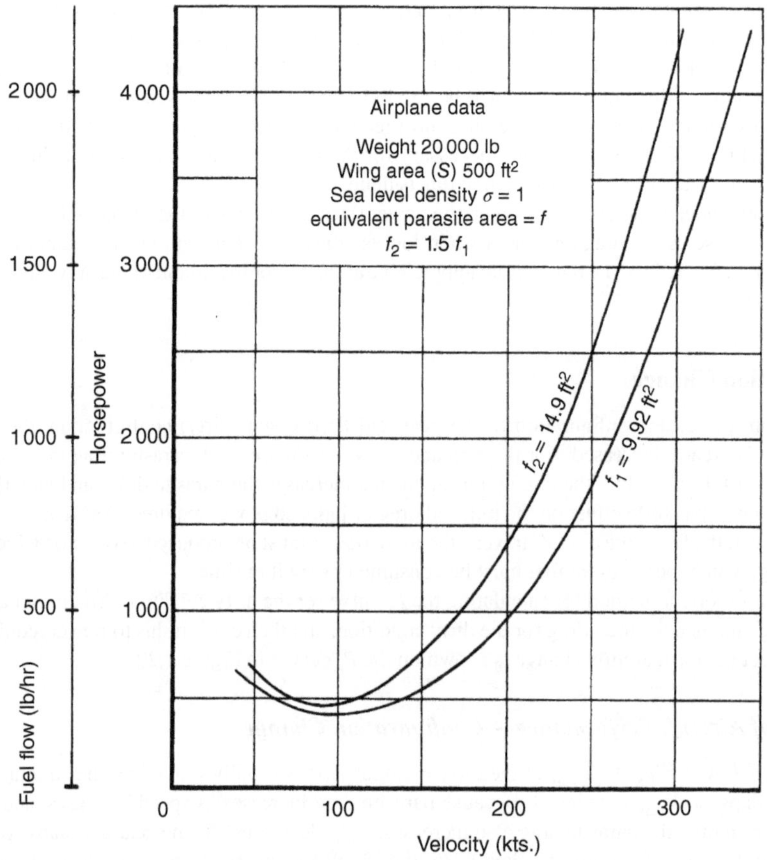

Figure 7.29. Effect of configuration on the P_r curve.

Rate of Climb Aircraft with gear and flaps deployed experience a reduction in the ROC performance, as shown in the angle of climb discussion above. In order to maximize the performance of an aircraft after takeoff, during missed approach in inclement weather, or in an emergency, retract the gear as soon as possible and retract the flaps as required.

Application 7.5

Using Figure 6.33, calculate the rate of climb for the jet aircraft at 200 kts. for both the "Clean Configuration" and "Full Flaps & Gear." Note the percent increase for a dirty configuration. Using Figure 7.29, calculate the rate of climb for the propeller aircraft at 200 kts. (assume $P_a = 3\,750$ HP) for both the "Clean Configuration (f_1)" and "Full Flaps & Gear (f_2)," note again the percent increase in the rate of climb for a dirty configuration.

Which aircraft suffered a greater percent decrease in the rate of climb between a clean configuration and a dirty configuration? Why?

Endurance Endurance is directly proportional to fuel flow: if drag is increased due to the deployment of gear and/or flaps, P_r increases, and thus fuel flow increases. When in a holding pattern, keep the gear retracted, deploy approach flaps as required, and slow down to reduce parasite drag. Again, comparing the change in fuel flow (assume fuel flow equals thrust required) at V_{BE} for a thrust-producing aircraft (Figure 6.33) in clean versus dirty configuration, and the change in fuel flow at V_{BE} for a propeller-driven aircraft (Figure 7.29) in clean and dirty configuration, the advantages of a propeller-driven aircraft at low speeds can be seen.

Specific Range As has been discussed, if it is necessary to fly in a dirty configuration, slow the airplane to reduce drag. If the $(L/D)_{max}$ AOA is known for the new configuration, then that AOA should be maintained to achieve V_{BR}.

Altitude Changes

In our discussion of the effect of altitude on the drag of an aircraft, we saw that the drag of the aircraft was unaffected by altitude, but that the TAS at which the drag occurred did change:

$$D_2 = D_1 \quad \text{and} \quad V_2 = \frac{V_1}{\sqrt{\sigma_2}}$$

Subscript 1 is at sea level ($\sigma_1 = 1$) and subscript 2 is at altitude. Applying the above equations to Eq. 1.7 gives

$$\frac{P_{r_2}}{P_{r_1}} = \frac{1}{\sqrt{\sigma_2}} \rightarrow P_{r_2} = P_{r_1}\frac{1}{\sqrt{\sigma_2}} \tag{7.10}$$

The drag does not change with altitude but the P_r does. The velocity changes by the same amount:

$$\frac{V_2}{V_1} = \frac{1}{\sqrt{\sigma_2}} \rightarrow V_2 = V_1\frac{1}{\sqrt{\sigma_2}} \tag{7.11}$$

In plotting new curves for a change in altitude, note that both Eqs. 7.10 and 7.11 must be applied. All points on the sea level curve move to the right and upward, by the same amount, when correcting for altitude changes. This is shown in Figure 7.30. Note that the line drawn from the origin, which is tangent to the curve and locates the $(L/D)_{max}$ point, remains tangent to the curve at all altitudes. The angle remains the same at all altitudes.

Another point of interest in the altitude curves is that the curves move farther apart at higher velocities. Each point on the sea level curve is moved to the right and also moved upward by the same multiplier, $1/\sqrt{\sigma_2}$, so the change is greater when the velocity or power required is greater. Thus, points on the right are changed more than those on the left. Even though the drag at sea level is equal to the drag at altitude, proportionally greater power is required because of the higher TAS.

Variation of Aircraft Performance – Altitude Change

Straight and Level Flight The power available at altitude is affected by the type of engine and/or propeller. If turbocharged reciprocating engines with constant speed props are flown below their critical altitude, they can develop as much power as at sea level. In this case, the V_{max} TAS will be increased significantly. Non-turbocharged reciprocating engines with fixed props, on the other hand, will suffer a power loss as altitude increases and may experience either an increase or a decrease in V_{max}, depending on the intersection point of the P_a and P_r curves. Similar reasoning can be applied to turboprop engines.

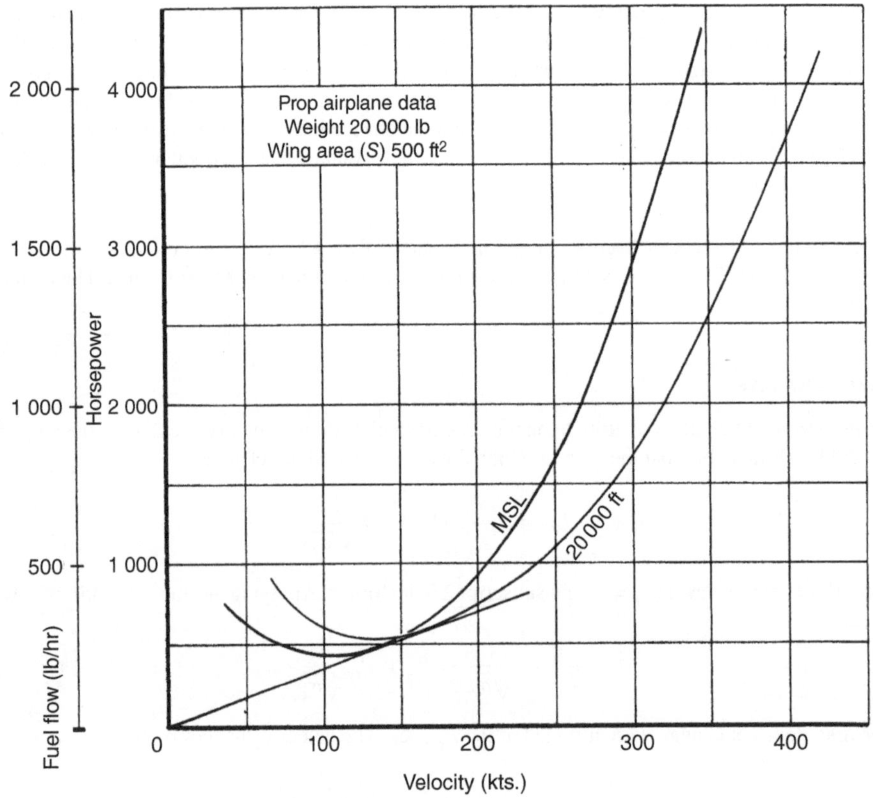

Figure 7.30. Effect of altitude on a P_r curve.

The engine loses power with increased altitude, but the P_r curve moves to the right, so the intersection of the curves will determine V_{max}. The indicated airspeed (IAS) will be lower in all cases.

Angle of Climb/Rate of Climb The climb performance of a power producer depends on excess thrust for the best angle of climb (V_X), and on excess power for the best ROC (V_Y). When altitude is increased, P_r increases, and depending on the type of power-producing engine, the P_a may decrease as well. From Eq. 7.7, it can also be seen that the increase in P_r at altitude will reduce the angle of climb, even if P_a is not reduced. As altitude is increased for a given aircraft, V_X will increase and V_Y will decrease. When the aircraft is unable to climb at a rate of 100 fpm, then the *service ceiling* has been reached. When the aircraft has no excess power and produces zero climb rate, an *absolute ceiling* has been reached. Figure 7.31 is a standard plot of V_X and V_Y versus altitude. Note that at the absolute ceiling, $V_X = V_Y$.

Application 7.6

Note that Figure 7.31 is based on indicated airspeed (IAS). As altitude increases, the IAS for V_X increases at a much smaller rate than the IAS for V_Y decreases.

How might this look if we substituted TAS instead of IAS? Why?

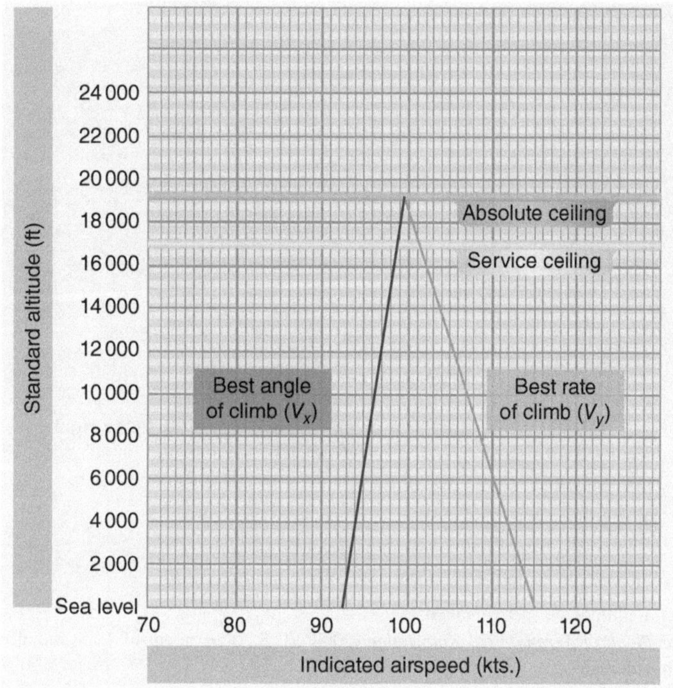

Figure 7.31. V_X versus V_Y with altitude.
Source: Pilot's Handbook of Aeronautical Knowledge (2008a)/U.S. Department of Transportation Federal Aviation Administration/Public domain.

Endurance With a turboprop, the reduction in fuel flow at altitude will be more than compensate for the increase in P_r, but if it is necessary to climb to altitude, more fuel may be required than if the airplane had remained at the lower altitude. For a reciprocating engine, the endurance at altitude is reduced, but only if the fuel/air mixture is leaned appropriately. So, be aware of the manufacturer's recommendations when manually adjusting the mixture.

Specific Range In our earlier discussion, we pointed out that an altitude increase changed both the P_r and the velocity by the same amount. We also saw that the tangent line to the curves remained the same for all altitudes. In Figure 7.25, we saw that the intersection of the tangent line and the curve indicated the maximum SR point. Thus, as far as the airframe is concerned, altitude has no effect on the SR (Figure 7.32). The ratio of fuel flow to velocity is constant with altitude. Propeller-driven airplanes do not show large increases in range with increasing altitude, unlike jet airplanes.

We must consider at what altitude the engine–propeller combination is most efficient. First, let us look at the non-turbocharged reciprocating engine with a fixed-pitch propeller. In general, these engines can develop sea level power to about 5 000 ft. This seems to be about the altitude where the maximum range is obtained. For the turbocharged, reciprocating engine with a constant speed propeller, about 10 000 ft is best. The turbine engine of the turboprop aircraft favors high altitudes and operates best at about 25 000 ft.

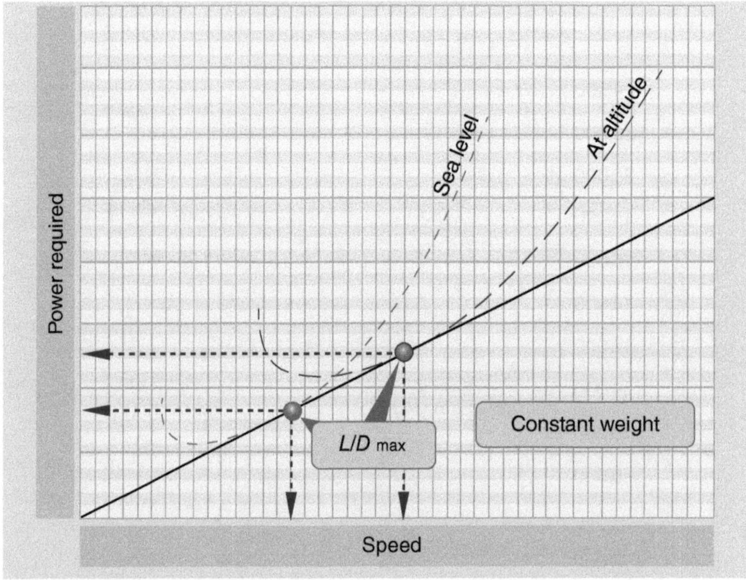

Figure 7.32. Effect of altitude on specific range.
Source: Pilot's Handbook of Aeronautical Knowledge (2008a)/U.S. Department of Transportation Federal Aviation Administration/Public domain.

Propeller Performance Summary

All aircraft in flight must produce thrust to overcome the drag of the aircraft. In thrust-producing aircraft, thrust is produced directly from the engine, but in aircraft that have propellers (or rotors), the engine does not produce thrust directly. These aircraft are called *power producers* because the engine produces power, turning the propeller, and developing an aerodynamic force as it turns through the air. The resultant force is thrust.

Propeller efficiency is the ratio of the power output to the power input. This efficiency will never be 100%. This loss in efficiency is why P_a is always less than BHP. When reviewing P_r curves to overcome total drag, two important items were realized:

1. The induced power required to overcome induced drag varies as the inverse of the velocity ratio to the first power.
2. The parasite power required to overcome parasite drag varies directly as the cube of the velocity ratio.

For a power-producing aircraft, we have seen that two items of aircraft performance occur at $(L/D)_{max}$:

1. Maximum engine-out glide ratio (minimum drag)
2. Maximum range

Changing the weight of an aircraft changes the induced drag much more than the parasite drag.

- The effect on the power-required curve with an increase in weight is that all points are moved up and to the right.

- The value of $(L/D)_{max}$ does not vary with a change in weight, but changing the weight changes the drag proportionately.

Changes in airplane configuration from a clean configuration to a dirty configuration increase the parasite drag but have little effect on induced drag.

- The faster the aircraft travels, the more power must be produced to overcome the increased parasite drag, which means more fuel must be consumed per unit of time.

With an altitude increase, all points on the power-required curves move upward and to the right, as well as farther apart at higher velocities.

- Even though the drag at sea level is equal to the drag at altitude, proportionally greater power is required because of the higher TAS.

SYMBOLS

c_b	Brake-specific fuel consumption (lb fuel/BHP)
P_a	Power available (horsepower units)
P_r	Power required (horsepower units)
P_{ri}	Power required, induced (horsepower units)
P_{r_0}	Power required, profile (horsepower units)
P_{rp}	Power required, parasite (horsepower units)
BHP	Brake horsepower (horsepower units)
SHP	Shaft horsepower (horsepower units)
THP	Thrust horsepower (horsepower units)
ESHP	Equivalent SHP (horsepower units)
V_X	Best angle of climb speed (kts. IAS)
V_Y	Best rate of climb speed (kts. IAS)
η (eta)	Propeller efficiency (%/100)

EQUATIONS

7.1 $\text{ESHP} = \text{SHP} + \frac{TV}{325\eta}$

7.2 $\eta = \frac{\text{THP}}{\text{BHP (or SHP)}}$

7.3 $P_{ri_2} = P_{ri_1}\frac{V_1}{V_2}$

7.4 $P_{r_{p2}} = P_{r_{r1}}\left(\frac{V_2}{V_1}\right)^3$

7.5 $P_a = (\eta_p)(\text{BHP})$

7.6 $c_b = \frac{\text{FF}}{\text{BHP}}$

7.7 $\sin\gamma = \frac{325(P_a - P_r)}{V_k W}$

7.8 $\text{ROC} = \frac{33\,000(P_a - P_r)}{W}$

7.9 $P_{r_2} = P_{r_1}\left(\frac{W_2}{W_1}\right)^{3/2}$

7.10 $P_{r_2} = P_{r_1} \frac{1}{\sqrt{\sigma_2}}$

7.11 $V_2 = V_1 \frac{1}{\sqrt{\sigma_2}}$

KEY TERMS

Angle of climb
Brake horsepower
Brake-specific fuel consumption
Constant-speed propeller
Effective pitch
Fixed-pitch propeller
Geometric pitch
Power available curve
Power producers
Power-required curve
Propeller efficiency
Rate of climb
Reciprocating engine
Shaft horsepower
Thrust horsepower
Turbocharged engine
Turboprop engine

PROBLEMS

Note: Answers to the problems in the first two sections are given at the end of the book.

Application of Chapter Theory

1. In the formula HP $= \frac{TV_k}{325}$
   a. power required is in horsepower units.
   b. power required is in ft-lb/sec.

2. Power is
   a. (Force × Velocity)/Time.
   b. Work/Time.
   c. (Force × Distance)/Time.
   d. Both (b) and (c).

3. The horsepower lost between indicated and brake horsepower is known as:
   a. Friction horsepower
   b. Shaft horsepower
   c. Thrust horsepower
   d. Useable horsepower

4. Propeller blade angle and propeller pitch are considered one and the same.
   a. True
   b. False

5. During level off after a climb with a fixed pitch propeller, if the throttle is not repositioned, the rpm of the propeller will increase.
   a. True
   b. False

6. The power required to overcome induced drag varies
   a. inversely with V^2.
   b. inversely with V^3.
   c. inversely with V.
   d. directly with V.

7. The power required to overcome parasite drag varies
   a. directly with V^2.
   b. directly with V^3.
   c. directly with V.
   d. inversely with V^2.

8. Maximum rate of climb for a propeller airplane occurs
   a. at $(L/D)_{max}$.
   b. at $P_{r(min)}$.
   c. at $C_{L(max)}$.
   d. at $(P_a - P_r)_{max}$.

9. The lowest point on the P_r curve is $(L/D)_{max}$.
   a. True
   b. False

10. Propeller aircraft are more efficient than jet aircraft because
    a. they do not go so fast.
    b. they process more air and do not accelerate it as much.
    c. they use gasoline instead of jet fuel.
    d. V_1 is less, so propulsive efficiency is greater.

11. Turboprop aircraft are classified as power producers because
    a. nearly all of the engine output goes to the propeller.
    b. the engine is a turbine engine.
    c. the fuel flow is proportional to the power produced.
    d. Both (a) and (c).

12. Propeller aircraft use V_X to climb
    a. when a climb is needed for maximum height to minimum horizontal distance.
    b. when a large excess of power is available.
    c. when flying at $(L/D)_{max}$.
    d. when a climb is needed for maximum height to minimum time.

13. Propeller aircraft get the highest angle of climb at $(L/D)_{max}$.
    a. True
    b. False

14. A lightly loaded propeller airplane will be able to glide _______when it is heavily loaded.
a. farther than
b. less far than
c. the same distance as

15. To obtain maximum glide distance, a heavily loaded airplane must be flown at a higher airspeed than if it is lightly loaded.
a. True
b. False

16. 16. To maximize range on a propeller-driven airplane at high altitude, true airspeed should be ________.
a. decreased
b. increased
c. not changed

17. In Figure 7.30, the specific range for a propeller airplane is
a. less at altitude than at sea level.
b. more at altitude than at sea level.
c. the same at altitude as at sea level.

18. The pilot of a propeller airplane is flying at the speed for the best range under no-wind conditions. A headwind is encountered. To obtain the best range, the pilot now must
a. speed up by the amount of the wind speed.
b. slow down by the amount of the wind speed.
c. speed up by less than the wind speed.
d. slow down by more than the wind speed.

19. As a propeller airplane burns up fuel, to fly for maximum range, the airspeed must
a. remain the same.
b. be allowed to speed up.
c. be slowed down.

20. The turboprop aircraft has its lowest specific fuel consumption at about 25 000 ft altitude because
a. the turbine engine wants to fly at high altitudes.
b. the propeller has higher efficiency at lower altitudes.
c. This is a compromise between (a) and (b) above.

21. The power-required curves for an increase in altitude show that the
a. P_r remains the same as altitude increases.
b. P_r increases by the same amount as the velocity.
c. P_r increases but the velocity does not.

22. A propeller aircraft in the dirty condition shows that the P_r moves up and to the left over the clean configuration. This is because
a. the increase in the induced P_r is more at low speed.
b. the increase is mostly due to parasite P_r.
c. both induced P_r and parasite P_r increase by the same amount.

23. The increase in the P_r curves for a weight increase is greater at low speeds than at high speeds because the increase in
a. induced P_r is greatest.
b. parasite P_r is greatest.
c. profile P_r is greatest.

Calculations Using Chapter Principles

24. $P_a - P_r$ curves for a propeller airplane are shown in Figure 7.24. Find

a. V_{max} at full power.
b. Climb angle at V_s.
c. Climb angle at $(L/D)_{max}$.
d. Rate of climb at V_s.
e. Rate of climb at 150 kts.
f. Velocity for maximum endurance.
g. Velocity for maximum range.
h. Velocity for maximum range with 50-kts. headwind.

25. In Chapter 5, Problem 20, you calculated the drag for a 10 000-lb turbojet airplane for sea level standard day conditions. The values that you found are repeated below. Calculate the values for horsepower required and plot them on graph paper.

V_k	Drag	P_r
125	1 233	
150	1 058	
172	1 020	
200	1 067	
300	1 720	
400	2 860	

26. Using Figure 7.27, find the velocity for the best range for the airplane at 30 000 and 20 000 lb gross weight (GW).

27. Using Figure 7.27, calculate the specific range for the airplane at both weights if it is flying at the best range of airspeed.

28. Using Figure 7.27, calculate the specific range for the 20 000-lb airplane if the throttle is not retarded and the airspeed is allowed to increase as weight is reduced. Compare your answer to that of Problem 7.27 and state your conclusions.

29. Using Figure 7.30, calculate the specific range for this airplane if it is flying at the best range airspeed at mean sea level (MSL) and at 20 000 ft altitude.

30. A 20 000-lb propeller airplane at sea level standard day has the following data. Calculate its P_r and plot the curve on the graph paper.

V	Drag	P_r
150	2 074	
182.7	1 923	
200	1 954	
300	2 948	
400	4 809	

31. The airplane in Problem 7.30 now has its GW increased to 30 000 lb. Calculate the new values and plot the curve on the graph paper.

$V_{20\,000\text{ lb}}$	$P_{r(20\,000\text{ lb})}$	$V_{30\,000\text{ lb}}$	$P_{r(30\,000\text{ lb})}$
150			
182.7			
200			
300			
400			

32. Calculate new values for the 20 000-lb airplane in Problem 7.30, but now flying at 20 000 ft density altitude. Plot the values and draw the curve on the graph paper.

V_{SL}	$P_{r(SL)}$	$V_{(20000\text{ ft})}$	$P_{r(20000\text{ ft})}$
150			
182.7			
200			
300			
400			

Correlation Through Scenario-Based Learning

Aircraft – *Diamond DA50RG*

1. Using the Diamond DA50RG aircraft flight manual (AFM) as found on the Diamond Aircraft Technical Publication's website, find the following:

a. Engine manufacturer and engine type
b. Air-cooled or liquid cooled
c. Are intercoolers included as part of the turbocharger system?
d. Takeoff power (HP) and takeoff propeller rpm
e. Time restriction for takeoff power
f. Maximum continuous power (HP) and propeller rpm
g. Propeller manufacturer and propeller type
h. Propeller construction material
i. Constant-speed or fixed-pitch propeller
j. What are the various load percentages used as reference points throughout the cruise performance charts of the AFM? Why do they change with altitude?
k. Why does this aircraft have the option for ground idle and flight idle?

2. What purpose does the engine gearbox serve? Why is it paramount to check the gearbox oil before every flight?

3. Research FADEC and explain why the DA50RG does not have a "prop control" or mixture control. How do these systems work on this airplane versus a *conventional* propeller system? What is the purpose of the dual-engine control units?

4. Reviewing the notes in Section 3 of the AFM regarding propeller overspeed, when in this condition the propeller now acts as a ______________.

5. Reference the Propeller Setpoint Curve figure in Section 7 of the AFM under the Propeller Control section, what is the difference in propeller rpm between 90% rpm and 45% rpm? What rpm can be expected when the engine is set at 80% load?

6. Calculate the thrust being produced when the aircraft is traveling 180 kts. and an assumed 272 Hp.

7. During the climb, the load is set to 90% with a result of 2 300 rpm, as the DA50 levels at altitude with the same 90% load setting, the airspeed increased, why did the rpm remain at 2 300?

8. Using equation 7.6, calculate c_b when the fuel flow is 14 gal/hr and the engine maintains the HP published while at 90% load.

9. Calculate the climb angle of the DA50 based on the following:

Standard day, sea level

Maximum takeoff weight

$P_a - P_r = 150\,\text{HP}$

TAS = 110 kts

10. Calculate the rate of climb of the DA50 based on the following information.

Standard day, sea level

Maximum takeoff weight

$P_a - P_r = 150\,\text{HP}$

11. Calculate the endurance (hours/minutes) of the DA50 based on the following information.

Maximum usable fuel of Jet A

Average fuel consumption rate = 12.3 gal/hr

12. Using the Cruise Performance chart for 4 407 lb in Section 5 of the DA50 AFM, calculate the SR_{max} (lb/nm) of the DA50 based on the following two scenarios:

Scenario 1:	Scenario 2:
Pressure altitude = 6 000 ft	Pressure altitude = 6 000 ft
Load 90%	Load 60%
ISA	ISA

13. Using the Cruise Performance chart for 4 407 lb in Section 5 of the DA50 AFM, now calculate the SR_{max} (lb/nm) of the DA50 based on the same two scenarios as the previous problem, except change the pressure altitude to 16 000 ft.

14. In which of the two previous problems, and the two scenarios, will the SR_{max} be most desirable (assuming no other environmental impacts and a constant weight of 4 407 lb)? Based on the chapter reading, what should happen to SR_{max} as weight decreases with fuel consumption?

8 Takeoff Performance

CHAPTER OBJECTIVES

After completing this chapter, you should be able to:

- Identify the considerations that should be made before any takeoff, and summarize the factors that are included in the calculation of takeoff performance.
- Describe the procedures used for crosswind takeoff, water takeoff, and performance takeoffs.
- Identify and characterize the performance speeds associated with a normal takeoff.
- Explain the dangers involved with improper takeoffs and illustrate how an improper takeoff may develop.
- Describe when a rejected takeoff (RTO) may occur, and summarize the performance speeds and definitions associated with the calculation of RTO performance data.
- Correlate the linear motion equations presented in Chapter 1 with those used for takeoff calculations.
- Examine the factors that affect takeoff performance, and calculate the change in takeoff distance as the factors change.

During takeoff, the aircraft starts from zero velocity, accelerates to takeoff velocity, and becomes airborne. The relationships among acceleration, velocity, takeoff distance, and time are not familiar to the average pilot and should be understood as well as experienced. This chapter discusses the factors involved in attaining takeoff velocity.

The following considerations are very important factors in evaluating takeoff performance for a given aircraft, location, mission, and pilot technique:

- Takeoff velocity, which could be a function of stall speed, minimum control speed, or the thrust (or power) available, thrust (or power) required relationship.

Flight Theory and Aerodynamics: A Practical Guide for Operational Safety, Fifth Edition.
Brian A. Johnson and Philip R. Fittante.

Companion website: www.wiley.com/go/flighttheoryandaerodynamics5e

- The acceleration during takeoff (we will assume constant acceleration).
- The distance required to complete the takeoff.
- Obstacle clearance after takeoff.
- Regulatory guidance and/or company operation specifications.

There are many factors that affect the calculation of takeoff performance:

1. Aircraft gross weight
2. Thrust
3. Temperature
4. Pressure altitude
5. Wind direction and velocity
6. Runway slope
7. Runway surface

The manufacturer of the aircraft publishes operating specifications, as well as performance charts and tables for the pilot(s) to follow. As the seven factors above change, and one or more of them will be on every flight, pilot(s) must adjust accordingly to remain inside the operating envelope of the aircraft. An important reminder is warranted here before reviewing any performance or weight and balance chart: **Before using any chart, table, or graph, ALWAYS review the Notes and Conditions as provided by the manufacturer.**

Figure 8.1 is a typical takeoff distance graph used to calculate takeoff ground roll and takeoff distance over a 50 ft obstacle for a small, retractable gear aircraft.

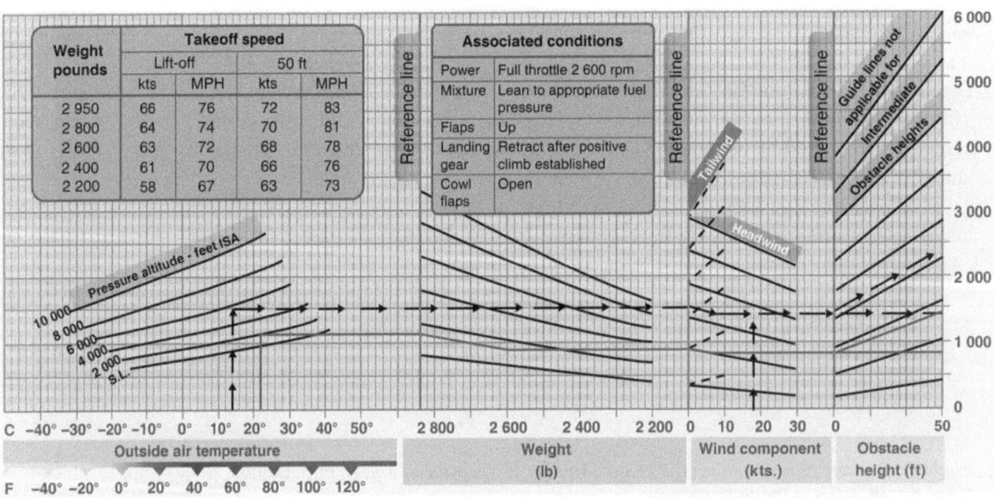

Figure 8.1. Takeoff distance graph.
Source: Pilot's Handbook of Aeronautical Knowledge (2023c)/U.S. Department of Transportation Federal Aviation Administration/Public domain.

EXAMPLE

Using Figure 8.1, calculate the takeoff ground roll and takeoff distance over a 50 ft obstacle with the given information: pressure altitude = 2 000 ft, outside air temperature (OAT) = 22 °C, takeoff weight = 2 600 lb, headwind component = 6 kts. Starting at 22 °C, follow the solid vertical line up to 2 000 ft pressure altitude, then follow the solid horizontal line to the first reference line. Then follow the solid line downward, paralleling the provided diagonal reference lines until the vertical line which represents 2 600 lbs is reached.

Continue horizontally until the reference line for the wind component, using the same technique for the weight, parallel the diagonal reference lines until the vertical line representing 6 kts. is reached, then move horizontally to the final reference line.

Finally, note the split of the solid line after the final vertical reference line. Move horizontally to the final vertical axis for the ground roll (~700 ft), and parallel the closest diagonal reference line to calculate the total takeoff distance over a 50 ft obstacle (~1 400 ft).

Figure 8.2 is another format of a typical takeoff distance chart used to calculate takeoff distance for a small, retractable gear aircraft. Due to the format of this chart in contrast to Figure 8.1, interpolation is required to calculate the takeoff distance. The example of how to use Figure 8.2 is as follows, to calculate takeoff distance use the following given information: pressure altitude of 2 500 ft and an outside air temperature (OAT) of 20 °C. Using the 20 °C column, interpolate the ground roll of the aircraft at 2 500 ft as indicated, which results in a ground roll distance of 1 173 ft.

Application 8.1

Using Figure 8.1, calculate the takeoff ground roll and takeoff distance over a 50 ft obstacle using the same criteria provided in the example, except reduce the OAT by 10 °C. Then, using the same criteria as in the example, calculate both takeoff performance values, except change the 6 kts. headwind into a 6 kts. tailwind, as if the pilot selected the incorrect runway to takeoff.

With the decrease in temperature, what happened to the takeoff ground roll and takeoff distance over a 50 ft obstacle? Why? What happened to the takeoff performance values when the pilot selected the incorrect runway and the 6 kts. of headwind became 6 kts. of tailwind?

Conditions: Flaps 10°, Full throttle prior to brake release, Paved level runway, Zero wind				Takeoff distance Maximum weight 2 400 LB									
Weight (lb)	Takeoff speed KIAS		Press ALT (ft)	0 °C		10 °C		20 °C		30 °C		40 °C	
	Lift off	AT 50 ft		Grnd roll (ft)	Total feet to clear 50 ft OBS	Grnd roll (ft)	Total feet to clear 50 ft OBS	Grnd roll (ft)	Total feet to clear 50 ft OBS	Grnd roll (ft)	Total feet to clear 50 ft OBS	Grnd roll (ft)	Total feet to clear 50 ft OBS
2 400	51	56	S.L.	795	1 460	860	1 570	925	1 685	995	1 810	1 065	1 945
			1 000	875	1 605	940	1 725	1 015	1 860	1 090	2 000	1 170	2 155
			2 000	960	1 770	1 035	1 910	1 115	2 060	1 200	2 220	1 290	2 395
			3 000	1 055	1 960	1 140	2 120	1 230	2 295	1 325	2 480	1 425	2 685
			4 000	1 165	2 185	1 260	2 365	1 355	2 570	1 465	2 790	1 575	3 030
			5 000	1 285	2 445	1 390	2 660	1 500	2 895	1 620	3 160	1 745	3 455
			6 000	1 425	2 755	1 540	3 015	1 665	3 300	1 800	3 620	1 940	3 990
			7 000	1 580	3 140	1 710	3 450	1 850	3 805	2 000	4 220	---	---
			8 000	1 755	3 615	1 905	4 015	2 060	4 480	---	---	---	---

To find the takeoff distance for a pressure altitude of 2 500 ft at 20 °C, average the ground roll for 2 000 ft and 3 000 ft.

$$\frac{1115 + 1230}{2} = 1173 \text{ ft}$$

Figure 8.2. Using interpolation to calculate takeoff distance.
Source: Pilot's Handbook of Aeronautical Knowledge (2023c)/U.S. Department of Transportation Federal Aviation Administration/Public domain.

NORMAL TAKEOFF

Proper takeoff roll and liftoff are essential to achieving the takeoff performance numbers calculated before departure. During a normal takeoff, the airplane uses the runway that offers the greatest headwind component, with comfortable safety margins for takeoff distance and obstacle clearance during the initial climb. The runway has been inspected for debris and for obstructions such as animals or other aircraft. As power is applied, the rudder (usually the right rudder in a single-engine airplane) is used to keep the aircraft on the runway centerline as the takeoff roll begins. In a crosswind, the ailerons are used to assist the rudders in maintaining the centerline and to keep the upwind wing from rising. The airplane is allowed to accelerate normally to the planned rotation speed and lifts off without over-rotating, finally establishing a climb at V_X or V_Y.

Figure 8.3 demonstrates a typical takeoff into the wind and a climb to a safe altitude. In Position 1, the airplane accelerates down the runway until proper rotation speed is attained. In Position 2, the pilot deflects the elevator upward by pulling back on the yoke/stick, giving the horizontal stabilizer an increased angle of attack (AOA), producing a tail-down force (because the lift from the horizontal stabilizer/elevator is downward), thus raising the nose. In Position 3, the increased AOA of the wing produces an increase in lift, and this unbalanced vertical force produces the vertical acceleration needed for liftoff. After liftoff, if the rate of climb is constant, the vertical forces are once again balanced. As the velocity is increased to

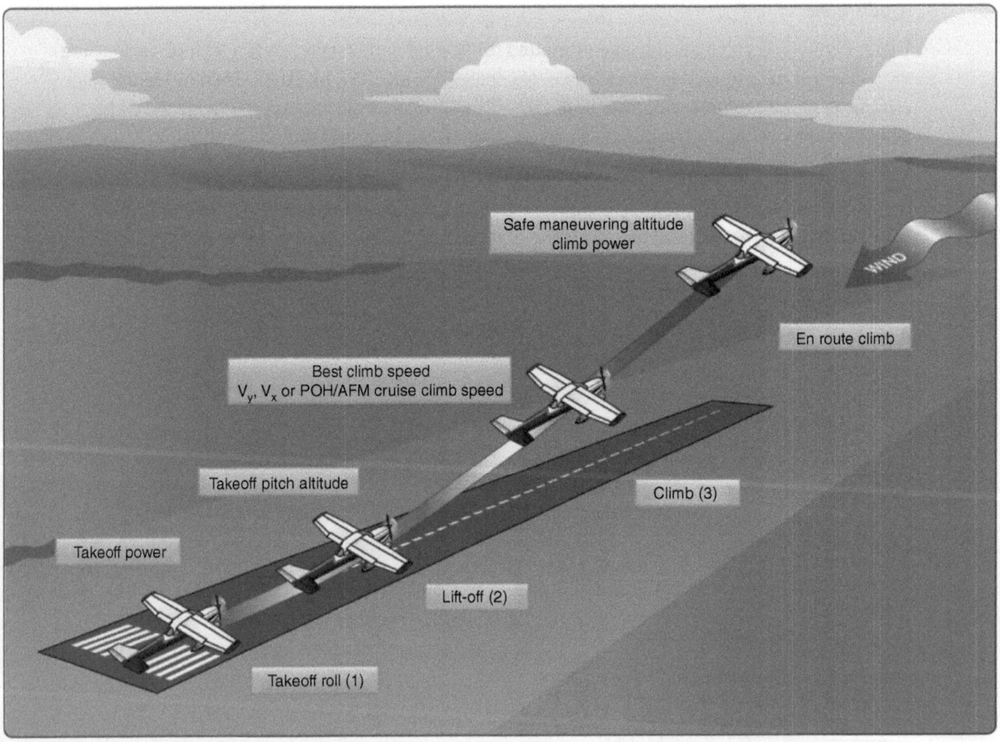

Figure 8.3. Normal takeoff and climb.
Source: Airplane Flying Handbook (2016a)/U.S. Department of Transportation Federal Aviation Administration/Public domain.

V_X or V_Y, the airspeed increases, and as we calculated in Chapter 4, as velocity increases so does lift, for a given AOA.

Crosswind Takeoff

Though a direct headwind is always preferred for takeoff, a headwind component without a crosswind component is rarely experienced. During training in single-engine propeller aircraft, pilots should practice takeoffs with varying wind velocities and crosswind components to better understand the wind effect on the aircraft during takeoff roll. Once the takeoff roll begins, the pilot should position the yoke/stick as if banking into the wind. A downward force is created on the upwind wing, keeping the wing from lifting due to the effect of the crosswind. Directional control should be maintained with a rudder, to keep the aircraft on the runway centerline. Depending on the aircraft type and rpm setting, the pilot may have to correct for the left-turning tendency of the aircraft, which will require additional right rudder that produces a larger AOA and resulting side force on the left side of the vertical stabilizer.

As speed increases, the aerodynamic forces acting on the ailerons also increase, making them more effective. Thus, aileron pressure should be gradually reduced as speed increases. Depending on the strength of the crosswind component and on the speed of the aircraft, partial windward aileron deflection may need to be continued throughout the takeoff roll and liftoff in order to keep the upwind wing from rising. If the pilot attempts a crosswind takeoff with insufficient windward aileron deflection, the aircraft may begin "skipping" toward the downwind edge of the runway (Figure 8.4). Not only is this poor technique, but depending on the strength of the crosswind component, the aircraft could exit the side of the runway or

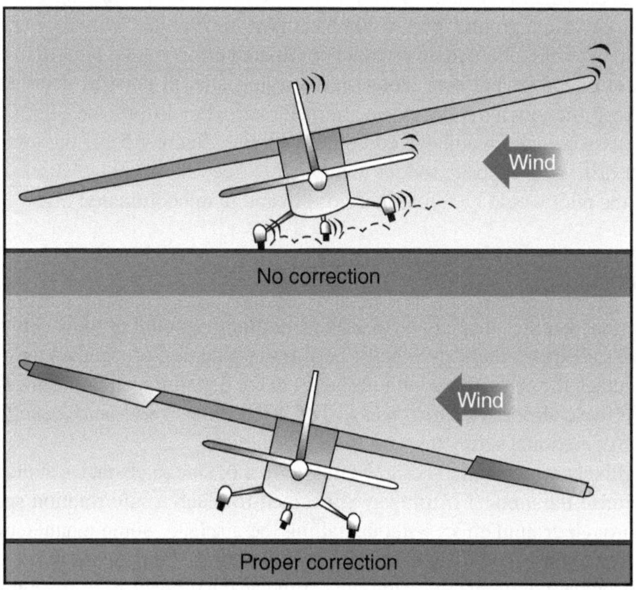

Figure 8.4. Crosswind takeoff.
Source: Airplane Flying Handbook (2004a)/U.S. Department of Transportation Federal Aviation Administration/Public domain.

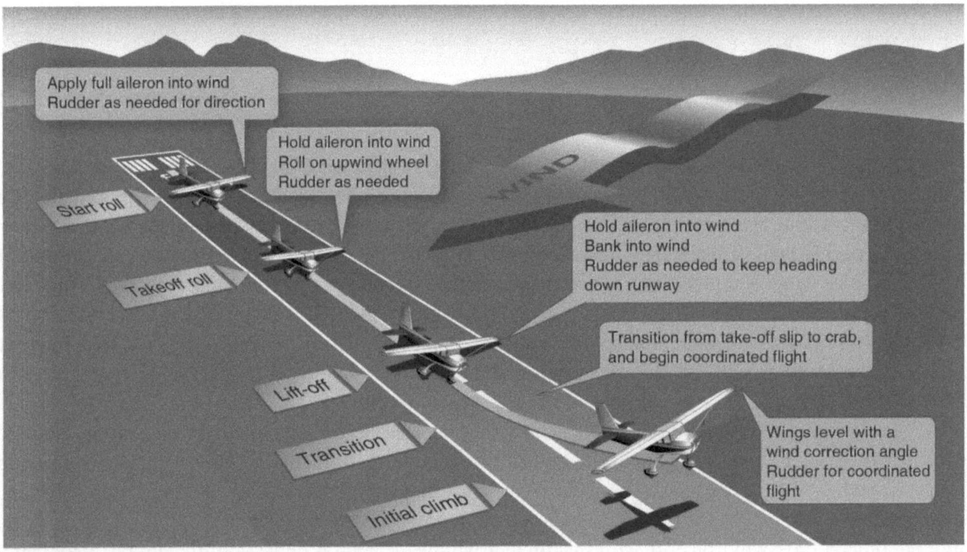

Figure 8.5. Coordinated use of flight controls during a crosswind takeoff and climb.
Source: Airplane Flying Handbook (2021)/U.S. Department of Transportation Federal Aviation Administration/Public domain.

experience structural damage to the landing gear, or even collapse of the landing gear, from the sideward force. In a tailwheel aircraft, a ground loop would be almost inevitable.

Proper crosswind takeoff technique involves a coordinated effort by the pilot to maintain runway centerline by proper use of rudder pressure along with varying aileron pressure from takeoff roll through liftoff in order to keep the upwind wing down. Immediately after liftoff, the pilot transitions to a crab to counteract the crosswind, and maintains coordinated flight (Figure 8.5). If the ailerons are positioned to prevent left/right drift, and opposite rudder is applied to keep the fuselage aligned with the extended runway centerline, the pilot would be flying in a sideslip and in uncoordinated flight.

Performance Takeoffs

While almost all normal takeoffs are planned to use the minimum amount of runway needed to reach a safe rotation speed above stall speed, there are specific performance takeoff procedures for operating at airports with shorter than average runways, those with obstacles in the departure path, or runways with suboptimal runway surfaces. At these airports, a *short-field takeoff* procedure or *soft-field takeoff* procedure may be required to meet the operational safety margins during departure.

A basic short-field takeoff and climb is shown in Figure 8.6. Due to obstacles at the end of the runway, the pilot must minimize the amount of runway consumed to reach a safe rotation speed and then must initiate the climb through ground effect without stalling the airplane. Some manufacturers recommend a takeoff flap setting for a short-field takeoff; as we have discussed, the increase in the C_L with the first or second flap setting provides additional lift to minimize time on the runway and outweighs the small increase in induced drag (and slight parasite drag) that is produced. The initial V-speed during climb-out will be V_X; as discussed in Chapter 7, this speed produces the greatest gain in altitude in the smallest horizontal distance. Climbing at an indicated air speed (IAS) of V_X gives the pilot the best chance of clearing an

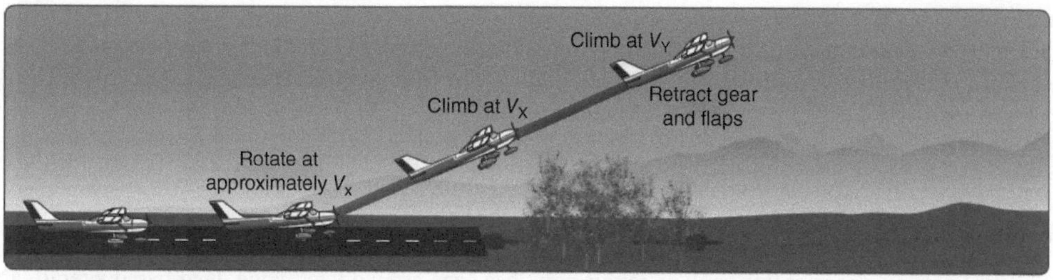

Figure 8.6. Short-field takeoff.
Source: Airplane Flying Handbook (2016a)/U.S. Department of Transportation Federal Aviation Administration/Public domain.

obstacle off the departure end of the runway. As also discussed in Chapter 7, if a climb is attempted at a speed below V_X the aircraft may stall, and if a speed above V_X is maintained then the aircraft may not clear the obstacle as it is not climbing at the best AOA. A headwind will further increase the amount of altitude gained over a given horizontal distance, while a tailwind will degrade our calculated obstacle clearance since ground speed will increase if the same IAS is maintained. Once the obstacle has been cleared, the AOA is decreased by lowering the aircraft's nose, and the airspeed is increased until either V_Y or a cruise climb airspeed is attained.

In a soft-field (or rough field) takeoff, as with a short takeoff, the pilot's goal is to minimize the takeoff roll and subsequent time with wheels on the ground. However, runway conditions require different techniques in a soft-field takeoff as opposed to a short-field takeoff. In order to minimize the takeoff roll in a soft-field takeoff, the pilot needs to use lift to offset the weight of the aircraft even *before* liftoff, to reduce the friction from a rough, grassy, sandy, or muddy runway. Figure 8.7 shows the basic technique to accomplish a proper soft-field takeoff to minimize the takeoff roll. The goal during the takeoff roll in a soft-field takeoff is to transfer the weight of the aircraft from the wheels to the wings as soon as safely practical. The takeoff roll usually starts with the yoke/stick in the aft position to minimize nosewheel pressure on the ground in a tricycle-gear airplane. As speed increases during the takeoff roll, full back pressure results in an excessive AOA, which if continued would keep the wing in a stalled condition, preventing liftoff. To prevent this condition, and to take advantage of the ground effect discussed in Chapter 5, the AOA must be reduced as speed increases during the takeoff roll, to maximize lift. This lift reduces the friction of the runway on the tires, which minimizes the takeoff roll by allowing the aircraft to accelerate to flying speed in the shortest distance. After liftoff, the procedure is the same as for a short-field takeoff: reduce pitch to

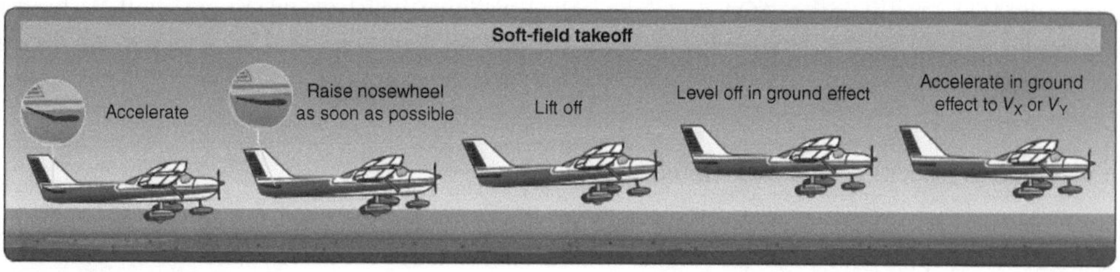

Figure 8.7. Soft-field takeoff.
Source: Airplane Flying Handbook (2016a)/U.S. Department of Transportation Federal Aviation Administration/Public domain.

accelerate to V_X in ground effect, then increase pitch to climb out of ground effect at V_X till clear of the obstacle, then reduce pitch to climb at V_Y or a cruise-climb airspeed. In Figure 8.7, note the position of the elevator to reduce tail-down force as the aircraft levels off in ground effect. Once the required airspeed is reached, usually V_X or V_Y, back pressure is once again applied to the yoke, and the AOA is increased to establish an aircraft climb out-of-ground effect.

Water Takeoffs

While the primary focus of this book up to this point has been around fixed-wing, land-based aircraft, an introduction to fixed-wing seaplane aerodynamics during takeoff allows for an introduction to the unique flight dynamics experienced when taking off from water. A full discussion involving all the environmental and man-made conditions that could be experienced when taking off from water is not warranted here, and many sources are available to expand beyond this discussion.

A seaplane on floats initiating a takeoff from water experiences many of the same forces discussed in Chapters 1–7, though these forces, such as lift and drag, occur in different areas during aircraft acceleration and must be well understood to avoid an accident. Because the floats are on the water, the amount of weight results in varying displacement in the water: the heavier the aircraft, the farther down in the water the floats are sitting (wetted area) as more water must be displaced to carry the weight. As we experienced during a soft-field takeoff from land, the yoke/control stick must also be held back during taxi and when initiating takeoff to provide a continual tail-down force (resulting in aircraft nose-up).

As a seaplane gains speed during the initial application of power, the movement of the floats through the water produces a large amount of drag. The floats are designed to be vertically narrower in the rear, so as they move horizontally through the water, the water is deflected downward which results in an upward lifting force; as discussed in Chapter 1, this is Newton's third law of motion in action. This upward force is known as *hydrodynamic lift*.

As the velocity continues to increase, the hydrodynamic lift, combined with the tail-down force due to the pilot's back pressure on the yoke/control stick, raises the front of the floats and as a result places the rear of the floats farther down in the water. This increase in drag can result in increased takeoff distance, minimizing the time spent in this position until moving into the next phase of takeoff. The area of greatest resistance is referred to as the hump and must be overcome to move into the next phase. Figure 8.8 shows how propeller thrust decreases with increasing airspeed, as discussed in Chapter 7.

Once the airspeed increases and hydrodynamic lift increases enough to raise the rear of the floats out of the water, the seaplane is *on the step* and finally drag is minimized enough that acceleration to takeoff speed is possible. Figure 8.9 shows a seaplane on the step. The aft portions of the floats are out of the water and most of the aircraft's weight is supported by hydrodynamic lift and aerodynamic lift, and not by the buoyancy of the floats. Back pressure on the yoke/control stick is reduced at this point to avoid placing the AOA near the critical AOA, creating a stalled condition, which would prevent liftoff. As further acceleration takes place, the aerodynamic lift takes over and the flight controls react as they would during the takeoff from land, eventually leading to the seaplane becoming airborne.

Performance Speeds – Normal Takeoff

There are many performance *V-speeds* that are important to a pilot during a normal takeoff and climb. Using these speeds allows the pilot to operate the aircraft as recommended by the manufacturer while maintaining regulatory compliance. The definitions below are not a complete listing of all the speeds/definitions that

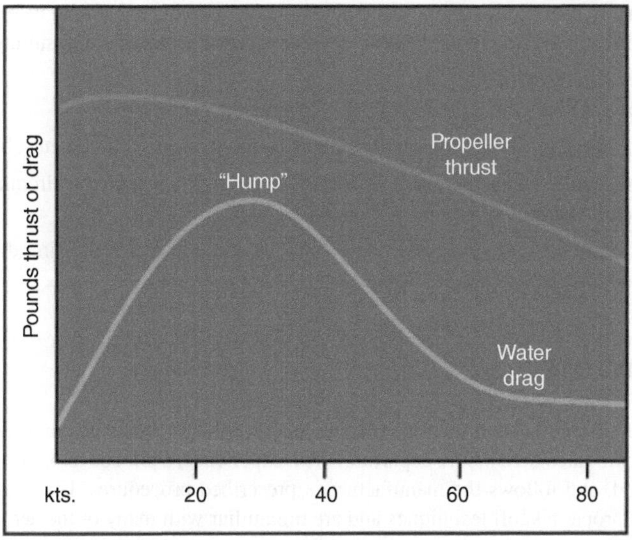

Figure 8.8. Water drag and propeller thrust on takeoff.
Source: Seaplane, Skiplane, and Float/Ski Equipped Helicopter Operations (2004b)/U.S. Department of Transportation Federal Aviation Administration/Public domain.

Figure 8.9. Hydrodynamic lift while on the step.
Source: Seaplane, *Skiplane, and Float/Ski Equipped Helicopter Operations* (2004b)/U.S. Department of Transportation Federal Aviation Administration/Public domain.

may be encountered but included are the more common terms used by most smaller fixed-wing aircraft during normal takeoff planning.

V_{FE} **– flaps extended speed:** Maximum speed at which the wing flaps can be in a given extended position due to air loads per the aircraft manufacturer.

V_{LE} **– gear extended speed:** Maximum speed at which the aircraft should fly with the retractable landing gear extended and locked.

V_{LO} – gear operating speed: Maximum speed at which the landing gear should be extended or retracted and involves the actual moving of the landing gear up or down during flight, resulting in air loads on the operating mechanism; usually less than V_{LE}.

V_{LOF} – liftoff: The speed at which the aircraft first departs the runway.

V_R – rotation: The speed at which the aircraft rotates to initiate takeoff attitude, may be the same as V_{LOF}.

V_X – best angle of climb: As we discussed previously, the speed at which the aircraft will gain the most altitude in a given *distance*.

V_Y – best rate of climb: As we discussed previously, the speed at which the aircraft will gain the most altitude in a given *time*.

IMPROPER LIFTOFF

As discussed in the Normal Takeoff section, a proper takeoff roll and liftoff are paramount to achieving the performance numbers calculated before departure. With experience, a pilot develops a smooth, coordinated takeoff, on airspeed, and follows the manufacturer's prescribed procedures. Unfortunately, many pilots have not practiced proper takeoff techniques and are unfamiliar with many of the aerodynamic principles discussed. Forcing the aircraft into premature liftoff before the airplane is ready to fly outside of ground effect can lead to longer takeoff distances and climb gradients that cannot be met. High pitch angles lead to increased rates of induced drag, and when combined with the effect of increased pitch angle when climbing out of ground effect the consequences can be deadly.

Premature Takeoff in Ground Effect

In Chapter 5, we discussed the effects of ground effect on landing and takeoff, and the characteristics that the aircraft exhibited when entering and departing ground effect. Ground effect will also allow an aircraft to depart the runway and remain in the air before "normal" flying speed has been reached due to the reduction in induced drag, as long as the aircraft remains one wingspan or less above the ground. If the pilot continues to pitch up with an increasing AOA, drag is increased and the aircraft may settle back to the runway, greatly increasing takeoff distance and possibly preventing stopping on the remaining runway (Figure 8.10). Training maneuvers such as soft-field takeoff procedures maximize the benefit of ground effect while allowing the pilot to gain the feel of an aircraft at minimum flying speed within one wingspan of the ground.

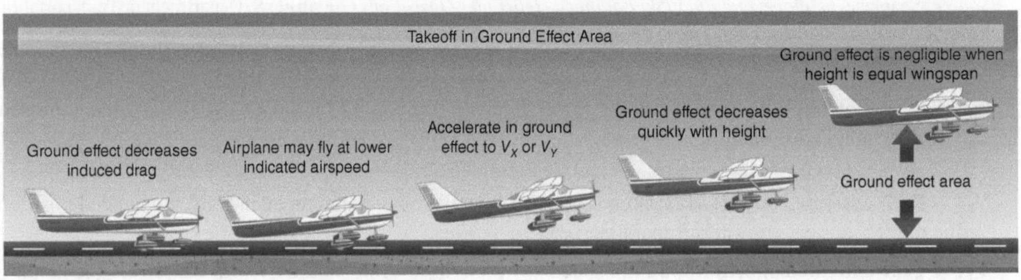

Figure 8.10. Premature takeoff due to ground effect.
Source: Airplane Flying Handbook (2021)/U.S. Department of Transportation Federal Aviation Administration/Public domain.

Dangers of Over-Rotation

The dangers of over-rotation during takeoff are twofold. First, high AOA results in high induced drag and possible loss of lift (if $C_{L(max)}$ is exceeded). Second, once the aircraft is airborne, it may be impossible to climb out of ground effect. When the aircraft departs ground effect, the downwash is unhampered by the ground and is noticeably increased, causing a large increase in induced drag. The increased downwash may envelop the horizontal tail and cause a pitch-up moment, which adds to the high AOA condition.

A widely publicized example of an accident initiated by over-rotation occurred in 1956 at Edwards Air Force Base when an F-100 fighter attempted an emergency landing. This accident has been nicknamed the "Sabre Dance." It is a classic example of the danger of over-rotation. The aircraft had a misaligned nosewheel and the pilot was attempting an emergency landing. The runway was partially foamed to lessen the chance of fire. The pilot thought he was over-running the foamed portion and decided to go around for another approach. To prevent the damaged nosewheel from contacting the runway, the pilot purposely held the nose high as he added a throttle for the go-around. The airplane remained airborne but could not accelerate or fly away from the runway (out of ground effect) because of the high induced drag. The nose pitched up further as the center of pressure shifted forward and the wings rocked due to roll-yaw coupling. Several thousand feet down the runway, the aircraft lost directional control and crashed.

REJECTED TAKEOFFS

Any takeoff may need to be rejected (formerly referred to as "aborted"); so, each takeoff should include the proper planning, preflight briefing, and expectation that an abnormal situation will occur during the takeoff roll. A good mindset for a pilot to develop is, "Every takeoff is an emergency."

Reasons for a rejected takeoff (RTO) range from the extreme, such as an engine failure or a runway incursion by another aircraft, to more benign reasons as an engine abnormality, bird strike, or air traffic control (ATC) directive. Whether a single-engine or multi-engine aircraft, performance speeds and published performance charts by the manufacturer should be followed to achieve operational safety margins and to avoid violating regulations. As presented in Chapter 1, as the kinetic energy (KE) of the aircraft increases, more runway distance will be consumed during an RTO and the stopping of the aircraft on the runway. Because KE varies as the square of the velocity but only the first power of the mass, an increase in ground speed will have a much greater effect on stopping distance than an increase in weight. A 10% increase in takeoff weight produces roughly a 10% increase in KE, while a 10% increase in speed results in a 21% increase in KE.

Many factors affect stopping distance, and a detailed discussion is beyond the scope of this textbook. Factors such as runway friction, runway contaminants, density altitude, flap configuration, aircraft tire condition, inoperative aircraft equipment, and pilot technique all influence whether the performance data calculated before takeoff can actually be achieved in the event of an RTO. Pilots should prepare for "low-speed" and "high-speed" regimes and should decide before takeoff for each regime what necessitates conducting an RTO or continuing with the takeoff and handling the abnormality in the air.

RTO V-Speeds and Definitions

As discussed earlier in this chapter, there are many performance *V-speeds* that are important to a pilot during takeoff to operate the aircraft safely and as recommended by the manufacturer, and to avoid violating regulations. The definitions and V-speeds below can be found in Title 14 of the Code of Federal Regulations (CFR) parts 25, 91, 121, and 135. These federal aviation regulations (FAR) govern aircraft operations, whether at a small, nontowered airport or the largest part 139 regulated airports. Many of

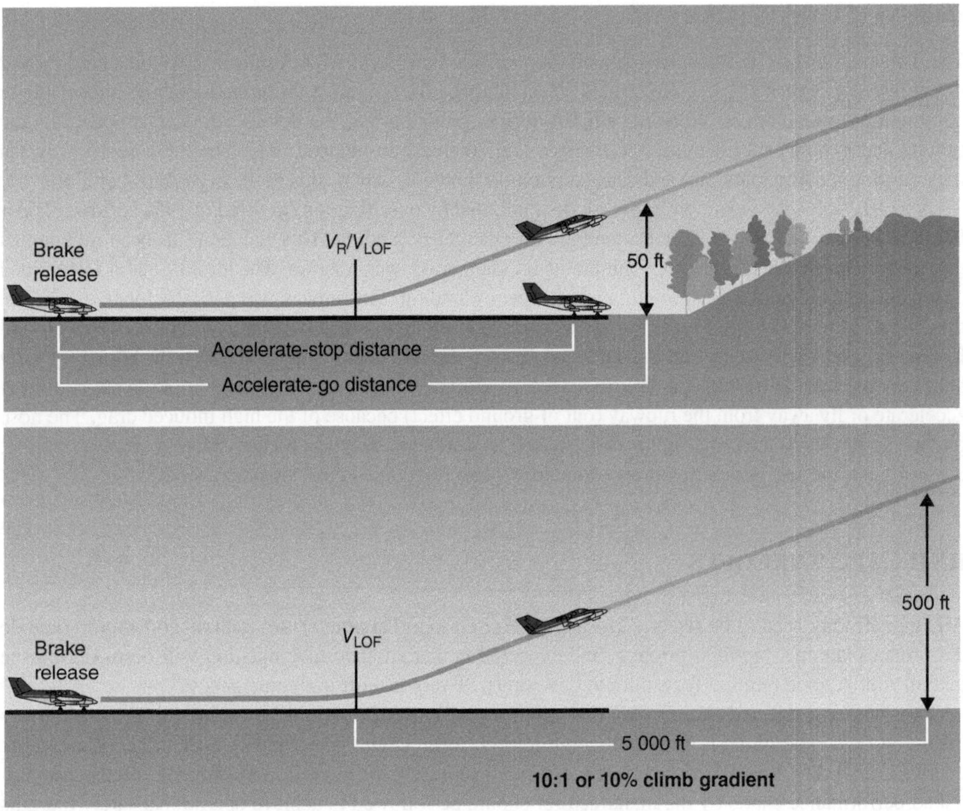

Figure 8.11. Accelerate-stop distance, accelerate-go distance, and climb gradient.
Source: Airplane Flying Handbook (2004a)/U.S. Department of Transportation Federal Aviation Administration/Public domain.

the following definitions and V-speeds were acquired from the Federal Aviation Administration's (FAA's) Federal Aviation Regulations/Aeronautical Information Manual (FAR/AIM), the FAA's Advisory Circular (AC) 120–62 *Takeoff Safety Training Aid*, or the FAA's *Airplane Flying Handbook*, and these documents should be consulted for more detailed information:

Accelerate-go distance: The distance required to accelerate to a specified speed and takeoff and climb to 50 ft, assuming an engine failure at V_R or V_{LOF}.

Accelerate-stop distance: The distance required to accelerate to a specified speed, V_R or V_{LOF}, experience an engine failure, and bring the aircraft to a stop (Figure 8.11). A more expanded definition can be found at 14 CFR part 25, §25.109.

Balanced field length (BFL): The runway length (or runway plus clearway and/or stopway) where, for the takeoff weight, the engine out accelerate-go distance equals the accelerate-stop distance (Figure 8.12). For a safe takeoff, the BFL must be no greater than the runway available.

Critical field length (CFL): The minimum runway length required for a specific takeoff weight. This distance may be the longer of the balanced field length (BFL) or 115% of the all-engine takeoff distance. The critical field length (CFL) should not be greater than the runway available.

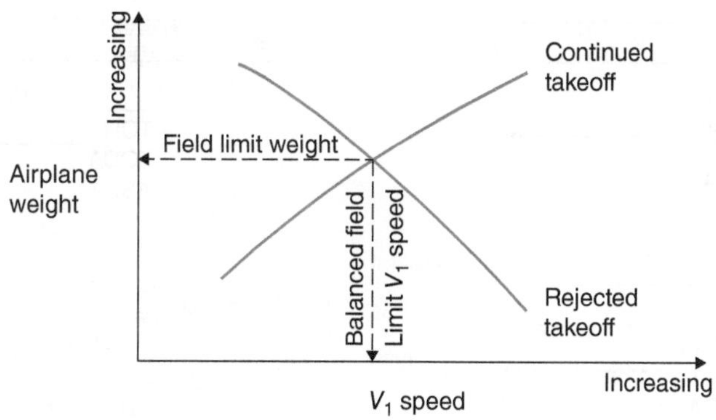

Figure 8.12. Balanced field length.
Source: AC 120–62 – Takeoff Training Aid (1994)/U.S. Department of Transportation Federal Aviation Administration/Public domain.

Takeoff distance: The distance required to complete an all-engines operative takeoff to a 35 ft height, and should be at least 15% less than the distance required for a one-engine inoperative takeoff distance. A more expanded definition can be found at 14 CFR part 25, §25.113.

V_{EF} **– critical engine failure speed:** The speed at which the critical engine is assumed to fail during takeoff.

V_{MBE} **– maximum braking energy speed:** The highest speed from which the airplane may be brought to a stop without exceeding the maximum energy absorption capacity of the brakes.

V_{MCG} **– ground minimum control speed:** The minimum airspeed at which the airplane, while on the ground, can lose an outboard engine and maintain directional control. V_{MCG} does not apply to single-engine aircraft.

V_1 **– takeoff decision speed:** The maximum speed in the takeoff at which the pilot must take the first action (e.g. apply brakes, reduce thrust, and deploy speed brakes) to stop the airplane within the accelerate-stop distance. V_1 also means the minimum speed in the takeoff run at which, following a failure of the critical engine at V_{EF}, the pilot can continue the takeoff and achieve the required height above the takeoff surface within the takeoff distance. The time interval between V_{EF} and V_1 is the longer of the flight demonstration time or one second. **Due to delayed pilot reaction time, the Go/No-Go decision needs to be made before V_1**, to allow for the RTO procedures to be initiated at V_1 to meet the published performance for the aircraft. An expanded discussion is located later in this chapter in the RTO Aerodynamics section.

V_2 **– takeoff safety speed:** The speed to which the airplane must be accelerated at or before reaching 35 ft above the end of the runway. Used primarily for transport category aircraft during instrument departures after experiencing failure of an engine during takeoff, and when a desired climb gradient must be met.

***Do not confuse V_1 and V_2 as defined here, with V_1 and V_2 as we have used many times within formulas in this textbook to describe an initial velocity (V_1) and subsequent velocity (V_2) after a specified condition.**

V_{YSE} **– best rate of climb single engine:** with one engine inoperative, the speed at which the aircraft will gain the most altitude in a given time. Marked with a blue line on most airspeed indicators. Discussed in more detail in Chapter 13.

Transport category aircraft fall under parts 25 and 29 of Title 14 CFR, and must display high levels of performance with built-in safety margins. In addition to adherence to many of the V-speeds mentioned

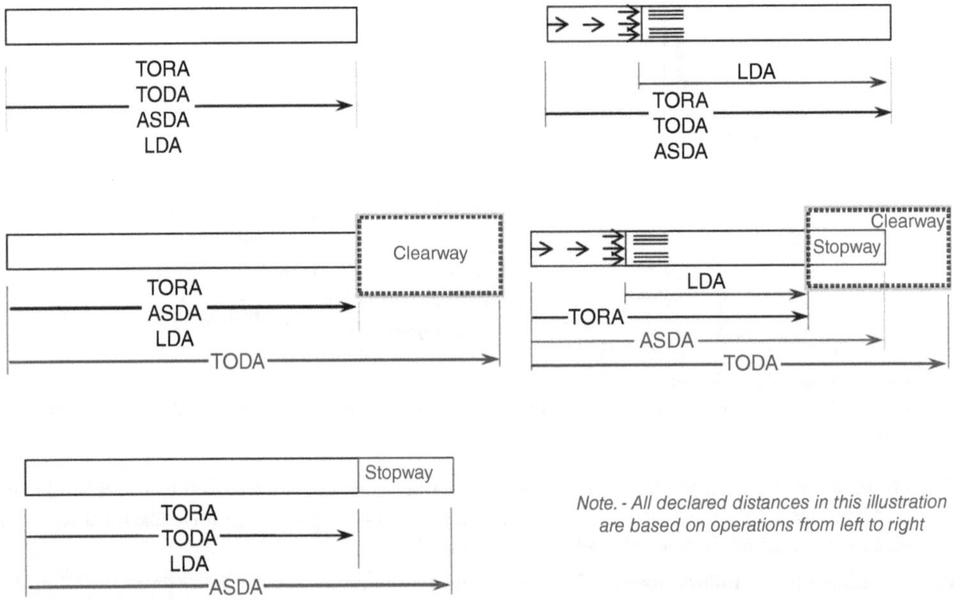

Figure 8.13. FAA declared distances.
Source: Aeronautical Information Manual – AIM (2024)/U.S. Department of Transportation Federal Aviation Administration/Public domain.

above, additional performance criteria must be calculated and followed during all phases of flight in transport category aircraft.

Declared Distances

Per the FAA FAR/AIM, declared distances for a runway represent the maximum distances available and are suitable for meeting takeoff and landing distance performance requirements. We have already introduced *accelerate-stop distance available* (ASDA) and accelerate-go distance available, and as we expand into single and multi-engine takeoff considerations, we must understand the forces at work and why there are limitations set by the FAA (Figure 8.13). The *takeoff run available* (TORA) is the runway length declared available and suitable for the ground run of an airplane taking off. The *takeoff distance available* (TODA) is the TORA plus the length of any remaining runway or clearway beyond the far end of the TORA (Figure 8.14). The landing distance available (LDA) will be discussed in Chapter 9. Stopway and clearway definitions are beyond the scope of this textbook but are instrumental to maintaining safe operations when flying high-performance aircraft.

Single-Engine Airplane

A graphic representation of the refused takeoff situation for a single-engine airplane is shown in Figure 8.15. At the brake release point, the aircraft begins to accelerate, and the increase in speed during the takeoff run is shown on the takeoff acceleration profile. The deceleration profile shows the decrease in speed as the airplane is braked to a stop at the end of the runway; this is the accelerate-stop distance discussed previously, applied in this case to a single-engine airplane. The refusal speed is determined

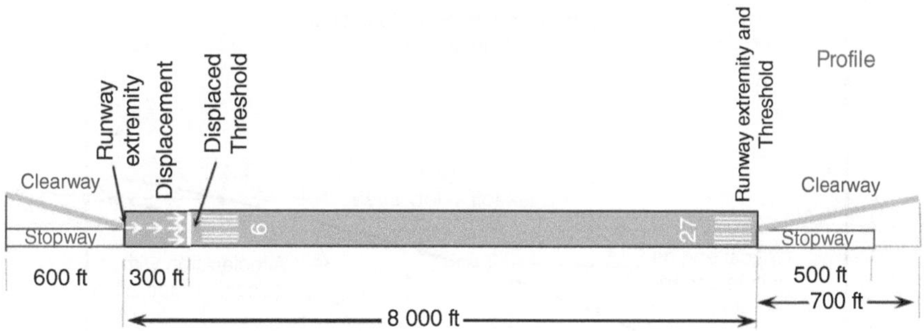

Runway	Length (feet)	TORA	ASDA	TODA	LDA
9	8 000	8 000	8 500	8 700	7 700
27		8 000	8 600	8 600	8 000

Figure 8.14. Runway application of declared distances.
Source: Aeronautical Information Manual – AIM (2024)/U.S. Department of Transportation Federal Aviation Administration/Public domain.

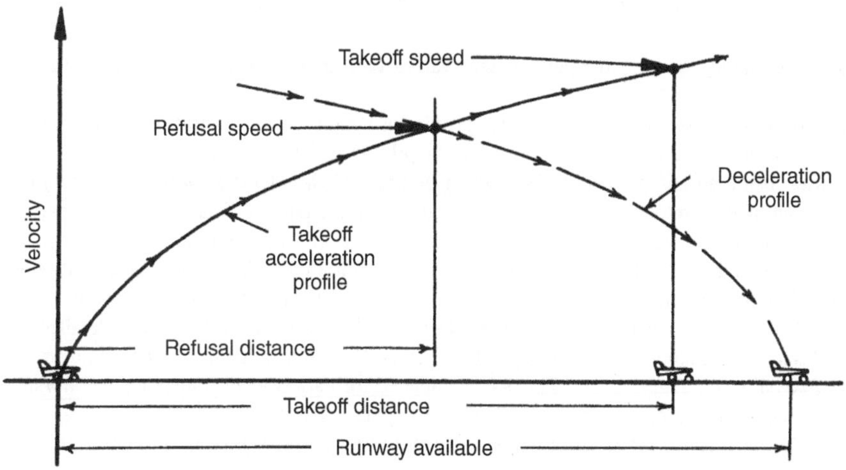

Figure 8.15. Single-engine velocity–distance profiles.

by the intersection of the acceleration and deceleration profiles. Once the refusal speed is exceeded, the aircraft cannot be brought to a stop on the runway remaining. If a single-engine airplane loses its engine above refusal speed, it will end up in the overrun.

Multi-engine Airplane

With the addition of multiple engines on an aircraft, a more detailed look at the acceleration and deceleration profiles during takeoff is warranted. Depending on whether the multi-engine aircraft is a light, twin-engine aircraft at maximum takeoff weight (MTOW), or a four-engine air transport category aircraft

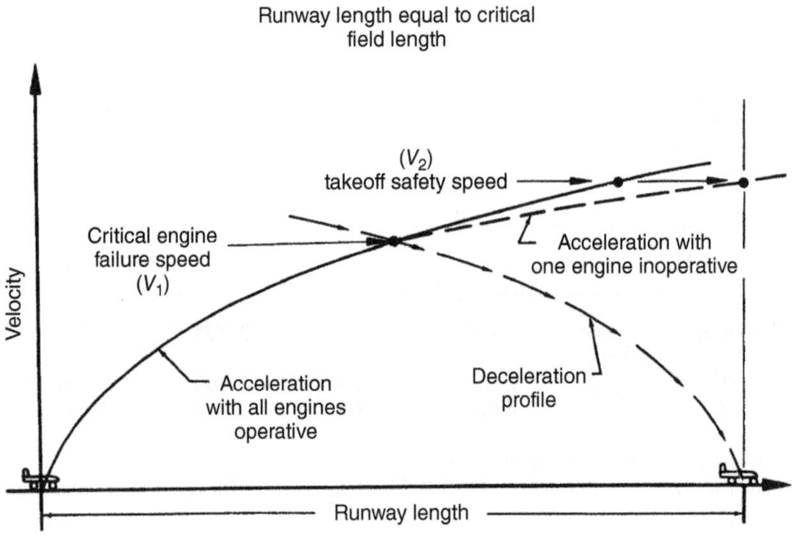

Figure 8.16. Multi-engine velocity–distance profiles.

with expansive safety margins, the performance values vary greatly, as do the options available to assist with an RTO.

A generic velocity versus runway distance profile for a multi-engine aircraft is shown in Figure 8.16. Note the position of V_1 and V_2. At or above V_1, the pilot of the multi-engine airplane must continue the takeoff after losing an engine. As can be seen in Figure 8.16, acceleration will be less with the loss of an engine, but takeoff velocity, V_2, will be the same. Thus, the takeoff distance will be greater. In some cases, the remaining runway may not be long enough to allow the airplane to accelerate to V_2. Thus, BFL becomes important because it assures that this will not happen. Figure 8.16 shows the condition where the actual field length is exactly equal to BFL and V_2 is reached at the exact end of the runway.

Figure 8.16 illustrates the performance items discussed thus far for an RTO. A more detailed discussion of takeoff safety, including but going beyond the items discussed thus far, can be found in FAA's AC 120–62 *Takeoff Safety Training Aid*. This AC is intended for Part 121 (air transport) operators, but much of its guidance is applicable to Parts 91, 129, and 135 operators as well.

Three additional RTO performance items pertinent to transport category aircraft are as follows:

1. **All-engine go distance:** 115% of the actual distance required to accelerate, then liftoff and reach a point 35 ft above the runway with all engines operating. This is the top image of Figure 8.17.
2. **Engine-out accelerate-go distance:** The distance required to accelerate with all engines operating, have an engine fail at V_{EF}, at least one second below V_1, continue the takeoff, liftoff, and reach a point 35 ft above the runway surface at V_2 speed. This is shown in the middle image of Figure 8.17.
3. **Engine-out accelerate-stop distance:** The distance required to accelerate with all engines operating, have an engine fail at V_{EF}, at least one second below V_1, recognize the failure and reconfigure for stopping and bring the airplane to a stop using maximum wheel braking with the speed-brakes extended. Reverse thrust is not considered in determining accelerate-stop distance. This is shown in the bottom image of Figure 8.17.

The takeoff field length to meet regulatory guidance is the most limiting of the three criteria.

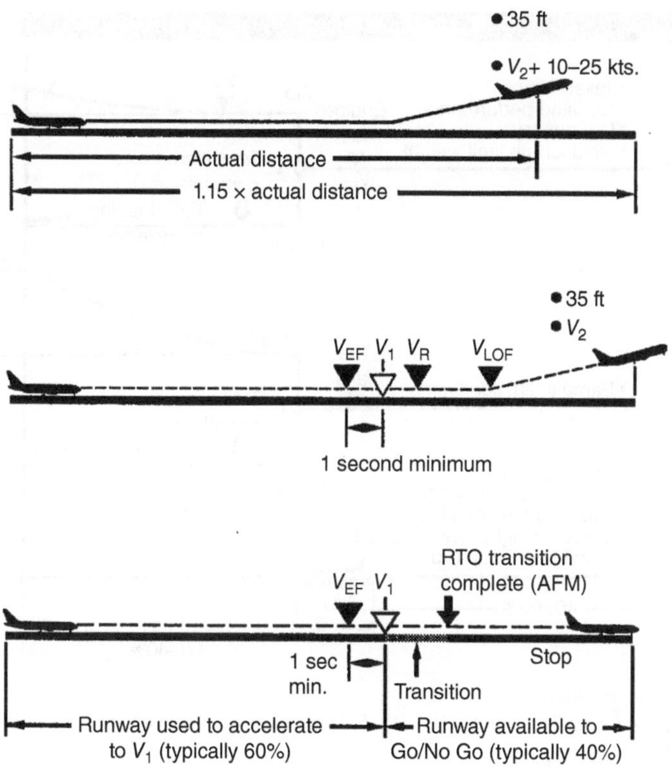

Figure 8.17. FAR takeoff field length.
Source: AC 120–62 – Takeoff Training Aid (1994)/U.S. Department of Transportation Federal Aviation Administration/Public domain.

Other considerations such as tire failure during the takeoff roll must be briefed prior to takeoff; in certain situations, it may be better to continue the takeoff roll if there is a tire failure even though V_1 has not been reached due to the loss of braking force from one or more wheels. Even more critical is the danger of a high-speed abort with directional control problems that could lead to a runway excursion at high gross weight. This is why many aircraft will specify a go-no-go speed for tire failure that is below V_1.

As can be seen in Figure 8.18, tire failure near V_1 could lead to an overrun at the end of the runway if an RTO is attempted and the calculated performance numbers cannot be met. A continued takeoff with the failed tire(s) would be preferred in this situation, followed by an emergency landing once the fuel is reduced, with full runway length available for landing.

Rejected Takeoff Aerodynamics

As noted previously, the calculation for the stopping distance after an RTO depends on the immediate deployment of speedbrakes as the stopping distance calculated does not include thrust reversers because the failure of an engine may have been the reason for initiating an RTO. Aircraft accidents have happened because pilots failed to deploy the speedbrakes during an RTO event or during landing. The FAA Pilot

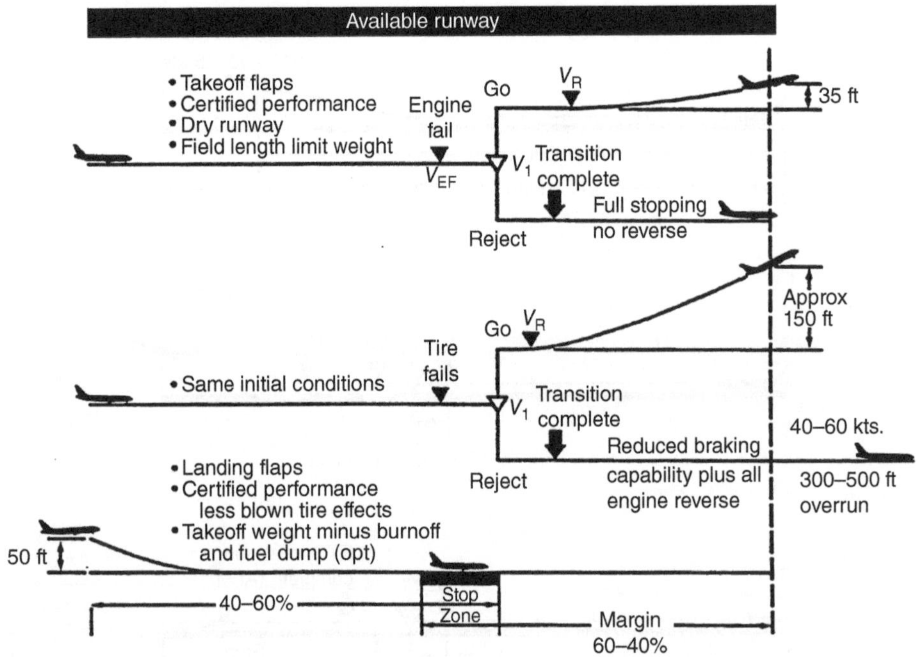

Figure 8.18. RTO and tire failure.
Source: AC 120-62 – Takeoff Training Aid (1994)/U.S. Department of Transportation Federal Aviation Administration/Public domain.

Guide to Takeoff Safety states there is one RTO for every 3 000 takeoffs, with 75% of those initiated at 80 kts. or less. About 2% of RTOs are initiated at speeds above 120 kts. and of course, most likely to result in an overrun. Approximately 21% of the accidents/incidents reviewed were because of an RTO initiated by an engine malfunction (Figure 8.19).

In 2004, an updated calculation time was used to calculate the transition time for the Airplane Flight Manual (AFM). As noted earlier in this chapter V_1 is not the "decision speed" where a pilot must begin the thought process to decide whether to continue the takeoff run or perform an RTO, the decision must be made by the time the aircraft reaches V_1. When calculating transition times during flight test, an increase in time is added to calculate the certified accelerate-stop distances as specified in the AFM. The goal is to allow sufficient time for the average pilot to transition from the takeoff mode to the stopping mode. Figure 8.20 shows the updated certification standards as of 2004 with the respective transition times. No deceleration credit is given during the 2-second time after speedbrake deployment which means retardation forces resulting from brake application and speed brake deployment are not applied. As a general rule, accelerate-stop distances calculated for every takeoff in the AFM are 400–600 ft longer than flight test accelerate-stop distance.

Figure 8.21 shows the effect of speedbrakes on the stopping capability of a typical mid-size two-engine transport that weighs 193 000 lb at brake release. The assumption is V_1 is 137 kts. and the main gear load would have decreased by 63 000 lb due to wing lift reducing the load on the gear. Notice the difference in the lift with the speedbrake position, *Up* represents the speedbrakes are deployed. As intended, with speedbrakes deployed drag is significantly more, and lift is significantly reduced, contributing greatly to the stopping distance. The total stopping force is 34% with the speedbrakes deployed.

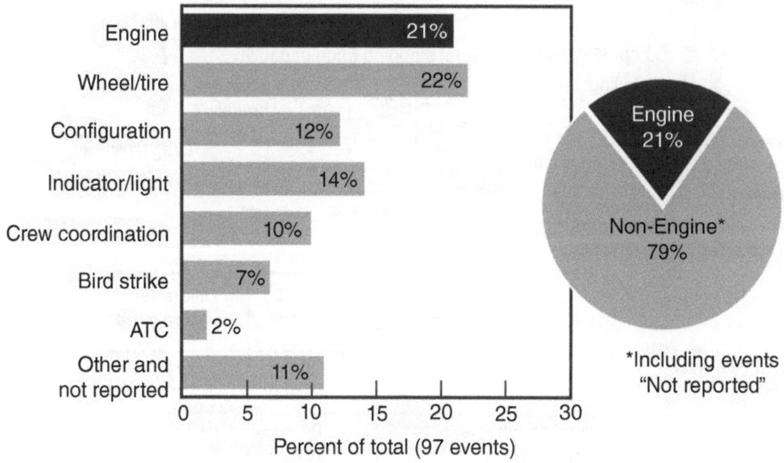

Figure 8.19. RTO initiation events leading to accident/incident (new image).
Source: Takeoff Training Safety Aid (2004c)/U.S. Department of Transportation Federal Aviation Administration/Public domain.

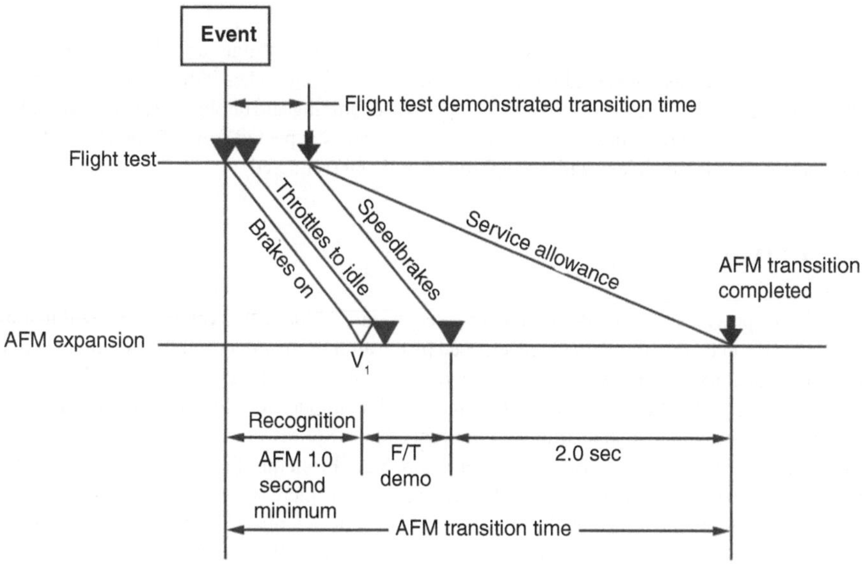

Figure 8.20. Establishing AFM transition time during an RTO.
Source: Takeoff Training Safety Aid (2004c)/U.S. Department of Transportation Federal Aviation Administration/Public domain.

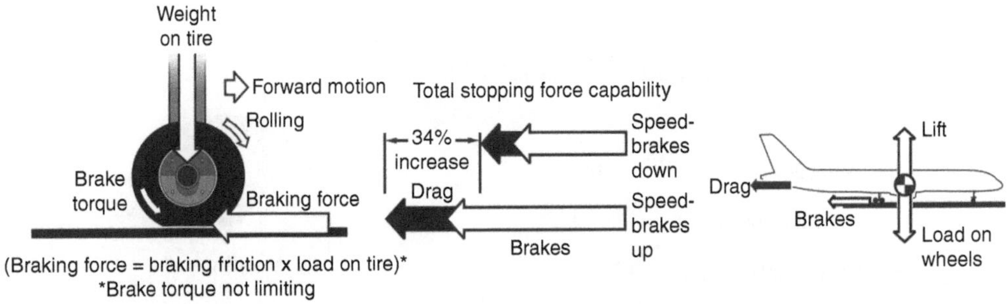

	Speedbrake position		Differene speedbrake up
	Down	Up	
Drag Lift	8 500 lbs 52 000 lbs	14 700 lbs −1 200	+73% −102%
Net load on wheels Max. braking force	141 600 75 900	194 800 98 000	+38% +29%
Max. stopping force (brakes and drag)	84 400	112 700	+34%

Figure 8.21. Effect of speedbrake usage during RTO.
Source: Takeoff Training Safety Aid (2004c)/U.S. Department of Transportation Federal Aviation Administration/Public domain.

Figure 8.22 represents the forces on the airplane when the pilots initiate an RTO. Note there are two separate situations, one has the speedbrakes deployed (Up) while the other indicates the speedbrakes are down. The pilots close the thrust levers, begin to apply wheel braking and deploy the speedbrakes. With the speedbrakes deployed the retarding force is much greater, and as expected from Chapters 4 and 5 reading the aerodynamic drag is much less with the speedbrakes down (retracted).

INITIAL CLIMB

As discussed in a previous section of this chapter, once the aircraft has lifted off, the pilot will maintain an IAS of V_X, V_Y, or some other recommended airspeed by the aircraft manufacturer or operator as required to meet any operational requirements. Depending on the size of the aircraft and on the type of FAA certificate under which the aircraft is being operated, the aircraft will continue to follow a climb profile as dictated by environmental conditions or standard operating procedures (SOP). An example of a basic performance chart for a light, single-engine aircraft is shown in Figure 8.23. From our discussion earlier in this chapter, as well as in Chapter 2, takeoff distance will increase with an increase in density altitude, or with an increase in weight. From the image on the left of Figure 8.23, the decrease in the rate of climb can be seen as the density altitude increases, a clear degradation in performance. In areas that experience high-density altitudes, pilots must be very familiar with the use of performance charts to calculate climb performance and obstacle clearance during instrument departure procedures (DPs).

Figure 8.24 shows a typical takeoff and climb profile of a corporate jet. Once takeoff thrust is set and the aircraft begins to accelerate, the rudder becomes more effective and nosewheel steering is accomplished with the rudder pedals. Airspeed is cross-checked around 80 kts., the V_1 call-out is made at V_1 (**V_1 will always be less than V_R**), and at V_R smooth backpressure is applied to raise the aircraft's nose and increase

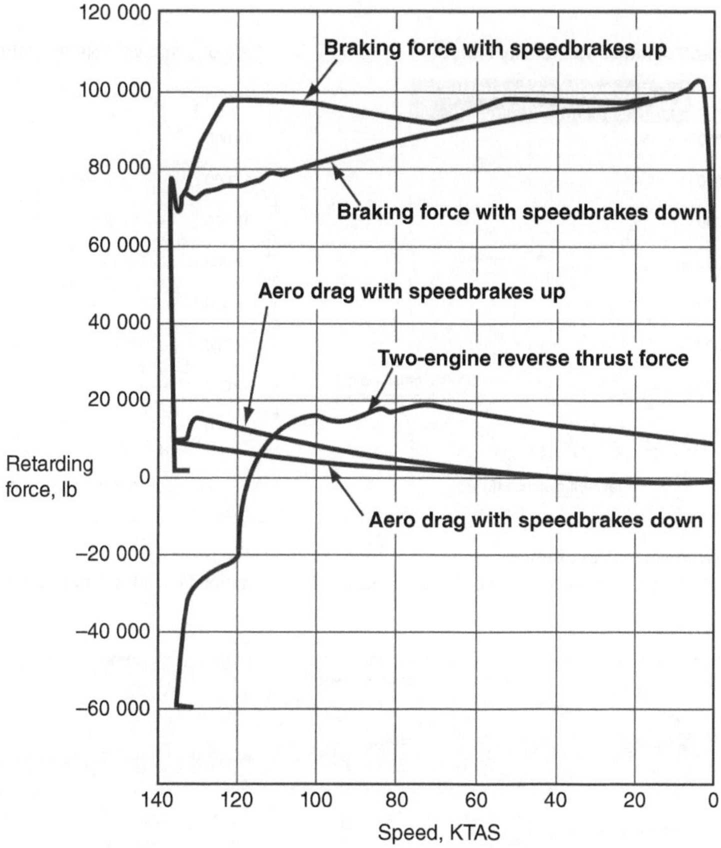

Figure 8.22. Retarding forces during RTO.
Source: Takeoff Training Safety Aid (2004c)/U.S. Department of Transportation Federal Aviation Administration/Public domain.

the AOA. Once airborne with a positive rate climb, the gear is retracted and the climb thrust is set along with flap retraction at the appropriate times based on manufacturer and operational considerations.

Figure 8.25 shows an example of the strict safety margins required of transport category aircraft during the climb segment breakdown with one engine inoperative. Note the above ground level altitudes and respective segments for gear and flap retraction.

LINEAR MOTION

Takeoff and landing distance computation is a complex subject, and the takeoff charts in the aircraft flight manual should be closely studied. This book makes no effort to analyze the problems in detail but simply explains how changes in weight, altitude, runway slope, headwind/tailwind, and other factors influence the aircraft's performance. The problem is simplified if we assume that the aircraft's acceleration is constant during the takeoff roll.

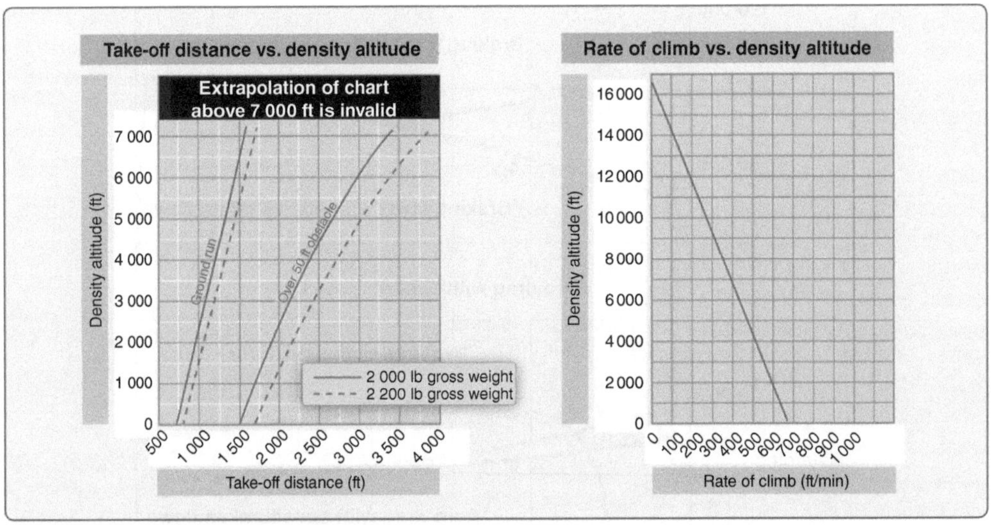

Figure 8.23. Performance chart examples.
Source: Airplane Flying Handbook (2016a)/U.S. Department of Transportation Federal Aviation Administration/Public domain.

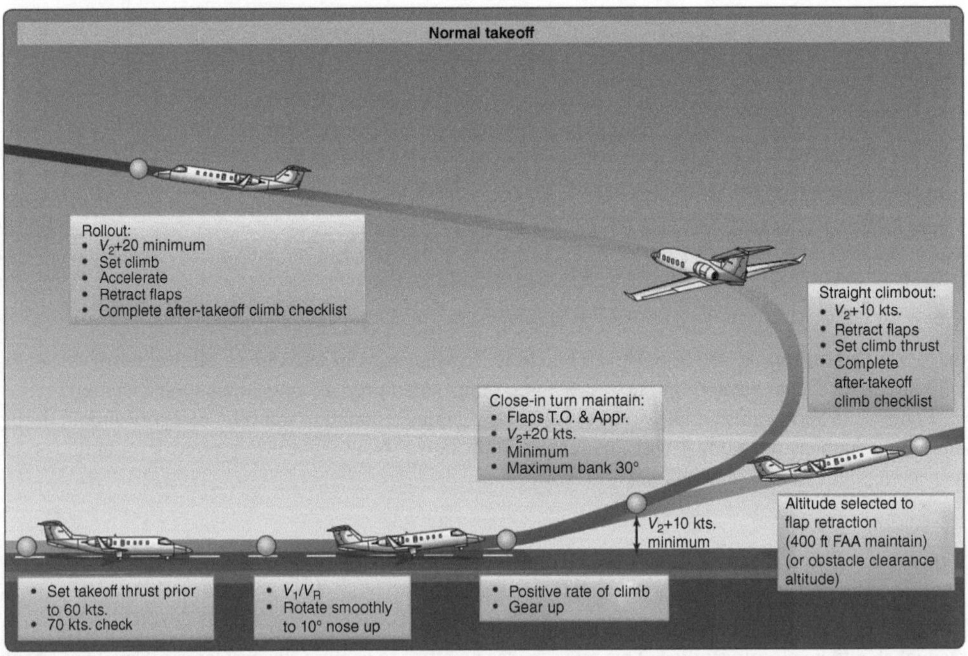

Figure 8.24. Jet takeoff and departure profile.
Source: Airplane Flying Handbook (2016a)/U.S. Department of Transportation Federal Aviation Administration/Public domain.

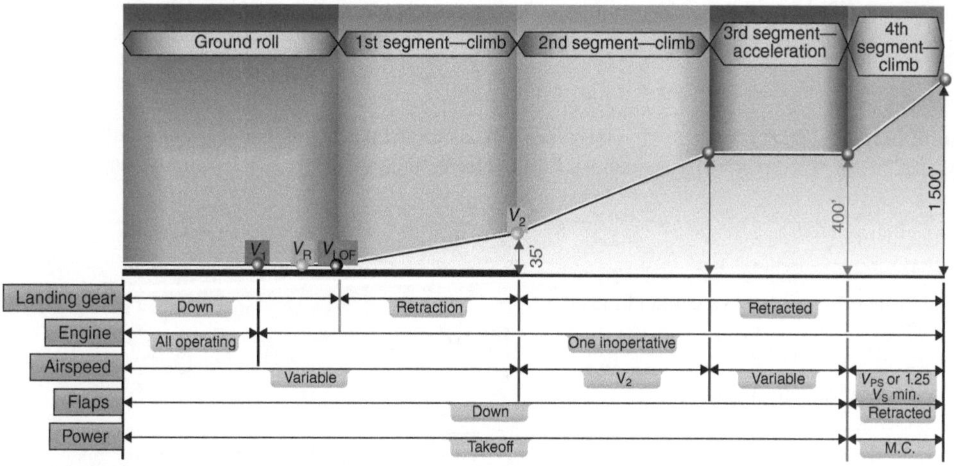

Items	1st T/O segment	2nd T/O segment	Transition (acceleration)	Final T/O segment
2 Engine	Positive	2.4%	Positive	1.2%
* 3 Engine	3.0%	2.7%	Positive	1.5%
4 Engine	5.0%	3.0%	Positive	1.7%
Wing flaps	T.O.	T.O.	T.O.	Up
Landing gear	Down	Up	Up	Up
Engines	1 Out	1 Out	1 Out	1 Out
Power	T.O.	T.O.	T.O.	M.C.
Air speed	$V_{LOF} \rightarrow V_2$	V_2	$V_2 \rightarrow 1.25\ V_S$(Min)	$1.25\ V_S$(Min)

* Required absolute minimum gradient of flightpath

M.C. = Maximum continuous
V_1 = Critical-engine-failure speed
V_2 = Takeoff safety speed
V_S = Calibrated stalling speed, or minimum steady flight speed at which the aircraft is controllable
V_R = Speed at which aircraft can start safely raising nose wheel off surface (rotation speed)
V_{LOF} = Speed at point where airplane lifts off

Figure 8.25. Segmented one-engine climb graph.
Source: Pilot's Handbook of Aeronautical Knowledge (2008a)/U.S. Department of Transportation Federal Aviation Administration/Public domain.

Jet transport aircraft have been found to normally accelerate at about 6 kts./s. This makes the stop or takeoff (Go/No-Go) decision at V_1 particularly important. The V_1 speed allows a one-second delay for the line pilot to react (i.e. initiate the stop or continue the takeoff). Any further delay defaults to the takeoff decision.

Newton's laws of motion express relationships among force, mass, and acceleration, but they stop short of discussing velocity, time, and distance. These items are covered here and will involve a brief review of Chapter 1 and the basic formulas introduced at the beginning of the textbook. As before, the formulas that are derived here are valid only for *constant acceleration*. From Chapter 1:

$$\text{Acceleration } a = \frac{\text{Change in velocity}}{\text{Change in time}} \rightarrow \frac{\Delta V}{\Delta t} \rightarrow \frac{V_f - V_i}{t_f - t_i} \text{ or } \frac{V_f - V_i}{t}$$

where

Δ (cap delta) means "change in"

V_f = final velocity at time t_f

V_i = initial velocity at time t_i.

If we start the time at $t_i = 0$ and rearrange the above, then Eq. 1.9:

$$V_f = V_i + at$$

If we start the time at $t_i = 0$ and $V_i = 0$ (brakes locked before takeoff roll) and rearrange the above where V_f can be any velocity given, for example, liftoff velocity, then we derive Eq. 1.10 from Chapter 1:

$$t = \frac{V_f}{a}$$

The distance s traveled in a certain time is

$$s = V_{av}t$$

The average velocity V_{av} is

$$V_{av} = \frac{V_i + V_f}{2}$$

And incorporating Eq.1.9, and substituting for V_f, we get

$$s = \left(\frac{((V_i + at) + V_i)}{2}\right)(t)$$

Therefore, Eq. 1.11:

$$s = V_i t + \frac{1}{2}at^2$$

Solving Eqs. 1.9 and 1.11 simultaneously and eliminating t, we derived Eq. 1.12 to solve for distance:

$$s = \frac{V_f^2 - V_i^2}{2a}$$

In applying these equations to the takeoff problem, where the aircraft starts from a "brakes locked" position on the runway, V_i is zero. The equations are then simplified and Eq. 1.12 becomes:

$$s = \frac{V_f^2}{2a} \tag{8.1}$$

where

s = distance (ft)

V_f = takeoff velocity (ft/s)

a = acceleration (ft/s^2).

The acceleration of an aircraft during takeoff is determined by applying Newton's second law: $F = ma$. The force that provides the acceleration for takeoff is the unbalanced force acting on the aircraft during the takeoff roll. This is called the net accelerating force and is illustrated in Figure 8.26.

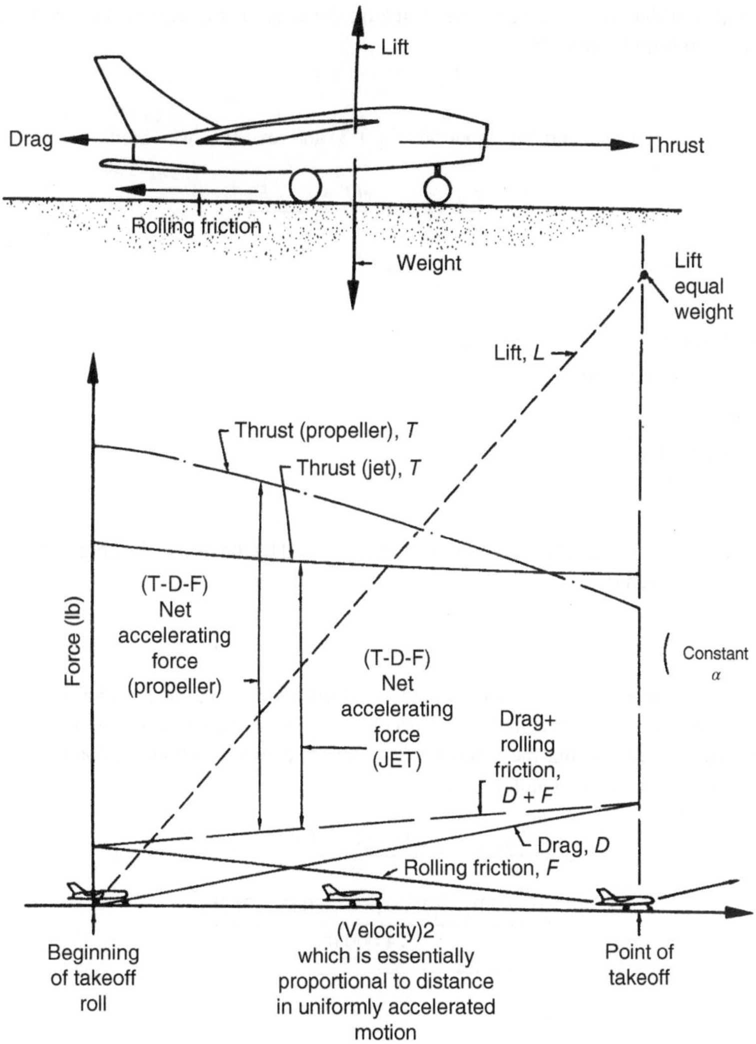

Figure 8.26. Forces on an airplane during takeoff.

Two assumptions are made in constructing Figure 8.26. The first is that no lift is being developed as the aircraft gathers speed. This is true for many of the higher-performance aircraft that have the wing positioned on the aircraft at an angle of incidence that produces the least drag rather than one that produces lift. These aircraft must "rotate" at takeoff speed to develop lift. Most transport aircraft and military aircraft fit into this category. If no lift is developed during the takeoff run, rolling friction remains constant until rotation. The second assumption is that the thrust increases during acceleration and that the net accelerating force is constant throughout the takeoff run, allowing us to use Eq. 8.1.

Figure 8.26 shows that the net accelerating force on the aircraft, F_N, equals the thrust, T, minus the drag, D, minus the rolling friction, F:

$$F_N = T - D - F$$

By substituting in Newton's second law equation (Eq. 1.3) and rewriting we obtain

$$a = \frac{F_N}{m} \rightarrow a = \frac{g(T - D - F)}{W} \tag{8.2}$$

where

a = acceleration (fps^2)

F_N = net accelerating force (lb)

W = weight (lb)

g = gravitational acceleration (=32 fps^2)

m = mass (slugs) (= W/g).

Equation 8.2 shows that the acceleration of the aircraft is affected by the thrust, the drag, the rolling friction, and the weight of the aircraft.

EXAMPLE

An airplane with a single, turbojet engine weighs 14 000 lb and accelerates down a runway with the following given set of conditions, calculate the acceleration (a) assuming constant acceleration. Given: T_a = 4 500 lb, average drag during the takeoff roll = 350 lb, average rolling friction during the takeoff roll = 75 lb, sea level, and standard conditions.

$$a = \frac{F_N}{m} \rightarrow a = \frac{g(T - D - F)}{W}$$

$$a = \frac{32.2\ \text{ft/s}^2(4500\ \text{lb} - 350\ \text{lb} - 75\ \text{lb})}{14\,000\ \text{lb}}$$

$$a = 9.4\ \text{ft/s}^2$$

FACTORS AFFECTING TAKEOFF PERFORMANCE

We saw at the beginning of this chapter that multiple factors affected takeoff performance. To see how these factors affect takeoff distance(s), we use Eq. 8.1 to make a ratio. A subscript 1 denotes a known set of conditions, and a subscript 2 applies to a new set of conditions:

$$\frac{s_2}{s_1} = \left(\frac{V_2}{V_1}\right)^2 \left(\frac{a_1}{a_2}\right) \tag{8.3}$$

The effects of weight change, altitude change, and wind conditions on takeoff distance can be found by applying Eq. 8.3.

Weight Change

Increasing the gross weight of an aircraft has a threefold effect on takeoff performance: (i) takeoff velocity is increased, (ii) there is more mass to accelerate, and (iii) there is more rolling friction. Liftoff speed will have to be higher to compensate for the increase in weight; this results in a longer takeoff distance. In general, a 10% increase in takeoff gross weight may result in a 21% or higher increase in takeoff distance. A 10% increase in takeoff gross weight would result in a 9% decrease in the rate of acceleration. Takeoff velocity varies as the square of the weight. This was discussed in Chapter 4 and shown in Eq. 4.7:

$$\frac{V_2}{V_1} = \sqrt{\frac{W_2}{W_1}}$$

Squaring this equation gives

$$\left(\frac{V_2}{V_1}\right)^2 = \frac{W_2}{W_1}$$

The effect of increasing weight on an aircraft's acceleration is twofold: There is more mass to be accelerated, and there is more rolling friction. For a dry concrete runway, the coefficient of rolling friction is about 0.02, so the increase in rolling friction due to an increase in weight of 1 000 lb means a loss in net accelerating force of 20 lb. This change is small, so the principal effect on acceleration is due to the change in the mass of the aircraft.

Acceleration is inversely proportional to the mass (or weight) of the aircraft. Therefore, if the rolling friction increase is ignored, the acceleration term is

$$\frac{a_1}{a_2} = \frac{W_2}{W_1}$$

If we substitute values from these last two equations into Eq. 8.3, the effect of a weight change on takeoff distance is

$$\frac{s_2}{s_1} = \left(\frac{V_2}{V_1}\right)^2 \left(\frac{a_1}{a_2}\right) \text{ substitute } \left(\frac{V_2}{V_1}\right)^2 = \frac{W_2}{W_1} \text{ and } \frac{a_1}{a_2} = \frac{W_2}{W_1}, \text{thus: } \frac{s_2}{s_1} = \left(\frac{W_2}{W_1}\right)^2 \rightarrow s_2 = s_1\left(\frac{W_2}{W_1}\right)^2 \tag{8.4}$$

EXAMPLE

The takeoff distance for a turbojet airplane, at sea level on a standard day, is 2 400 ft when at a weight of 8 000 lb. If the airplane weighs 500 lb less during the next takeoff, assuming all other conditions are the same, what is the new minimum takeoff distance required?

$$\frac{s_2}{s_1} = \left(\frac{W_2}{W_1}\right)^2 \rightarrow s_2 = 2\,400\text{ ft}\left(\frac{7\,500\text{ lb}}{8\,000\text{ lb}}\right)^2$$

$$s_2 = 2\,109\text{ ft}$$

Altitude

Runway density altitude is runway pressure altitude corrected for runway temperature. Runway density altitude may be considerably higher than the field density altitude because runway temperatures can be higher than the official field temperature. As we learned in Chapter 2, an increase in density altitude has a twofold effect on takeoff performance: (i) a higher takeoff TASis required, and (ii) for non-turbocharged reciprocating engines and for turbine engines, less thrust is available, so the net accelerating force is reduced. As a result, if there is an increase in pressure altitude or density altitude, an aircraft will require a longer takeoff distance (Figure 8.27).

To produce lift equal to the weight of the aircraft at a given AOA, the dynamic pressure must be the same, regardless of the altitude. Thus, the airplane will takeoff at the same equivalent airspeed (EAS) at altitude as at sea level. Because of the reduced density of the air, however, the required true airspeed (TAS) will be greater at altitude.

From basic aerodynamics, we learned that the relationship between true and equivalent airspeed was expressed by Eq. 2.12:

$$\text{TAS} = \frac{\text{EAS}}{\sqrt{\sigma}}$$

Thus, the takeoff airspeed at an airfield with a density altitude above sea level (V_2), as compared to the takeoff airspeed at sea level on a standard day (V_1), is

$$\frac{V_2}{V_1} = \frac{1}{\sqrt{\sigma_2}} \quad \text{or} \quad \left(\frac{V_2}{V_1}\right)^2 = \frac{1}{\sigma_2}$$

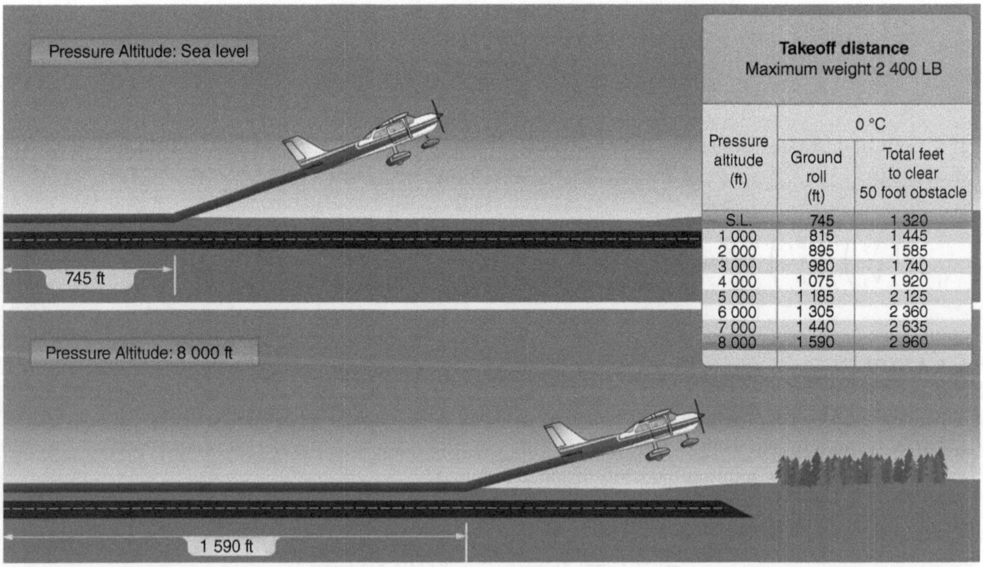

Figure 8.27. Takeoff distance with altitude change.
Source: Pilot's Handbook of Aeronautical Knowledge (2023c)/U.S. Department of Transportation Federal Aviation Administration/Public domain.

For non-turbocharged reciprocating engines and turbine engines, the power or thrust available decreases approximately linearly with the air density (or density ratio). As thrust is decreased, acceleration (a_2) is also decreased as the following:

$$\frac{a_1}{a_2} = \frac{1}{\sigma_2}$$

Substituting values from these last two equations into Eq. 8.3 gives us the effect of altitude on takeoff distance for **non-turbocharged reciprocating or turbine engine aircraft**:

$$\frac{s_2}{s_1} = \left(\frac{V_2}{V_1}\right)^2\left(\frac{a_1}{a_2}\right) \quad \text{substitute} \quad \left(\frac{V_2}{V_1}\right)^2 = \frac{1}{\sigma_2} \quad \text{and} \quad \frac{a_1}{a_2} = \frac{1}{\sigma_2}, \quad \text{thus:}$$

$$\frac{s_2}{s_1} = \left(\frac{1}{\sigma_2}\right)^2 \rightarrow s_2 = s_1\left(\frac{1}{\sigma_2}\right)^2 \tag{8.5}$$

Turbocharged reciprocating engines can deliver sea level power up to their critical altitude, so no decrease in thrust or acceleration occurs below that altitude. Thus, substituting values from these last two equations into Eq. 8.3,

$$\frac{s_2}{s_1} = \left(\frac{V_2}{V_1}\right)^2\left(\frac{a_1}{a_2}\right) \quad \text{substitute} \quad \left(\frac{V_2}{V_1}\right)^2 = \frac{1}{\sigma_2} \quad \text{and} \quad \frac{a_1}{a_2} = 1, \text{thus:}$$

$$\frac{s_2}{s_1} = \frac{1}{\sigma_2} \rightarrow s_2 = s_1\left(\frac{1}{\sigma_2}\right) \tag{8.6}$$

where

s_1 = standard sea level takeoff distance

s_2 = altitude takeoff distance

σ_2 = altitude density ratio.

EXAMPLE

The takeoff distance for a turbojet airplane, at sea level on a standard day, is 2 400 ft when at a weight of 8 000 lb. If the same airplane departs an airport with a density ratio of 0.8617, assuming all other conditions are the same, what is the new minimum takeoff distance required?

$$\frac{s_2}{s_1} = \left(\frac{1}{\sigma_2}\right)^2 \rightarrow s_2 = 2\,400\text{ ft}\left(\frac{1}{0.8617}\right)^2$$

$$s_2 = 3\,232\text{ ft}$$

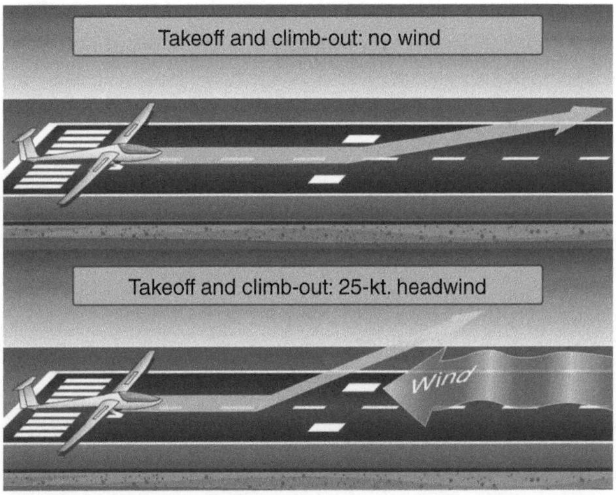

Figure 8.28. Takeoff distance reduced by headwind.
Source: Pilot's Handbook of Aeronautical Knowledge (2023c)/U.S. Department of Transportation Federal Aviation Administration/Public domain.

Wind

Takeoff into a headwind is preferred and allows the aircraft to reach takeoff velocity at a lower ground speed than for a no-wind condition, and results in a shorter takeoff distance (Figure 8.28). A headwind that is 10% of the takeoff airspeed will reduce takeoff distance by approximately 20% while taking off with a tailwind that is 10% of the takeoff airspeed will increase the takeoff distance by close to the same amount (Figure 8.29). The FAA's *Pilot's Handbook for Aeronautical Knowledge* states that if a headwind component is 50% of the takeoff velocity, the takeoff distance would only be 25% of the zero-wind takeoff distance, which is a 75% reduction. The takeoff velocity, V_2, with a headwind, in terms of the no-wind takeoff velocity, V_1, and the velocity of the headwind or tailwind component V_w, is

$$V_2 = V_1 - V_w$$

Dividing both sides of this equation by V_1 and squaring gives

$$\left(\frac{V_2}{V_1}\right)^2 = \left(\frac{V_1 - V_w}{V_1}\right)^2 \rightarrow \left(1 - \frac{V_w}{V_1}\right)^2$$

Although the acceleration over the ground is reduced by a headwind, the acceleration through the air mass is not affected. Therefore,

$$a_2 = a_1 \quad \text{or} \quad \frac{a_1}{a_2} = 1$$

Substituting the values of these last two equations into Eq. 8.3, we find that the **effect of a headwind** on the takeoff distance is

$$\frac{s_2}{s_1} = \left(\frac{V_2}{V_1}\right)^2\left(\frac{a_1}{a_2}\right) \text{ substitute } \left(\frac{V_2}{V_1}\right)^2 = \left(1 - \frac{V_w}{V_1}\right)^2 \text{ and } \frac{a_1}{a_2} = 1, \text{ thus:}$$

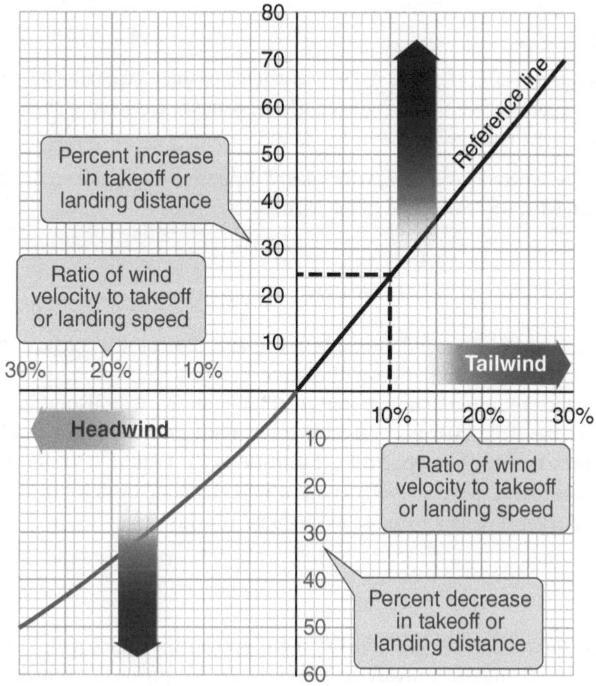

Figure 8.29. Effect of wind on takeoff.
Source: Pilot's Handbook of Aeronautical Knowledge (2008a)/U.S. Department of Transportation Federal Aviation Administration/Public domain.

$$\frac{s_2}{s_1} = \left(1 - \frac{V_w}{V_1}\right)^2 \rightarrow s_2 = s_1\left(1 - \frac{V_w}{V_1}\right)^2 \tag{8.7}$$

Takeoff with a tailwind simply means that the ground speed must be increased over the no-wind ground speed by the amount of the **tailwind**:

$$\frac{s_2}{s_1} = \left(1 + \frac{V_w}{V_1}\right)^2 \rightarrow s_2 = s_1\left(1 + \frac{V_w}{V_1}\right)^2 \tag{8.8}$$

EXAMPLE

The no-wind takeoff distance for a turbojet airplane, at sea level on a standard day, is 2 400 ft when at a weight of 8 000 lb and a takeoff velocity of 112 kts. If the same airplane departs an airport with a tailwind component of 20 kts., assuming all other conditions are the same, what is the new minimum takeoff distance required?

$$\frac{s_2}{s_1} = \left(1 + \frac{V_w}{V_1}\right)^2 \rightarrow s_2 = 2\,400\text{ ft}\left(1 + \frac{33.8\text{ ft/s}}{189.3\text{ ft/s}}\right)^2$$

$$s_2 = 3\,333.6\text{ ft}$$

Application 8.2

Compare the increase in takeoff distance that was calculated due to the tailwind component in the previous example problem, with the calculated takeoff distance expected with the same tailwind component using only Figure 8.29.

Does the approximate percentage increase in takeoff distance with the tailwind as determined by Figure 8.29 match what was calculated in the example problem?

Runway Surface

Most performance charts are for a paved, smooth, level, and dry runway surface; of course, that is not always possible. Any surface contamination will often decrease acceleration, delay takeoff velocity, and extend the takeoff distance required. These "abnormal" conditions will also affect performance speeds and change the value of the BFL. **Always consider the warnings, notes, and conditions published on the performance charts or in other operating limitations.** In Figure 8.30, the takeoff distance chart states that a 15% increase in the "paved, level, dry runway" calculated distance is required to meet the new short-field performance takeoff distance due to a dry, grass runway.

Runway Slope

Most runways have a slight slope (or gradient) to them even if they appear flat to the human eye. The gradient is the amount of change in runway height over the length of the runway and is expressed as a percentage. If a runway has a 2% gradient uphill, then for every 100 ft of runway distance the height would increase by 2 ft. This may seem small, but to an aircraft attempting to accelerate uphill for takeoff, or attempting to stop when landing downhill, the takeoff performance numbers can greatly be affected, especially when combined with any runway contamination. The component of the weight that is parallel to

Takeoff distance
Maximum weight 3 800 lbs

Short field

CONDITIONS:
Flaps 10°
2 850 RPM, Full Throttle and Mixture Set at Placard Fuel Flow Prior to Brake Release
Cowl Flaps Open
Paved, Level, Dry Runway
Zero Wind

Mixture setting	
Press ALT	PPH
S.L.	144
2 000	138
4 000	132
6 000	126
8 000	120

NOTES:
1. Short field technique as specified in Section 4.
2. Landing gear extended until takeoff obstacle is cleared.
3. Where distance value has been deleted, climb performance after lift-off is less than 150 fpm. Rate of climb is based on landing gear extended and flaps 10° at takeoff speed.
4. Decrease distances 10% for each 10 knots headwind. For operation with tailwinds up to 10 knots, increase distances by 10% for each 2.5 kts.
5. For operation on a dry, grass runway, increase distance by 15% of the "ground roll" figure.

Weight LBS	Takeoff speed kias		Press ALT FT	0°C		10°C		20°C		30°C		40°C	
	LIFT	AT		GRND	Total to clear	GRND	Total to clear	GRND	Total to clear	GRND	Total to clear	GRND	Total to clear

Figure 8.30. Takeoff distance chart with runway surface adjustment.
Source: U.S. Department of Transportation Federal Aviation Administration (2016b)/Public Domain.

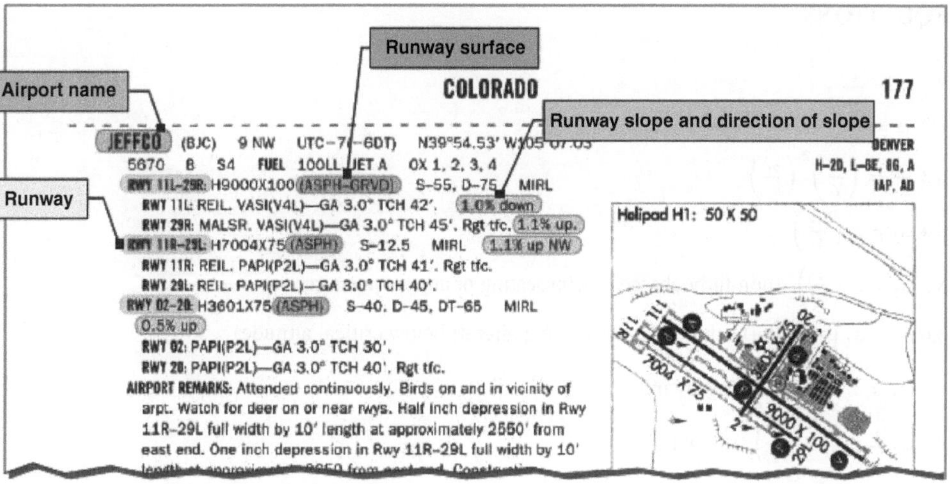

Figure 8.31. U.S. Chart Supplement information.
Source: U.S. Department of Transportation Federal Aviation Administration (2008b)/Public Domain.

the runway surface will reduce the net accelerating force, in case of an upslope, or increase the accelerating force, in case of a downslope:

$$\text{Accelerating force} = \pm W \text{ sine slope}$$

This results in an increase in takeoff distance if attempting to takeoff uphill, and a decrease in takeoff distance if taking off downhill. The takeoff distance charts in the aircraft flight manual usually include runway slope calculations. A rule of thumb is a *five percent increase in takeoff distance for each percent of uphill slope*. Figure 8.31 shows an example of a common source of runway surface and slope information for planning a flight.

SYMBOLS

F	Rolling friction force (lb)
F_N	Net accelerating force
V_1	Takeoff decision speed (kts.)
V_2	Takeoff speed (kts.)
V_{EF}	Critical engine failure speed (kts.)
V_{FE}	Flaps extended speed (kts.)
V_{LE}	Gear extended speed (kts.)
V_{LO}	Gear operating speed (kts.)
V_{LOF}	Liftoff (kts.)
V_{MBE}	Maximum braking speed (kts.)
V_{MCG}	Ground minimum control speed (kts.)
V_R	Rotation (kts.)
V_W	Wind speed (kts.)

EQUATIONS

8.1 $s = \frac{V_f^2}{2a}$

8.2 $a = \frac{F_N}{m} \rightarrow a = \frac{g(T-D-F)}{W}$

8.3 $\frac{s_2}{s_1} = \left(\frac{V_2}{V_1}\right)^2 \left(\frac{a_1}{a_2}\right)$

8.4 $s_2 = s_1\left(\frac{W_2}{W_1}\right)^2$

8.5 $s_2 = s_1\left(\frac{1}{\sigma_2}\right)^2$ (non-turbocharged reciprocating or any turbine aircraft)

8.6 $s_2 = s_1\left(\frac{1}{\sigma_2}\right)$ (turbocharged reciprocating aircraft below critical altitude)

8.7 $s_2 = s_1\left(1 - \frac{V_w}{V_1}\right)^2$ (headwind)

8.8 $s_2 = s_1\left(1 + \frac{V_w}{V_1}\right)^2$ (tailwind)

KEY TERMS

Accelerate-go distance
Accelerate-stop distance available (ASDA)
Balanced field length
Critical engine failure speed (V_{EF})
Crosswind takeoff
Declared distances
Hydrodynamic lift
Landing distance available (LDA)
On the step
Rejected takeoff (RTO)
Rotation (V_R)
Short-field takeoff
Soft-field takeoff
Takeoff decision speed (V_1)
Takeoff distance available (TODA)
Takeoff field length
Takeoff safety speed (V_2)
Takeoff run available (TORA)

PROBLEMS

Note: Answers to the problems in the first two sections are given at the end of the book.

Application of Chapter Theory

1. Before using a performance chart always be sure to review the notes and conditions.
   a. True
   b. False

2. V_1 should always be at or above V_R:
   a. True
   b. False

3. Takeoff distance is a function of
   a. takeoff velocity.
   b. acceleration.
   c. Both (a) and (b).

4. Takeoff distance is
   a. directly proportional to the acceleration.
   b. inversely proportional to the velocity squared.
   c. inversely proportional to the acceleration.
   d. directly proportional to the velocity squared.
   e. Both (c) and (d).

5. After completing a short-field takeoff, the following IAS should be maintained to maintain obstacle clearance:
   a. V_X
   b. V_Y
   c. V_{LE}
   d. V_R

6. The speed at which the aircraft first departs the runway:
   a. V_R
   b. V_{LO}
   c. V_{LE}
   d. V_{LOF}

7. A 10% increase in takeoff weight produces roughly a 10% increase in kinetic energy, while a 10% increase in speed results in a ____% increase in kinetic energy.
   a. 10
   b. 21
   c. 15
   d. 25

8. The maximum speed in the takeoff at which the pilot must take the first action (e.g. apply brakes, reduce thrust, deploy speed brakes) to stop the airplane within the accelerate-stop distance:
   a. V_{EF}
   b. V_{MCG}
   c. V_1
   d. V_2

9. Which of the following applies to declared distance information:
   a. A stopway is calculated into the TORA.
   b. If a clearway is available, then ASDA will always be longer than TORA.
   c. If only a stopway is available, then TODA will always be longer than ASDA.
   d. If a clearway is available, then TODA will always be longer than TORA.

10. As weight increases, the takeoff distance increases because
   a. there is less thrust and thus less acceleration.
   b. there is more mass and thus less acceleration.

c. the takeoff velocity is higher.
d. Both (b) and (c).

11. As weight (W_2) increases, the takeoff distance increases
a. directly as the square of W_2/W_1.
b. directly as W_2/W_1.
c. inversely as the square of W_2/W_1.
d. inversely as W_2/W_1.

12. For jet aircraft, takeoff distance at altitude is greater than at sea level because
a. the thrust available is less.
b. the EAS for takeoff is greater.
c. the TAS for takeoff is higher.
d. Both (a) and (c).

13. Which of the aircraft below will pay the greatest penalty in high-altitude takeoff distance?
a. One with a turbocharged reciprocating engine.
b. One with a turbine engine.
c. One with a non-turbocharged reciprocating engine.
d. Both (b) and (c).

14. Takeoff acceleration through the air mass is not reduced by a headwind.
a. True
b. False

15. Takeoff with a tailwind requires that the takeoff ground speed be _____ by the amount of the tailwind.
a. increased
b. decreased
c. not affected

16. Which of the following is not calculated into stopping distance after an RTO?
a. RTO corrective procedures before V_1
b. speedbrakes
c. thrust reversers

Calculations Using Chapter Principles

17. If the net accelerating force (F_N) on an aircraft traveling down a runway is 5 000 lb, what is the rolling friction (F) given the following information: Thrust = 7 000 lb, Drag = 1 560 lb
a. 2 000 lb
b. 440 lb
c. 5 440 lb
d. 3 440 lb

18. Using Figure 8.1, calculate the takeoff ground roll given the following information:
Temperature = 10 °C
Pressure altitude = 8 000 ft
Weight = 2 400 lb
Tailwind component = 5 kts.

a. 2 400 ft
b. 900 ft
c. 1 500 ft
d. 1 100 ft

19. Calculate the airplane's no-wind takeoff distance from a brakes-locked position given the following information:

Liftoff speed = 110 kts.

Sea level, standard day

Takeoff constant acceleration = 10 ft/s^2

20. The takeoff distance for a turbojet airplane, at sea level on a standard day, is 3 500 ft when at a weight of 9 000 lb. If the airplane weighs 1 000 lb less during the subsequent takeoff, assuming all other conditions are the same, calculate the new minimum takeoff distance required.

21. The takeoff distance for a turbojet airplane, at sea level on a standard day, is 2 800 ft when at a weight of 6 000 lb. If the same airplane departs an airport with a density ratio of 0.8881, assuming all other conditions are the same, what is the new minimum takeoff distance required?

Aircraft data for Problems 22–27:

Thrust available during takeoff = 3 000 lb

Turbojet engines

Average drag = 400 lb

Average rolling friction = 100 lb

Gross weight = 10 000 lb

Liftoff speed = 120 kts.

Sea level standard conditions

22. Calculate the airplane's acceleration.

23. What should the airspeed be at the 1 000-ft runway marker?

24. Calculate the no-wind takeoff distance.

25. If the weight increases to 15 000 lb, calculate the no-wind takeoff distance.

26. Calculate the takeoff distance for the 10 000 lb airplane if there is a 12-kts. headwind.

27. Calculate the no-wind takeoff distance for the 10 000 lb airplane operating from an airfield where the density ratio is 0.8.

Correlation Through Scenario-Based Learning

Aircraft – *Diamond DA50RG* / Section 5 AFM

1. Read the takeoff distance chart warnings, notes, cautions, and conditions at the beginning of Section 5, DA50 AFM.

2. Using the Take-off Distance – Normal Procedure chart for 4 407 lb (imperial units), calculate the takeoff distance over a 50 ft obstacle and the ground roll for the DA50 based on the following:

   ATIS Pressure = 29.80

   Airport elevation = 880 ft MSL

   OAT = 21 °C

3. Using your answers from Problem #2, calculate your takeoff distances for a 10-kt. *headwind* component.

4. Using your answers from Problem #2, calculate your takeoff distances for a 5-kt. *tailwind* component and when using a dry, *grass* runway (2 in. long grass).

5. Using the Take-off Distance – Normal Procedure chart for 3 968 lb (imperial units), calculate the takeoff distance over a 50 ft obstacle and the ground roll for the DA50 based on the following:

   ATIS pressure = 29.80″

   Airport elevation = 880 ft MSL

   OAT = ISA

6. Using the Climb Performance Takeoff Climb chart for 4 407 lb, calculate the rate of climb based on the following:

   Pressure altitude = 5 000 ft

   OAT = ISA

7. Using the Climb Performance Cruise Climb chart for 4 407 lb, calculate the rate of climb based on the following:

   Pressure altitude = 5 000 ft

   OAT = ISA

8. Which chart from Problems #6 and #7 provided the highest ROC? Why? What is the distinguishing difference between the two charts? Is your answer supported by the chapter discussion?

9. If an instrument flight rules departure procedure (DP) required a 500 ft/nm climb gradient, and the airspeed stated for the respective charts in Problems #6 and #7 were ground speed, would either scenario meet the required climb gradient for the DP? How would a headwind or tailwind component impact your answer if the airspeeds were IAS?

10. Using the Time, Fuel, and Distance to Climb chart for 3 968 lb, calculate the fuel, time, and distance to climb based on the following:

    Departing airport = KCLT

    Departing airport ATIS pressure = 29.67″

    Cruise altitude = 15 000 ft pressure altitude

    OAT = ISA

11. Using the maximum takeoff weight of 3 968 lb, calculate the mass of the DA50RG at the beginning of the takeoff roll.

12. Using your answer from Problem #11, and an assumed average net force during the takeoff roll of 800 lb, calculate the average acceleration of the DA50RG.

13. Using only the calculations and discussions provided in this textbook, calculate the takeoff distance if the aircraft is started from a brakes-locked position and rotates at 70 kts., with the average acceleration as found in Problem #12.

14. Using your average acceleration from Problem #12 and the rotation speed from Problem #13, calculate the time in seconds from brakes release to rotation.

15. Using your ground roll answer from Problem #5, calculate the takeoff distance using Eq. 8.4 if the aircraft weight was increased to 4 407 lb. Now, using the Take-off Distance – Normal Procedure chart for 4 407 lb (imperial units), use the chart to interpolate the takeoff ground roll for the same conditions in Problem #5. How does your answer from the manufacturer's performance chart for 4 407 lb compare to the answer you calculated using Eq. 8.4?

9 Landing Performance

CHAPTER OBJECTIVES

After completing this chapter, you should be able to:

- Identify the considerations that should be made before any landing, and summarize the factors that are included in the calculation of landing performance.
- Describe the forces on an airplane during gliding flight, and explain the importance of glide angle at $(L/D)_{max}$ to the safety of flight.
- Identify and characterize the performance speeds associated with the landing approach.
- Define a stabilized approach and summarize approach path options during the approach to landing.
- Describe the procedures used for crosswind and performance landings.
- Explain the dangers involved with improper landing procedures, provide examples of improper landing situations, and illustrate how an improper landing may develop.
- Identify the three types of hydroplaning and summarize the development of each type.
- Distinguish the forces on an airplane during landing and explain what measures can be taken to minimize landing distance.
- Correlate the linear motion equations used to calculate takeoff performance with those used to calculate landing performance.

Landing performance involves decelerated motion. During the landing phase of flight, the aircraft touches down at a certain velocity, and then decelerates to zero velocity. This chapter discusses the approach to a landing and the process of speed reduction once the aircraft has touched down. The following factors must be considered when evaluating landing performance. Note that some of these factors will be more important than others for different aircraft, locations, missions, and pilot techniques:

- Approach paths/speeds, including obstacle clearance.
- Deceleration during landing.
- The distance required to stop the aircraft.

Flight Theory and Aerodynamics: A Practical Guide for Operational Safety, Fifth Edition.
Brian A. Johnson and Philip R. Fittante.

Companion website: www.wiley.com/go/flighttheoryandaerodynamics5e

- Regulatory guidance and/or company operation specifications.
- Hazards of hydroplaning.
- Go-around/missed approach.

As with takeoff performance, there are many factors that affect the calculation of landing performance:

1. Aircraft gross weight
2. Thrust
3. Temperature
4. Pressure altitude
5. Wind direction and velocity
6. Runway slope
7. Runway surface

As discussed in Chapter 8, and demonstrated in Figures 8.1 and 8.2, the manufacturer of the aircraft publishes operating specifications, as well as performance charts and tables for the pilot to follow. One or more of the seven factors above will change on every flight, and the pilot must adjust accordingly to remain inside the operating envelope of the aircraft. Figure 9.1 is a typical landing distance graph used to calculate landing ground roll and landing distance over a 50 ft obstacle. ALWAYS review the notes and conditions of the performance charts before beginning the calculations.

Figure 9.2 depicts a typical landing distance table used by aircraft manufacturers to determine landing distance instead of the graph method. As before, review the notes and conditions before beginning the calculation for landing distance. Figure 9.2 indicates that at a gross weight of 1 600 lb with an approach speed of 60 mph, the landing distance can be interpolated if the pressure altitude and outside air temperature

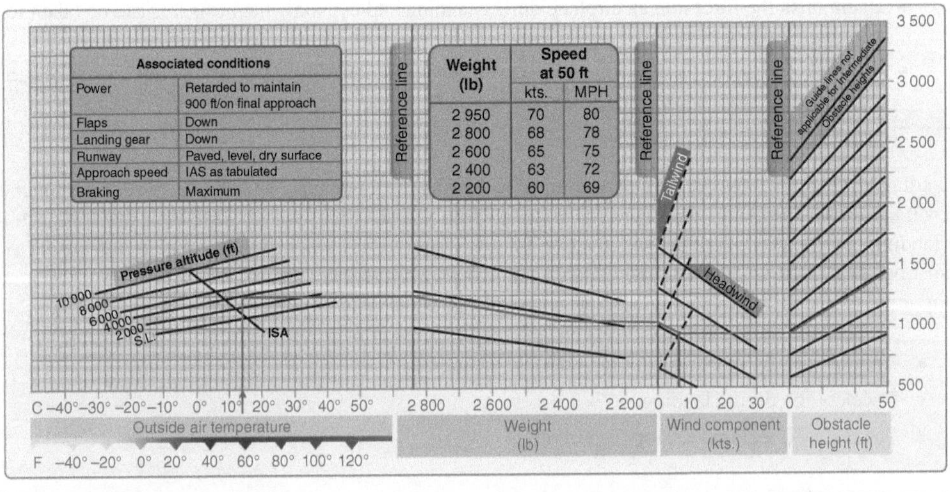

Figure 9.1. Landing distance graph.
Source: *Pilot's Handbook of Aeronautical Knowledge* (2016b)/U.S. Department of Transportation Federal Aviation Administration/Public domain.

Landing distance

Conditions: Flaps lowered to 40°; Power off; Hard surface runway; Zero wind

Gross weight lb	Approach speed IAS, MPH	At sea level and 59 °F		At 2 500 ft and 50 °F		At 5 000 ft and 41 °F		At 7 500 ft and 32 °F	
		Ground roll	Total to clear 50 ft OBS	Ground roll	Total to clear 50 ft OBS	Ground roll	Total to clear 50 ft OBS	Ground roll	Total to clear 50 ft OBS
1 600	60	445	1 075	470	1 135	495	1 195	520	1 255

Note:
1. Decrease the distances shown by 10% for each 4 kts. of headwind.
2. Increase the distance by 10% for each 60 °F temperature increase above standard.
3. For operation on a dry, grass runway, increase distance (both "ground roll" and "total to clear 50 ft obstacle") by 20% of the "total to clear 50 ft obstacle" figure.

Figure 9.2. Using interpolation to calculate landing distance.
Source: *Pilot's Handbook of Aeronautical Knowledge* (2023c)/U.S. Department of Transportation Federal Aviation Administration/Public domain.

are known. In Figure 9.2, if we were departing from an airport with a pressure altitude of 1 250 ft, then we would find the 50% point between the highlighted column for sea level and the highlighted column for 2 500 ft. For this example, the landing distance over a 50 ft obstacle would be 1 105 ft with an airport pressure altitude of 1 250 ft. Finally, note within the Notes section of Figure 9.2 for discussion later in the chapter: A headwind decreases landing distance, warmer air increases landing distance, and a dry, grass runway increases landing distance.

As discussed in Chapter 8, this is a complex subject and the simplified discussion presented here should not be used to supersede the information presented in the aircraft flight manual.

EXAMPLE

Using Figure 9.1, calculate the landing ground roll and landing distance over a 50 ft obstacle with the given information: pressure altitude = 4 000 ft, outside air temperature (OAT) = 1 °C, landing weight = 2 400 lb, and headwind component = 6 kts.

As shown in the image, starting at 14 °C, follow the vertical line up to 4 000 ft pressure altitude, then move horizontally to the first reference line. Note in the given example, the line then parallels the closest diagonal reference line downward until the given weight of 2 400 lb is met. Then, move horizontally to the next vertical reference line. After considering the 6 kts. headwind component, continue horizontally to the final vertical axis for the ground roll (~975 ft), and parallel the closest diagonal reference line to read the total distance over a 50 ft obstacle (~1 500 ft).

Application 9.1

Using Figure 9.1, calculate the landing ground roll and landing distance over a 50 ft obstacle using the same criteria provided in the example, except reduce the pressure altitude to sea level. Note the associated conditions table as an inset in the landing distance graph.

With the decrease in pressure altitude, what happened to the landing ground roll and landing distance over a 50 ft obstacle? Why? Using these same set of conditions, and based on your reading in Chapter 8, what change in performance values would you expect if the aircraft were landing on a paved, level, grass surface? Why?

PRELANDING PERFORMANCE

Gliding Flight

One of the most important concepts of aerodynamics is the equilibrium of forces, which was discussed in Chapter 1. An airplane that is flying at a constant velocity is in equilibrium, as shown in Figure 9.3. If the engine(s) should fail, thrust is reduced to zero, lift is decreased, and the forces on the aircraft are no longer balanced. This means that the aircraft will not be able to maintain altitude and will start to descend. If this occurs, the pilot would want to know several things:

1. How far can the airplane glide?
2. How far away is the nearest suitable landing site?
3. What will the sink rate be?

An understanding of the basic aerodynamics of gliding flight should help the pilot make an intelligent estimate of the performance of the airplane in a power-off glide. Once the engine quits and the aircraft starts to descend, the forces acting on it are shown in Figure 9.4. The figure shows that the forces of lift and drag act along the vertical and longitudinal axes of the aircraft, but the weight acts toward the center of the earth. To better analyze the reaction of these forces, it is desirable to replace the weight force by two component forces that act in the same rectangular coordinate system as the lift and drag forces.

The glide path makes a glide angle with the horizontal, which we will call γ (gamma). The component of the weight that acts perpendicular to the flight path is $W \cos \gamma$. It acts at the aircraft's center of gravity (CG) and is balanced by the lift, L. The component of weight that acts along the flight path is $W \sin \gamma$, and this force balances the drag D. Thus, the overall net force on the aircraft is zero, so there is no acceleration. In Figure 9.4, the component $W \sin \gamma$ is shown at some distance below the CG. This component actually acts at the CG and is shown in the figure merely to explain the right triangle relationship of the vectors.

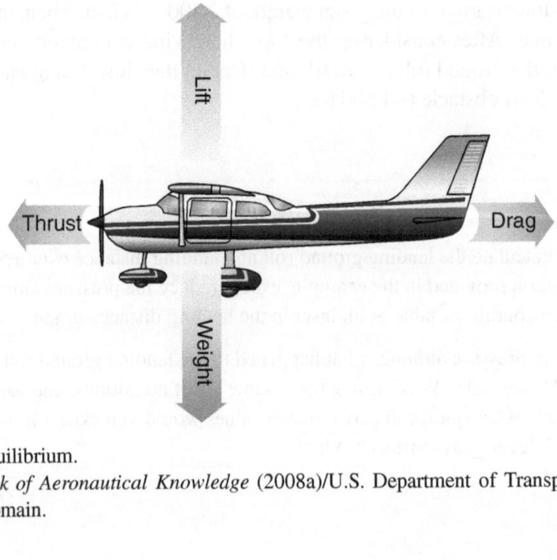

Figure 9.3. Forces in equilibrium.
Source: *Pilot's Handbook of Aeronautical Knowledge* (2008a)/U.S. Department of Transportation Federal Aviation Administration/Public domain.

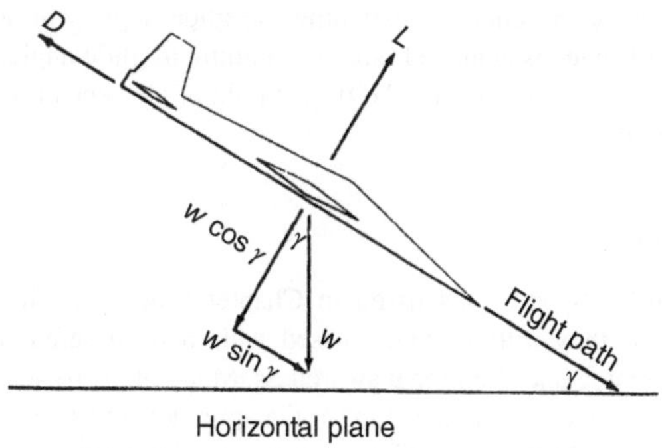

Figure 9.4. Forces acting in a power-off glide.

Once glide has been established and velocity is constant, equilibrium again exists, and the force equations are

$$L = W\cos\gamma \tag{9.1}$$

$$D = W\sin\gamma \tag{9.2}$$

As discussed in Chapter 5, the pilot is concerned about how to achieve the maximum glide ratio, meaning the smallest altitude loss per horizontal distance traveled. Flying at the airspeed that gives maximum glide ratio means that maximum glide range will be attained. The ratio of horizontal distance to vertical distance (altitude) is the glide ratio. To cover the most distance over the ground, the pilot wants to have the maximum glide ratio. The flight path angle must be a minimum to achieve this.

Dividing Eq. 9.1 by Eq. 9.2 gives

$$\frac{L}{D} \rightarrow \frac{W\cos\gamma}{W\sin\gamma} \rightarrow \frac{1}{\tan\gamma} \quad \text{or} \quad \tan\gamma = \frac{D}{L} \rightarrow \frac{1}{\mathrm{L/D}} \tag{9.3}$$

Equation 9.3 shows *that the minimum glide angle is obtained at* $(L/D)_{max}$ and, therefore, the airplane must be flown at the airspeed corresponding to the $(L/D)_{max}$ angle of attack (AOA). The glide ratio vector diagram is shown in Figure 9.5. The tangent of the glide angle in Figure 9.5 is equal to the opposite side divided by the adjacent side of the right angle shown in the figure. From Eq. 9.3, the tangent of the glide angle is D/L, so the opposite side of the triangle can be labeled as drag and the adjacent side can be labeled as lift. The vertical velocity $-V_V$ and the horizontal velocity $+V_H$ also represent the opposite side and adjacent side of the triangle, respectively; *thus, the numerical value of the L/D ratio is equal to the glide ratio.*

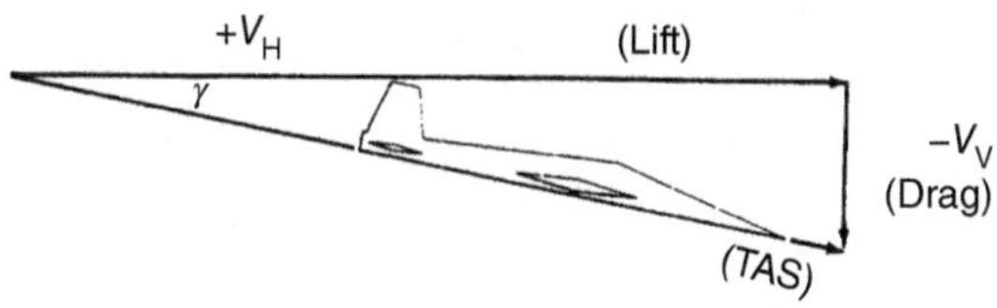

Figure 9.5. Glide ratio vector diagram.

A pilot who tries to stretch the glide by flattening the glide angle will actually decrease the glide distance. Maximum glide distance is achieved only at a minimum glide angle, and this occurs when the plane flies at the airspeed that corresponds to $(L/D)_{max}$. Flight at airspeeds faster or slower than this will decrease the range of the glide.

The Landing Approach

Many of the performance "V-Speeds" discussed in Chapter 8 apply to the landing approach and to deceleration after landing, so they will not be discussed again here. A safe landing approach and touchdown starts with proper planning and situational awareness. Many of the basic principles of aerodynamics discussed up to this point are involved in the setup for the approach and stabilized decent. The pilot must have a thorough understanding of lift, drag, high C_L devices, and proper coordination of the flight controls in order to end an approach at the proper velocity and position over the runway threshold.

A typical approach and landing profile for a jet can be seen in Figure 9.6. The definitions below are not a complete listing of all the speeds that may be encountered, but include some of the terms that can be found for Title 14 of the Code of Federal Regulations (CFR) pertaining to Part 25, 91, 121, or 135 aircraft operations:

Approach climb: The speed that gives the best climb performance in the approach configuration, with one engine inoperative, landing gear retracted, and maximum takeoff power on the operating engine(s); used for missed approach calculations to meet a minimum climb gradient.

Landing climb: The speed giving the best performance in the full landing configuration (gear down) with maximum takeoff power on all engines, used for missed approach calculations to meet a minimum climb gradient.

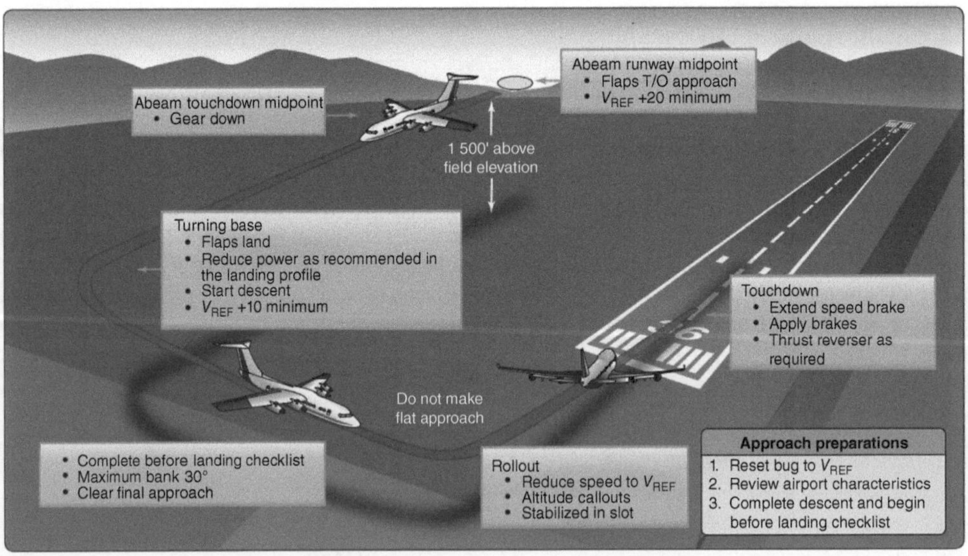

Figure 9.6. Jet approach and landing profile.
Source: Airplane Flying Handbook (2016a)/U.S. Department of Transportation Federal Aviation Administration/Public domain.

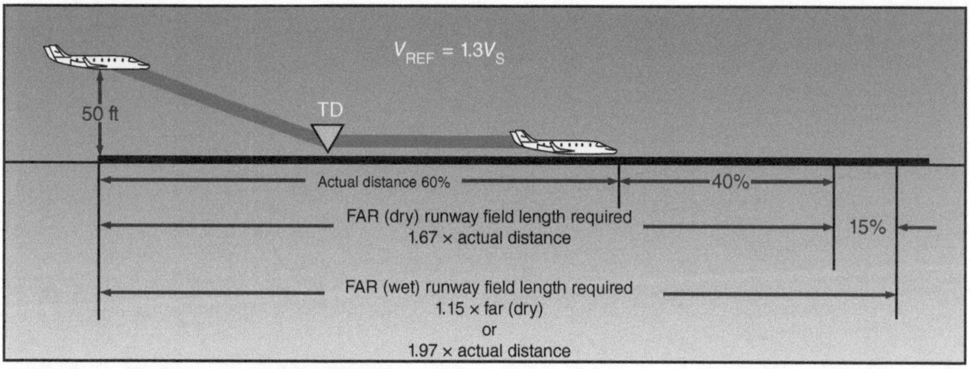

Figure 9.7. FAR landing field length required.
Source: Airplane Flying Handbook (2004a)/U.S. Department of Transportation Federal Aviation Administration/Public domain.

V_{REF} **– reference speed:** 1.3 times the stalling speed in the landing configuration; often the required speed at the 50 ft height above the threshold on the approach end of the runway (Figure 9.7).

V_{S0} **– power-off stalling speed (landing configuration):** The power-off stalling speed or minimum steady flight speed at which the aircraft is controllable in the landing configuration.

V_{S1} **– power-off stalling speed (specified configuration):** The speed at which the aircraft is controllable in a specified configuration, published by the manufacturer.

Stabilized Approaches

Precise flying is required to ensure a stabilized, steady flight path to touchdown. The approach speed specified for the weight of the aircraft is given in the airplane flight manual and provides a margin of safety above the minimum safe airspeed for the aircraft. The minimum airspeed may depend on the aircraft's stall speed, minimum control speed, or the speed at which $P_a = P_r$ (or $T_a = T_r$).

A fairly long approach path allows the pilot to stabilize the aircraft's airspeed, descent rate, and alignment with the runway centerline. Once these are brought under control, a smooth flight path can be maintained and only small control forces will be needed to make the actual landing. Steep turns should be avoided during the alignment of the aircraft with the runway. Steep turns increase the induced drag, increase the stall speed, and will bleed off the airspeed, possibly leading to a cross-controlled stall. Once proper alignment to the runway has been established, only minor banking for drift corrections will be required.

Figure 9.8 shows a typical stabilized approach for a jet aircraft. At 1 000 ft above ground level (AGL), the aircraft is established on glide path with a constant rate of descent and is configured for landing. The Federal Aviation Administration (FAA)'s *Airplane Flying Handbook* (2016a) describes five basic elements of a stabilized approach:

1. The aircraft should be in a landing configuration early in the approach.
2. Before descending below 1 000 ft AGL, the aircraft should be on the briefed profile, and usually a glide path of 2.5°–3° with configuration and trim set for landing.

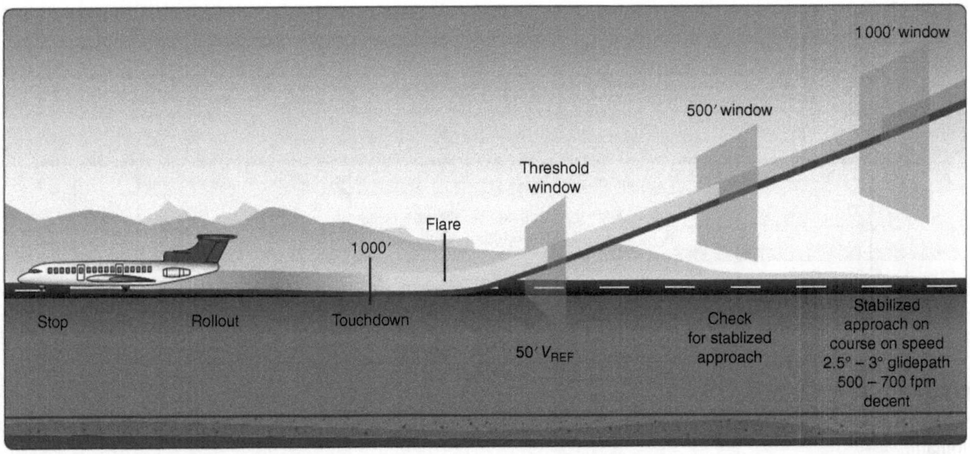

Figure 9.8. Stabilized approach for a jet aircraft.
Source: Airplane Flying Handbook (2016a)/U.S. Department of Transportation Federal Aviation Administration/Public domain.

3. The indicated airspeed (IAS) should be within 10 kts. of the target airspeed. As discussed in Chapters 6 and 7, thrust producers and propeller-driven aircraft (P_r versus T_r) operate differently in the slower flight regimes as induced drag comprises the majority of the total drag.
4. Based on ground speed (GS), the descent rate should be between 500 and 700 fpm, and should never exceed 1 000 fpm.
5. Whether a thrust producer or power producer, the engine speed should be set to allow for a rapid throttle increase and go-around.

Approach Glide Paths

A stabilized glide path makes a safe landing within the aircraft's published performance parameters much more likely. During visual approaches, an aiming point on the runway is chosen to assist in maintaining a stabilized approach. Various approach glide paths are shown in Figure 9.9, and the respective views from the flight deck are shown in Figure 9.10.

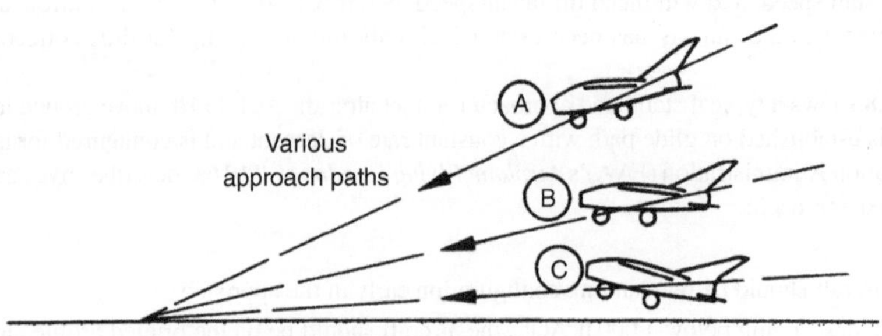

Figure 9.9. Approach glide paths.

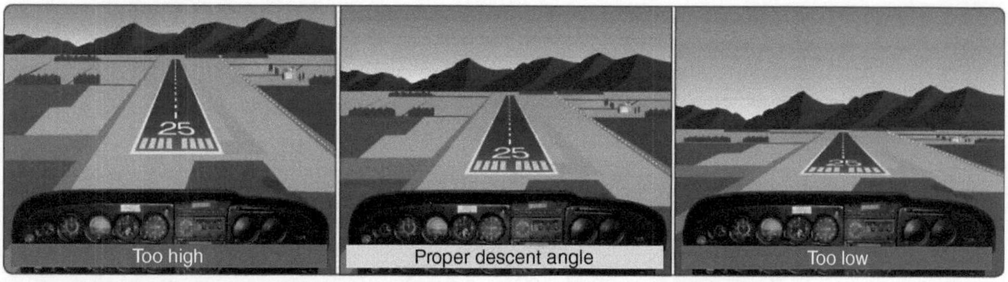

Figure 9.10. Approach glide path views from the flight deck.
Source: Airplane Flying Handbook (2016a)/U.S. Department of Transportation Federal Aviation Administration/Public domain.

Approach Path A

Within Figure 9.9, approach path A shows a steep, low-power approach. The far left viewpoint from the flight deck in Figure 9.10 shows what the pilot would see if on a steep, low-power approach; an aiming point within the touchdown zone would lead to an abrupt round out and flare with an extended touchdown distance. Such an approach usually results from making the turn from the downwind leg too soon and not allowing for a long enough final approach aligned with the runway centerline. The pilot finds the plane too close to the touchdown point and too high. There are three ways to correct this:

1. Cut the throttle and attempt to lose altitude rapidly. If the approach is made at idle throttle, the pilot must dive (or "forward slip") the airplane to lose altitude.
2. Continue the normal approach and thus land long on the runway.
3. Execute a go-around and make a new approach.

The first option results in a high rate of descent and/or an increase in airspeed. Pilots of light aircraft may not have difficulty in reducing the high rate of descent by either flaring the airplane or adding throttle. In either case, the plane will probably either make a hard landing or float excessively. The high rate of descent is much more serious for heavier airplanes. Recovery involves more than merely pulling the nose up to flare the plane. Such a recovery technique will add greatly to the induced drag but will provide little increase in lift. A large increase in thrust is required to overcome the high rate of descent. In some cases, there is simply not enough thrust available, and a hard, short landing results.

Landing long may succeed if the runway is long enough. However, *the third way – making a go-around – is recommended*. Although it may be hard on the pilot's ego, it is far easier on the airplane. If the aircraft is not properly lined up on the final approach, it is certainly better to perform a go-around rather than to "press on" and hope to salvage a good landing from a poor approach. A poor approach usually leads to a poor landing.

Approach Path C

Approach path C in Figure 9.9 shows a long, shallow approach. This is a result of getting too low during the approach and then having to "drag it in" using high thrust and high AOA. In Figure 9.10, the image on the far right shows an example of a flat approach from the perspective of the pilot. In some aircraft, the viewpoint would show a much higher deck angle, as depicted in approach path C of Figure 9.9. Such approaches may be successful in propeller aircraft (remember P_r in the low-velocity regime), but they

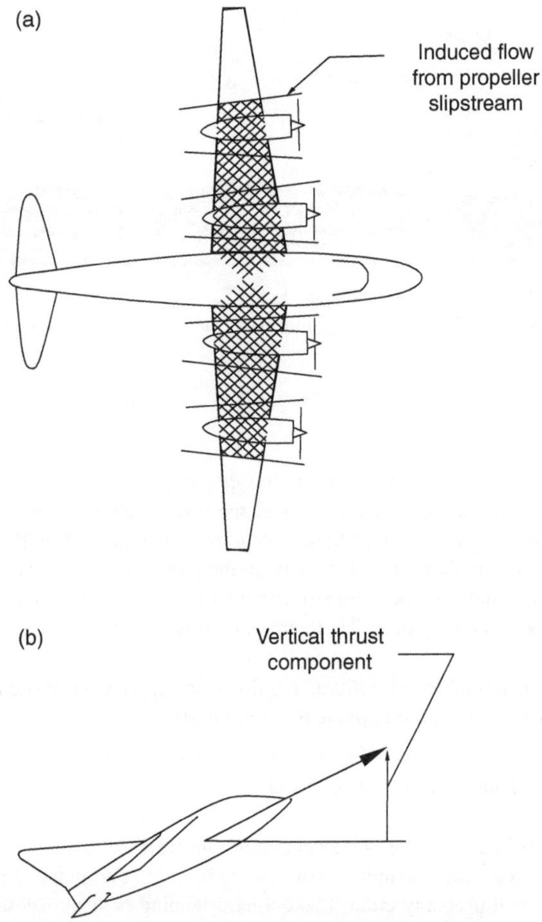

Figure 9.11. Lift from (a) propellers and (b) turbojets.

should be avoided in all airplanes. Propeller aircraft derive much more lift at high power settings due to the air being blown over the wing area behind the propellers, as shown in Figure 9.11a.

The only advantage of making a high-thrust approach in a conventional jet aircraft is the generation of lift from the vertical component of thrust, as shown in Figure 9.11b. The large increase in drag that results from the high AOA for the jet aircraft far exceeds the benefits of the lift.

Approach Path B

Propeller aircraft are power producers, and they produce high thrust at low airspeeds. Propeller aircraft do not suffer from a thrust deficiency at low airspeeds as do jet aircraft. Even with the advantages of blowing air and high thrust available, propeller aircraft should not be flown in a low, flat, high-powered approach. Engine failure under this type of approach can be disastrous. Thus, the intermediate approach B is by far the most desirable of the three options. It does not require a high rate of sink, excessive speed, or the extreme flare that the steep approach requires. The intermediate path also does not require the high thrust and AOA that the low path does. The middle image within Figure 9.9 shows the flight deck view of an aircraft on a stabilized approach at the proper descent angle.

As discussed in Chapter 4, the use of flaps will change the angle of attack, and the viewpoint from the flight deck, during a stabilized approach. With flap deployment, a change in the pitch attitude of the aircraft will be noticed, usually an initial nose-up movement due to a change in camber and higher C_L when flaps are <15°, then a nose-down pitching moment due to the change in center of pressure. With >15° of flaps, more drag than lift is produced, and pitch must be decreased to maintain desired airspeed. If flaps are not used, the angle of attack will have to be increased (higher C_L) to generate the lift lost as velocity decreases. As flaps are deployed incrementally, the nose of the aircraft will have to be lowered to maintain proper V_{REF} due to increasing C_L with trailing edge flaps. The critical angle of attack is now smaller, and the wing generates additional lift at a lower angle of attack (Figure 9.12). Figure 9.13 shows the effects of flaps on the approach angle and landing point for a single-engine propeller aircraft. Full flaps allow a steeper angle of descent for the same IAS during the final approach, while deploying flaps during the approach will result in a touchdown point closer to the landing threshold if airspeed and power are not changed.

An aircraft in a stabilized approach – at the correct airspeed, at the correct descent rate, and aligned with the runway centerline, has the best chance of avoiding common pitfalls during the landing phase of the flight. An unstabilized approach may lead to an improper flare during landing, or floating, ballooning, porpoising, and bouncing on the runway. All of these items lead to longer landing distances, and, in extreme cases, may damage the aircraft. Once established on the proper approach path and aircraft heading, the pilot

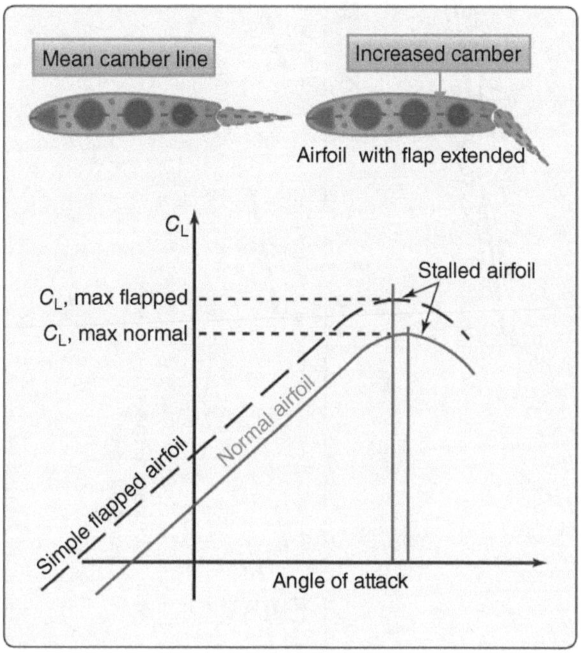

Figure 9.12. Coefficient of lift comparison for flap extended and retracted positions.
Source: Airplane Flying Handbook (2016a)/U.S. Department of Transportation Federal Aviation Administration/Public domain.

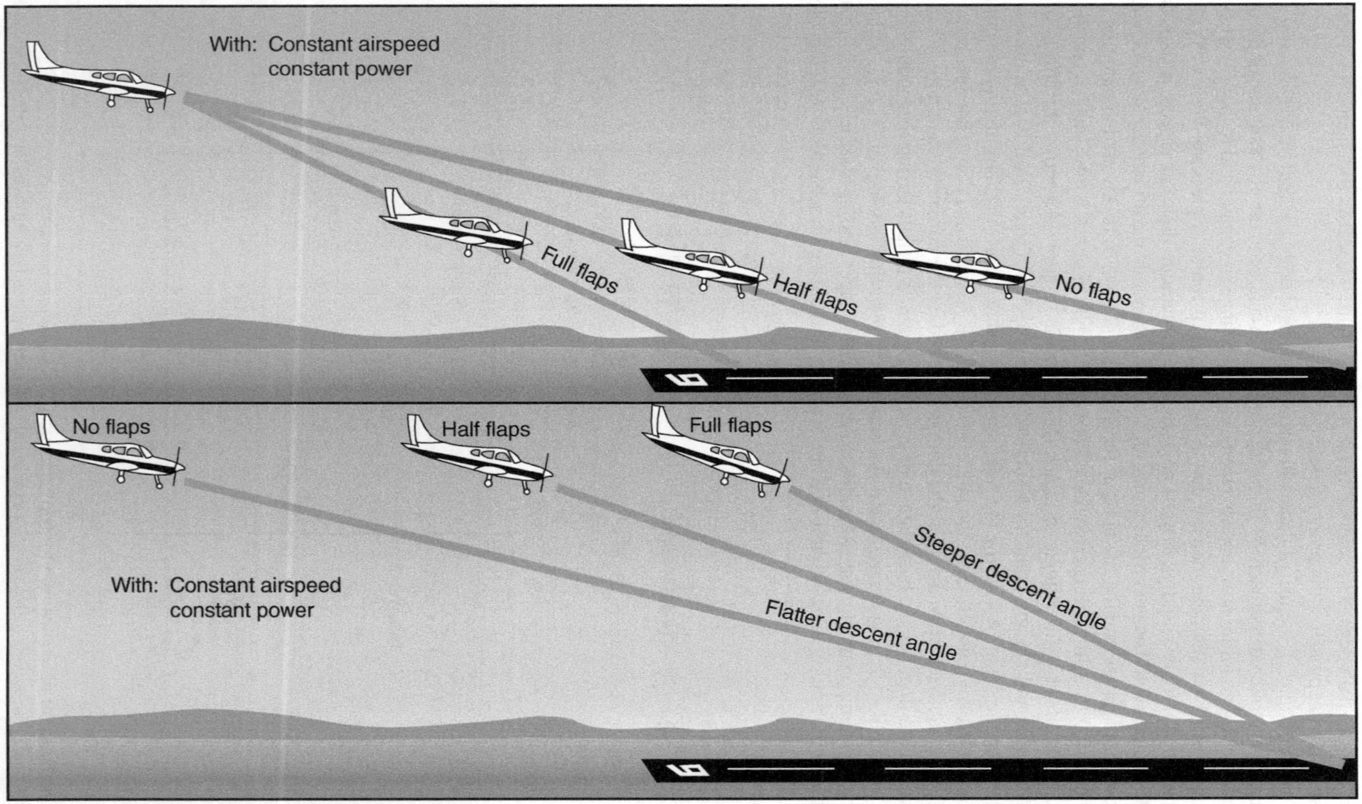

Figure 9.13. Effect of flaps on final approach.
Source: Airplane Flying Handbook (2004a)/U.S. Department of Transportation Federal Aviation Administration/Public domain.

must control the airspeed and altitude. The proper AOA will produce the desired airspeed. Lowering the AOA will increase the airspeed, and increasing the AOA will lower the airspeed. As we have emphasized several times, *the stick controls the airspeed.* Once proper airspeed has been established, *the primary control of the rate of descent will be the throttle.* Pitch for airspeed, power for altitude.

NORMAL LANDING

A normal landing begins with a proper round out and flare, the anticipation of and correction for ground effect, with a touchdown at the minimum safe speed above stall speed. Depending on the mission of the aircraft and on operating regulations, the pilot may need to land in the shortest possible distance, or the landing may finish with a landing roll and excess runway available. The forces on an aircraft during deceleration and the procedures to minimize landing rollout will be discussed in the Forces on the Aircraft During Landing section.

Figure 9.14 shows the change in the angle of attack of a typical round-out and touchdown for a small single-engine aircraft. As the airplane reaches the appropriate height above the ground, the pitch attitude is increased to increase the angle of attack. This action increases the lift and slows the rate of descent. As lift increases, so does the drag, so velocity steadily decreases. As the airspeed continues to decrease, further back pressure is required on the yoke/control stick to continue to increase the angle of attack and thus lift, resulting in an almost leveling attitude within the flare (ground effect).

Crosswind Landing

For most runway landings, the wind will not be perfectly aligned with the runway, so there is usually a crosswind component. Crosswind landings are usually a challenging obstacle to new pilots, and therefore limits are set by instructors, flight schools, employer operating specifications, or the manufacturer. Improper crosswind landing (and takeoff) technique can lead to unwanted side loads on landing gear, structural damage to the aircraft including wingtips, and at the very least a blow to the pilot's confidence. A proper understanding of crosswind limitations, pilot technique, and the aerodynamic forces at play greatly influence the positive outcome of a crosswind landing. The techniques discussed in this section are primarily for smaller general aviation aircraft, as larger transport category and military

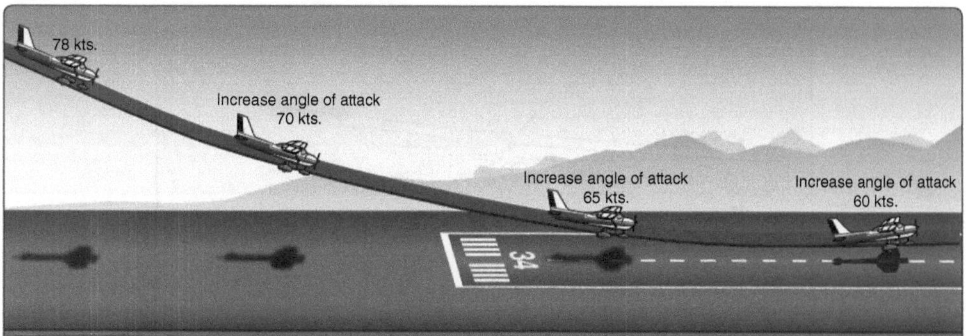

Figure 9.14. Changing angle of attack during landing.
Source: Airplane Flying Handbook (2016a)/U.S. Department of Transportation Federal Aviation Administration/Public domain.

aircraft may require different techniques based on gear configuration, wingspan, and the manufacturer's recommendations.

When a pilot transitions from a "crabbing" landing approach into the flare, the aircraft's longitudinal axis should be transitioned so it is aligned parallel to the runway centerline, and the aircraft should be over the center of the runway without drifting in either direction. In Figure 9.15, the wind is from the pilot's left, and improper crosswind correction is being used. The aircraft is drifting to the pilot's right as the wind is pushing on the left side of the aircraft and underneath the wing. This aerodynamic force to the pilot's right moves the aircraft to the right and raises the upwind (left) wing, so that when the aircraft touches down it will do so on the right main wheel and drift to the right. This places a significant side load on the landing gear and creates a torque that raises the upwind wing even more. This could cause the right wingtip to strike the ground in a low-wing aircraft, or in extreme cases could cause a ground loop (especially with conventional gear aircraft).

In addition to the negative side effects discussed above, quite often the aircraft will weathervane into the wind, resulting in the longitudinal axis of the airplane becoming misaligned with the runway centerline. As the aircraft touches down on the downwind main landing gear, the aircraft may pivot more into the wind complicating the landing or eventual go-around.

After using a crabbing approach with the aircraft weathervaning into the wind so the aircraft's ground track remains aligned with the centerline of the runway, a common technique to counteract the forces due to crosswind component is to transition to sideslip correction when in the flare (Figure 9.16). In this case, the rudder is used to keep the longitudinal axis aligned with the centerline, and windward aileron is used to lower the upwind wing. This tilts the lift vector to windward, and the horizontal (windward) component of lift now balances the force of the crosswind. The aircraft is now in a sideslip, with no drift to the left or the right of the centerline, and with its longitudinal axis aligned with the centerline. The aircraft should touchdown on the upwind main gear first, then the downwind main gear, and finally the nosewheel. As the aircraft slows down in the flare and after landing, aerodynamic forces on the ailerons and rudder are decreasing, making them less effective. Thus, more control input is required from the pilot to remain aligned with the centerline and to prevent drift. During rollout, full aileron may be needed to keep the upwind wing on the ground while less rudder will be required once the nosewheel is on the ground. Manufacturers publish the maximum demonstrated crosswind component in the aircraft's flight manual; this should be treated as a number to never exceed, requiring the selection of an alternate runway or airport should the crosswind component exceed this value. A discussion on sideslips and forward slips is found in Chapter 13.

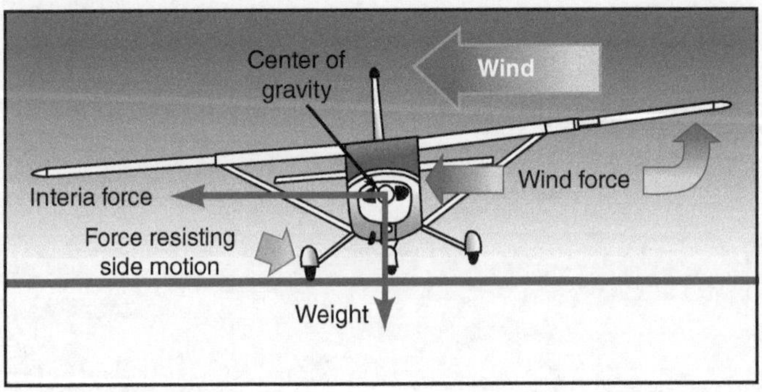

Figure 9.15. Improper drift correction.
Source: Airplane Flying Handbook (2004a)/U.S. Department of Transportation Federal Aviation Administration/Public domain.

Figure 9.16. Sideslip during crosswind.
Source: Airplane Flying Handbook (2004a)/U.S. Department of Transportation Federal Aviation Administration/Public domain.

Performance Landings

As discussed in the Performance Takeoffs section in Chapter 8, there are situations that require a pilot to land on a runway with an obstacle at the arrival end, on a runway with minimal runway length for rollout, or on a runway that is contaminated or is not a hard surface. When these situations occur, a short field or soft-field approach and landing procedure may be required. Because this textbook is not an instruction manual on these techniques, they will briefly be introduced and only those aspects of the landing related to our previous discussions, and the forces at work, will be discussed here.

A *short-field landing* is usually required when an obstacle along the approach path limits the runway length available for landing and stopping (effective runway length), or the runway is of minimal length without an obstacle and almost equal to the landing distance required for the aircraft to land and stop on the available runway (Figure 9.17). When this occurs, the pilot must fly the airplane at a minimal speed

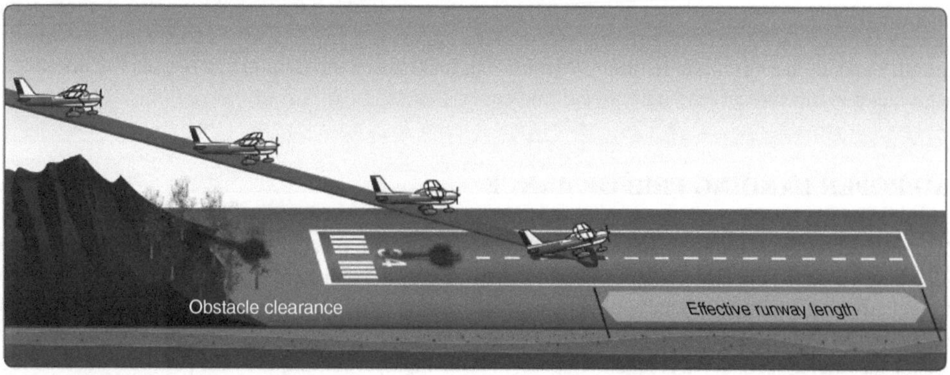

Figure 9.17. Short-field approach and landing over an obstacle.
Source: Airplane Flying Handbook (2016a)/U.S. Department of Transportation Federal Aviation Administration/Public domain.

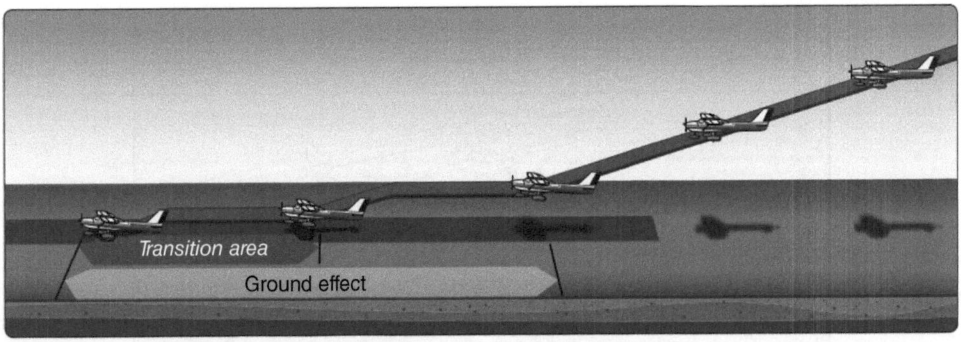

Figure 9.18. Soft/rough-field approach and landing.
Source: Airplane Flying Handbook (2016a)/U.S. Department of Transportation Federal Aviation Administration/Public domain.

that allows for positive control of the aircraft in the air with a safe margin above V_{SO} but also allows for little or no floating while in the landing flare and stopping in the shortest distance possible. As discussed in Chapter 1 with the kinetic energy formula, because the velocity is squared, any unnecessary airspeed on final will translate into excessive runway length being consumed. Most manufacturers require full flap deployment, which results in a high C_L value and high drag, so more power will be required to maintain V_{REF}. A stabilized approach is required as deviations from a constant decent rate and positive airspeed control will result in excessive runway usage or improper flare and touchdown. Because the aircraft will most likely be operating in the region of reversed command (discussed in Chapter 10), the proper go-around procedure must be briefed before the approach is started and if required, initiated as soon as possible.

As with a soft/rough-field takeoff, a *soft/rough-field landing* involves an approach and landing technique to minimize time on the runway at high velocities due to the extra drag and contaminants involved. During a soft/rough-field landing, the weight of the airplane should remain on the wings as long as possible, and the airplane should touchdown at the slowest possible flying speed without stalling (Figure 9.18). Power may be utilized during the landing flare while in ground effect to minimize the forces on touchdown that could result in damage to the nose-gear or nose-over tendency. Because the approach speed will be very similar to a short-field approach, the angle of attack will be relatively high and power will be required to overcome the increase in drag experienced at low velocity. Braking force will be minimal once on the runway as frictional forces on the ground will slow the aircraft and any excessive braking may damage the aircraft or result in a nose-over tendency. Maximizing brake force and minimizing the landing distance is not a factor in this type of landing.

IMPROPER LANDING PERFORMANCE

The final descent and landing of any aircraft should be stabilized, on airspeed, and on centerline to allow a landing within the touchdown zone. We previously discussed the hazards of not preparing properly for ground effect and the consequences of approaching a runway in a steep descent with idle thrust, or when low and slow with high thrust settings. Unstabilized approaches commonly lead to unstabilized landings outside the briefed touchdown area, off airspeed, or with improper technique by the pilot to try and correct for a deteriorating situation. In almost all cases, the pilot should execute a go-around or missed approach. Better to attempt another approach than force an unstabilized approach and landing.

Improper Round Out

To most passengers, a smooth approach and landing seem like a normal and stress-free act by the pilot in concert with the aircraft. In fact, a smooth and stable approach and landing occur only because a pilot applies the proper technique and feel for the aircraft acquired over hundreds, if not thousands, of landings. Understanding energy management, and when to begin the round-out and flare is necessary for a successful landing, and for achieving the performance for stopping distance calculated before landing. A high round-out commonly leads to the slowing of the aircraft well above the runway, and well into the region of reversed command; within seconds, the airspeed begins to slow and approaches stall speed. If the pilot maintains back pressure on the stick, the angle of attack, already high, may become critical, and the risk of a stall above the runway is high. If the pitch angle is high and the airspeed is low, power should be added while level pitch is maintained until the aircraft is able to re-establish the proper landing attitude (Figure 9.19). In most cases, a go-around should be initiated.

Delaying the round-out can be just as detrimental as rounding out too high. If the pilot suddenly decides that impact is imminent, pulling back too aggressively on the stick can lead to an accelerated stall due to high stall speed and critical angle of attack, complicating an already unstable situation. If the pilot is unable to raise the nose in time, then the nosewheel of the aircraft may strike the runway. This could cause a prop strike, but even if that does not occur, it may damage the nose landing gear and may result in the aircraft bouncing back into ground effect. In either case, it is advisable to apply full power, pitch for airspeed, initiate a go-around, and retract flaps and gear as applicable once a positive climb rate has been established.

Floating during round-out is a common occurrence for student pilots, or when pilots are transitioning to new aircraft and are unsure of proper approach airspeed. Sometimes, excessive floating is blamed on ground effect, when in reality it is a combination of ground effect and poor approach technique by the pilot, in particular, beginning the round out with too much airspeed. If too much velocity is carried into the landing flare ($KE = 1/2mV^2$), or the pilot dives to the runway to achieve a prescribed landing point, extra airspeed will be carried into the landing flare. If the runway is short, there may be insufficient landing distance to stop the aircraft once a touchdown has been made, or the pilot may attempt to force the aircraft on the runway, further complicating the situation. If sufficient runway remains, the pilot should decrease power, maintain a level pitch attitude, and let the airspeed bleed off in the flare. As the airspeed decreases,

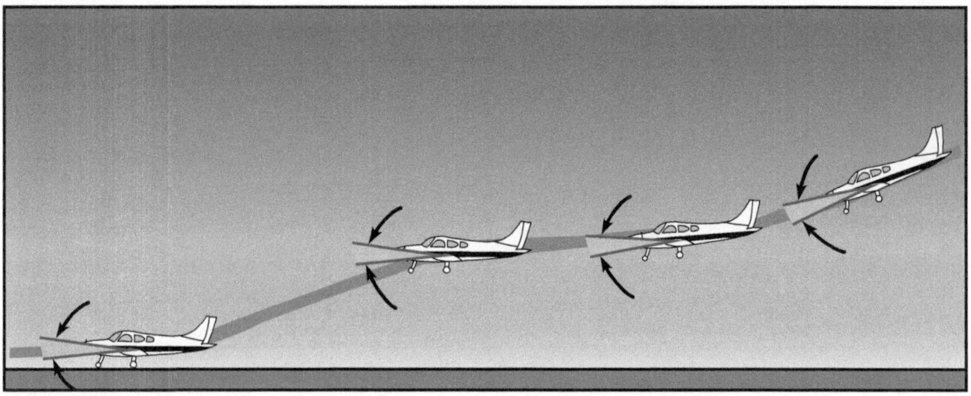

Figure 9.19. High round out.
Source: Airplane Flying Handbook (2004a)/U.S. Department of Transportation Federal Aviation Administration/Public domain.

the angle of attack will have to increase in order to maintain level flight over the runway, eventually touching down on the main landing gear allowing aerodynamic braking to slow the aircraft.

Bouncing and Ballooning

In the previous section, we discussed how leveling too high or leveling too low when initiating the landing flare could result in a stall above the runway or a nosewheel-first touchdown. This section and the following section address these and other consequences of an improper round-out in more detail, and describe what the pilot should and should not do when a landing is not going as planned.

When a pilot feels that the descent rate is excessive, he or she may instinctively pull back on the stick. This causes a tail-down force, raising the nose as well as the AOA, thus increasing the lift. The sudden increase in lift may cause the aircraft to *balloon* away from the runway and possibly out of ground effect. The airspeed will now be decreasing, and because of the nose-high attitude, the wing will be closer to the critical angle of attack ($C_{L(max)}$). If the aircraft is propeller driven and the power setting is high, the left turning tendency may cause the longitudinal axis to be diverted left of the centerline, and should any crosswind exist the aircraft is subject to drifting left or right of the centerline. If the aircraft does balloon out of ground effect, the nose may pitch up further, and if this does occur, induced drag will increase also. Experience will dictate what corrective action is best for each situation, but usually, the best procedure is to reduce the angle of attack to minimize airspeed loss, apply full power, and initiate a go-around. If the balloon is managed properly, runway is available, and the aircraft is a safe distance above the runway, the pilot may choose to reestablish a proper angle of attack and add power as needed to cushion the landing, using the rudder to keep the aircraft's longitudinal axis aligned with the runway centerline and touchdown on the main gear.

Bouncing during touchdown results from similar problems with pilot technique as ballooning, but whereas in ballooning the aircraft does not touchdown, in a bounce the pilot lets the aircraft touchdown and bounce at least once into the air, if not multiple times (Figure 9.20). Due to a pilot's instinctive reaction of pulling back on the stick, and to the downward movement of the tail when the main gear strikes the runway, the nose goes up and the AOA is increased. Again, C_L is increased and approaching $C_{L(max)}$, especially if airspeed is low and the pilot is slow to react. If the aircraft exits ground effect, the nose will tend to pitch up further, induced drag will increase, and the same deteriorating situation as ballooning may occur. Corrective action should be the same; in most cases, it is best to initiate a

Figure 9.20. Bouncing during touchdown.
Source: Airplane Flying Handbook (2016a)/U.S. Department of Transportation Federal Aviation Administration/Public domain.

go-around, maintaining the proper AOA and airspeed for the climb and "cleaning up" the aircraft by retracting gear and/or flaps when a positive rate of climb has been established. If a bounce is not severe and if sufficient runway remains, a proper landing attitude can be reestablished, and a normal touchdown can be made as long as the rudder is used to align the longitudinal axis with the centerline and correction for any crosswind-induced drift has been made with ailerons.

Porpoising

Excessive floating during the round-out and flare always increases landing distance. When in doubt as to whether the aircraft can safely be brought to a stop without a runway overrun, a go-around is recommended.

Many pilots during training, or when transitioning to a new aircraft, attempt to force the airplane onto the runway when enough lift still is being developed for the aircraft to fly. As a result, the aircraft begins to *porpoise* similar to the way a true porpoise moves in and out of the water. These oscillations by the pilot can become worse during the flare if the pilot jockeys the throttle, or overcorrects as the angle of attack of the wing is varied by the constantly moving elevator and loss of tail down force in ground effect. Sometimes during these oscillations, the nosewheel is allowed to strike the runway, forcing the aircraft back in the air and in a nose-high, increasing angle of attack situation. If a go-around is not selected, it may be possible to salvage the landing by establishing a proper landing pitch attitude, increasing power as required to cushion the landing, and using aerodynamic braking on the runway (having reduced power to idle) without the aircraft climbing back into ground effect (Figure 9.21).

If a pilot lands when the aircraft is still producing enough lift for the aircraft to fly and then adds forward stick pressure, a situation known as wheelbarrowing may occur. Due to the deflection of the elevator or stabilator with a forward stick, the lift on the tail is now an upward force, and more pressure is placed on the nosewheel. This situation can also occur when a pilot is on takeoff roll and continues to force the aircraft on the runway when the wing is producing sufficient lift to fly. The proper corrective technique is to re-establish aft pressure on the stick, and for a landing verify that the throttle is set in the idle position. Depending on aircraft manufacturer recommendations, as well as the operating rules or the pilot, flaps can be retracted, placing more weight on wheels and lowering the C_L of the wing.

Rejected Landings (Go-Arounds)

When a "normal landing" as described in the Normal Landing section cannot be assured, a go-around should be initiated. Whether a simple go-around during training in visual meteorological conditions (VMC), or when executing a low-visibility instrument approach, all go-around procedures should be

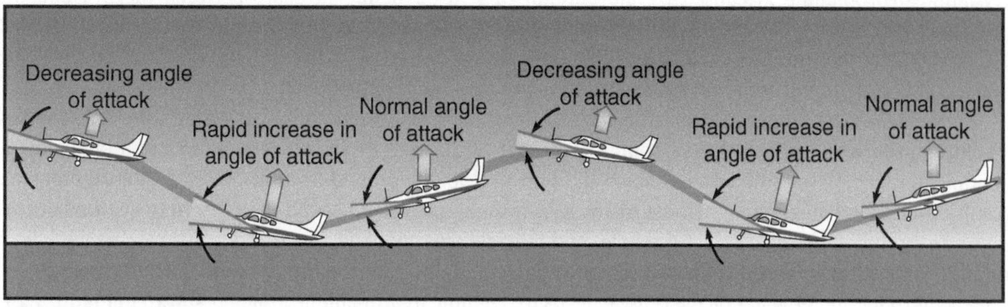

Figure 9.21. Porpoising during round out.
Source: Airplane Flying Handbook (2004a)/U.S. Department of Transportation Federal Aviation Administration/Public domain.

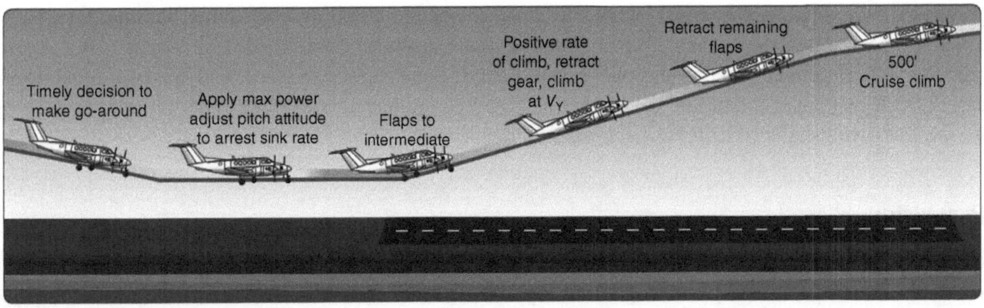

Figure 9.22. Go-around procedure.
Source: Airplane Flying Handbook (2016a)/U.S. Department of Transportation Federal Aviation Administration/Public domain.

briefed and planned before beginning the initial landing approach. Power should be increased smoothly to the maximum allowable power. Once the airspeed is safely above stall speed, the pitch attitude/angle of attack should be increased to gain altitude as soon as practical using V_X. As discussed in Chapter 4, increasing the pitch attitude too quickly will cause an abrupt increase in the angle of attack. If the AOA increases to the critical angle of attack, a stall will occur, and recovery from a stall at a low altitude is unlikely. Based on our Chapter 5 discussion, if applicable the landing gear should be retracted as soon as a positive rate of climb is established, and flaps retracted incrementally until fully retracted and V_Y is established (Figure 9.22).

HAZARDS OF HYDROPLANING

Hydroplaning occurs when a tire loses contact with the runway surface due to a buildup of water or slush between the tire and the ground. Besides the obvious dangers you may already associate with hydroplaning in an automobile, other adverse effects may occur in aircraft, such as ground controllability and weathervaning into a strong crosswind. NASA has researched this problem and identified three forms of hydroplaning: dynamic, viscous, and reverted rubber.

Dynamic Hydroplaning

Dynamic hydroplaning is caused by the buildup of hydrodynamic pressure between the tire and the pavement and usually occurs at relatively high speeds. The water pressure creates an upward force that effectively lifts the tire off the surface. When complete, separation of the tire and pavement occurs, the condition is called total dynamic hydroplaning, and wheel rotation will stop.

Figure 9.23 shows the forces on a tire without the presence of water on the runway. Figure 9.23a shows a standing tire with the only forces being the weight of the aircraft and the ground reaction opposing it. As the aircraft moves to the left, as shown in Figure 9.23b, the ground friction causes a spin-up moment, resulting in wheel rotation. When the tire is rolling freely at a fixed speed on a dry runway, the vertical ground reaction shifts forward of the axle, and a spin-down moment that offers resistance to the wheel rotation is developed. When these two moments are equal, the wheel is turning at a constant rpm.

The introduction of water on the runway leads to dynamic hydroplaning, which is shown in Figure 9.24. Deep fluid on the runway creates additional drag on the tire when it is displaced from the tire path, and a high-spray pattern is produced, as shown in Figure 9.24a. As the forward speed of the aircraft is increased

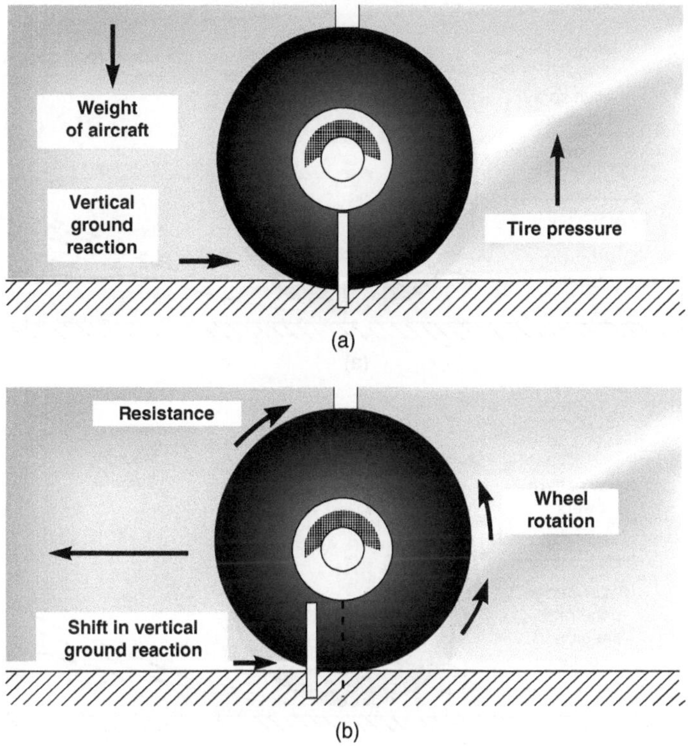

Figure 9.23. Forces on the tire: (a) static condition, (b) rolling tire.

(Figure 9.24b), the spray pattern thrown up by the tire lowers and the wedge of water penetrates the tire–ground contact area and produces a hydrodynamic lift force on the tire. This is *partial hydroplaning*. As the speed increases, the spray pattern becomes flatter, and the wedge of fluid penetrates farther into the ground contact area until at some high forward speed complete separation of the tire and runway takes place and total hydroplaning occurs, as shown in Figure 9.24c.

The ground friction is progressively reduced as the wedge of water penetrates beneath the tire. It approaches zero at total hydroplaning, and the spin-down moment causes the tire to stop the wheel rotation. Obviously, no braking action is available when the wheel is not making contact with the runway.

Total dynamic hydroplaning is more of a landing than a takeoff problem, but crosswind takeoffs are dangerous under these conditions. Within dynamic hydroplaning, Horne's formula can predict the approximate speed at which total dynamic hydroplaning occurs:

$$V_{\mathrm{H}} = 9\sqrt{P} \tag{9.4}$$

where

V_{H} = predicted speed at which hydroplaning begins (kts.)

P = tire inflation pressure (psi)

9 = constant (may vary 6–9 depending on the type of tire).

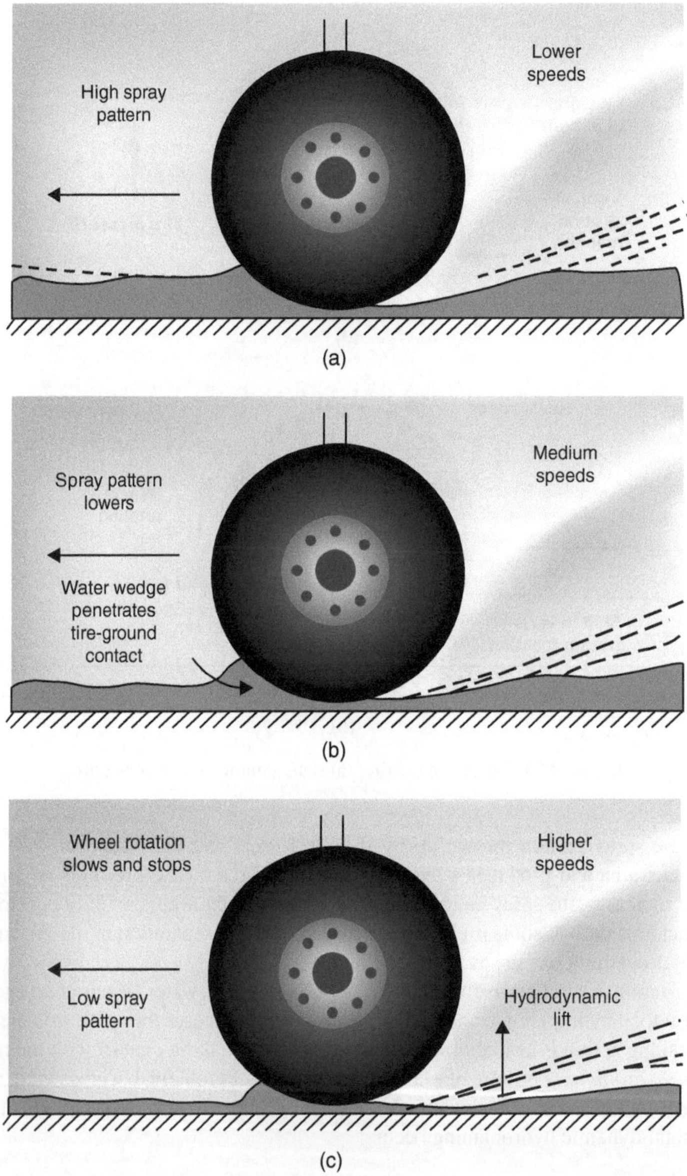

Figure 9.24. Hydroplaning forces on the tire: (a) low speed, (b) medium speed, and (c) high speed.

Total dynamic hydroplaning usually does not occur unless a severe rain shower is in progress. There must be a minimum water depth present on the runway to support the tire. The exact depth cannot be predicted since other factors, such as runway smoothness and tire tread, influence dynamic hydroplaning. Both smooth runway surfaces and smooth tread tires will induce hydroplaning with lower water depths. While the exact depth of water required for hydroplaning has not been accurately determined, a conservative estimate for an "average" runway is that water depths in excess of 0.1 in. may induce full hydroplaning.

EXAMPLE

A Cirrus SR22T lands on a runway contaminated with water from a recent thunderstorm, calculate the potential hydroplaning speed during landing if the main gear tire pressure is 63 psi.

$$V_{\mathrm{H}} = 9\sqrt{P}$$

$$V_{\mathrm{H}} = 9\sqrt{63\ \text{psi}} \rightarrow V_{\mathrm{H}} = 9(8)$$

$$V_{\mathrm{H}} = 72\ \text{kts.}$$

Viscous Hydroplaning

Viscous hydroplaning is more common than dynamic hydroplaning. Viscous hydroplaning may occur at lower speeds and at lower water depths than dynamic hydroplaning. Viscous hydroplaning occurs when the pavement surface is lubricated by a thin film of water. The tire is unable to penetrate this film, and contact with the pavement is partially lost. Viscous hydroplaning often occurs on smooth runway pavements or where rubber deposits are present, usually in the touchdown area where a thin water film can significantly reduce the coefficient of friction.

Reverted Rubber Hydroplaning

The third type of hydroplaning is known as *reverted rubber (steam) hydroplaning*. This may occur if the pilot reacts to dynamic hydroplaning by locking the brakes. As the aircraft gradually slows down, the wheels eventually make contact with the runway. With the wheels still locked, the friction between the tires and the runway melts the rubber of the tires. White streaks on the runway are an indication that this type of hydroplaning has occurred. Examination of the aircraft tire will show an elliptically shaped tacky or melted rubber condition. This condition occurs when the heat that is generated during a locked wheel skid reverts the rubber to a softer, slippery state.

Methods to prevent hydroplaning include transverse runway grooving, frequent removal of rubber deposits from the touchdown areas, maximum use of aerodynamic braking, and pilot education.

LANDING DECELERATION, VELOCITY, AND DISTANCE

Forces on the Aircraft During Landing

Transport and other large aircraft use a different landing technique than many military aircraft. Nearly all modern transports have anti-skid brakes, ground spoilers, and thrust reversers. They land in a three-point attitude and deploy spoilers, thrust reversers, and brakes soon after touchdown (Figure 9.25). The forces acting on any aircraft during the landing roll are shown in Figure 9.26, note that thrust reverser forces are not shown in this figure.

Many military and similar high-performance aircraft use aerodynamic drag braking as well as brake friction to slow the airplane; even small, single-engine general aviation airplanes can benefit from some aerodynamic drag at higher speeds. The use of aerodynamic drag is most efficient to assist deceleration when the speed of the aircraft on the runway is at least 60–70% of the touchdown speed and can save on wear and tear on tires and brakes. These aerodynamic forces are shown in Figure 9.27. It is assumed that:

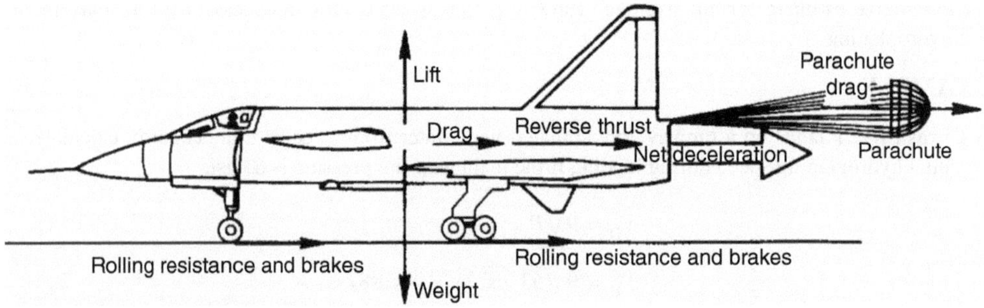

Figure 9.25. Net forces on a decelerating military aircraft.
Source: NASA/NASA-SP-367/Public Domain.

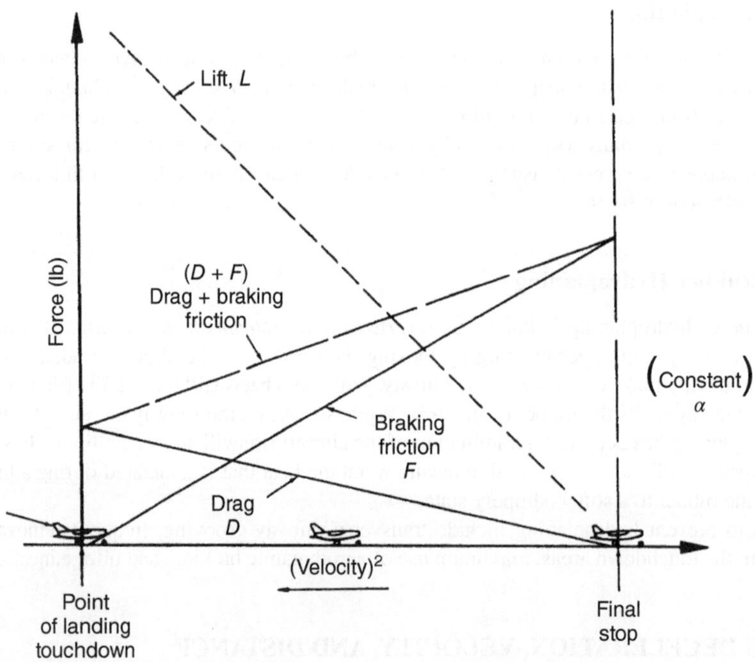

Figure 9.26. Forces acting on an airplane during landing.

1. The aircraft has a tricycle landing gear and lands on the main gear and the pilot holds the nose up to take advantage of the aerodynamic drag.
2. Foot brakes are not applied until the pilot can no longer keep the nosewheel off the runway.
3. The aircraft does not develop lift in the three-point position and the nosewheel is on the runway.

The braking action of the wheel brakes involves many factors, including:

- Tire material.
- Tread design and condition.
- Runway surface material and condition.

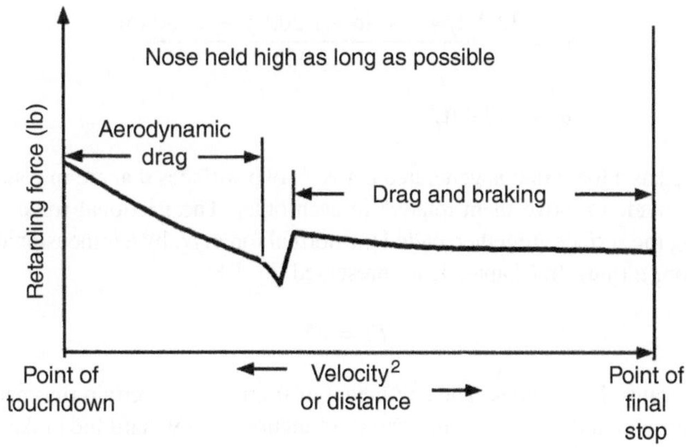

Figure 9.27. Aerodynamic braking and wheel braking.

- Amount of braking applied (wheel slippage).
- Amount of normal (squeezing) force between tires and the runway.
- Pilot technique.

After touchdown, an aircraft has negative acceleration ($-a$), or deceleration. Similar to Figure 8.26, Figure 9.26 shows that the net accelerating (or decelerating) force on the aircraft, F_N, equals the thrust T, minus the drag, D, minus the rolling friction, F. Assuming constant acceleration (or deceleration), we apply Newton's second law (Eq. 1.3), from which Eq. 8.2 is derived. Thus:

$$a = \frac{F_N}{m} \rightarrow a = \frac{g(T - D - F)}{W}$$

where

a acceleration (or deceleration) (fps^2)

F_N = net accelerating force (lb)

W = weight (lb)

g = gravitational acceleration (= 32 fps^2)

m = mass (slugs) (= W/g).

Equation 8.2 shows that the acceleration (deceleration) of the aircraft is affected by the thrust, the drag, the rolling friction, and the weight of the aircraft.

EXAMPLE

An airplane with a single turbojet engine lands on a runway. Calculate the net acceleration (a) (assuming acceleration is constant). Given: aircraft weight = 14 000 lb, T_a (average thrust during landing roll) = 700 lb, average drag during the landing roll = 1 200 lb, average braking friction during the landing roll = 5 100 lb, sea level, standard conditions.

$$a = \frac{F_N}{m} \rightarrow a = \frac{g(T - D - F)}{W}$$

$$a = \frac{32.2\ \text{ft/s}^2(700\ \text{lb} - 1\,200\ \text{lb} - 5\,100\ \text{lb})}{14\,000\ \text{lb}}$$

$$a = -12.9\ \text{ft/s}^2$$

Friction is defined as a force that is generated between two surfaces that are in contact with each other when an attempt is made to move them relative to each other. The frictional force, F, is related to the force that is pressing the surfaces together, called the normal force, N, by a dimensionless factor called the coefficient of friction, μ (mu). In Chapter 1, we presented Eq. 1.8:

$$F_b = \mu N$$

Figure 9.28 illustrates these forces. The coefficient of friction for a certain tire material, tread design, and tire wear operating on a dry concrete runway is a function of how hard the brakes are being applied: the harder the brakes are applied, the more slip there is between the tires and the runway. A rolling tire with no braking action does not slip. A locked wheel that skids without turning is defined as 100% slip.

A plot of the coefficient of friction versus wheel slippage is shown in Figure 9.29. The maximum braking coefficient of friction can be seen to be about 0.7. It occurs when wheel slip is about 10%. Locking the wheels reduces the value to about 0.5, causes the tires to wear unevenly, and may cause tire blowout. Water, snow, or ice on the runway causes a large reduction in the coefficient of friction, as shown in Figure 9.30.

Application 9.2

Using Figure 9.30 and Eq. 1.8, review the difference in the value of μ for dry concrete and smooth, clear ice.

Calculate the braking force of an aircraft for both conditions if the weight of the aircraft is 7 000 lb and we assume 100% of the weight is available for braking (no lift being generated by the wings) as the normal force (N). What is the difference in F_b for each condition? Consider an aircraft landing on a runway for each of the conditions, what would be the impact on the runway length requirement for the icy runway?

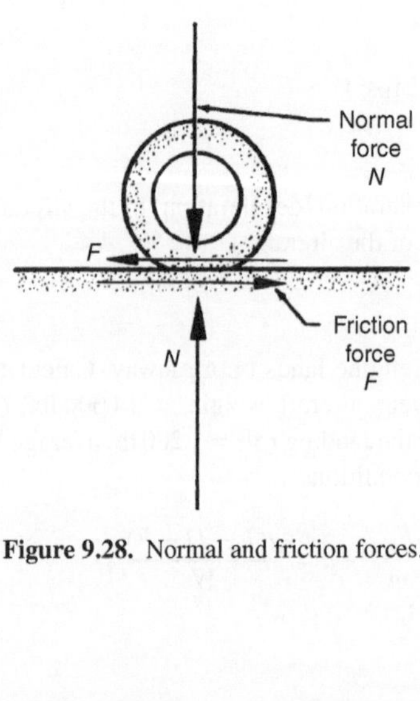

Figure 9.28. Normal and friction forces.

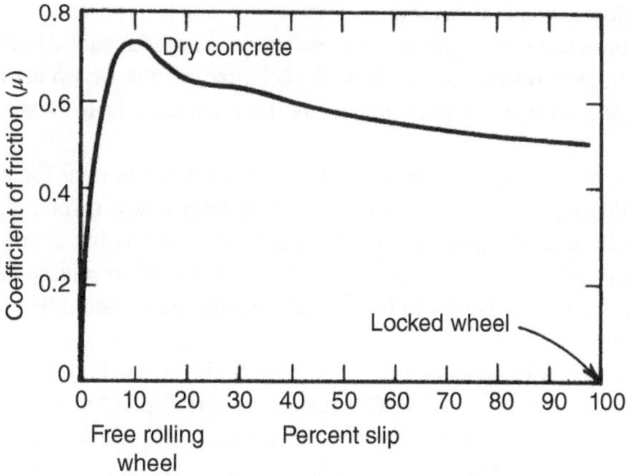

Figure 9.29. Coefficient of friction versus wheel slippage.

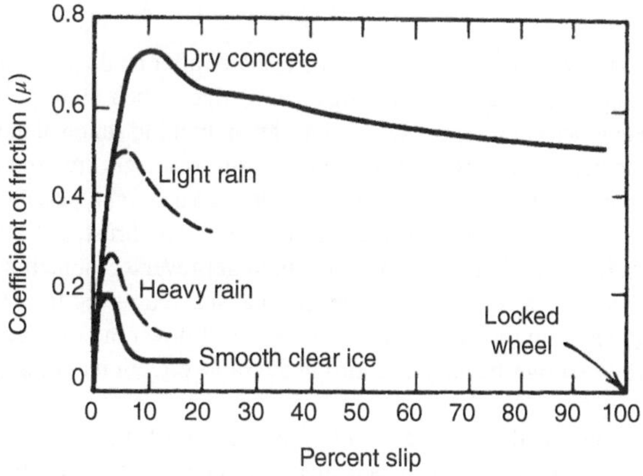

Figure 9.30. Effect of runway condition on coefficient of friction.

Braking Techniques

The technique of using braking devices may vary with aircraft type and model, but general principles can be stated for aircraft using aerodynamic braking. During the initial landing roll, the velocity and dynamic pressure are high, and aerodynamic braking will be most effective, especially when above 60% of the touchdown speed. Holding the nose of a tricycle landing gear airplane high will increase the parasite drag and is an effective method of slowing the aircraft, though the pilot must be careful to keep the aircraft from becoming airborne again, raising the nose too high causing a tail scrape, using wheel brakes while aerobraking and letting the nose fall too quickly.

At some point during the landing roll, the reduced airspeed will cause the elevator to lose effectiveness, and the nose of the aircraft can no longer be held off the runway. Usually, this is the point where wheel braking is initiated and becomes the principal method of stopping the aircraft (in combination with thrust reversers if applicable). It is debatable whether it is better to retract the flaps, thus increasing the weight on the wheels, or to leave them down, thus preserving the aerodynamic braking effect. Inadvertent retraction of the landing gear instead of the flaps has occurred, so the best practice is to leave the flaps down until taxiing off the runway, unless required by the manufacturer to obtain maximum short-field performance.

Wheel braking force for all aircraft will be a maximum when the normal force on the braked wheels is greatest. When landing on a contaminated runway, especially if icy, reverse thrust and aerodynamic braking may be most efficient at higher speeds. For small, tricycle landing gear aircraft, this will occur when the control stick is held in the back position. Many pilots tend to release back pressure when the nosewheel contacts the runway. This should be avoided, and the stick should be held full back as long as braking action is desired.

If an aircraft is equipped with spoilers, they should be deployed as soon as the aircraft is "weight on wheels" during touchdown. As discussed in Chapter 3, spoilers "spoil" most of the lift over the wing, allowing more of the weight of the airplane to be on the wheels. The FAA's *Airplane Flying Handbook* states that spoilers may increase drag by up to 60% if applied as soon as touchdown occurs, and they may increase wheel loading by as much as 200% in the landing flap configuration. Many modern aircraft automatically deploy the inboard spoilers once weight on wheels is sensed.

Thrust Reversers

Many jet aircraft use thrust reversers to reduce stopping distance after landing, or during a rejected takeoff. Because of the kinetic energy of these aircraft during landing (due to their high speed and large mass), plus the fact at flight idle the engines are still producing some thrust, braking action alone is insufficient to bring these aircraft to stop at an acceptable distance. Therefore, thrust reversers are used to assist in slowing the aircraft down after landing. Thrust reversers are grouped into two general types, *mechanical-blockage* and *aerodynamic-blockage*. Figure 9.31 shows an example of two types of thrust reversers that are incorporated as part of the turbine engines we discussed in Chapter 6, target reverser (sometimes known as clamshell or bucket) and cascade reverser. A component of the jet exhaust is redirected from the normal flow out the back of the engine to an angle in the direction of aircraft travel down the runway (Figure 9.32). This of course slows the aircraft down significantly, as thrust in the same direction as drag is now used to slow the aircraft.

Target reversers have clamshell doors that rotate from the rear of the engine and redirect most of the exhaust forward. This is known as a type of mechanical-blockage system. This type of system is usually found with ducted turbofan engines where the fan and core flow mix in a common nozzle.

Cascade reversers are common on unducted turbofan engines and redirect the bypass fan air for the reverse component. This type of reverser system usually consists of blocker doors, a translating cowl, and cascade vanes to redirect the fan flow to slow down the aircraft. Because the fan may produce up to 80% of the engine's thrust, by redirecting the fan airflow through the cascade vanes forward (instead of aft) in the direction of aircraft travel, the aerodynamic force quickly slows the airplane and minimizes wear on the braking system.

Either type of thrust reverser should be applied as soon as possible after the nosewheel touches down. This will allow the kinetic energy of the airplane to be dissipated faster because the speed is greater. Some aircraft pitch up when thrust reversers are applied, especially when combined with the effect of spoilers, so careful monitoring of touchdown speed is important.

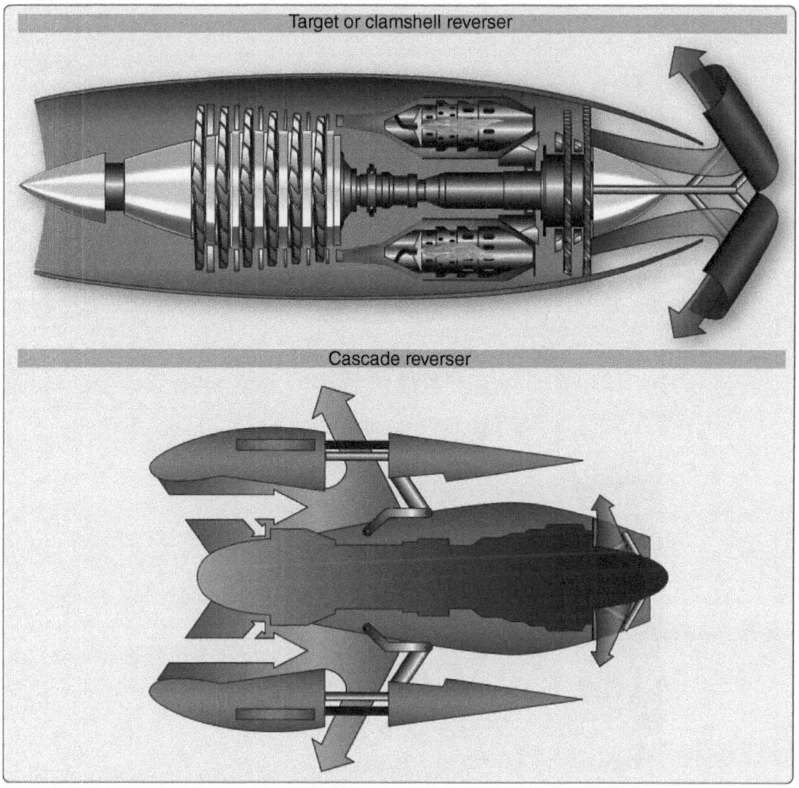

Figure 9.31. Thrust reversers.
Source: Airplane Flying Handbook (2016a)/U.S. Department of Transportation Federal Aviation Administration/Public domain.

LANDING EQUATIONS

General Equation

The approach and landing procedures discussed in this chapter, and the use of mechanical devices available to minimize stopping distance, must take into consideration the weight changes and environmental conditions that change with each separate flight. Of these factors, the change in gross weight with each flight may have the largest effect on landing distance as seen in Figure 9.1. As the density altitude increases or decreases for a given location, so will the landing distance. If the runway configuration does not allow for an optimum headwind component, or the runway slope or surface is not a level, dry, and paved surface, the landing distance will change. Using the same linear equations first introduced in Chapter 1, and then applied to changing conditions in Chapter 8, we can calculate the new landing distance with a change in weight or environmental conditions.

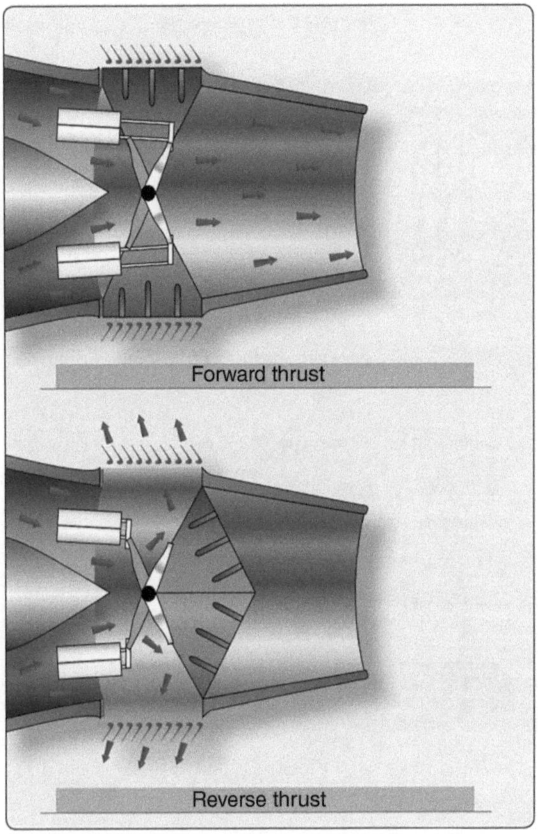

Figure 9.32. Thrust reverser activation.
Source: Aviation Maintenance Technician Handbook (2023d)/U.S. Department of Transportation Federal Aviation Administration/Public domain.

The calculation of landing distance, s, in terms of touchdown velocity, V_i, and deceleration, $-a$, is similar to the takeoff case, and when combined with Eq. 1.12 we find:

$$s = \frac{V_f^2 - V_i^2}{2(-a)}, \text{ and if landing to a full stop, then :}$$

$$s = \frac{-V_i^2}{2(-a)}$$

$$s = \frac{V_i^2}{2a} \tag{9.5}$$

To determine how the landing distances are affected by variations in weight, altitude, and wind conditions, a ratio is established like that used for the takeoff case (Eq. 8.3):

$$\frac{s_2}{s_1} = \left(\frac{V_2}{V_1}\right)^2 \left(\frac{a_1}{a_2}\right)$$

Effect of Weight Change

Changing the gross weight of a landing aircraft affects the landing speed in the same way the takeoff speed is affected. For an increase in gross weight, a greater speed will be required due to the increase in lift needed for the increase in weight. A 21% increase in landing weight requires a 10% increase in landing speed. The landing speed must be changed by using Eq. 4.7:

$$\frac{V_2}{V_1} = \sqrt{\frac{W_2}{W_1}} \text{ or } \left(\frac{V_2}{V_1}\right)^2 = \frac{W_2}{W_1}$$

Changing the weight of the aircraft has minimal effect on the deceleration of the aircraft caused by braking action. At first, this statement seems counterintuitive, but the heavier the aircraft, the more weight there will be on the braking wheels. Therefore, more braking force is available to counteract the greater mass to be decelerated. For the braked phase of the landing,

$$\frac{a_1}{a_2} = 1$$

Substituting these last two equations into Eq. 8.3 gives the effect of a weight change on landing distance. An increase in gross weight translates to an increase in landing distance, as shown in Eq. 9.6:

$$\frac{s_2}{s_1} = \left(\frac{V_2}{V_1}\right)^2 \left(\frac{a_1}{a_2}\right) \text{ substitute } \left(\frac{V_2}{V_1}\right)^2 = \frac{W_2}{W_1} \text{ and } \frac{a_1}{a_2} = 1, \text{then :}$$

$$\frac{s_2}{s_1} = \frac{W_2}{W_1} \text{ or,}$$

$$s_2 = s_1 \left(\frac{W_2}{W_1}\right) \tag{9.6}$$

EXAMPLE

The landing distance for a turbojet airplane, at sea level on a standard day, is 2 400 ft when at a weight of 8 000 lb. If the airplane consumes 500 lb of fuel before the next landing attempt, assuming all other conditions are the same, what is the new minimum landing distance required?

$$\frac{s_2}{s_1} = \left(\frac{W_2}{W_1}\right) \rightarrow s_2 = 2\,400 \text{ ft} \left(\frac{7\,500 \text{ lb}}{8\,000 \text{ lb}}\right)$$

$$s_2 = 2\,250 \text{ ft}$$

Effect of Altitude

Airspeed is affected by the same factors at altitude whether the aircraft is taking off or landing; as pressure and temperature change, so does the density of the air and the performance of the aircraft. An increase in density altitude increases the landing speed, IAS will remain the same but true air speed (TAS) will increase. The FAA's *Airplane Flying Handbook* (2016a) states that a 10% increase in landing speed requires a 21% increase or more in landing distance. In mountain flying operations, this higher-than-normal TAS can lead to problems on landing, as the GS will be higher. This higher GS may cause the pilot to inadvertently slow the aircraft to near stall speed as he or she tries to make the approach speed closure rate look "normal." *Fly by the numbers* IAS to remain safe.

The equation showing the velocity change at altitude in Chapter 8 holds for the landing phase as well:

$$\left(\frac{V_2}{V_1}\right)^2 = \frac{1}{\sigma_2}$$

Unless the aircraft is equipped with thrust reversers, it does not rely on the engine performance to decelerate the aircraft as it did for takeoff acceleration. So,

$$\frac{a_1}{a_2} = 1$$

Substituting the values from these last two equations into Eq. 8.4 gives us the *effect of altitude on landing distance for all non-thrust-reverser aircraft*:

$$\frac{s_2}{s_1} = \left(\frac{V_2}{V_1}\right)^2 \left(\frac{a_1}{a_2}\right) \text{ substitute } \left(\frac{V_2}{V_1}\right)^2 = \frac{1}{\sigma_2} \text{ and } \frac{a_1}{a_2} = 1\text{, then:}$$

$$\frac{s_2}{s_1} = \frac{1}{\sigma_2} \text{ or,}$$

$$s_2 = s_1 \left(\frac{1}{\sigma_2}\right) \tag{9.7}$$

where

s_1 = standard sea level landing distance

s_2 = new altitude landing distance

σ_2 = new altitude density ratio.

EXAMPLE

The landing distance for a turbojet airplane, at sea level on a standard day, is 2 400 ft when at a weight of 8 000 lb. If the same airplane lands at an airport with a density ratio of 0.8617, assuming all other conditions are the same, what is the new minimum landing distance?

$$\frac{s_2}{s_1} = \left(\frac{1}{\sigma_2}\right) \rightarrow s_2 = 2\,400\text{ lb}\left(\frac{1}{0.8617}\right)$$

$$s_2 = 2\,785\text{ ft}$$

Application 9.3

Using Figure 9.1, calculate your sea level, standard day (standard temperature) landing distance ground roll with your own derived landing weight, and zero headwind/tailwind. Then, recalculate your landing distance ground roll for a runway at 5 000 ft pressure altitude at standard temperature (Hint: Table 2.1 has standard day temperatures for pressure altitudes).

Did the landing distance ground roll increase or decrease with an increase in altitude? Why? Using Eq. 9.7, calculate s_2 using your derived s_1 landing distance and density ratio for 5 000 ft pressure altitude. How does it compare with your Figure 9.1 result?

Effect of Wind

A headwind will decrease the landing distance and a tailwind will increase the landing distance due to the change in touchdown GS, as we saw in Chapter 8 for takeoff performance calculations. Figure 9.33 shows us once again the effect of landing with a headwind or tailwind. Figure 9.34 shows the effect on the touchdown point when gliding into a headwind, the touchdown point will be closer to the threshold of the runway because the GS will be less than if there was not a headwind given the same V_{REF} speed on final. A pilot in a powered aircraft may either need to fly a base leg closer to the runway or adjust power settings to land at the desired point during the final approach. Figure 9.35 also shows the consequence of landing with a tailwind, although V_{REF} may be the same, landing with a tailwind leads to a higher GS and a touchdown point farther down the runway than desired.

Headwinds and tailwinds affect takeoff and landing performance in exactly the same way, so Eqs. 8.7 and 8.8 apply to the landing situation also. We rewrite them with new equation numbers:

$$s_2 = s_1\left(1 - \frac{V_w}{V_1}\right)^2 \text{(Headwind)} \tag{9.8}$$

$$s_2 = s_1\left(1 + \frac{V_w}{V_1}\right)^2 \text{(Tailwind)} \tag{9.9}$$

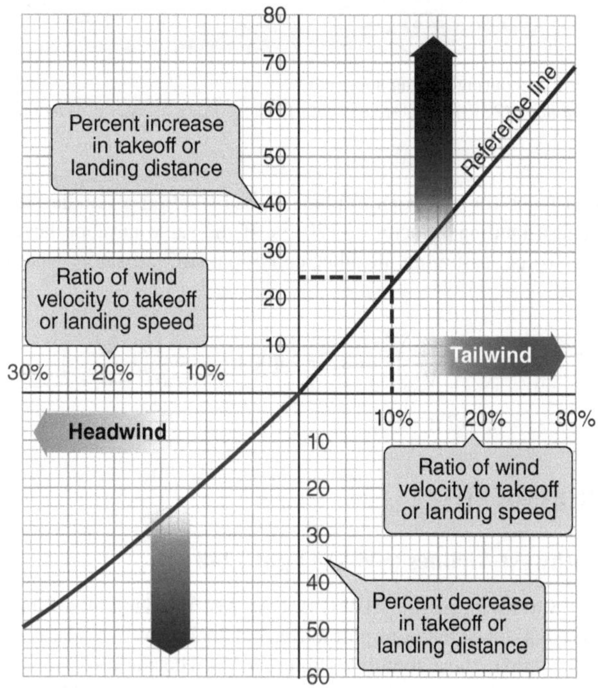

Figure 9.33. Effect of wind on landing.
Source: Pilot's Handbook of Aeronautical Knowledge (2008a)/U.S. Department of Transportation Federal Aviation Administration/Public domain.

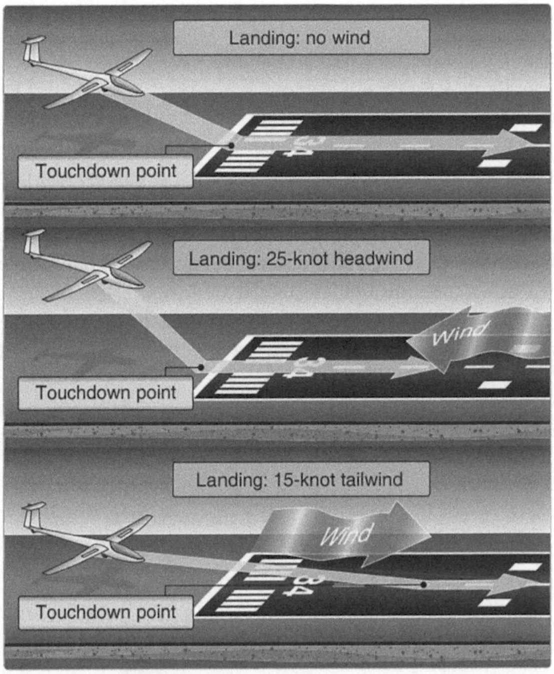

Figure 9.34. Effect of headwind during landing approach.
Source: Glider Flying Handbook (2013)/U.S. Department of Transportation Federal Aviation Administration.

EXAMPLE

The no-wind landing distance for a turbojet airplane, at sea level on a standard day, is 2 400 ft when at a weight of 8 000 lb and a landing velocity of 120 kts. If the same airplane lands at an airport with a tailwind component of 10 kts., assuming all other conditions are the same, what is the new minimum landing distance required?

$$\frac{s_2}{s_1} = \left(1 + \frac{V_w}{V_1}\right)^2 \rightarrow s_2 = 2\,400\,\text{ft}\left(1 + \frac{16.9\,\text{ft/s}}{202.8\,\text{ft/s}}\right)^2$$

$$s_2 = 2\,817\,\text{ft}$$

LANDING ENVIRONMENT

Accident Brief: Southwest Airlines (SWA) Flight 1248

On 8 December 2005, a Southwest Airline (SWA) Boeing 737-700, landing at Chicago Midway International Airport, rolled through a blast fence, airport fence, and finally onto a roadway striking an automobile, resulting in the death of a child in the automobile (Figure 9.35). The National Transportation Safety Board (NTSB) determined the probable cause to be the pilots' failure to use available reverse thrust in a timely manner, most likely due to the pilots' lack of familiarity with the airplane's autobrake system, which caused a distraction away from proper thrust reverser operation. Contributing to the accident were reports of poor braking action on the runway and a tailwind component greater than 5 kts. Specifically, at the time of the

Figure 9.35. Southwest Airlines Flight 1248 post-accident.
Source: National Transportation Safety Board/Public domain.

accident, the pilots reported they received runway condition/braking action reports for runway 31C, good or fair braking action for the first half of the runway, and poor braking action for the second half of the runway.

Upon landing, SWA Flight 1248 fully deployed ground spoilers, and the autobraking system was engaged. As discussed earlier in this chapter and demonstrated in Figure 9.25, drag and friction forces were at work to slow the Boeing 737 as normal. Unfortunately, the autobrakes were deactivated after 12 seconds and the first indication of thrust reverser activation was not recorded until 15 seconds after touchdown, without full deployment until 18 seconds after touchdown. Parametric data below from five flights that landed 25 minutes before the accident flight show similar aircraft and expected performance values to the accident aircraft, except for the time to full thrust reverser deployment which resulted in a faster speed at the 40-kts. mark (Table 9.1). Combine this delayed response for thrust reverser activation, inactivation of the autobrakes, a contaminated runway, a tailwind, and finally a shorter-than-average runway for a Part 139 airport, and you have the increased chance of an overrun.

Runway Surface

We discussed in Chapter 8 that most performance charts are for a paved, smooth, level, and dry runway surface. When calculating landing distance, many of the same concerns become factors that must be incorporated into the manufacturer's performance charts. Not only will contamination on the runway impact friction and deceleration, but braking action may be impacted as well, and the required runway length for stopping increased. Poor approach procedures, higher V_{REF} speeds in icing conditions, and contamination on the runway can lead to large increases in landing distance.

Table 9.1. Parametric data for Southwest Airlines Flight 1248.

Airline and flight number	Airplane type	Time	Touchdown ground speed in knots	Vertical load (*G*)	Gross weight (lbs)	Seconds to brake use	Seconds to full thrust reverser deployment	Seconds to 40 kts.
United Airlines 1446	A320	1849	128	1.33	113 830	3.5[a]	6	18
SWA 2920	737	1852	140	N/A[b]	114 560	0.5	4	21
SWA 321	737	1901	121	1.5	103 320	6.0	4	25
SWA 2947	737	1902	132	1.95	110 320	8.0	4	26
SWA 1830	737	1904	126	1.35	105 520	2.0	4	24
Accident flight	737	1913	131	1.41	118 280	3.3[a]	18	33

Source: National Transportation Safety Board/Public domain.

[a]Note: The autobrake system initially applied the brakes for United Airlines flight 1446 and the accident flight.

[b]N/A in this table indicates not available.

As a result of the SWA Flight 1248 accident investigation, the NTSB's study found that the airplane's braking action ability was five times worse than would be expected on a bare and dry runway. A broader problem was found throughout the aviation industry that different methods and terminologies were used to define the relationship between runway surface condition and an airplane's braking ability. The European Aviation Safety Agency (EASA) and the Joint Aviation Authorities (JAA) used a conservative style approach that correlated the effective braking of an anti-skid braked wheel and contaminant type and depth (in the absence of any direct evidence). Boeing and subsequently SWA calculated operational airplane landing performance based on correlations between braking action reports and the airplane's braking ability. The International Civil Aviation Organization (ICAO) defines the relationship between runway surface conditions and an airplane's braking action (Table 9.2). Note the correlation between the Mu (μ) values in Table 9.2 and Figure 9.30. Finally, Transport Canada, in their *Aeronautical Information Manual*, publishes the Canadian Runway Friction Index (CRFI) that provides guidance to correlate ground surface vehicle friction survey measurements to airplane performance on certain winter-contaminant surfaces (Table 9.3). The CRFI uses a very detailed method of measuring and reporting friction on contaminated runways to determine landing distances. These friction values were obtained from decelerometers and are reported in the surface condition reports within Notice to Airmen (NOTAMS) as runway friction values.

Table 9.2. ICAO correlation of the measured runway braking action to estimated braking action.

Measured runway braking action (MU reading)	Estimated braking action
0.40 and above	Good
0.39 to 0.36	Medium to good
0.35 to 0.30	Medium
0.29 to 0.26	Medium to poor
0.25 and below	Poor

Source: National Transportation Safety Board/Public domain.

Table 9.3. CRFI index values published by Transport Canada.

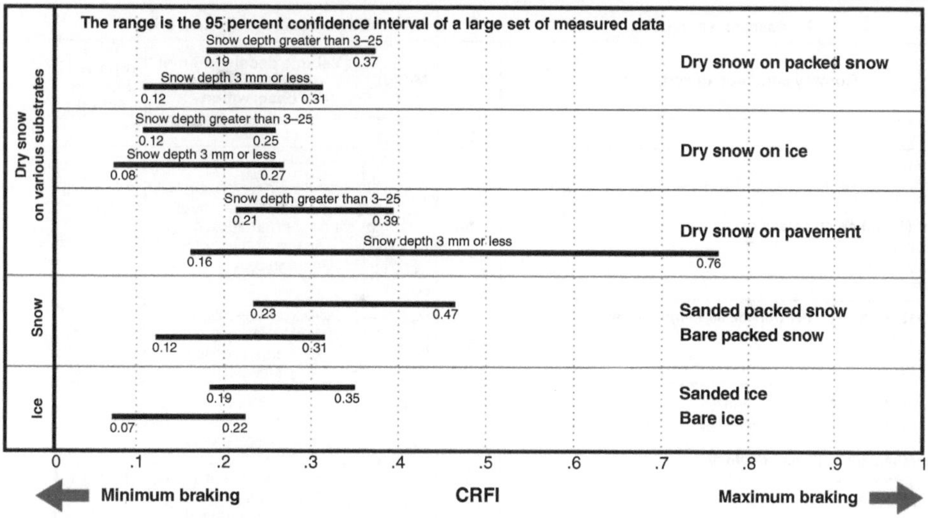

Source: Transport Canada/Public domain.

Through the issuance of AC 150/5200-30D, the FAA has developed the Runway Condition Assessment Matrix (RCAM) by which an airport operator reports a runway surface assessment (paved runways only) when contaminants are present. Upon completion of the assessment, the RCAM defines the format for the airport operator report, and a runway condition "code" is generated via the NOTAM System (Table 9.4). Finally, a pilot interprets the runway condition in terms related to airplane performance. With the Mu decimal removed, note again the Mu values in Table 9.4 compared to braking action Mu reports in Table 9.2 and Figure 9.30.

Runway Slope

As with takeoff performance, the slope or gradient of the runway can greatly impact stopping distance. Landing uphill reduces landing rollout while landing downhill will increase landing rollout. A rule of thumb when landing on a runway with a downslope is to account for a 10% increase in landing distance for every 1% of downslope. If your runway is 2 000 ft long and has a 3% downslope in the direction you are landing, then the *effective* runway length is only 1 400 ft. If there is a strong headwind then the headwind component may counteract the decrease in performance due to the down sloping runway. For every 1% of downslope, the equivalent landing distance would be equal to ~3 kts. of tailwind, so choosing to land on the down sloping runway would only be warranted if there were obstacles on the opposite approach end to the upsloping runway, or there was a strong headwind that would outweigh the negative impact of the down sloping runway. As in all performance matters, consult the aircraft flight manual (AFM) and add a comfort factor to the calculations to allow for mistakes.

Due to possible visual illusions for upsloping and down sloping runways, improper approach flight paths can lead to increased landing speeds. Always consider the warnings, notes, and conditions published on the performance charts, and be mindful of your personal limitations.

Table 9.4. Braking action assessment using RCAM.

Assessment criteria		Downgrade assessment criteria		
Runway condition description	**Code**	**Mu (μ)** [1]	**Vehicle deceleration or directional control observation**	**Pilot reported braking action**
• Dry	6	40 or Higher	---	---
• Frost • Wet (Includes Damp and 1/8 inch depth or less of water) **1/8 *in.* (3 mm) depth or less of:** • Slush • Dry Snow • Wet Snow	5	40 or Higher	Braking deceleration is normal for the wheel braking effort applied AND directional control is normal.	Good
5 ° *F* (–15 °C) and *Colder outside air temperature:* • Compacted Snow	4	39 to 30	Braking deceleration OR directional control is between Good and Medium.	Good to medium
• Slippery When Wet (wet runway) • Dry Snow or Wet Snow (Any depth) over Compacted Snow ***Greater than* 1/8 *in.* (3 mm) *depth of*:** • Dry Snow • Wet Snow ***Warmer than* 5 ° *F* (–15 °C) *outside air temperature:*** • Compacted Snow	3	39 to 30	Braking deceleration is noticeably reduced for the wheel braking effort applied OR directional control is noticeably reduced.	Medium
Greater than* 1/8 (3 *mm*) *inch depth of: • Water • Slush	2	29 to 21	Braking deceleration OR directional control is between Medium and Poor.	medium to poor
• Ice [2]	1	29 to 21	Braking deceleration is significantly reduced for the wheel braking effort applied OR directional control is significantly reduced.	Poor
• Wet Ice [2] • Slush over Ice [2] • Water over Compacted Snow [2] • Dry Snow or Wet Snow over Ice [2]	0	20 or Lower	Braking deceleration is minimal to non-existent for the wheel braking effort applied OR directional control is uncertain.	Nil

Source: Federal Aviation Administration (2020)/U.S. Department of Transportation/Public Domain.

SYMBOLS

N	Normal force (lb)
P	Tire inflation pressure (psi)
V_H	Hydroplaning speed (kts.)
V_{S0}	Power-off stalling speed (landing configuration) (kts.)
V_{S1}	Power-off stalling speed (specified configuration) (kts.)
V_{REF}	Reference speed (kts.)
γ (gamma)	Glide angle (degrees)
μ (mu)	Coefficient of friction

EQUATIONS

9.1 $L = W \cos \gamma$

9.2 $D = W \sin \gamma$

9.3 $\tan \gamma = \frac{1}{\text{L/D}}$

9.4 $V_{\text{H}} = 9\sqrt{P}$

9.5 $s = \frac{V_i^2}{2a}$

9.6 $s_2 = s_1 \frac{W_2}{W_1}$

9.7 $s_2 = s_1 \frac{1}{\sigma_2}$

9.8 $s_2 = s_1 \left(1 - \frac{V_{\text{w}}}{V_1}\right)^2$ (Headwind)

9.9 $s_2 = s_1 \left(1 + \frac{V_{\text{w}}}{V_1}\right)^2$ (Tailwind)

KEY TERMS

Approach climb
Ballooning
Bouncing
Canadian Runway Friction Index (CRFI)
Crosswind landing
Deceleration
Dynamic hydroplaning
Effective runway length
Flare
Friction
Glide path
Hydroplaning
Landing climb
Landing field length
Porpoising
Power-off stalling speed (landing configuration)
Power-off stalling speed (specified configuration)
Reference speed
Reverted rubber hydroplaning
Round out
Runway Condition Assessment Matrix (RCAM)
Short-field landing
Sideslip
Soft/rough-field landing
Stabilized approach
Thrust reversers
Weathervane
Viscous hydroplaning

PROBLEMS

Note: Answers to the problems in the first two sections are given at the end of the book.

Application of Chapter Theory

1. An airplane is in trimmed flight and its velocity is constant.
   a. It has no unbalanced forces acting on it.
   b. It has no unbalanced moments acting on it.
   c. It is in a state of equilibrium.
   d. All of the above are true.

2. An airplane is in a constant velocity engine out glide. Its altitude is decreasing. It is not in a state of equilibrium.
   a. True
   b. False

3. At $(L/D)_{max}$ an airplane will
   a. be at minimum glide angle.
   b. be achieving maximum glide distance.
   c. have a glide ratio equal to the numerical value of $(L/D)_{max}$.
   d. All of the above.

4. The power-off stalling speed or minimum steady flight speed at which the aircraft is controllable in the landing configuration:
   a. V_{SO}
   b. V_{REF}
   c. V_{S1}
   d. V_{LOF}

5. The speed at which the aircraft is controllable in a specified configuration, published by the manufacturer:
   a. V_{SO}
   b. V_{REF}
   c. V_{S1}
   d. V_{S2}

6. When landing in an airplane equipped with tricycle-gear while experiencing a right crosswind, with proper crosswind control technique applied, the first wheel to make contact with the runway should be the:
   a. Nosewheel.
   b. Left main.
   c. Right main.
   d. Left or right main.

7. When conducting a short-field approach, a stabilized approach is not recommended in order to clear the required obstacle and/or to stop in the shortened runway available.
   a. True
   b. False

8. Once on the ground, the recommended braking technique for a short-field and a soft/rough-field landing are the same.
   a. True
   b. False

9. A steep, low-power approach is more dangerous for heavy airplanes than light airplanes because
   a. recovery from a high rate of descent involves a great increase in power (or thrust).
   b. the aircraft will float down the runway and possibly overshoot the runway.
   c. flaring the aircraft to decrease the rate of descent increases induced drag.
   d. Both (a) and (c).

10. A low-angle, high-power (or thrust) approach
   a. may be used for propeller aircraft if a short-field landing is required.
   b. is dangerous for propeller aircraft if an engine fails.
   c. should be avoided at any cost for jet airplanes as high induced drag results.
   d. All of the above.

11. Aerodynamic drag is more effective than wheel braking during the first part of a landing.
   a. True
   b. False

12. As a general rule, when the nose of a tricycle landing gear airplane can no longer be held off the runway, it is time to start applying wheel brakes.
   a. True
   b. False

13. In applying wheel brakes, it is not a good idea to apply full brake pressure because:
   a. you might cause a blowout.
   b. you get lower coefficients of friction when wheel slippage exceeds about 15%.
   c. Both (a) and (b).

14. In general, if maximum wheel braking of a tricycle landing gear airplane is desired, the pilot should keep the stick full aft even after the nose is on the runway.
   a. True
   b. False

15. Headwinds and tailwinds affect the landing distance by the same amount as they affect the takeoff distance.
   a. True
   b. False

Calculations Using Chapter Principles

16. Using Figure 9.1, calculate the landing ground roll given the following information:

   Temperature = 10 °C

   Pressure altitude = 8 000 ft

   Weight = 2 400 lb

   Tailwind component = 5 kts.

   a. 2 000 ft
   b. 1 400 ft

c. 1 250 ft
d. 2 100 ft

17. Using Figure 9.1, calculate the landing distance over a 50 ft obstacle given the following information:

Temperature = 20 °F

Pressure altitude = 4 000 ft

Weight = 2 600 lb

Headwind component = 5 kts.

a. 2 000 ft
b. 1 750 ft
c. 1 250 ft
d. 1 500 ft

18. If your tire pressure is 140 psi, calculate the approximate speed in knots where total dynamic hydroplaning may occur.

a. 106 kts.
b. 110 kts.
c. 12 kts.
d. 68 kts.

19. Calculate the deceleration of an aircraft during landing rollout given the following conditions:

T_a (average) = 575 lb

Average drag = 1 300 lb

Average braking/rolling friction = 4 200 lb

Aircraft weight = 12 000 lb

Assumptions = constant deceleration, standard day, sea level conditions.

a. 10.2 fps
b. 16.3 fps
c. 11.3 fps
d. 13.2 fps

20. Calculate your landing distance if your initial approach speed at touchdown is 140 kts. and your average acceleration during landing rollout is −19.6 fps.

a. 1 428 ft
b. 500 ft
c. 549 ft
d. 1 639 ft

Aircraft data for Problems 21–25:

Landing speed = 100 kts.

Gross weight = 8 000 lb

Average retarding force = 2 000 lb (brakes + thrust reversers if applicable)

Sea level standard conditions.

21. Calculate the deceleration.

22. Calculate the no-wind stopping distance.

23. If the weight is increased to 13 000 lb, calculate the no-wind stopping distance. Assume the same landing speed (100 kts.).

24. If the 8 000 lb airplane lands with 10 kts. of tailwind, calculate the stopping distance.

25. If the aircraft is operating from an airfield where the density ratio is 0.8, calculate the no-wind stopping distance.

Correlation Through Scenario-Based Learning

Aircraft – *Diamond DA50RG*/Section 5 AFM

1. Read the landing distance chart warnings, notes, cautions, and conditions at the beginning of the Landing Distance section within Section 5, DA50 AFM.

2. Based on the chapter reading and the DA50 AFM, calculate the potential hydroplaning speed of the main gear tires. What about the nose wheel tire?

3. Using the Landing Distance – Flaps LDG/4 407 lb (imperial units) chart, calculate the landing distance over a 50 ft obstacle and the ground roll for the DA50 based on the following:

ATIS pressure = 30.20″

Airport elevation = 1 280 ft MSL

OAT = 10 °C

4. Using your answers from Problem #3, calculate your landing distances for a 20-kt. *headwind* component.

5. Using your answer from Problem #3, calculate your landing distances for a 6-kt. *tailwind* component and when using a *wet*, paved runway.

6. Using the Landing Distance – Flaps LDG//3 968 lb (imperial units) chart, calculate the landing distance over a 50 ft obstacle and the ground roll for the DA50 based on the following:

Landing airport – KSEZ, landing Runway 21 (review the runway length and slope % within the FAA digital chart supplement)

Pressure altitude = 5 000 ft

OAT = 32 °C

7. Using the runway information and the DA50 downhill slope % correction, does the aircraft meet the performance criteria to land on Runway 21, and is it safe to do so?

8. Review the latest Diamond Aircraft *Ice Protection System for Flight into Known Icing* supplement online for the DA50, and answer the following:

- **a.** Define freezing rain versus freezing drizzle per the DA50 AFM
- **b.** Define FIKI
- **c.** The operation of a TKS system versus deicing boots or a "hot wing"
- **d.** Define icing conditions per the DA50 Ice Protection supplement
- **e.** Maximum and minimum airspeed for Flaps T/O when in icing conditions
- **f.** V_{REF} when landing in icing conditions for *Flaps UP* and *Flaps T/O*

9. Using your ground roll answer from Problem #3, calculate the landing distance using Eq. 9.6 if the aircraft weight was decreased to 3 527 lb. Now, using the Landing Distance – Flaps LDG/3 527 lb (imperial units) chart, interpolate the landing ground roll for the same conditions in Problem #3. How does your answer from the manufacturer's performance chart for 3 527 lb compare to the answer you calculated using Eq. 9.6?

10 Slow-Speed Flight

CHAPTER OBJECTIVES

After completing this chapter, you should be able to:

- Define the region of reversed command and summarize aircraft performance when operating in this region.
- Compare and contrast the performance of thrust producers and power producers when operating in the region of reversed command.
- Explain how a stall develops and correlates stall patterns to wing shapes.
- Describe various types of stall entry and explain the importance of stall recovery procedures on the safety of flight.
- Summarize the development of a spin and identify the recovery procedures.
- Compare the aerodynamic characteristics of a spin between a straight-wing and sweptwing aircraft.
- Describe low-level wind shear and its impact on aircraft takeoff and landing performance.
- Describe wake turbulence and its impact on aircraft takeoff and landing performance.

Throughout the history of aviation, the slow-speed region of flight has been the most hazardous, sometimes leading to loss of control (LOC) accidents. During takeoff, landing, instrument approaches, and go-arounds, the aircraft is forced to move in "slow flight" (meaning flight that is deliberately slower than cruising speed, sometimes slow enough that stalling is a risk). Recognizing the hazards during this time is essential to understanding aircraft control characteristics at these slower speeds. In addition to takeoffs, landings, instrument approaches, and go-arounds, the aircraft may be maneuvered deliberately at airspeeds 5–10 kts. above the $1G$ stall speed during level, climbing, or descending flight; this controlled maneuvering is also referred to as slow flight. During flight training, applicants are required to maneuver in slow flight to demonstrate mastery of aircraft control at these slow airspeeds. If an aircraft slows to minimum controllable airspeed, any additional increase in the angle of attack or load factor, or any reduction in airspeed, will result in a stall. Failure to understand how an aircraft responds to control inputs at these

Flight Theory and Aerodynamics: A Practical Guide for Operational Safety, Fifth Edition.
Brian A. Johnson and Philip R. Fittante.

Companion website: www.wiley.com/go/flighttheoryandaerodynamics5e

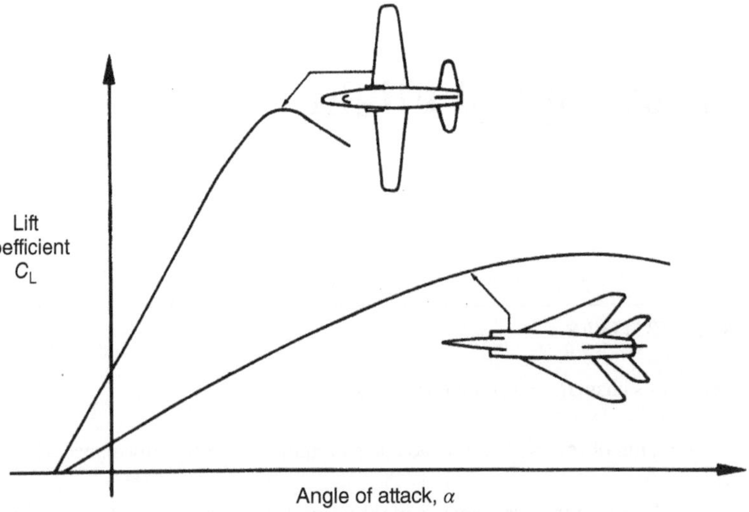

Figure 10.1. Effect of sweepback on C_L–α curves.

slower speeds, and improper recognition of and recovery from inadvertent spins, continues to cause far too many fatal aircraft accidents.

Different aircraft respond differently to slow flight; the reasons for this are quite different for straight-wing aircraft and for sweptwing aircraft. Examination of the coefficient of lift curves for each type of aircraft will help explain some of these differences.

Figure 10.1 shows the lift versus angle of attack (AOA) curves for a straight-wing, propeller-driven aircraft, and a sweptwing, turbojet aircraft. Four differences can be seen from these curves:

1. The straight-wing aircraft has a higher value of C_L. Recall the basic lift equation, Eq. 4.3:

$$L = \frac{C_L \sigma V^2 S}{295}$$

   It can be seen that if the lift equals the aircraft weight, then stall speed, V_S, will be a minimum when the value of the lift coefficient is at its maximum, $C_{L(max)}$. The higher the value of $C_{L(max)}$, the lower the stall speed will be. This means that the straight-wing aircraft has a lower stall speed for the same weight and wing area.
2. The sweptwing aircraft must fly at a higher AOA to achieve maximum lift.
3. There is a sudden reduction of CL for the straight-wing aircraft at stall, but not for the sweptwing aircraft.
4. The straight-wing aircraft is more sensitive to AOA changes.

REGION OF REVERSED COMMAND

Understanding the impact of drag on aircraft at low airspeeds is vital to safe flight in this flight regime, especially in jet aircraft, so a discussion of the *region of reversed command* and of the thrust required for various airspeeds below $(L/D)_{max}$ is warranted. Most pilots are familiar with this flight regime as the "backside of the thrust curve" or, incorrectly, as the "backside of the power curve." This last expression is correct only for propeller aircraft. Level flight below the airspeeds for $(L/D)_{max}$ is sometimes referred to as an area of "speed instability." If an aircraft experiences an action by the pilot, such as an oversight to add throttle during level out from a descent, the airspeed will continue to decrease unless a power or pitch correction is made by the pilot.

Thrust Producers

A typical thrust-required curve for a turbojet aircraft in the dirty configuration is shown in Figure 10.2. Minimum drag occurs at the $(L/D)_{max}$ point. For this aircraft, these figures are 2 400 lb thrust required at 160 kts. true air speed (TAS). All airspeeds greater than that for minimum drag are said to be in the region of *normal command*. During climb, cruise, and descent, the aircraft is operating in the region of normal command. In this region, added thrust must be supplied if greater airspeed is desired. Figure 10.2 shows that to increase the airspeed to 200 kts., the pilot must add throttle until the engine(s) produce 2 750 lb of thrust.

In the region of normal command, thrust is directly related to velocity. All airspeeds below the minimum drag airspeed, which occurs at $(L/D)_{max}$, are said to be in the region of reversed command. In this region, the thrust required to fly the aircraft is inversely related to the airspeed. The slower the aircraft flies, the greater the thrust required to maintain level flight. At an airspeed of 140 kts., for instance, Figure 10.2 shows that the T_r is 2 750 lb, the same thrust that was required at 200 kts.

It can be reasoned that "if more thrust is required to fly slower, then to slow down the pilot should add throttle." This, of course, is not true. The fallacy of this statement lies in the fact that, at any stabilized airspeed, the thrust available equals the thrust required, and adding throttle produces a thrust excess. This

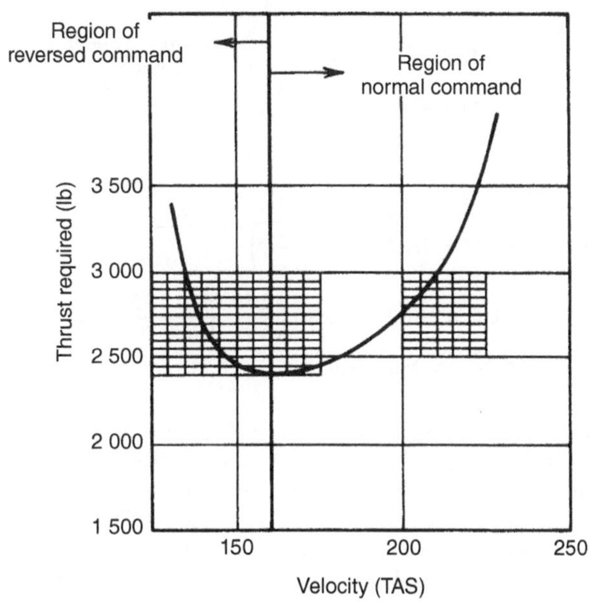

Figure 10.2. Regions of normal and reversed command.

excess will cause the aircraft either to climb or to accelerate. The correct statement should be: *If for any reason, the aircraft slows down, more thrust is required to maintain altitude at that airspeed.*

Take the case of an aircraft flying at minimum drag airspeed. This is 160 kts. for the aircraft in Figure 10.2. The T_r is 2 400 lb, then T_a must also be 2 400 lb for stabilized flight. There is no excess thrust, so the aircraft cannot climb under these conditions. To slow the aircraft, the pilot has several alternatives: (i) increase the drag by using speed brakes, extending the flaps, or lowering the landing gear; (ii) retard the throttle, thus reducing thrust; and (iii) increase the AOA by back pressure on the control stick.

Let us examine these alternatives in more detail. For the first alternative, increasing the drag is certainly a viable method of slowing the aircraft. That is why speed brakes were invented. However, speed brakes, flaps, and lowering the gear usually result in rapid changes in airspeed, so let us examine how we could slowly and more gradually reduce the airspeed.

For the second alternative, reducing the throttle results in a thrust deficiency. At any stabilized flight airspeed, a thrust deficiency results in a rate of sink (ROS). This can be seen by an adaptation of Eq. 6.6, where T_a is now subtracted from D (equivalent to T_r) to provide a positive ROS, which is the same as a negative rate of climb (ROC). The adaptation of Eq. 6.6 will now be Eq. 10.1:

$$\text{Rate of sink} = 101.3\ V_k \left(\frac{D - T_a}{W}\right)\ \text{(fpm)} \tag{10.1}$$

EXAMPLE

A 10 000 lb airplane has the T_r curve shown in Figure 10.2. The T_a is 2 500 lb and it is the same for all airspeeds. The airplane is flying at a speed of 140 kts. in 1*G* flight. Find the ROS using Eq. 10.1 and ROC using Eq. 6.6.

$$\text{Rate of sink} = 101.3\ V_k \left(\frac{D - T_a}{W}\right)$$

$$\text{ROS} = 101.3\ (140\ \text{kts.}) \left(\frac{2\,750\ \text{lb} - 2\,500\ \text{lb}}{10\,000\ \text{lb}}\right)$$

$$\text{ROC} = 354.6\ \text{fpm}$$

or using Eq. 6.6;

$$\text{ROC} = 101.3\ V_k \left(\frac{T_a - T_r}{W}\right)$$

$$\text{ROC} = 101.3\ (140\ \text{kts.}) \left(\frac{2\,500\ \text{lb} - 2\,750\ \text{lb}}{10\,000\ \text{lb}}\right)$$

$$\text{ROC} = -354.6\ \text{fpm}$$

The pilot who elects to slow the aircraft by only using a reduction in throttle cannot maintain altitude. Any attempt to increase AOA to compensate for less thrust will further decrease the airspeed and increase the drag. This results in an even greater thrust deficiency and a greater ROS.

The third method of slowing the aircraft is to increase the AOA. In Chapter 4, we saw that the basic lift equation shows that the AOA controls the value of the lift coefficient, and this in turn controls the airspeed. We concluded that the *angle of attack is the primary control of airspeed in level flight.*

For example, suppose the pilot pulls back on the control stick and thus increases AOA. The lift coefficient is increased, airspeed is reduced, and drag is increased. Thrust is no longer sufficient to produce

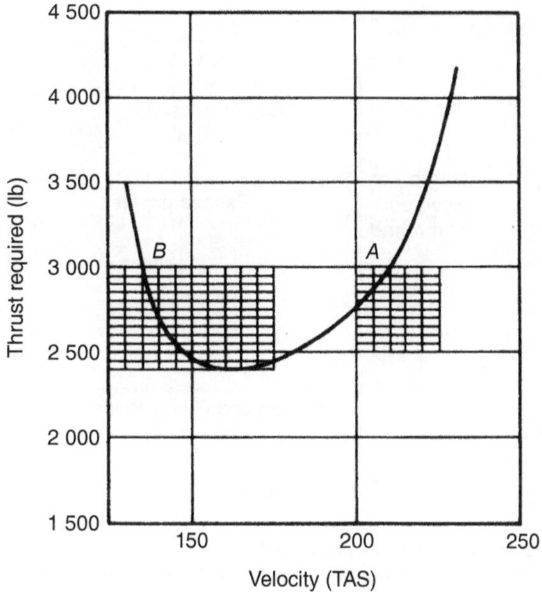

Figure 10.3. Constant airspeed climb. Stick or throttle?

enough lift to maintain altitude, and the aircraft will descend. So, the pilot must add power to maintain altitude. This can be a smooth coordinated method of slowing the aircraft without loss of altitude and without excessive "throttle jockeying."

Another example of using AOA to control airspeed is illustrated in Figure 10.3. Suppose a pilot is flying an airplane having the T_r curve shown in Figure 10.3 and that the throttle is set to produce 3 000 lb of thrust. For stabilized flight, the airplane must be at point *A* or at point *B*. At point *A*, the airspeed is 210 kts., and at point *B* it is 135 kts. To climb the airplane from point *A*, the pilot can pull back on the control stick and increase the AOA. This will increase the C_L and decrease airspeed. If this maneuver slows the aircraft to 200 kts., the thrust available is still 3 000 lb, but the thrust required has dropped to 2 750 lb. The aircraft now has an excess thrust of 250 lb and will be able to climb.

Now, suppose the plane is at point *B*. To climb the aircraft, the pilot pulls back on the control stick and increases the AOA. The airspeed will be reduced again, but this time the thrust required will be increased. If the airspeed is decreased to 130 kts., the thrust required will be about 3 500 lb, a thrust deficiency of 500 lb. The climb method that worked in the region of normal command will not work in the region of reversed command.

Power Producers

With power producers as with thrust producers, the region of reversed command occurs at low airspeeds, below the airspeed corresponding to the bottom of the P_r curve. But, as we saw in Chapter 7, for power-producing aircraft, the bottom of the power-required curve is no longer $(L/D)_{max}$. Instead, it is the point at which the least power is required to maintain level flight. As we see in Figure 10.4, at airspeeds below the lowest point of the P_r curve, more power is needed to maintain level flight when airspeed is decreased. In this regime, if a high sink rate should develop, simply adding power will not be sufficient to stop the descent. To slow or stop the descent, a combination of reducing the pitch attitude and adding

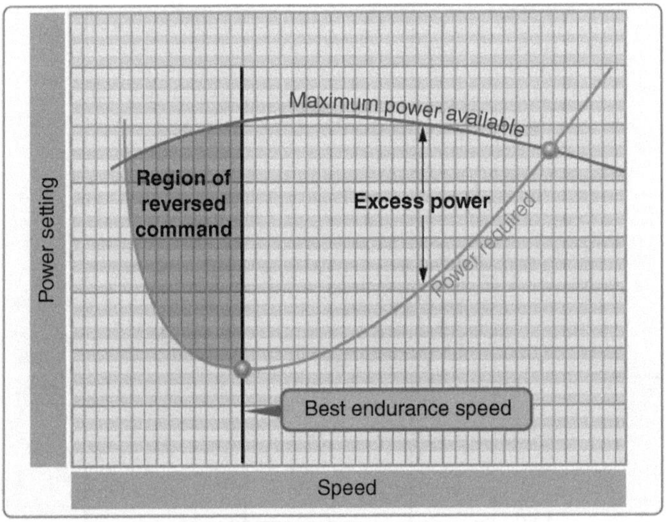

Figure 10.4. Region of reversed command for a power producer.
Source: Pilot's Handbook of Aeronautical Knowledge (2016b)/U.S. Department of Transportation Federal Aviation Administration/Public domain.

power will be needed. In smaller, single-engine aircraft, full power is required to stop the descent and initiate a climb if a go-around is required on a bad approach.

A method of climbing the aircraft that works at all airspeeds is to control the airspeed with the stick. If the airspeed is held constant, the thrust or power required will be constant. The rate of climb is then controlled by the excess thrust (power), and the thrust or power available is determined by the throttle setting. So, the *throttle setting controls the rate of climb or descent in steady flight.*

Flight in the region of reversed command cannot be avoided. Every takeoff and landing require flight in this region. Pilots must be careful not to paint themselves into a corner, where drag is so high that enough thrust may not be available to overcome it. When the drag (T_r or P_r) exceeds the thrust available (T_a) or power available (P_a), there is only one way to go – *down.*

STALLS

As explained in Chapter 4, a stall occurs when the boundary layer separates from the wing, the critical angle of attack has been reached, and the wing is no longer producing enough lift to sustain level flight. Pilots practice intentional stalls to recognize the conditions that produce stalls, the characteristics of an imminent stall, and proper recovery technique to avoid causing a secondary stall (or, even worse, a spin) and to minimize altitude loss while maintaining positive aircraft control.

However, an aircraft does not need to be in slow flight to enter a stall, as stalls can occur at any airspeed, altitude, and configuration. The pilot of an aircraft in a steep dive can pull abruptly back on the stick and exceed the critical angle of attack at a speed well above V_{S_0}. The critical AOA does not change for a given aircraft, regardless of airspeed, pitch attitude, angle of bank, weight, or altitude. As we will see in Chapter 11, the stall speed also increases once an aircraft in flight enters a level turn. The aircraft will stall at the same AOA regardless of an increase in load factor. Back pressure is applied to the stick or yoke during a level turn, lift increases, as does the wing AOA, and eventually a stall will be produced if back pressure continues to increase. Once an aircraft stalls, the wing design as well as the placement of the center of gravity (CG) greatly influences the recovery technique, and the altitude loss suffered.

Regardless of the type of aircraft, type of wing planform, or type of stall, an imminent stall will always display certain characteristics. Pilots must recognize these characteristics to initiate recovery procedures before the stall develops and before altitude has been lost. Colgan Flight 3407 (preoccupation with automation), Air France Flight 447 (incorrect response to instrumentation error and stall indications), and many other high-profile accidents, in addition to hundreds of general aviation accidents, continue to show that pilots are still not responding correctly to indications of an imminent stall. While improved automation and artificial stall indicators do raise awareness of an imminent stall, the pilot still needs to pay attention to the sight picture through the windshield, sounds including wind noise and engine noise, and sensations such as control wheel forces, all of which can offer important clues that an aircraft is nearing a stall. During slow flight, feel continues to be the best indicator that the aircraft is approaching a stall, and corrective action needs to be taken immediately. The pilot may sense mushy controls and a change in the direction of the aircraft. The pilot may feel vibration in the controls or a feeling of buffeting in the flight controls. Hearing also plays an important role. In a fixed-pitch propeller aircraft, there is an audible change in rpm when additional load is placed on the propeller during a climb, and the sound of the air flowing around the aircraft decreases as the aircraft slows toward a stall. In each of these examples, a pilot's experience plays a vital role in how soon the indicators of an impending stall are recognized, and in how quickly the correct recovery techniques are begun to transition smoothly back to normal flight.

Stall Patterns

While the basic principles of the stall, as introduced in Chapter 4, are the same for all wings, different wings behave differently in a stall. Here we will discuss the stall patterns of various wing designs, as well as the advantages and disadvantages of generic wing shapes.

Elliptical Wing

The lift distribution in the spanwise direction of an ellipse is shown in Figure 10.5. The lift is zero at the wingtips and increases elliptically, with the maximum lift occurring at the wing root. If the wing area is exactly proportional to the lift at all points on the wing, the local wing loading will be a constant value everywhere on the wing. Stall of a wing will occur where the ratio of the local lift coefficient, C_l, to the wing lift coefficient, C_L, is highest. If the wing planform area is also elliptical, the local wing loading will be constant. This is shown by the value of $C_l/C_L = 1.0$ in Figure 10.6. Thus, the entire trailing edge of an elliptical planform wing will stall at the same time. The elliptical wing is the most efficient planform and was used with much success on the British Spitfire during World War II. Structurally, the elliptical wing is difficult to manufacture, so it is not widely used.

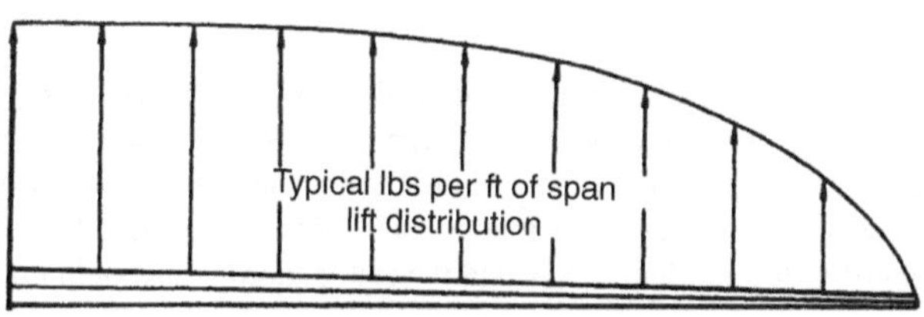

Figure 10.5. Spanwise lift distribution.

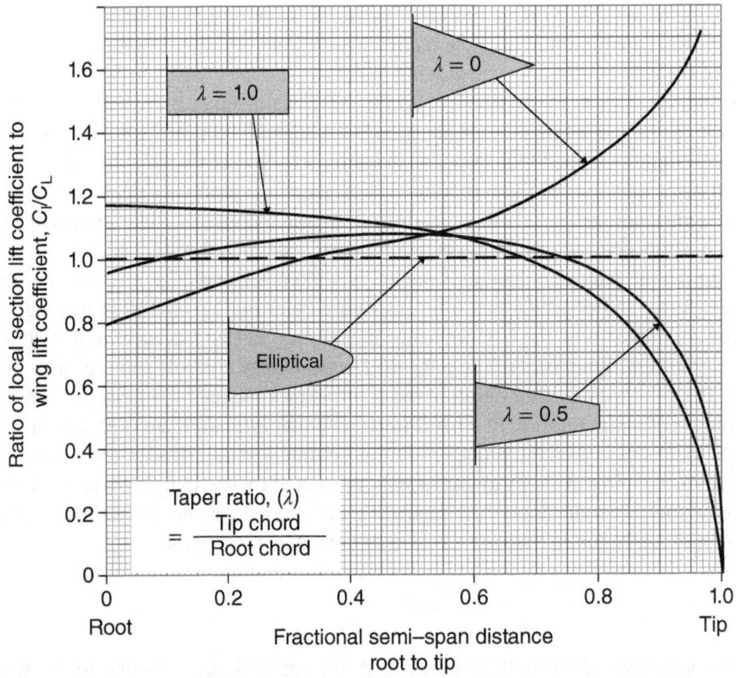

Figure 10.6. Wing spanwise lift distribution.

Straight/Rectangular Wing

The straight wing is usually rectangular or may have a slight taper. The wing area near the wingtip is large compared to the lift in that area, and the wing loading is, therefore, low. Figure 10.6 shows that the value of C_1/C_L approaches zero near the wingtip for the rectangular wing. The maximum wing loading is at the wing root. Therefore, the stall will start at the trailing edge of the wing root and then spread outward and forward as the stall progresses, as shown in Figure 10.7.

Straight-wing aircraft, whose wings stall starting at the roots, generally have better stall characteristics than sweptwing aircraft. First, there is a better stall warning, caused by the separated air buffeting the fuselage. Second, because the ailerons of the straight-wing aircraft are not installed air until the stall has progressed outboard from the wing roots, they remain effective longer in a stall than those of a sweptwing aircraft, in which the ailerons are compromised from the beginning of a stall.

Swept Wing

Swept wings have more taper, with the extreme being the delta wing, which has a taper ratio of zero. Both suffer from spanwise flow that leads to early airflow separation at the wingtips. The delta wing's area at the tip is very small, thus the wing loading is very high. The stall starts where the wing loading is highest, so the stall starts at the wingtip and progresses inboard and forward. The high loading at the wingtip can be seen from the values of C_1/C_L in Figure 10.6, and the stall pattern is shown in Figure 10.7.

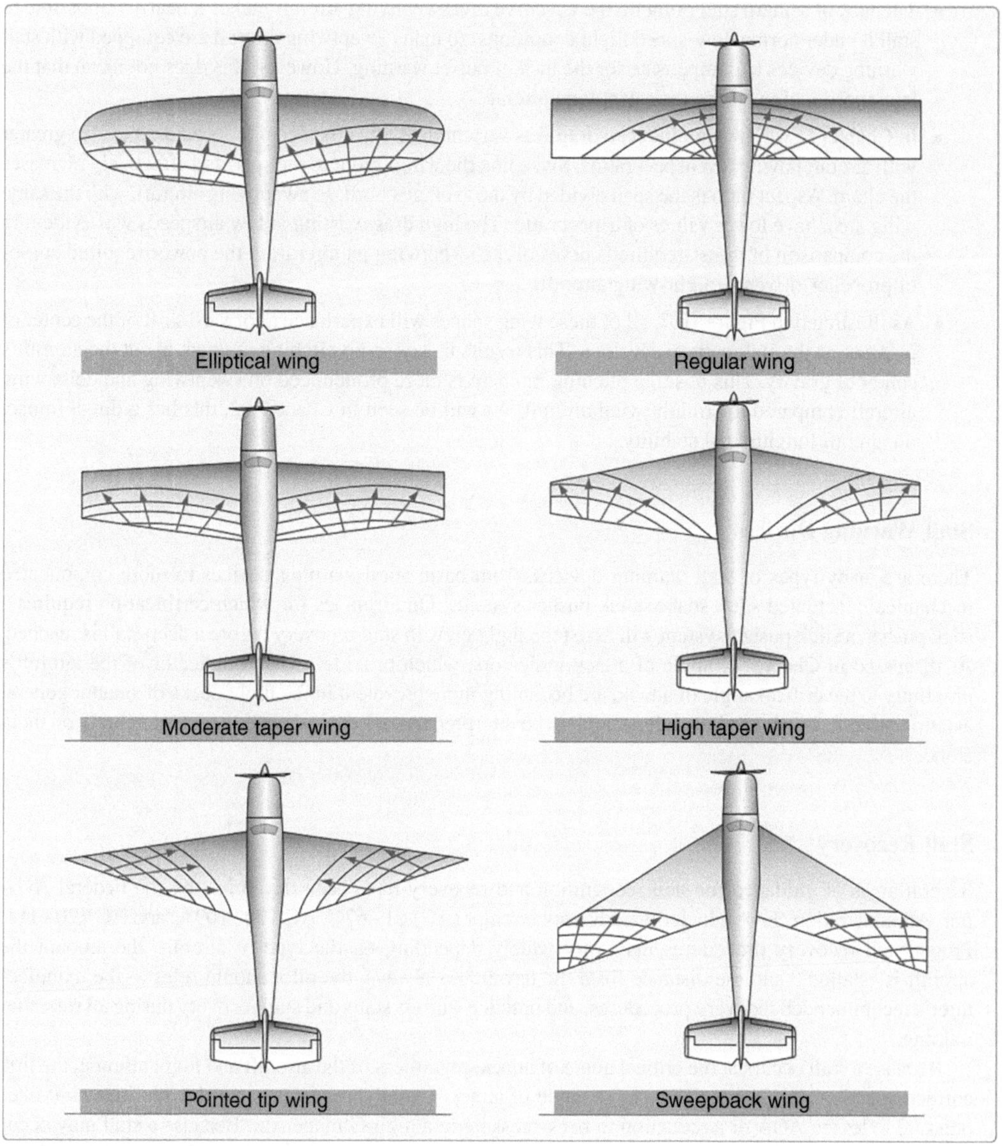

Figure 10.7. Stall patterns.
Source: Airplane Flying Handbook (2021)/U.S. Department of Transportation Federal Aviation Administration/Public domain.

- The lack of a sharp stall point on the C_L curve of a sweptwing aircraft makes it nearly impossible to stall it under normal low-speed flight conditions, so many sweptwing aircraft are equipped with stall warning devices to compensate for the lack of buffet warning. However, this does not mean that the low-speed region is safe for sweptwing aircraft.
- In Chapter 5, we saw that induced drag was very high at low airspeeds. Induced drag is also greater with aircraft having low aspect ratios. Sweeping the wings reduces the span and effectively increases the chord. Aspect ratio is the span divided by the average chord, so sweptwing aircraft, with the same wing area, have lower values of aspect ratio. The high drag existing at low airspeeds was evident in the comparison of thrust-required curves of the sweptwing jet aircraft to the power-required curves of propeller-driven straight-wing aircraft.
- As illustrated in Figure 10.7, all of these wing shapes will experience a forward shift of the center of pressure as the stall pattern develops. This results in a nose-up pitching moment about the aircraft's center of gravity. This nose-up pitching moment is more pronounced on sweptwing and delta wing aircraft compared to straight wing aircraft. As will be seen in Chapter 12, this has a direct impact on aircraft longitudinal stability.

Stall Warning Devices

There are many types of stall warning devices, from basic aural warning devices to more complicated mechanically actuated stick shaker/stick pusher systems. On airplanes for which certification requires a stick pusher, a stick pusher system will assist the flight crew in stall recovery before a deep stall is reached. As discussed in Chapter 4, angle of attack indicators, which provide a visual indicator of the aircraft's proximity to the critical angle of attack, are becoming more prevalent in the flight decks of smaller general aviation aircraft. Pilots need to understand how to interpret and use these devices but must not rely on them alone.

Stall Recovery

Several areas of guidance for stall recognition and recovery have been developed by the Federal Aviation Administration (FAA), including advisory circular (AC) 61–67C, AC 120–109A, and AC 120–111. Proper stall recovery procedures may vary widely depending on the type of aircraft, the amount the aircraft is "stalled," and the distance from the terrain. As always, the pilot should refer to the manufacturer's recommended recovery procedures, and practice various stalls and stall recovery during all refresher training.

Because a stall occurs at the critical angle of attack, regardless of the aircraft and flight attitude, the first corrective action should be to reduce the angle of attack. Usually, this means a reduction in stick (yoke) pressure to lessen AOA or a reduction in back pressure in a high-G maneuver. Because a stall may occur in any attitude, "lowering the nose" is not the correct response unless the aircraft is in an upright attitude in relation to the ground. In some cases, forward pressure may be required, but only enough to reduce the excessive angle of attack and not enough to produce an excessive nose-low attitude (during normal upright attitude recoveries) and a negative G-load on the wings. If too much forward stick pressure is used, altitude loss may be extreme, jeopardizing recovery before impact with terrain. Figure 10.8 is a typical stall recovery procedure for an "upright" stall recovery scenario.

As discussed in the first section of this chapter, while reducing the angle of attack will begin the process of recovery from a stall, power may be required to minimize altitude loss as the pilot reduces the angle of attack. If the aircraft is already at full power, in most cases, the power should be maintained to assist

Stall recovery template	
1. Wing leveler or autopilot	1. Disconnect
2. (a) Pitch nose-down	2. (a) Apply until impending stall indications are eliminated
(b) Trim nose-down pitch	(b) As needed
3. Bank	3. Wings Level
4. Thrust/Power	4. As needed
5. Speed brakes/spoilers	5. Retract
6. Return to the desired flight path	

Figure 10.8. Stall recovery template.
Source: Airplane Flying Handbook (2016a)/U.S. Department of Transportation Federal Aviation Administration/Public domain.

with recovery, but managed appropriately so that the aircraft does not overspeed or exceed the load factor limitations. Depending on the altitude and on the type of stall experienced, the pilot should then regain straight-and-level flight at normal airspeed until the pilot has verified that the aircraft is under control.

During the stall and stall recovery, care should be taken to avoid excessive roll and yaw. As discussed earlier, sweptwing aircraft stall first at the wingtip, and the stall progresses inward and forward. If the ailerons are in aerodynamically dead air (as they are on a sweptwing aircraft in a stall), the pilot will have limited roll control, and the aircraft may become unstable about the longitudinal axis. During practice stalls in a single-engine propeller aircraft, on the other hand, banking during stalls is commonplace, and even advisable during training to build the pilot's confidence. Due to the rectangular wing planform, the stall begins at the wing root and progresses outward and forward, allowing the pilot positive roll control during the stall and recovery. During stalls in a single-engine aircraft with high power settings, right rudder application will be required to maintain coordinated flight. Left aileron may also be needed to prevent overbanking, so aileron control during this time is very important. If the stall is allowed to continue until a "full break," a wing drop may occur. If the pilot attempts recovery by rolling in the opposite direction, the downward deflecting aileron on the low wing would produce a greater angle of attack, thus increasing the drag on that wing and aggravating the situation. The outboard portion of the wing may now become stalled. If this occurs, the aircraft will begin a yaw in the direction of the stalled (lower) wing, and a spin could develop if the angle of attack is not reduced, and yaw is not corrected properly (discussed in detail in the Uncoordinated Flight section). If a spin develops, much more altitude will be lost than in a proper stall recovery, and if a spin occurs at a low altitude, there may not be sufficient altitude to recover.

Power-Off Stall

Power-off stalls are usually practiced to simulate an impending stall during the approach to a landing. The aircraft is configured for landing with flaps and landing gear deployed, and normal V_{REF} speed is achieved. After the nose has been lowered to a normal approach attitude, it is raised to an attitude, which if maintained will induce a stall. The wings should be held level, and directional control maintained by coordinated use of the rudder and aileron. When the critical angle of attack has been reached, the nose should be lowered to reduce the angle of attack, and the recovery procedure should be initiated immediately. On single-engine propeller aircraft, the right rudder will be needed to maintain coordination as power is applied and the nose is raised to minimize altitude loss. Gear and flaps are retracted per the manufacturer's procedures, and the airplane is trimmed for straight-and-level flight (Figure 10.9).

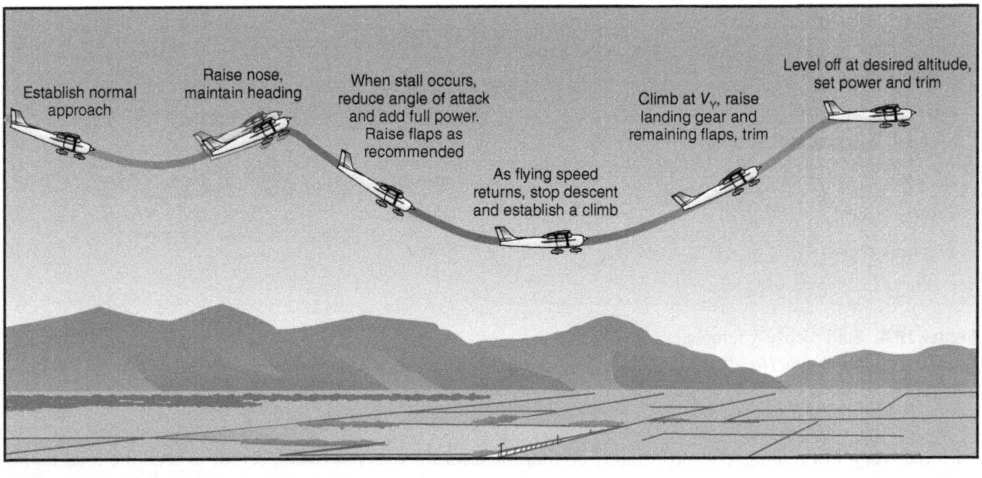

Figure 10.9. Power-off stall and recovery.
Source: Airplane Flying Handbook (2004a)/U.S. Department of Transportation Federal Aviation Administration/Public domain.

Power-On Stall

A power-on stall is usually practiced simulating an inadvertent stall during takeoff or climb. Takeoff gear and flap settings are used, and full power may be used throughout the power-on stall in smaller training aircraft, though in higher performance aircraft a percentage of full power is more commonly used to avoid excessive nose-high attitudes. To begin the maneuver, the aircraft slows to lift-off speed while in level flight. Once the target airspeed has been reached, the power is increased to the power setting appropriate for that aircraft, and the nose is smoothly raised with an aft stick until the critical angle of attack is reached. Care must be taken when demonstrating this maneuver in a single-engine propeller aircraft, as increased right rudder must be used to compensate for the left-turning tendency experienced at high power, and high angle of attack attitudes (left-turning tendency is discussed in Chapter 13). If proper directional control is maintained with the rudder, the aircraft will stall wings level, and the proper recovery technique should be initiated. Usually, once the nose is lowered to reduce the angle of attack to less than the critical AOA, the aircraft will recover, and the nose can once again be raised carefully to minimize altitude loss. As with a power-off stall, if the aircraft experiences a wing drop at the moment of full stall, opposite aileron should *not* be applied because the stall may become aggravated, and a spin could develop. The rudder opposite the wing drop must be applied as soon as possible: for example, if the nose of the aircraft begins to yaw to the left and the left-wing drops, immediate application of right rudder is required. *Use the opposite rudder to correct any wing drop.*

Often power-on stalls are practiced only until the stall is imminent and the pilot is allowed to recover when one of the stall characteristics is recognized. If at any time during the maneuver, an unsafe condition should develop, one of the pilots should call "knock-it-off" and recovery procedures should be initiated.

A variation of the power-on stall can occur during an *elevator trim stall*. If an aircraft is trimmed for landing, or for best glide speed during a simulated emergency descent, the nose may pitch to an extreme nose-high attitude when full power is applied for a go-around or missed approach. Heavy stick forces may be encountered as the pilot tries to maintain a safe pitch attitude below the critical angle of attack until the trim forces can be reduced.

During training events for power-on and power-off stalls, pilots should periodically practice secondary stalls to familiarize themselves with the conditions that cause a secondary stall, and with the proper control

Figure 10.10. Secondary stall due to improper stall recovery.
Source: Airplane Flying Handbook (2004a)/U.S. Department of Transportation Federal Aviation Administration/Public domain.

actions to recover and prevent additional secondary stalls. A secondary stall results when the pilot pulls the stick back too abruptly as the nose drops during stall recovery (Figure 10.10). The aircraft will once again approach the critical angle of attack and experience another stall, which in extreme cases can lead to multiple secondary stalls. Multiple secondary stalls are sometimes referred to as "falling leaf stalls," in which an excessive angle of attack is maintained with aft stick pressure, and the rudder is used to maintain coordination and prevent excessive wing drop.

Accelerated Stall

An accelerated stall occurs at a higher airspeed ($>1G$ stall speed) than is usually experienced by a pilot during "normal" stall training. Accelerated stalls may occur when the pilot makes an abrupt control movement, resulting in the wing exceeding the critical angle of attack at higher than the usual published stall speeds. Accelerated stalls are in essence a "G-induced" stall, as they occur at load factors greater than 1, which is the load factor for stalls during normal student training in light aircraft. During steep turns, a stall will occur at airspeeds much higher than the published stall speed, often surprising the inexperienced pilot. Corrective actions should be immediately applied, usually beginning with reducing aft stick pressure and leveling the wings. Accelerated stalls can be more insidious than power-off and power-on stalls and occur with much less warning. Often pilots are caught unaware of the impending accelerated stall, resulting in loss of directional control and improper yaw correction, and in extreme cases, a spin can occur. Pilots should practice accelerated stalls during training and recognize the importance of design maneuvering speed and load factor. More on this topic will be presented in Chapter 11. Figure 10.11 demonstrates the impact of an increase in stall speed due to the angle of bank during an accelerated stall; note the effect of 40° versus 0° of flaps.

Cross-Control Stall

A cross-control stall is a type of accelerated stall that deserves special attention because it usually occurs in traffic patterns, at altitudes that may not allow recovery. In civilian pilot training, only flight instructor

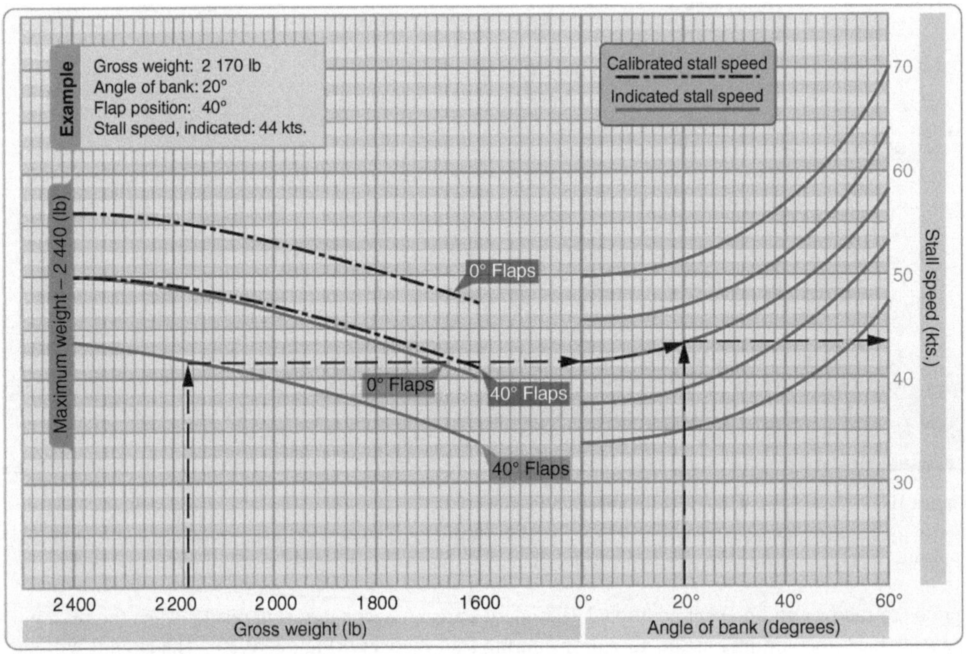

Figure 10.11. Stall speed chart.
Source: Pilot's Handbook of Aeronautical Knowledge (2016b)/U.S. Department of Transportation Federal Aviation Administration/Public domain.

candidates may be required to demonstrate a cross-control stall during a practical test, but all fixed-wing pilots should be aware of the conditions that may cause a cross-control stall. Because of the insidious nature of this stall, there may be little or no warning of the oncoming stall. The aircraft may enter a spin immediately due to excessive yaw, with little time for the unsuspecting pilot to recover. A common situation begins when the pilot overshoots the base-to-final turn and realizes that the aircraft is not aligned with the runway centerline. Unwilling to steepen the bank angle of the aircraft, the pilot moves the longitudinal axis of the aircraft toward the runway with the rudder, increasing the rate of the turn. As the speed of the outer wing increases, its lift increases, and the aircraft begins to bank in the direction of the skidding turn. To counteract this, the pilot applies opposite aileron, which as we have discussed exacerbates the situation because the downward deflecting aileron on the low wing increases its drag, producing a yaw toward the low wing. Because this action lowers the nose even more toward the ground, the pilot instinctively applies back pressure on the yoke/control stick, raising the nose. The indicated airspeed (IAS) now begins to decrease due to the increase in drag; simultaneously, the stall speed begins to increase and within seconds, the IAS and stall speed meet with little warning.

Due to the excessive yaw, the wing may drop violently and unexpectedly, causing the aircraft to roll inverted. Because this occurs close to the ground, even with proper recovery techniques there may not be sufficient time and altitude to recover. Avoidance of the cross-control stall is the only reasonable course of action, and this requires that pilots be trained to understand how this stall occurs, so they can avoid situations in which a cross-control is likely to occur.

Uncoordinated Flight

The previous sections on stalls and stall recovery emphasized the importance of understanding coordinated flight during slow flight. There are times when uncoordinated flight is necessary, as was introduced in Chapter 9 during crosswind landing techniques, and will be expanded upon in Chapter 13 during the discussion on forward slips. But a review of proper rudder coordination and yaw is warranted as we transition into spin entry and spin recovery, as each year the general aviation industry experiences accidents based on loss of control (LOC) due to inadvertent uncoordinated flight during stalls. In the next chapter we will introduce accelerated stalls, again proper coordinated flight is paramount to avoid a possible LOC event, or at the very least, exceeding a limitation on an aircraft component.

Figure 10.12 shows an aircraft in a standard rate turn to the right, this means the aircraft is turning at a rate of 3°/second, and is visually shown by the fact the wing of the airplane in the turn coordinator is aligned with the lower, white mark above the *R*. The "ball" within the inclinometer indicates whether the turn is coordinated or not as shown by the headers for *Skid*, *Coordinated Turn*, and *Slip*. Above the turn coordinator, there is a depiction of a digital attitude indicator, also indicating a turn to the right. The "coordination" of the turn on a digital indicator is shown by the white mark below the triangle, when the triangle is centered with the white bottom then the aircraft is in coordinated flight.

A *skid* represents uncoordinated flight. In Figure 10.12, the aircraft is banking to the right with too much right rudder applied by the pilot, so the ball in the inclinometer is pushed outside of the turn as the aircraft "skids" into the turn. The corrective action would be for the pilot to release enough of the right

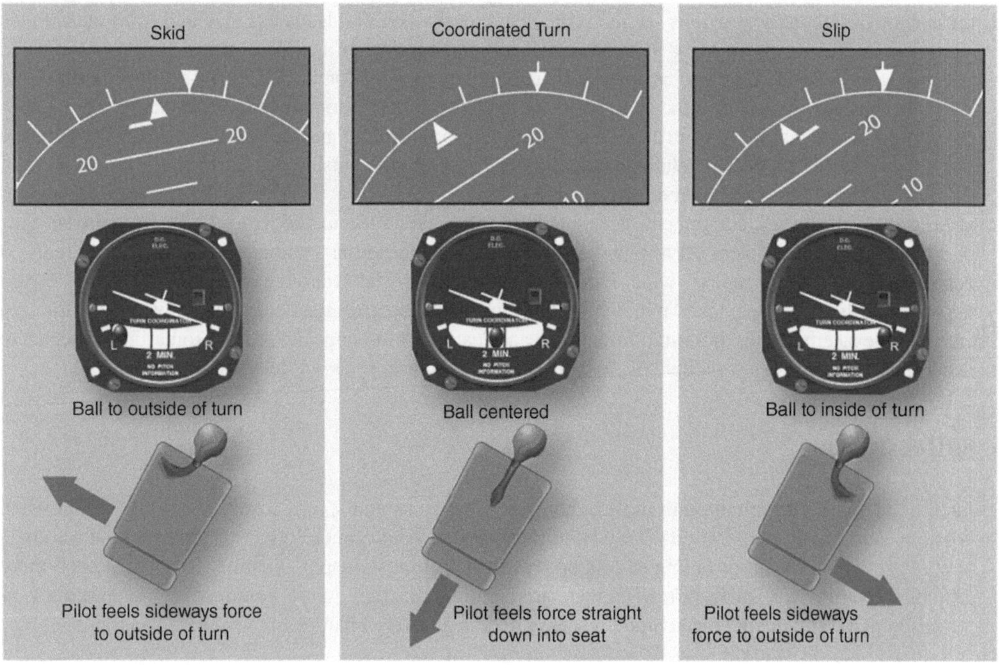

Figure 10.12. Skid, slip, and coordinated flight.
Source: Pilot's Handbook of Aeronautical Knowledge (2016b)/U.S. Department of Transportation Federal Aviation Administration/Public domain.

rudder, so the ball returns to the center of the inclinometer. A *slip* also represents uncoordinated flight. In Figure 10.12, the aircraft once again banks to the right, but this time the pilot adds too much left rudder (or not enough right rudder), and both the pilot and inclinometer ball are pushed to the inside of the turn. The expression "flying by the seat of your pants" derives from an experienced pilot being able to feel when the aircraft is not coordinated. Uncoordinated flight is commonplace early in a pilot's flight training, and many new pilots are familiar with the directive, "step on the ball" to maintain coordinated flight. But, as shown by accident reports and anecdotal discussions with many pilots, the importance of coordinated flight and their impact on spin entry and spin recovery is not always truly understood, even by flight instructors.

Both slips and skids lead to uncoordinated flight, and if this occurs during a moment of stall, then an incipient spin may develop (introduced in the next section). So, understanding how the aircraft is placed in a slip or skid, and knowing what to expect should uncoordinated flight occur during a stall, can help prevent inadvertent spin entry. During a slip, when the nose of the aircraft yaws to the outside of the turn, the outside wing has a higher angle of attack. As that wing stalls and drops, the pilot may be banking in the opposite direction resulting in a lowered aileron on the wing, exacerbating the stall. When the aircraft fully stalls the wings would be closer to level before the aircraft yaws and enters a spin. This can be visualized by looking at Figure 10.12 (Slip), as the aircraft approaches a stall, the outside wing is "more stalled" than the right wing, so at the moment of a full stall, the left wing will drop, and the airplane will begin to level. Often the angle of attack is decreased at this point and no longer exceeds the critical angle of attack, and in many general aviation training aircraft this is all that is needed to mitigate spin entry with little corrective action by the pilot.

A skid entry at the moment of a full stall is more insidious and less desirable than a slip entry. The cross-control stall in the previous section is an example of a skid that leads to a stall and spins close to the ground, with little or no warning. During a skid, too much rudder is applied in the direction of the bank, which results in a nose drop and an overbanking tendency by the pilot. As the bank becomes unsettling to the pilot, they attempt to move the controls in the opposite direction to reduce the bank angle. This lowers the aileron on the inside wing further, contributing to more lift/drag and a higher AOA than the opposite wing. Also, during a skid, the relative wind approaching from the outside wing toward the inside wing of the turn is partially blocked by the fuselage and the relative wind that does strike the leading edge strikes at an angle leading to a component of the spanwise flow down the inside wing. Because spanwise flow does not produce as efficient lift as relative wind perpendicular to the leading edge, lift is further reduced contributing to a "more stalled" wing. Finally, the inside wing stalls, and the aircraft begins a roll toward the inside of the turn. A panicked pilot who further moves the controls to attempt to roll away from the spin entry only exacerbates the situation. Immediate application of the opposite rudder, sometimes referred to in aerobatic circles as "top rudder," is the only option to limit the yaw while both wings are stalled.

SPINS

Intentional spins are prohibited in most light aircraft but spins and spin recovery should be taught when possible as part of all pilot training. The statistics continue to show that pilots are not heeding the warnings of stall recognition and are failing to properly recover from a stall/spin situation. Spins are much more difficult to control in a jet aircraft, so they are not common practice. Actual spin recovery techniques vary for different aircraft, so the discussion presented here is general in nature.

Spin Warning

For an aircraft to spin, it must be near an aerodynamic stall and in a yawed state. Therefore, stall warnings must be heeded if spins are to be avoided. For most general aviation, light training aircraft spins are considered to have four stages (Figure 10.13.): entry, incipient, developed, and recovery.

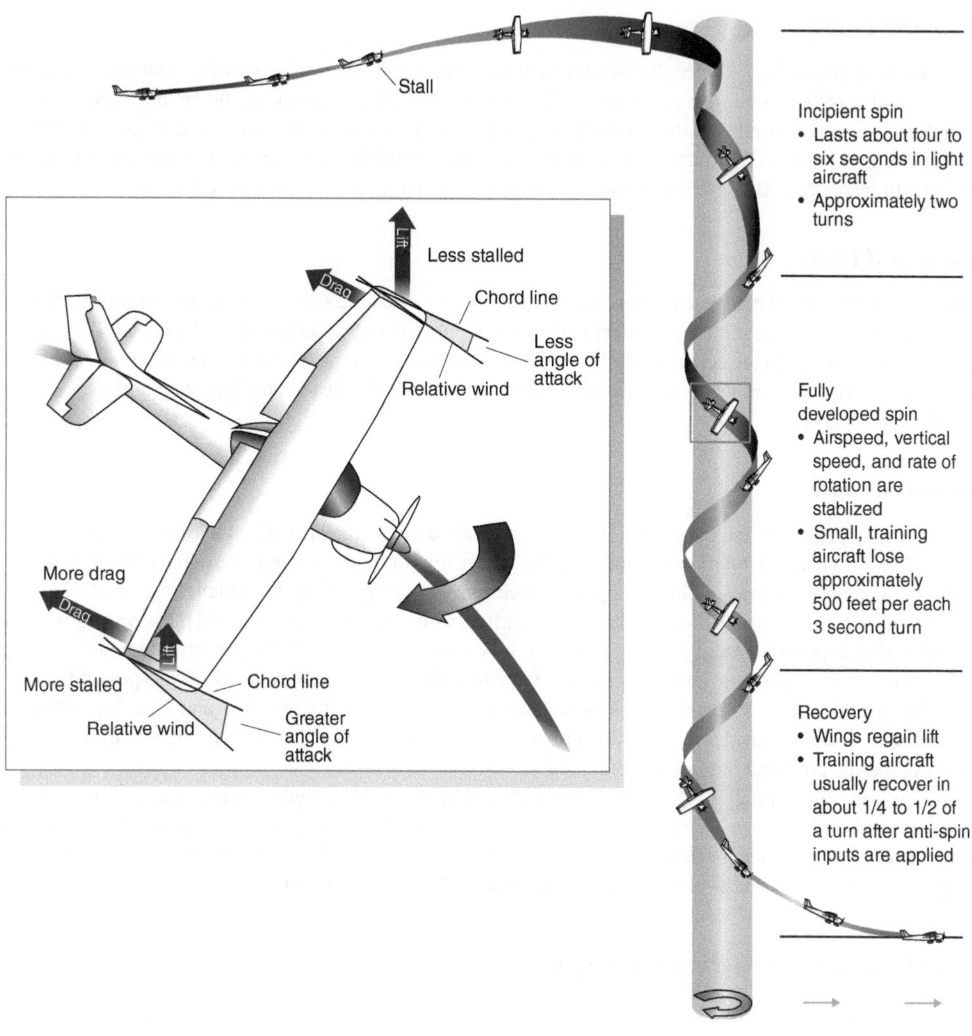

Figure 10.13. Spin entry and recovery.
Source: Airplane Flying Handbook (2004a)/U.S. Department of Transportation Federal Aviation Administration/Public domain.

Entry

During the entry stage, the pilot introduces the conditions for stalled flight, whether accidentally or intentionally. The critical angle of attack has been exceeded, and due to the unequal angle of attack on each wing, one wing is more stalled than the other and a yawed state exists. If steps are not taken immediately to reduce the angle of attack and restore coordinated flight, a spin will begin.

Incipient Phase

The incipient phase begins once the aircraft begins rotation and usually lasts two to four turns until the inertial and aerodynamic forces come into balance. During the incipient phase, the airspeed will begin to stabilize at a low airspeed, and the symbolic airplane of the turn coordinator will indicate the direction of spin. As a general rule, departures that occur at high airspeeds (high kinetic energy) require more time and altitude to reach the fully developed phase than departures that occur at low kinetic energy.

Developed Phase

Once in equilibrium, the airspeed, rotation rate, and vertical speed are stabilized, and the aircraft continues to move downward in a spiraling vertical path (helical path) as indicated in Figure 10.13. Unlike a spiral dive, in which the airplane is not in equilibrium, in a stabilized spin the airplane's aerodynamic forces are in balance. Keep in mind that some aircraft will never reach a fully developed spin state and may remain oscillatory in nature.

Recovery Phase

The recovery process is initiated when the AOA is reduced below the critical angle of attack and rotation stops. The spin equilibrium is disrupted, and the wings are no longer stalled. Care must be taken at this point to avoid entry into a spin in the opposite direction, or overspeeding the airplane.

In our discussion of stalls in the Stalls section earlier in this chapter, we mentioned the advantages of the straight wing in providing stall warning to the pilot. When pilots of straight-wing aircraft feel the buffeting that warns of an impending stall, they merely have to reduce back pressure to avoid the possibility of a spin. Figure 10.14 shows how a straight-wing aircraft stalls at the wing root first and how the stalled air envelops the tail and provides a buffet warning to the pilot. At this point, the ailerons are still experiencing airflow, and the pilot still has good lateral (roll) control.

Sweptwing aircraft stall at the wingtips first, rather than at the roots (Figure 10.15). The stall warning for sweptwing aircraft is much less than for straight-wing aircraft, and in addition, the ailerons are enveloped in the stalled air and become ineffective early in the stall–spin sequence. Stall warning devices are especially helpful in anticipating and avoiding an incipient spin in sweptwing aircraft.

Aerodynamic Characteristics of a Spin

Wing/Fuselage Characteristics

The aerodynamic characteristics of a spin for a straight-wing aircraft and those for a sweptwing aircraft are quite different. The straight-wing aircraft has C_L and C_D curves as shown in Figure 10.16. If this type of aircraft is flying at an AOA at or above that for $C_{L(max)}$ and the aircraft is rolled to the right (and/or the right wing drops), the right wing is flying with a higher local AOA and becomes even more stalled. The left wing, on the other hand, moves upward and is flying with a lower local AOA and becomes less stalled, thus producing more lift. The right rolling motion of the aircraft is aided by the relative increase in lift of the left wing, and a yawing moment to the right is produced. These moments result in autorotation. This means that the aircraft will continue to roll and yaw in the same direction without any input from the pilot. In addition to the aerodynamic forces acting on the aircraft, there are also gyroscopic effects that complicate the problem.

The straight-wing aircraft's spin is characterized primarily by rolling motion with a moderate yaw. The nose-down attitude in the spin is about 40°. Whereas a straight-wing aircraft must stall before a spin can develop, in sweptwing aircraft a spin does not have to be preceded by a stall. The sweptwing aircraft has

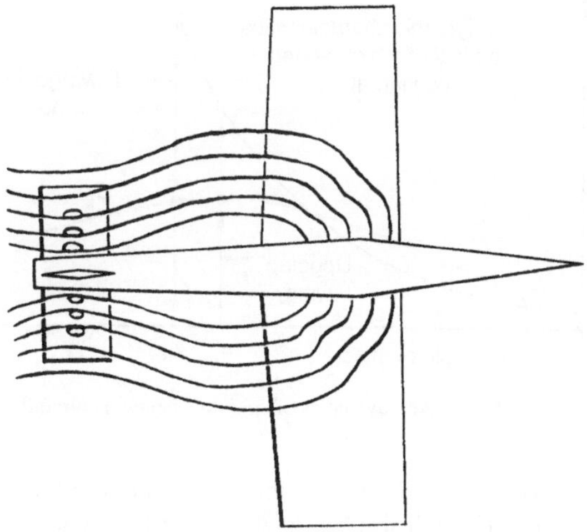

Figure 10.14. Stall hitting the horizontal tail.

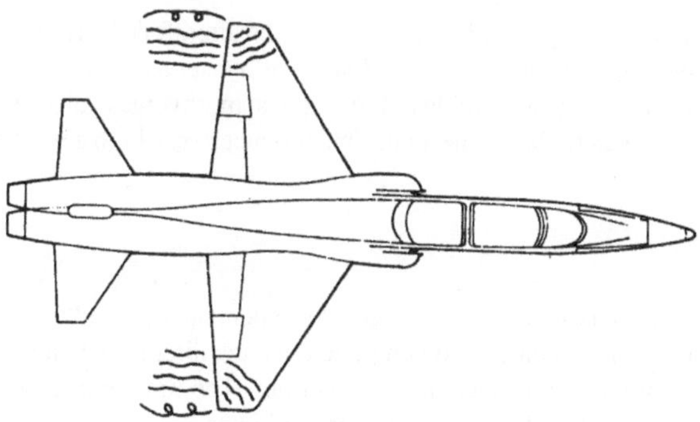

Figure 10.15. Swept wings stall at tips first.

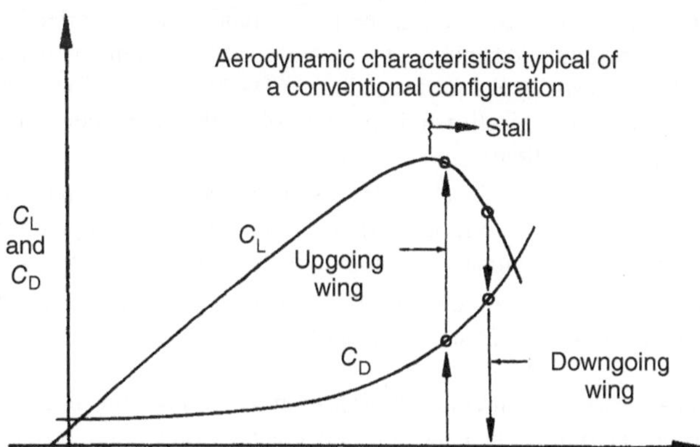

Figure 10.16. Aerodynamics of spin for straight-wing aircraft.

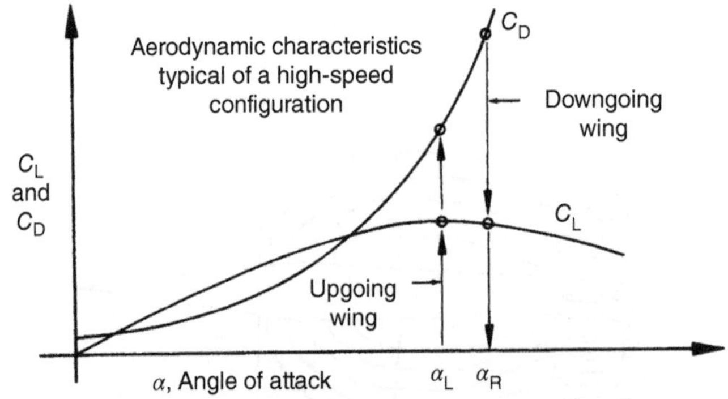

Figure 10.17. Aerodynamics of spin for sweptwing aircraft.

C_L and C_D curves as shown in Figure 10.17. Note that the C_L curve does not have a well-defined maximum lift point. When this type of aircraft is rolled at high angles of attack, only small changes in C_L take place. In the sweptwing aircraft, there is no definite stall, and the wing autorotation contribution will be quite weak. The change in C_D that occurs between the two wings, however, is substantial, and a strong yawing moment is produced.

Sweptwing aircraft have low aspect ratios. The mass of the aircraft is distributed along the longitudinal axis of the aircraft rather than in the wings. This is known as a *fuselage-loaded* aircraft, as represented by the T-38 in Figure 10.15. As the yaw develops, during the spin, this mass distribution contributes to the rotational momentum and tends to flatten the spin. This produces very high angles of attack and high sink rates.

Recovery

As indicated above, spin recovery procedures are quite dependent on the aerodynamic design of the aircraft. However, recovery from a spin is considered complete once the aircraft is below its stall angle of attack and there are no significant un-commanded angular motions. The time to achieve this recovered state is measured from the time the pilot initiates recovery control inputs.

There are three general actions for spin recovery:

1. **Flight control inputs:** For most aircraft, the use of rudder is most effective to recover from a developed spin. The resulting reduction in yaw rate also reduces coupling in the pitch axis, thereby reducing AOA. Forward stick/yoke further reduces AOA below stall. For fuselage-loaded aircraft, the ailerons may be the most effective anti-spin control. In this case, aileron in the direction of spin will reduce roll/yaw accelerations.
2. **Engine power:** Generally, power is reduced in a spin recovery to avoid nose-up pitching moments and reduce propeller gyroscopic effects. Turbine aircraft engines also produce gyroscopic effects though these are considered minimal in relation to the aerodynamic moments.
3. **Emergency devices:** If equipped, anti-spin parachutes or anti-spin rockets may be employed to recover from a spin, particularly in aircraft under test.

Once autorotation has ceased, the aircraft will transition into a nose-low dive. It is important to neutralize the rudder to avoid entering a spin in the opposite direction. Add power as required to recover from the dive without initiating a secondary stall.

This discussion of spins is a simplified general overview. Each model of airplane spins differently, and a given aircraft will spin differently when the configuration of the aircraft is changed. For those who accidentally enter a spin condition, disorientation and confusion in the direction of rotation can occur, especially if in instrument meteorological conditions (IMC) and flying solely by reference to instruments. Determining whether you are in a spin or spiral by looking at your airspeed indicator is paramount to begin recovery procedures. In any aircraft, it is best to follow the manufacturer's recommended spin recovery procedure for that aircraft. Within modern general aviation, single-engine training aircraft, once the proper spin recovery procedures have been initiated, the aircraft may still spin from one quarter of a turn to up to several turns, depending partly on how quickly and correctly the recovery procedure is initiated.

In the absence of a published procedure, the following technique is widely accepted as a typical spin recovery procedure in small, fixed-wing, general aviation training aircraft:

1. Reduce the throttle to idle.
2. Position the ailerons to neutral.
3. Apply the opposite rudder to the direction of rotation.
4. Apply stick (yoke) forward beyond the neutral position and hold (decrease AOA).
5. Neutralize the rudder after rotation stops.
6. Apply back pressure on the stick (yoke) until level flight (caution of secondary stall or excessive loads).

Many fixed-wing general aviation flight instructors teach the PARE method: (i) **P**ower to Idle; (ii) **A**ilerons Neutral. (iii) **R**udder Opposite Direction of Spin. (iv) **E**levator Forward.

Pilots must be familiar with the spin characteristics of their airplanes as described in the operator's handbook; and if spins are approved by the manufacturer, pilots should **strongly** consider taking spin training from an experienced flight instructor, and learn proper spin **avoidance** and recovery, firsthand.

Inverted Spins

Unlike an erect spin, the AOA in an inverted spin is always negative. Also, while roll and yaw are in the same direction in an erect spin, in an inverted spin, the roll rate is always the opposite of the yaw rate. Recovery from an inverted spin is generally easier than from an erect spin. In fact, many aircraft will recover from an inverted spin as soon as the controls are neutralized. However, it is important to initiate the proper control inputs based on the specific aircraft.

Impact of Weight and Balance During a Spin

The weight of a loaded aircraft is not as important as the position of the various "weights" within an aircraft when considering spin entry and recovery. An aft CG tends to increase the aircraft's pitch attitude (and thus AOA). If the AOA is already high (as in a climb, or in slow flight, for example), this further increase in AOA may precipitate a stall. Because this type of stall usually occurs as a surprise to the pilot, the controls may be uncoordinated, and the yawing of the aircraft leads to spin entry. Depending on the amount of weight, and how far aft the CG is, recovery of the stalled condition may be difficult or impossible, and recovery from the spin may be delayed or impossible. An aircraft that is approved for spins under the FAA's utility category may not be safe to spin when loaded within the normal category. A more detailed discussion about longitudinal stability is found in Chapter 12.

Application 10.1

This textbook presents a generic spin recovery procedure that will facilitate recovery from an incipient spin in most general aviation airplanes.

Consider the spin recovery procedure if the same aircraft was in an inverted, developed spin. Would the sequence of the recovery procedures be any different? How can a spin be detected if an aircraft is in instrument meteorological conditions (IMC)? What are the spin recovery procedures for the T-6 Texan, T-38 Talon, and the T-45 Goshawk?

HAZARDS DURING SLOW-SPEED FLIGHT – LOW-LEVEL WIND SHEAR

Wind shear can be defined as a sudden and prolonged change in wind direction and/or speed, relative to an airplane's ability to accelerate. While wind shear can occur at any altitude, it is most hazardous below 2 000 ft above ground level (AGL) during approach, landing, and takeoff operations, so this discussion is limited to low-level wind shear. Under unusual weather conditions, wind-direction changes of 180° and wind-speed changes greater than 50 kts. have been observed.

We have been taught that wind cannot affect an aircraft in flight except for ground speed and drift. This is true for steady winds or winds that change gradually, but it is not true when rapid wind changes occur. When the air velocity changes, the mass of the aircraft must be accelerated or decelerated, and this takes time. Slow-accelerating airplanes will have more trouble recovering from wind shear.

Common causes of wind shear are thunderstorm activity, weather fronts, terrain, and temperature inversions with strong winds above. Figure 10.18 illustrates a wind-shear situation caused by a downdraft. At position 1, the aircraft encounters the outflow, and then a degradation in performance occurs at position 2 when the headwind shifts to a tailwind and the downdraft becomes stronger. In position 3, the aircraft descends because of the direct effect of the increasing downdraft itself in addition to the loss of lift caused

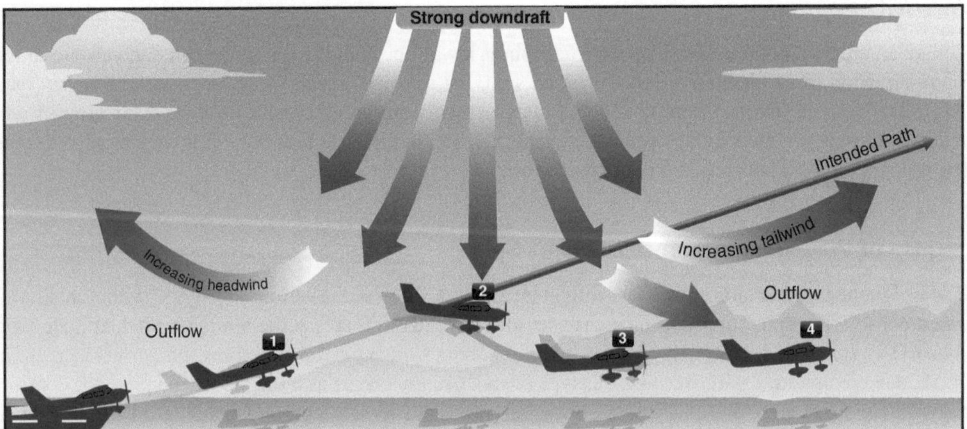

Figure 10.18. Wind shear caused by a downdraft.
Source: Pilot's Handbook of Aeronautical Knowledge (2008a)/U.S. Department of Transportation Federal Aviation Administration/Public domain.

by the strong tailwind, leading to position 4 when the aircraft may or may not clear terrain as the tailwind continues. The downward airflow in this situation, commonly associated with a thunderstorm, is called a *downburst*, and when its diameter is 2.5 mi or less it is more commonly referred to as a *microburst*. Studies have indicated that downdrafts of more than 6 000 fpm were encountered in strong microbursts. As a headwind shifts direction to become a tailwind, the aircraft suffers a loss of performance and may not be able to out-climb the strong downdraft. A downburst, and specifically a microburst, is the most dangerous type of wind shear because it can force the airplane into the ground.

When a downburst hits the ground, it is diverted outward. This creates strong horizontal winds known as *headwind bursts, tailwind bursts*, and *crosswind bursts*, depending on the position of the aircraft relative to the center of the thunderstorm. These are shown in Figure 10.19, as viewed from above. The performance of an aircraft flying through these bursts is discussed in the next section.

Another hazard associated with thunderstorms is called the gust front, which may precede a thunderstorm by as much as 15 mi. This is shown in Figure 10.20. The slope of the gust front makes it impossible to determine when an aircraft will fly through it, especially when it is in visual meteorological conditions (VMC) outside of precipitation. Aircraft trying to stay ahead of a storm into or out of an airport may encounter the unexpected shear of a gust front.

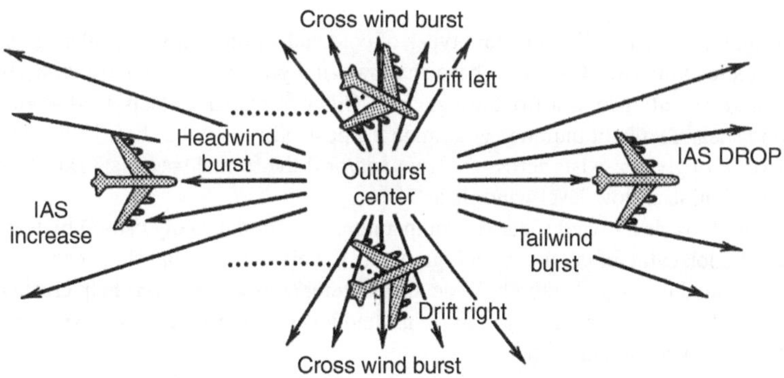

Figure 10.19. "Bursts" caused by a thunderstorm.

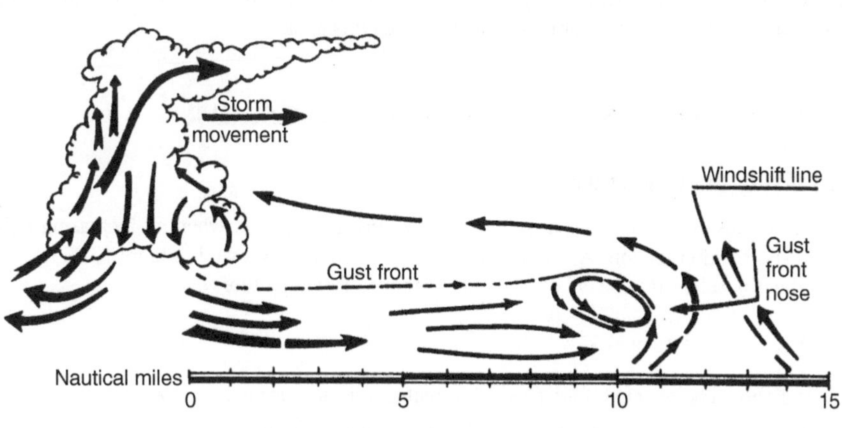

Figure 10.20. Thunderstorm gust front.

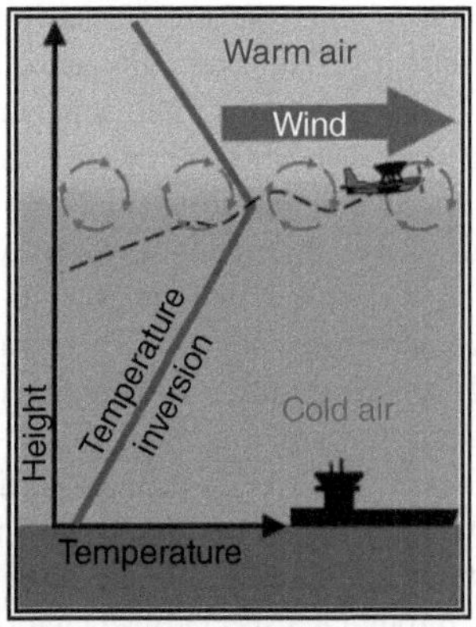

Figure 10.21. Temperature inversion LLWS.
Source: Airplane Flying Handbook (2016a)/U.S. Department of Transportation Federal Aviation Administration/Public domain.

Wind shear may also be present in certain types of cold and warm fronts. Turbulence may or may not exist in wind-shear conditions. Most fronts have shallow wind gradients, so in most cases, the changes in wind direction and velocity across a front are gradual. Other fronts have steep wind gradients and large amounts of wind shear. A rule of thumb to determine the possibility of wind shear is this: If a temperature differential of 10 °F or greater exists across the front and/or if the frontal speed is 30 kts. or more, there is the possibility of significant low-level wind shear.

A third source of low-level wind shear is a temperature inversion, usually caused by warm air moving above a pocket of cool, calm air, as shown in Figure 10.21. This type of wind shear can be found on clear nights when the ground is dry and smooth. Wide diurnal temperature variations help create this situation. The horizontal winds in the boundary layer above ground level have been found to be as much as 70 kts., while the surface winds remained calm.

Little or no turbulence, or vertical wind velocity, is usually found at the boundary of the warm and cold air in a temperature inversion. When an aircraft is flying at a constant altitude, no significant wind shear will occur across the boundary. However, large variations in headwind, tailwind, and crosswind components can be experienced when climbing or descending through this area.

AIRCRAFT PERFORMANCE IN LOW-LEVEL WIND SHEAR

During Takeoff and Departure

The most hazardous condition caused by wind shear during takeoff is when an increasing tailwind is encountered shortly after takeoff. A simplified sequence of events is shown in Figure 10.22.

If wind shear is believed to be encountered during takeoff or departure:

Before V_1:

1. Takeoff should be rejected if airspeed abnormalities occur during the initial takeoff roll.
2. If airspeed abnormalities occur before V_1, but within the high-speed regime of the takeoff roll, a takeoff should be continued if the runway may be too short to abort.

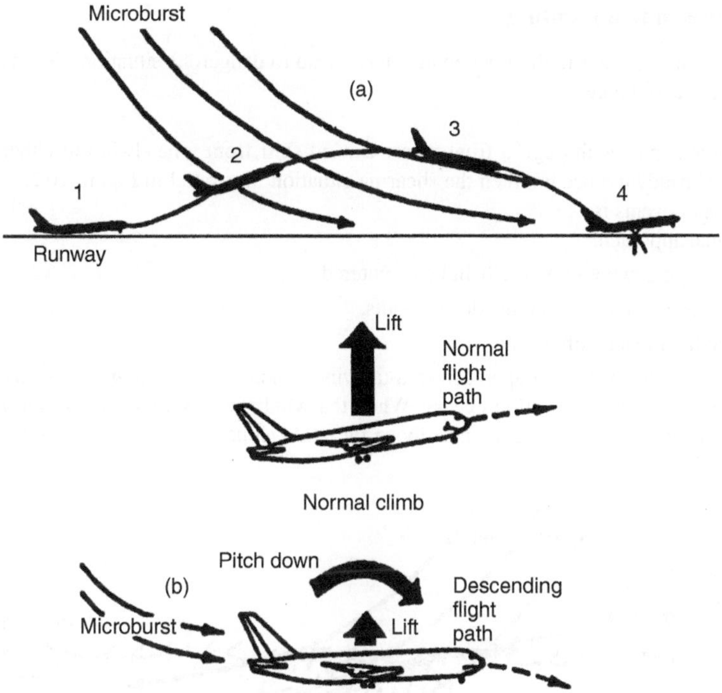

Figure 10.22. Tailwind wind shear encountered on takeoff: (a) flight path, (b) forces and moments.

After V_1:

1. Maintain or set throttles to maximum thrust.
2. Rotate at V_r as normal.
3. Set the required pitch for maximum climb as referenced by the manufacturer or follow the flight director (FD) pitch command if wind-shear guidance is provided; takeoff/go-around (TOGA) may be an option on the FD in some aircraft.

Initial climb:

1. Maintain or set throttles to maximum thrust.
2. Continue to pitch for maximum climb as referenced by the manufacturer or follow the FD pitch command if wind-shear guidance is provided.
3. If a turn is not required for obstacle clearance, level the wings to maximize the climb rate.
4. Continue to monitor the airspeed for indications of wind shear, or for indications that the situation has been cleared. Airspeed near or at the intermittent speed for the stick shaker may be required.
5. Do not change the flap or landing gear configuration until out of the wind shear, unless specific guidance from the manufacturer is given.
6. Once clear of wind shear, maintain proper airspeed and normal climb profile, and configure as recommended by the aircraft checklist.

During Approach to a Landing

Wind shear encountered during the final approach can lead to dangerous situations. The two most likely situations are discussed here.

1. If an aircraft passes through a front where a wind shift from a headwind to either a tailwind or a reduced headwind occurs, then the shearing situation illustrated in Figure 10.23 may exist. The sequence of events is
   a. Normal approach.
   b. Increasing downdraft and tailwind encountered.
   c. Airspeed decreases and pitch down results.
   d. Potential impact with terrain.
2. When an aircraft makes an approach in a tailwind condition that changes to a headwind or calm condition, the reverse situation occurs. When the wind shear is encountered, the IAS increases, the aircraft pitches up, and it climbs above the glide slope. If thrust is not reduced, the aircraft

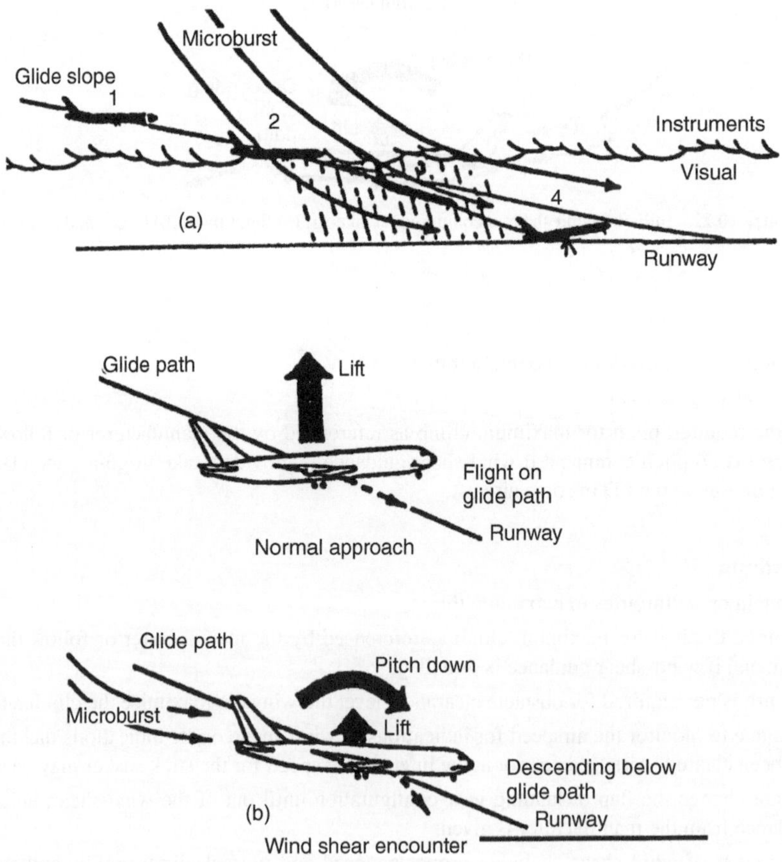

Figure 10.23. Tailwind shear encountered in landing approach: (a) flight path, (b) forces and moments.

climbs above the flight path, and a high and fast approach results in the possibility of overshooting the runway. If the pilot reduces thrust after going high on the glide slope, then thrust must be re-established as the aircraft returns to the glide slope or the airplane will go below the flight path, and a short, hard landing will result.

If wind shear is believed to be encountered during the final approach to landing, the following go-around/recovery procedures should be initiated:

1. Set maximum go-around thrust and pitch for maximum climb as recommended by the aircraft manufacturer or set TOGA if able.
2. If the FD provides guidance and the aircraft is connected to autopilot, keep the autopilot engaged.
3. If a turn is not required for obstacle clearance, level the wings to maximize the climb rate.
4. Continue to monitor the airspeed for indications of wind shear, or for indications that the situation has been cleared. Airspeed near or at the intermittent speed for the stick shaker may be required.
5. Do not change the flap or landing gear configuration until out of the wind shear, unless specific guidance from the manufacturer is given.
6. Once clear of wind shear, maintain proper airspeed and normal climb profile, configure as indicated by the aircraft checklist.

Vertical Shear Level

The previous section introduced an overall analysis of the impact of horizontal wind shear on a landing aircraft, but the vertical level of the shear also must be reviewed for a true understanding of wind shear detection and proper avoidance. The closer the shear level to the runway, the less time the pilots have to react and initiate proper recovery procedures. In the top image of Figure 10.24, the landing aircraft encounters wind shear at the outer marker (OM) of an instrument landing system (ILS) at glideslope intercept, usually around 2 000 ft AGL. In this example, the shear is a headwind to a tailwind which results in a decrease in performance. As a loss of lift is experienced, to maintain the V_{REF}-indicated airspeed that was briefed, the nose drops, and altitude is lost as the aircraft descends below the glideslope. Power (thrust) must be added to correct for altitude and pitch is reset for the proper airspeed. Descent on a glideslope is a constant rate descent maneuver, the greater the ground speed, the greater the descent rate. At this AGL altitude recovery from the shear is usually attainable and the glideslope recaptured for a normal approach and landing.

The middle image in Figure 10.24 depicts a shear level closer to the runway, for this example, we will use 1 000 AGL. As above, when the shear level is encountered the headwind once again shifts to a tailwind, resulting in a loss of performance. If the headwind was 15 kts. and transitioned to a 15 kts. tailwind, the shear would be a difference of 30 kts., quite substantial for any aircraft. Once again, the aircraft descends below the glideslope, additional power is applied, and the pitch is increased to regain the glideslope. If the pilots overcorrect, the aircraft climbs above the glideslope and is now high and potentially above V_{REF}. Due to the lower altitude, the pilots may land long and fast, using up more runway than was calculated. In this situation, it would be best operating practice to initiate a go-around upon the first indication of wind shear and reestablish a second approach or holding procedures until the reason for the shear has moved from the approach area.

In the bottom image of Figure 10.24, the shear level moves even closer to the runway and may result in a high sink rate, and hard landing scenario. As before, when the headwind shifts to a tailwind, the nose drops, and the aircraft descends below the glideslope. Aware of the low AGL altitude, the pilots

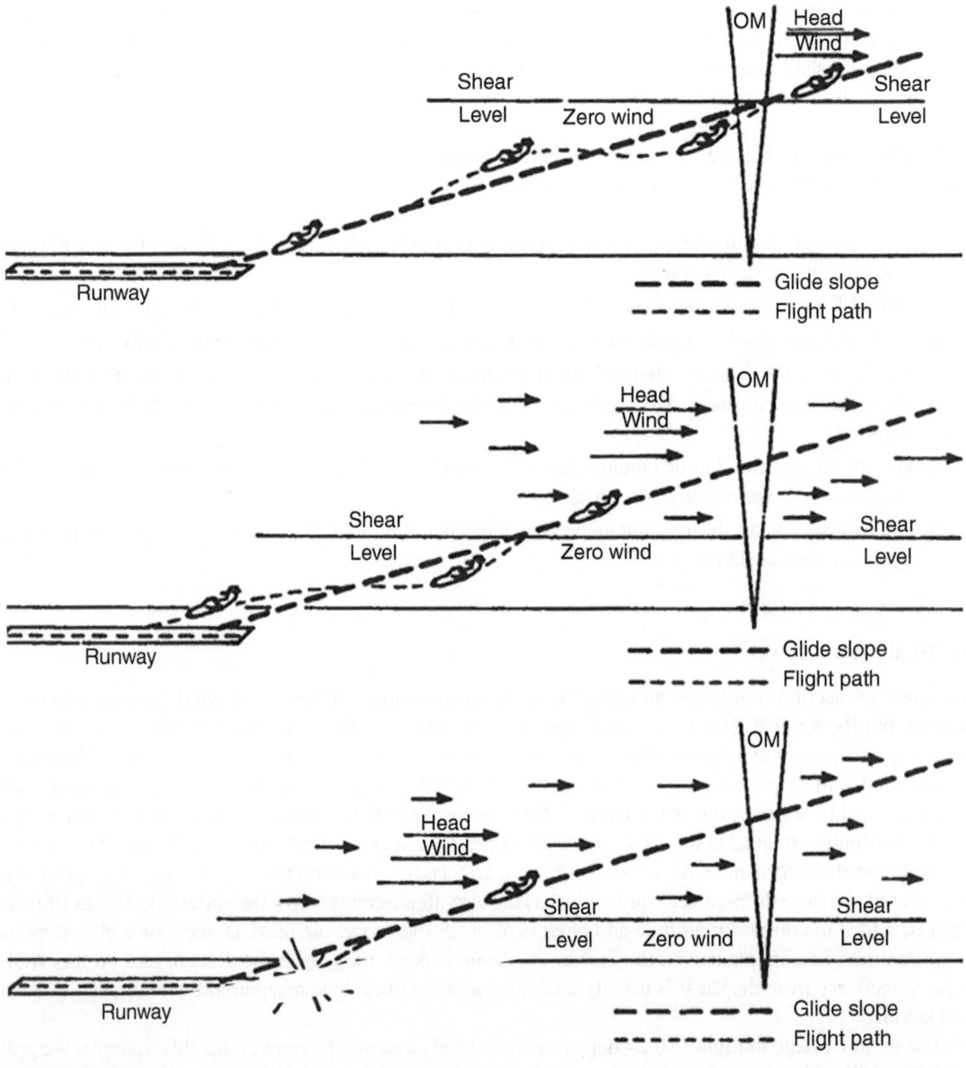

Figure 10.24. Impact of shear level altitude on landing aircraft.
Source: Wind Shear FAA-P-8740-40 (2008b)/U.S. Department of Transportation Federal Aviation Administration/Public domain.

once again add power and increase the pitch, which results in a higher angle of attack. As discussed in Chapter 5, drag increases and the pilots must react with proper pitch and power to avoid a high descent rate, with high power, as will be discussed in the next chapter as the aircraft is in the region of reversed command. A go-around is the only option when encountering shear close to the ground. Depending on spool up time of the engines, the aircraft will sink below the glideslope and only proper wind shear avoidance procedures and quick action will prevent the aircraft from striking the ground short of the runway.

Crosswind Burst Response

A crosswind shear burst causes the airplane to roll and/or yaw. Large crosswind shears may require large and rapid aileron inputs. Wind shear increases the pilot's workload and causes distraction. Horizontal vortices may also be encountered, which can cause severe roll forces and require full aileron movement to maintain control of the aircraft.

Heavy Rain

Wind shear is often accompanied by heavy rain due to the precipitation-induced downdraft. Earlier studies of rain effects at altitude concluded that rain had a significant effect on performance, but, due to sufficient altitude, would not force the aircraft to the ground. In the approach or post takeoff flight phase, however, conditions are much more critical, and the performance margin of the aircraft is considerably reduced.

Rain affects the aircraft in several ways:

1. The momentum of the raindrops impacts the aircraft in a downward and backward direction.
2. The increased weight of the rainwater film affects the gross weight of the aircraft.
3. The water film is roughened by the impact of the raindrops and the aerodynamic properties of the wing are adversely affected.
4. The raindrops hit the airplane unevenly, causing possible pitching or rolling moments.

Of these effects, the roughened airfoil and resulting loss of aerodynamic properties are the most important. Recent studies have indicated that as much as a 30% increase in C_D and a loss of lift of more than 30% can occur due to this roughness. The AOA for $C_{L(max)}$ can also be reduced by 2–6°. At least nine major commercial aircraft accidents investigated and reported by the National Transportation Safety Board (NTSB) have involved heavy rain during takeoffs and landings.

HAZARDS DURING SLOW-SPEED FLIGHT – TURBULENCE

Turbulence Effects

Turbulence differs from wind shear in that it is sporadic. Airspeed may fluctuate in an unpredictable pattern, and in extreme cases, injury to the passengers or damage to the aircraft may occur. Moderate or greater turbulence can make changes in airspeed difficult to detect and thus may delay recognition of the presence of wind shear.

Wake Turbulence

Introduced in Chapter 5, wake turbulence is caused by wingtip vortices generated when a wing is producing lift. The greater the lift being produced, the greater the induced drag, and the more intense the wingtip vortices will be.

A *vortex* is a highly developed rotational mass of disturbed, high-energy air created by the wing of an aircraft as it produces lift. An aircraft creates two such rotating vortices, one from each wingtip, as shown in Figure 10.25. These vortices travel outward from the wingtips and sink toward the ground. As an aircraft that produces lift passes through the air, there is a differential pressure on the bottom and the top of the wing, resulting in a movement of air around the wingtips from high to low pressure. The air rolls into two distinct vortices, one at each wingtip.

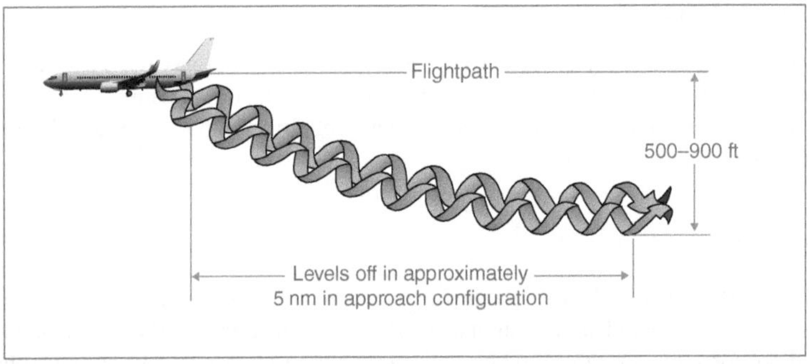

Figure 10.25. Wingtip vortices behind an aircraft.
Source: Aircraft Wake Turbulence AC 90-23G (2014)/U.S. Department of Transportation Federal Aviation Administration/Public domain.

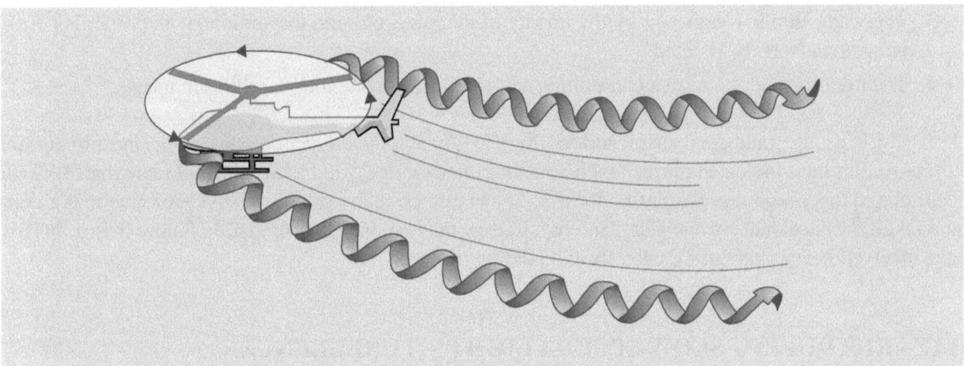

Figure 10.26. Helicopter vortices.
Source: Aircraft Wake Turbulence AC 90-23G (2014)/U.S. Department of Transportation Federal Aviation Administration/Public domain.

Because these wingtip vortices are more intense as more lift is being produced, they are strongest when an aircraft is heavy, clean, and slow. This rotational energy is directly related to the weight and to the lift being generated, and inversely related to the airspeed, so the vortices are strongest during takeoff and landing at high gross weights. During slow-speed flight, the aircraft is at a high angle of attack, resulting in a greater pressure differential between the upper and lower surfaces of the wing. As we saw in Chapter 4, as weight increases, so does the lift required, and thus, so does the intensity of wingtip vortices. The rotating process is normally completed in about two to four times the wingspan (200–600 ft) behind the aircraft.

Vortices are also developed by helicopters and trail along behind the aircraft in the same manner as those of fixed-wing aircraft (Figure 10.26).

The force of wingtip vortices trailing behind a heavy aircraft can easily exceed the aileron control capacity or climb rate of light aircraft. These forces can cause some light aircraft to roll completely over and to be forced into the ground. A typical mishap is a case where, just before touchdown, a wing drops,

Figure 10.27. Wake turbulence avoidance.
Source: Pilot's Handbook of Aeronautical Knowledge (2008a)/U.S. Department of Transportation Federal Aviation Administration/Public domain.

and contact is made with the runway while the aircraft is in a wing-down attitude. Another possible result, at altitude, is the inadvertent entry into an unusual attitude and resultant spin.

Avoidance of wake turbulence during takeoffs and landings is accomplished by requiring time intervals between aircraft depending on size, usually between two and four minutes. In the takeoff pattern, the following aircraft should become airborne before the lead aircraft and should climb above the turbulence. In the landing pattern, the following aircraft should touch down past the point where the lead aircraft has touched down (Figure 10.27).

Figure 10.28 shows various methods for avoiding wake turbulence caused by another aircraft during slow-speed flight in the airport environment. In Figure 10.28, Image (a), the landing aircraft touches down after the touchdown point of the preceding aircraft. The quartering headwind blows the upwind vortex into the runway, but closer to the approach end, which the landing aircraft has already cleared.

Lateral movement of tip vortices due to crosswinds can present a hazard to takeoffs and landings when aircraft are operating from airports with parallel runways that are less than 2 500 ft apart. This is illustrated in Figure 10.28, Image (b). The no-wind movement of the vortices is away from the centerline of the runway, but if a quartering headwind from the left in this example is present, the right vortex will move to the right and will present a hazard to aircraft operating on a parallel runway.

In Figure 10.28, Image (c), an aircraft in slow-speed flight during an approach to land is crossing behind the glide path of another landing aircraft on an intersecting runway. The position of the crossing traffic must be noted, and though the wingtip vortices from the crossing traffic will usually sink, local winds and thermals may cause the vortices to rise into the glide path of the landing traffic.

Understanding when wake turbulence occurs, what aircraft are more likely to produce the greatest strength vortices (i.e. heavy, clean, or slow), and the regulations that exist to help pilots avoid wake turbulence will all help improve aviation's safety record. Remember: *if an aircraft is producing lift, it is producing wake turbulence.*

Mechanical Turbulence

Mechanical turbulence is most hazardous to aircraft during the takeoff and landing phases of flight. As the wind flows around man-made obstructions and terrain around an airport, the direction and velocity of the wind can change without warning, complicating the crosswind takeoff and landing procedures that were discussed in Chapters 8 and 9 (Figure 10.29). The strength/magnitude of mechanical turbulence depends on many factors: the speed of the wind, the angle at which the wind strikes the object, and most importantly the stability of the air. The intensity of the turbulence associated with the ground obstructions varies with

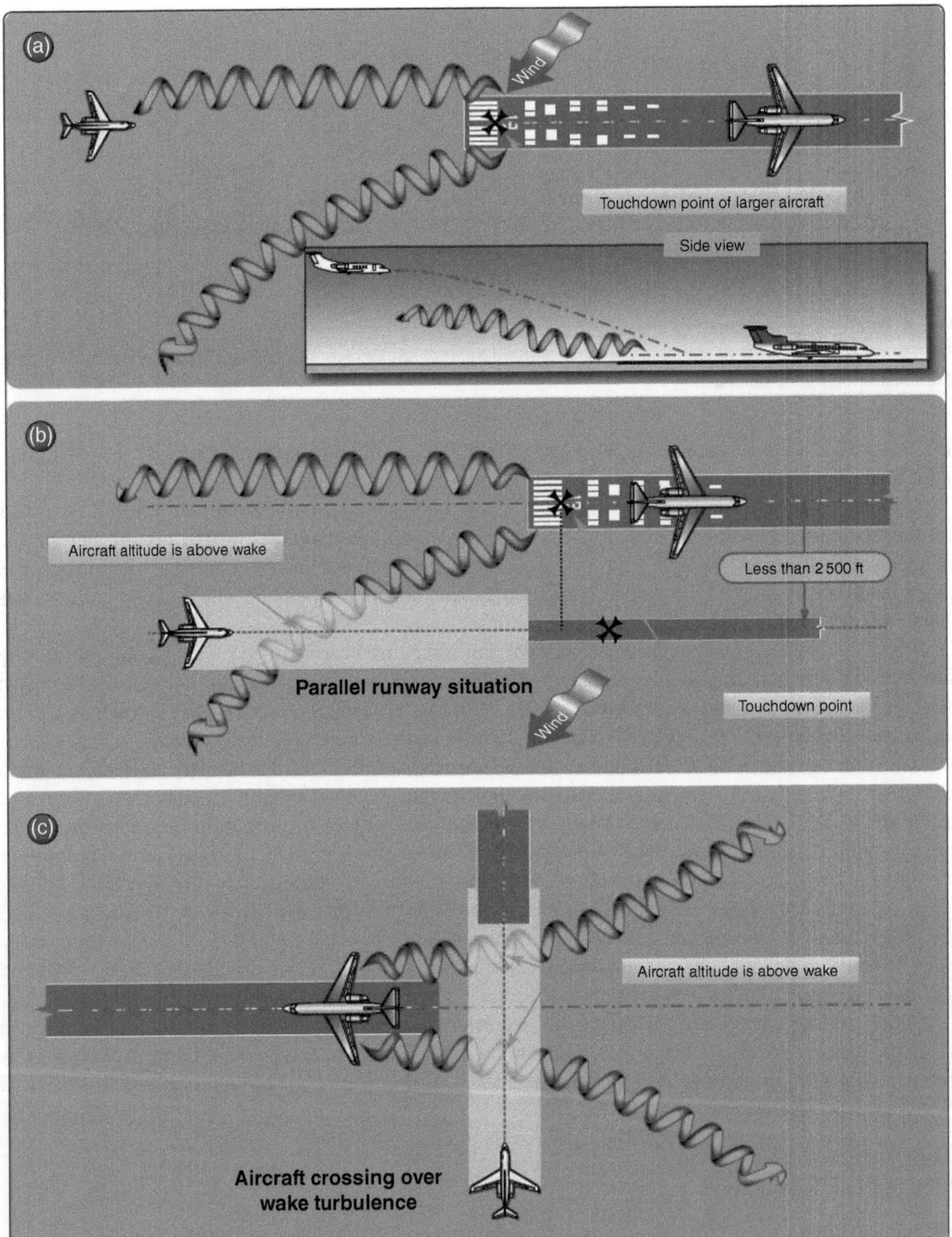

Figure 10.28. Wake turbulence avoidance procedures. (a) Upwind vortices of landing aircraft blown onto the runway, but closer to the approach end of the runway and not impacting the landing aircraft if the same touchdown point is used. (b) Vortices created by the landing aircraft on one runway are blown by a crosswind onto the touchdown zone of a parallel runway, possibly impacting landing aircraft. (c) Landing aircraft on one runway may be impacted by wingtip vortices from landing traffic on a crossing runway.

Source: Pilot's Handbook of Aeronautical Knowledge (2016b)/U.S. Department of Transportation Federal Aviation Administration/Public domain.

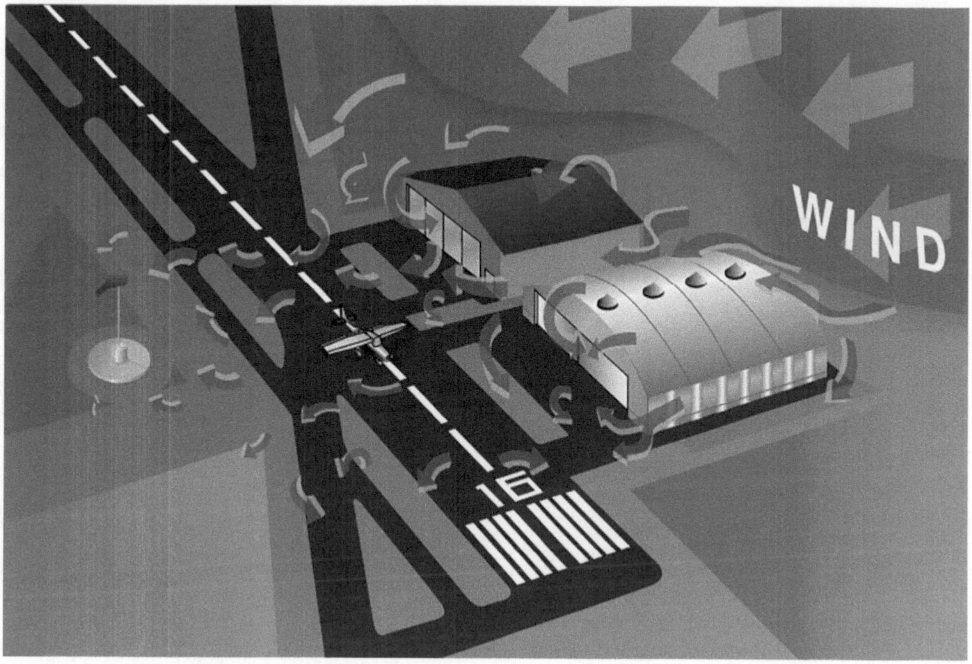

Figure 10.29. Mechanical turbulence.
Source: Pilot's Handbook of Aeronautical Knowledge (2023c)/U.S. Department of Transportation Federal Aviation Administration/Public domain.

the size and density of the structures around the airport, and airspeed gains and losses due to changing wind direction (wind shear) can occur when least expected, resulting in loss of altitude or directional control. During the landing phase of flight, if the approach airspeed is allowed to degrade too far, and the shear due to mechanical turbulence results in a sudden loss of headwind, the critical AOA could be exceeded and a landing short of the runway may occur.

Convection and Mountain Wave Turbulence

In addition to the horizontal movement of air discussed in Chapters 8 and 9 affecting takeoff and landing performance, vertical movement of the air due to thermal changes and uneven heating of the earth's surface also affects the glide path of an aircraft. Figure 10.30 shows an aircraft on a stabilized approach to the runway when the constant rate of descent becomes disrupted due to rising and sinking air on the final approach. Due to the man-made parking lot and building, according to Stefan–Boltzmann's law of radiation, the darker object that has absorbed more of the sun's energy is emitting more energy warming the air over the structures, resulting in the rising of the less dense air. Over the cooler water/vegetation, the air is denser and may result in descending air currents. As a light aircraft descends on final during a stabilized approach, pitch and power may need to be adjusted to maintain a proper glide path due to the varying vertical movement of the air.

Slow-flight or takeoffs and landings around higher terrain offer additional risks depending on the wind velocity and stability of the air. Figure 10.31 shows the impact of predominantly downdrafts on the leeward

Figure 10.30. Thermal changes due to varying surface conditions.
Source: Pilot's Handbook of Aeronautical Knowledge (2023c)/U.S. Department of Transportation Federal Aviation Administration/Public domain.

Figure 10.31. Low-level Mountain wave turbulence.
Source: Balloon Flying Handbook (2008c)/U.S. Department of Transportation Federal Aviation Administration/Public domain.

side of higher terrain. Many normally aspirated engines used by general aviation aircraft lack the extra power to clear terrain high enough to avoid mountain wave turbulence. Close to the surface, violent rotors may exceed 3 000 ft/min, far beyond the maximum rate of climb for many aircraft. Combined with varied terrain heights and higher density altitude, a pilot may quickly exceed the critical AOA while trying to evade low-level mountain turbulence. Gravity waves, characterized by successive wavelengths of updrafts and downdrafts two to ten miles wide may extend far beyond the terrain, and at the cruising altitudes of many general aviation aircraft.

EQUATION

10.1 Rate of sink (ROS) $= 101.3\, V_k \left(\frac{D - T_a}{W}\right)$ (fpm)

KEY TERMS

Accelerated stall
Cross-control stall
Downburst
Microburst
Normal command
Power-off stall
Power-on stall
Rate of sink (ROS)
Region of reversed command
Skid
Slip
Speed instability
Spin
Temperature inversion
Wake turbulence
Wind shear

PROBLEMS

Note: Answers to the problems in the first two sections are given at the end of the book.

Application of Chapter Theory

1. The main difference between the $C_L - \alpha$ curves for straight-wing aircraft is that
   a. the sweptwing aircraft has a lower value of $C_{L(max)}$.
   b. the straight-wing aircraft does not fly at as high an AOA for $C_{L(max)}$.
   c. the sweptwing aircraft does not have an abrupt loss of lift at $C_{L(max)}$.
   d. All of the above.

2. Low-speed stall will start at the wing's trailing edge
   a. where C_l/C_L is a minimum.
   b. Where C_l/C_L is a maximum.
   c. at the root for a swept wing.
   d. at the wingtip for a straight rectangular wing.

3. The region of reversed command for a jet aircraft is also correctly known as
   a. the backside of the thrust-required curve.
   b. The backside of the power curve.
   c. the backside of the drag curve.
   d. Both (a) and (c) above.

4. An aircraft should never be flown in the region of reversed command.
 a. True
 b. False

5. Pulling back on the control stick (or yoke) will cause the airplane to climb if the plane is flying at
 a. low speed.
 b. high speed.
 c. any speed.

6. A method of climbing that will work at any speed is to control the airspeed with the stick and add throttle to climb.
 a. True
 b. False

7. An airplane is making a final approach for landing and encounters a horizontal wind shear. Which of the types of bursts below is the most dangerous?
 a. Headwind burst.
 b. Crosswind burst.
 c. Tailwind burst.

8. An aircraft in which of the following situations is most likely to create the most intense wake turbulence?
 a. Heavy, dirty, and slow.
 b. Heavy, clean, and slow.
 c. Light, clean, and slow.
 d. Light, dirty, and slow.

9. Wake turbulence can cause an airplane to be turned completely upside down. To escape wake turbulence, a pilot should avoid
 a. flying behind and below a large airplane.
 b. taking off or landing behind a large airplane.
 c. taking off or landing on a parallel runway that is downwind from one being used by large airplanes.
 d. All of the above.

10. Spin recovery consists of
 a. stopping spin rotation with a rudder.
 b. reducing the critical AOA.
 c. applying ailerons into the spin direction (sweptwing airplanes).
 d. All of the above.

11. An aircraft loaded within weight and balance limits, but with a CG at the forward limit, will experience a higher stall airspeed.
 a. True
 b. False

12. To avoid an accidental spin, a pilot should avoid which phase of a spin:
 a. Entry
 b. Incipient
 c. Developed
 d. Recovery

13. A cross-control stall setup is most likely to occur
    a. during departure while clearing an obstacle.
    b. In the traffic pattern.
    c. on a straight-in final approach in IMC.
    d. When experiencing wake turbulence.

14. Which wing planform is the most aerodynamically efficient?
    a. Elliptical
    b. Rectangular
    c. Sweptwing
    d. Tapered

15. Which of the following statements is not accurate based on the Figure 10.7 discussion?
    a. The center of pressure will shift forward on all wing shapes as the stall pattern develops.
    b. The center of pressure will shift forward on sweptwing and delta wing aircraft but remain the same on elliptical wings and tapered wings as the stall pattern develops.
    c. Each wing will develop a nose-up pitching moment as the stall pattern develops
    d. The nose-up pitching moment is more pronounced on sweptwing and delta wing aircraft.

16. When an aircraft is banked to the left and in a skid, the corrective action should be to apply the right rudder to correct.
    a. True
    b. False

17. During a stall and spin entry to the left, the left wing "drops" as the left wing is "more stalled" than the right wing.
    a. True
    b. False

18. If a microburst is encountered on takeoff, what change in performance should be expected when going from a tailwind to a headwind?
    a. No change in performance.
    b. Decrease in performance.
    c. Increase in performance.

Calculations Using Chapter Principles

19. According to Figure 10.11, regardless of the angle of the bank, anytime 0° flaps are used instead of a 40° flap setting, the accelerated stall speed will always be higher for a given weight.
    a. True
    b. False

20. Using Figure 10.11, calculate the IAS accelerated stall speed given the following information:

    Gross weight: 2 000 lb

    Angle of bank: 40°

    Flap position: 40°

    a. 52 kts.
    b. 55 kts.

c. 47 kts.
d. 40 kts.

21. A 10 000 lb airplane has the T_r curve shown in Figure 10.3. The T_a is 3 000 lb and it is the same for all airspeeds. The airplane is flying at 210 kts. and the pilot decides to climb and pulls back on the stick, thus reducing the speed to 200 kts. Find the ROC.

22. Consider the same situation as in Problem 21, except the airspeed at the start of the maneuver is 135 kts. and after the pilot pulls back on the stick, the speed is 130 kts. Find the ROC.

23. A 14 000 lb airplane has the T_r curve shown in Figure 10.3. The T_a is 3 000 lb and it is the same for all airspeeds. The airplane is flying at 200 kts. when the pilot pulls back on the stick, resulting in a speed of 175 kts. Find the ROC at the new speed.

Correlation Through Scenario-Based Learning

Aircraft – *Diamond DA50RG*

1. Review the *Stalling Speeds* section in Section 5 of the DA50 aircraft flight manual (AFM), and find the following:
  a. Position of the throttle and center of gravity (CG) for the listed speeds.
  b. What is the *Note* about IAS values? Why does this happen? What is the relationship between IAS and calibrated airspeed (CAS) as flaps are deployed?
  c. Based on the chapter reading, for any of the respective listed stall speeds, if the CG was not full forward, would the IAS for stall be higher or lower than what is published? Why?
  d. For any specific bank angle, does the stall speed increase or decrease as flaps are deployed?
  e. For any specific bank angle and gear/flap setting, how does the IAS stall speed change with a change in weight? Why?
  f. What does the position of the gear have to do with the published stall speeds?

2. Review the *Stalling Speeds* section in Section 5 of the DA50 *Ice Protection System for FIKI* **supplement**, and find the following:
  a. Position of the throttle and center of gravity (CG) for the listed speeds.
  b. Does the same *Note* about IAS values exist when compared to the AFM in the previous problem?
  c. According to the supplement, in what environmental conditions do the supplement stalling speeds apply?
  d. Overall, how do the stalling speeds in the supplement compare to the published stalling speeds from the AFM? Why is there a difference?
  e. Why is the row *Flaps LDG* for the published stalling speeds not included?

3. Review the *Stall Warning System Description* section in Section 7 of the DA50 AFM, and answer the following:
  a. How is a stall warning recognized mechanically on the DA50, and how is that warning translated to the pilot?
  b. What is the preventative measure to prevent the stall warning system from icing over?
  c. How is the pilot notified if there is a failure in the stall warning system?

4. Review the *Recovery from Spin* procedures in Section 3 of the DA50 AFM. How closely do the procedures parallel the PARE recovery procedures as discussed in the chapter reading? At what point does the manufacturer recommend the flaps be retracted if deployed at the beginning of the spin?

5. What does the *Caution* state regarding the order of the initial spin recovery steps?

11 Maneuvering Performance

CHAPTER OBJECTIVES

After completing this chapter, you should be able to:

- Identify the differences in shallow, medium, and steep turns, and explain the aerodynamic forces on an airplane during constant altitude, turning flight.
- Define load factor and calculate the load factor for varying angles of bank.
- Summarize the relationship between bank angle, load factor, and stall speed.
- Explain the purpose of a V–G diagram and locate the maneuvering speed and limit load factor performance values.
- Calculate maneuvering speed and correlate with aircraft performance load factor limitations.
- Summarize the relationship between radius of turn and rate of turn with aircraft bank angle and airspeed.

For our discussion, maneuvering performance will be divided into two main categories: *general turning performance* and *energy maneuverability*. General turning performance is common to all aircraft. It involves constant altitude turns and constant speed (but not constant velocity) conditions. Remember, velocity is a vector, having speed and direction. Energy maneuvering is relevant to tactical military missions, such as air-to-air combat maneuvers and air-to-ground maneuvers.

The discussion in this chapter will primarily cover General Turning Performance, but Energy Management will be introduced at the end of the chapter as an understanding of the basics of energy management is critical to all pilots.

All turns discussed in this chapter will be accomplished while maintaining a constant altitude.

GENERAL TURNING PERFORMANCE

Introduced in Chapter 3, the ailerons are used to initiate a turn. If the airplane is banked to the left, the aileron on the left wing will be raised, while simultaneously the aileron on the right wing will be lowered. At moderate bank angles, the airplane may continue in the turn as the ailerons are neutralized, and opposite

Flight Theory and Aerodynamics: A Practical Guide for Operational Safety, Fifth Edition.
Brian A. Johnson and Philip R. Fittante.

Companion website: www.wiley.com/go/flighttheoryandaerodynamics5e

deflection of the ailerons will be needed to level the airplane. Coordinated use of the rudder will be required in the direction of turn in most aircraft to counteract *adverse yaw*, which will be discussed in more detail in Chapter 13. As a review from Chapter 10, Figure 11.1 once again shows the proper use of the rudder to keep the ball in the center of the inclinometer, but now there is another tool used to visualize a coordinated turn, a "yaw string." When the yaw string is centered, the banked turn is coordinated. In the top image, the pilot is banking to the left but is applying too much right rudder for the turn, so the aircraft is in a *slip*. The relative wind striking the aircraft can be visualized by the yaw string. In the bottom image, the pilot continues in a left, banked turn, but the pilot is applying too much left rudder, so the airplane is considered in a *skid*. The image in the center of Figure 11.1 depicts an aircraft in level coordinated flight, the ball is in the center of the inclinometer, and the yaw string is vertical in the windscreen to indicate proper flow of the relative wind around the aircraft. To maintain a constant altitude, back pressure may be required as the bank angle is increased, and power/thrust will be added to maintain the same IAS as when the airplane was in level flight.

As an introduction to level flight, the Federal Aviation Administration (FAA)'s *Airplane Flying Handbook* (2016a) introduces three categories of turns: shallow, medium, and steep. Before the forces that initiate and maintain a turn can be discussed, a brief review of these three categories is warranted:

1. **Shallow:** with a bank angle of <20°, the lateral stability incorporated into the design of most aircraft results in the pilot having to maintain yoke/stick pressure in the direction of turn to prevent the aircraft from leveling. Minimal back pressure on the yoke/stick is required to maintain level flight, and little power/thrust is needed to maintain the original IAS. Standard rate turns in instrument meteorological conditions (IMC) are normally in this category for slower aircraft.
2. **Medium:** medium banked turns have bank angles ≥20° to <45°. Once the pilot neutralizes the ailerons, the airplane will most likely remain in the banked state, and will not inherently level on its own, or continue into an overbanking tendency. Turning flight while in the airport traffic pattern commonly uses medium banked turns.
3. **Steep:** turns are considered steep when the bank angle during level flight is ≥45°. Coordinated rudder in the direction of turn, increased back pressure on the yoke/stick, and increased power/thrust will be needed to maintain level flight, constant altitude, and original indicated airspeed (IAS) throughout a steep turn. The airplane may continue banking into an overbanking tendency as shown in Figure 11.2. Unless opposite control pressures are utilized to maintain the required bank angle. Satisfactory demonstration of steep turns is required for the private pilot and commercial pilot FAA certification standards.

Forces in Turns

In turning flight, an airplane is not in a state of equilibrium: if the forces on the airplane were balanced, the airplane would not deviate from its straight path. As we saw in Chapter 4, the unbalanced force that initiates the turn and keeps the plane turning is the horizontal component of lift. As the angle of bank increases in a level turn, so will the unbalanced force. Unlike an automobile, an airplane must be banked in order to turn, and to maintain level flight the angle of attack must be increased.

First, let us review acceleration. In Chapter 1, acceleration was defined as the rate of change of velocity. Velocity was defined as a vector quantity that involves both the speed of an object and the direction of the object's motion. Any change in the direction of a turning aircraft is therefore a change in its velocity, even though its speed remains constant. An aircraft in a turn is subjected to an unbalanced force acting toward the center of rotation. Newton's first law of motion, the Law of Inertia, states that a body at rest will remain at rest, and a body in motion will remain in motion, in a straight line, unless acted on by an unbalanced force. Newton's second law then states that an unbalanced force will accelerate a body in the direction

Figure 11.1. Slip, skid, and coordinated flight.
Source: U.S. Department of Transportation (2013)/Public domain.

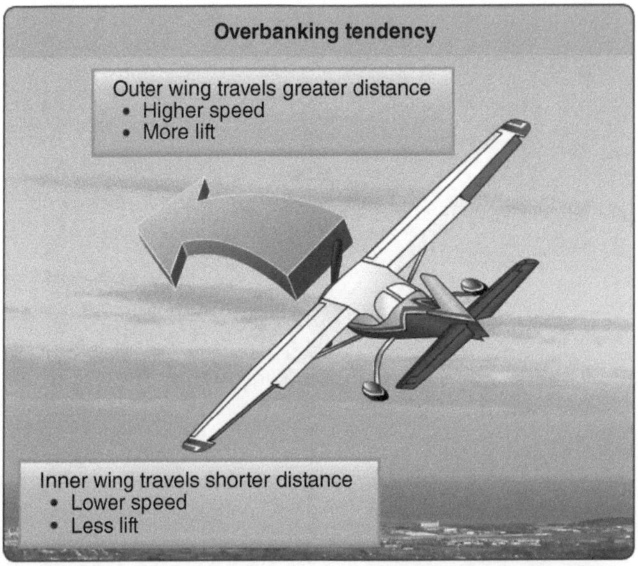

Figure 11.2. Overbanking tendency.
Source: Airplane Flying Handbook (2016a)/U.S. Department of Transportation Federal Aviation Administration/Public domain.

of that force. This unbalanced force is called the centripetal force, which in the case of a turning aircraft is the horizontal component of lift, and it produces an acceleration toward the center of rotation known as centripetal acceleration (Figure 11.3). The magnitude of the centripetal force (horizontal component of lift) is directly related to the angle of bank: as the angle of bank increases, so does the magnitude of the centripetal force.

During a banked turn, the pilot will interpret a force in the opposite direction of centripetal force, toward the outside of the turn. Some FAA figures and discussions still reference the term, *centrifugal force*, or inertial force. Inertia is not a force, it is just a measure of a body's resistance to change in its current motion; it has no units, but it is all about the body's mass. The force you feel in a turn that feels like it is pulling you toward the outside of the turn is the aircraft exerting centripetal force on your body. Like the airplane, your body was moving in a straight line before the turn. A centripetal force on your body is required to cause your body to now move in a curved path. The seat and the door exert that force on you.

The forces acting on an aircraft in a coordinated, constant altitude turn are shown in Figure 11.4. In wings level, constant altitude flight, the lift equals the weight of the airplane and is opposite to it in direction. But in turning flight, the lift is not opposite, in direction, to the weight. Only the vertical component of the lift, called effective lift, is available to offset the weight. Thus, if constant altitude is to be maintained, the total lift must be increased until the effective lift equals weight. *An aircraft in a ~45° bank, maintaining level flight, has roughly 80% of the original effective lift*. Until the remaining 20% loss of vertical lift is replaced, the aircraft will not be able to maintain level flight. As introduced in Chapter 4, as lift is increased so is induced drag. If sufficient thrust is not added to compensate for the increase in drag, then the aircraft will slowdown. In the next section, we will see how this increase in bank angle and a corresponding increase in lift also cause the stall speed to increase, increasing the risk of an accelerated stall, as introduced in Chapter 10.

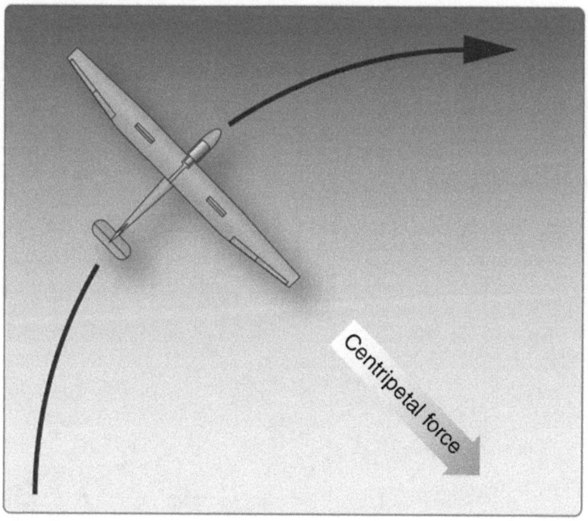

Figure 11.3. Turning airplane and centripetal force.
Source: Glider Flying Handbook (2013)/U.S. Department of Transportation Federal Aviation Administration/Public Domain.

LOAD FACTOR

The *load factor*, commonly referred to in FAA publications by the symbol *"G"*, is defined as the ratio of the lift required of an aircraft to maintain level altitude, to the total weight of the aircraft, or lift/weight. The letter "*n*" is more commonly used in aerodynamics to denote load factor. The load factor is a unit of force equal to the force exerted by gravity on a body at rest and signifies the force to which the body is subjected when the aircraft is accelerated. This is felt by the pilot as the aircraft is deflected from its straight-line path during a level altitude, banked turn. A load factor of 2 means the total load on an aircraft's structure is two times its weight and is expressed as 2 *G*s.

Understanding load factors and the ability of an aircraft (and pilot) to withstand high *G* flight is vital for the safety of flight. The structure of an aircraft is only designed to withstand a certain amount of stress in flight, and it is imperative that pilots understand the structural capabilities of their aircraft as well as the individual components of the aircraft. An increased load factor (beyond 1*G*) increases the stalling speed as we discussed in Chapter 10, and an accelerated stall may occur to an unsuspecting pilot at an airspeed much higher than V_{S0}. This could result in a stall/spin loss of control incident with little to no warning. This chapter will focus primarily on load factor as it relates to turning flight, but it is important to note that when an aircraft makes an abrupt elevator/stabilator input to pull out of a dive, the aircraft is subjected to an increased load factor and a possible accelerated stall. The pilot will be pressed down in the seat as they are subjected to two or three times the force equal to their weight, as will the components of the aircraft.

The centripetal force causing the airplane to turn is found by solving the triangle in Figure 11.4:

$$\sin\varphi = \frac{\text{Centripetal force}}{\text{Total lift }(L)}, \quad \text{thus:}$$

$$\text{Centripetal force} = L\sin\varphi \tag{11.1}$$

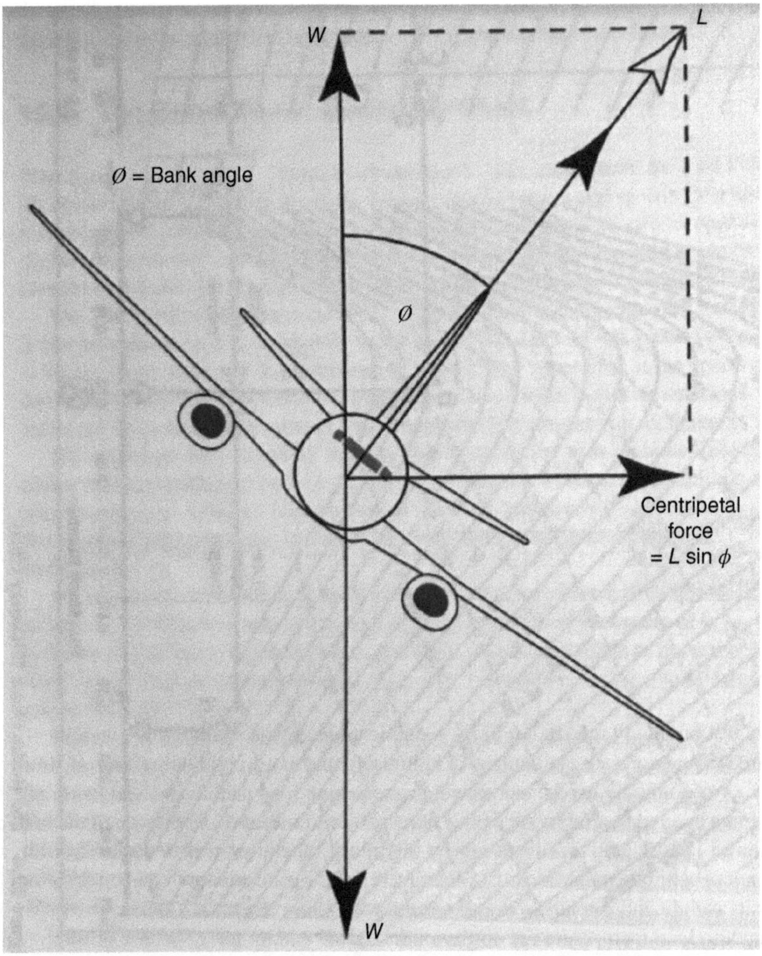

Figure 11.4. Forces on an aircraft in a coordinated level turn.

Solving the triangle in Figure 11.4 also gives

$$\cos \varphi = \frac{\text{Effective lift}}{\text{Total lift}} = \frac{W}{L} \tag{11.2}$$

Equation 11.3 shows the calculation of the load factor in terms of the required lift to maintain level altitude, compared to the aircraft weight:

$$G = \frac{L}{W} \tag{11.3}$$

From Eq. 11.3, the total lift, L, can be replaced by its equivalent, GW, as now $L = GW$.

Inverting Eq. 11.2 and substituting the value of L/W into Eq. 11.3, we obtain

$$G = \frac{1}{\cos \varphi} \tag{11.4}$$

Load Factors on an Aircraft in a Coordinated Turn

Equation 11.4 tells us that the load factor experienced by an aircraft in a level-coordinated turn is determined by the bank angle alone. The type of aircraft, airspeed, or other factors have no influence on the load factor. Any aircraft maintaining level flight, in a 60° bank will have a load factor of $2G$. If the total gross weight of the aircraft is 2 500 lb, the wings must produce 5 000 lb of lift to maintain level flight in a 60° banked turn. Figure 11.5 depicts the load factors required at various bank angles. Note that for bank angles greater than 45°, the load factor increases at a much faster rate. This dramatic increase in load factor is why significant back pressure may be necessary to maintain altitude in a steep turn. During a medium banked level turn, the load factor increases only slightly, and for a shallow banked level turn the increase in load factor is negligible.

Note that the load factors required to fly at constant altitudes in turning flight were determined by assuming that the wings of the aircraft provided all the lift of the aircraft. According to this analysis, the load factor at a 90° bank will be infinite and thus impossible to attain. But we have all seen aircraft fly an eight-point roll, with the roll stopped at the 90° position. The secret is that lift is provided by parts of the aircraft other than the wings. If the nose of the aircraft is held above the horizontal, the vertical component of the thrust will act as lift (Figure 11.6). The fuselage is also operating at some effective AOA and will develop lift. The vertical tail and rudder will also develop lift.

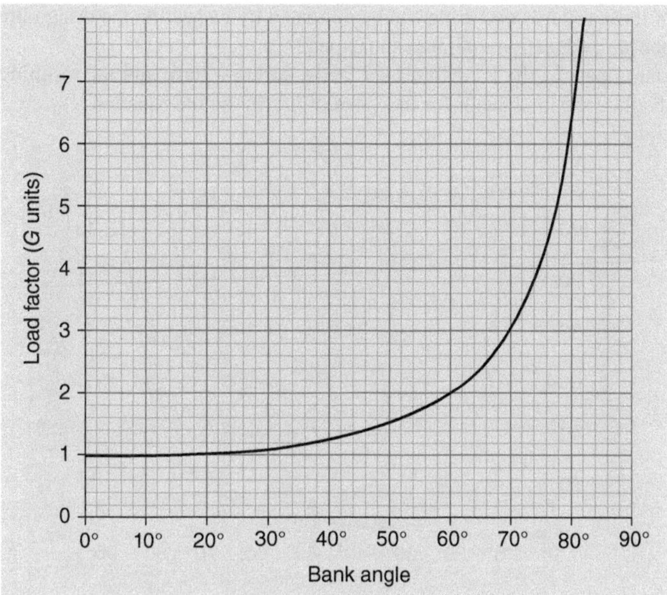

Figure 11.5. Load factors at various bank angles.
Source: Pilot's Handbook of Aeronautical Knowledge (2008a)/U.S. Department of Transportation Federal Aviation Administration/Public domain.

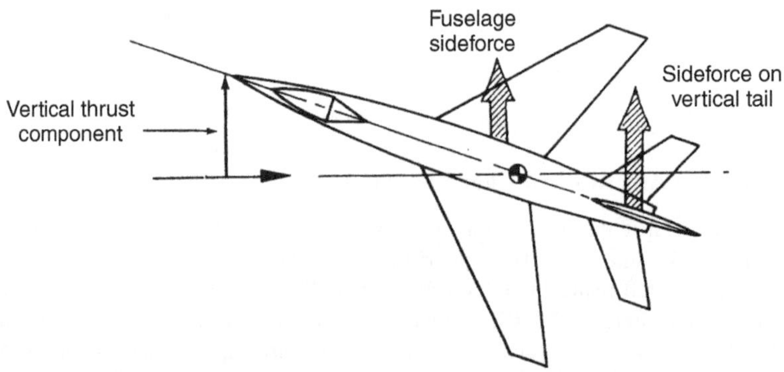

Figure 11.6. Forces on an aircraft during a 90° roll.

Effect of a Coordinated Banked Turn on Stall Speed

Any aircraft can be stalled at any airspeed within its design limits. In basic aerodynamics, we learned that the aircraft always develops its slow speed stall at the critical AOA. At this AOA, the value of the lift coefficient is a maximum, $C_{L(max)}$. So, V_S occurs at $C_{L(max)}$. As the load factor is increased, the stall speed increases. Figure 11.7 shows the impact of increasing bank and load factor on stall speed. Note again in Figure 11.7, the rapid increase in stall speed once the 45° angle has been surpassed during a constant altitude turn. The hazard can be found when a pilot inadvertently stalls the aircraft while in a steep turn or stalls the aircraft above its designed structural limits when the aircraft is subjected to an excessive load factor. The risk of these hazards occurring can be mitigated by respecting the design limits of an aircraft, and through proper pilot education, training, and experience.

To calculate the impact of the increased load factor during a turn, and the resulting increase in stall speed, we can substitute the load factor for the value of lift, in the lift equation.

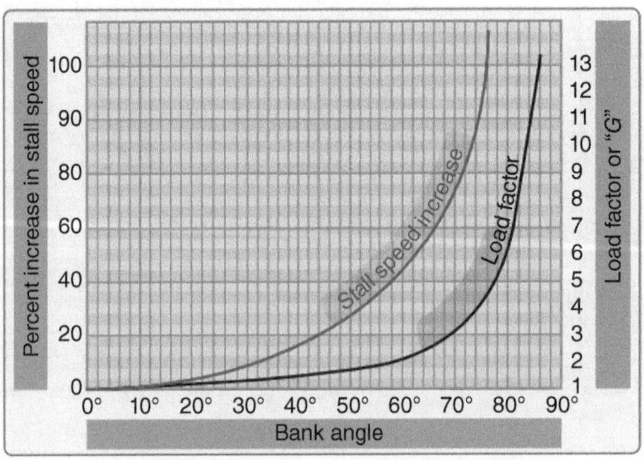

Figure 11.7. Increase in stall speed with load factor.
Source: Airplane Flying Handbook (2016a)/U.S. Department of Transportation Federal Aviation Administration/Public domain.

The basic lift Eq. 4.3 can be rewritten as

$$V_S = \sqrt{\frac{295L}{C_{L(max)}\sigma S}}$$

and since $L = GW$,

$$V_S = \sqrt{\frac{295GW}{C_{L(max)}\sigma S}}$$

All other factors in the equation are constant for a given altitude, leaving only the square root of the load factor to calculate the accelerated stall speed. Thus, stall speed depends only on the square root of the G loading, *when maintaining a constant altitude*. Therefore,

$$\frac{V_{s_2}}{V_{s_1}} = \sqrt{G} \text{ or } V_{s_2} = V_{s_1}\sqrt{G} \tag{11.5}$$

where

V_{s_1} = stall speed under 1G flight
V_{s_2} = stall speed under other than 1G flight
G = load factor for condition 2.

EXAMPLE

Calculate the new stall speed (V_{s_2}) of an airplane when the V_{s_1} is 47 kts., and the airplane is experiencing a load factor of 3.8G during a level altitude, banked turn.

$$V_{s_2} = V_{s_1}\sqrt{G}$$
$$V_{s_2} = 47 \text{ kts.} \cdot \sqrt{3.8G}$$
$$V_{s_2} = 91.6 \text{ kts.}$$

To calculate the stall speed when given a known bank angle, we substitute the value of G from Eq. 11.4:

$$V_{s_2} = V_{s_1}\sqrt{\frac{1}{\cos\varphi}} \tag{11.6}$$

EXAMPLE

Calculate the new stall speed (V_{s_2}) of an airplane when the V_{s_1} is 47 kts., and the airplane is experiencing a bank angle of 78° during a level altitude, banked turn.

$$V_{s_2} = V_{s_1}\sqrt{\frac{1}{\cos\varphi}}$$

$$V_{s_2} = 47 \text{ kts.}\sqrt{\frac{1}{\cos 78°}}$$

$$V_{s_2} = 103 \text{ kts.}$$

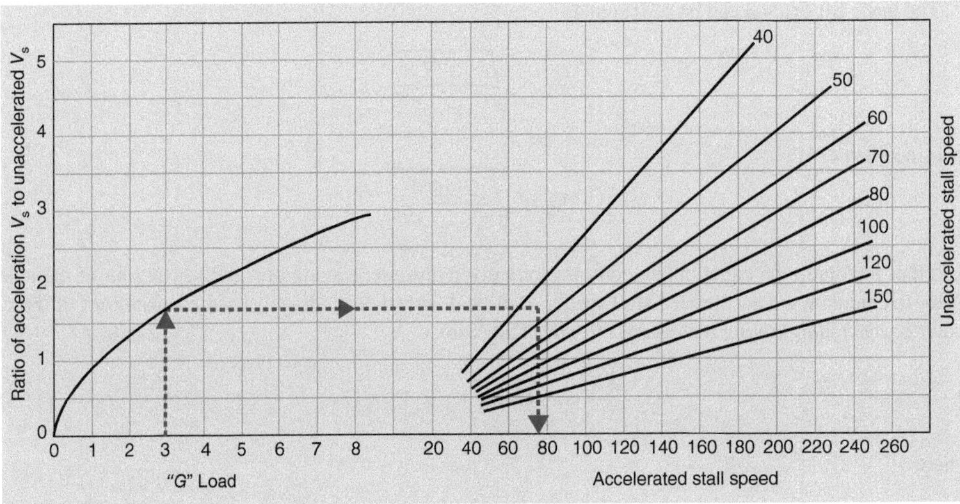

Figure 11.8. Load factor and stall speed.
Source: Pilot's Handbook of Aeronautical Knowledge (2008a)/U.S. Department of Transportation Federal Aviation Administration/Public domain.

From Eq. 11.5 above, an aircraft with a V_{SO} of 50 kts. will stall at 100 kts. if experiencing a load factor of $4G$. Because the stalling speed doubles due to the squared load factor, pilots can impose excessive loads on the structure of an aircraft very easily, especially when stalling intentionally (or unintentionally) at higher speeds. Pilots need to be aware that in a constant altitude, and increasing bank turn, the stall speed is increasing due to increased load factor, and the indicated airspeed is decreasing due to increased drag from the increase in lift. Figure 11.8 can be used to calculate the increased stall speed due to the increased load factor when the aircraft is experiencing a greater than $1G$ load factor. In Figure 11.8, the example aircraft shown is experiencing a load factor of $3G$ and has a given of 45 kts. Due to the increased load factor, the increased stall speed with the new condition increases to ~75 kts. From Figure 11.5, we find that the bank angle required to produce a $3G$ load factor on an airplane in a constant altitude, level turn, is 72°. Should the example aircraft experience a 72° bank angle during a level altitude turn, an airspeed of >75 kts. will need to be maintained to avoid stalling the aircraft.

Application 11.1

Using Figure 11.8, calculate the new stall speed of an airplane when the $1G$ stall speed is 47 kts., and the airplane is experiencing a load factor of $3.8G$ during a level altitude, banked turn.

Has your stall speed increased due to the increased load factor? How did your answer compare to the same example given in this section when calculated using Eq. 11.5?

THE *V–G* DIAGRAM (FLIGHT ENVELOPE)

Equation 11.5 shows how the stall speed increases as greater than $1G$ loading is applied to an aircraft. If an aircraft is under a $4G$ load, the stall speed is twice the $1G$ stall speed (because stall speed varies as

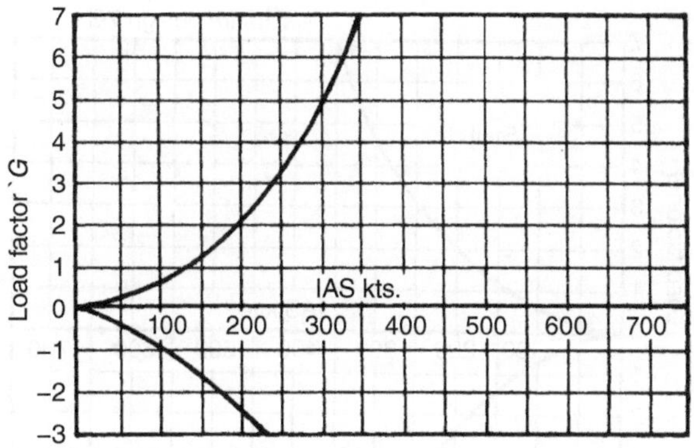

Figure 11.9. First-stage construction of a V–G diagram.

the square root of load factor). At zero G the stall speed is zero because no lift is being developed. This equation cannot be applied when an aircraft is developing negative load factors (square root of a negative number is not a real number), but an aircraft can be stalled under negative G loading.

Figure 11.9 shows the first construction lines of the V–G diagram (flight envelope), and the plot of the stall speed at various G loadings. This velocity versus G load diagram is valid for a specific aircraft, with a given weight and altitude. These curved lines can also be thought of as the number of Gs that can be applied to the aircraft before it stalls at a given airspeed. They are called the aerodynamic limits of the aircraft. It is impossible to fly to the left of these curves because the aircraft is stalled in this region. From Figure 11.9, the $1G$ stall speed of the example aircraft can be interpolated to be ~125 kts. IAS, while the $2G$ stall speed can be interpolated to be 200 kts. IAS. The values for negative lift flight will be similar to those experienced during positive G flight for each respective bank angle, but the airspeed necessary to produce a given negative load factor will be greater.

Equation 11.4 relates the G loading to the bank angle. Rearranging this equation, we see that the bank angle is the angle whose cosine is $1/G$. Mathematically, this is Eq. 11.7:

$$\varphi = \text{arc cos}\left(\frac{1}{G}\right) \tag{11.7}$$

Table 11.1 shows the bank angles that produce the load factors shown in Figure 11.9.

Table 11.1. Load factors at various bank angles.

Load factor, G	Bank angle, $\theta°$
1	0
2	60
3	70.5
4	75.5
5	78.5
6	80.4
7	81.8

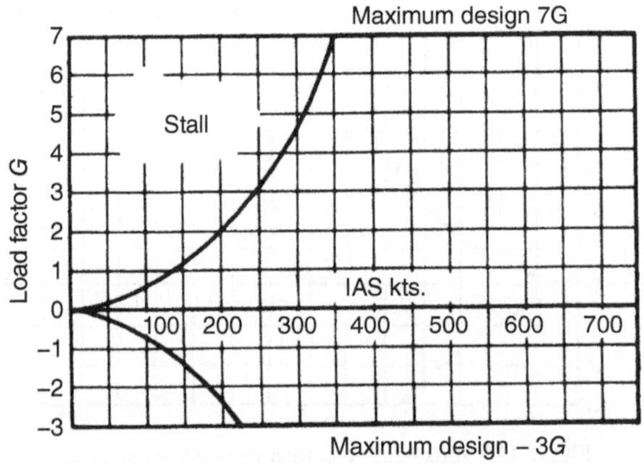

Figure 11.10. Second-stage construction of $V-G$ diagram.

The second stage of construction of the $V-G$ diagram consists of drawing horizontal lines at the positive and negative *limit load factors* (LLFs). These limits are specified for each model aircraft and are shown in Figure 11.10 as LLF $+7G$ and as LLF $-3G$. The FAA has designed a category system for many general aviation aircraft. A detailed examination of how LLFs are developed is beyond the scope of this textbook, but a review of AC 23-19A and Code of Federal Regulations (CFR) 25.337 should be a place to start as these sources are based on regulatory guidance. A placard on the flight deck as well as warnings in the Airplane Flight Manual (AFM) specify what category (or categories) that aircraft may be operated in which apply to the following LLFs. Figure 11.11 shows that some smaller general aviation aircraft can be operated in two different categories based on the loaded aircraft weight and the position of the aircraft's center of gravity (CG). Note that the Utility category requires much stricter parameters of operation than the Normal category. Examples of positive and negative LLF set by the FAA for aircraft up to 4 000 lbs. are as follows:

- **Normal category:** +3.8 to −1.52
- **Utility category:** +4.4 to −1.76
- **Acrobatic category:** +6.0 to −3.00

The structural LLFs show the maximum *G*s that may be imposed on the aircraft in flight without damaging the aircraft structure. Pilots should be aware of the load factors imposed on the aircraft they are flying at all times; accidents involving stalls initiated from steep turns in addition to structural failures as the result of poorly executed acrobatics may occur. With experience, a pilot may be able to judge the amount of load factor imposed on an aircraft, but more accurate measures can be determined by installing an accelerometer. Though not often found in non-aerobatic general aviation aircraft, accelerometers are becoming more common.

Several important facts about load factors should be understood:

1. The limits are for a certain gross weight, and they change if the gross weight differs from that specified.
2. The limits are for symmetrical loading only.
3. The assumption is made that there is no corrosion, metal fatigue, or other damage to the aircraft.
4. Up-gusts and turbulent air can add to pilot-imposed load factors.

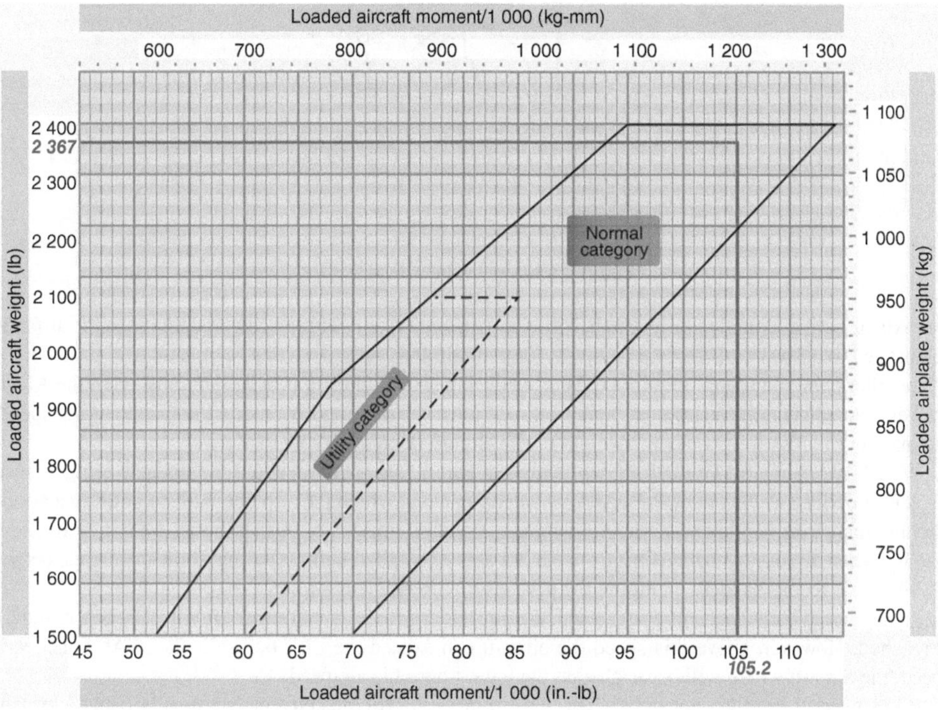

Figure 11.11. Normal versus utility category.
Source: Airplane Flying Handbook (2021)/U.S. Department of Transportation Federal Aviation Administration/Public domain.

The aircraft designer designs a certain strength into the airframe. *AC 23-19A Airframe Guide for Certification of Part 23 Airplanes* is a beneficial resource for the tolerances that must be met. The allowable total strength divided by the weight of the aircraft determines the LLF. If the design weight is exceeded, then the number of *G*s that can be tolerated by the airframe must decrease. This is an inverse proportion as shown in Eq. 11.8:

$$\frac{\text{LLF}_2}{\text{LLF}_1} = \frac{W_1}{W_2} \tag{11.8}$$

During unsymmetrical maneuvers, such as a rolling pullout, the wing that is rising will have more lift on it, and thus a greater load factor, than the downgoing wing (Figure 11.12). The accelerometer is located on the centerline of the aircraft and will measure the average load factor. In Figure 11.12, the pilot reads 5*G*, but the upgoing wing will be subjected to 7*G*.

Maneuver Speed and Limit Load Factor

An interesting point on the *V–G* diagram is the intersection of the aerodynamic limit line and the structural limit line. The aircraft's speed at this point is called the *design maneuver speed*, V_A, or simply, the *maneuvering speed*. The FAA defines maneuvering speed in the *Pilot's Handbook of Aeronautical Knowledge* (2016b) as, "the speed below which you can move a single flight control, one time, to its full deflection, for

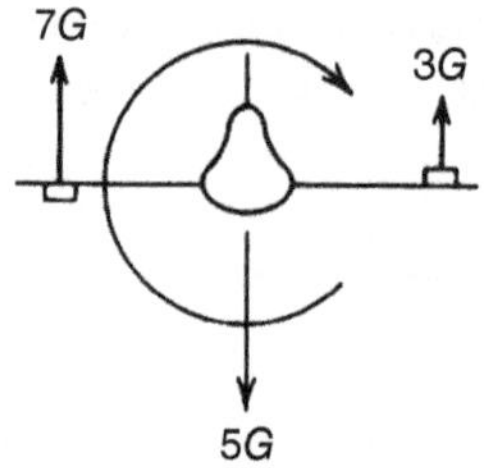

Figure 11.12. Antisymmetrical loading.

one axis of airplane rotation only (pitch, roll, yaw), in smooth air, without risk or damage to the airplane." Pilots need to be aware of what this speed is for their airplane on a given day, so it is usually posted in IAS as are other common speeds required to be used during each flight. V_A is usually published in the AFM as well as placarded on the instrument panel for maximum visibility and is usually the maneuvering speed at the maximum gross weight. As the weight of the aircraft decreases, so will V_A.

AC 23-19A Airframe Guide for Certification of Part 23 also adds an additional maneuvering airspeed, V_o, and is defined as the *operating maneuvering airspeed* at which the airplane will stall during a nose-up pitching maneuver before exceeding the airplane structural limits. For this textbook, we will only emphasize V_A when discussing maneuvering speed, but a more detailed understanding of the difference between V_A and V_o is recommended, including the updates of Amendment §23.45.

At any speed below maneuvering speed, the aircraft cannot be overstressed: it will stall before the LLF is reached. However, above this speed the aircraft can exceed the LLF before it stalls. At maneuvering speed, the aircraft's LLF will be reached at the lowest possible airspeed.

Pilots need to heed V_A not only when performing acrobatic maneuvers and steeply banked turns but should also slow the aircraft to an airspeed below V_A when entering turbulent air. Due to gust factors and unexpected turbulence, especially around thunderstorms and when crossing fronts, adherence to design maneuvering speed is imperative. Clear air turbulence, often encountered close to the tropopause or jet stream, but not uncommon down to 15 000 ft, may also require a reduction in speed below V_A.

The maneuver airspeed, V_A, is shown in Figure 11.13. It can be calculated using Eq. 11.9:

$$V_A = V_s\sqrt{\text{LLF}} \tag{11.9}$$

where

V_A = maneuver speed (kts.)

V_s = stall speed (kts.)

LLF = limit load factor.

EXAMPLE

Calculate the maneuvering speed (V_A) if the V_s of the aircraft is 50 kts., and the aircraft is at the positive LLF for a normal category aircraft.

$$V_A = V_s\sqrt{\text{LLF}}$$

$$V_A = 50\text{ kts.}\sqrt{3.8\,G}$$

$$V_A = 97.5\text{ kts.}$$

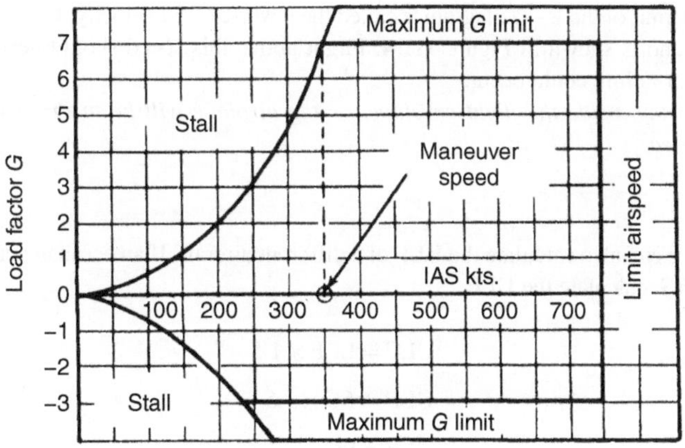

Figure 11.13. Maneuver speed.

All missions that an aircraft is designed to perform can be accomplished without exceeding the LLFs, but on occasion, a pilot will cause the aircraft to exceed these load factors. If this happens, the aircraft will probably suffer permanent deformation (aircraft components do not return to their original shape), resulting in costly repairs, but it will not necessarily result in failure of the primary structure.

However, there is a load factor which, if exceeded, may lead to catastrophic failure of the airframe. This load factor is called the *ultimate load factor* (ULF). Numerically, the positive and negative ULFs are 1.5 times the LLFs, essentially a 50% safety margin. These are shown in Figure 11.14 and are labeled as structural failures.

To calculate the ULF, we would use Eq. 11.10:

$$\text{ULF} = \text{LLF} \times 1.5 \tag{11.10}$$

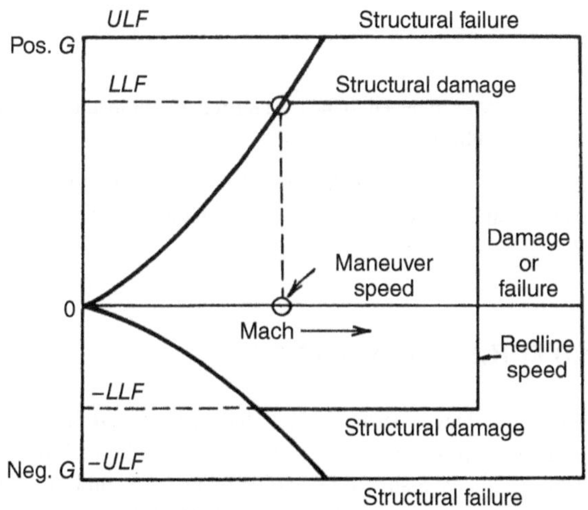

Figure 11.14. Ultimate load factors.

The high-speed limit of the $V-G$ diagram is called the never-exceed velocity, V_{NE}, or, more commonly, the redline speed, and is shown in Figure 11.14. Flight above this speed should never be attempted, as structural damage or failure could occur.

The pilot must stay inside the flight envelope, or the airplane will be in the structural damage or structural failure area.

EXAMPLE

Using Eq. 11.10, we can also calculate the ULF of a different aircraft. If an acrobatic category aircraft has a positive LLF of 6*G*, calculate the ULF.

$$\text{ULF} = \text{LLF} \times 1.5$$

$$\text{ULF} = 6G \times 1.5$$

$$\text{ULF} = 9G$$

If the aircraft weighed 3 000 lb, the limit load (LL) would be 18 000 lb, and the ultimate limit load (ULL) would be 27 000 lb.

$$\text{LL} = \text{Weight} \times 6G$$

$$\text{LL} = 3\,000\ \text{lb} \times 6G$$

$$\text{LL} = 18\,000\ \text{lb}$$

$$\text{ULL} = \text{Weight} \times 9G$$

$$\text{ULL} = 3\,000\ \text{lb} \times 9G$$

$$\text{ULL} = 27\,000\ \text{lb}$$

Application 11.2

Consider an airplane that has been tested to have an ultimate limit load of 18 000 lb, and a maximum gross weight of 3 000 lb.

How would you find the LLF (maximum *G* loading) for this aircraft? What is the limit load for this aircraft? What is the value of the maximum *G*s this aircraft is able to withstand before structural damage?

LOAD FACTOR AND FLIGHT MANEUVERS

Unless the aircraft is in straight and level, unaccelerated flight with a load factor of 1*G*, the pilot should be concerned about the load factors imposed on the aircraft's structure in flight. Full application of the primary flight controls as discussed in Chapter 3 should be avoided above V_A, and rapid movements of the flight controls at any speed should be avoided, especially if they involve full movement from one extreme to the other.

As discussed in the Load Factor section, maneuvering above 1G of flight will result in increased load factors, as well as increased stall speeds. If a stall is encountered and the pilot aggressively pushes the nose over during recovery, then aggressively recovers from the dive by applying back pressure, overloading of the primary control surfaces may occur or damage to the rigging system connecting the control surfaces could be aggravated. All control movements should be made with smooth application, with adherence to the V_A IAS. During higher speed stalls, there is added stress to the aircraft structure, most notably the wings and empennage. Aggressive movements at these speeds, especially when lightly loaded during acrobatic maneuvers if the airplane is not in the acrobatic category, leave very little room for extra limit loading before structural damage or failure.

In a typical spin for a single-engine aircraft loaded within the normal category or utility category weight/CG limits, exceeding the LLF is unlikely. When recovering from a spin as discussed in Chapter 10, care must be taken not to commit one of the common errors of spin recovery because the aircraft may be subjected to excessive structural loads. If immediate, abrupt back pressure is applied to the yoke/control stick during spin recovery after the rotation has stopped, it is possible to exceed the LLF and overstress the elevator/stabilator. If power was not reduced to idle before spin recovery or applied too early after spin recovery before attaining a level flight attitude, the LLF could be exceeded, in addition to V_{NE}.

Figure 11.15 shows the shift in the P_r curve as additional load factor is applied to a maneuvering aircraft. For example, assume a pilot in level flight banks the aircraft to initiate a turn. As the pilot pulls back on the yoke (stick) to maintain altitude (remember total lift required increases when bank increases), the angle of attack is increased, and thus, so does the total drag. The P_r to maintain the original speed increases as the P_r is needed to overcome the increase in total drag (Chapter 7). For our example the P_a does not change,

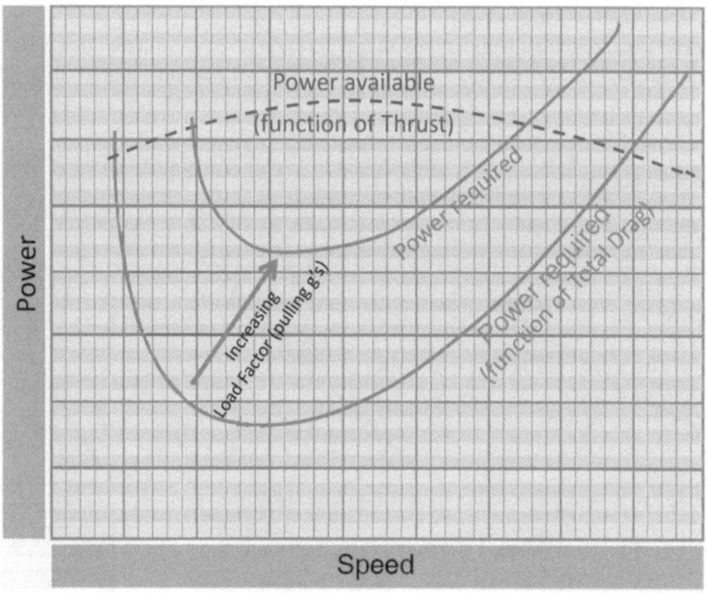

Figure 11.15. Power required curve and load factor.
Source: *Airplane Flying Handbook* (2021)/U.S. Department of Transportation Federal Aviation Administration/Public domain.

so the difference in P_a and P_r for a given speed decreases as load factor, *G*s, increases. Again, before any flight maneuver is started, proper calculation of the weight and balance, awareness of normal CG limits, and recognition of V_A and all structural limitations should be briefed and monitored.

Radius of Turn

The *radius* of a turn is critical to the success of a maneuver, whether it be air-to-air combat or simply turning base to final. As we have discussed, an increased bank angle during a coordinated turn in level flight decreases the turn radius. When an aircraft increases its bank angle to limit the turn radius in level flight, stall speed increases. A balance must be found between the bank angle necessary to perform the maneuver and the need to maintain a safe margin above stall speed without exceeding V_A.

Figure 11.16 shows the varying bank angle, stall speed, and turn radius of a glider. Note the following items:

1. As the glider's bank angle increases, stall speed increases.
2. For a constant bank angle, the turn radius is directly proportional to true airspeed (TAS). As the TAS increases, so does the radius.
3. For a constant TAS, the turn radius is inversely proportional to the bank angle. As the bank angle increases, the radius decreases.

The radius of a turn and the rate of a turn are directly linked, regardless of the aircraft. Figure 11.17 shows the radius of a turn around a fixed object at a constant altitude, for example, during ground reference maneuvers for private pilot certification. The radius again increases in two situations; constant angle of bank with increasing airspeed, and constant airspeed with decreasing angle of bank. The radius of a turn decreases when the situations are reversed.

A body traveling in a circular path undergoes an acceleration toward the center of rotation. This is called centripetal acceleration, a_r, and is usually measured in meters per second squared (m/s^2). It is a function

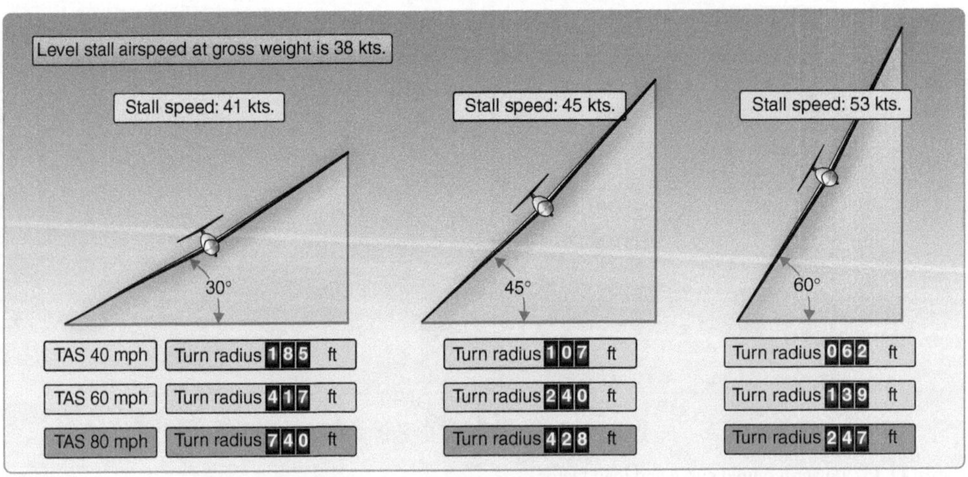

Figure 11.16. Stall speed and turn radius with varying angle of bank.
Source: Glider Flying Handbook (2013)/U.S. Department of Transportation Federal Aviation Administration.

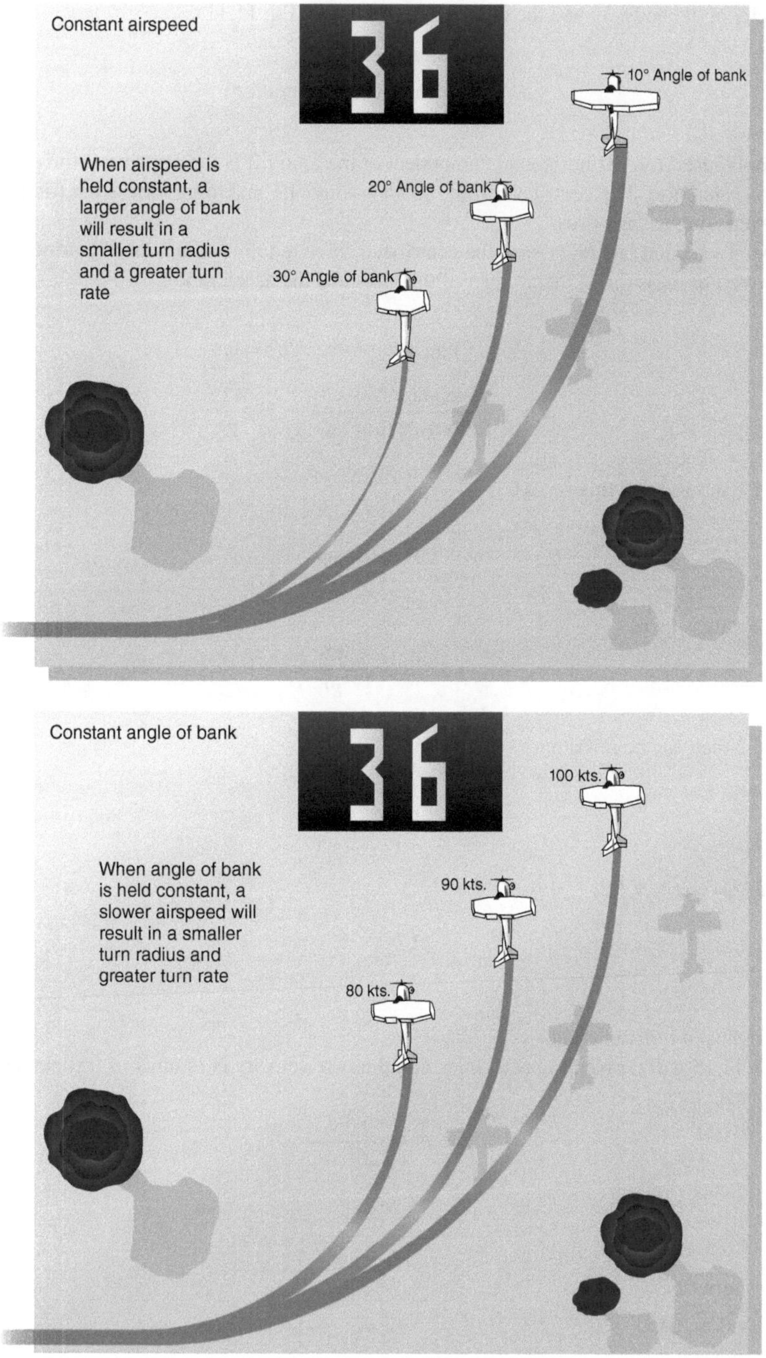

Figure 11.17. Rate and radius of a turn.
Source: Airplane Flying Handbook (2004a)/U.S. Department of Transportation Federal Aviation Administration/Public domain.

of the velocity of the body, V, and the radius, r, of the circle (Eq. 11.11):

$$a_r = \frac{V^2}{r}\text{(centripetal acceleration)} \tag{11.11}$$

As shown in Figure 11.4, the horizontal component of the total lift is the centripetal force that causes the centripetal acceleration. The centripetal force in an automobile making a turn is the frictional force the pavement exerts on the car's tires.

By Newton's second law ($F = ma$), the centripetal force required to produce centripetal acceleration (a) in an aircraft of mass (m) is the product of the mass and the acceleration:

$$\text{Centripetal force (CF)} = ma_r$$

$$\text{CF} = \frac{(\text{Weight} \times \text{Velocity}^2)}{(\text{gravity} \times \text{turn radius})} = \frac{WV^2}{gr} \tag{11.12}$$

Equation 11.1 showed that this force is

$$\text{CF} = L \sin \varphi$$

Equating CF values gives,

$$L \sin \varphi = \frac{WV^2}{gr} \tag{11.13}$$

Equation 11.2 then can be rewritten as

$$L \cos \varphi = W \tag{11.14}$$

By dividing Eq. 11.13 by Eq. 11.14, we obtain

$$\tan \varphi = \frac{V^2}{gr} \text{or, } r = \frac{V^2}{g \tan \varphi} \tag{11.15}$$

This is the radius of turn equation.

Equation 11.15 was derived in basic units, and thus the velocity is in units of feet per second. If V is measured in kts., then

$$r = \frac{V_k^2}{11.26 \tan \varphi} \tag{11.16}$$

where

r = radius of turn (ft)

V_k = velocity (kts. TAS)

φ = bank angle (degrees)

11.26 = conversion factor from g and conversion to kts.

EXAMPLE

Calculate the radius of a turn of an airplane with a TAS of 200 kt., and a bank angle of 50° while maintaining a constant altitude.

$$r = \frac{V_k^2}{11.26 \tan \varphi} \rightarrow r = \frac{(200\text{ kts.})^2}{11.26\,(\tan 50°\,)} \rightarrow r = 2\,981\text{ ft}$$

Rate of Turn

The *rate of turn* (ROT) is primarily used by most pilots to determine a standard turn rate, and is the change in heading of the aircraft per unit of time:

$$\text{ROT} = \frac{g \tan \varphi}{V_{\text{fps}}}\ (\text{rad/s})$$

If velocity is in knots, and ROT is measured in degrees per second, then

$$\text{ROT} = \frac{1\,091 \tan \varphi}{V_k} \quad (\text{deg /s}) \tag{11.17}$$

EXAMPLE

Calculate the rate of turn of an airplane with a TAS of 200 kt, and a bank angle of 50° while maintaining a constant altitude.

$$\text{ROT} = \frac{1\,091\ \tan \varphi}{V_k} \rightarrow \text{ROT} = \frac{1\,091\ \tan 50°}{200\text{ kts.}} \rightarrow \text{ROT} = 6.5\ \text{deg /s}$$

The standard rate turn is 3 deg/s and is the rate of turn used while banking for most small, general aviation aircraft operating in IMC. Faster aircraft often use a slower turn rate of 1.5 deg/s because turning 3 deg/s would require an excessive bank angle. A rough approximation of the angle of bank that can be expected while maintaining a 3 deg/s standard rate turn, can be determined by dividing the TAS (kts.) by 10 and then adding 5. So, with a TAS of 110 kts., the bank angle is ~16°. As the TAS increases, so does the bank angle required to maintain the desired turn rate.

The importance of bank angle and velocity can is found in Eqs. 11.16 and 11.17. High bank angles and slower airspeeds produce small turn radii and high ROT. Figure 11.17 once again shows us the relationship between ROT and bank angle. ROT increases when a constant airspeed is maintained and the angle of bank increases, or airspeed is decreased while maintaining a constant angle of bank. As before, the ROT decreases when the situations are reversed.

V_A is the speed at which the highest bank angle can be achieved at minimum airspeed. The minimum turn radius and maximum rate of turn will be realized at this speed. Figure 11.18 is a chart that shows turn radii and rates of turn for various bank angles and airspeeds for an aircraft making a coordinated, constant altitude turn.

In our discussion of induced drag in Chapter 5, we saw that the rearward component of the total lift vector was induced drag and that the vertical component was effective lift. These are shown again in Figure 11.19. We see that the effective lift must equal the weight of the plane if constant altitude is to be maintained while turning. We also see that more thrust is required to overcome this increase in induced

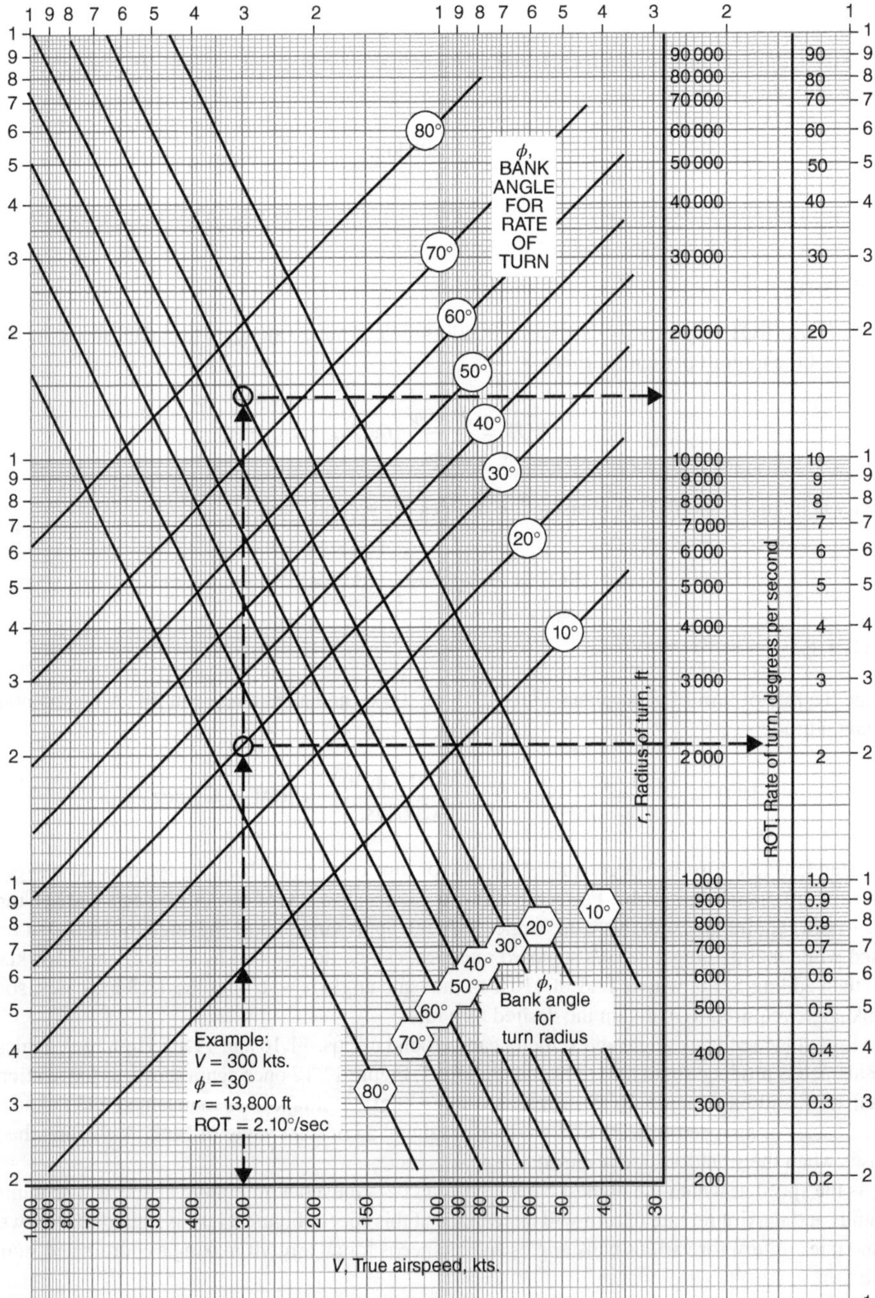

Figure 11.18. Constant altitude turn performance.

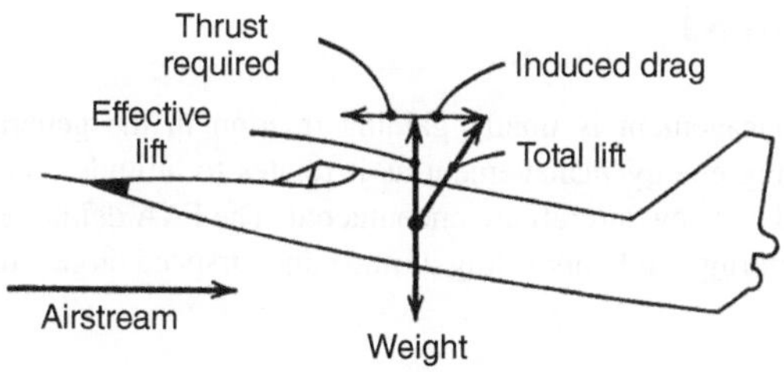

Figure 11.19. Forces on the complete aircraft.

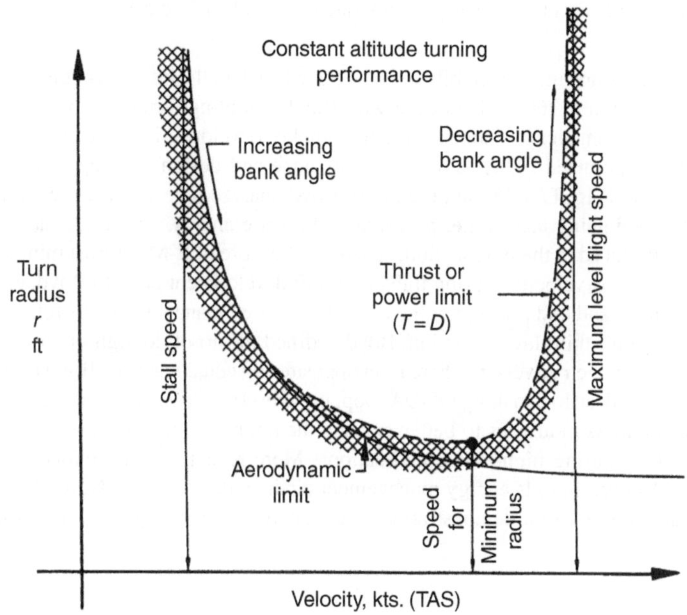

Figure 11.20. Thrust-limited turn radius.

drag if airspeed is to be maintained. Thus, it may be that the turn radius may be limited by thrust available instead of structural limitations.

Figure 11.20 shows a thrust-limited aircraft's turn radius plotted against true airspeed. The aerodynamic limits are the same as described in Figure 11.13. The thrust limit curve is determined by a combination of bank angle and airspeed.

Application 11.3

Review the example problems that were calculated for the Radius of Turn and Rate of Turn sections.

Using the given bank angle (50°), and the given TAS (200 kts.), confirm the radius and rate of turn answers from the examples using Figure 11.18.

ENERGY MANAGEMENT

The subject of energy management is finally gaining traction in the general aviation flight training environment. Understanding energy management as it relates to altitude and airspeed is vital for new pilots as habits are formed and new aircraft are encountered. The FAA defines energy management as the process of planning, monitoring, and controlling altitude and airspeed targets in relation to the airplane's energy state in order to:

1. Attain and maintain desired vertical flightpath-airspeed profiles.
2. Detect, correct, and prevent unintentional altitude-airspeed deviations from the desired energy state.
3. Prevent irreversible deceleration and/or sink rate that results in a crash.

The subject of energy and maneuverability can be traced back to the infamous fighter pilot and military strategist John Boyd. In the 1960s, what began with Boyd scribbling formulas on an Air Force officers' club tablecloth for mathematician Thomas Christie, quickly developed into the energy-maneuverability (E-M) model which optimized fighter aircraft performance and design, leading to the development of fighter aircraft like the F-16, F/A-18, and the A-10. Boyd analyzed the various energy states of aircraft as they related to dogfighting tactics, i.e. potential and kinetic energy, speed, and acceleration. Boyd's simple equation, introduced in the next section, stated that the core of E-M is thrust minus drag divided by weight, then multiplied by velocity. During fighter aircraft development prior to Boyd, engineers focused on wing loading, range, and endurance, not the subject of energy management. As related to our Chapter 1 discussion on Newton's third law of motion, Boyd codified the art of dogfighting by analyzing that for every action by a pilot or their adversary, there is an opposite and equal reaction. Boyd is also credited with the development of the decision-making OODA loop, or "Observe-Orient-Decide-Act." The OODA loop has been used in countless industries to better manage the human–human or human–machine interface when decision-making during high-pressure situations. More recently, many pilots may remember the late Robert "Bob" Hoover and his energy management airshow demos in the North American Rockwell Shrike Commander, a true visual demonstration of the energy balance equation that is presented in the next section.

The Airplane as an Energy System

As introduced in Chapter 1, the mechanical energy of an airplane in flight is the sum of the kinetic energy from airspeed and the potential energy in altitude. Once again, we can see the relationship with Eq. 1.6:

$$\mathrm{TE} = \mathrm{PE} + \mathrm{KE}\ (\text{ft-lb})$$

An aircraft in flight is an open system, fuel is used to gain energy while drag imposes a loss of energy. In flight, the aircraft gains energy from the force of the engine and thrust (T) and loses energy from drag (D). So as before, the net change in energy is the difference between thrust and drag ($T - D$). Mechanical energy, stored as altitude and airspeed, either increases, decreases, or remains unchanged. If ($T - D > 0$) then the aircraft can climb, increase speed, or a combination of both. If ($T - D < 0$) then the aircraft's mechanical energy decreases, stored energy as altitude can be used as a decent, or stored energy as airspeed can be depleted as an aircraft maintains altitude. When ($T - D = 0$), the mechanical energy gained equals the energy that was lost, all thrust spent on drag, altitude and airspeed are now back at a constant again.

John Boyd, and now the FAA's *Airplane Flying Handbook*, defines the rate of energy change as *Specific Excess Power* (P_s), and quantified as the net energy flow as the direct function of the difference between thrust and drag:

$$P_s = \frac{(T - D)\,V}{W} \tag{11.18}$$

where

P_s = Specific excess power, rate of change in distance or height (usually in fpm)

T = Thrust

D = Drag

V = Velocity (airspeed)

W = Aircraft weight.

The relationship between the net energy flow and the changes in energy stored as altitude and airspeed can best be shown with the airplane's energy balance equation (Figure 11.21). The left side of the equation is always equal to the right side of the equation, although the rate of energy change (P_s) may be positive, negative, or zero, if there is a net energy change then altitude and/or airspeed must change as well to balance the equation.

Energy System Control

Earlier in this textbook we emphasized the importance of "pitch for airspeed and power for altitude," a common directive during primary flight training. Within the context of energy management, an energy-centered approach details the roles of the engine and the flight controls. Using coordinated throttle and elevator inputs, we can now restate the popular adage by saying "pitch plus power controls energy state." During flight the throttle setting becomes the *total energy controller*, if the throttle is increased so thrust is greater than drag, then the mechanical energy has increased. If the throttle is set so that drag eventually is greater than thrust, then the mechanical energy has decreased. The elevator on the other hand becomes the *energy distribution controller*, by adjusting the pitch attitude with the elevator, changes in total energy are exchanged between the vertical flight path and the airspeed. Through the coordinated

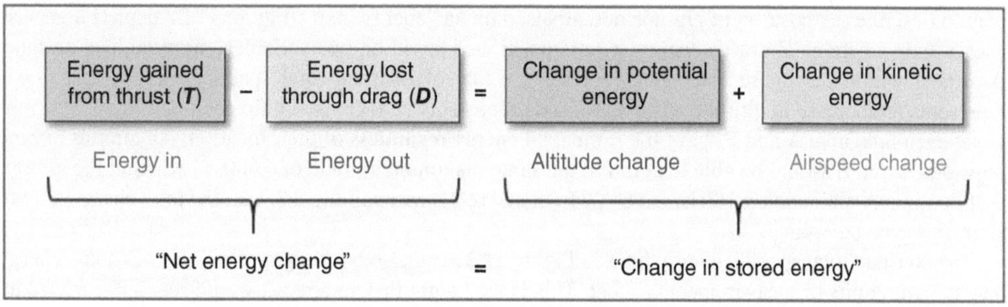

Figure 11.21. Energy balance equation.
Source: Airplane Flying Handbook (2021)/U.S. Department of Transportation Federal Aviation Administration/Public domain.

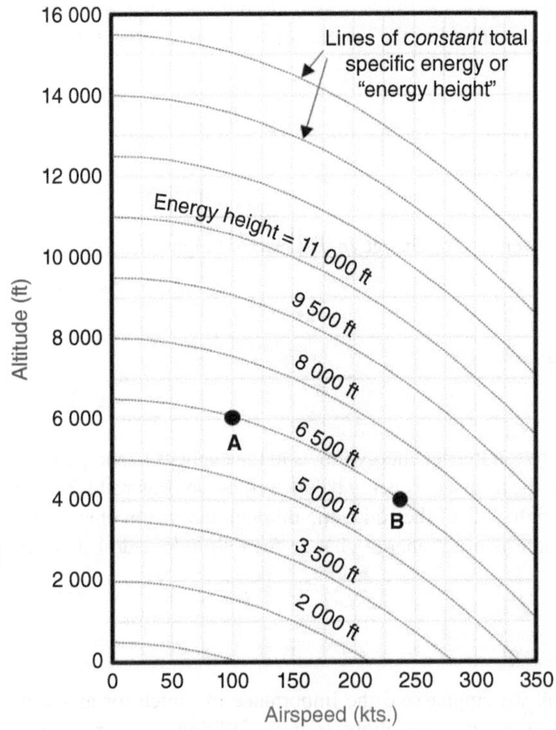

Figure 11.22. Energy state map.
Source: Airplane Flying Handbook (2021)/U.S. Department of Transportation Federal Aviation Administration/Public domain.

use of both the throttle setting and the elevator, we move beyond only using pitch for airspeed and power for altitude, and both pitch (elevator) and power (throttle) are intricately related to control the total energy state.

The FAA's *Airplane Flying Handbook* uses the analogy of latitude and longitude on an aeronautical map to describe the position of altitude and airspeed on an "energy map." Figure 11.22 depicts a generic energy map where an aircraft's energy state is positioned based on two variables, airspeed, and altitude. As stated within the figure, each line represents a line of constant total specific energy. The Y-axis represents an increase in altitude while the X-axis represents an increase in airspeed. Different positions along each line, note A and B, have the same total energy regardless of their location. An aircraft in both positions A and B would be able to zoom to the same maximum altitude of 6 500 ft., whether the aircraft is at Position A maintaining 100 kts. at 6 000 ft, or at 4 000 ft maintaining 240 kts., the total energy of both aircraft remains the same.

The constant total specific energy lines in Figure 11.22 can also be thought of as lines of equal "energy height," with units or measurements in feet. This is the height that an aircraft would reach if it traded all its available airspeed into height. We can calculate the energy height by using the formula for total specific energy, which we will symbolize with an E_s. E_s is the sum of the potential and kinetic energies divided by the aircraft's weight. So as a review of Chapter 1, if we incorporate Eqs. 1.1, 1.4, and 1.5 into Eq. 1.6, then divide by the weight, we arrive at Eq. 11.19:

$$E_s = h + \frac{V^2}{2g} \tag{11.19}$$

where

E_s = Total specific energy (ft)

h = Height (altitude)

V = Velocity (airspeed)

g = Gravitational constant

W = Aircraft weight.

Application 11.4

Calculate the energy height (in ft) of both positions A and B in Figure 11.22 using Eq. 11.19. As a reminder, Position A aircraft is at 6 000 ft and maintains 100 kts., while Position B aircraft is at 4 000 ft and maintains 240 kts. Does the energy height calculated match the energy height given in Figure 11.22? Do not forget to convert kts. to ft/s!

If the same aircraft is at an altitude of 4 000 ft, and located on the total specific energy state line for 5 000 ft, does the airspeed you calculate using Eq. 11.19 match the airspeed from Figure 11.22?

Application of the Energy System

Now that a foundation of energy management has been created, understating how to apply the principles of altitude and airspeed management through proper thrust and flight control movement for the desired flight condition is the final step. We will divide the application of the energy management system into three scenarios that are required of all basic flight maneuvers: the addition of more total energy, the new energy state that requires less total energy, and finally the move to a new energy state that requires no change in total energy.

If the desired flight condition requires more total energy, then the throttle will have to increase so that $T > D$ and total energy is increased. The elevator will be used to adjust the pitch attitude to distribute the energy state, either by pitching up to maintain the desired airspeed or by pitching down to maintain altitude as the airspeed increases (remember the lift formula and Eq. 4.3).

If the new energy state requires less total energy, then the throttle is reduced so that $T < D$, now the total energy is less. Again, the elevator is used to distribute the total energy, either by pitching down to maintain a desired speed and/or descent rate, that is, Instrument Landing System (ILS) approach, Global Positioning System (GPS) vertical navigation, or visual glideslope.

Finally, if the new energy state requires no change in total energy, no throttle change is needed unless it is used to compensate for a change in drag depending on the maneuver. The elevator is used to trade airspeed for additional altitude by pitching up or to trade altitude for increased speed by pitching down.

Altitude \ Airspeed	Slower	Desired airspeed	Faster
Higher	(1) Total Energy: OK Potential energy: high Kinetic Energy: low	(4) Total Energy: high Potential energy: high Kinetic Energy: OK	(7) Total Energy: very high Potential energy: high Kinetic Energy: high
Desired altitude	(2) Total Energy: low Potential energy: OK Kinetic Energy: low	**(5) Desired Energy State** Total Energy: OK Potential energy: OK Kinetic Energy: OK	(8) Total Energy: high Potential energy: OK Kinetic Energy: high
Lower	(3) Total Energy: low Potential energy: low Kinetic Energy: low	(6) Total Energy: low Potential energy: low Kinetic Energy: OK	(9) Total Energy: OK Potential energy: low Kinetic Energy: high

Figure 11.23. Energy state matrix.
Source: Airplane Flying Handbook (2021)/U.S. Department of Transportation Federal Aviation Administration/Public domain.

Figure 11.23 can be used as a generalized energy state matrix for pilots to understand their state of energy management. In combination with Figure 11.23, Figure 11.24 can be used to "solve" the possible energy error found within the energy state matrix. In Positions 2, 3, and 6, the aircraft has potentially too little energy and a correction must be made. Figure 11.24 states the corrective action for these positions begins with the addition of throttle to increase total energy, then proper adjustment of the elevator for the desired pitch attitude. For an aircraft in Positions 4, 7, and 8, the energy may be too high, and once again using Figure 11.24 we see an adjustment in the throttle setting to change the total energy, except this time the throttle needs to be reduced so that the total energy is reduced. Elevator movement at this point will be adjusted for the proper airspeed to maintain the desired altitude or to maintain the proper airspeed for the descent. Within Positions 1 and 9, the total energy is correct but there are energy distribution errors, Figure 11.24 states how the exchange of energy can be transferred by proper use of the elevator. Finally, Position 5 indicates the desired energy state where the pilot should maintain both the throttle and elevator settings. Figure 11.24 also includes a column for reference with the aircraft being "very slow" at different altitudes, note the same directive is given for elevator position as was introduced in Chapter 10, lower the elevator to increase airspeed and consider full power to increase altitude.

Application 11.5

Consider a box canyon scenario where an aircraft lacks the climb performance to clear an approaching ridgeline. What is the best method to maneuver the aircraft to execute a turn reversal? What airspeed will you use to minimize turn radius and maximize turn rate? What considerations do you have for wind direction and your own aircraft's flight envelope? Consider the energy state matrix in your analysis.

Altitude \ Airspeed	Cautions when very slow	Slower	Desired speed	Faster
Higher	Relatively safe. Surplus altitude is available to gain speed by pushing the elevator forward.	(1) Exchange energy by pushing the elevator forward to accelerate and descent simultaneously. Maintain the throttle setting.	(4) Reduce throttle setting to reduce total energy. Use elevator to maintain correct airspeed and allow the airplane to descend.	(7) Reduce throttle setting significantly to decrease total energy. Pull back on elevator gradually to decelerate to correct airspeed and then descend.
Desired altitude	Risky. Consider gaining speed at the expense of some altitude initially to improve climb performance with full throttle.	(2) Increase throttle setting to gain total energy by accelerating. Use elevator to maintain the desired altitude.	(5) DESIRED ENERGY STATE Maintain both throttle and elevator (trim) settings.	(8) Reduce throttle setting to decelerate. Use elevator to maintain the desired altitude.
Lower	Dangerous! Apply full throttle to resolve this condition. Avoid pitching up, which would increase drag and reduce or impede climb performance when on the backside of the power required curve.*	(3) Increase throttle setting significantly to gain total energy. Push elevator forward gradually to accelerate to correct airspeed and then climb.	(6) Increase throttle setting to gain altitude and pull back on elevator to maintain correct airspeed.	(9) Exchange energy by pulling the elevator back to climb and decelerate simultaneously. Maintain the throttle setting.

* Depending on aircraft type, as full throttle is applied at the start of correction maneuver, slight forward or aft elevator pressure may be needed to maintain a constant pitch attitude. As the airplane gains total energy, use the elevator to accelerate to correct airspeed and then climb.

Figure 11.24. Corrective actions for the energy state matrix.
Source: Airplane Flying Handbook (2021)/U.S. Department of Transportation Federal Aviation Administration/Public domain.

SYMBOLS

E_s	Total specific energy
P_s	Specific excess power
V_A	Design maneuvering speed (kts.)
V_o	Operating maneuvering speed (kts.)
V_{NE}	Never-exceed speed (kts.)

EQUATIONS

11.1 $\mathrm{CF} = L \sin \phi$

11.2 $\cos \phi = \frac{W}{L}$

11.3 $G = \frac{L}{W}$

11.4 $G = \frac{1}{\cos \phi}$

11.5 $\frac{V_{s_2}}{V_{s_1}} = \sqrt{G}$ *or* $V_{s_2} = V_{s_1} \sqrt{G}$

11.6 $V_{s_2} = V_{s_1} \sqrt{\frac{1}{\cos \phi}}$

11.7 $\phi = \arccos \left(\frac{1}{G} \right)$

11.8 $\frac{\mathrm{LLF}_2}{\mathrm{LLF}_1} = \frac{W_1}{W_2}$

11.9 $V_\mathrm{A} = V_\mathrm{s} \sqrt{\mathrm{LLF}}$

11.10 $\mathrm{ULF} = \mathrm{LLF} \times 1.5$

11.11 $a_\mathrm{r} = \frac{V^2}{r}$ (centripetal acceleration)

11.12 $\mathrm{CF} = \frac{WV^2}{gr}$ (centripetal force)

11.13 $L \sin \phi = \frac{WV^2}{gr}$

11.14 $L \cos \phi = W$

11.15 $r = \frac{V^2}{g \tan \phi}$

11.16 $r = \frac{V_\mathrm{k}^2}{11.26 \tan \phi}$ (ft)

11.17 $\mathrm{ROT}\ (\mathrm{deg/s}) = \frac{1\,091 \tan \phi}{V_\mathrm{k}\ (\mathrm{kts.})}$

11.18 $P_\mathrm{s} = \frac{(T-D)\,V}{W}$

11.19 $E_\mathrm{s} = h + \frac{V^2}{2g}$

KEY TERMS

Centripetal force
Design maneuvering speed (V_A) (kts.)
Energy map
Energy state
Inertia
Load factor
Limit load factor (LLF)
Operating maneuvering speed (V_o) (kts.)
Radius of turn (ft)
Rate of turn (ROT) (deg/s)

Specific excess power
Total energy controller
Ultimate load factor (ULF)
V–G diagram

PROBLEMS

Note: Answers to the problems in the first two sections are given at the end of the book.

Application of Chapter Theory

1. In a constant altitude and constant airspeed turn, an aircraft is in a state of equilibrium.
   a. True
   b. False
2. The *G* forces on an airplane in a level turn depend on
   a. airspeed.
   b. type of airplane (jet or prop).
   c. bank angle.
   d. All of the above.
3. An airplane performing an 8-point roll can stop in the 90° position because
   a. the wings are not supporting the weight.
   b. the fuselage and vertical tail support much of the weight.
   c. the pilot pulls the nose above the horizontal, thus some of the thrust helps support the weight.
   d. All of the above.
4. When an airplane is in a constant altitude bank, the stall speed
   a. remains the same as in level flight.
   b. increases as the square root of the *G*s.
   c. increases as the square root of $1/\cos \varphi$.
   d. Both (b) and (c).
5. What can we find from the aerodynamic limit line on a *V–G* diagram?
   a. It shows the maximum *G*s that can be pulled at any airspeed below the maneuver speed.
   b. It shows the stall speed when *G*s are being pulled.
   c. It is impossible to fly to the left of this line because the airplane will stall.
   d. All of the above.
6. The limit load factor (LLF) is also called maximum design *G*.
   a. True
   b. False
7. If the weight of an airplane, W_2, is increased above the design gross weight, W_1, the limit load factor
   a. remains the same.
   b. increases by the weight ratio, W_1/W_2.
   c. decreases by the weight ratio, W_1/W_2.
   d. None of the above.

8. If an airplane is flying in symmetrical flight at its maneuver speed
a. it cannot be overstressed.
b. it can make the smallest turn radius.
c. it can make the highest rate of turn.
d. All of the above.

9. The constant altitude turn performance chart (Figure 11.18)
a. can be used for only one particular model of airplane.
b. does not show if the airplane is overstressed.
c. is good for any model airplane.
d. Both (b) and (c).

10. When an airplane makes a level banked turn, the airspeed will decrease (unless throttle is added) because
a. parasite drag is increased the most.
b. induced drag is increased the most.
c. profile drag is increased the most.

Calculations Using Chapter Principles

11. Calculate the new stall speed (V_{s_2}) of an airplane when the V_{s_1} is 60 kts., and the airplane is experiencing a load factor of 4.4*G* during a level altitude, banked turn.
a. 78.1 kts.
b. 116.9 kts.
c. 125.9 kts.
d. 103.9 kts.

12. For the airplane in Problem 11, calculate the new stall speed (V_{s_2}) of an airplane when the airplane is experiencing a bank angle of 60° during a level altitude, banked turn.
a. 84.9 kts.
b. 131.6 kts.
c. 64.5 kts.
d. 78.0 kts.

13. Calculate the maneuvering speed (V_A) if the V_s of the aircraft is 78 kts., and the aircraft is at the positive LLF for a utility category aircraft.
a. 163.6 kts.
b. 152.1 kts.
c. 104.9 kts.
d. 191.1 kts.

14. Calculate the radius of a turn of an airplane with a TAS of 162 kt, and a bank angle of 45° while maintaining a constant altitude.
a. 1 956 ft
b. 2 331 ft
c. 3 553 ft
d. 2 778 ft

15. Calculate the rate of turn of an airplane with a TAS of 90 kt, and a bank angle of 30° while maintaining a constant altitude.

a. 3.1 deg/s
b. 12.1 deg/s
c. 14.5 deg/s
d. 7.0 deg/s

16. A twin-engine airplane has an engine failure shortly after takeoff. The pilot tries to turn back toward the field by making a left turn but crashes after making a 180° turn. The wreckage diagram shows that the turn radius was 6 000 ft. The airspeed during the turn is estimated to be 200 kts. TAS. Calculate the bank angle of the airplane.

17. Using Figure 11.18, verify the answer to Problem 16.

18. Assuming a level turn, find the *G*s on the airplane in Problem 16 during the turn.

19. For the airplane in Problem 16, calculate the rate of turn and time to make the 180° turn.

20. Using Figure 11.18, verify the ROT you calculated in Problem 19.

Correlation Through Scenario-Based Learning

Aircraft – *Diamond DA50RG*

1. Review Section 2 of the DA50 AFM and find the following: Z
a. V_A or V_o
b. V_{FE}
c. V_{NO} and V_{NE}
d. V_{LO} and V_{LE}
e. Locate and review *Airspeed Indicator Markings* (table)

2. Explain why there are different maneuvering speeds published for different aircraft weights. How does this compare to the chapter discussion? How would you calculate maneuvering speed if the aircraft's weight was between one of the weights that are listed?

3. Review the *Center of Gravity (CG)* section within Section 2 of the AFM, in Figure 1, what happens to the maximum takeoff weight as the CG decreases (moves toward the front of the aircraft)? Why?

4. Review the *Approved Maneuvers* and *Maneuvering Load Factors* sections within Section 2 of the AFM and answer the following questions:
a. What does the note regarding 60° of bank emphasize?
b. In which category as discussed in this chapter is the DA50 certified?
c. What are the approved maneuvers for the DA50?
d. Do the positive and negative LLF at V_o match what was discussed in the chapter?
e. How do the LLFs change with flaps deployed? Why? What about at V_{NE}?

5. Review the *Emergency Descent* section within Section 3 of the AFM. What is the recommended flap position for the descent? What is the warning that is listed for the descent? What scenarios would result in a pilot having to use the emergency descent procedure? Based on the chapter reading and additional research, would banking the aircraft during the emergency descent aid in increasing the rate of descent or be detrimental to aircraft safety?

6. Review the *Stalling Speeds Description* section within Section 5 of the AFM and answer the following questions:
a. For aircraft weight 3 527 lb, gear and flaps up, what happens to the IAS for stall speed as the bank angle increases? Does this verify with the chapter discussion?

b. Compare your answer to the same row for an aircraft weight of 4 407 lb, how have the stall speeds changed for the same respective bank angle? Why have the stall speeds changed with weight? Are these speeds based on coordinated or uncoordinated flight? Do you think these speeds are based on level altitude flight, or does it matter?

c. Why is there a bigger discrepancy between the IAS and calibrated airspeed (CAS) published stall speeds at 60° angle of bank versus 0° bank?

7. Using Eq. 11.5, choose one weight and one configuration row to compare CAS stall speeds at a 60° angle of bank (how many *G*s?) and 0° bank. Does your numerical calculation using CAS support what is published in the table?

8. Using Eq. 11.6, choose one weight and one configuration row to compare CAS for any bank angle. Does your numerical calculation using CAS support what is published in the table?

12 Longitudinal Stability and Control

CHAPTER OBJECTIVES

After completing this chapter, you should be able to:

- Explain the difference between static stability and dynamic stability and provide examples of each.
- Define weight and balance terminology, explain the importance of weight and balance to the safety of flight, and calculate the center of gravity of an airplane.
- Summarize the relationship between aircraft weight and balance and aircraft performance.
- Identify the three reference axes of an airplane and recognize which axis relates to pitch stability.
- Describe the contribution of aircraft components to static pitch stability.
- Identify the two types of oscillatory motion associated with dynamic longitudinal stability.
- Compare and contrast the pitching tendencies in a stall between a low-tailed airplane and a T-tail airplane.

In addition to adequate performance, an aircraft must have satisfactory handling qualities. An aircraft must be able to maintain stable flight and be able to recover from the effects of disturbances such as gusts. This property is called the *stability* of an aircraft. Stability reduces the pilot's workload. In some cases, such as helicopter flight, it is necessary to provide artificial stabilization by using *automatic stabilization equipment* (ASE). The mission of the aircraft is very important to the manufacturer when considering the design of the aircraft: a single-engine Cessna trainer is by design much more stable than an F/A-18 Super Hornet.

The stability of an aircraft impacts two areas of aircraft design, *maneuverability* and *controllability*. The *controllability* of an aircraft refers to the response of an aircraft in the flight path and attitude to the control input of a pilot, specifically when maneuvering the aircraft. For an aircraft to respond to the controls, its stability must be overcome. Stability and control are often compared to a seesaw, with stability at one end and control at the other. The more stability an aircraft has, the less controllability it has, and vice versa.

Stability also affects the maneuverability of an aircraft. Maneuverability allows an aircraft to be maneuvered easily and withstand the stresses we discussed in Chapter 11. This design characteristic is

Flight Theory and Aerodynamics: A Practical Guide for Operational Safety, Fifth Edition.
Brian A. Johnson and Philip R. Fittante.

Companion website: www.wiley.com/go/flighttheoryandaerodynamics5e

a function of the weight, structural strength, flight control design, and powerplant, among other items. Modern complex, high-performance aircraft have complicated stability problems that are beyond the scope of this book. Our Chapter 12 discussion is basic, and some simplifying assumptions will be made.

DEFINITIONS

Equilibrium

An aircraft is said to be in a state of equilibrium if the sum of all moments and forces at its center of gravity (CG) is zero. This means that there is no pitching, yawing, or rolling and that velocity (not just speed) is constant.

Static Stability

Static stability is the *initial tendency* of a body to move once it has been displaced from its equilibrium position. If a body has the initial tendency to return to its equilibrium position, it is said to have positive static stability. Suppose a properly trimmed aircraft in level flight is disrupted by thermal turbulence. If, without control input from the pilot, the aircraft initially tends to return to its original airspeed and altitude, it is said to have positive static stability. The location of the center of lift relative to the CG, the distance of the horizontal tail surfaces from the CG, and the *tail volume* (defined later) of the airplane all help determine whether an aircraft has positive static *longitudinal* stability.

The three conditions of static stability can be illustrated by the ball/cone in Figure 12.1:

1. **Positive static stability:** The bottom of the bowl is the equilibrium position of the ball in the image on the far left of Figure 12.1. If the ball is moved to one side, its initial tendency is to return toward the bottom: it has positive static stability. Note that for the ball to have positive static stability, it is

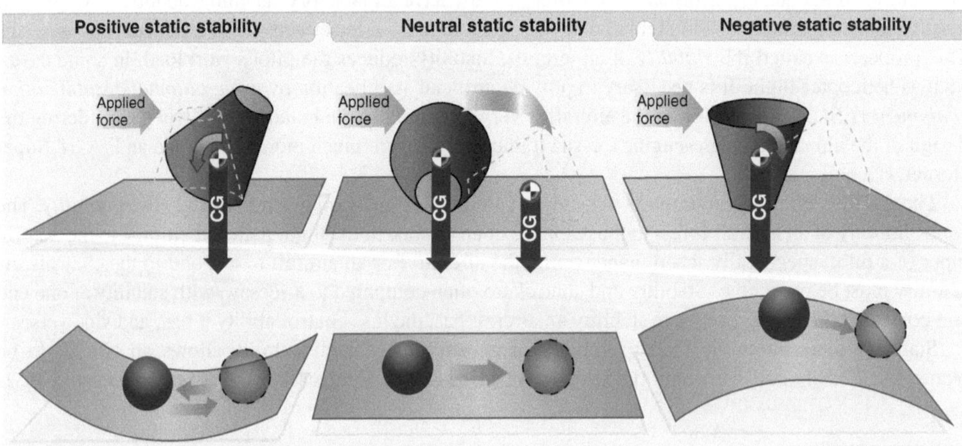

Figure 12.1. Types of static stability.
Source: Pilot's Handbook of Aeronautical Knowledge (2008a)/U.S. Department of Transportation Federal Aviation Administration/Public domain.

not important that the ball return to the bottom, only that it has the tendency to return. Similarly for the cone on the far left in Figure 12.1: a force is applied, and the natural tendency of the cone is to return to its original state. Suppose an aircraft in straight and level flight encountered a downdraft, causing a brief, almost instantaneous descent. If without control input from the pilot the aircraft began a climb back toward the original altitude, it would be said to have positive static stability.

2. **Negative static stability:** If an aircraft that has been disturbed from its equilibrium condition has the initial tendency to move farther away from equilibrium, it is said to have negative static stability. This is illustrated by the ball/cone on the far right of the image in Figure 12.1. The ball has been displaced from its equilibrium position on the left (at the top of the bowl) and moved to the right by a force. The initial tendency of the ball is to move farther from its original position, so the ball has negative static stability. The cone on the right of the image exhibits the same natural tendency as the ball when a force is applied; it moves farther away from the original state of equilibrium. Consider again the aircraft in the example above. If without control input from the pilot the aircraft were then to begin a descent with an increasing rate of descent, it would be said to have negative static stability.
3. **Neutral static stability:** If an aircraft is disturbed from its equilibrium position and has the tendency neither to return nor to move farther away from its equilibrium position, it is said to have *neutral static stability*. This is illustrated by the ball/cone on a flat surface, as shown in the middle of the image in Figure 12.1. If the ball/cone is displaced by a force, it does not have a tendency to return to its original position, nor does it have a tendency to move away from its original position. This is neutral static stability.

Dynamic Stability

Dynamic stability is the response of a body *over time* to a disruption of equilibrium, after its initial response to the disruption. With respect to pitch, roll, and yaw, an aircraft can be statically stable or unstable, but it can only be dynamically stable if the aircraft is statically stable, producing the oscillations shown in Figure 12.2. Over time, a statically unstable aircraft would continuously diverge in the direction of

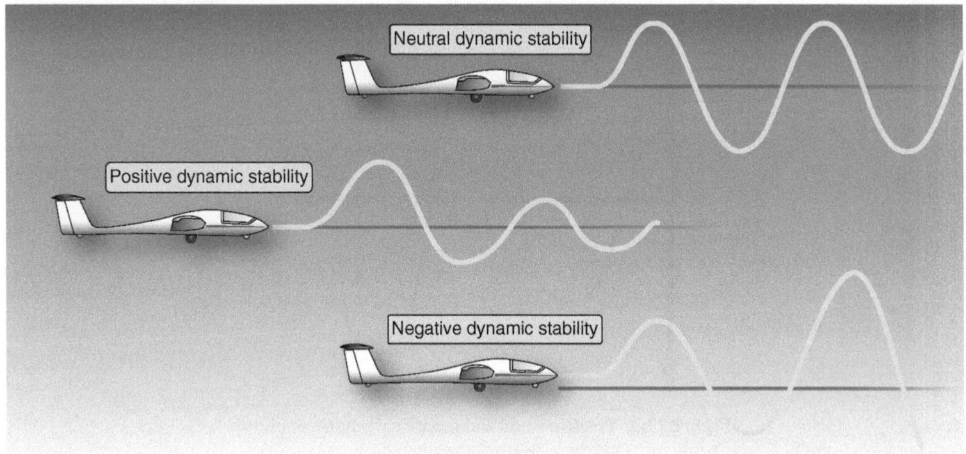

Figure 12.2. Dynamic stability.
Source: Glider Flying Handbook (2013)/U.S. Department of Transportation Federal Aviation Administration/Public Domain.

disruption. Like static stability, dynamic stability can be positive, negative, or neutral. Therefore, our review of dynamic stability around the axes of flight for Chapters 12 and 13 will focus on:

1. **Positive dynamic stability:** If an aircraft is disturbed from its equilibrium state and the maximum displacement decreases with time, the aircraft is said to have positive dynamic stability.
2. **Negative dynamic stability:** If the maximum displacement increases with time, and becomes more divergent, the aircraft is said to have negative dynamic stability.
3. **Neutral dynamic stability:** If the displacement remains constant with time, the aircraft is said to have neutral dynamic stability.

OSCILLATORY MOTION

The equilibrium of an aircraft in flight is constantly being disrupted (e.g. by updrafts, downdrafts, and intentional or inadvertent control inputs by the pilot). If the aircraft did not have positive static stability, it would be difficult to control (by a human pilot), so all aircraft must have some degree of positive *static* stability. However, this does not ensure that the aircraft will have positive *dynamic* stability as well.

In Figure 12.3, the initial reaction to the displacement is to return toward the equilibrium position. This is positive static stability. The subsequent oscillations, however, are divergent, so the dynamic stability is negative. An aircraft with negative dynamic stability would be difficult to control (by a human pilot), so this would be an unacceptable design feature.

Figure 12.4 illustrates positive static stability and neutral dynamic stability. No damping of the oscillations occurs.

The desired combination of positive static and positive dynamic stability produces a damped oscillation. An airplane with positive static and dynamic stability will return to its equilibrium condition when its equilibrium has been disrupted. This is shown in Figure 12.5.

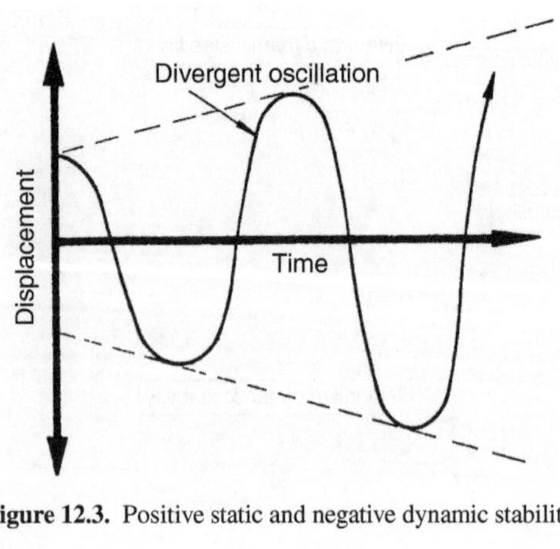

Figure 12.3. Positive static and negative dynamic stability.

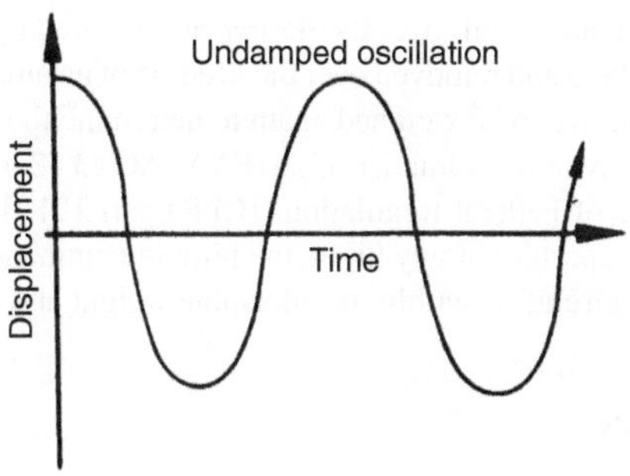

Figure 12.4. Positive static and neutral dynamic stability.

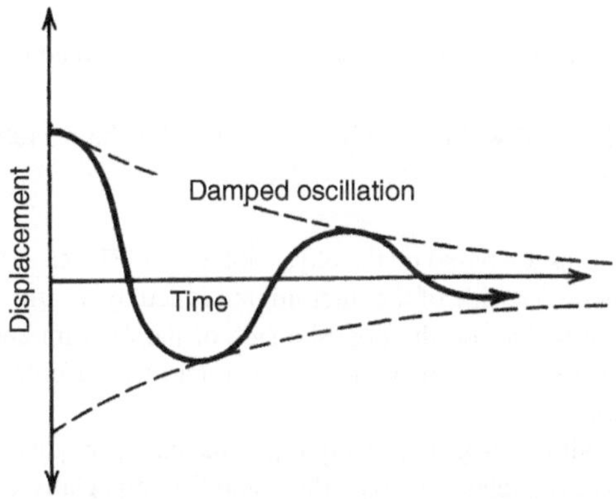

Figure 12.5. Positive static and positive dynamic stability.

WEIGHT AND BALANCE

A discussion of stability would not be complete without an introduction to aircraft weight and balance and its effect on performance. The weight and balance of an aircraft must be calculated and evaluated before and during every flight. The importance of the impact of weight and balance on the stability of an aircraft, limit load factor, and overall safety of flight cannot be overstated.

Aircraft designers and engineers set the maximum weight based on the amount of lift the wings or rotors can provide during normal operations, as well as the structural strength of the aircraft. The CG of the aircraft is also determined, along with the operational range of the CG, based on the mission of the aircraft. Additional certification information for small aircraft can be found by researching Civil Air Regulation (CAR) 3 certification or the General Aviation Manufacturers Association (GAMA) published information for part 23 and part 27 aircraft. Additionally, for larger part 25 aircraft certification, the most current version of FAA Advisory Circular (AC) 120-27, *Aircraft Weight and Balance Control* should be reviewed.

Any repairs or alternations to an aircraft over its lifetime are required by federal aviation regulations to be recorded. As equipment is added and removed over the lifetime of an aircraft, it is also must be recorded, and quite often the aircraft will have to be weighed again to determine the most up-to-date weight per the most current edition of Federal Aviation Administration (FAA) AC 43.13-1 or per the applicable operating certificate/authority of 14 Code of Federal Regulations (CFR) part 121, 125, and 135 operators. During the preflight inspection before departure of any flight, the pilot in command (PIC) has the moral and legal responsibility to verify that the aircraft is within the allowable weight and CG range.

Weight and Balance Theory

All passenger-carrying aircraft certificated by the FAA will possess CG information located within the Type Certificate Data Sheet (TCDS). Additional weight and balance information in the TCDS may include maximum weights, number of seats, fuel capacity, oil capacity, maximum baggage, and the datum location. Two general conditions must be met when calculating the weight and balance before flight:

1. The total ramp weight (and takeoff weight) must be no greater than the maximum allowed by the FAA.
2. The CG must be maintained within the allowable range for the operational weight of the aircraft from takeoff to landing.

Weight and balance problems are based on the physical law of the lever. A lever is balanced, much like an airplane, when the weight on one side of the fulcrum multiplied by its arm (distance from the fulcrum) is equal to the product of the weight on the opposite side of the fulcrum and its arm. At this point, as discussed in Chapter 1, the sum of the moments around the fulcrum (and aircraft CG) is zero, and the lever (or aircraft) is in equilibrium.

A detailed discussion on all the aspects of weight and balance is beyond the scope of this textbook, but certain terminology and principles directly affect stability, especially along the longitudinal axis, and warrant further evaluation. Figure 12.6 shows the location of many of the terms commonly used in the calculation of aircraft weight and balance. The *datum* is an imaginary vertical line set by the aircraft manufacturer at an arbitrary location, from which all moment arms are measured. The *arm* is the distance between a certain object (the pilot, fuel in the main tanks, baggage behind rear seats, etc.) in the aircraft and the datum, often measured in inches. When the aircraft manufacturer places the arm to the left of the datum the values are negative, and when the arms are placed to the right of the datum the values are positive (Figure 12.6). The moment is the product of the arm and the weight, expressed in pound-inches (lb-in.), and could be positive or negative. The calculation of the moment is shown in Eq. 12.1:

$$\text{Moment (lb-in.)} = \text{Weight (lb)} \times \text{Arm (in.)}, \quad \text{or}$$
$$W \times A = M \text{ (lb-in.)} \tag{12.1}$$

Once all of the weights of the aircraft are totaled, including the empty weight of the aircraft itself, and the individual moments are calculated, the final "balance point" (total CG) of the aircraft can be calculated. To determine this point, we will use Eq. 12.2:

$$\text{CG (in.)} = \frac{\text{Total moment (lb-in.)}}{\text{Total weight (lb)}} \tag{12.2}$$

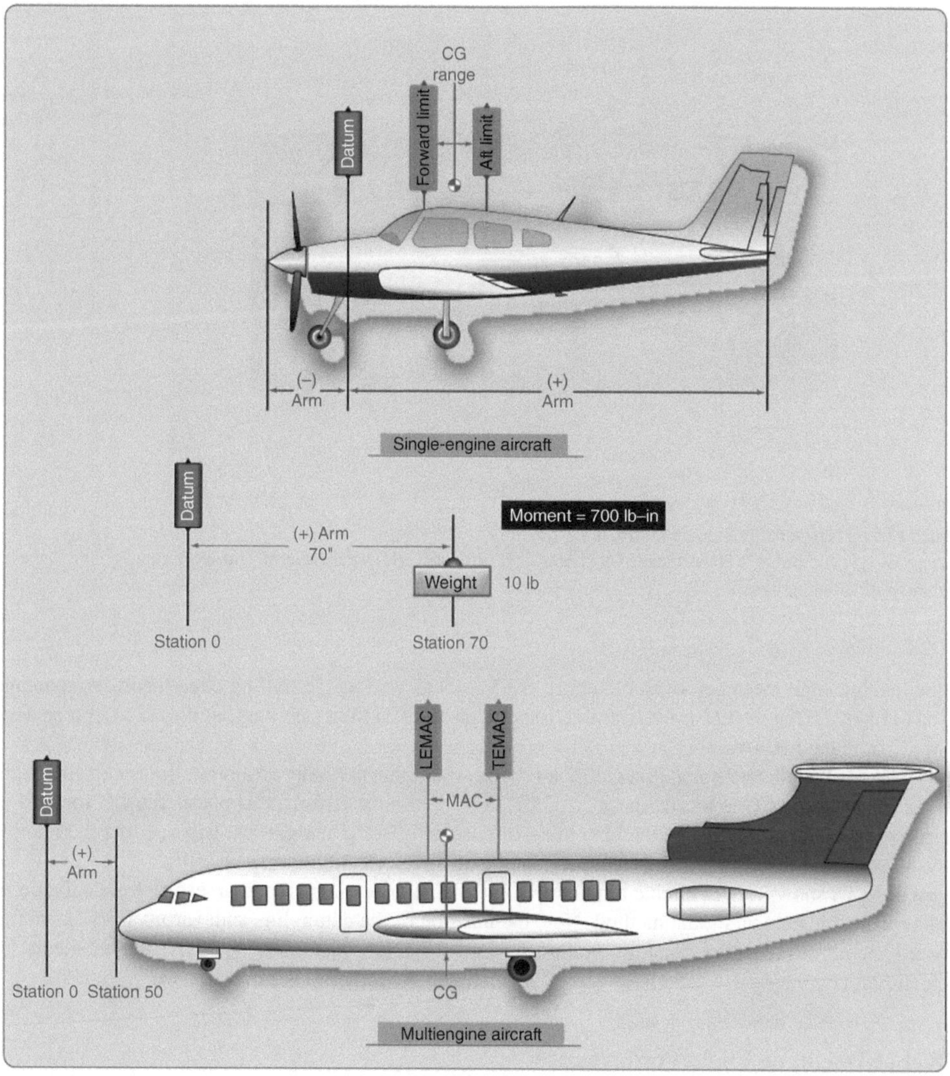

Figure 12.6. Key weight and balance locations on an aircraft.
Source: Weight and Balance Handbook FAA-H-8083-1B (2016d)/U.S. Department of Transportation Federal Aviation Administration/Public domain.

Mean Aerodynamic Chord

For most general aviation aircraft, the location of the CG is expressed as a distance from the datum. Many larger aircraft, from corporate jets to air transport category aircraft and military aircraft, express the CG position as a percentage of the length of the mean aerodynamic chord (MAC), or "percent MAC" (Figure 12.7). These more advanced aircraft often incorporate tapered, swept, and blended wing designs that necessitate a complex calculation of MAC by the manufacturer. This calculation includes the position

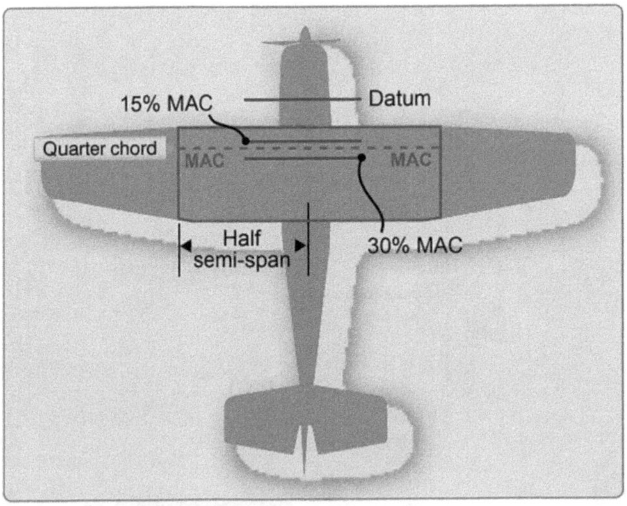

Figure 12.7. Mean aerodynamic chord (MAC).
Source: Weight and Balance Handbook FAA-H-8083-1B (2016d)/U.S. Department of Transportation Federal Aviation Administration/Public domain.

of the leading edge mean aerodynamic chord (LEMAC) as well as the trailing edge mean aerodynamic chord (TEMAC) (Figure 12.6). An aircraft's forward and aft CG limits are then expressed as a percentage of MAC from the LEMAC.

The FAA's Weight and Balance Handbook (2016d) states that normally an aircraft has acceptable flight characteristics if the CG is located near 20–25% MAC. For most airfoils, this places the CG forward of the aerodynamic center (AC). In the upcoming discussion of static margin, we will see that the relative positions of CG and AC are critical to understanding an aircraft's longitudinal stability.

Figure 12.8 shows a twin-engine airplane weight and balance diagram and an example calculation of loading using the computational method. Note the position of the datum line and ±arm values extending forward and aft of the datum point, as well as the total CG limit range of 32.0–43.6″. The moments for each aircraft component are calculated using Eq. 12.1, and the total CG is calculated using Eq. 12.2.

Effect of Weight on Flight Performance

Both the *total weight* of an aircraft and the *location* of the weight have profound effects on the performance of the aircraft. An aircraft that is loaded beyond the allowable range may not be as stable and controllable as when loaded within limits. As discussed in Chapters 8 and 9, increased weight adversely affects takeoff and landing performance, and the location of the CG affects the stability of the aircraft not only on takeoff and landing but throughout the flight. In extreme cases, shifting individual weights inside the aircraft during abrupt flight maneuvers, or during climbs and descents, compromises the safety margins of an aircraft by moving the CG location. As discussed in Chapter 2, higher-density altitude calculations degrade aircraft performance and may further impact the flight performance envelope for certain aircraft. For many light, single- and multi-engine aircraft, it may not be possible to load the aircraft as desired with passengers and cargo, or they may need to be loaded in specific configurations due to cargo requirements and increased density altitude.

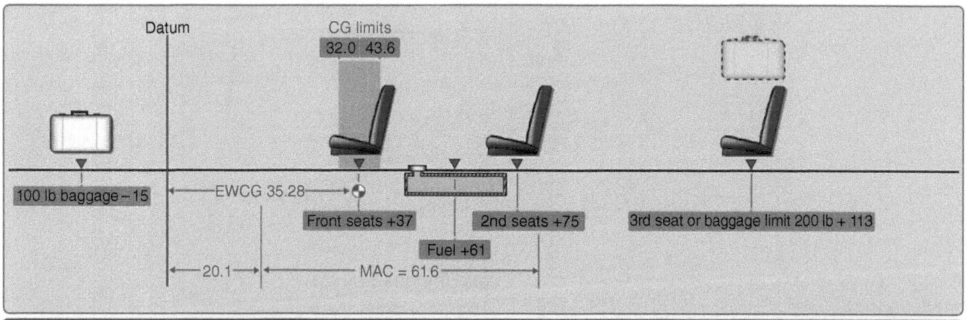

Item	Weight (lb) (5 200 max.) ×	Arm (in) =	Moment (lb-in)	CG
Airplane	3 404	35.28	120 093	
Fuel (140 gal)	840	61.0	51 240	
Front seat	320	37.0	11 840	
Row 2 seats	310	75.0	23 250	
Forward baggage	100	−15.0	−1 500	
Aft baggage	90	113.0	10 170	
Total	5 064		215 093	42.47

Figure 12.8. Weight and balance diagram and computational method calculation.
Source: Weight and Balance Handbook FAA-H-8083-1B (2016d)/U.S. Department of Transportation Federal Aviation Administration/Public domain.

As discussed in Chapter 11, the weight of an aircraft greatly impacts the load factor and structural integrity of the aircraft. The V–G diagram shows that a change in weight changes the position of the various limitations and that the accelerated stall speed increases as weight increases. A minor oversight by a pilot of loading a small, single-engine airplane with 50 lb of extra weight may not seem like much, but if the aircraft is loaded within the utility category and experiences the maximum positive load limit of 4.4G, that 50 lb overload results in a 220 lb effective load, with possible catastrophic effects.

Effect of Weight on Load Distribution

The location of an aircraft's CG relative to the CG limits, and the respective position of the CG for the various stations within an aircraft, impact the flight performance and stability of the aircraft. Although the lateral location of the CG can be important for larger aircraft, and for smaller aircraft with a lateral fuel imbalance, our discussion will focus on the fore and aft distribution of weight along the longitudinal axis and on its effect on the aircraft's longitudinal stability. Figure 12.9 shows the effect of fore and aft load distribution within an aircraft.

During level cruise flight, the amount of tail down force on the horizontal stabilizer must be added to the total lift produced by the aircraft to remain in level flight. For example, if the weight of the aircraft is 2 000 lb and the tail down force required to maintain level flight is 50 lb, the wings must generate 2 050 lb of lift. As reviewed in Chapter 4, the greater the lift required, the greater the drag produced; thus, many of the aircraft's performance airspeeds are adversely affected. As the CG moves forward, more tail down force must be generated to counteract the nose-down pitching moment caused by the forward CG. If the tail down force required is now 100 lb, the wings must now generate 2 100 lb of lift. Note that there has been no change in the weight of the aircraft, only a change in the location of a portion of the weight.

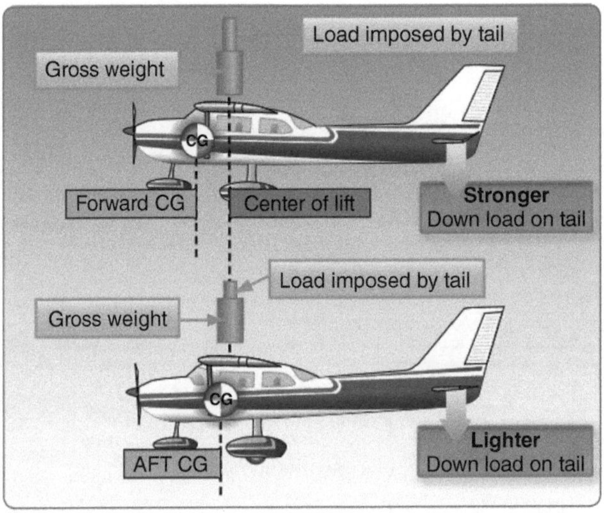

Figure 12.9. Effect of load distribution on balance.
Source: Pilot's Handbook of Aeronautical Knowledge (2016b)/U.S. Department of Transportation Federal Aviation Administration/Public domain.

To generate this additional lift at a given airspeed, the wings require a higher coefficient of lift (C_L) and thus a higher angle of attack AOA. Because the AOA is now closer to the critical AOA, the stall speed is increased, and because the wing is generating more lift, induced drag is increased as well, adversely affecting performance. A forward CG also results in heavier flight controls upon landing: as airspeed decreases, airflow over the horizontal stabilizer decreases, and so the tail down force it generates decreases as well, requiring progressively increasing back pressure on the stick/yoke to keep the nose-up. In extreme situations, the airplane may have to land at higher ground speeds to allow adequate elevator effectiveness, and a proper flare during landing rollout may be impossible. As the CG moves aft, there is less wing loading due to the decrease in downward force on the tail, allowing for increased performance, lower stall speed, and a faster cruise speed. As many aircraft operators have realized, a reduction in total lift required equals a reduction in fuel required, resulting in improved fuel efficiency and lower operating costs. Positioning heavier payload, or fuel, by the aircraft operators toward the fore and aft, or lateral limits, could result in less responsive controls and reduced maneuverability.

The faster cruise speed, improved fuel economy, and decreased stall speed associated with an aft CG are all positive performance values, so without further evaluation, it appears the aircraft should be loaded with the CG as far aft as possible. However, there are other factors to consider. As the CG moves aft, longitudinal stability decreases. And while the stall speed may decrease with an aft CG, an aft CG out of limits may affect the ability of an airplane to recover from a stalled condition, or from a spin. The airplane that was certificated to be recoverable from an incipient spin may no longer have that ability when the airplane is loading incorrectly. In an extreme example, as we saw in Chapter 10 with a swept-wing airplane, an aft CG out of limits could result in a flat spin for a straight-wing airplane, which may be unrecoverable. With an aft CG out of limits, flight controls during normal maneuvers such as chandelles and lazy eights may be more sensitive, and stall recovery procedures during advanced stall demonstrations may result in additional recovery time and increased loss of altitude.

Application 12.1

Refer to Figure 12.8, consider the situation where the 90 lb suitcase currently located at a station with an arm of +113 is moved to the second row seats.

What impact would this move have on the total weight of the aircraft and on the location of the CG? What methods are available to efficiently and accurately calculate the new total CG due to this shift in weight? What effect does this change in total CG have on the flying qualities of the airplane? Could the suitcase be moved to the forward baggage area without exceeding any limitations?

AIRPLANE REFERENCE AXES

To better envision the forces and moments acting on an airplane, we assign three mutually perpendicular reference axes, all intersecting at the CG. These are shown in Figure 12.10. The longitudinal axis is

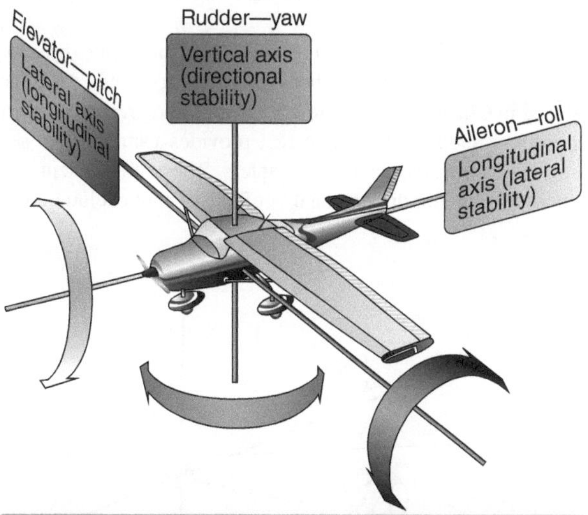

Primary control surface	Airplane movement	Axes of rotation	Type of stability
Aileron	Roll	Longitudinal	Lateral
Elevator/ Stabilator	Pitch	Lateral	Longitudinal
Rudder	Yaw	Vertical	Directional

Figure 12.10. Airplane reference axes.
Source: Pilot's Handbook of Aeronautical Knowledge (2008a)/U.S. Department of Transportation Federal Aviation Administration/Public domain.

assigned to the letter X, and the aircraft experiences rolling motion around this axis using the ailerons. The lateral axis is the Y axis and can be considered for our discussion as the axis from wingtip to wingtip through the CG. The aircraft experiences pitching moments around the lateral axis by using either the elevator or stabilator primary flight controls. The vertical axis is the Z-axis, and the aircraft experiences yawing motion around the vertical axis by using the rudder. Again, all three imaginary axes pass through the aircraft's CG.

As seen in Figure 12.10, the positive direction of the axes is determined by positioning the thumb, index finger, and middle finger of the right hand so that they are at right angles to each other. Point the thumb in the forward direction of the X-axis, and the other fingers point to the positive directions of the Y- and Z-axes.

Three moments are possible about the three axes. These are given three identifying letters that occur in sequence in the alphabet. The rolling moment is L (to avoid confusion with lift we will call this L'), the pitching moment is M, and the yawing moment is N. The positive direction of these moments is determined by using the right-hand rule, shown in Figure 12.11. The right thumb points to the positive axis and the curvature of the fingers shows the direction of the positive moments. The completed axes and moment illustration are shown in Figure 12.12.

Longitudinal stability and control refer to the behavior of the airplane in pitch, that is, movements of the longitudinal axis about the lateral axis as controlled by the elevator or stabilator. This movement is illustrated in Figure 12.13. In the pure pitching case, no rolling or yawing of the aircraft takes place.

The roll, pitch, and yaw moments are a function of roll rate (p), pitch rate (q), and yaw rate (r) as well as roll angle (ϕ), AOA (α), and sideslip (β) among other factors. Do not confuse the symbol "q" for pitch rate with "q" that was introduced in Chapter 2. These relationships are represented by stability derivatives that help define an aircraft's motions and stability. Table 12.1 provides a summary of selected aircraft stability derivatives that will be referenced in the next few chapters. These are presented in the table as dimensional derivatives as opposed to the nondimensional coefficients that exclude consideration of mass and inertia.

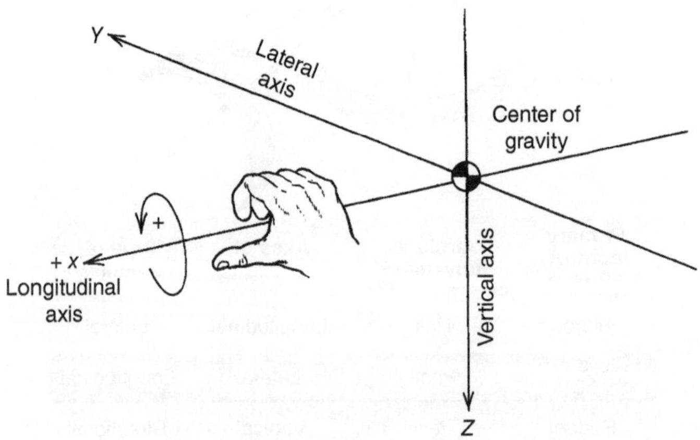

Figure 12.11. Establishing positive moment direction.

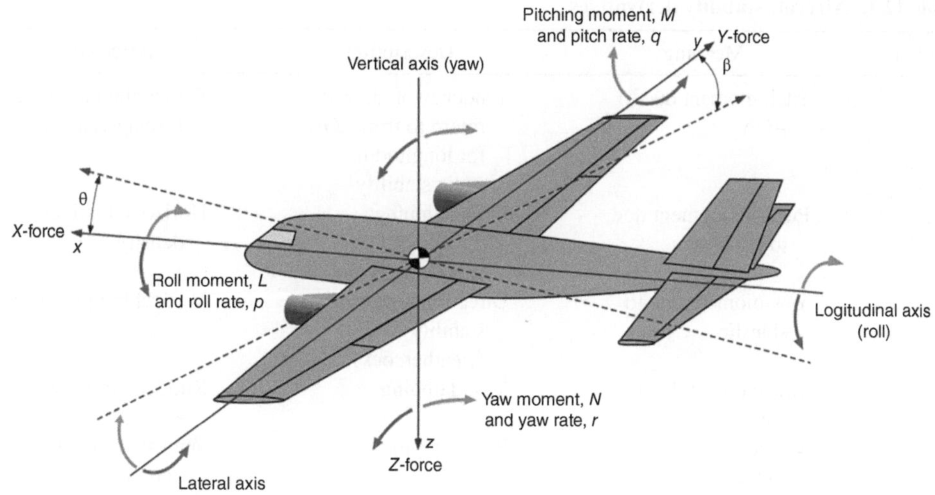

Figure 12.12. Airplane axes and moment directions.
Source: Introduction to Aerospace Flight Vehicles Section 52 (2023)/reproduced with permission of J. Gordon Leishman.

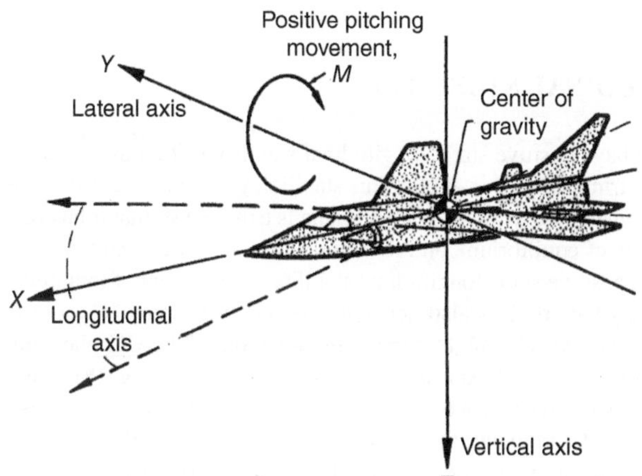

Figure 12.13. Movement of the longitudinal axis in pitch.

Table 12.1. Aircraft stability derivatives.

Symbol	Meaning	Description	Affected by
M_{α}	Pitch moment due to AOA	Tendency of aircraft to return to trim AOA for longitudinal static stability	Horizontal tail, wing, fuselage/nacelles
M_q	Pitching moment due to pitch rate	Pitch damping, damping of short-period motion	Horizontal tail, air density
N_{β}	Yaw moment due to sideslip	Directional static stability (weathercock)	Vertical tail primarily
N_r	Yaw moment due to yaw rate	Yaw damping	Rudder, air density
L_p	Roll moment due to roll rate	Roll damping	Wingspan and area, vertical tail
L_{β}	Roll moment due to sideslip	Dihedral effect, Dutch roll and spiral stability effects	Wing dihedral, sweepback and wing location, vertical tail
L_r	Roll moment due to yaw rate	Dutch roll and spiral stability effects	Wingspan, vertical tail

STATIC LONGITUDINAL STABILITY

An airplane is said to have positive static longitudinal stability if it tends to return to a constant AOA in flight once it has been trimmed to that angle. This stabilizing pitch moment is represented by $C_{M\alpha}$, or the coefficient of pitch moment due to AOA. If an airplane is trimmed so that it has zero pitching moments at some AOA, is in a state of equilibrium, and is then disturbed in pitch and tends to return to the trimmed AOA, it is said to have positive static longitudinal stability. The amount of longitudinal stability an aircraft has depends upon many factors. It is also dependent on pitch damping (C_{Mq}) or the coefficient of pitch moment due to pitch rate. As you might expect, the position of the CG plays an important role in the aircraft's stability around the lateral axis, as does the location of the wing (the center of lift) and of the tail relative to the aircraft's CG. Because we are dealing with moments, not only does the location of the tail (i.e. its distance from the CG) influence the longitudinal stability, but so does the magnitude (and direction) of the vertical force on the tail, which depends partly on the area of the horizontal tail surfaces.

An airplane that has negative static longitudinal stability, on the other hand, will continue to pitch away from the trimmed AOA. If an airplane has neutral static longitudinal stability, it will remain at whatever AOA the disturbance has caused.

It is important for an airplane to have positive static longitudinal stability. A stable airplane is easy and safe to fly. It can be trimmed at any desired speed and will tend to maintain that speed. An airplane with negative static longitudinal stability will be impossible to trim and will always require the pilot's attention. A tendency to dive or initiate a climb when upset is possible, even to the point of a stall, which in the case of an excessive aft CG location may be unrecoverable. Of course, an aircraft with excessive positive static longitudinal stability will be difficult to maneuver due to the heavy control forces. Therefore, it is important to consider the mission of the aircraft when designing for stability.

The Pitching Moment Equation

The pitching moment about the aircraft CG is shown in Eq. 12.3:

$$M_{CG} = C_{M(CG)}\, qSc \tag{12.3}$$

where:

M_{CG} = pitching moment about the CG (ft-lb)
$C_{M(CG)}$ = coefficient of pitching moment about CG
q = dynamic pressure (psf)
S = wing area (ft^2)
c = mean aerodynamic chord, MAC (ft).

Rearranging Eq. 12.3 gives

$$C_{M(CG)} = \frac{M_{CG}}{qSc} \tag{12.4}$$

Because the values of q, S, and c are always positive, it follows that for a nose-up (+) pitching moment, the value of $C_{M(CG)}$ must also be positive; for a nose-down (−) pitching moment, the value of $C_{M(CG)}$ must be negative.

Graphic Representation of Static Longitudinal Stability

A plot of the variation of $C_{M(CG)}$ at different values of C_L (different AOAs) for an airplane with positive static longitudinal stability is shown in Figure 12.14. The trim point is the value of C_L where the aircraft has no pitching moment. At all values of C_L above the trim point, such as point y, the aircraft will have a nose-down (−) pitching moment. If the aircraft is disturbed by an up-gust, the AOA (and C_L) will be increased. For stability, a nose-down pitching moment is required, so this is a stabilizing condition.

If the airplane is disturbed by a down-gust, the AOA and value of C_L will be reduced. This is represented by point X in Figure 12.14.

The value of $C_{M(CG)}$ at point X is positive, and a nose-up pitching moment results. This is what is required for static stability. Thus, a negative slope on this graph represents an aircraft with positive static

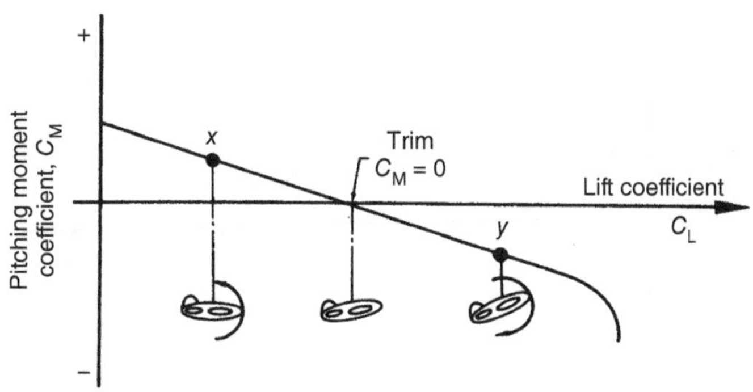

Figure 12.14. Positive static longitudinal stability.

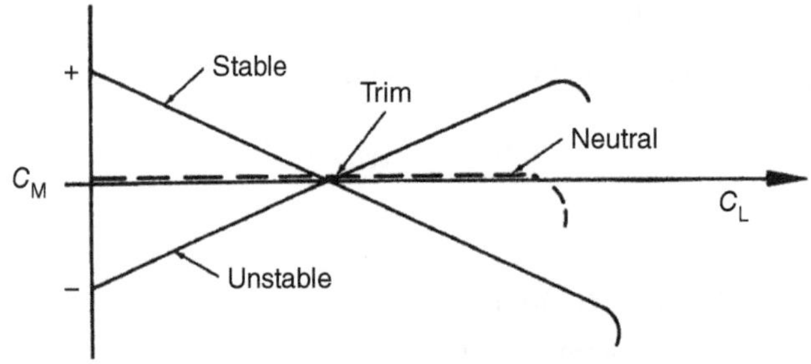

Figure 12.15. Types of static longitudinal stability.

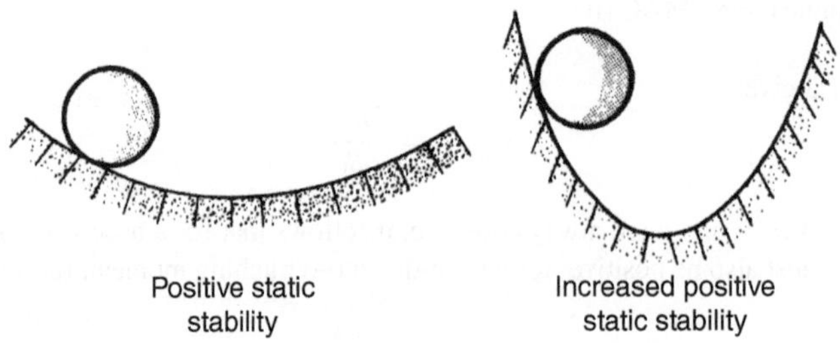

Figure 12.16. Degrees of positive static stability.

stability. Conversely, a positive slope would indicate an unstable aircraft and a zero slope would represent an aircraft with neutral stability. These are shown in Figure 12.15.

There are different degrees of stability. Some aircraft tend to return to their equilibrium positions faster than others. Again, using the analogy of the ball/cone in the curved container, the degree of stability can be illustrated as shown in Figure 12.16. Degrees of stability are shown on the $C_{M(CG)} - C_L$ plot by the slope of the curve (Figure 12.17). The steeper slope of the stable curve shows more stability, and the steeper slope of the unstable curve shows more instability. An airplane can be stable at lower angles of attack but may be unstable at higher angles of attack. Instability at high AOA would increase the risk of reaching critical AOA, resulting in a stall.

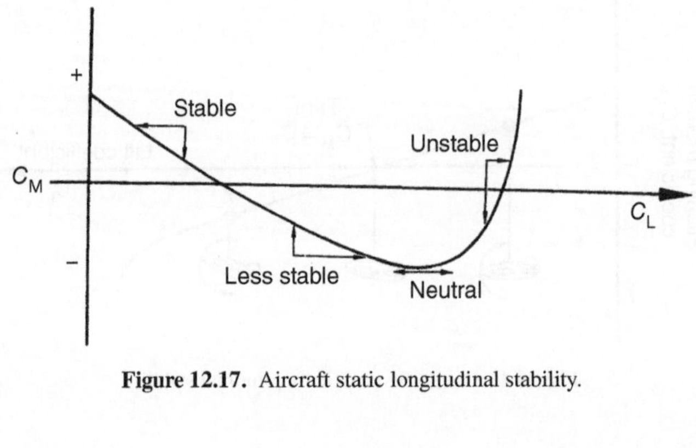

Figure 12.17. Aircraft static longitudinal stability.

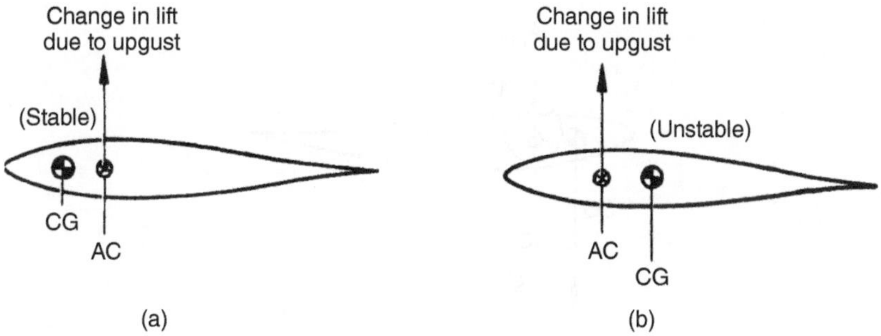

Figure 12.18. Effect of CG and AC location on static longitudinal stability. (a) CG is located forward of the AC. (b) CG is located aft of the AC.

Contribution of Aircraft Components to Pitch Stability

Wings

The static stability contribution of the wings depends on the relative position of the AC and the CG of the airplane. Consider a tailless (flying wing) airplane with a symmetrical airfoil section as shown in Figure 12.18. We will consider two possibilities of CG and AC location:

1. CG is located forward of the AC (Figure 12.18a).
2. CG is located aft of the AC (Figure 12.18b).

If the airplane in Figure 12.18a experiences an up-gust, the AOA will be increased and the lift at the AC will increase. The airplane will then rotate about the CG (a free body will rotate around the CG), and this nose-down moment will tend to rotate the airplane and return it to its equilibrium AOA. *This is positive static stability*. The greater the distance between the AC and the CG, the greater the nose-down moment in response to an up-gust.

If the airplane in Figure 12.18b experiences an up-gust, the increase in lift at the AC will rotate the airplane about its CG creating a nose-up pitching moment, and rotate the airplane away from its equilibrium position. *This is negative static stability*. The CG must be forward of the AC for stable flight, in this case.

The airplane in Figure 12.18a is stable in pitch, but it is not in equilibrium. The conditions for equilibrium require that there be no unbalanced forces or unbalanced moments acting on the airplane. The first requirement is easily satisfied by reducing the AOA so that lift equals weight. However, the second requirement has not been met. The forces of lift and weight create a nose-down pitching moment because the AC is behind the CG and the aircraft naturally rotates around the CG. Therefore, the unbalanced force must be canceled out by an opposing nose-up pitching moment to restore equilibrium (and stable flight).

In our discussion of pressure distribution on airfoils in Chapter 3, we saw that the forces acting on the top and bottom surfaces of a symmetrical airfoil were in the same position along the chord and that no pitching moment about the center of pressure (CP) resulted from their location (see Figure 3.30). The pressure location on cambered airfoils, on the other hand, was found not to be located at the same chordwise position, and so a nose-down pitching moment resulted from positive camber (Figure 3.30). A negatively cambered airfoil, again with the AC positioned behind the CG, produces a nose-up pitching moment. Aircraft manufacturers not only design an airfoil with the correct position of AC to give the correct amount of nose-up moment when in cruise flight, but also design the airfoil in certain cases with a given negative AOA to provide the necessary nose-up force. The correct amount of force for a given aircraft is

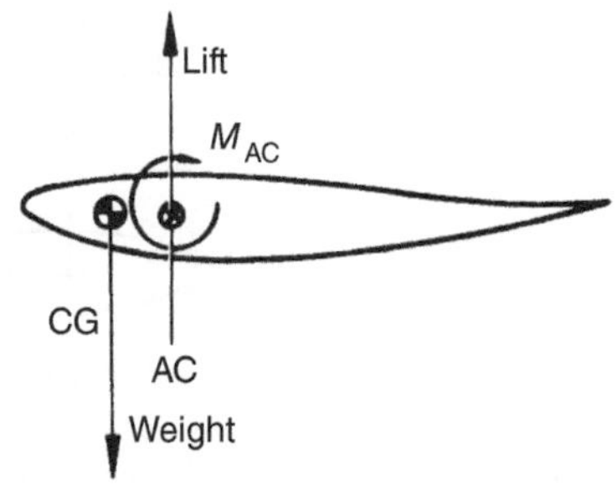

Figure 12.19. Static longitudinally stable flying wing in equilibrium.

what we need to cancel the nose-down moment caused by the lift and weight vectors. This is shown in Figure 12.19. Delta-wing airplanes use reflexed (negatively cambered) trailing edges to create the nose-up pitching moments required for equilibrium.

In summary, the overall contribution of the wings of an aircraft to the static longitudinal stability depends on the location of the AC and the CG of the entire aircraft. If the AC is behind the CG, the contribution is stabilizing. *If the AC is ahead of the CG, the contribution is destabilizing*. In the case of the statically stable symmetric flying wing aircraft (Figure 12.18a), a trailing edge negative camber was necessary to balance the natural nose-down pitching moment and provide equilibrium in flight (Figure 12.19). When designing a conventional airplane with separate wing and horizontal tail sections, we must still consider the AC of the entire aircraft (wings, fuselage, and tail) in relation to the CG for flight equilibrium. Consider the vertical forces acting on the aircraft if the AC is located behind the CG as shown in Figure 12.20. In order to maintain flight in equilibrium, the tail must provide enough downforce to compensate for the rotation around the CG. This tail-down force must be accounted for by the total lift.

It is assumed that this airplane has a symmetrical airfoil, so it has no pitching moment due to the pressure distribution on the airfoil. For balance, the airplane must have a download on the tail ($LIFT_T$). The lift on the wing ($LIFT_W$) must equal the weight of the aircraft plus the lift on the tail. Thus, it must operate at a higher AOA than if the lift on the tail acted upward. Higher angles of attack produce more drag.

One of the newer concepts in aircraft design is to reduce static stability deliberately by moving the CG backward, and thus reduce the drag. This may require automatic stabilization devices such as fly-by-wire, as the aircraft may not be controllable by a human pilot. In conventional aircraft, rearward CG can also reduce drag and increase airspeed, but caution is warranted: if the CG is too far aft, even a full down elevator may be insufficient to overcome the nose-up pitching moment, and an aft CG may make stall/spin recovery difficult or impossible.

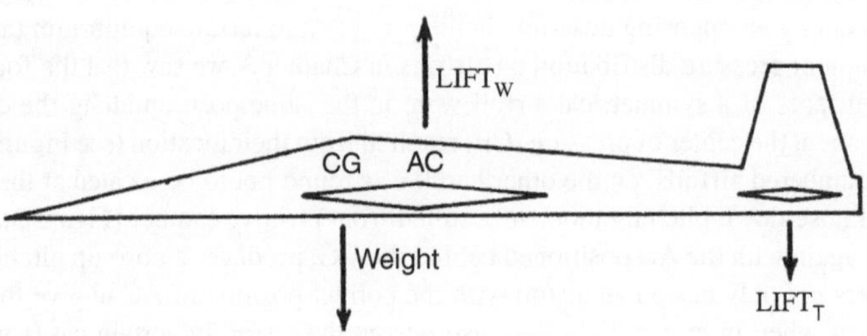

Figure 12.20. Airplane with static longitudinal stability.

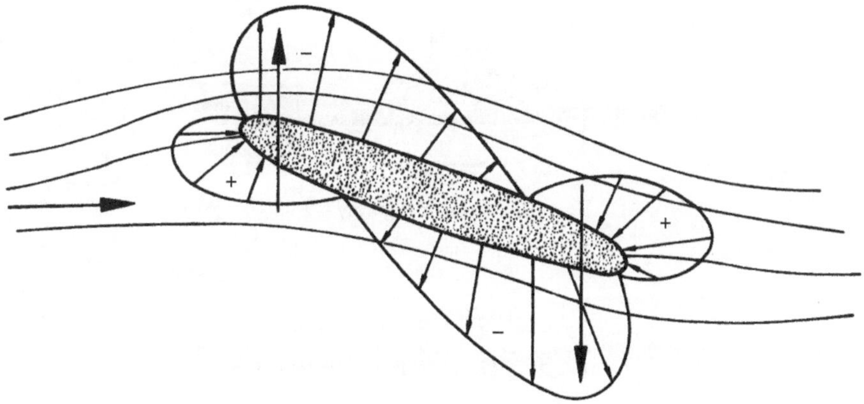

Figure 12.21. Pressure distribution about a body of revolution.

Fuselage

A streamlined fuselage has a pressure distribution similar to that of a body of revolution when placed in an airstream. The pressure distribution about such a body is shown in Figure 12.21. It can be seen that no net lift is developed by the pressure distribution, but a nose-up pitching moment is developed by an up-gust, so the pitching moment is destabilizing. Thus, *the fuselage is a destabilizing component.*

Engine and Engine Nacelles

The amount of power and position of the thrust line also can impact longitudinal stability. A high thrust line can be seen in Figure 12.22 to rotate the aircraft around the CG in a nose-down moment, whereas a low thrust line below the CG introduces a nose-up moment. Power changes in our single-engine aircraft example also influence longitudinal stability (Figure 12.23). If full power is applied, a nose-up rotation around the CG is produced, and vice versa. At cruise power, with the trim set properly, there should be no longitudinal pitching moment. Pilots transitioning to certain types of seaplanes may experience differences in longitudinal stability as compared to aircraft previously flown during thrust changes due to the exaggerated new position of the thrust line (Figure 12.24). Note the position of the horizontal stabilizer and the role of the changing lift forces as thrust is increased and decreased.

The direction of the airflow through a propeller disk or through a turbojet engine is not changed if the axis of the engine is in line with the aircraft's flight path. If the engine axis is at an angle with the relative wind, however, the airflow is turned so that it flows in the direction of the engine axis. When this happens, a side force is exerted on the propeller shaft or on the side of the jet engine intake, in accordance with Newton's third law. This is shown on the aircraft in Figure 12.25 and is a destabilizing force when in front of the CG.

The longitudinal axis of the aircraft is at a positive angle to the relative wind and the resulting force creates a nose-up pitching moment. This is a destabilizing moment. If the engines are mounted so that the propellers or jet intakes are behind the CG and the axis of the engine makes a positive angle with the relative wind, the downwardly directed thrust produces a nose-down moment. This is a stabilizing moment.

Propellers or jet engine intakes located forward of the CG are destabilizing components, and propellers or jet engine intakes located behind the CG are stabilizing components (tail-mounted).

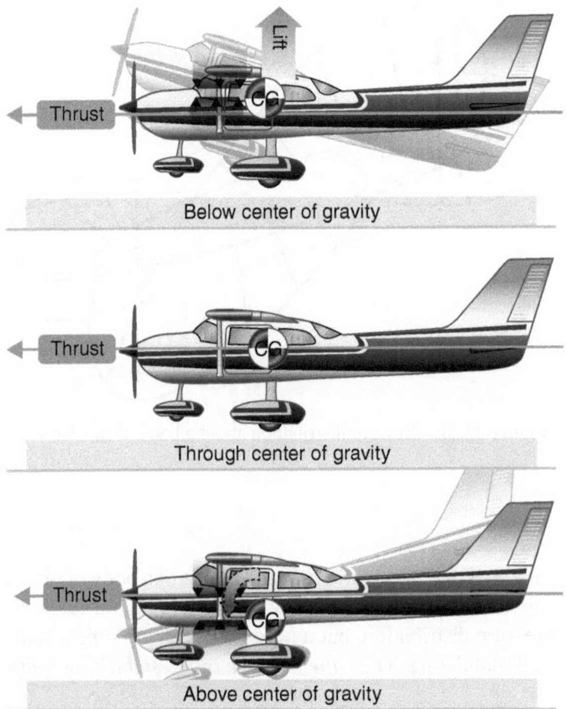

Figure 12.22. Thrust line and longitudinal stability.
Source: Pilot's Handbook of Aeronautical Knowledge (2008a)/U.S. Department of Transportation Federal Aviation Administration/Public domain.

Boeing's Maneuvering Characteristics Augmentation System (MCAS)

On 29 October 2018, Lion Air Flight 610 crashed 13 minutes after takeoff into the Java Sea off the coast of Indonesia, killing all 189 people on board. On 10 March 2019, Ethiopian Airlines Flight 302 crashed six minutes after takeoff from Addis Ababa, killing all 157 people on board. Both accidents involved Boeing's latest 737 version, the Boeing 737 MAX. It soon became apparent that Boeing's new software, used to mitigate pitch changes during certain configuration changes, may have been a contributing factor, especially when additional components on the aircraft experienced failure, , such as the AOA sensor.

The redesign of the 737 NG to the 737 MAX required larger engines, which then resulted in the position of the larger engines being mounted farther forward and higher leading to a change in the CG. Previous variants of the 737 had the CG well forward of the center of lift, so as previously discussed, with neutral control inputs the plane would nose-down and recover on its own when encountering a stall. Due to the engine size and position change with the MAX variant, the 737 tended to pitch up under certain conditions, especially when at high angles of attack when body lift from the large engine nacelles mounted ahead of the wings created a strong nose-up force. To mitigate this tendency (and to meet FAA airworthiness criteria for stability), Boeing attempted to replicate the "feel" of previous 737 versions by developing software called the Maneuvering Characteristics Augmentation System (MCAS), which the company defines as "flight control law implemented on the 737 MAX to improve aircraft handling characteristics and decrease pitch-up tendency at elevated angles of attack." The flight control law, or MCAS, was enhanced so the pilot

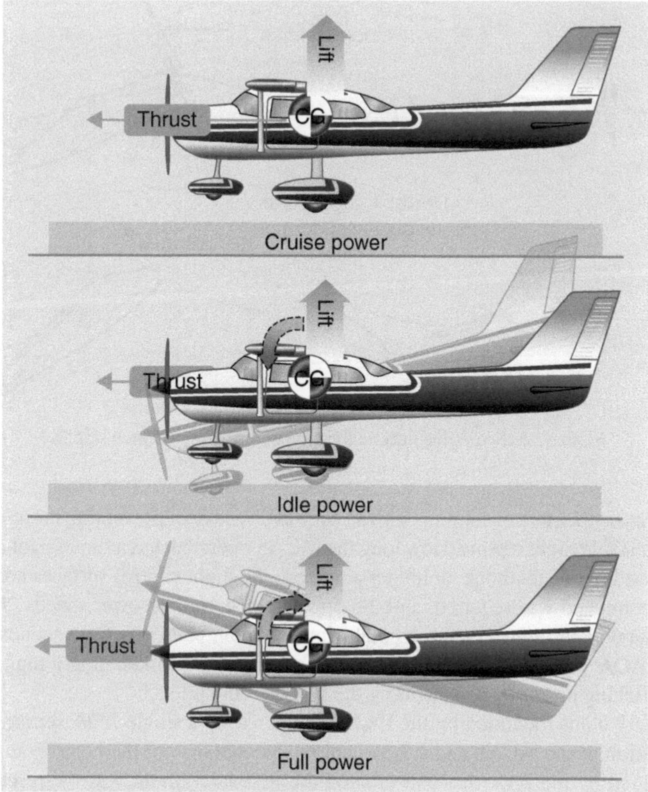

Figure 12.23. Power changes and longitudinal stability.
Source: Pilot's Handbook of Aeronautical Knowledge (2008a)/U.S. Department of Transportation Federal Aviation Administration/Public domain.

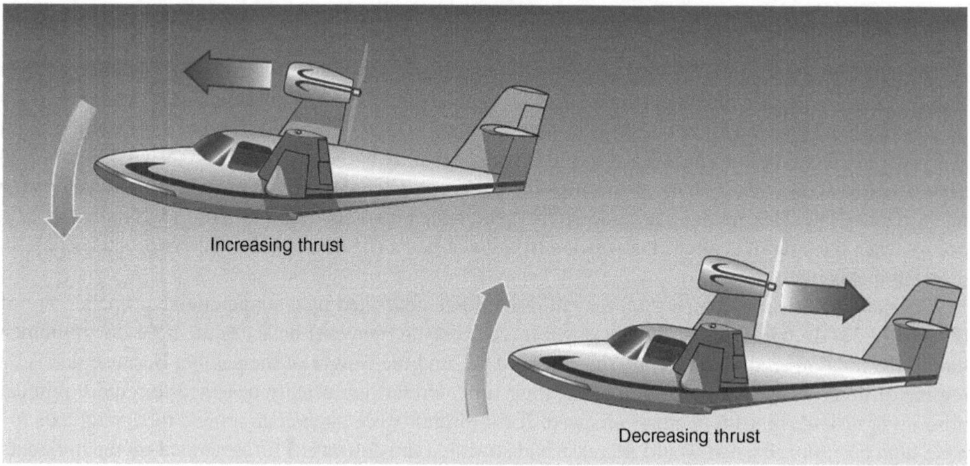

Figure 12.24. Pitching forces of a high thrust line.
Source: Wind Shear FAA-P-8740-40 (2004b)/U.S. Department of Transportation Federal Aviation Administration/Public domain.

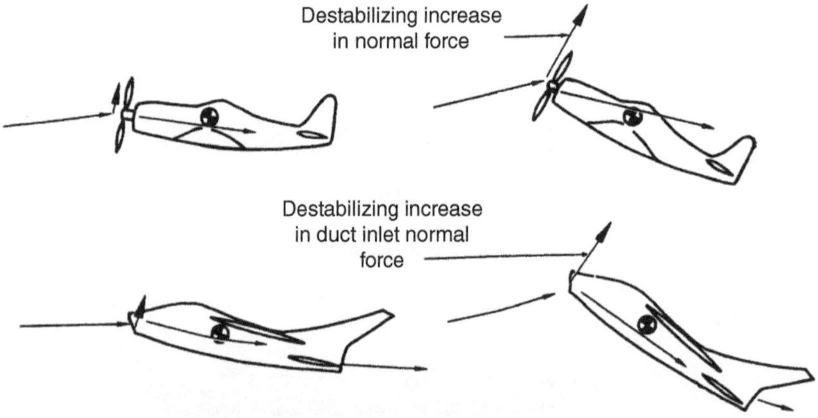

Figure 12.25. Engine nacelle location contribution to pitch stability.

could feel the forces on the control column, with an overall directive of preventing the aircraft from entering a hazardous and unstable flight regime. How long the MCAS operated was a function of airspeed and AOA, as we have discussed in this textbook, at higher airspeeds the flight control surfaces are more effective so minimal MCAS command was required, and the opposite occurred at lower speeds. When in operation, the MCAS command would move the horizontal trim stabilizer, pause for five seconds, and then operate again if the high AOA was still sensed. If the AOA sensor was still indicating a high AOA, the aircraft would continue pitching nose-down, overriding the inputs of the pilot.

One of the safety items identified by the FAA was the use of a single AOA sensor, which may result in incorrect activation of the MCAS and subsequent nose-down trim of the horizontal stabilizer. Another safety item identified by the FAA that exacerbated the first safety item was the reset of MCAS by the pilot's use of the electric trim, thus resulting in repetitive MCAS commands and additional movement of the horizontal stabilizer toward a nose-down position. An overall analysis of the original MCAS design is that there were several single-points-of-failure built into the system as a result of cost-cutting measures.

Horizontal Stabilizer

As the name implies, the horizontal stabilizer is a strongly stabilizing influence on longitudinal static stability. It is usually a symmetrical airfoil because it must produce both upward and downward lift, and it may be in a fixed position as designed by the manufacturer to provide the best stability during cruise flight. The contribution of the horizontal stabilizer to the pitch stability of the aircraft can be seen in Figure 12.26. Figure 12.26a shows that if an up-gust causes the aircraft to pitch up, then an upward lift is developed by the horizontal tail. This creates a nose-down moment, which is stabilizing. In Figure 12.26b, the opposite occurs when the aircraft is pitched downward by a down-gust. The horizontal stabilizer produces nose-up moment in this case.

Consider the effect of airspeed on the tail-down force generated by a single-engine propeller aircraft (Figure 12.27). In cruise flight, the tail-down force (nose-up moment) balances the nose-down moment due to the rotation around the CG by the rearward AC and the weight of the engine. Because this is the regime of flight where pilots spend most of their time, aircraft are usually designed to require minimal nose-up or nose-down trim at cruise airspeed. If the aircraft were pitched up around the lateral axis to a nose-high position, airspeed would decrease and so would the downward lift generated by the horizontal tail surfaces. As a result, the nose would fall, the airspeed would increase, and so would the speed of the

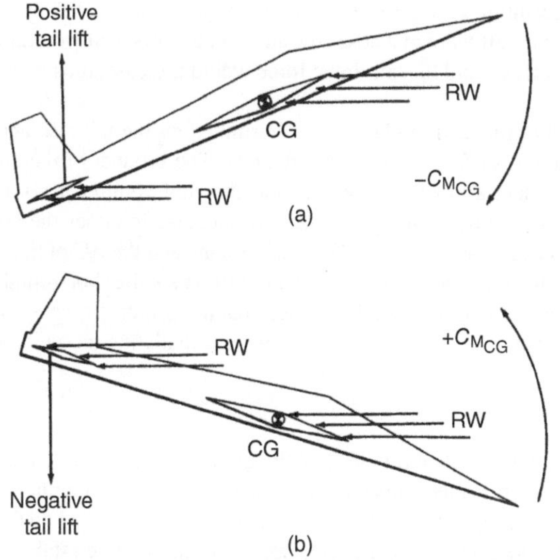

Figure 12.26. Lift of the horizontal stabilizer produces a stabilizing moment. (a) Up-gust causes the aircraft to pitch up, resulting in an upward lift developed by the horizontal tail, creating a nose-down stabilizing moment. (b) The opposite occurs when the aircraft is pitched downward by a down-gust as the horizontal stabilizer produces a stabilizing nose-up moment.

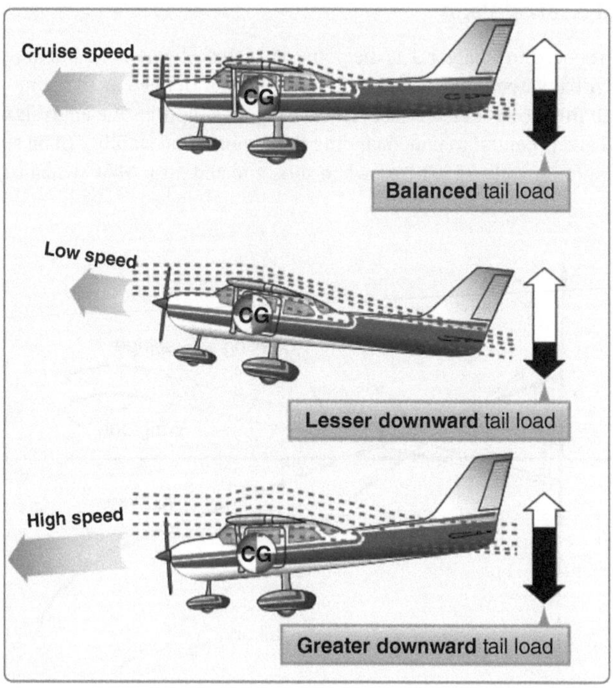

Figure 12.27. Effect of speed on tail-down force.
Source: Pilot's Handbook of Aeronautical Knowledge (2016b)/U.S. Department of Transportation Federal Aviation Administration/Public domain.

air going over the tail, resulting in a nose-up force. Eventually, the aircraft would return to the original trimmed airspeed. If the aircraft were in a nose-low attitude, the airspeed would increase, the speed of the air over the tail would increase, and the tail-down force would increase, thus raising the nose back toward equilibrium flight.

The degree of stability produced by the tail is determined by the tail size and the moment arm from the AC of the horizontal tail surfaces to the aircraft's CG. The product of the area of the horizontal tail surfaces (ft^2) and the moment arm (ft) is referred to as the "tail volume" (ft^3). It is an indication of the stabilizing effectiveness of the horizontal stabilizer. An increase in either the size of the horizontal tail control surfaces or the distance between the CG of the airplane and the AC of the horizontal stabilizer will increase the tail volume and thus the stability of the aircraft. A smaller horizontal tail area located farther from the CG can have the same effect as a larger horizontal tail area closer to the CG. Of course, many other factors in aircraft design must be accounted for when calculating tail volume.

Total Airplane

As previously described, when evaluating the static longitudinal stability of an aircraft, we must consider the contributions of all aircraft components including the wing, fuselage, engines, and horizontal stabilizer (tail). Figure 12.28 shows the typical contributions of each component based on the relationship of AC and CG for each (Figure 12.28 does not consider engine impact on aircraft pitch stability). The wing alone, with AC forward of CG, is a destabilizing force as represented by a positive slope on the graph, as are the wing and fuselage together. The horizontal stabilizer provides enough stability to overcome the negative stability of the wing and fuselage. This is represented by the overall negative slope for the aircraft as a whole.

Static Margin and Neutral Point

The term AC was introduced in Chapter 3 as the point on an airfoil where the pitching moment is constant with changing AOA if the velocity is constant. It can be further defined as the point along the chord line of the airfoil where all lift is considered to act. The use of AC simplifies the analysis of pitching moments about the aircraft CG and is central to understanding the longitudinal stability of an aircraft. The AC of an airfoil is located at approximately 25% MAC when subsonic and 50% MAC when supersonic.

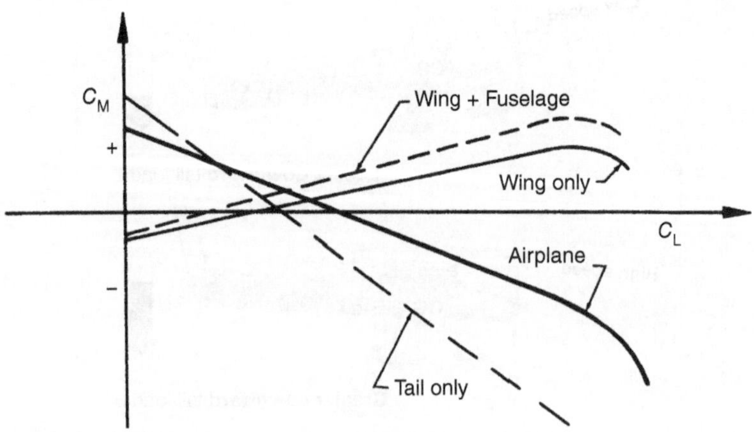

Figure 12.28. Typical buildup of aircraft components.

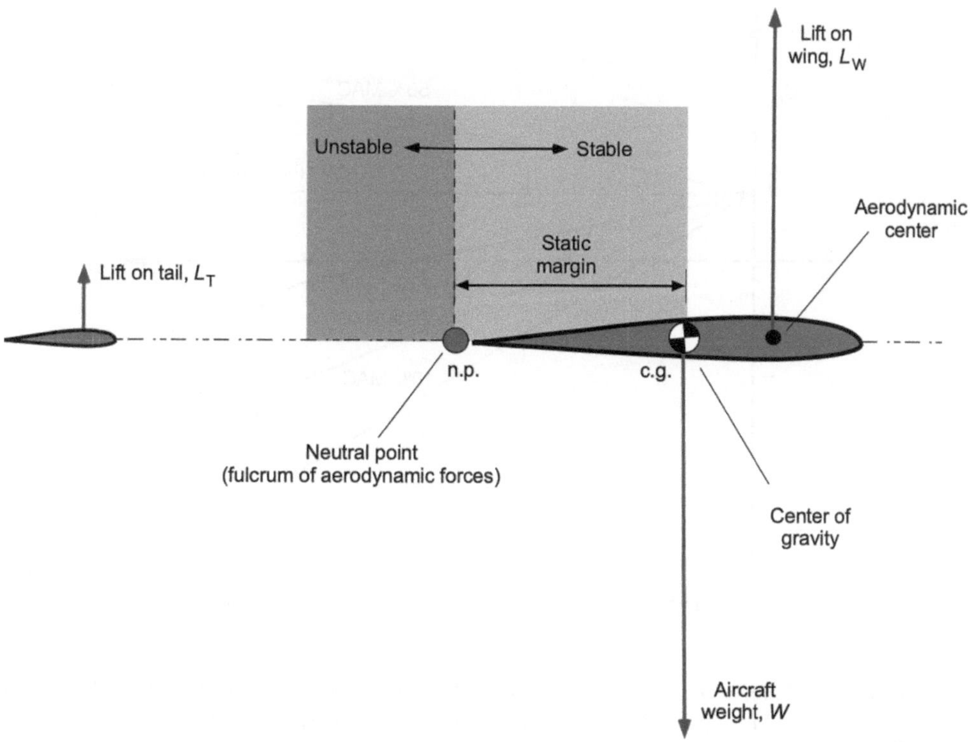

Figure 12.29. Neutral point and static margin.
Source: Introduction to Aerospace Flight Vehicles Section 52 (2023)/reproduced with permission of J. Gordon Leishman.

Aircraft static longitudinal stability must consider the impact of all aircraft components (wings, fuselage, and tail) on the pitching moments. The point on an airplane where all nose-up and nose-down pitching moments are balanced is referred to as the *neutral point* (NP) of the aircraft. The NP may also be considered the AC of the entire aircraft.

The relative location of the NP and CG is critical to understanding an aircraft's longitudinal stability. The term *static margin* refers to the distance between an aircraft's CG and NP. As shown in Figure 12.29, if the CG is forward of the NP, the aircraft has a positive static margin and thus is statically stable. If the CG is aft of the NP, the aircraft is statically unstable.

Most aircraft are designed to have positive static longitudinal stability. However, some military aircraft such as the F-16 are designed to be statically unstable (CG aft of the NP) to provide greater maneuverability. Again, the mission of the aircraft will influence the amount of static margin. Swatton (2011, p. 253) states that conventional transport airplanes have a static margin between 10% and 30% of the MAC ahead of the NP, which is a few feet in length; whereas for small airplanes, the margin ranges between 5% and 15% of the MAC and will only be a few inches.

As discussed in the Wings section, and as shown in Figure 12.30, varying the CG position has a great effect on the longitudinal static stability of an airplane. Positive longitudinal stability is increased by moving the CG forward of the NP; again, the greater the distance the greater the positive stability. Forward positioning of the CG results in increased pitch stability, as was explained earlier and is shown by the

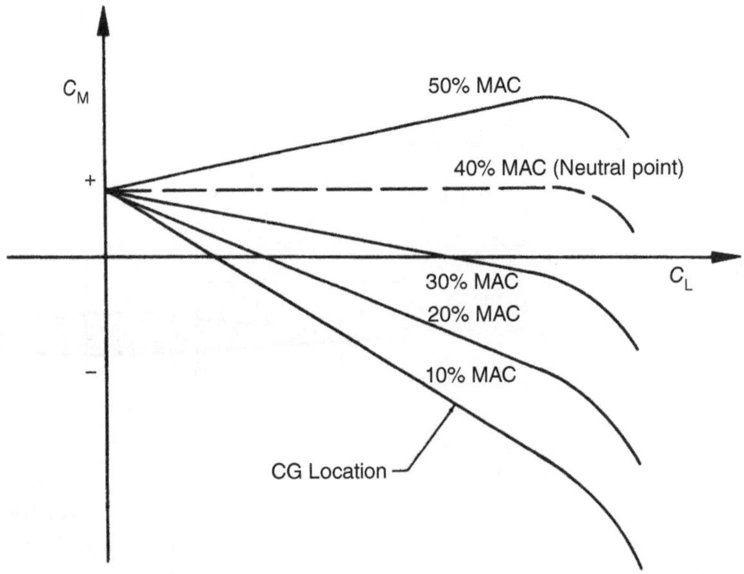

Figure 12.30. Effect of CG location on static longitudinal stability.

slope of the $C_M - C_L$ curves in Figure 12.30. At 40% MAC, the slope is zero, and the aircraft has neutral stability at this point. Moving the CG behind this point results in an unstable condition.

Stick-Fixed Versus Stick-Free Stability

Stick-position stability refers to the amount of elevator deflection needed at various speeds to obtain the desired longitudinal stability. In most aircraft at low speeds, it is preferred that aft movement of the yoke/control stick results in an increase in AOA, while at high speeds forward movement of the yoke/control stick results in decreasing AOA. The terms stick-fixed stability and stick-free stability are used to describe the impact on longitudinal static stability based on free-floating flight controls or non-free floating flight controls. If the elevators (or other control surfaces) are allowed to float free, they will be moved from their neutral positions by outside disturbances, and the total stabilizing surface will be reduced. The stability of an aircraft with "free-floating" control surfaces is known as *stick-free stability*. An aircraft with the controls held in a fixed position by a powered irreversible control system will be more stable, and its stability is referred to as *stick-fixed stability*. If you can manipulate the control surfaces of the aircraft while standing outside the aircraft it is stick-free. Pilots can create a "stick-fixed" condition in flight by preventing rudder movement with their feet and by holding the stick/yoke in one position.

Figure 12.31 shows the $C_M - C_L$ curves for both conditions. The slope of the stick-fixed condition is greater; hence, the stability is greater for the stick-fixed case. The further the yoke/control stick has to be moved to initiate a pitch change, the more stable the airplane. As the CG moves aft, the control stick movement required to produce the same change in airspeed or C_L decreases, and when the stick position movement needed to maintain a change of airspeed is zero, then the CG is at the stick-fixed NP. Unless the airplane is equipped with irreversible powered controls, better stability will be realized if the controls are held in the neutral position, simulating stick-fixed conditions. In an aircraft with free-floating controls

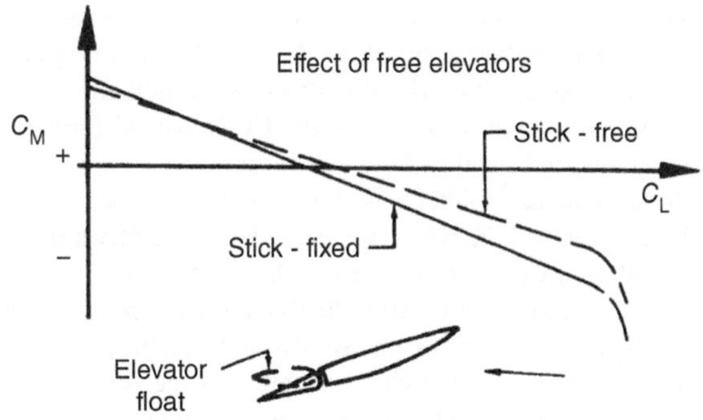

Figure 12.31. Stick free–stick fixed stability.

(stick-free), the NP will be forward of the stick-fixed NP, thus the greater the pilot's ability to simulate stick-fixed conditions, the closer the NPs will be together and improved longitudinal static stability.

Accident Brief: B-1A Static Margin Test

In 1984, a B-1A test aircraft (forerunner to the B-1B) crashed during a flight test mission at Edwards Air Force Base, CA. The mission included testing the aircraft's CG management system for this variable wing sweep aircraft, which ranges from a wing-sweep angle of 15° during full forward sweep, to 67.5° when full aft (Figure 12.32). The aircraft was flying with the wings swept to a full aft position for high-speed flight.

Figure 12.32. B-1A Lancer with T-38 escort.
Source: NASA/Public domain.

The CG position was manually set by the crew to an aft position for the test point to check the proper movement of aircraft fuel to maintain the desired CG. The B-1 has a fuel system that can manually or automatically move the fuel position to adjust the aircraft's CG during flight. The aft wing sweep position resulted in an aft aircraft AC with an aft CG, though the CG remained forward of the AC to ensure a positive static margin for longitudinal stability.

After completion of the aft wing test point, the next test point called for a forward wing sweep position for slow speed flight. The crew moved the wing sweep control to the forward position with the selected CG still set to manually hold an aft CG location. Soon, the aircraft's CG was perilously positioned aft because fuel had not been transferred forward. Also, the B-1 was equipped with a full-stabilizer-authority variable-gain stability control augmentation system (SCAS), and with the pitch SCAS command-path gain set at maximum, the pilots experienced very light stick forces as the AC shifted. As the wings traveled forward, the aircraft AC continuously shifted forward as well. Eventually, the AC moved forward of the CG and the aircraft went into an uncontrolled pitch up as the static margin decreased and became negative. The aircrew did not have enough control authority to arrest the pitch up so the aircraft stalled and departed controlled flight. The crew successfully ejected in the B-1A capsule, though one pilot was killed due to a failure that resulted in an improper capsule angle at ground impact.

This accident illustrates the importance of static margin to aircraft stability. Though the B-1's CG control was normally automatic, with any aircraft (and particularly a variable wing sweep aircraft) it is critical to always understand the relative positions of AC and CG to ensure proper static margin and thus aircraft stability.

DYNAMIC LONGITUDINAL STABILITY

Stability considerations that we have discussed up to now have dealt with static stability. *Dynamic stability* refers to the response of an aircraft to disturbances over a period of time. Dynamic stability exists when the amplitude of these disturbances dampens out with time, a desirable feature in an aircraft.

The reaction of an aircraft to disturbances differs, depending on whether the controls are in the stick-free or the stick-fixed configuration. For our discussion, let us consider the stick-free or reversible-controls aircraft. Most light general aviation aircraft are in this category. Remember, if a control surface is capable of being manually moved from outside the aircraft, it is considered stick-free. Two types of dynamic oscillations are possible, the long-period mode (referred to as the phugoid mode), and the short-period mode. They are summarized as follows:

1. **Long period mode (phugoid):** The long period mode (function of true airspeed (TAS) is characterized by widely varying airspeed, pitch, and altitude of the airplane with poor damping oscillation, but the AOA remains nearly constant (Figure 12.33). During a phugoid oscillation which usually lasts only two minutes or less, the airspeed may vary by ± 5 kts. The motion is so slow that the effects of damping forces are very low, and sometimes the oscillation may not even be noticed. The whole phugoid oscillation can be thought of as a slow interchange between kinetic and potential energy. If an aircraft is trimmed for an AOA and then power is changed, the aircraft will move through a series of pitch, altitude, and airspeed changes to maintain the original trimmed AOA condition.
2. **Short period mode:** The second mode is a short period mode, as shown in Figure 12.34, found in all non-augmented aircraft. The short period mode in the stick-free configuration is caused during an attitude change of the airplane, a first-order response. How the airplane reacts in the second order (oscillatory response) determines the magnitude of the short period. A typical recovery mode may have a period of 0.3–1.5 seconds and has heavy damping of the oscillations, typical in light,

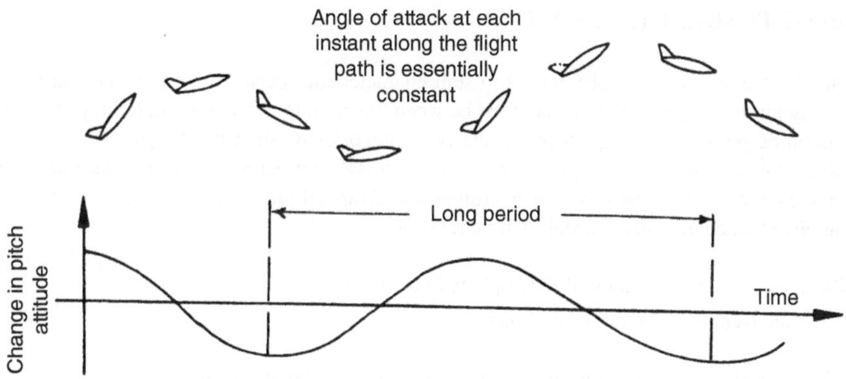

Figure 12.33. Phugoid longitudinal dynamic mode.

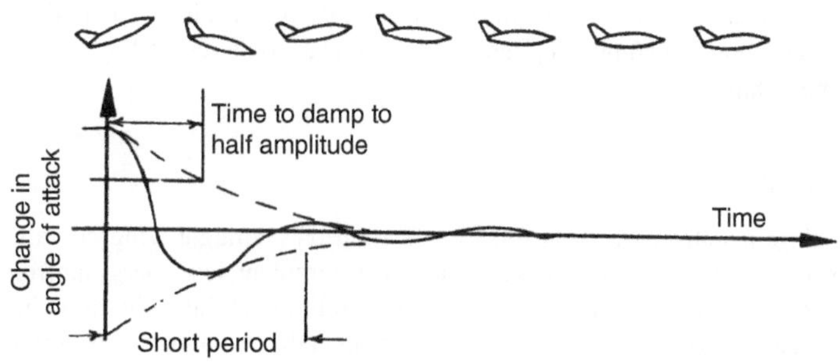

Figure 12.34. Short-period dynamic mode.

general aviation airplanes. Recovery from this type of pitching can be accomplished by releasing the controls or, more rapidly, by holding the controls in their neutral positions. This has the effect of putting the airplane into the stick-fixed configuration. A quick deliberate input by the pilot of forward and aft yoke/control stick motion (referred to as a "doublet") after which the controls are immediately returned to the neutral position can be used to initiate the short period response of an airplane demonstrating longitudinal stability.

Pilots must be careful not to try to dampen out the oscillations. Pilot reaction time is close to the natural period of the oscillations, and inadvertent reinforcement of the pitching moment may result. While this may only result in a rough ride for a relatively slow light airplane, it can prove disastrous for a high-performance jet aircraft. If this reinforcement occurs in a high-performance jet, it is called a *pilot-induced oscillation* (PIO). A PIO is especially dangerous during a transition to landing and has often resulted in aircraft accidents.

A similar dynamic stability problem, called *collective bounce*, exists in helicopter flight for the short period mode. If an up-gust is encountered and the helicopter is forced upward, the pilot is forced downward in the seat. If the pilot has hold of the collective stick, this will also be forced downward, and the AOA of all the rotors will be decreased, causing the helicopter to descend. Again, this results in the pilot being forced upward, the collective stick is raised, and the process is repeated.

PITCHING TENDENCIES IN A STALL

In Chapter 10, we reviewed the differences in stall characteristics between a swept-wing airplane and a straight-wing airplane, specifically the contrast between the buildup of C_L with increasing angle of attack and the defining point of $(L/D)_{max}$ between the two categories of aircraft. Our discussion in regard to longitudinal stability and resultant pitch tendencies in a stall will start with a summary of the stalling characteristics of a T-tail airplane versus a straight-wing/low-tail plane (non T-tail). Two areas of focus regarding pitch tendencies during a stall will be reviewed:

1. The static pitching tendency of the airplane in the stall.
2. The effectiveness of the tail in stall recovery.

The forces in the pitching plane for a T-tailed aircraft are shown in Figure 12.35. Assume that the aircraft is trimmed in straight and level flight and that the aircraft's CG is forward of the AC (as shown). A downward balancing force on the tail is required to maintain equilibrium. In the case of an aft CG location, an upward force on the tail is required. If the speed is reduced, a gentle up-deflection of the elevator is required to maintain altitude, and the downward load on the tail is increased. The static stability of the aircraft causes a nose-down pitching tendency, which has to be resisted by a further-up elevator to keep the nose from dropping.

Low-Tailed Aircraft

Figure 12.36 displays a stall progression for a low-set tail on a straight-wing aircraft. Initially, as the aircraft slows down, the low-set tail becomes engulfed in the turbulent, low-energy air from the wing wake, reducing the efficiency of the tail. Pilots should heed the buffeting at this point as an imminent stall and begin immediate recovery procedures. At the stall, two distinct things happen. First, the airplane responds with a nose-down pitch. Back pressure on the yoke/control stick by the pilot will raise the nose, decreasing the speed even more, but upon release the nose once again pitches down; this demonstrates positive static stability. Second, at the moment of stall, the wing wake passes straight aft and goes above the low-set tail. This leaves the tail in undisturbed, high-energy air, and it now is at a high positive AOA, causing an upward lift on the tail. This lift increases the nose-down pitching tendency and assists with reducing the AOA.

T-Tail Aircraft

Again, assume that an aircraft is trimmed in straight and level flight. As the aircraft is slowed toward a stall, the handling characteristics are much the same as for the low-tailed aircraft, except that the high tail remains

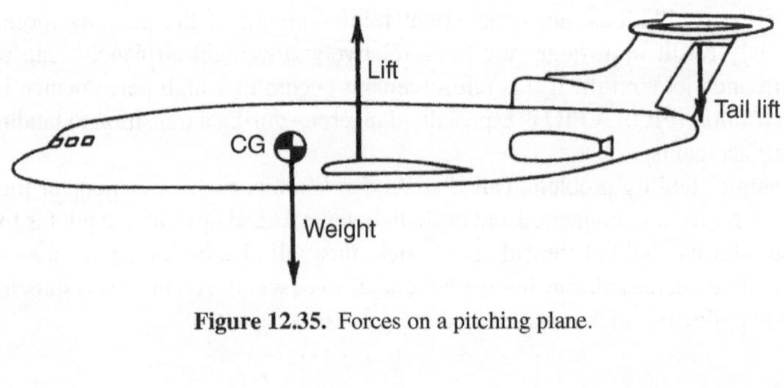

Figure 12.35. Forces on a pitching plane.

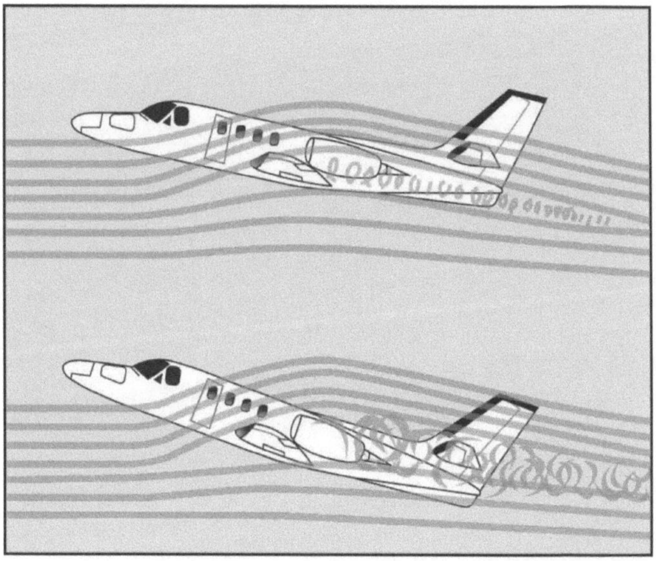

Figure 12.36. Wing wake influences on a low-tail aircraft.
Source: Airplane Flying Handbook (2004a)/U.S. Department of Transportation Federal Aviation Administration/Public domain.

clear of the wing wake and retains its effectiveness, allowing the pilot to continue into a stall without pre-stall warning. Continued speed reduction is, therefore, more efficient than in a low-tailed aircraft.

At the stall, two distinct things happen. First, the swept-wing, high-tail airplane tends to suffer a marked nose-up pitch after the stall. Second, the wing wake, which has now become low-energy turbulent air, passes straight aft and immerses the T-tail, which is now in the right position to catch it. This greatly reduces the tail's effectiveness and makes it incapable of combating the nose-up pitch and so the airplane continues to pitch up. The great reduction in lift and the increase in drag cause a rapidly increasing descent path. Thus, the angle of attack is increased and the pitch-up problem is worsened. The airplane is well on its way to extreme angles of attack and deep stall, as is shown in Figure 12.37.

Once in a deep stall, recovery may be difficult to impossible. As reviewed in Chapter 10, some airplanes have stick shakers/stick pushers installed in the aircraft to assist with stall prevention and stall recovery. Once the airplane begins to descend, if a constant pitch attitude is maintained, the AOA increases rapidly as the flightpath becomes deflected downward, and the stall deepens. Several thousand feet of altitude loss are possible; pitch controls AOA and airspeed, and throttle controls altitude. Initial response and timely recovery procedures are key to recovering from a deep stall.

Explanation of Nose-Up Pitch Following Stall in Swept-Wing Aircraft

Three features affect the pitching tendency:

1. **Wing section characteristics:** The section characteristics of most wings are such as to cause a nose-down pitch, providing that the whole wing could be stalled at the same instant (spanwise). The pressure distribution is rather flat in normal flight, as shown in Figure 12.38. As the stall is approached, the pressure distribution changes to an increasingly leading-edge peaked pattern

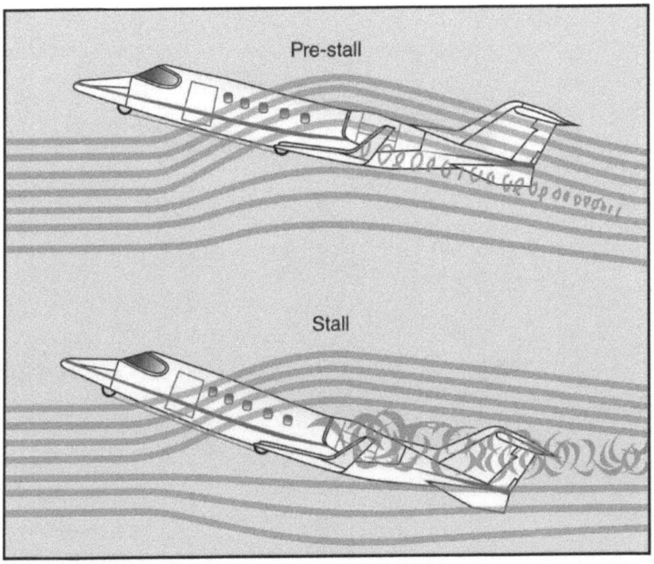

Figure 12.37. Wing wake influences on a sweptwing T-tail aircraft.
Source: Airplane Flying Handbook (2004a)/U.S. Department of Transportation Federal Aviation Administration/Public domain.

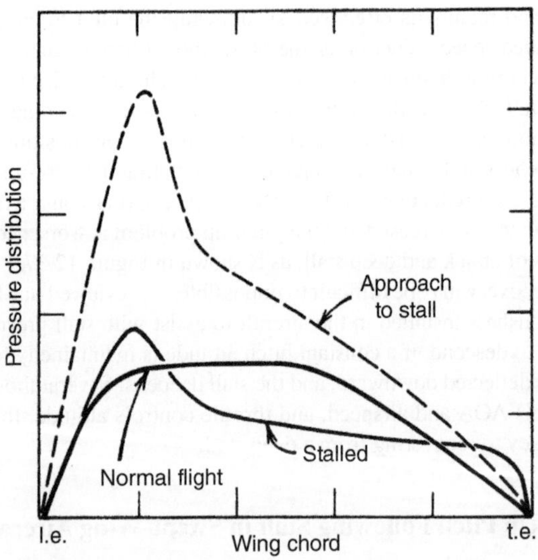

Figure 12.38. Change in pressure distribution at stall.

because of the enormous low-pressure area developed by the nose profile. At the stall this peak collapses, the pressure distribution pattern changes and the result is usually a nose-down pitch.

2. **Wing sweep:** In practice, the whole wing does not stall at the same instant. In Chapter 10, we discussed the fact that swept and tapered wings tend to stall at the tips first because of the high wing loading at the tips (Figure 12.39). This of course is a problem that affects roll control should

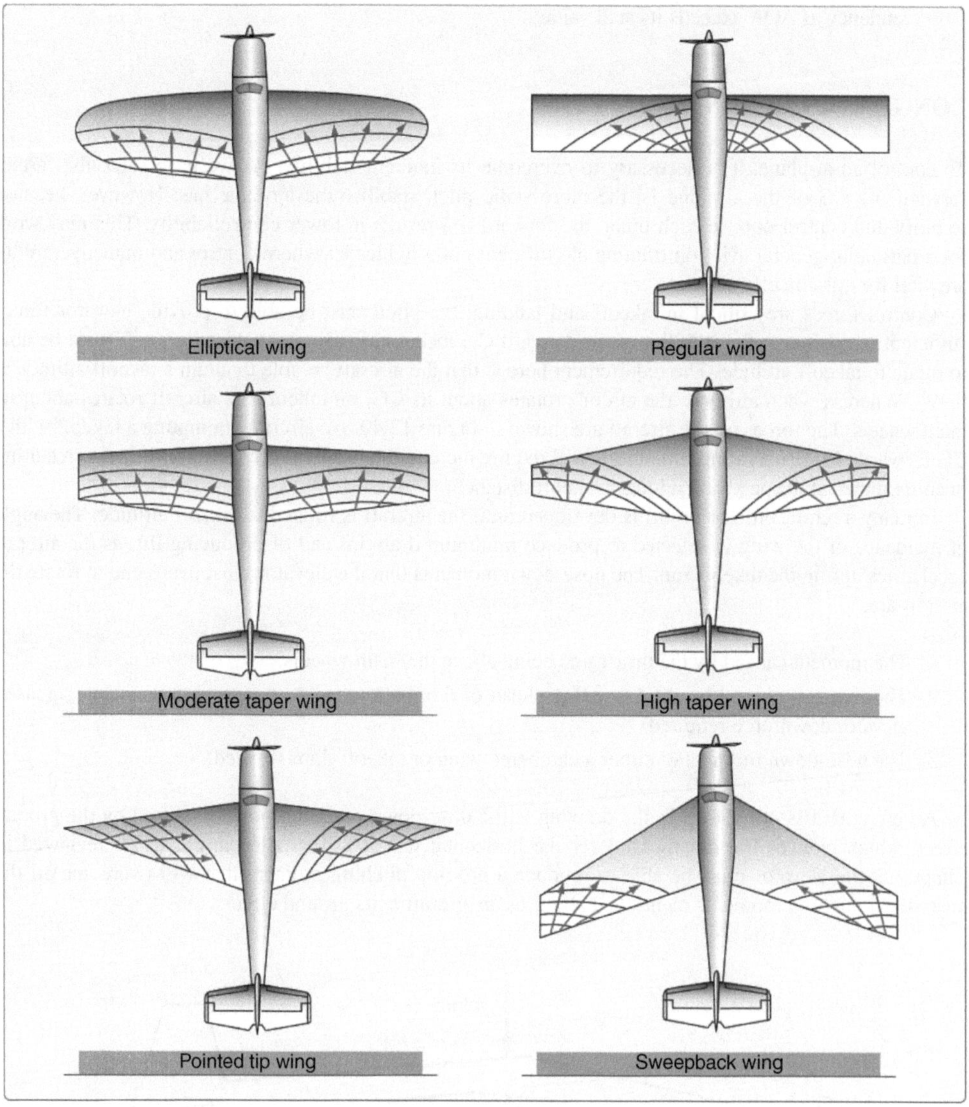

Figure 12.39. Sweptwing stall characteristics.
Source: Pilot's Handbook of Aeronautical Knowledge (2016b)/U.S. Department of Transportation Federal Aviation Administration/Public domain.

the ailerons lose their effectiveness. The boundary layer outflow also resulting from sweep slows the airflow, reduces the lift near the tips, and worsens the situation. This loss of lift outboard (and therefore aft) causes the CP to move forward, which augments the pitch-up tendency as shown in Figure 12.39.

3. **The fuselage:** On a modern aircraft, the forward fuselage overhangs the wing to a greater extent than on the older straight-wing models. As we have seen, the fuselage is a destabilizing factor and an even greater one with the longer overhang. Thus, the fuselage contributes to the nose-up pitching tendency as AOA exceeds its stall value.

LONGITUDINAL CONTROL

To control an airplane, it is necessary to overcome its inherent stability. As we have seen, the farther forward the CG of the airplane is, the more static pitch stability the airplane has. However, because stability and control oppose each other, the forward CG results in lower controllability. This may work for a particular general aviation training aircraft, but not a fighter jet where control and maneuverability are vital for mission effectiveness.

Control forces are critical in takeoff and landing: the pilot must be able to provide elevator forces sufficient to overcome the stability of the forward CG location. During takeoff, an aircraft must be able to rotate to takeoff attitude. The requirement here is that the aircraft be able to attain a takeoff attitude at $0.9V_s$. Whereas when airborne the aircraft rotates about its CG, on takeoff the aircraft rotates about its main wheels. The forces on the aircraft are shown in Figure 12.40. An aircraft attempting a takeoff with a CG forward of safe operating limitations will require increased tail-down force to raise the nose, resulting in an increased stall speed and a longer takeoff distance.

In many aircraft, little or no lift is developed until the aircraft is rotated to takeoff attitude. The angle of incidence of the wing is selected to produce minimum drag, instead of producing lift, as the aircraft accelerates during the takeoff run. The nose-down moments that the elevator must overcome to rotate the aircraft are:

1. The moment caused by the thrust line being above the main wheels.
2. The moment caused by the CG being ahead of the main wheels (increased moment = increased elevator downforce required).
3. The nose-down moment of either a cambered wing or takeoff flaps (if used).

An effect that is not shown in the drawing is the nose-down pitching moment caused by the ground effect, which reduces the downwash over the horizontal tail of a low-tailed airplane. As reviewed in Chapter 5, the elevator must be able to produce a nose-up pitching moment that will overcome all the nose-down pitching moments mentioned above as an aircraft exits ground effect.

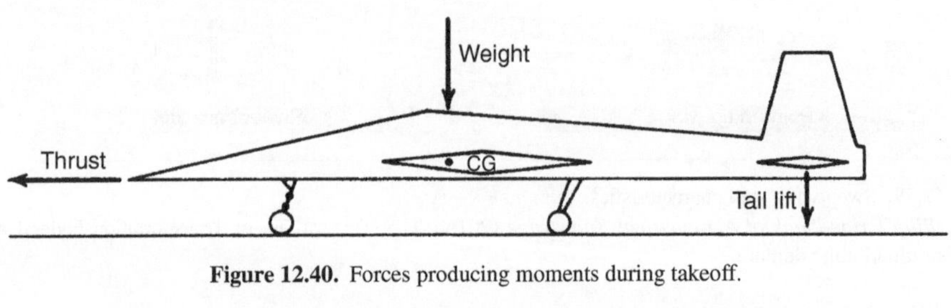

Figure 12.40. Forces producing moments during takeoff.

Landing requirements include flaring the aircraft prior to touchdown and overcoming the nose-down moment caused by the airplane entering ground effect. If the airplane has enough elevator authority for takeoff, it will usually have enough elevator control for landings.

SYMBOLS

$C_{M(CG)}$	Pitching moment coefficient (dimensionless)
L'	Rolling moment (ft-lb)
L_β	Roll moment due to sideslip
L_p	Roll moment due to roll rate
L_r	Roll moment due to yaw rate
M_α	Pitch moment due to AOA
M_{CG}	Pitching moment about CG
M_q	Pitching moment due to pitch rate
N	Yawing moment
N_β	Yaw moment due to sideslip
N_r	Yaw moment due to yaw rate
PIO	Pilot-induced oscillation
X	Longitudinal axis of aircraft
Y	Lateral axis of aircraft
Z	Vertical axis of aircraft

EQUATIONS

12.1 $W \times A = M$ (lb - in.)

12.2 $\text{CG (in.)} = \frac{\text{Total Moment (lb-in.)}}{\text{Total Weight (lb)}}$

12.3 $M_{CG} = C_{M(CG)} qSc$

12.4 $C_{M(CG)} = \frac{M_{CG}}{qSc}$

KEY TERMS

Arm
Controllability
Datum
Doublet
Dynamic stability
Leading edge mean aerodynamic chord (LEMAC)
Longitudinal stability
Maneuverability
Mean aerodynamic chord (MAC)
Moment
Neutral point (NP)
Phugoid
Trailing edge mean aerodynamic chord (TEMAC)
Short period

Static margin
Static stability
Stick-fixed
Stick-free

PROBLEMS

Note: Answers to the problems in the first two sections are given at the end of the book.

Application of Chapter Theory

1. The more stability an airplane has, the easier it is to control.
   a. True
   b. False

2. Static stability of an airplane is:
   a. the ability to return to its equilibrium position once it has been disturbed.
   b. the long-time reaction to a disturbance.
   c. the initial tendency to move toward the equilibrium position after a disturbance.
   d. None of the above.

3. Dynamic stability of an airplane is _____ to a disturbance.
   a. the immediate reaction
   b. the long-time reaction
   c. oscillatory in nature.
   d. Both (b) and (c).

4. A damped oscillation is one that:
   a. shows positive static stability.
   b. shows positive dynamic stability.
   c. Both (a) and (b).
   d. Neither (a) nor (b).

5. Most light, single-engine training airplanes are designed with the following stability characteristics:
   a. Positive static, positive dynamic.
   b. Positive static, negative dynamic.
   c. Negative static, negative dynamic.
   d. Negative static, positive dynamic.

6. Weight and balance of airplanes is important because:
   a. CG affects the stability and control of the aircraft.
   b. if the CG is moved forward, the static pitch stability is increased.
   c. if the CG is too far forward, the plane may not respond to elevator commands.
   d. All the above.

7. Which of the following conditions is not associated with an aft CG:
   a. Higher cruise speed.
   b. Nose-heavy tendency in the landing flare.
   c. Lower stall speed.
   d. Lighter flight control forces.

8. Longitudinal stability refers to the behavior of an airplane in pitch, which involves rotation around the ____________ axis.
   a. vertical
   b. longitudinal
   c. lateral

9. The farther back the wings are located, the more static pitch stability an airplane has.
   a. True
   b. False

10. Placing the jet engines at the rear of an airplane:
    a. Contributes to a stabilizing moment for pitch stability.
    b. Contributes to a destabilizing moment for pitch stability.
    c. Does not create a stabilizing or destabilizing moment regarding pitch stability.

11. Phugoid oscillation can best be described as an example of static stability.
    a. True
    b. False

12. Phugoid oscillation is dangerous because a pilot reaction can lead to PIO.
    a. True
    b. False

13. In a phugoid oscillation:
    a. oscillations are damped in one to three seconds.
    b. the aircraft is considered unstable.
    c. damping is weak.
    d. All the above.

14. The greater the static margin of an airplane, the greater the impact on longitudinal stability.
    a. True
    b. False

15. The most critical condition for pitch control is that the elevators must be able to
    a. rotate the airplane for takeoff.
    b. flare the airplane for landing.
    c. overcome a forward CG location.
    d. All the above.

16. Consider a small, single-engine general aviation airplane with two rows of seating and an aft baggage compartment. If a 50 lb bag were moved from the aft baggage compartment to the front seat, what overall effect would the movement have on aircraft weight and balance?
    a. total CG moves forward, total weight decreases.
    b. total CG moves aft, total weight remains the same.
    c. total CG moves forward, total weight remains the same.
    d. total CG moves aft, total weight increases.

Calculations Using Chapter Principles

17. Calculate the moment of a rear seat passenger in an aircraft with the following given information:

    Passenger weight = 182 lb

    Arm (rear seat) = 121.0″

18. Calculate the CG of an aircraft with the following given information:

Total weight (including aircraft) = 3 270 lb

Total moment = 277 680 lb-in.

19. Calculate the total moment of an aircraft with the following given information:

Aircraft basic empty weight (Arm 78.0) = 2 125 lb

Front seat occupant weight (Arm 83.0) = 320 lb

Rear seat occupant weight (Arm 118.0) = 250 lb

Fuel weight (Arm 77.0) = 210 lb

Baggage Area 1 (Arm 148.0) = 50 lb

20. Calculate the total CG for the aircraft in Problem 19.

Correlation Through Scenario-Based Learning

Aircraft – *Diamond DA50RG*

1. Review the *Mass (Weight)* section within Section 2 of DA50 aircraft flight manual (AFM), and find the following:
   - **a.** Maximum takeoff mass.
   - **b.** Does the note regarding the *beginning of flight* apply to aircraft operating under FAA regulations? If so, what is the maximum ramp weight?
   - **c.** Maximum zero fuel mass. How is this defined in the AFM?
   - **d.** The maximum landing weight compared to the maximum takeoff weight.
   - **e.** The baggage area weight restrictions.
2. Review the *Center of Gravity* (CG) section within Section 2 of DA50 AFM, and find the following:
   - **a.** The location of the datum plane on the DA50.
   - **b.** Maximum takeoff weight change as the CG is shifted toward the datum plane
   - **c.** Locate and review the placards specifically for baggage areas A-D. What are the weight restrictions for each area? What is the difference in the "position" of baggage areas A & B versus C & D?
3. Review the *Mass and Balance* (Section 6) within the DA50 AFM, and find the following:
   - **a.** Review Figure 1 and the datum plane description.
   - **b.** Where is the actual empty weight of each individual DA50 located in the aircraft?
   - **c.** Aircraft operating fluids are included in the empty mass.
   - **d.** Weight and volume of fluids that are included in the empty mass.
   - **e.** The type of fuel is used to operate the DA50. What is the weight of each gallon of fuel?
   - **f.** Unusable fuel for both the left and right tanks.
   - **g.** Usable fuel capacity in gallons and weight (lb.) for the DA50.
   - **h.** Locate and review the *Moment Arms* table.
   - **i.** Locate and review the *Example Calculation of Loading Condition* table.
4. Find the *Calculation of Loading Condition (imperial)* table in Section 6, and use the provided information to calculate the total mass (weight) and moment. Then, use the Permissible Center of Gravity Range figure to confirm the aircraft is within balance. Based on the information given below, is the aircraft in legal weight and balance? If not, what are the options to correct either weight and/or balance?

Component	Weight (lb)	Moment (lb-in.)
A/C empty weight	3 352.25	315 495.47
Front seats	405	
Rear pax seats	0	
Baggage area A	10	
Baggage area B	50	
Baggage area C	10	
Baggage area D	0	
Fuel	Full fuel	

5. Find the *Calculation of Loading Condition (imperial)* table in Section 6, and use the provided information to calculate the total mass (weight) and moment. Then, use the Permissible Center of Gravity Range figure to confirm the aircraft is within balance. Based on the information given below, is the aircraft in legal weight and balance? If not, what are the options to correct either weight and/or balance?

Component	Weight (lb)	Moment (lb-in.)
A/C empty weight	3 352.25	315 495.47
Front seats	200	
Rear pax seats	324	
Baggage area A	50	
Baggage area B	77	
Baggage area C	20	
Baggage area D	0	
Fuel	Full fuel	

6. Research the FAA's *Weight and Balance Handbook FAA-H-8083-1B* to find the weight shift formula. What are the consequences to the safety of flight if an aircraft is loaded forward or aft of the acceptable CG range?

7. Does the DA50 have a conventional tail, V-tail, or a T-tail design?

8. Research *AC91-74B Pilot Guide: Flight in Icing Conditions*, specifically the sections on tailplane icing. What are the dangers involved with tailplane icing related to longitudinal stability? How is the flap position involved with the dangers of a tailplane stall?

9. Research and review the NTSB accident report for Simmons Airlines, d.b.a. American Eagle Flight 4184 regarding the accident flight on 31 October 1994. Review the probable cause statement, including the contributing factors and associated findings and analysis.

13 Directional and Lateral Stability

CHAPTER OBJECTIVES

After completing this chapter, you should be able to:

- Identify three types of static directional stability and provide examples of each.
- Describe the contribution of aircraft components to static yaw stability.
- Identify and describe six critical aspects of flight where the rudder is instrumental to maintaining directional control.
- Summarize multi-engine flight principles, define performance items related to multi-engine flight on one engine, and explain the importance of directional control during one-engine inoperative flight.
- Define static lateral stability and illustrate its importance concerning the safety of flight.
- Describe the contribution of aircraft components to roll stability.
- Identify the types of motion that can result from the interaction of yaw and roll.

In this chapter, we will discuss directional and lateral aircraft stability. Directional stability refers to the behavior of the airplane in yaw while lateral stability refers to the behavior of the aircraft in roll. We will examine these aircraft motions individually with regard to static and dynamic stability as was done in Chapter 12. However, unlike longitudinal motion that is restricted to a single axis (pitch), directional and lateral motion appear in two axes (yaw and roll). These two motions are often coupled together, resulting in complex aircraft motions such as Dutch roll. These coupled effects are discussed in the section Types of Motion Resulting from Coupled Effects at the end of this chapter.

STATIC DIRECTIONAL STABILITY

Directional stability and control refers to the behavior of the airplane in yaw, that is, movements of the longitudinal axis when it is rotated about the vertical axis. The area of the vertical stabilizer and the position

Flight Theory and Aerodynamics: A Practical Guide for Operational Safety, Fifth Edition.
Brian A. Johnson and Philip R. Fittante.

Companion website: www.wiley.com/go/flighttheoryandaerodynamics5e

of the center of gravity (CG) relative to the vertical stabilizer are important considerations when evaluating directional stability.

Yawing moments ($C_{N\beta}$ – coefficient of yaw moment due to sideslip) result in rotation around the vertical axis as illustrated in Figure 13.1. By convention, a positive yawing moment is to the right of the aircraft's initial heading and a negative yawing moment is to the left of the aircraft's initial heading. In the pure yawing case, there is no pitching or rolling of the aircraft.

When the longitudinal axis is not aligned with the relative wind, the aircraft is said to be in a sideslip. The *sideslip angle,* β (beta), is the angle between the relative wind and the airplane's longitudinal axis. When the relative wind is from the right, the sideslip is positive in value. Figure 13.2 shows a positive sideslip angle.

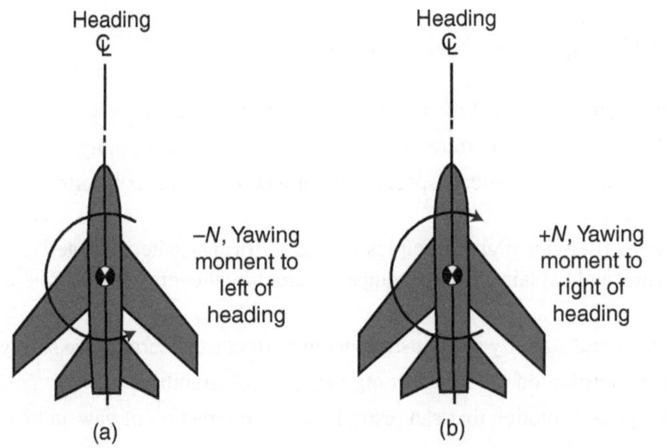

Figure 13.1. (a) Negative yawing moment and (b) positive yawing moment.

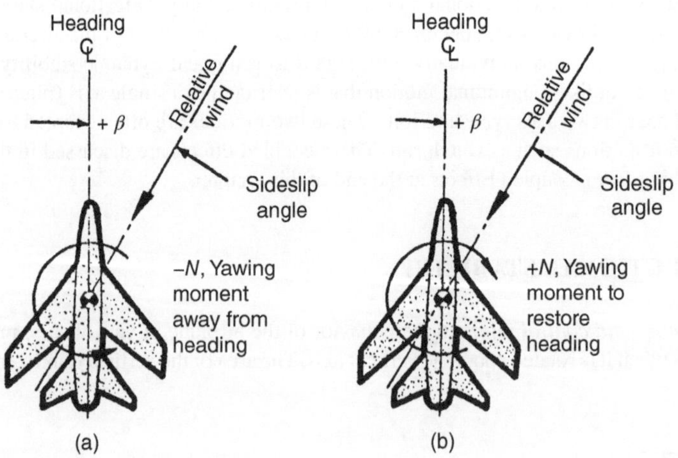

Figure 13.2. (a) Unstable and (b) stable in yaw.

Directional stability is considered to be positive, negative, or neutral:

1. **Positive directional stability:** An airplane is said to have positive directional stability if when trimmed for non-sideslip flight and then subjected to a sideslip condition, it reacts by turning into the new relative wind so that the sideslip angle is reduced to zero. This is the preferred design and flight characteristic of an aircraft.
2. **Negative directional stability:** If the airplane reacts to the sideslip by turning away from the new relative wind, thus increasing the sideslip angle, the airplane has negative directional stability.
3. **Neutral directional stability:** If the airplane continues in the sideslip, maintaining the same angle to the relative wind, the airplane is said to have neutral directional stability.

Figure 13.2a shows an airplane with negative static directional stability, while Figure 13.2b shows an airplane with positive static directional stability. With positive static directional stability, the vertical tail provides a "weathervane" effect ($C_{N\beta}$) that turns the airplane into the relative wind. With negative static directional stability, an airplane tends to turn away from the relative wind. This leads to sideward flight that is often called "swapping ends," a clearly unacceptable condition.

The Yawing Moment Equation

The yawing moment about the aircraft CG is

$$N_{\text{CG}} = C_{N_{\text{CG}}} qSb \tag{13.1}$$

where

N_{CG} = yawing moment about CG (ft-lb)

$C_{N_{\text{CG}}}$ = coefficient of yawing moment about CG

q = dynamic pressure (psf)

S = wing area (ft^2)

b = wing span (ft).

Rearranging Eq. 13.1 gives

$$C_{N_{\text{CG}}} = \frac{N_{\text{CG}}}{qSb} \tag{13.2}$$

Because the values of q, S, and b are always positive, it follows that for a nose-right (+) yawing moment, the value of the yawing coefficient must also be positive. Similarly, nose-left (−) yawing moments require the value of the yawing coefficient to be negative.

Graphic Representation of Static Directional Stability

A plot of the variation of $C_{N_{\text{CG}}}$ at different sideslip angles for an aircraft with positive static directional stability is shown in Figure 13.3. It has a positive slope, which is exactly opposite to the slope for static pitch stability (see Figure 12.14).

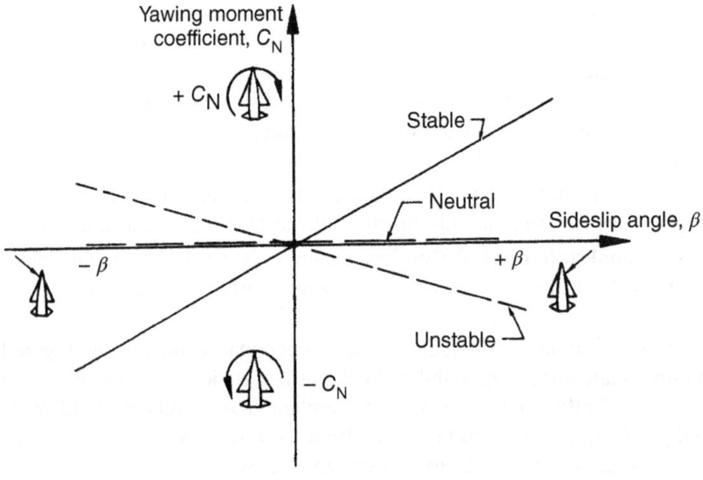

Figure 13.3. Static directional stability.

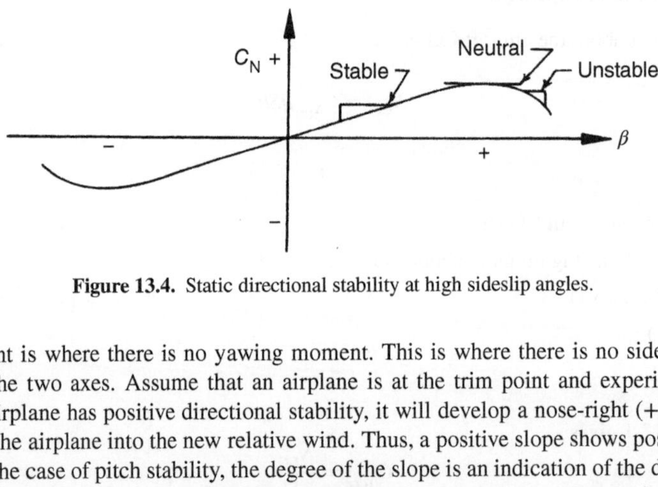

Figure 13.4. Static directional stability at high sideslip angles.

The trim point is where there is no yawing moment. This is where there is no sideslip angle, at the intersection of the two axes. Assume that an airplane is at the trim point and experiences a right (+) sideslip. If the airplane has positive directional stability, it will develop a nose-right (+) yaw coefficient, which will yaw the airplane into the new relative wind. Thus, a positive slope shows positive stability.

Again, as in the case of pitch stability, the degree of the slope is an indication of the degree of stability. The steeper slope means increased stability (or instability). It is not unusual for an airplane to be stable at small sideslip angles but unstable at high sideslip angles; this is shown in Figure 13.4. The unstable condition must not exist at angles of sideslip encountered in ordinary flight conditions.

Contribution of Aircraft Components to Yaw Stability

Wings

The wing contribution to positive static directional stability is small but increases with the amount of sweepback (leading edge slopes backward). In Figure 13.5, the sideslip is to the right, placing more of the leading edge of the right wing into the relative wind; hence the right wing now produces more drag than the left wing, resulting in a stabilizing influence as the aircraft is brought back to its original path. The right wing also produces more lift, but this is a roll factor, which is discussed in the section on Lateral Control.

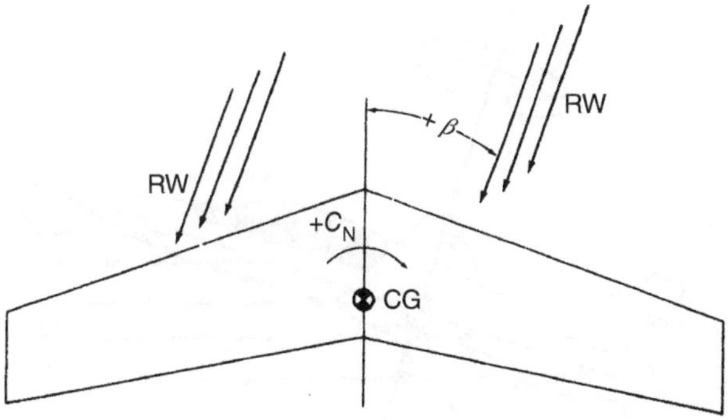

Figure 13.5. Effect of wing sweepback on directional stability.

Engine Nacelles

The contribution of the propeller or jet engine inlets to the directional stability of an aircraft depends on their fore and aft location with respect to the aircraft's CG. If the aircraft is sideslipped, the relative wind must be changed to a direction parallel to the axis of the engine. This is similar to the discussion of pitch stability (see Figure 12.23). A sideward force is developed when the direction of the airflow is changed, and this will be a destabilizing factor if the propeller or engine inlet is forward of the CG. Aft engines (e.g. pusher props) are stabilizing in both pitch and yaw.

Fuselage

In our discussion of pitch stability, we discovered that the fuselage contribution was a destabilizing factor. A similar effect occurs with sideslip. The center of pressure is located near the quarter length of the fuselage, which is ahead of the CG and is destabilizing.

Another factor that contributes to the directional instability of the fuselage is that if the side area of the fuselage forward of the CG is greater than that behind it, the relative wind hitting this area creates a destabilizing yawing moment. Figure 13.6 shows the fuselage effect.

Vertical Tail

The vertical stabilizer is a strongly stabilizing factor, like the feathers on an arrow. It is located behind the CG, and a sideslip creates a horizontal tail force on the aerodynamic shape of the tail. This produces a stabilizing yawing moment about the CG, as shown in Figure 13.7. If a gust moves the nose of the airplane to the right, the relative wind increases on the left side of the vertical fin (and fuselage), and tail side force moves the tail back to the right around the CG, which brings the nose back in line with the original flight path. Note that the distance from the CG to the vertical stabilizer is larger than the distance from the CG to the nose of the airplane, this creates positive static directional stability.

Tail stability can be enhanced by the addition of a dorsal fin (extension of the tail), which does not increase the parasite drag as much as increasing the size of the vertical tail would. Another advantage of the dorsal fin is that it delays the tendency of the vertical tail to stall at high sideslip angles. Figure 13.8 shows both the reduction of parasite drag (sketches) and the increased yaw stability (graph) resulting from the addition of a dorsal fin.

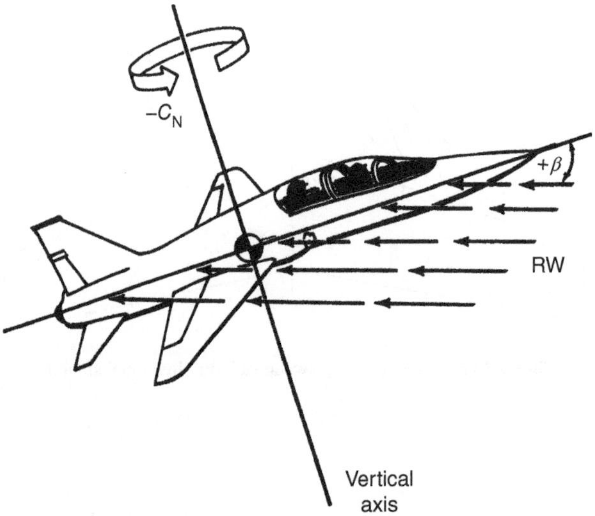

Figure 13.6. Directional instability of fuselage.

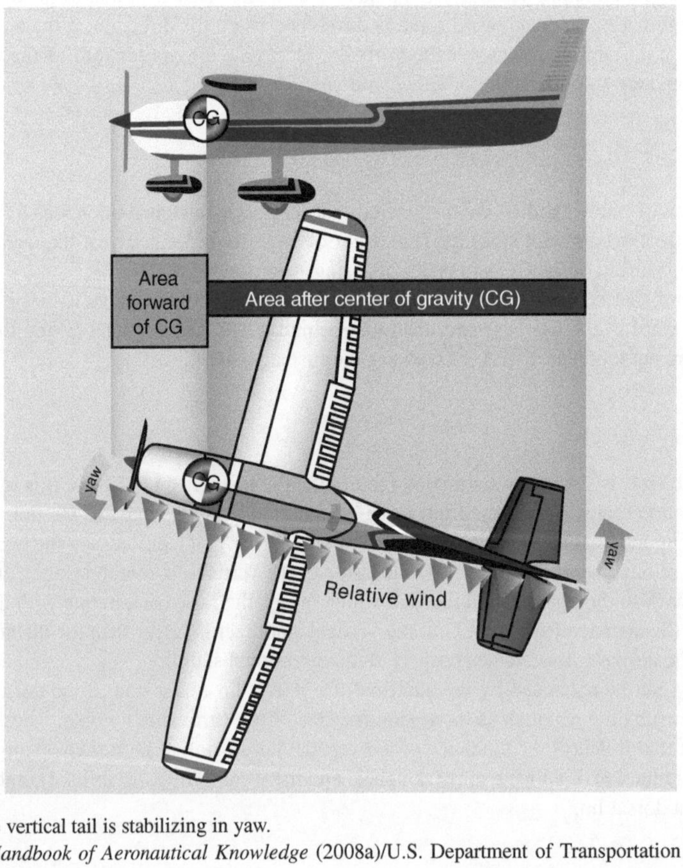

Figure 13.7. The vertical tail is stabilizing in yaw.
Source: Pilot's Handbook of Aeronautical Knowledge (2008a)/U.S. Department of Transportation Federal Aviation Administration/Public domain.

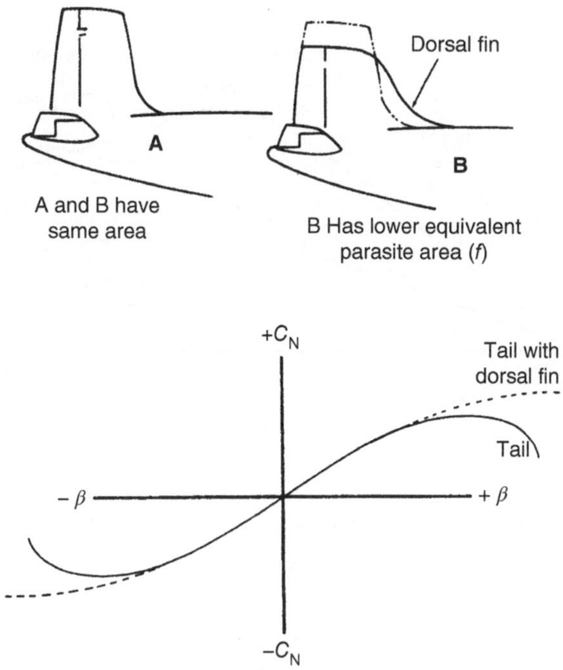

Figure 13.8. The dorsal fin decreases drag and increases stability.

Some aircraft incorporate a *ventral fin* or *strake* to aid in directional stability or decrease lateral stability in airplanes that have too much lateral stability (some T-tail airplanes). These elongated fins are most commonly found along the empennage and by definition on the "belly" of the airplane. Some strakes are found on jet engine nacelles to aid in energizing the air going over the top of the wing. Strakes located on the rear of the empennage improve directional stability during takeoff, improve stability and handling at low airspeeds and high angles of attack, reduce yaw, and in some multi-engine aircraft reduce V_{MC}.

Total Airplane

Figure 13.9 shows a typical buildup of the fuselage, vertical tail, and dorsal fin as they affect static directional stability. Wing and engine effects are not shown. The wing effect is small, and the engine effect depends on location, as discussed earlier in this section.

Rudder-Fixed–Rudder-Free Stability

In Chapter 12, the concept of stick-fixed–stick-free stability was introduced to demonstrate the increased stability when the elevator was fixed in place and not allowed to float by either the pilot holding the rudder in place or by a powered irreversible control system. In the same way, fixing the rudder in the neutral position will prevent "rudder float" and will effectively increase the vertical tail area and thus increase the directional static stability. For aircraft with conventional, reversible controls, increased directional stability will result if the pilot keeps both feet on the pedals and holds the rudder in the neutral position, as shown in Figure 13.10.

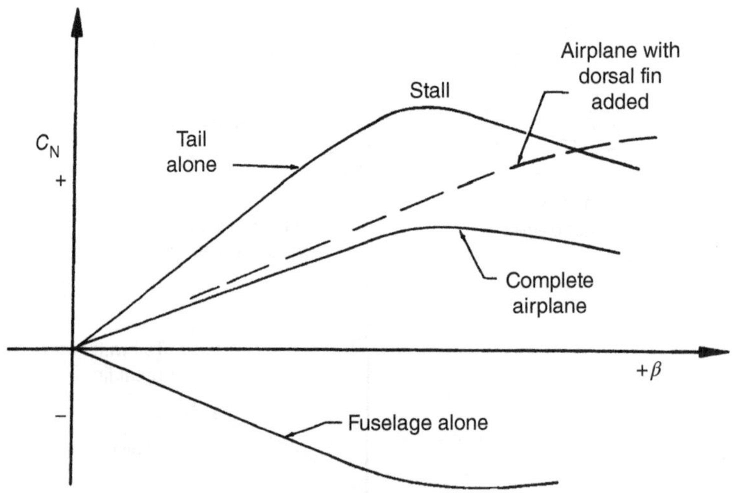

Figure 13.9. Typical buildup of component effects on static directional stability.

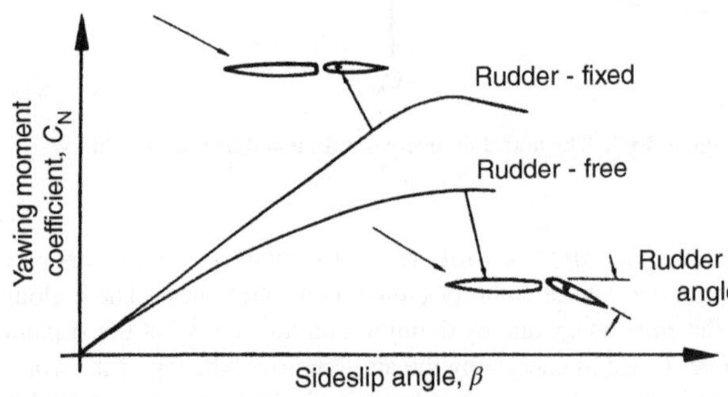

Figure 13.10. Rudder-fixed–rudder-free yaw stability.

Effect of High Angle of Attack

If the vertical tail is engulfed in stalled air from the wings at high angles of attack, it will not be effective in developing sideward forces, and static directional stability will deteriorate (Figure 13.11). The decay in stability will have a strong adverse effect on the ability to recover from spins and unusual attitudes.

Yaw Damper

A yaw damper in its most basic form is an augmentation system consisting of a servo that adjusts rudder movement in response to inputs from a gyroscope or accelerometer to minimize undesired yaw around the vertical axis. Yaw dampers are also utilized on swept-wing airplanes to mitigate Dutch roll characteristics and will be further discussed in the final section of this chapter, Types of Motion Resulting From Coupled Effects. During a coordinated turn in an airplane with a yaw damper installed, the pilot will not need to add rudder input due to the automatic nature of the yaw damping system, alleviating the problems as introduced

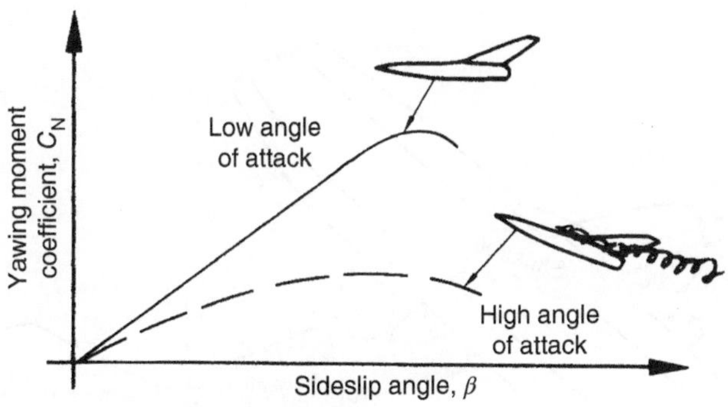

Figure 13.11. Loss of directional stability at high AOA.

in the slip/skid discussion earlier in this textbook. In addition to assisting in turns and mitigating adverse yaw, a yaw damper system also increases stability in turbulence. A NASA (1970) study found that a yaw damper significantly reduced the short-period lateral-directional motions in turbulent air, resulting in a more comfortable ride for passengers and improved flying qualities of the airplane. Yaw dampers have also been found to increase the time available to apply corrective action due to asymmetrical thrust following an engine failure in multi-engine airplanes. Most airplanes with a yaw damper installed incorporate the ability to manually control the activation of the yaw damper, and when the autopilot is engaged the yaw damper will automatically activate. On some newer aircraft, such as some versions of the Cirrus SR22, the yaw damper engages automatically after takeoff and disengages before landing.

DIRECTIONAL CONTROL

Directional control forces exerted by the rudder are critical in six aspects of flight. The type of airplane and its mission will determine which of these is the most important.

1. Spin recovery (discussed in Chapter 10).
2. Adverse yaw (discussed in the Adverse Yaw section of this chapter).
3. Slipstream rotation.
4. Crosswind takeoff and landing.
5. Asymmetrical loading/thrust.
6. Intentional slips.

Slipstream Rotation

Slipstream rotation is predominant in single-engine, propeller aircraft. Its effect is most pronounced at high-power and low airspeed, a combination always encountered during takeoff and landing, as well as in approaches to power-on stalls. The slipstream from the propeller rotates about the fuselage of the aircraft as shown in Figure 13.12. If it strikes the left side of the vertical stabilizer, it will cause a nose-left yawing moment around the vertical axis that must be overcome with right rudder to maintain directional control. As the airplane increases speed, the effect of the spiraling slipstream decreases as the spiral elongates.

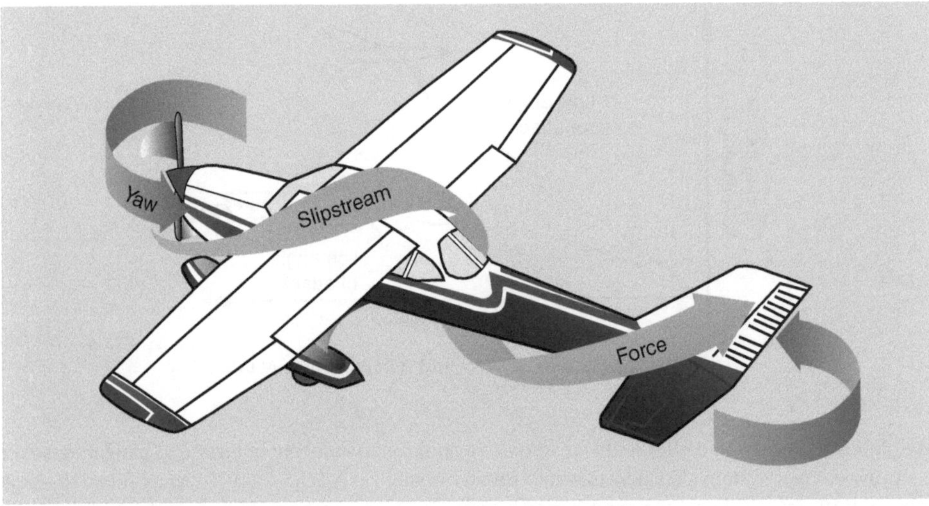

Figure 13.12. Slipstream rotation causes yaw.
Source: Pilot's Handbook of Aeronautical Knowledge (2008a)/U.S. Department of Transportation Federal Aviation Administration/Public domain.

Crosswind Takeoff and Landing

As we discussed during takeoff and landings, the aircraft rudder forces must be great enough to maintain a track down the runway when a crosswind exists. In effect, the airplane is sideslipping with respect to the wind. The positive static yaw stability of the airplane will try to turn it into the crosswind. This is undesirable during takeoffs and landings, and the rudder must be able to overcome this tendency. Rudder force may be required to realign the longitudinal axis of the aircraft with the runway centerline. For example, with a crosswind from the right the nose will weathervane to the right, so left rudder must be applied to "straighten" the nose. Because the wind will also be blowing the airplane to the left in this example, right wing down will be required to counteract drift (see the Intentional Slips section).

Asymmetrical Loading/Thrust

Also known as P-factor, asymmetrical loading occurs when the downward-moving blade of a propeller engine creates more lift than the upward-moving blade. This occurs when the longitudinal axis of the aircraft makes a positive angle with the relative wind – most commonly when the aircraft is in a pitch up attitude, such as in a climb. In effect, the right half of the propeller disc is generating more thrust than the left half, for reasons discussed below. This unbalanced force generates a moment that causes a left-turning tendency (for aircraft with clockwise propeller rotation as seen from the cockpit) for a single-engine propeller aircraft when at a high fuselage pitch angle and especially at high power settings. It should be noted that the left-turning tendency for single-engine aircraft is not due to P-factor alone but is actually the sum of four forces: P-factor, torque, gyroscopic effect, and slipstream effect. In the past, all these forces were lumped together and simply called "torque."

The lift (thrust) generated by a propeller blade, like any airfoil, is determined by its angle of attack and by the speed of the relative wind the blade encounters. Both (all) blades of the propeller have the same pitch, and the relative wind the *aircraft* encounters is the same for all blades of the propeller in level flight, but at high angles of attack there is a forward component of the velocity of the downward moving blade(s), which does two things: it increases the speed of the relative wind that blade encounters, and it increases the angle of attack of that blade. The effect of both of these is to increase the lift (thrust) produced by the downgoing blade(s). Similarly, a component of the velocity of the upgoing blade(s) is aft. This reduces the speed of the relative wind the upgoing blade encounters, and it reduces its angle of attack. Both of these effects reduce the lift (thrust) produced by the upgoing blade(s). Because the downgoing blades are on the right half of the propeller disc and the upgoing blades on the left, the result is that the right half of the propeller disc produces more thrust than the left half, resulting in a tendency to turn the aircraft to the left (Figure 13.13). It may help to imagine turning the single-engine airplane on its tail while maintaining forward velocity, like a helicopter rotor. The blade moving into the relative wind produces more lift than the blade moving with the relative wind, opposite the direction of flight. Due to asymmetric thrust, this imaginary vertical aircraft would tend to rotate counterclockwise as seen from behind.

The left-turning tendency is not theoretical – it is a very real phenomenon that pilots encounter on every takeoff and in every climb. It matters not just for directional control, but for the safety of flight: if while climbing out a pilot neglected to compensate for the left-turning tendency and allowed a stall to occur with the controls uncoordinated (insufficient right rudder), the stall could become a spin, which at low altitude would likely be unrecoverable.

In summary, the greater the angle of attack, horsepower of the engine, and rpm setting, the more right rudder that will be required to prevent a yaw to the left around the vertical axis. There have been limited experiments with counter-rotating propellers on single-engine aircraft, often known as contra-rotating propellers (Figure 13.14). The contra-rotating propellers counter the P-factor of each descending blade, and when set correctly the aft blade can take advantage of slipstream from the forward propeller. Due to more complex mechanical design and additional weight, this design has not been used in wide-scale production.

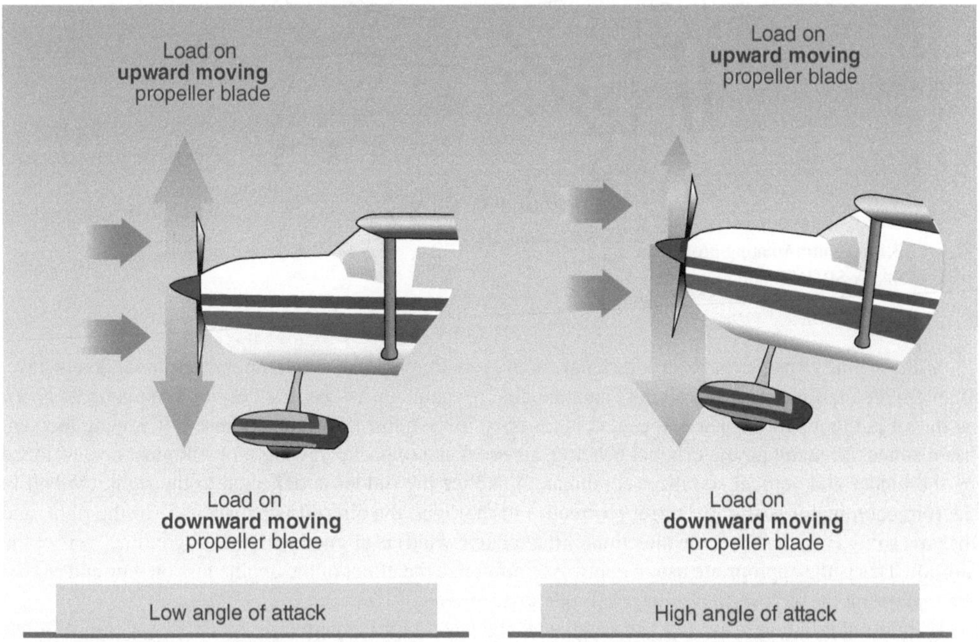

Figure 13.13. Asymmetrical loading.
Source: Pilot's Handbook of Aeronautical Knowledge (2008a)/U.S. Department of Transportation Federal Aviation Administration/Public domain.

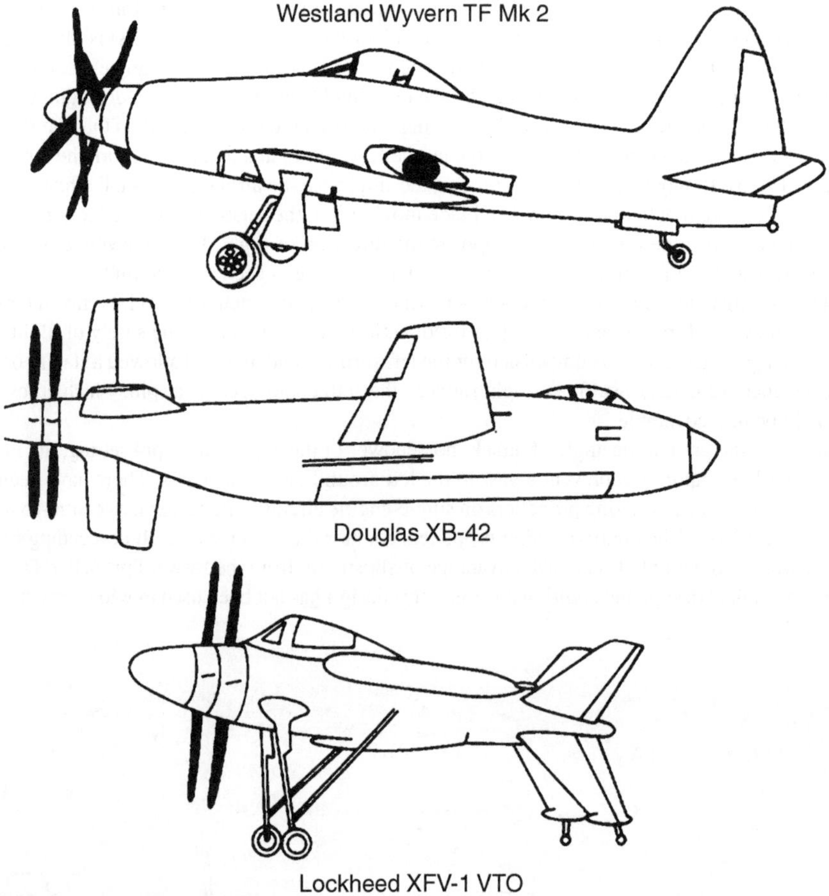

Figure 13.14. Contra-rotating propellers.
Source: NASA-SP-367/Public Domain.

Multi-engine airplanes have an additional control requirement that single-engine airplanes do not have. If an engine malfunctions, the aircraft must be able to maintain its heading despite *asymmetrical thrust* as shown in Figure 13.15. The left engine is assumed to have lost thrust, and a nose-left yawing moment has resulted. To maintain the original heading, an equal and opposite yawing moment must be developed by the rudder and vertical stabilizer combination. Notice the rudder is deflected to the right, the ball in the turn coordinator is slightly offset (correctly) to the right, the aircraft is banked ~5° to the right, and the yaw string (which shows the direction of the relative wind) is aligned with the longitudinal axis of the aircraft. This is the appropriate use of controls to minimize the effect of the engine loss on yaw and reduce drag, allowing the aircraft to minimize altitude loss.

If right rudder alone were used to counteract the yaw due to asymmetric thrust in the case of a left engine failure, there would be an imbalance in sideward forces acting on the airplane. This would cause a sideslip to the left unless the right wing was dropped slightly as indicated. This bank toward the operating engine reduces the amount of right rudder required, reducing the rudder's induced drag, and reducing form

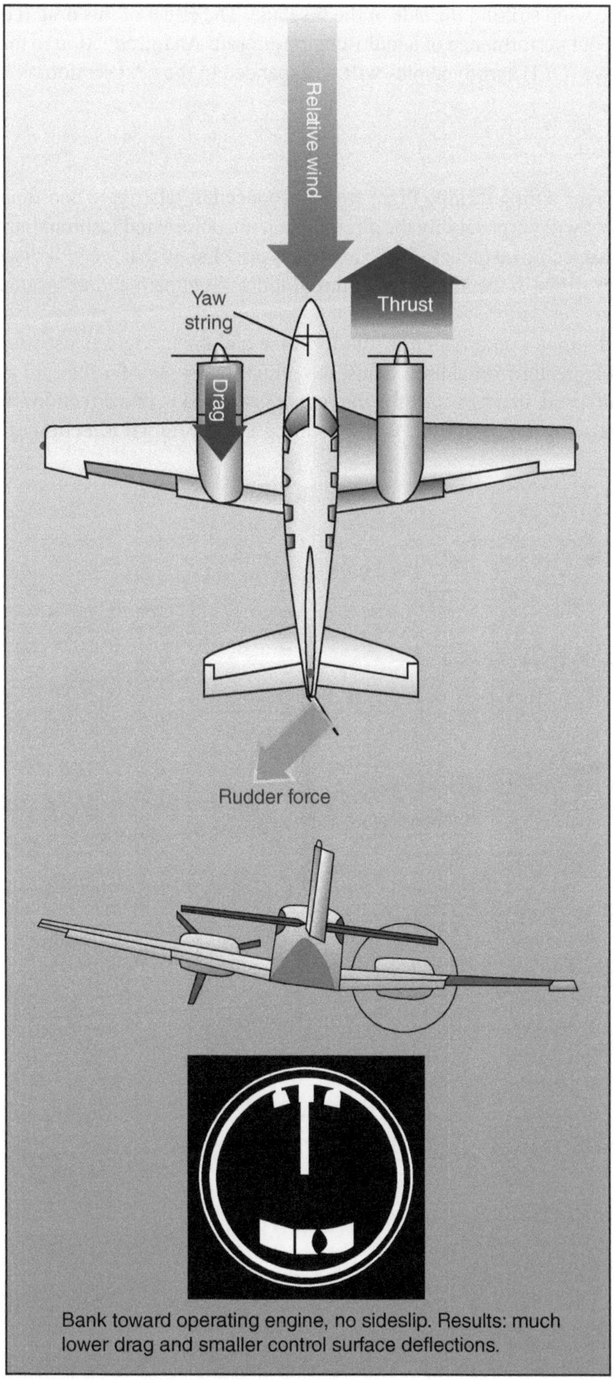

Figure 13.15. Yawing moment due to asymmetrical thrust.
Source: Airplane Flying Handbook (2004a)/U.S. Department of Transportation Federal Aviation Administration/Public domain.

drag from the relative wind striking the side of the fuselage. The effect of this drag reduction is to improve markedly the engine-out performance of a multi-engine aircraft. An introduction to multi-engine flight and one-engine inoperative (OEI) aerodynamics will be expanded in the next section.

Intentional Slips

In Chapter 11, the forces during turning flight were introduced in relation to coordinated flight. There are instances where a pilot will purposely fly the aircraft in an uncoordinated fashion, introducing a sideslip to control the aircraft in specific instances. There are two types of slips that we will discuss, the forward slip and the sideslip. They differ in the amount of control inputs, flight path and/or ground track, and fuselage alignment.

During a forward slip, a combination of rudder force and bank angle is used to lose altitude due to increased drag from a greater frontal area. Thus, the primary purpose of a forward slip is to lose altitude without increasing airspeed. In a forward slip, the rudder is applied in one direction while ailerons are used to lower the opposite wing so the aircraft is not pointing in the original direction of travel. Figure 13.16

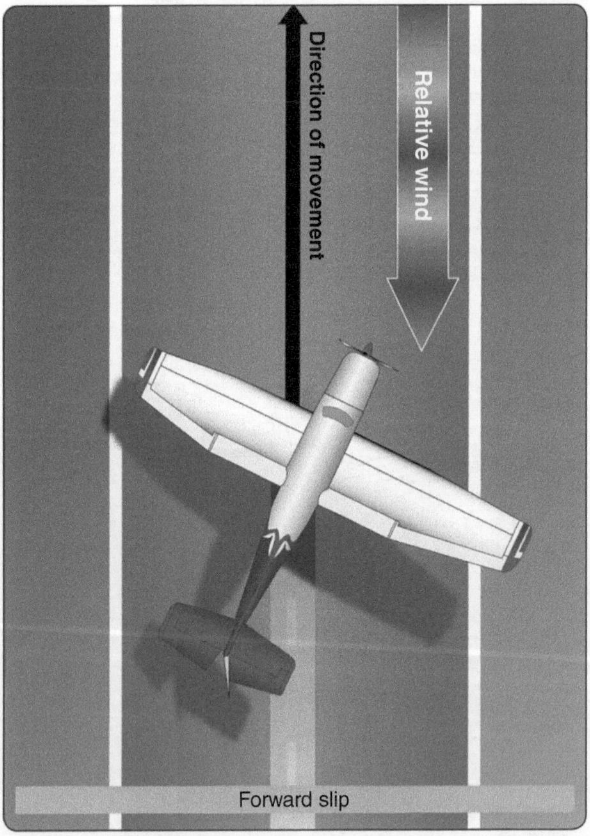

Figure 13.16. Forward slip during landing.
Source: Airplane Flying Handbook (2021)/U.S. Department of Transportation Federal Aviation Administration/Public domain.

shows a forward slip with the right rudder and left wing down, the original direction of movement (ground track) is maintained but the longitudinal axis is positioned to the right. Because more fuselage is presented to the relative wind, drag increases markedly and the aircraft loses altitude. Once full rudder is required to maintain the ground track, the slip limit has been reached. During a slip, the stall speed increases, while the increase in drag reduces the indicated speed, so the two speeds can "meet" very quickly, and the pilot may introduce a cross-control stall and subsequent spin at a very low altitude. As we have discussed several times, pitch controls airspeed; as the forward slip is initiated, the yoke/control stick may need to be pushed forward to lower the nose of the airplane to avoid an accelerated stall.

A sideslip was introduced during the discussion on crosswind landings. It should be noted that in the sideslip, the bank into the wind is used to offset the force of the wind. But when the lift vector is tilted for this offset, the aircraft wants to turn. Opposite rudder is applied to keep the fuselage from turning so it remains aligned with the runway centerline, and upon landing the wheels touch while being aligned with the runway. This does not mean they touch simultaneously: in a crosswind landing, the upwind wheel touches down first, followed by the downwind wheel as airspeed decreases. During a *forward* slip, the longitudinal axis of the airplane is not aligned with the ground track, while during a *side*slip the longitudinal axis is aligned with the ground track (Figure 13.17).

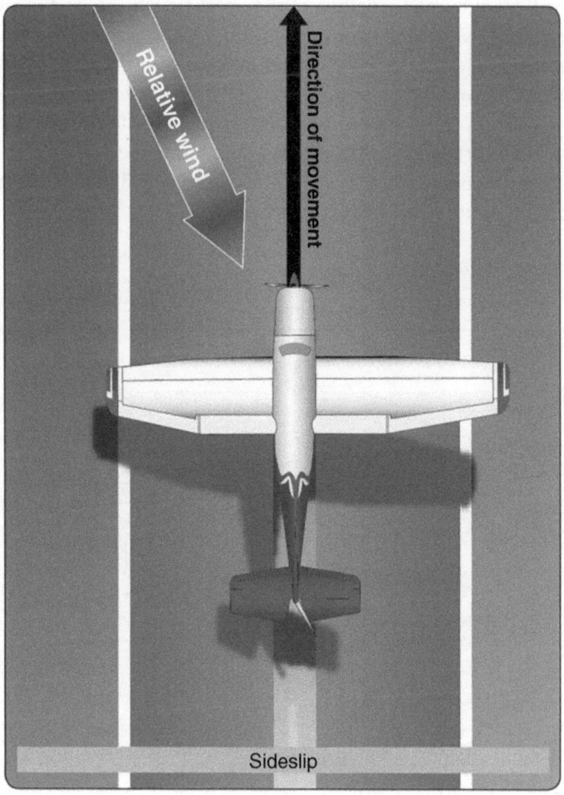

Figure 13.17. Sideslip during landing.
Source: Airplane Flying Handbook (2021)/U.S. Department of Transportation Federal Aviation Administration/Public domain.

MULTI-ENGINE FLIGHT PRINCIPLES

The dangers associated with single-engine operation in a twin-engine airplane go far beyond losing half of the power available. Losing 50% of your engines in a light, twin-engine may seem bad enough, but this loss usually translates into a loss of 80% or more of climb performance, and in some cases, an aircraft may not be able to climb at all and instead may be forced into a gradual descent (drift down). Though a second engine in this case may seemingly offer many advantages in safety over a single-engine aircraft, without the proper initial and recurrent training the second engine could very well result in a reduction in safety if an emergency is handled improperly.

Introduction to Single-Engine Flight

The minimum airspeed at which the rudder can produce the necessary yawing moment to maintain heading, if an engine fails, is called the in-flight minimum control speed, V_{MC}, or sometimes V_{MCA}; safe operating practice requires that V_{MC} not be greater than $1.2V_s$.

Figure 13.18 shows the effect of asymmetrical thrust and respective rudder input when dealing with a multi-engine propeller airplane that does not employ counter-rotating engines. If the left engine fails, more rudder authority will be required to maintain the correct sideslip since a longer arm exists as shown by D_2. P-factor (asymmetrical loading) is stronger on the descending blade of both engines; so, when the left engine fails, the yaw is greater because the arm of the thrust on the right engine is greater, and more rudder force will be required. For both reasons (among others), the left engine is considered the critical engine. Counter-rotating engines on a twin-engine airplane (the left engine rotates clockwise and the right engine rotates counter-clockwise) equalize the effect of the failure of either engine, so no longer is there a critical engine. This is to differentiate from an aircraft such as the P-38 where both engines were critical. If either engine fails, the forces are equal but opposite.

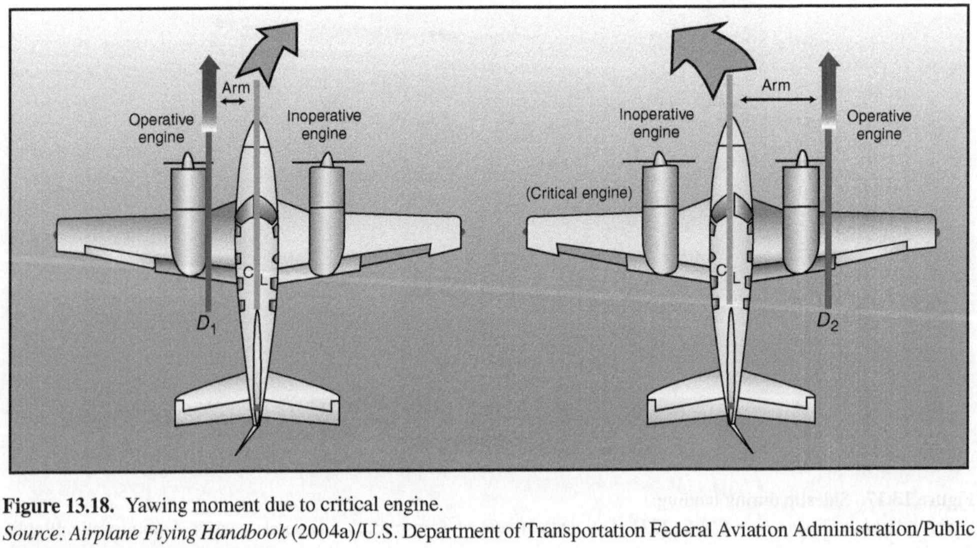

Figure 13.18. Yawing moment due to critical engine.
Source: Airplane Flying Handbook (2004a)/U.S. Department of Transportation Federal Aviation Administration/Public domain.

In addition to learning the limiting airspeeds that pertain to all aircraft, the multi-engine pilot must learn a new set of airspeeds that pertain specifically to multi-engine aircraft. These include V_{XSE} and V_{YSE}, which are the best angle-of-climb and best rate-of-climb speed, respectively, with one engine inoperative. V_{YSE} is usually marked with a blue line on most airspeed indicators and is of vital importance during the loss of an engine. $V_{MC,}$ which is defined as the minimum controllable airspeed with the critical engine inoperative, will be discussed later in this section. V_{MC} is usually marked with a red radial line on most airspeed indicators. V_{SSE} is the safe, intentional OEI speed, better known as the safe single-engine speed to most pilots. Knowledge of what these speeds are for a given aircraft is imperative to the safe operation of a multi-engine aircraft. These airspeeds must be considered during performance calculations as well as in the handling of emergencies.

Since there are a wide variety of aircraft with two engines or more flying today, with varied weights, sizes, and engine types, our discussion will center on twin-engine propeller-driven aircraft. Many pilots who desire to fly heavier, more complicated aircraft will develop their skills and habit patterns in light, twin-engine aircraft.

Application 13.1

There are other methods that have been employed on multi-engine aircraft to minimize the effects of torque and P-factor. What method is employed by the Russian Tu-95 bomber? What was different about the P-38 Lightning's use of counter-rotating propellers and how did it impact the critical engine?

Flight Principles with One Engine Inoperative

In the previous section, we discussed that a 50% reduction in power results in an even larger loss in climb performance. This should be considered very carefully whenever a pilot is considering taking off in a twin-engine propeller aircraft that is near maximum gross weight, and/or operating at a high-density altitude. For example, a light twin-engine aircraft may have a rated horsepower of 180 thrust horsepower on each engine. For normal cruise level flight, only 150 hp is required, which leaves a reserve horsepower of 210 for maneuvering, climbing, and increased forward airspeed. With the loss of one engine, 180 hp is now unavailable, so there is only 30 hp in excess of the 150 hp required for level flight. As we will see, the 30 hp excess may further be degraded when drag is increased due to the aerodynamic forces required to maintain level flight, and because of the windmilling propeller on the failed engine. Not only in this example will the aircraft more than likely not be able to climb, but it may also only be able to maintain a slow descent rate.

When an engine fails, the propeller usually continues windmilling (spinning without producing thrust) due to the angle of attack (low blade angle) of the propeller blade with the oncoming relative wind. While this windmilling does not produce thrust, it does increase the drag on an aircraft markedly due to the force required to turn the propeller on the dead engine. This windmill drag results in additional asymmetrical thrust yawing the aircraft in the direction of the failed engine, requiring additional rudder input from the pilot to maintain coordinated flight. This extra control input further increases parasite drag, and overcoming this additional drag uses more of the remaining thrust from the good engine. Figure 13.19 depicts a twin-engine airplane that has a failure of the left engine and is slipping in the direction of the "dead" engine resulting in uncoordinated flight.

Multi-engine propeller aircraft have the ability to feather their propellers, which means the propeller of the failed engine goes to a high blade angle of approximately 90°. This streamlines the propeller with the relative wind, minimizing parasite drag and in most cases allowing the propeller to stop windmilling.

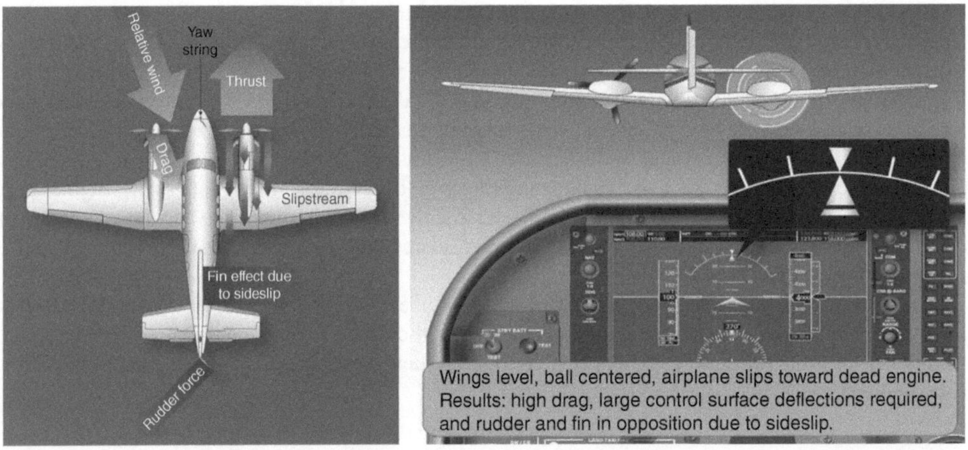

Figure 13.19. Sideslip into the failed engine.
Source: Airplane Flying Handbook (2021)/U.S. Department of Transportation Federal Aviation Administration/Public domain.

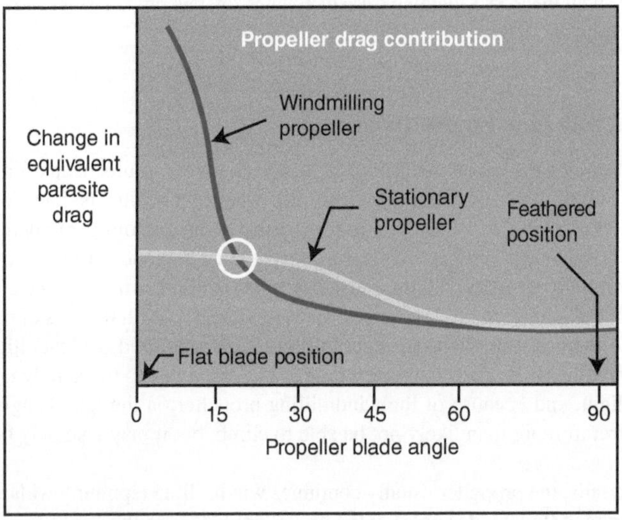

Figure 13.20. Propeller drag contribution.
Source: Airplane Flying Handbook (2004a)/U.S. Department of Transportation Federal Aviation Administration/Public domain.

Figure 13.20 shows the reduction in equivalent parasite drag when a propeller has been feathered versus one that has been allowed to continue windmilling.

Once a pilot has established that an engine has failed, adjusting the pitch for V_{YSE} should be the next step in order to maximize performance and minimize altitude loss. At this point (unless already accomplished), the pilot should initiate procedures to counteract the asymmetric thrust. Pilots of multi-engine aircraft must remain current and proficient regarding procedures to maintain coordinated flight in the event of an engine

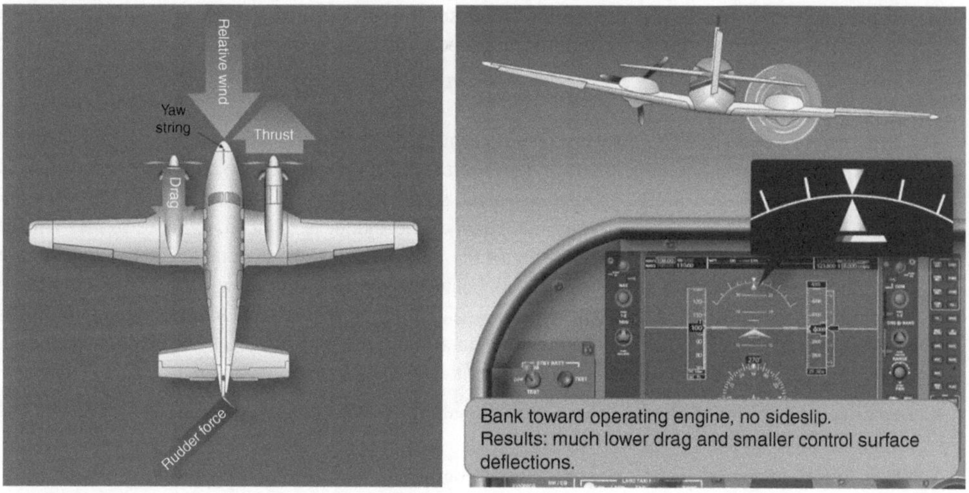

Figure 13.21. Zero sideslip with proper control application.
Source: Airplane Flying Handbook (2021)/U.S. Department of Transportation Federal Aviation Administration/Public domain.

failure. In addition to yaw correction from the rudder, the bank angle should be between 2° and 5° into the good engine, which allows the horizontal component of lift to assist in counterbalancing the asymmetrical thrust, thus preventing sideslip. Both rudder and bank angle need to be used in coordination as each one alone will detract from the desired single-engine performance of a multi-engine (Figure 13.21).

V_{MC} Considerations

Although we discussed the meaning of V_{MC} earlier in this section, the derivation, application, and conditions of V_{MC} are so important to understanding the aerodynamic principles during one-engine flight in a twin-engine reciprocating aircraft that further discussion is required. Pilots need to have an understanding of V_{MC} beyond knowing that V_{MC} is simply the red radial line on the airspeed indicator and believing that as long as the indicated airspeed remains above this line, they will be safe. Failure to fully comprehend that V_{MC} is *not* a fixed airspeed and that it *does* increase above the red radial line, or to understand the proper recovery procedure should the aircraft fall below V_{MC} during OEI flight, often results in an aircraft becoming uncontrollable.

During aircraft certification, V_{MC} is the calibrated airspeed at which at sea level the critical engine suddenly becomes inoperative, and the pilot is able to maintain *directional* control with no greater than 5° bank angle into the good engine. During flight testing, published V_{MC} is usually determined for all the aircraft circumstances that would result in the highest V_{MC}, thus allowing for real-world flight scenarios when an engine fails on an unsuspecting pilot. The aircraft manufacturer publishes V_{MC} in the aircraft flight manual, and all multi-engine pilots should be aware of the following 14 Code of Federal Regulations (CFR) part 23, section 23.149 certification conditions:

- **Maximum available takeoff power initially on each engine:** Because V_{MC} increases with takeoff power, and decreases with altitude, the highest V_{MC} is found at sea level with maximum available takeoff power. In a turbocharged aircraft, V_{MC} remains the same until the critical altitude has been reached, then decreases as altitude increases.

- **All propeller controls in the recommended takeoff position throughout V_{MC} determination:** The highest V_{MC} is found with a windmilling propeller in the takeoff position on the critical engine unless the engine has an auto-feathering system. Once the propeller of the inoperative engine is feathered, V_{MC} is reduced.
- **Unfavorable weight and CG position:** The determination of the most unfavorable weight is decided by the individual manufacturer, but V_{MC} does increase as the weight of the aircraft is *reduced*. V_{MC} is also higher the farther aft the CG is positioned, so the farthest aft CG results in the highest V_{MC}. As the CG is shifted farther aft, the moment arm of the rudder is reduced, so more rudder will be needed to counteract the asymmetric thrust of the operating engine, as depicted in the right image in Figure 13.22. When full rudder travel has been reached, the aircraft will no longer be able to maintain its current heading and will turn toward the inoperative engine. In addition, as the CG is moved toward the rear of the aircraft, the moment arm of the operating propeller blade is increased, increasing the yawing effect of asymmetric thrust.
- **Landing gear retracted:** Extended landing gear adds directional stability, so V_{MC} is decreased, thus retracted landing gear leads to higher V_{MC}.
- **Flaps in takeoff position:** For most light twin-engine aircraft, the takeoff flap setting will be 0°.
- **Maximum 5° angle of bank into the operating engine:** Because bank angle works to the manufacturer's favor in lowering V_{MC}, no more than 5° of bank is permitted during the certification of V_{MC} (though lower bank angles around 2° may yield the lowest V_{MC}). Remember that the horizontal component of lift while banked counteracts the asymmetric thrust of the operative engine when used in conjunction with proper rudder input. V_{MC} increases significantly when bank angle is used improperly and can increase by as much as 15–20 kts. when the wings are held in the level position and the rudder alone is used to counteract yaw.
- **Aircraft airborne and out of ground effect:** V_{MC} would be lower when in ground effect, so V_{MC} is determined when an aircraft is operating outside of the benefits of ground effect.
- **Aircraft trimmed for takeoff.**
- **Cowl flaps in the takeoff position.**

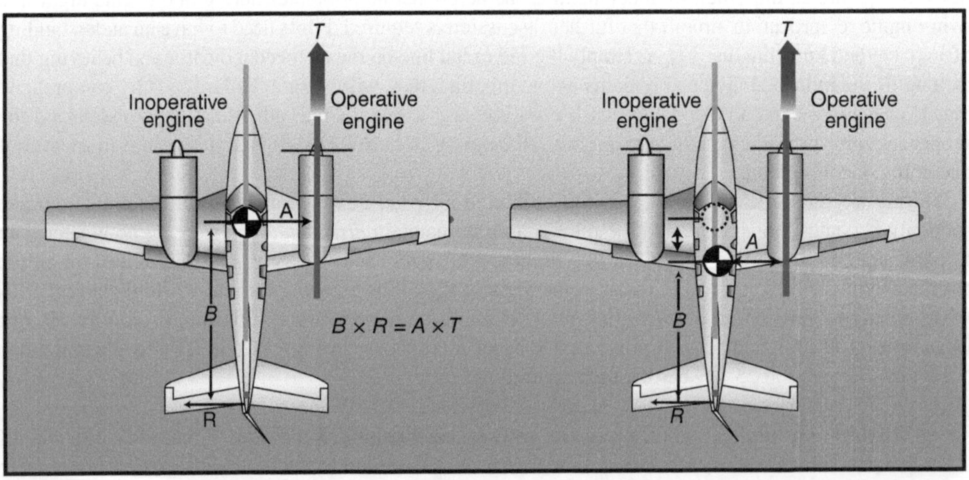

Figure 13.22. Effect of rearward CG on yaw.
Source: Airplane Flying Handbook (2004a)/U.S. Department of Transportation Federal Aviation Administration/Public domain.

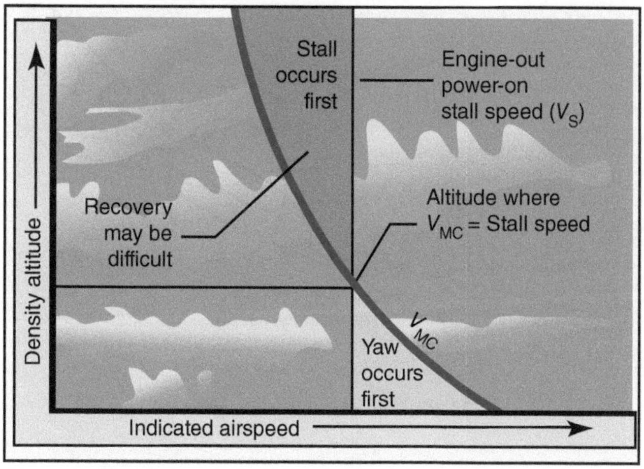

Figure 13.23. Relationship of V_{MC} to V_S.
Source: Airplane Flying Handbook (2004a)/U.S. Department of Transportation Federal Aviation Administration/Public domain.

The pilot of a multi-engine aircraft must also consider the relationship of V_{MC} to V_S. V_{MC} is usually higher than V_S. For a normally aspirated engine, V_{MC} decreases with altitude, and for a turbocharged engine, V_{MC} decreases above the critical altitude. V_S (indicated airspeed [IAS]), however, remains the same, so as V_{MC} decreases with altitude and V_S stays the same, eventually the two speeds meet. At high-density altitudes, they are close together from the start (Figure 13.23). If the pilot is operating at full power on the operating engine and reaches V_S at a density altitude at which V_{MC} is close to V_S, a sudden spin will most likely develop. In this case, regardless of the rudder direction, the spin will develop toward the inoperative engine. Multi-engine aircraft are not certificated for spins or required to be able to recover from spins. A spin entry during this situation will most likely involve a disoriented or nonproficient pilot and result in an unrecoverable and thus fatal outcome.

LATERAL STABILITY AND CONTROL

Lateral stability and control refer to the behavior of the airplane in roll, that is, movement of the lateral axis when it is rotated about the longitudinal axis using the ailerons. Roll results when a rolling moment (L') acts on the airplane. This rolling moment can be caused either by the pilot activating the ailerons (or spoilers) or by the aircraft being subjected to a sideslip angle. From the stability standpoint, we are more interested in the reaction of the airplane to a sideslip angle. Several factors impact the amount of lateral stability that is designed into an aircraft, including the vertical wing position, dihedral, and sweepback.

In the next section, we discuss static lateral stability and lateral control. Dynamic lateral stability is coupled with dynamic directional stability, and they are discussed in the final section of this chapter, *Types of Motion Resulting from Coupled Effects.*

STATIC LATERAL STABILITY

When considering the static lateral stability of an aircraft it is necessary to also consider the directional motions previously discussed. Roll stability is not a direct function of the aircraft bank, but rather is

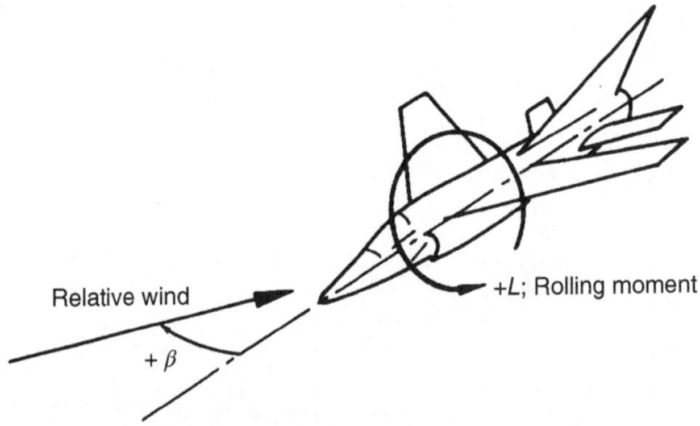

Figure 13.24. Rolling moment caused by sideslip.

dependent on the aircraft's response to the sideslip (β) that is produced in a banked turn due to adverse yaw. This sideslip produces a roll moment ($C_{L'\beta}$ – coefficient of roll moment due to sideslip) that may act as a stable, unstable, or neutrally stable force. This roll moment due to sideslip is known as the *dihedral effect*.

Figure 13.24 shows an aircraft that is in a right sideslip ("wind in right ear"). If this sideslip was due to a roll to the right and the aircraft continues to sideslip and roll to the right, then it is displaying unstable static lateral stability. This is known as *spiral instability* and will be discussed further at the end of this chapter. As we discussed with directional stability, the desired effect in this example would be for the right wing to rise, the left wing to drop, and the aircraft to return to stabilized flight.

For static stability in roll, we need a wings level rolling moment to be developed if a wing drops and sideslip results. Figure 13.25 shows the three possibilities. In Figure 13.25a, the airplane is in a right sideslip and right roll (aircraft flying toward you with $+\beta$). To have positive static roll stability, a left rolling moment ($-L'$) is required to level the wings. In Figure 13.25b, the airplane is sideslipping to its right (the viewer's left, since this view is looking aft, and the aircraft is flying toward the viewer) and is experiencing a + sideslip angle. To level the wings and have positive static roll stability, a left-wing-down (−) rolling moment is required. In Figure 13.25b, there is no rolling moment developed when a right sideslip is experienced, so this airplane has neutral static roll stability. Figure 13.25c shows an unstable airplane (the same situation shown in Figure 13.24).

The Rolling Moment Equation

The rolling moment equation about the aircraft CG is

$$L'_{CG} = C_{L'(CG)}\, qSb \tag{13.3}$$

where

L'_{CG} = rolling moment about the CG (ft-lb)
$C_{L'(CG)}$ = coefficient of rolling moment about CG
q = dynamic pressure (psf)

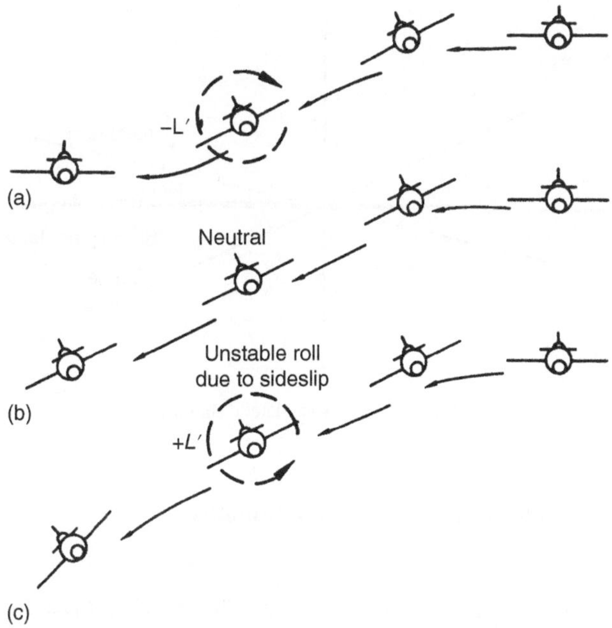

Figure 13.25. (a) Stable, (b) neutral, and (c) unstable static lateral stability.

S = wing area (ft^2)

b = wing span (ft).

Rearranging Eq. 13.3 gives

$$C_{L'(\mathrm{CG})} = \frac{L'_{\mathrm{CG}}}{qSb} \tag{13.4}$$

For a right-wing-down, +, rolling moment, $C_{L'}$ must be positive. For a left-wing-down, −, rolling moment, $C_{L'}$ must be negative.

Graphic Representation of Static Lateral Stability

A plot of the variation of $C_{L'(\mathrm{CG})}$ at different sideslip angles for an aircraft with positive static lateral stability is shown in Figure 13.26. The plot has a negative slope. This slope is the same as that for static longitudinal stability, but it is opposite to that for static directional stability (see Figures 12.14 and 13.3). The trim point is the point where there is no rolling moment. This occurs where there is no sideslip angle, at the intersection of the two axes.

Assume that an airplane is trimmed for zero roll and experiences a right, +, sideslip. If the airplane has positive lateral stability, it will develop a negative rolling moment coefficient, $-C_{L'}$, which raises the right wing. This is stabilizing. Thus, a negative slope shows positive stability. The magnitude of the slope is an indication of the degree of the stability (or instability) just as it was for both pitch stability and yaw stability.

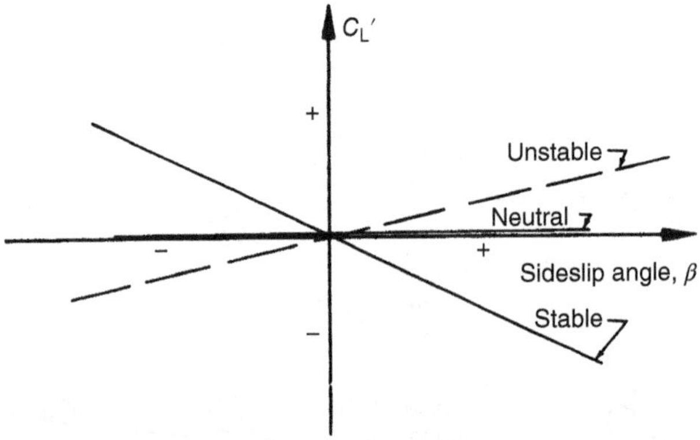

Figure 13.26. Static lateral stability.

Contributions of Aircraft Components to Roll Stability

Wing Dihedral

There are several aspects of aircraft design that impact the dihedral effect ($C_{L'\beta}$). As was seen in Chapter 12 concerning pitch stability, there are some design characteristics that provide a stabilizing influence and some that are destabilizing. We will focus on the primary factors of wing dihedral, vertical wing position, wing sweep, and the vertical tail. Together, these design factors may serve to offset individual contributions to stability and instability. The important thing to remember is that there is not an optimum amount of dihedral effect that applies to all aircraft. It will depend on the mission of the aircraft.

Dihedral is defined as the spanwise inclination of a wing or horizontal stabilizer to the horizontal, or to a plane equivalent to the horizontal. An upward inclination is a positive dihedral, and a downward inclination is a negative dihedral (commonly called *anhedral*). Dihedral is shown in Figure 13.27 and usually consists of an upward angle of 1–3°.

It is difficult to illustrate in a two-dimensional drawing the relative wind as it hits the aircraft in a sideslip, but an attempt is made in Figure 13.28. The glider rolls the aircraft to the left, which initiates a sideslip to the left. The angle of attack (AOA) of the left wing is greater than that of the right wing because of the dihedral of the wings. Therefore, more lift, L is produced on the left wing, and the positive coefficient

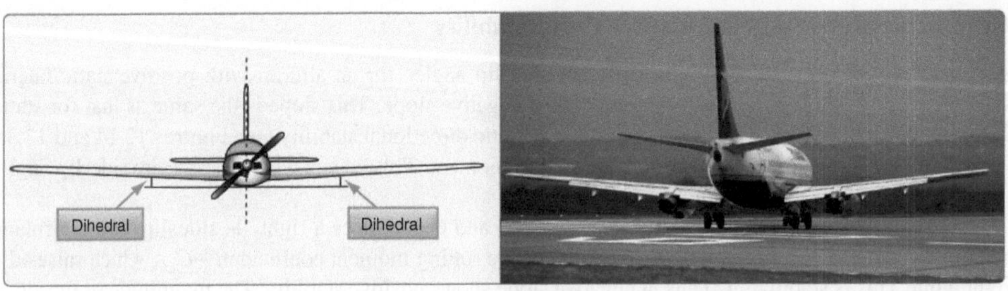

Figure 13.27. Dihedral angle.
Source: U.S. Department of Transportation (2016b)/Public domain.

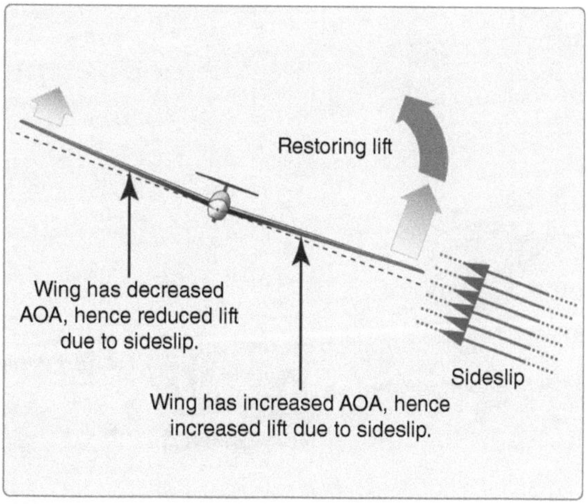

Figure 13.28. Dihedral producing static lateral stability.
Source: Pilot's Handbook of Aeronautical Knowledge (2023c)/U.S. Department of Transportation Federal Aviation Administration/Public domain.

of the rolling moment, $+C_{L'}$, is generated, and a stabilizing rolling moment results. Too much "dihedral effect" may result in increased Dutch roll tendencies.

An anhedral wing design, where there is a downward slant from the wing root to the wing tip, is incorporated into many aircraft such as large military cargo aircraft. The C-17 in Figure 13.29 has a high wing that is swept back with a T-tail. The strong dihedral effect of the wings and tail would make this aircraft too stable in roll. Therefore, the wings have anhedral to reduce the dihedral effect, thus increasing aircraft maneuverability.

Vertical Wing Position

The position of the wing impacts the flow pattern of air around the aircraft fuselage when it is in a sideslip. This in turn changes the AOA of the wing near the root. For example, a high wing will cause the wind striking the fuselage to flow upward over the wing, increasing the AOA at the wing root, and causing a roll away from the sideslip. This enhances the dihedral effect leading to greater roll stability. Conversely, a low-wing aircraft will have a reduced AOA at the wing root, decreasing the dihedral effect. This is why low-wing aircraft typically have greater wing dihedral than high-wing aircraft due to this natural dihedral effect of the high wing.

A high wing position also places the CG of an airplane below the center of pressure of the wing, often referred to as the keel or "pendulum effect." This pendulum effect is equivalent to 1–3° of positive effective dihedral. Conversely, a low-mounted wing with the CG of the airplane above the center of pressure of the wing will be unstable and is equivalent to 1–3° of negative effective dihedral. This is another reason why low-wing aircraft are designed with more dihedral than high-wing aircraft.

Wing Sweepback

The contribution of swept wings in producing positive static directional stability was discussed previously (see Figure 13.5). We saw that if a swept-wing airplane was sideslipped to the right, then the wing on the right had more drag than the wing on the left. The right wing also has more lift than the left wing

Figure 13.29. C-17 Globemaster III.
Source: NASA/Public domain.

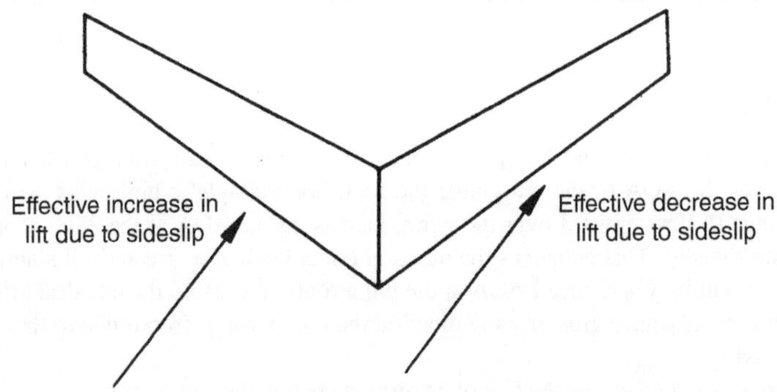

Figure 13.30. Dihedral effect of sweepback.

due to more of the leading edge being exposed to the relative wind at a more favorable angle. This is shown in Figure 13.30. This is stabilizing, much in the same way that dihedral is stabilizing; in fact, it is called the *dihedral effect of sweepback*. The Federal Aviation Administration (FAA)'s Pilot's Handbook of Aeronautical Knowledge states, "10° of sweepback on a wing provides about 1° of effective dihedral, while a high wing configuration can provide about 5° of effective dihedral over a low wing configuration."

Forward Swept Wing

In contrast to wing sweepback, there have been several aircraft designed with forward-swept wings, notably NASA's X-29 and the Russian Sukhoi Su-47. These designs have never materialized into mass military or civilian production. Figure 13.31 introduces several of the unique aerodynamic features of the X-29 that incorporate a forward-swept wing design. Several design features are noted here that were introduced in Chapter 3. The canards contribute to the lift of the aircraft along with the wings, and wing flaperons (combination flaps and ailerons) are used to change the camber of the wing and serve as ailerons for roll control. With a forward-swept wing design, the wings are mounted farther back along the fuselage, and instead of a horizontal tail, the aircraft has canards mounted in front of the wings. In contrast to our discussions in Chapters 4 and 5 regarding airflow around rearward-swept wings, the air flowing over forward-swept wing flows inward toward the wing root and not toward the wing tip; this means the wing tips (ailerons) do not stall first at high angles of attack. Despite positive design aspects such as this, as well as improved overall maneuverability and delayed onset of critical Mach (Chapter 14), these wing designs resulted in unstable flight characteristics, and fly-by-wire controls were required to safely fly the aircraft.

Vertical Tail

Side forces on the vertical tail will be a stabilizing influence since the tail is above the CG. The effect of the vertical tail on positive static lateral stability is shown in Figure 13.32. Aircraft with large vertical tails can be a major contributor to the dihedral effect. A T-tail configuration produces an even stronger dihedral effect, similar to the natural dihedral effect of a high wing.

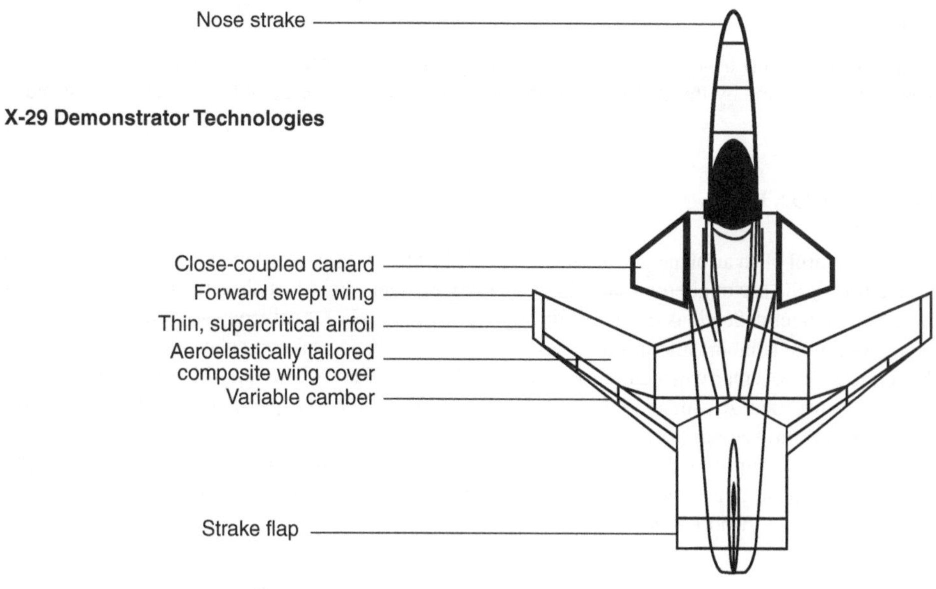

Figure 13.31. NASA X-29 aerodynamic features.
Source: NASA (2003)/Public Domain.

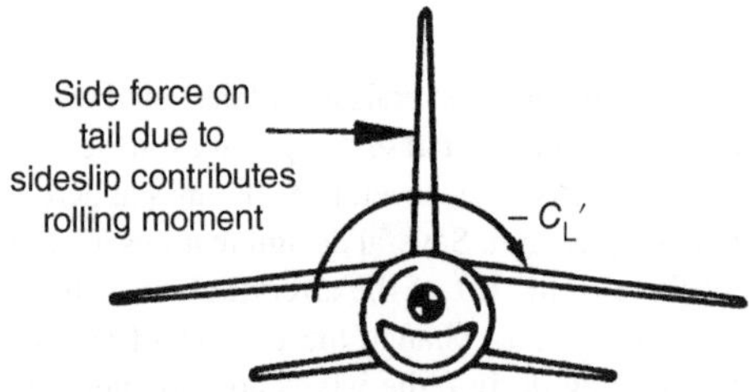

Figure 13.32. Vertical tail effect on lateral stability.

Engine

Torque reaction occurs on a single-engine propeller airplane due to Newton's third law. On most aircraft built in the United States, the propeller rotates clockwise as viewed from the cockpit. As the engine exerts a torque that rotates the propeller clockwise, the propeller exerts an equal and opposite torque on the engine which rotates the engine (and everything attached to the engine – namely the rest of the aircraft) counterclockwise. This torque produces a rolling tendency to the left, and this rolling tendency is greater at high power settings. It also occurs during cruise, and design compensation, such as offsetting the thrust line of the engine, allows for normal cruise flight with little correction from the pilot. Often, torque reaction is discussed as a left-turning tendency because on takeoff the torque imparts more force on the left wheel, resulting in a moment around the left main gear that must be corrected with the rudder. The horsepower of the engine, size of the propeller, and rpm setting all impact the amount of torque reaction created.

Complete Airplane

The total airplane must have positive lateral stability. Some components may have negative stability, but this must be overcome by the stabilizing moments of other components, and the entire aircraft must be laterally stable.

LATERAL CONTROL

The lateral control of an airplane is typically accomplished by providing differential lift on the wings, usually obtained by the use of ailerons and spoilers. However, some aircraft such as the B-1B use differential horizontal stabilizer deflections (split tail) during high-speed flight for roll control. Delta wing and flying wing aircraft often combine ailerons and elevators into a single control unit called elevons. Both left and right surfaces act together when elevator action is called for. They act in opposition to each other when roll motion is required. A combination of pitch and roll response is also possible. Pilot inputs are similar to those for conventional controls.

Critical lateral control conditions are takeoffs and landings in crosswind situations. Military fighters and acrobatic airplanes may need high roll rates and therefore may desire less lateral stability and more lateral control than transport aircraft.

DYNAMIC DIRECTIONAL AND LATERAL COUPLED EFFECTS

In our discussion of static directional and lateral stability, we saw that static stability depended on the aircraft's reaction to an imposed sideslip angle. Both yawing and rolling of an aircraft produce sideslip.

Conversely, sideslip produces both yawing and rolling moments. These two moments interact and result in coupled effects that determine the dynamic stability of an aircraft in yaw and roll. Several effects of coupled yaw and roll stability are described below.

Roll Due to Yawing

The normal way to produce rolling moments is by use of ailerons, then application of the rudder only to maintain a coordinated turn. However, yawing can also produce roll. If the pilot applies right rudder, the aircraft yaws to the right. In rotating about the CG on the vertical axis, to the right, the left wing moves faster than the right wing. The left wing then develops more lift, then rises, which lowers the right wing, and the aircraft rolls to the right.

This ties into our previous discussion of static roll stability. The yaw to the right creates a negative sideslip angle (relative wind from the left). This causes a positive rolling moment (right wing down), and the airplane rolls to the right.

Adverse Yaw

An airplane that enters a bank will yaw in the direction opposite the roll (e.g. if banked to the right, the airplane will yaw to the left) if proper coordination with the rudder is not initiated by the pilot or augmentation system at the beginning of aileron application. This tendency to yaw in the direction opposite the roll is referred to as *adverse yaw*. If an airplane yaws in the direction of the turn, the reaction is referred to as *proverse yaw*. In general, ailerons usually produce adverse yaw while spoilers usually produce proverse yaw. As presented in Chapter 5, this is in addition to the greater downwash angle of the upgoing wing causing an aft shift of the lift vector, resulting in induced drag that produces adverse yaw.

First, let us examine the lift forces on each wing when an airplane is rolled. Figure 13.33 shows an airplane that is making an aileron roll to the left. Consider the lift generated by each wing when an airplane rolls. Recall that the lift generated by any wing is a function of the velocity (speed and direction) of the relative wind the wing encounters. The "free stream" relative wind the entire aircraft encounters is the same for both wings. This is one component of the relative wind each wing encounters, but as the aircraft is rolling the motion of each wing (one moving upward, the other downward) creates another component of the relative wind of each wing, the effect of which is that the two wings encounter relative winds which, while equal in speed, differ in direction (Figure 13.34). Specifically, the upgoing wing creates a downward relative wind component, and the vector sum of this and the free stream relative wind is a relative wind that is directed slightly downward with respect to the free stream relative wind. Similarly, the downward movement of the downgoing wing creates a small upward relative wind component, so that the relative wind encountered by that wing is directed slightly upward with respect to the free stream relative wind. Because the lift any wing generates is perpendicular to its relative wind, the fact that the relative wind of the upgoing wing is directed slightly downward means that the lift of the upgoing wing is not vertically upward but canted slightly aft. This rearward component of that wing's lift is induced drag, which pulls that wing aft, contributing to adverse yaw. Likewise, because the relative wind of the downgoing wing is directed slightly upward with respect to the free stream relative wind, that wing's lift (which again is perpendicular to its relative wind) is canted slightly forward, and the small forward component of that wing's lift pulls that wing forward, again contributing to adverse yaw.

The rearward direction of the lift vector on the upgoing wing and the forward direction of the lift vector on the downgoing wing both oppose the yaw of the airplane in the intended direction of the turn. In this example, they try to turn the airplane to the right, hence the term adverse yaw.

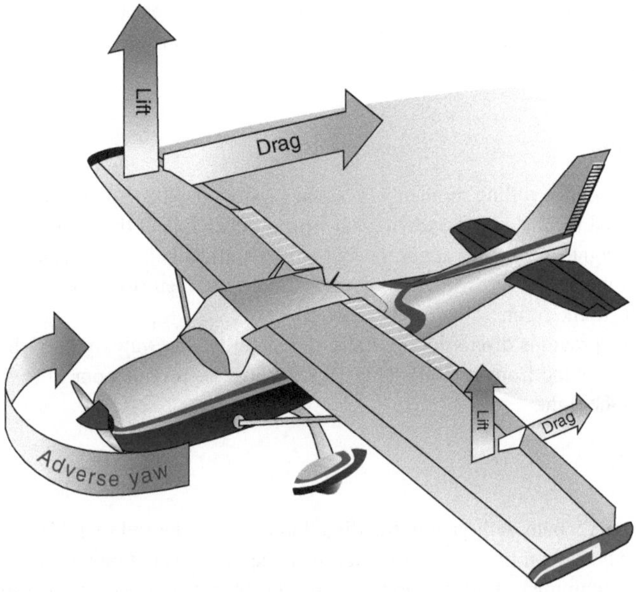

Figure 13.33. Adverse yaw.
Source: Pilot's Handbook of Aeronautical Knowledge (2008a)/U.S. Department of Transportation Federal Aviation Administration/Public domain.

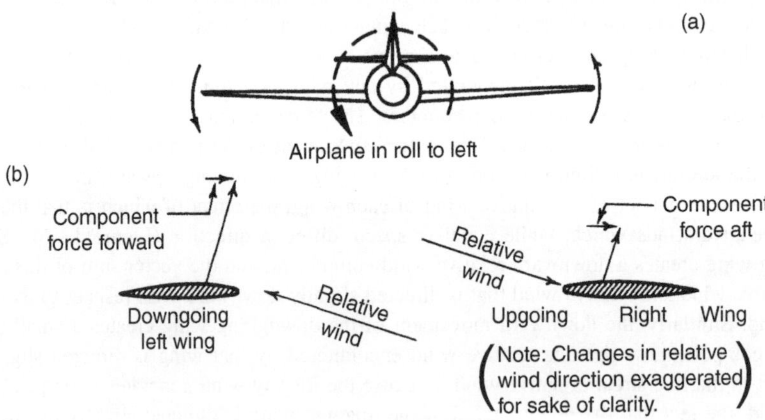

Figure 13.34. High AOA: (a) upgoing wing: (b) downgoing wing.

Adverse yaw is amplified at high angles of attack because airspeed is lower and aerodynamic pressure over the control surfaces is lower, thus requiring greater rudder deflection to counteract adverse yaw. As sailplane pilots are aware, adverse yaw is also more pronounced in aircraft with longer wingspans due to the longer moment arm of the force at the end of the wing.

Rudder in the direction of the bank is required to counteract adverse yaw. At slower airspeeds, more rudder will be required: as noted above, the reduction in aerodynamic pressure over the rudder at lower airspeeds means that the rudder will be less effective and will require greater control input (increased rudder pressure) to counteract adverse yaw.

To reduce the impact of adverse yaw, manufacturers have incorporated designs to assist the pilot during turns, including coupled ailerons and rudder, differential ailerons, and frise-type ailerons.

Differential ailerons usually reduce but do not completely eliminate the effects of adverse yaw. During a banked turn, the aileron that is deflected downward (on the upgoing wing) is deflected a smaller distance than the aileron deflected up (on the downgoing wing). The drag is now increased on the downgoing wing and reduced on the upgoing wing, reducing adverse yaw. With frise-type ailerons, the upward-deflected aileron (downgoing wing) is mechanically pivoted so the leading edge is exposed to the airflow. This increases the drag on the downgoing wing and assists in opposing the drag due to lift on the upgoing wing, thus reducing adverse yaw. Coupled ailerons and rudder have linked controls through interconnected springs that minimize adverse yaw by deflecting the rudder in the correct direction as the ailerons are used to initiate a banked turn (Figure 13.35). This was essentially the Wright Brothers' solution to counter adverse yaw from the use of wing warping for roll control.

Types of Motion Resulting from Coupled Effects

Three types of airplane motion can result from the interaction of yaw and roll:

1. *Spiral divergence (spiral instability)* results when the static directional stability ($C_{N\beta}$) is great in comparison to the static lateral stability ($C_{L'\beta}$). This is a weak dihedral effect. If a wing is lowered, the directional stability is greater than the roll stability so the aircraft will not readily sideslip. This means the wing will not raise back to a level position. As a result, the outside wing continues to travel faster than the inside wing. The faster outside wing generates more lift than the inside wing, which causes the bank angle to increase further even with neutral ailerons (this is the so-called "overbanking tendency"). The bank angle increases progressively, and the airplane enters an ever-tightening spiral dive commonly called a graveyard spiral.

   In most situations, the pilot recognizes the increasing spiral tendency and corrects the aircraft to normal flight. But for an unsuspecting pilot in instrument flight with improper scan technique, or during instrumentation failure, spiral divergence can lead to an unrecoverable dive at low altitude. Once the pilot exits the base of the clouds and regains visual contact with the ground, the instinctive, but incorrect, reaction is to pull back on the stick/yoke. Because airspeed has been building in the dive and the spiral has been "tightening," this can lead to a marked increase in load factor and structural failure. In extreme situations, the ultimate limit load factor may be exceeded, and components of the aircraft may separate from the aircraft.
2. *Directional divergence* results from negative directional stability, meaning that the aircraft is directionally unstable. The airplane with this design flaw, upon entering a sideslip after being disturbed either in roll or yaw, continues to yaw farther in the same direction. Once the yawing motion has started, it will continue until the airplane is broadside to the relative wind. This is incompatible with flight, so aircraft are designed so as not to exhibit directional divergence.
3. *Dutch roll* is a coupled directional and lateral oscillation that occurs when the dihedral effect is greater than the airplane's directional stability. The lateral/directional oscillation is usually dynamically stable but is undesirable due to its oscillatory nature. If a stable airplane is in a right sideslip

Figure 13.35. Coupled ailerons and rudder.
Source: Pilot's Handbook of Aeronautical Knowledge (2008a)/U.S. Department of Transportation Federal Aviation Administration/Public domain.

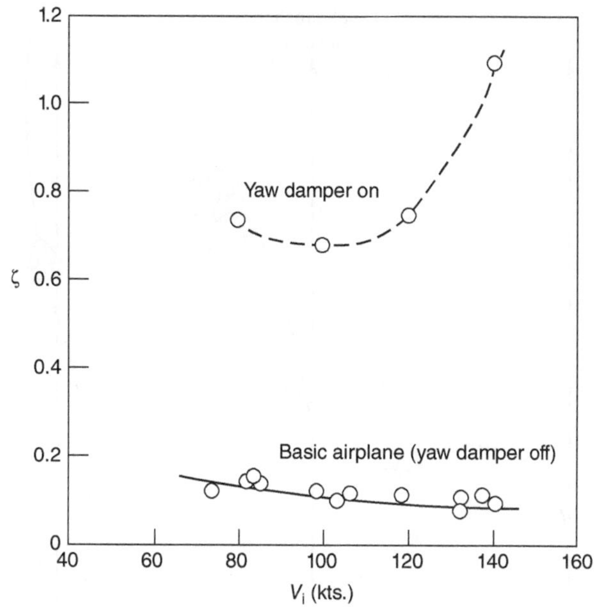

Figure 13.36. Yaw damper impact on Dutch roll characteristics.
Source: NASA (1970)/Public Domain.

(wind in the right ear), directional stability ($C_{N\beta}$) will start to reduce the sideslip. However, with strong lateral stability ($C_{L'\beta}$), the sideslip will cause a roll moment to the left providing a strong dihedral effect that results in spiral stability. If not controlled by the pilot or aircraft augmentation systems, adverse yaw will then cause a left sideslip. The whole process will be repeated continuously, and the aircraft will appear to make a figure eight on the horizon.

4. Aircraft designers must balance the desire to minimize Dutch roll through a reduced dihedral effect with the desire to control spiral instability. Again, aircraft mission will be a determining factor. In most non-swept-wing aircraft, the Dutch roll oscillations will dampen out on their own without pilot intervention, though the Dutch roll characteristics could be aggravated with an aft CG. A small amount of spiral divergence is tolerated because it is greatly preferred to Dutch roll.

As previously discussed in the Yaw Damper section, swept-wing aircraft may incorporate a yaw damper (or other types of stability augmentation systems) to minimize the effects of Dutch roll. A yaw damper incorporates a servo that moves the rudder in response to yaw, so as to minimize the motion around the vertical axis. Figure 13.36 shows the variation in Dutch roll damping with velocity with the yaw damper activated versus deactivated. The damping ratio (ζ) is indicated on the Y-axis while IAS is on the X-axis. The figure shows that the effect of Dutch roll characteristics is much less with the yaw damper activated. The damping ratio (ζ) is indicated on the Y-axis while IAS is on the X-axis.

Three possible flight paths due to coupled dynamic directional and lateral stability are shown in Figure 13.37.

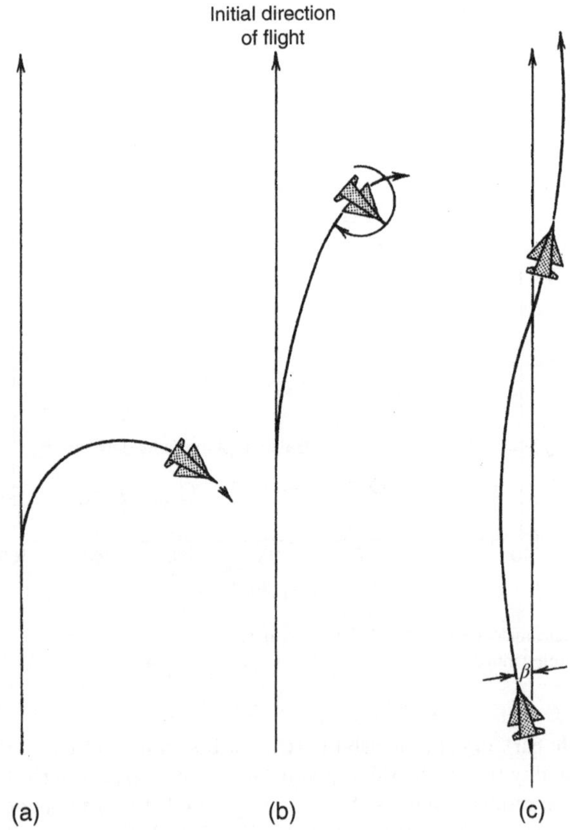

Figure 13.37. Flight paths due to coupled dynamic effects: (a) spiral divergence, (b) directional divergence, (c) Dutch roll.

SYMBOLS

ζ	Damping ratio
$C_{L'(CG)}$	Rolling moment coefficient about CG
$C_{N_{CG}}$	Yawing moment coefficient about CG
L'_{CG}	Rolling moment about CG (ft-lb)
N_{CG}	Yawing moment about CG
V_{MC}	Minimum controllable airspeed (critical engine inoperative) (kts.)
V_{SSE}	Intentional one-engine inop. speed (kts.)
V_{XSE}	Best angle of climb speed (one engine)
V_{YSE}	Best rate of climb speed (one engine)
β (beta)	Sideslip angle (degrees)

EQUATIONS

13.1 $N_{\text{CG}} = C_{N_{\text{CG}}} qSb$

13.2 $C_{N(\text{CG})} = \frac{N_{\text{CG}}}{qSb}$

13.3 $L'_{\text{CG}} = C_{L'(\text{CG})} qSb$

13.4 $C_{L'(\text{CG})} = \frac{L'_{\text{CG}}}{qSb}$

KEY TERMS

Adverse yaw
Anhedral
Asymmetrical thrust
Contra-rotating propellers
Dihedral
Directional divergence
Dutch roll
Forward slip
P-factor
Side slip
Spiral divergence
Spiraling slipstream
Strake
Ventral fin
Yaw damper

PROBLEMS

Note: Answers to the problems in the first section are given at the end of the book.

Application of Chapter Theory

1. For static directional stability, if an airplane sideslips to the left, the plane must
   a. keep its old heading
   b. yaw to the left
   c. yaw to the right

2. Directional stability and yaw refers to the behavior of the longitudinal axis to rotate around the __________ axis.
   a. vertical
   b. lateral
   c. longitudinal

3. The wings of an airplane contribute to static directional stability
   a. if they are forward of the CG
   b. if they are behind the CG
   c. either of the above

4. The fuselage is unstable in both static pitch and static yaw stability.
   a. True
   b. False

5. Engine nacelles with propellers or jet inlets behind the airplane's CG are stable in both static pitch and static yaw stability.
   a. True
   b. False

6. Which of the following aircraft components is a destabilizing component in regard to directional stability?
   a. Strake
   b. Ventral fin
   c. Tail-mounted engine
   d. Fuselage

7. Pilots can increase the stability of airplanes without powered controls by
   a. resting their feet on the rudders, even though the plane is trimmed
   b. holding the stick (or yoke) in the neutral position
   c. both of the above
   d. neither (a) nor (b)

8. Operating an airplane at a high AOA will result in a loss of yaw stability.
   a. True
   b. False

9. A yaw damper is a destabilizing component regarding directional stability, though it is still utilized in air transport category aircraft because it aids in mitigating the effects of turbulence.
   a. True
   b. False

10. The effects of the P-factor are most pronounced in which one of the following situations?
    a. High angles of attack and high power settings
    b. Low angles of attack and high power settings
    c. Low angles of attack and low power settings
    d. High angles of attack and low power settings

11. For a twin-engine airplane without counter-rotating propellers, which of the following is true regarding the critical engine?
    a. The right engine would be considered the critical engine
    b. The left engine would be considered the critical engine
    c. Either engine could be considered the critical engine
    d. Only single-engine aircraft have a critical engine

12. During a sideslip to land, the longitudinal axis of the airplane will be aligned with the ground track of the airplane.
    a. True
    b. False

13. For a twin-engine propeller airplane, a loss of one engine results in a reduction of 50% in performance available for maneuvering flight.
a. True
b. False

14. After an engine failure in a light twin-engine, propeller-driven airplane, what is the maximum recommended bank angle into the working engine to maintain during coordinated flight with proper rudder input?
a. 0°
b. 3°
c. 5°
d. 10°

15. Which of the following is not stated in the chapter as a certifying condition for the calculation of an airplane's V_{MC}?
a. Flaps in takeoff position
b. Landing gear extended
c. Maximum available takeoff power
d. Windmilling propeller

16. Which of these vertical wing locations will require the most wing dihedral for static roll stability?
a. High wing
b. Mid-wing
c. Low wing

17. Swept wings affect static roll stability more than straight wings by
a. increasing the stability
b. decreasing the stability
c. stability remains the same

18. Which term best characterizes the yaw of an airplane in the opposite direction of roll without coordinated rudder input?
a. Adverse yaw
b. Proverse yaw
c. Spiral instability
d. Dutch roll

19. If the pilot of a stable airplane applies the right rudder, the airplane will
a. roll to the left and yaw to the left
b. roll to the right and yaw to the right
c. roll to the left and yaw to the right
d. roll to the right and yaw to the left

20. Of the three types of coupled effects discussed, which of the following has the potential to be the most dangerous?
a. Spiral divergence
b. Directional divergence
c. Dutch roll

21. The yaw damper assists in the mitigation of which coupled effect of roll and yaw?
a. Spiral divergence
b. Directional divergence
c. Dutch roll

22. Which of the following airplane design situations may result in spiral divergence?
   a. Weak lateral stability, strong directional stability
   b. Strong lateral stability, weak directional stability
   c. Strong lateral stability, strong directional stability
   d. Weak lateral stability, weak directional stability

Correlation Through Scenario-Based Learning

Aircraft – *Diamond DA50RG*

1. Review the *Dimensions* section within Section 1 of DA50 aircraft flight manual (AFM) and find the following:
   a. Dihedral of the wing
   b. Leading edge sweep back of the wing
   c. Geometric twist of the wing
   d. Leading edge sweep back of the horizontal stabilizer
   e. Leading edge sweep back of the vertical stabilizer

2. What are the advantages of the sweepback for the horizontal and vertical stabilizers?

3. Where is the rudder trim located within the flight deck? How is it activated by the pilot? What specifically does a yellow RUD message on the G1000NXi represent?

4. Find the *Three-View Drawing* section within Section 1 of DA50 AFM. What is unique about the outer tips of the horizontal stabilizer? In terms of the chapter stability discussion, why do you think this design was incorporated into the tail?

5. What is the maximum lateral fuel imbalance (gal and lb) between the left- and right-wing fuel tanks?

14 High-Speed Flight

CHAPTER OBJECTIVES

After completing this chapter, you should be able to:

- Identify the regions of high-speed flight.
- Define the speed of sound and calculate the local speed of sound with a change in altitude.
- Identify and characterize the performance speeds associated with high-speed flight.
- Describe high-subsonic flight and explain critical Mach number and the development of shock waves.
- Describe transonic flight and identify the performance problems aircraft experience when operating in this region.
- Describe supersonic flight and correlate the development of shock waves in this region to other regions of high-speed flight.
- Explain the design features associated with aircraft components for aircraft operating in the supersonic region.

In this chapter, we discuss the airflow as the aircraft approaches the speed of sound (high subsonic), transonic flight, and supersonic flight. Hypersonic flight is not discussed.

In subsonic flight, the density change in the airflow is so small that it can be neglected in the flow equations without appreciable error. The airflow at these lower speeds (~250 kts.) can be compared to the flow of water and is called *incompressible flow*. However, at high speeds density changes take place in the airstream that are significant. Thus, this type of airflow is called *compressible flow*. High-subsonic, transonic, supersonic, and hypersonic flight all involve compressible flow.

Regions of high-speed flight have been arbitrarily named as follows:

- **High-subsonic:** Airspeeds where the airflow around the aircraft is completely below the speed of sound, the speed range for most civilian aircraft (Critical Mach number or less, usually below Mach 0.75).

Flight Theory and Aerodynamics: A Practical Guide for Operational Safety, Fifth Edition.
Brian A. Johnson and Philip R. Fittante.

Companion website: www.wiley.com/go/flighttheoryandaerodynamics5e

- **Transonic:** Aircraft speeds where the airflow around the aircraft is partially high-subsonic and partially supersonic (Critical Mach number 0.75 to 1.3).
- **Supersonic:** Aircraft speeds where the airflow around the aircraft is completely above the speed of sound but below hypersonic airspeed (from about Mach 1.3 to 5.0).
- **Hypersonic:** Aircraft speeds above Mach 5.0.

THE SPEED OF SOUND

The speed of sound is an important factor in the study of high-speed flight. Small pressure disturbances are caused by all parts of an aircraft as it moves through the air. These disturbances move outward from their source through the air at the speed of sound. A two-dimensional analogy is that of the ripples on a pond that result when a stone is thrown into the water. The behavior of these disturbances changes at airspeeds near, at, and above the speed of sound, and these changes have profound effects on the flow of air over the wings of aircraft operating at these speeds.

The speed of sound in air is a function of temperature alone, as can be shown in Eq. (14.1). The speed of sound varies from 661 kts. at sea level on a standard day (15 °C) to 574 kts. at the tropopause (−56.5 °C); see Table 2.1 for the speed of sound at various altitudes on a standard day.

$$a = a_0\sqrt{\theta} \tag{14.1}$$

where

a = speed of sound in air

a_0 = speed of sound in air at sea level, standard day (661 kts.)

θ = temperature ratio, $T/T_{0.}$

EXAMPLE

Calculate the speed of sound at a pressure altitude of 25 000 ft with an outside air temperature of −36 °C.

$$\theta = \frac{T}{T_0} \rightarrow \theta = \frac{(-36^\circ\text{C} + 273^\circ\text{K})}{288^\circ\text{K}} \rightarrow \theta = 0.823$$

$$\text{Eq. 14.1} \quad a = a_0\sqrt{\theta}$$

$$a = 661 \text{ kts.}\sqrt{.823}$$

$$a = 600 \text{ kts.}$$

Since Eq. 14.1 shows that the speed of sound is only a function of temperature, we can see that the higher in altitude the lower the speed of sound. Because the aircraft's speed in relation to the speed of sound is so important in high-speed flight, airspeeds are usually measured as Mach number. Mach number is the aircraft's true airspeed divided by the speed of sound (in the same atmospheric conditions, i.e. local speed of sound):

$$M = \frac{V}{a} \tag{14.2}$$

where

M = Mach number

V = true airspeed (kts.)

a = speed of sound (kts. local speed of sound).

EXAMPLE

Using the local speed of sound found in the last example, calculate the Mach number if that same airplane has a velocity of 375 kts. TAS.

$$M = \frac{V}{a}$$

$$M = \frac{375 \text{ kts.}}{600 \text{ kts.}}$$

$$M = 0.63$$

When an aircraft is flying below the speed of sound, the pressure disturbances will be moving faster than the airplane (ripples on a pond), and those disturbances that travel ahead of the aircraft influence or "warn" the approaching airflow. This "pressure warning" can be observed in a smoke wind tunnel as it causes the upwash well ahead of the wing. This is shown in Figure 14.1a.

If the aircraft is flying at a speed greater than the speed of sound, the airflow ahead of the aircraft is not influenced by the pressure field of the aircraft, because the speed of the pressure disturbances is less than the speed of the aircraft. The airflow ahead of the wing will have no warning of the approach of the wing and will not change its direction ahead of the wing's leading edge. Because the pressure waves are no longer "parting the way" for the oncoming wing, the air molecules compress in front of the wing and thus suddenly undergo extreme changes in pressure, density, and temperature. This is shown in Figure 14.1b. Two new "V-speeds" can now be introduced regarding the maximum operating speeds of jet aircraft:

V_{MO} – **Maximum operating speed:** Maximum operating speed in kts.

M_{MO} – **Maximum operating speed (Mach):** Maximum operating speed in terms of Mach (speed of sound).

Application 14.1

Consider Eqs. 14.1 and 14.2 in the following scenario. An aircraft maintains a constant true altitude of 25 000 ft, with a local outside air temperature of −30 °C and local area pressure of 30.00″ Hg (sea level).

Calculate the local speed of sound at an equivalent pressure altitude for the aircraft in the scenario above. If the aircraft was maintaining a calibrated airspeed (CAS) of 400 kts., what is the TAS? If the aircraft had a M_{MO} of 0.92 Mach, would the aircraft be exceeding a limitation? Hint: Review the airspeed discussion in Chapter 2.

Coffin Corner (Q Corner)

Pilots new to the world of high-speed flight are introduced to the concept of the "coffin corner," or "Q Corner," a situation occurring in flight at high speed and high altitude in which the stall speed of the

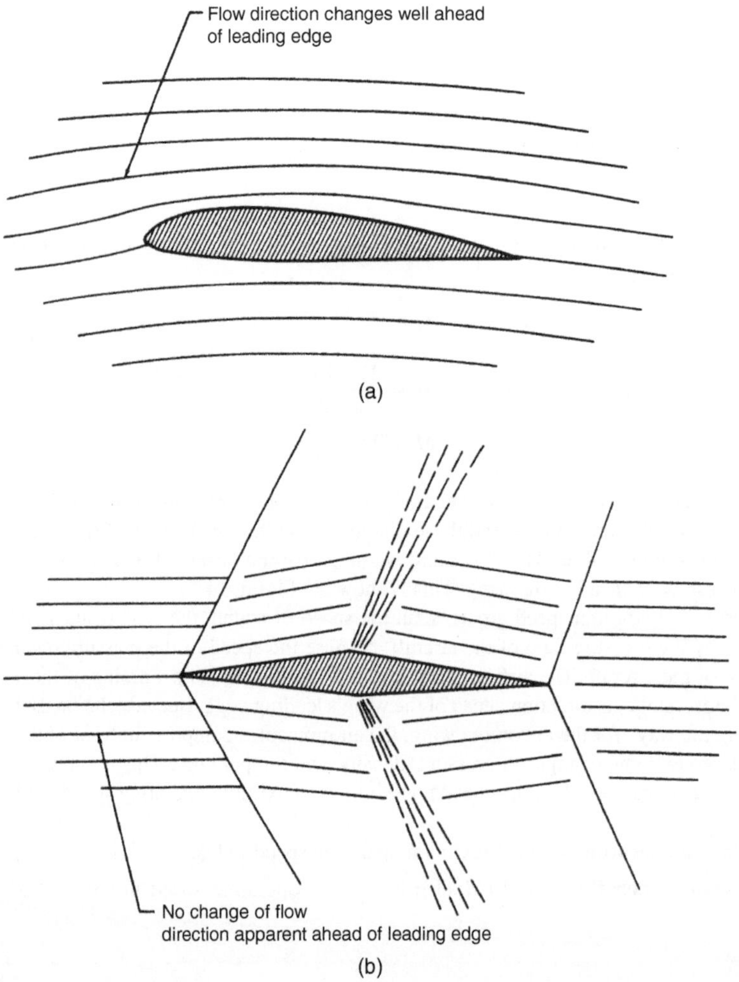

Figure 14.1. (a) Subsonic flow and (b) supersonic flow.

aircraft in Mach number equals the M_{MO} of the aircraft. In this situation, if the airspeed is decreased, the aircraft will stall, but if airspeed is increased, the maximum operating speed will be exceeded. As will be shown below, this is due to the change in density and temperature as an aircraft climbs in altitude; thus, it is important that pilots understand how airspeed varies with Mach number as an aircraft changes altitude.

If the aircraft climbs at a constant (CAS) (which is very close to indicated airspeed (IAS), and for purposes of this discussion will be considered to be identical), its TAS and Mach number will increase in the climb because as the air density decreases, the aircraft will need a higher TAS to obtain the same ram air pressure at the pitot tube. As we saw earlier in the textbook, the dynamic pressure at this faster TAS due to lower air density is equivalent to the dynamic pressure we experienced at sea level when the CAS and TAS were equal. Due to the decreasing temperature with altitude increase, the local CAS speed of sound is also decreasing.

Now consider how stall speed and stall Mach number are changing as an aircraft climbs. As discussed above, as altitude increases, the CAS stall speed remains the same, but the TAS stall speed increases. From Eq. 14.1, we know that the local speed of sound is decreasing in the climb. From Eq. 14.2, since the stall speed (as TAS) is increasing and the local speed of sound is decreasing, *the stall Mach number is increasing as the aircraft climbs.*

Equation 14.2 also says that if the Mach number is held constant in a climb, the TAS airspeed will decrease. Suppose an aircraft climbs at a constant Mach = M_{MO}. As it climbs at constant Mach, its TAS will be decreasing. Meanwhile, as discussed above, the stall speed TAS *increases* with altitude. There is an altitude at which the decreasing aircraft TAS equals the increasing stall TAS, and this is the coffin corner: you are already at M_{MO}, so if the airspeed is increased to avoid a stall, M_{MO} is exceeded, and structural damage could occur. But if airspeed is decreased to avoid exceeding M_{MO}, a stall will occur. There are no good options in the coffin corner – except to descend.

HIGH-SUBSONIC FLIGHT

Compressibility effects are not limited to aircraft that fly supersonically. A high-subsonic velocity will produce local supersonic flow on top of the wings, fuselage, and other parts of the aircraft, and these areas of local supersonic flow can strongly impact performance characteristics. Helicopters often experience compressibility effects on rotor tips; this can occur even when the helicopter is in a hover. The most important compressibility effect that occurs in high-subsonic aircraft is the formation of *normal* shock waves on the aircraft's wings or rotor blades. Once this occurs, the aircraft will be subject to an increase in drag, and possibly other negative forces the pilot may not have anticipated including buffeting and aircraft control difficulties.

Normal Shock Wave Formation on Wings

Air speeds up as it passes over an airfoil, so the local Mach number on top of the wing is always greater than the flight Mach number. For example, if an aircraft is flying at $M = 0.50$, the local airspeed might be $M = 0.70$ at the thickest point of the wing (depending on the thickness and/or camber of the airfoil). This is shown in Figure 14.2.

If the flight Mach number increases, the local velocity on top of the wing also increases. At some flight Mach number, the local maximum velocity reaches sonic speed, $M = 1.0$. This flight Mach number is called the critical Mach number, M_{crit}. For the example shown in Figure 14.2, M_{crit} is 0.72.

Once M_{crit} is exceeded, the aircraft is flying in the transonic speed range and a new set of drag and aerodynamic forces can be expected. Most jet aircraft will cruise just below their critical Mach number to avoid the negative effects of transonic flight. Supersonic airflow exists in the area of maximum thickness on top of the wing, but subsonic flow exists elsewhere. All pressure disturbances behind this sonic flow cannot be propagated forward because they run into sonic velocities traveling rearward. A type of *compression wave* now forms, called a *normal shock wave*, as the velocity of air over the wing exceeds M_{crit} (Figure 14.2). A normal shock wave is present when the air slows from supersonic to subsonic; this shock wave is perpendicular to the airflow. As the air flows through the normal shock wave, it undergoes rapid compression. The compression decreases the kinetic energy of the airstream and converts it into a static pressure as well as a temperature increase behind the shock wave. The heat rise behind the shock wave is either radiated to the atmosphere or absorbed by the wing surface, but in either case, it is lost. The lost energy in terms of pressure (dynamic pressure plus static pressure) must be continuously supplied by the engines. This creates a type of drag known as wave drag.

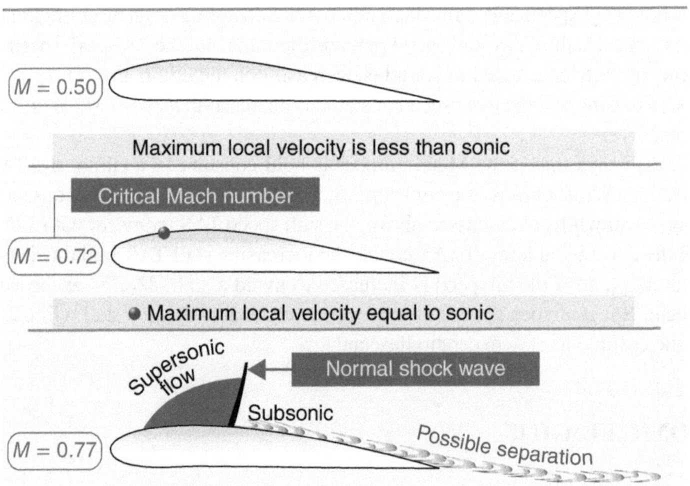

Figure 14.2. Airflow over a wing section.
Source: Pilot's Handbook of Aeronautical Knowledge (2008a)/U.S. Department of Transportation Federal Aviation Administration/Public domain.

The increase in static pressure caused by the normal shock wave has the same effect as the adverse pressure gradient that was discussed in basic aerodynamics. It slows the air in the boundary layer down to the point where faster-moving air from outside the boundary layer rushes in and reverse flow occurs, resulting in a higher pressure region behind the normal shock wave. This is shown as the "possible separation" area in Figure 14.2.

DESIGN FEATURES FOR HIGH-SUBSONIC FLIGHT

Subsonic jet aircraft can increase airspeed without encountering shock wave problems if the critical Mach number (M_{crit}) can be raised, so understanding when this occurs and designing airfoils to delay the onset of M_{crit} are of great importance. The larger the Mach number for M_{crit}, the better. Several design features help accomplish this:

- Thin airfoil sections
- Airfoil sections that have good high-subsonic Mach number characteristics
- Sweepback
- Vortex generators

Thin Airfoil Sections

The thinner the airfoil, the lower the increase in speed of the air over the top surface of the airfoil; thus, the airspeed must be higher in order to reach M_{crit}. One disadvantage of thin wings is that they do not have high values of $C_{L(max)}$, so higher takeoff and landing speeds are required. Another disadvantage is that it is difficult to design the structural strength and rigidity required in a thin wing. Many aircraft that fly close to, or exceed, the speed of sound also incorporate various high-speed flight controls that may be located on the wing, and thus, must be incorporated into the structure of the wing along with the necessary hydraulic control or fly-by-wire (FBW) controls. Finally, there is less room for fuel tanks in a thin wing as compared to a thicker wing of the same planform.

High-Speed Subsonic Airfoils

The laminar flow airfoil discussed in Chapter 5 was the first high-speed airfoil. Moving the maximum thickness backward from about 25% C to 40%–50% C did reduce drag and increase the critical Mach number, but one disadvantage still existed for this type of airfoil. The shock wave that develops on a laminar airfoil occurs in the adverse pressure gradient region. As the airflow is slowing up in this region, it has a tendency to separate (stall) easily.

The *supercritical airfoil* shown in Figure 14.3 introduced in Chapter 3 was designed in the 1970s by NASA to correct this deficiency and lower fuel costs. It currently is in use on corporate jets, military

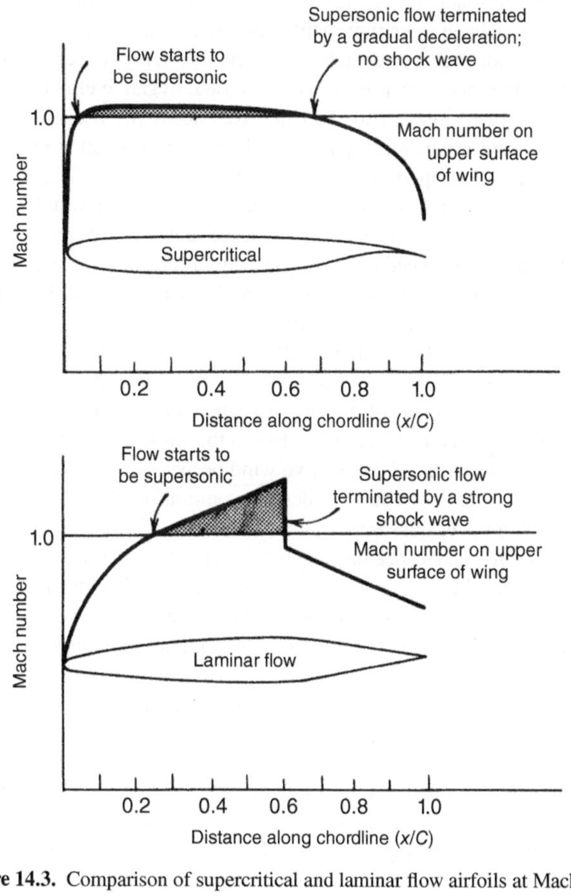

Figure 14.3. Comparison of supercritical and laminar flow airfoils at Mach 0.75.

aircraft, and transport category aircraft around the world to increase the cruise Mach number and improve fuel efficiency. Flatter on the top of the wing and rounded on the bottom with a reflex cambered surface near the trailing edge, it is shaped so that the normal shock will occur where the upper surface pressure gradient is favorable or zero. At that point, the boundary layer is able to encounter the pressure increase across the shock wave without separating. The surface curvature on the top surface of the supercritical airfoil is less than a conventional laminar airfoil, so the local Mach number will be less than that for the laminar flow airfoil delaying the onset of the critical Mach number enabling a higher cruise Mach number. If M_{crit} is exceeded, the airflow will be supersonic nearer the leading edge of the supercritical airfoil. However, the airflow will remain at a much lower Mach number, and the supersonic flow will be terminated by a gradual deceleration. Therefore, no shock wave will result, and drag will be less. Though some lift is lost due to less curvature of the top of the wing, the trailing edge is more curved than a conventional wing and a portion of the lift is restored. The reduced drag and increased M_{crit} of a supercritical wing allow higher airspeeds and/or better fuel efficiency than previous designs, making it easy to see why it is so widely used on both military and civilian jet aircraft.

Sweepback

We have already seen that swept-back wings offer better static directional stability and static lateral stability than straight wings. There is an even more important reason why swept-back wings are almost universally used in transport-category jet aircraft; however, wing sweepback delays the effects of compressibility and increases the critical Mach number. Because M_{crit} is increased, negative effects of exceeding the critical Mach, like force divergence Mach number and Mach tuck (defined in the Transonic Flight section), are also delayed, which allows the aircraft to travel faster before their onset. Instead of presenting the complicated conventional discussion of this, a simpler explanation is presented here.

Unlike a straight-wing aircraft, the leading edge of the swept-wing aircraft does not strike the oncoming air at 90°, so the full effect of compressibility is not felt until the swept-wing aircraft is flying at a higher speed compared to the same speed of a straight-wing aircraft. By sweeping the wings, the effective chord (parallel to the aircraft's longitudinal axis) has been lengthened, but the wing thickness has not been changed. Thus, the ratio of thickness/chord has been reduced. In effect, the wing thickness has been reduced, and a higher M_{crit} results. Swatton (2011, p. 363) shows that the approximate reduction in Mach number from a straight-wing design to a swept-wing design can be calculated by multiplying the straight-wing Mach number by the cosine of the wing-sweep angle. For example, if the acceleration of air over a straight-swing design resulted in Mach 0.85, then in the same airplane with a swept-wing angle of 40°, the "new" Mach number in terms of the relative wind perpendicular to the new leading edge would be Mach 0.65. This reduction due to swept-wing design would allow the aircraft to have a higher cruise Mach number before reaching the critical Mach number.

While the increase in stability and, especially, the increase in M_{crit} are great advantages of a swept wing, there are disadvantages to sweeping the wings. For example, since the wings in a swept-wing design are no longer at right angles to the relative wind, primary and secondary airfoils such as ailerons and flaps no longer have the same impact as if they were installed on a straight-wing design. As discussed in Chapter 10, swept-wing aircraft lose aileron effectiveness early in a stall, unlike straight-wing aircraft.

Also, as previously discussed, the coefficient of lift curve is affected by sweep in two ways (Figure 14.4). First, the value of $C_{L(max)}$ is reduced, so takeoff and landing speeds are increased. Second, the curve flattens and has no sharp stall angle of attack (AOA) point.

The lack of a definite stall point has caused some complacency in pilots who are making a transition from straight-wing to swept-wing aircraft. While it is true that the swept-wing aircraft does not show a clean stall, it is also true that sweeping the wings decreases the aspect ratio, and this means a large increase in the induced drag coefficient. This is particularly dangerous during takeoffs and landings, as

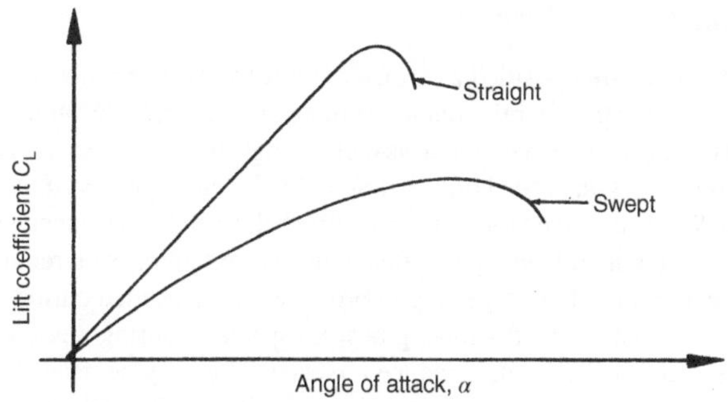

Figure 14.4. Effect of wing sweep on a $C_L - \alpha$ curve.

was emphasized earlier. If the same wing area and aspect ratio of a straight-wing design were required in a swept-wing design, the new wing span would have to be larger, and this would introduce twisting forces that are commonly experienced by high-speed flight in aircraft with longer wing spans. Wingtip stall is increased by swept-back wings, as was first presented in Chapter 10; the greater the wing sweep and wing taper, the greater the impact of wingtip stalls.

Vortex Generators

The use of vortex generators to energize the boundary layer and thus delay separation in the low-speed region of flight was discussed in Chapter 4. The same general principles apply to high-speed flight. Shock-induced separation occurs because the boundary layer does not have enough energy to overcome the adverse pressure gradient through a shock wave. Vortex generators mix higher kinetic energy air from outside the boundary layer with the slower air in the boundary layer and delay separation.

If a normal shock wave does develop, vortex generators are effective in breaking it up and are usually placed at about a 15° angle of attack to the airstream, well aft of the leading edge of the airfoil. The additional drag of the generators is small in comparison to the wave drag that they help dissipate. Figure 14.5 shows an example of vortex generators on a wing.

Figure 14.5. Vortex generators.
Source: U.S. Department of Transportation (2018)/Public domain.

High-Speed Subsonic Control Surfaces

Control surfaces on an aircraft are classified as primary and secondary controls, as reviewed in Chapter 3. The primary controls maneuver the aircraft around the three axes: lateral (elevator/stabilator), longitudinal (ailerons), and vertical (rudder). Secondary controls include high-lift devices such as flaps and slats, as well as other devices such as spoilers and speed brakes. Figure 14.6 includes many of the control surfaces used for high subsonic flight. Wings designed for high-speed flight also need to be maneuverable at lower speeds, especially when landing or taking off, and quite often to meet a certain mission requirement. Leading edge flaps and slats assist the aircraft at lower speeds and produce the C_L needed during slow-speed flight.

Due to the low drag of some aircraft, slowing down, especially during descent, can be problematic, so flight spoilers act as speed brakes to help reduce airspeed. This may be necessary to meet airspace or arrival route speed restrictions, for example. As introduced in Chapter 3, spoilers disrupt the air going over the top of the wings, and on the ground during landing or a rejected takeoff they markedly reduce lift and

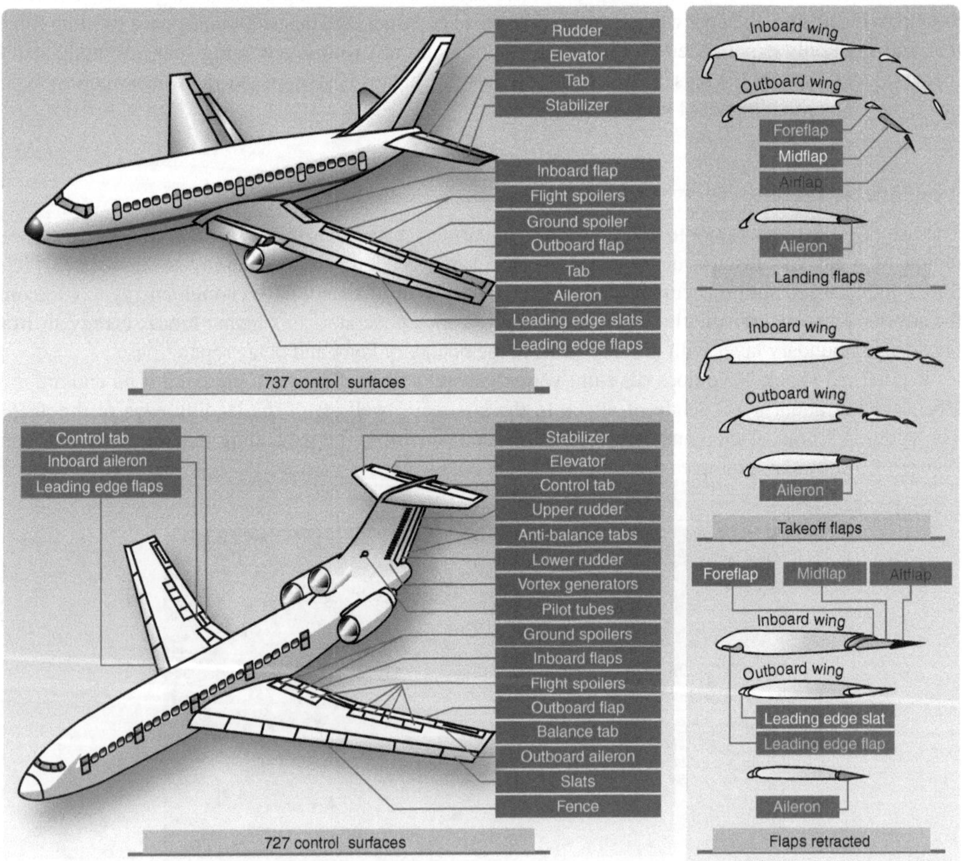

Figure 14.6. High-speed subsonic flight control surfaces.
Source: Pilot's Handbook of Aeronautical Knowledge (2016b)/U.S. Department of Transportation Federal Aviation Administration/Public domain.

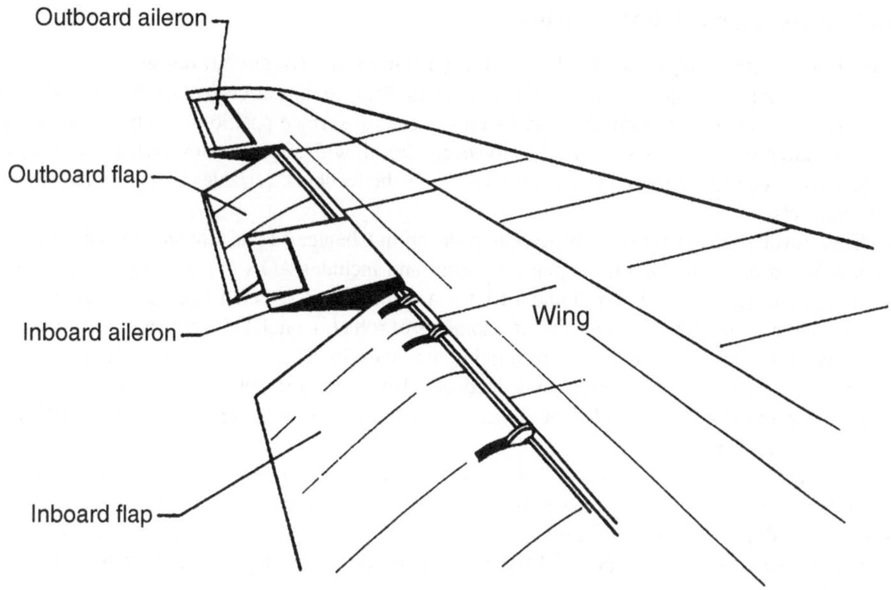

Figure 14.7. Inboard and outboard ailerons.
Source: NASA/NASA-SP-367/Public Domain.

place more weight on the wheels, increasing braking action. During flight in many jet aircraft, spoilers (roll spoilers) work together with the ailerons to provide roll control, reducing the twisting forces of the ailerons on the wing. For example, in a roll to the right, the ailerons remain neutral but the spoilers protrude slightly from the upper surface of the right wing, disrupting lift and causing the aircraft to roll to the right. Because spoilers also increase drag, they cause yaw in the direction of the bank, so the use of spoilers during turns also reduces adverse yaw.

It is also common in aircraft designed for high-speed flight to see two sets of ailerons on an aircraft: inboard high-speed ailerons and outboard low-speed ailerons (Figure 14.7). As we saw in Chapter 3, the outboard ailerons are near the end of the wing; as the aircraft turns during high-speed flight, the outboard aileron may also create a twisting or bending force on the wing due to the long moment arm during roll around the longitudinal axis. Often, these outboard ailerons are locked during high-speed flight and the inboard ailerons (high-speed ailerons) are utilized. Inboard ailerons are usually smaller and due to their position closer to the fuselage, the moment arm is smaller resulting in less twisting of the wing. Some aircraft, notably the Boeing 777, incorporate flaperons on the trailing edge of the wings. As their name suggests, these control surfaces combine the characteristics of both flaps and ailerons. Not only do they serve to control roll in the same way as conventional ailerons, but they also are symmetrically extended during landing to function as flaps.

As previously mentioned, many high-speed aircraft also incorporate tailplane designs with a variable incidence horizontal stabilizer. In addition to the multiple stability improvements incorporated by the vertical tail, a large aircraft that experiences higher aerodynamic loading due to higher velocities requires a higher range of trim response for varying pitch response. To minimize the amount, the elevator needs to travel during a wide range of weight and trim changes, the horizontal stabilizer moves through various fixed positions during the flight profile from takeoff to landing to minimize the range of movement required by the elevator.

Flight Control Augmentation Systems

Since the introduction of high-speed military jet aircraft in the 1950s, aircraft designers have employed augmentation systems to improve aircraft flying qualities. Prior to this, aircraft exhibited *open-loop* flight control systems where the pilot directly controlled the control surface positions. Such is still the case in light training aircraft such as the Cessna 172. However, aircraft with augmentation systems are considered *closed-loop* flight control systems where the pilot controls the feedback variables as opposed to the control surfaces themselves.

Feedback involves the stability derivatives introduced in Chapter 12 and Chapter 13. The appropriate feedback is based on actual aircraft response in flight and includes AOA (α), sideslip (β), roll rate (p), pitch rate (q), and yaw rate (r). See Table 12.1 for a summary of selected stability derivatives. These feedback variables impact the resultant aircraft moments of roll (L'), pitch (M), and yaw (N). For example, as introduced in Chapter 12 in reference to longitudinal stability, C_{mq} (coefficient of pitch moment due to pitch rate) provides pitch damping using feedback from the horizontal tail (see Figure 14.8). For lateral-directional motion, C_{Nr} (coefficient of yaw moment due to yaw rate) provides yaw damping using feedback from the rudder.

Yaw dampers are a common feature on many aircraft to control undesirable motions such as Dutch roll ($C_{\text{L}'\beta}$). Simple closed-loop feedback systems such as yaw dampers are known as *stability augmentation systems* (SAS). More advanced closed-loop systems are known as *control augmentation systems* CAS, do not confuse with calibrated airspeed. These augmentation systems may utilize mechanical linkages or digital linkages known as FBW.

The first production aircraft to incorporate an FBW flight control system was the F-16. By altering the feedback provided to individual control surfaces, aircraft such as the F-16 can provide tailored flying qualities dependent on the phase of flight. For example, in most phases of flight, the F-16 uses AOA (α) feedback for longitudinal stability. However, for specific tasks such as aerial refueling, the feedback changes to pitch angle (θ). This means that when aerial refueling, the aircraft will hold a particular pitch angle making position control behind the tanker easier when minimal maneuvering is required. The ramifications of this change in flight control laws are apparent if the F-16's aerial refueling door is opened in inverted flight (for demonstration purposes only). The aircraft (without pilot input) will hold the same pitch attitude, maintaining level flight at negative one G, instead of the nose tracking downward as most aircraft would do.

Many aircraft now incorporate multiple computers that control FBW flight controls, not only saving on weight as well as maintenance costs but also aiding in providing a degree of envelope protection to prevent the pilot from exceeding set limits. Whereas in conventional aircraft the flight controls are operated through mechanical linkages, in FBW aircraft, the flight controls are controlled through the use of electrical wiring and commands by the various flight computers. For example, in the Airbus family, the primary flight controls are usually controlled through inputs from the pilots using a side stick, using computers such as

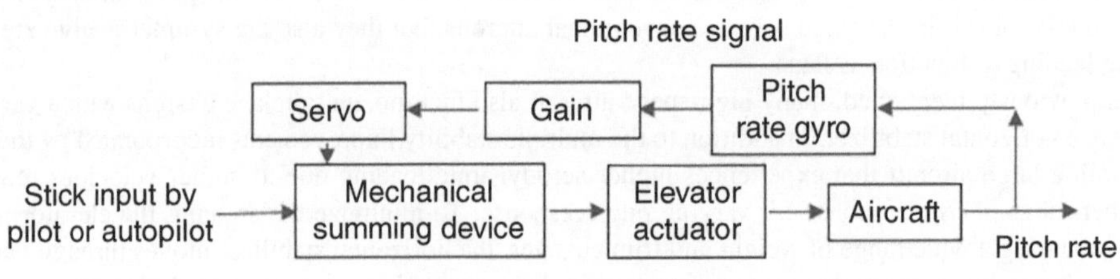

Figure 14.8. Closed-loop system.

the Elevator Aileron Computer (ELAC), Spoiler Elevator Computer (SEC), and the Flight Augmentation Computer (FAC). The elevator/aileron computers along with the spoiler/elevator computers control the movement of the elevators and ailerons and automatically compute the correct input of spoilers and yaw damping. FACs compute the maximum operating limits for the aircraft, including wind shear and speed information for the cockpit displays. In addition, FACs also control yaw damping and the rudder to maintain turn coordination. Further discussion of flight control design and the importance of stability derivatives is beyond the scope of this textbook.

TRANSONIC FLIGHT

Early attempts to "break the sound barrier" were not successful because of problems encountered in the transonic region. When these problems were better understood, several important design changes were made, and today supersonic aircraft have little difficulty in flying through this speed region. Subsonic aircraft do not incorporate these design changes and will experience difficulties if the critical Mach number is exceeded, hence the importance of delaying the onset of M_{crit}. This discussion also applies to subsonic aircraft and helicopters, if they venture into the transonic region of flight, as well as to supersonic aircraft while operating in the transonic region.

Wave Drag and Force Divergence

At airspeeds of about 5% above M_{crit}, the normal shock wave on the top of the wing causes the boundary layer to separate from the wing. This causes a change in the aerodynamic force coefficients, both C_L and C_D, as well as pitching moments. This airspeed is called the *force divergence Mach number* or the *drag divergence Mach number* (M_{DD}).

Figure 14.9 shows how C_D varies with Mach number (for a constant value of C_L). When the drag divergence Mach number is reached, the C_D increases rapidly. The drag increase is caused by the energy loss across the normal shock wave and by boundary layer separation and is called wave drag. Note that even though the value of C_D again decreases above Mach 1, the total drag continues to increase with airspeed. The rate of increase will be less, however, above Mach 1.

In addition to the drag increase behind the normal shock wave, the airflow separation also results in a loss of lift. Figure 14.10 shows a sudden lowering of the coefficient of lift at the force divergence Mach number. Shock-induced separation creates a local stall situation similar to the low-speed stall, except that it occurs behind the normal shock wave rather than at the trailing edge of the wing. This causes the center

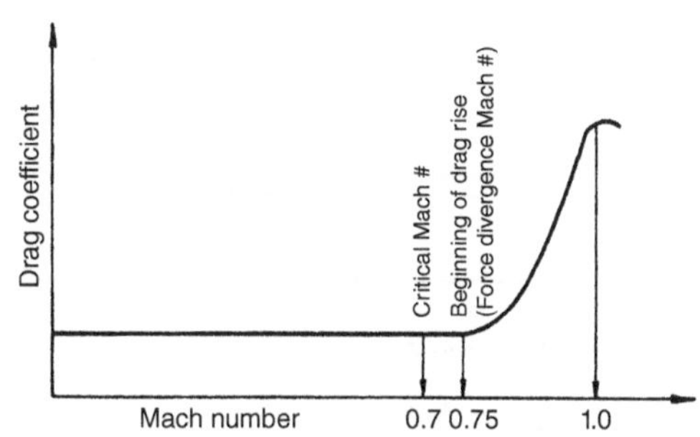

Figure 14.9. Force divergence effect on C_D.

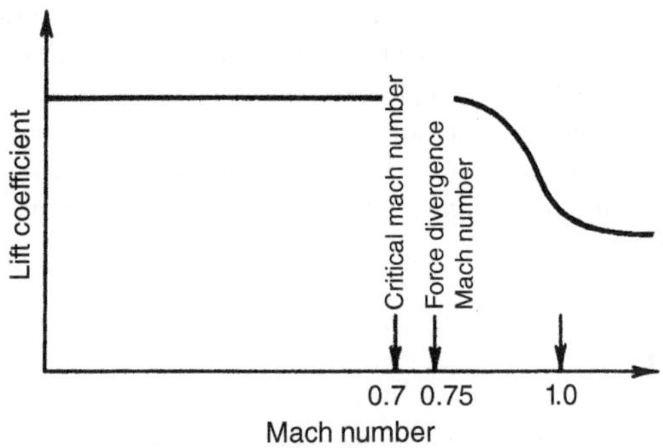

Figure 14.10. Force divergence effect on C_L.

of pressure to be shifted forward and produces a nose-up pitching moment. If one wing develops a normal shock before the other, a rolling moment can be produced toward the wing with the earlier shock. If a wing drops, the aircraft will tend to yaw in that direction, and a condition similar to Dutch roll may develop. Sweepback reduces the effects of exceeding the force divergence Mach number by lowering the magnitude of the changes in C_L, C_D, and pitching moments allowing the pilot more control. Figure 14.11 shows the impact of a thinner wing on the M_{DD}, note as the thickness-chord ratio (*t/c*) decreases, the Mach number where M_{DD} is reached increases.

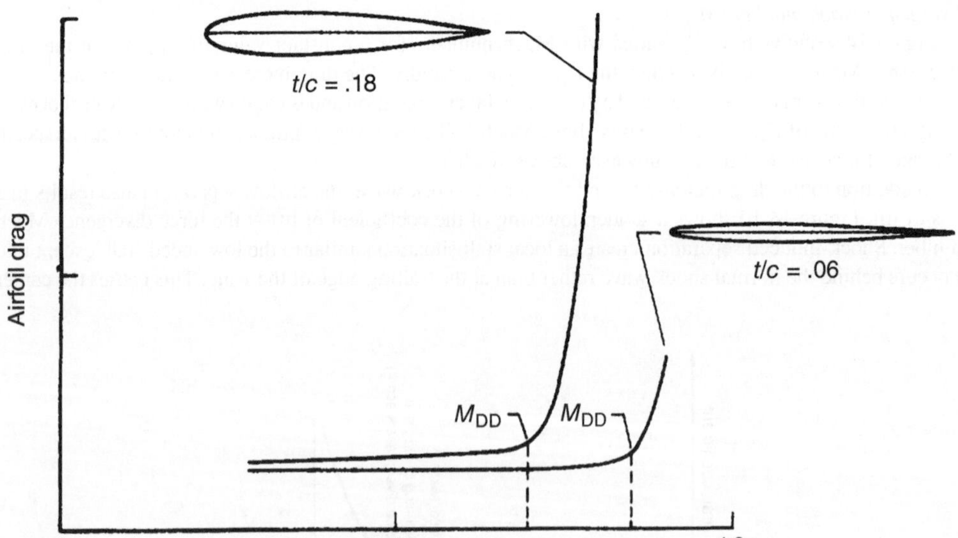

Figure 14.11. Thick airfoils and drag-divergence Mach.
Source: NASA/NASA-SP-367/Public Domain.

In summary, the effects of reaching force divergence Mach number are:

- An increase in C_D for a given value of C_L.
- A decrease in C_L for a given AOA.
- A change in pitching moment as the aerodynamic center shifts.

Mach Tuck

Once a shock wave forms on the wing, and separation of the boundary layer occurs, pitching moments change on an aircraft and the nose of the aircraft will drop. This is referred to as *Mach tuck* or *tuck under*. Depending on the position of the horizontal stabilizer (or stabilator), pitch authority may not be sufficient to compensate for the nose drop without significant input from the pilot. Three factors may contribute to a nose-down pitching moment in transonic flight:

1. At airspeeds above the force divergence Mach number, the normal shock wave on top of the wing increases in intensity and moves backward. A second normal shock wave then appears on the bottom of the wing (Figure 14.12).
2. These shock waves move aft as the top shock moves aft, the separation point also moves aft, and so does the center of pressure, thus adding to the Mach tuck tendency.
3. The aerodynamic center moves from the quarter chord point to the 50% chord point when the aircraft reaches supersonic flight, again influencing the pitching moment of the aircraft.

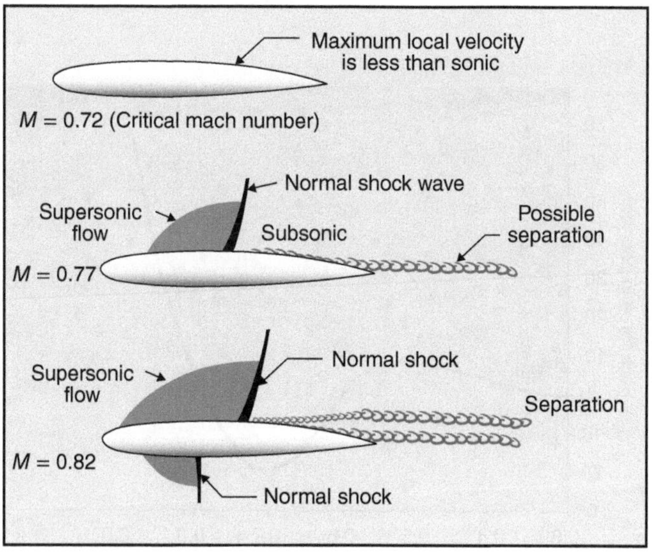

Figure 14.12. Normal shock wave on bottom of the wing.
Source: Airplane Flying Handbook (2004a)/U.S. Department of Transportation Federal Aviation Administration/Public domain.

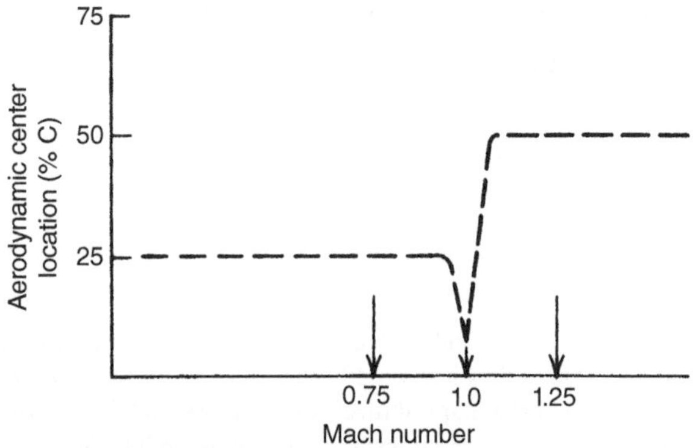

Figure 14.13. Aerodynamic center location shift.

All aircraft flying supersonically suffer some amount of nose-down pitching moment, though due to automatic flight augmentation systems compensating for the nose-down effect, the resultant impact felt by the pilot may be negligible. The shift of the aerodynamic center is not a smooth movement. Sharp leading edges make the shift smoother, but in many aircraft, the aerodynamic center moves forward before the eventual rearward shift (Figure 14.13). This movement can cause a pitch up moment in the early stages of transonic flight, but eventually, it will cause Mach tuck as the speed increases. Figure 14.14 shows the stick forces for a typical jet aircraft as the Mach number increases. As the critical Mach number is approached, lift increases, and stick push is required to maintain level flight as the nose rises. As the shock wave forms above Mach 0.70 in this example, the nose begins to "tuck" and aft stick (pull) is required to compensate.

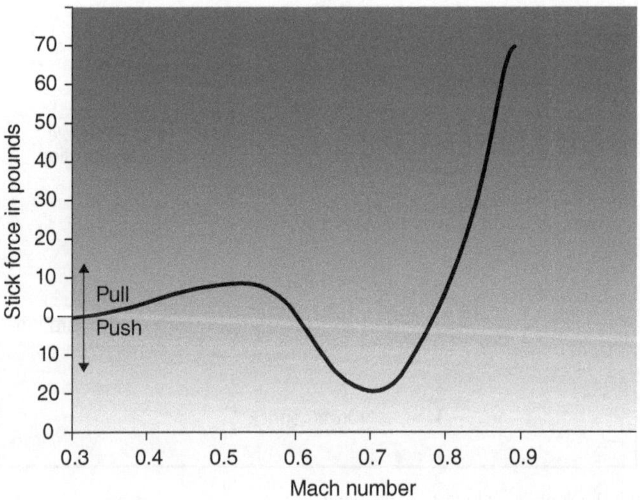

Figure 14.14. Stick forces versus Mach number.
Source: Airplane Flying Handbook (2004a)/U.S. Department of Transportation Federal Aviation Administration/Public domain.

High-Speed Mach Buffet

The violently turbulent separated air behind a normal shock wave often produces buffeting of the aircraft. Regardless of the aircraft speed, a high-speed buffet will occur when the AOA is such that boundary layer separation begins. The buffet is caused by the flow of turbulent air over the horizontal stabilizer, especially when the horizontal stabilizer is located lower on the tail in the wake of the separation from the wing. The T-tail configuration helps eliminate this problem by putting the horizontal tail above the wing wake, though as we saw in earlier chapters this T-tail design can adversely affect stall recovery.

At higher altitudes, where the density of the air is less, a higher AOA is required to maintain level flight, thus placing the AOA of the airplane closer to the critical AOA. As previously discussed, at a higher aircraft weight (especially at higher altitudes) an aircraft must produce a greater AOA to maintain level flight, so higher weight also puts the aircraft closer to the critical AOA and boundary layer separation. As an aircraft experiences higher G-loading due to level turns or turbulence, a higher AOA is required to maintain the same lift, and once again, the AOA is closer to the critical angle of attack and boundary layer separation.

For all of the reasons discussed above, flight in the transonic range is undesirable, so supersonic aircraft fly through this region as rapidly as possible.

Control Surface Buzz and Flutter

Another hazard to high-speed flight is *buzz* and *flutter*. Boundary layer separation acting on control surfaces often causes rapid oscillations called buzz or flutter, and in extreme situations can even cause twisting (torsion) of the wings in flight. As the aircraft continues to accelerate, and/or the vibration of the surface amplifies and develops into divergent oscillations, loss of control and/or structural damage could occur. Control surface buzz can very quickly lead to metal fatigue damage to hinge fittings and other parts of the control surfaces, and if prolonged or severe can cause structural failure. Flutter can occur on a variety of control surfaces, from the wing to the ailerons, to the elevators (or ruddervators). Aircraft are tested during certification for flutter, but flutter could occur if V_{NE} is exceeded, and slack control surface cables or other maintenance deficiencies may lead to control surface buzz or flutter at airspeeds well below V_{NE}.

SUPERSONIC FLIGHT

As the flight Mach number increases in the transonic speed range, the upper and lower normal shock waves increase in size and strength and move to the trailing edge of the wing, as shown in Figure 14.15.

When the flight Mach number is 1.0, the entire airfoil is supersonic, except the leading-edge stagnation area. The shock waves at the trailing edge are still normal shock waves, so the velocity behind them is still subsonic. A new shock wave now appears ahead of the airfoil. It is a compression wave just like the normal shock wave, and because the air at the leading edge is subsonic, the new wave, called a bow wave, must also be a normal shock wave, at least in the area directly ahead of the leading edge. The air ahead of the shock wave is moving faster than the speed of sound, and the air between the shock wave and the leading edge is subsonic due to stagnation, so a normal shock wave exists.

Figure 14.16 shows a bow wave ahead of the leading edge. Full supersonic flight exists when the bow wave becomes attached to the leading edge of the wings, nose of the aircraft, tail sections, and all other parts of the aircraft.

The bow wave may not attach easily if the leading edges are too blunt. Sharp leading edges are used on supersonic aircraft, but it still is necessary for the flight Mach number to be greater than 1.0 for the airflow over the entire wing to be supersonic. Thus, the supersonic range starts at about Mach 1.2–1.3, depending

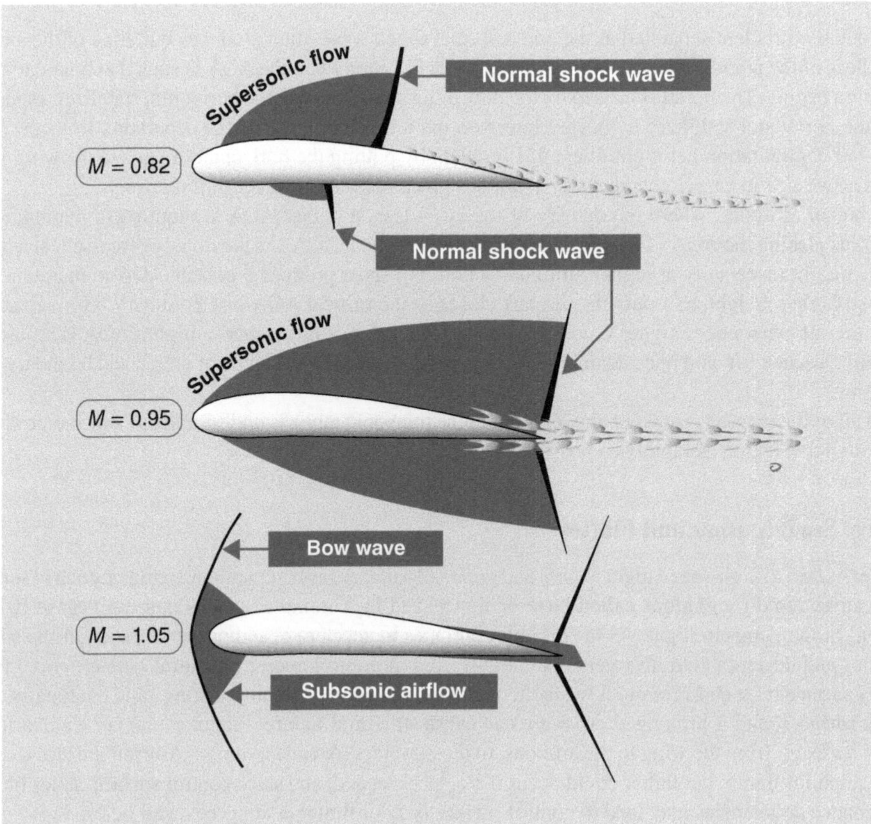

Figure 14.15. Normal shock waves move to trailing edge.
Source: Pilot's Handbook of Aeronautical Knowledge (2008a)/U.S. Department of Transportation Federal Aviation Administration/Public domain.

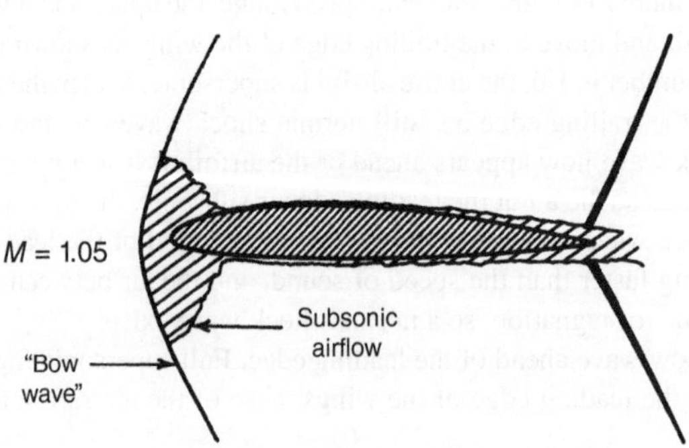

Figure 14.16. Unattached bow wave at transonic speed.

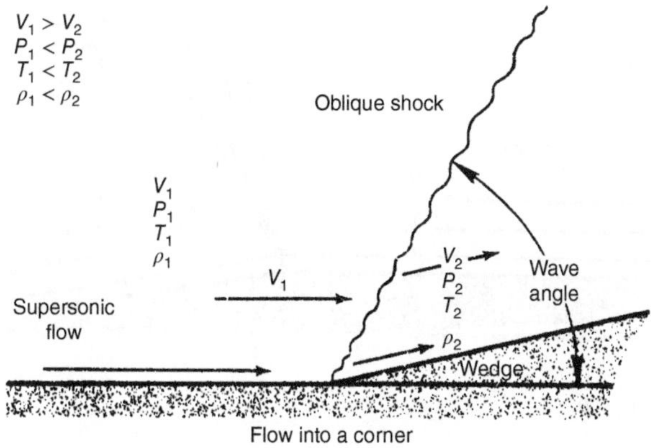

Figure 14.17. Formation of an oblique shock wave.

on individual aircraft design. The normal waves at the trailing edges also are deflected backward and now are called *oblique shock waves*.

Oblique Shock Waves

An oblique shock wave, like a normal shock wave, is a compression wave. It differs from the normal shock wave because the airflow direction changes as air flows through it and slows down but remains supersonic. To see how an oblique shock wave forms, consider a wedge-shaped object placed in a supersonic flow, as shown in Figure 14.17.

The airflow next to the horizontal surface must change direction when the wedge is encountered. It does this and assumes the direction of the arrow V_2. This sudden change in direction, as the air molecules turn into the path of other molecules, V_1, traveling in the horizontal direction, produces a shock wave at an oblique angle to the horizontal. The angle that the wave makes with the horizontal is called the wave angle. This is shown in Figure 14.17.

The wave angle depends on the Mach number of the approaching flow and the angle of the wedge. It is the sum of the Mach wave angle and the wedge angle (Eq. 14.3):

$$\sin \mu = \frac{1}{M} \tag{14.3}$$

where

μ(mu) = Mach wave angle (degrees).

EXAMPLE

$M = 2.0$ and the wedge angle is 10°, then

$$\text{Wave angle} = \text{arc } \sin 0.5 + 10^\circ = 30^\circ + 10^\circ = 40^\circ$$

The oblique shock wave is weaker than the normal shock wave, but it still is a shock wave and wastes energy. As air passes through an oblique shock wave, its density, pressure, and temperature all rise and

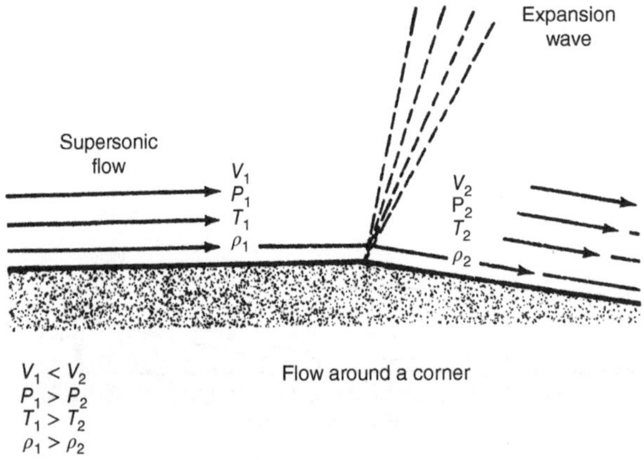

Figure 14.18. Formation of an expansion wave.

its velocity decreases. *The principal difference is that the air remains supersonic behind an oblique shock wave, but it is always slowed to subsonic behind a normal shock wave.*

Expansion Waves

The third kind of wave that occurs in supersonic flow is the expansion wave. Unlike the normal and oblique shock waves, the expansion wave is not a shock wave. In fact, the expansion wave is just the opposite of a compression wave. Density, pressure, and temperature all decrease as the air flows through an expansion wave. Air velocity increases through this wave and no energy is lost. Expansion waves form where the airstream turns around a convex corner, as on the trailing edge of some airfoils. This is shown in Figure 14.18. In flowing around the convex corner, the airstream makes an infinite number of minute direction changes until it has turned the corner and the flow direction is parallel to the downstream surface. The expansion wave takes place through a fan-shaped area, as shown by the dashed lines in Figure 14.18.

Shock waves, on the other hand, take place through a very narrow area. As the static pressure decreases through an expansion wave, lift is produced in this region. Subsonic flow could not negotiate a convex corner because the flow would separate from the wing. However, supersonic flow has a higher energy level and can turn these sharp corners by means of an expansion wave, without separation.

Characteristics of the waves are summarized in Figure 14.19.

Aerodynamic Forces in Supersonic Flight

Aerodynamic forces are produced by differential pressures acting on the surfaces of a wing in an airstream. This is true whether the airstream is subsonic or supersonic. In supersonic flight, sudden increases in pressure result when the airflow passes through shock waves. Static pressure drops when passing through an expansion wave, though gradually, while static pressure increases along with temperature when passing through oblique and normal shock waves. An expansion wave at a sharp corner produces a fairly rapid pressure drop, while an expansion over a curved surface will produce a gradual pressure drop.

Type of wave formation	Oblique shock wave	Normal shock wave	Expansion wave
	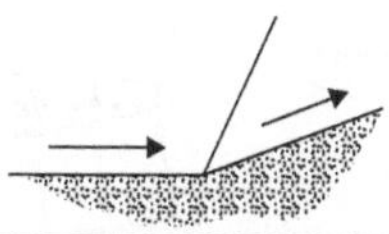	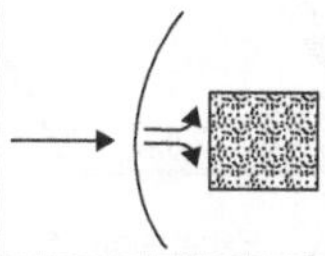	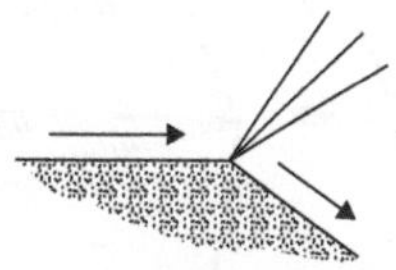
Flow direction change:	"Flow into a corner," turned to flow parallel to the surface	No change	"Flow around a corner," turned to flow parallel to the surface
Effect on velocity and Mach number:	Decreased but still supersonic	Decreased to subsonic	Increased to higher supersonic
Effect on static pressure and density:	Increase	Great increase	Decrease
Effect on energy or total pressure:	Decrease	Great decrease	No change (no shock)
Effect on temperature:	Increase	Great increase	Decrease

Figure 14.19. Summary of supersonic wave characteristics.

Supersonic Airfoils

One type of supersonic airfoil is the double-wedge airfoil shown in Figure 14.20 This airfoil cannot be used subsonically because the air will become separated when it reaches the convex angles at the top and bottom of the wedges. It does have use in supersonic missiles that are boosted through the subsonic flight region.

Figure 14.20a shows the wave pattern about a double-wedge airfoil at zero AOA and thus zero lift in supersonic flight. Oblique shock waves, of the same strength, form at the top and the bottom of the leading and trailing edges. Behind these shocks the pressure rises but no net lift results. At the sharp corners at the 50% chord point, expansion waves form, of the same strength. Lower pressures exist behind the expansion waves, but again the pressures on the top and bottom are equal in value and cancel each other. This is shown in Figure 14.20b. The rearward components of the pressure arrows represent "wave drag" that exists even when no lift is being developed.

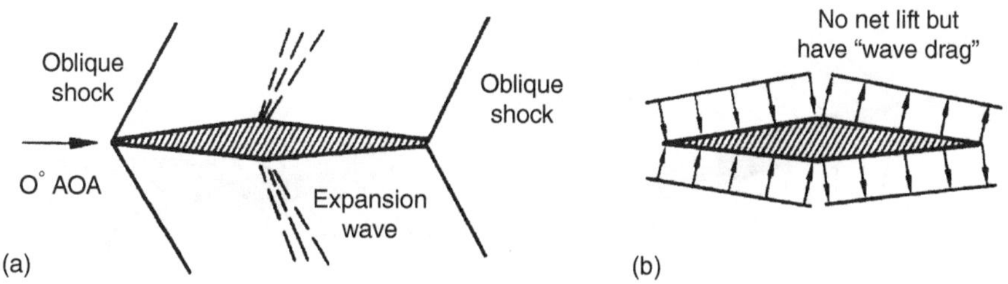

Figure 14.20. Double-wedge airfoil in supersonic airflow: (a) wave pattern and (b) pressure distribution.

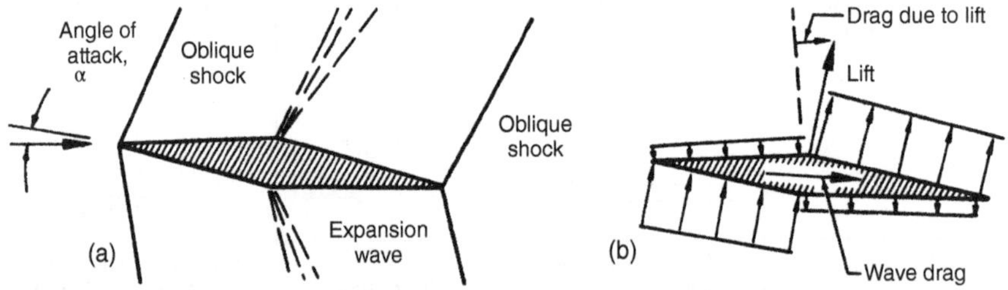

Figure 14.21. Double-wedge airfoil developing lift: (a) wave pattern and (b) pressure distribution.

Consider the double-wedge airfoil at a small positive AOA, as shown in Figure 14.21a. The oblique shock waves are now unequal in strength. The shock wave at the top of the leading edge does not have to turn at as high an angle as it did at zero AOA and is now weaker. The pressure rise through the top shock is now lessened.

The reverse is true for the bottom leading-edge shock. It is stronger and the pressure rise is greater than at zero AOA. The expansion waves at the midpoint corners are not of equal strength at a positive AOA. The top expansion is now greater because the angle to be turned is greater than it was at zero AOA. The pressure decrease is also greater on the top. The opposite reasoning applies to the bottom expansion wave, and the pressure decrease behind the bottom expansion wave is lessened. Net lift is now being developed. The pressure distribution is shown in Figure 14.21b.

To avoid the subsonic problems of the double-wedge airfoil, another supersonic airfoil section is used. This is called the circular arc or biconvex airfoil and uses two arcs of circles to define its shape. With sharp leading edges, the bow wave attaches easily. This airfoil develops shock waves at the leading and trailing edges, but because there are no sharp angles at the midchord point, no concentrated expansion wave occurs at that point. Instead, a continuous expansion wave, from the leading edge to the trailing edge, forms on the top and bottom airfoil surfaces. A symmetrical circular arc airfoil produces no net lift at zero AOA because the waves are symmetrical. But at positive AOA, the shock wave on the top leading edge is weaker and the top expansion wave is stronger than the corresponding waves on the bottom of the airfoil. Thus, lift is created as shown in Figure 14.22. The center of pressure and the aerodynamic center are both at the same position for both the double-wedge and the circular arc airfoils. This position is at the 50% chord

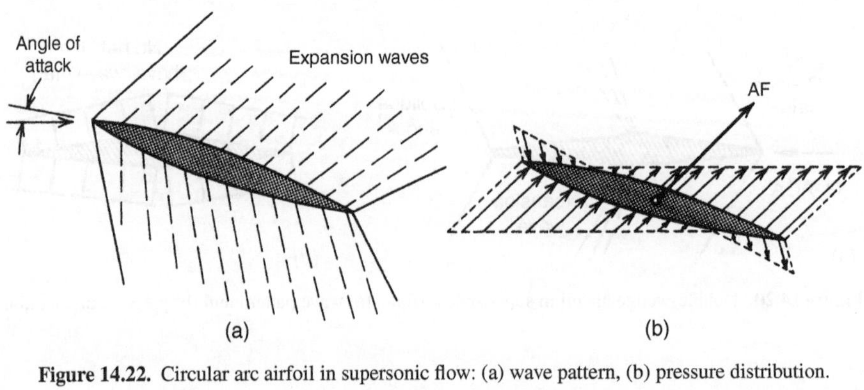

Figure 14.22. Circular arc airfoil in supersonic flow: (a) wave pattern, (b) pressure distribution.

point. The circular arc airfoil can be flown at subsonic speeds. However, like all supersonic airfoils, it is quite thin, so its $C_{L(\max)}$ is low. Thus, its performance at low airspeeds is poor, and high takeoff and landing speeds are required.

Wing Planform

In the Sweepback section of this chapter, we discussed the use of sweepback as a means of increasing M_{crit} for subsonic aircraft. Sweepback has other advantages for supersonic aircraft as well. In addition to increasing the force divergence Mach number, sweepback also reduces the increase in C_D. This is shown in Figure 14.23. The increase in drag coefficient is both delayed and reduced by sweepback.

It is also interesting to compare straight wings (0° sweep) with swept wings. Note that the straight-wing and the swept-wing curves cross at some Mach number greater than 1.0. For the 45° wing, they cross at about $M = 1.4$. This means that at this speed both wings would have the same drag and at higher speeds, the straight wing actually has less drag. Therefore, for high-supersonic aircraft, the straight-wing configuration may be desirable. Of course, at speeds lower than the crossover point, the advantage is with the swept-wing aircraft. Figure 14.24 shows the variation in aerodynamic efficiency using $L/D_{\max}$ to indicate the mission profile requirements of straight-wing and swept-wing design. Aging variable sweep aircraft designs are being replaced by modern designs that combine the advantages of both straight and swept-wing design along with improved lighter structural components, without the complexity and weight of variable sweep mechanisms.

In our discussion of the formation of oblique shock waves, we limited our analysis of Mach wave angle to two-dimensional flow. In reality, the pressure disturbances radiate outward in all directions, like expanding spheres. Thus, instead of a Mach wave, we have a Mach cone. This is shown in Figure 14.25.

Another advantage of sweep is that it places the wing within the Mach wave (cone) of the aircraft. This allows subsonic airfoil sections to be used for supersonic flight. Using subsonic airfoils is desirable because they develop higher $C_{L(\max)}$ values and thus reduce takeoff and landing speeds.

Consider a Mach wave that is generated at the wing root as in Figure 14.26. The air is supersonic ahead of the Mach wave and it is also supersonic in the direction of flight behind the Mach wave. The component of air behind the Mach wave that is perpendicular to the Mach wave will, however, be subsonic. If the wing is swept back so that its leading edge is behind the Mach wave, it will be flying in a subsonic flow and thus can use a subsonic airfoil section.

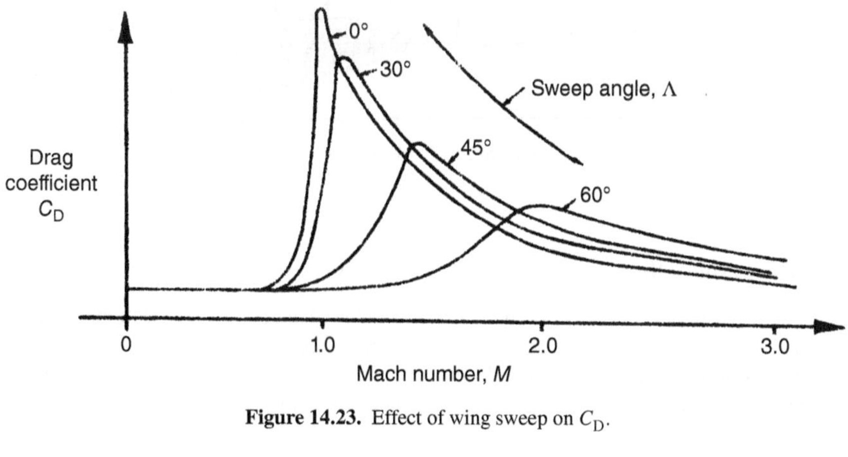

Figure 14.23. Effect of wing sweep on C_D.

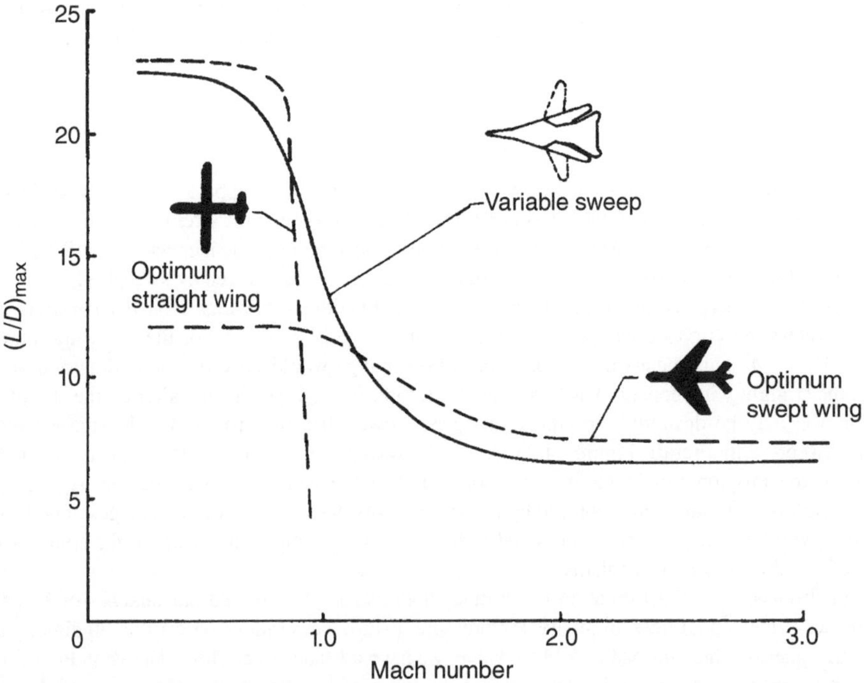

Figure 14.24. Variable sweep and drag-divergence Mach.
Source: Courtesy of NASA, NASA-SP-367.

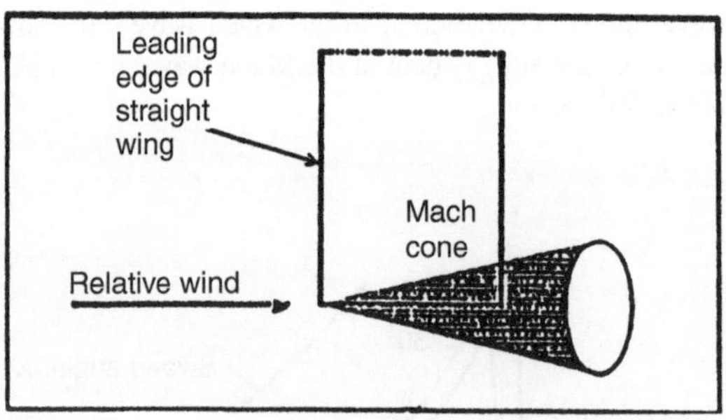

Figure 14.25. Mach cone.

Area Rule Drag Reduction

Experiments in wind tunnels showed that bodies with sudden changes in cross-sectional area produced more drag than similar bodies with gradual changes in area, especially at speeds approaching Mach 1; sometimes referred to as *Whitcomb's area rule* after the engineer who pioneered these early wind tunnel tests. This increase in primarily wave drag resulted in a need for additional thrust in order to push through

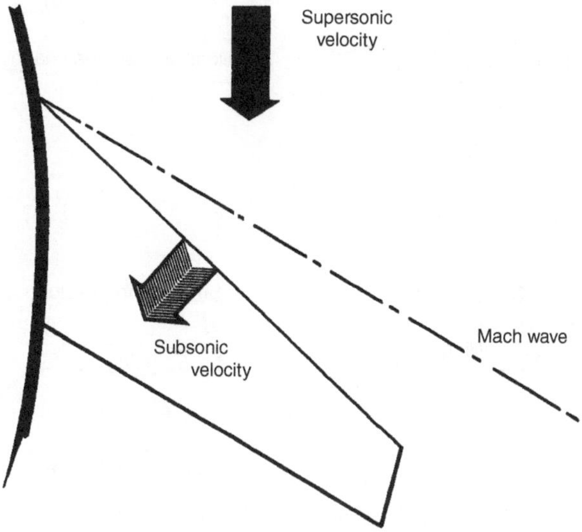

Figure 14.26. Swept wing in supersonic flight.

the transonic range and beyond. The frontal cross-sectional area of the entire airplane is considered in the aircraft design phase to minimize total drag; the fuselage, wings, and tail area are considered as a whole and not as individual components. A large increase in the cross-sectional area occurs at the wing root where the wings attach to the fuselage. By reducing the fuselage area at this point, the total area was enlarged gradually and drag was reduced. At the trailing edge of the wing, the area of the fuselage was again increased to compensate for the loss of wing area. This resulted in the "coke bottle," or wasp waist, shape of the fuselage, first used on early jet fighters. Although the distinct "coke bottle" shape is no longer used in modern aircraft design as more powerful engines now exist to overcome drag, the principle behind the area rule is still used to minimize drag on aircraft, especially in high-subsonic aircraft such as the design with the classic forward "hump" in Boeing's 747 as one example.

Figure 14.27 shows the ideal fuselage shape versus a modified fuselage shape with a "waist" to reduce the cross-sectional area. Note the cross-sectional area for both Figure 14.27a and b, the cross-sectional area for the fuselage with the "waist" is much less, resulting in a reduction in wave drag as well as in the magnitude of shock waves in high-speed flight.

Control Effectiveness

At the high speeds found in the transonic and supersonic ranges, shock-induced separation can reduce the effectiveness of control surfaces in two ways.

1. If the surface is located in a region of stalled air, it is operating in an "aerodynamically dead" air mass and cannot produce effective aerodynamic forces.
2. Control surfaces are effective if they change the airflow around the entire wing or stabilizer. When there is a shock wave ahead of the control surface, deflection of that surface cannot influence the airflow ahead of the shock wave, so what little aerodynamic force is developed is restricted to the control surface and area behind the shock wave.

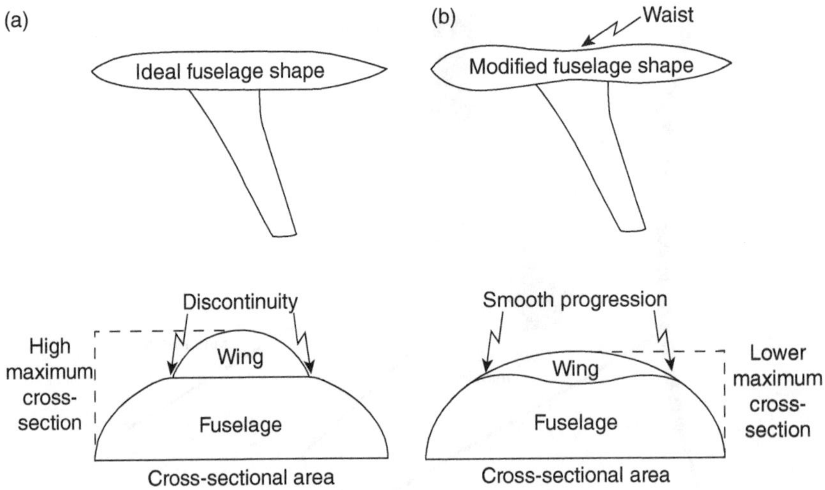

Figure 14.27. The area rule. (a) Cigar-shaped fuselage (b) Waisted fuselage.
Source: Swatton (2011)/reproduced with permission of John Wiley & Sons.

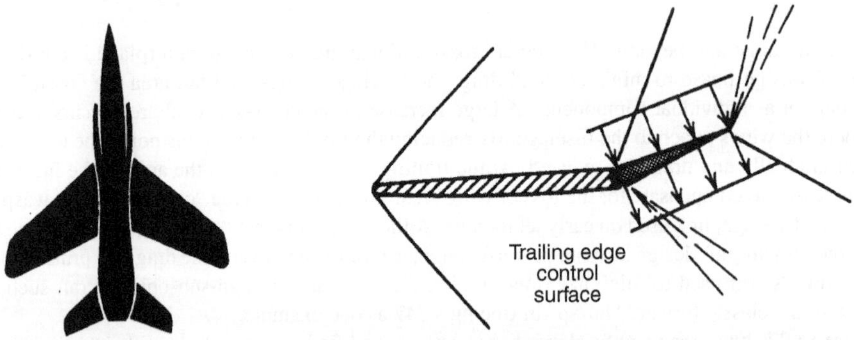

Figure 14.28. Subsonic control surface.

Because the design features that would later allow safe, supersonic flight had not yet been developed, early attempts to break the sound barrier met with disastrous results. To attain the required speed, the aircraft had to be put into a steep dive. When the aerodynamic center shifted to the 50% chord location, the aircraft's center of gravity was forward of the aerodynamic center, increasing the static margin. This created an overly stable dive condition that could not be overcome with the elevators, and often the aircraft continued in the dive until it hit the earth. The elevators were constructed in a conventional manner, as shown in Figure 14.28. The control forces were limited to the small elevator itself and could not influence the flow about the horizontal stabilizer due to the shock wave that formed at the junction of the horizontal stabilizer, and elevator as also shown in Figure 14.28. The amount of force that the pilot could exert was also limited because no power assist controls were available.

As the state of the technology progressed, all moving control surfaces, as shown in Figure 14.29, were developed, as were fully powered, irreversible control systems. Much more control was then available, and

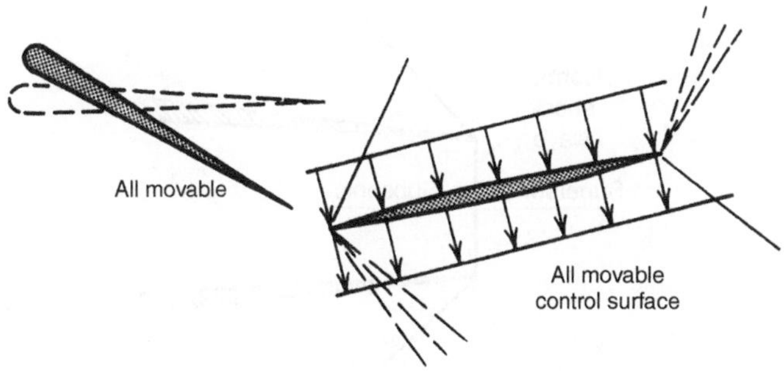

Figure 14.29. Supersonic control surface.

pitch control was no longer a problem. Many large transport-category aircraft today incorporate a variable incidence horizontal stabilizer to handle most of the trim control of the aircraft in pitch, with the smaller elevator taking care of the rest.

Modern aircraft traveling at supersonic speed still have stability and maneuvering concerns. As we have seen, the effectiveness of an airfoil decreases as the surface encounters supersonic flow. For example, this results in a reduction in aircraft weathercock stability ($C_{N\beta}$). The stabilizing yaw moment generated by the vertical tail is diminished, requiring larger vertical tails, the use of twin tails, or the addition of ventral fins. Another issue concerns the increased static margin when an aircraft is supersonic. As we saw in Chapter 12, increased static margin results in decreased aircraft maneuverability. Therefore, an aircraft such as the F-16 that is designed with a negative static margin (center of gravity aft of the neutral point) for enhanced maneuverability becomes less maneuverable when supersonic.

Application 14.2

Consider an aircraft such as the B-1B that enters a turn while supersonic, but transitions to subsonic flight while in the turn.

Would the amount of aft stick required to turn be different when supersonic as opposed to subsonic? Why? Since the aircraft is limited to 3G, what concern should the pilot have as the aircraft transitions to subsonic flight while in the turn?

Supersonic Engine Inlets

The compressor blades of a turbine engine can be made to operate either subsonically or supersonically, but not both. The necessity for slow-speed flight, as well as supersonic flight, rules out the use of supersonic flow in the compressor. Therefore, the flow must be slowed to subsonic when the aircraft is flying supersonically. This requires that the airflow pass through a normal shock wave to slow it down to subsonic speeds. Many fighter aircraft incorporate slots and slats as diverters to channel the airflow, reducing the turbulence of the air going into the engine, while other fighter aircraft have boundary layer diverters such as a splitter plate to reduce turbulence caused by boundary layer air at the engine inlet.

The Mach number behind a normal shock wave is approximately $1/M$, where M is the Mach number ahead of the shock. If the flight Mach is only slightly above 1.0, this is no problem, and a normal shock inlet such as shown in Figure 14.30 can be used. As aircraft speed increases, there must be uniform delivery of air to the compressor unit, with minimal turbulence and pressure change. With the increase in speed

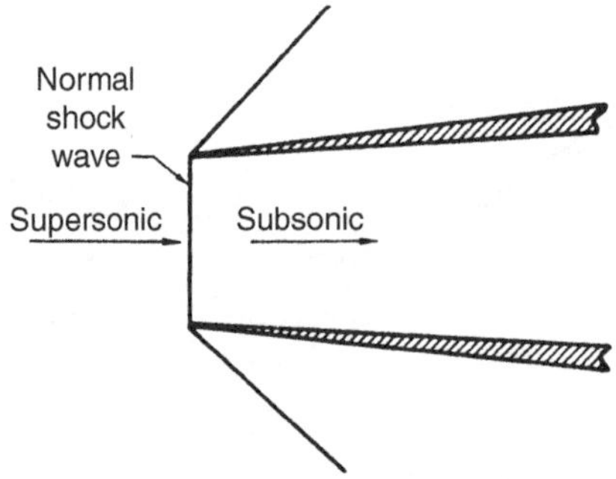

Figure 14.30. Normal shock engine inlet.

Figure 14.31. Divided-entrance inlet ducts.
Source: U.S. Department of Transportation (2023d)/Public domain.

comes increased compression in the inlet, which can compensate for any loss in thrust due to an increase in aircraft speed. This "recovery of pressure lost" is referred to as *total pressure recovery* or *ram recovery*. To recover as much of the total free airstream as possible, there are several designs of inlet ducts intended to maximize airflow and pressure to the engine.

The design and placement of the inlet ducts are based on the placement of the engine on the aircraft as well as the mission of the aircraft, that is, what airspeed, altitude, and attitude will the aircraft be operated. Like all inlet ducts, the *divided-entrance duct* is designed to function as a diffuser with a divergent shape, thus decreasing the velocity in the inlet and increasing the static pressure. Figure 14.31 is an example

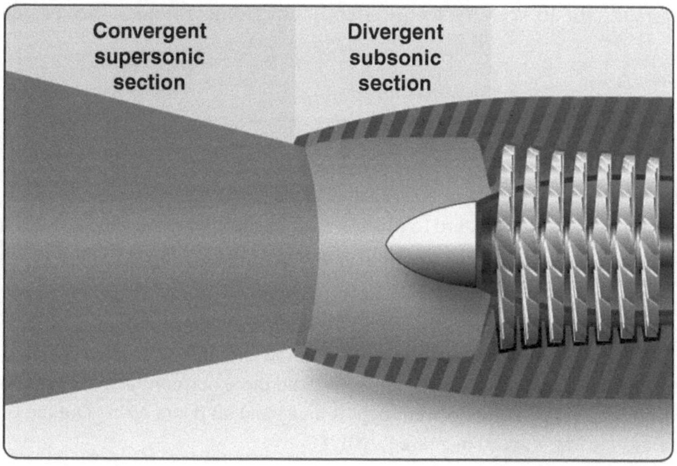

Figure 14.32. Inlet duct acting as a diffuser.
Source: Aviation Maintenance Technician Handbook (2023d)/U.S. Department of Transportation Federal Aviation Administration/Public domain.

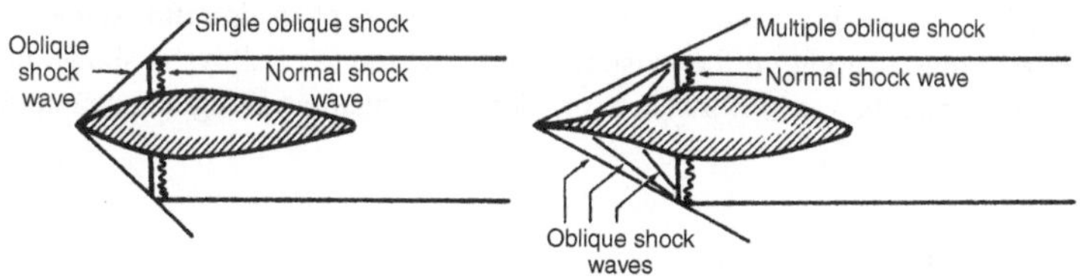

Figure 14.33. "Spike" oblique shock engine inlets.

of a divided-entrance duct and is used when the pilot sits low and near the front of the aircraft where a single-entrance duct would not be used. The balance with the divided-entrance duct is to take inlet air through ducts mounted in the wing root or through scoops, and evenly distribute to the compressor with minimal turbulence as well as drag due to inlet design.

A *variable-geometry duct* involves a general design with a diffuser that progressively decreases in area as the air moves through the duct, and then increases in area (Figure 14.32). High-speed military aircraft may use a variable-geometry duct with a mechanical device to diffuse the incoming air to lower than supersonic speeds as the airspeed increases or decreases (Figure 14.33). The Mach cone angle varies with the flight Mach number, so for proper position of the shock waves, the spike (or ramp) must be adjustable in flight.

In some high-speed military aircraft, there exists a combination of shock methods, the development of a shock wave, and the variable-geometry duct design to reduce the speed of the incoming air to subsonic levels. As the flight Mach number increases, the reduction in airspeed in the normal shock wave becomes greater, the intensity of the shock increases, and the energy loss in the shock wave is increased. To reduce this energy loss, an oblique shock inlet is used. This type of inlet may use a "spike" or a "ramp," which causes oblique shock waves to form. When air passes through an oblique shock wave, its velocity is reduced but remains supersonic. The energy lost in passing through an oblique shock is much less than that lost in passing through a normal shock. Often the air is slowed by a series of oblique shocks, each slowing the air until the air velocity is only slightly supersonic; the air is then passed through a weak normal shock.

Application 14.3

In the 1950s, the Pratt & Whitney division of the United Aircraft Corporation designed an engine that could power the SR-71 Blackbird to Mach 3+ and included many technological innovations. The aircraft was utilized for years by the United States Air Force and the Central Intelligence Agency for high-altitude reconnaissance.

What was the name of this engine? What led to the many metallurgical development problems during the design? Based on the altitude and speed of the aircraft, the engines operated as what two different types of engine design? At what speed did the shock cone achieve maximum forward movement, as well as the forward bypass doors and aft doors open? Did the spike move forward or aft as the Mach speed increased? Why?

Aerodynamic Heating

The temperature of the air in the stagnation area of the leading edge of the wing, at the nose of the aircraft, at the leading edges of the stabilizers, and at other stagnation points of an aircraft is shown in Figure 14.34. The effect of aerodynamic heating on the tensile strength of structural alloys is shown in Figure 14.35. These figures help illustrate the vast difference in problems of structural strength encountered in designing

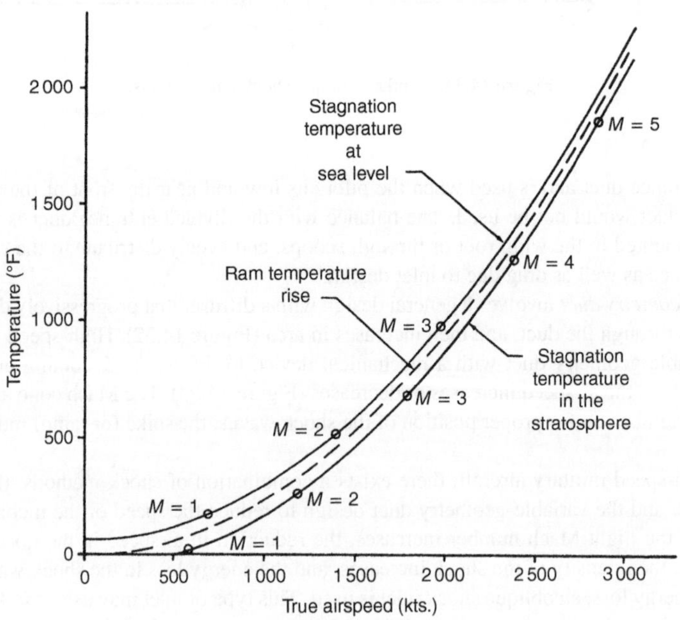

Figure 14.34. Stagnation temperatures.

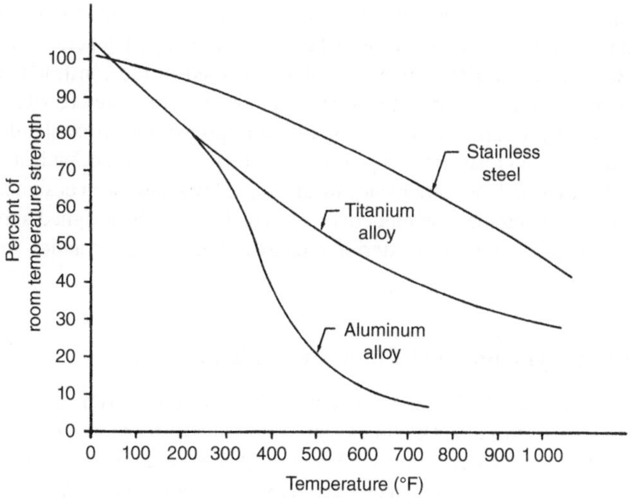

Figure 14.35. Effect of temperature on tensile strength of metals after half-hour exposure.

a Mach 3 supersonic transport (SST) aircraft as compared to a Mach 2 SST. The stagnation temperature in the stratosphere is about 250 °F for the Mach 2 aircraft and about 700 °F for the Mach 3 SST. Comparing the tensile strength of aluminum alloy at these temperatures shows that at 250 °F, about 80% of the room temperature tensile strength remains, but at 700 °F, less than 10% of the room temperature strength remains. Clearly, it is not feasible to make leading edges out of aluminum alloy for a Mach 3 aircraft, so other alloys are used.

Figure 14.36 depicts the specific tensile strength as a function of the use temperature of selected strength materials compared with Ti alloys and gamma Ti aluminides. These compounds are based on

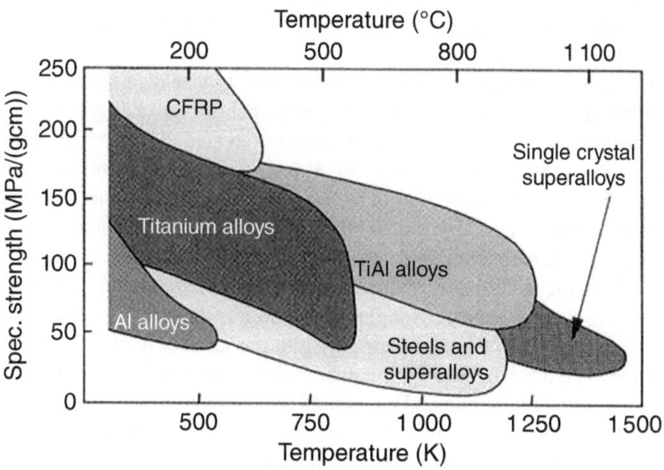

Figure 14.36. Specific tensile strength as a function of use temperature.
Source: Courtesy of NASA, Citation 20180001137.

titanium and aluminum which are light (low density), relatively stiff (high modulus) and have properties that allow them to be tensile and resist creep when exposed to high temperatures. Note the inclusion of CFRP, Carbon Reinforced Plastics, a popular composite used in the construction of most new aircraft. The use of ceramics, often referred to as ceramic-matrix composites (CMCs), has also become widespread in commercial and military aircraft, first used to protect the space shuttle. Ceramic components are extremely durable and resistant to repeated thermal cycling and mechanical wear, in addition to providing excellent electrical and radio frequency insulation. CMCs have been designed to withstand temperatures up to 2 700 °F, a requirement on modern engines that have been designed to run hotter and more efficiently. Often used as protection from exterior exhaust in fighter jets, ceramics are noncorrosive and lightweight.

Computational Fluid Dynamics and Computer-Aided Design

Computational fluid dynamics (CFD) and computer-aided design (CAD) are now standard design approaches to any new aircraft or aircraft component. Traditionally, CAD was not used until later in the design process, but as technology has evolved, CFD and CAD are used by designers in all phases of the design process. CFD is now used in high-speed aerodynamic design to account for flexibility in wings (aeroelastics), noise propagation models, vibration prediction, low-speed (high-lift system) analysis, and icing models. Safety should always be the driving force in aircraft design, followed of course by mission requirements, regulatory mandates, and production costs. CFD and CAD allow the modeling and testing of new aircraft and component designs before wind tunnel testing and flight testing are begun. By using computer fluid modeling and design, engineers are able to narrow down the best fuselage and wing shape for the aircraft mission, sometimes leading to direct comparison to flight test, alleviating additional rounds of wind tunnel testing. By allowing designers of supersonic aircraft to calculate the critical Mach number, as well as the effects of drag as the aircraft accelerates through transonic flight, CFD and CAD allow areas of inefficiency to be noted before expensive wind tunnel tests are initiated. An aircraft that is able to exceed Mach 1 may also need to be highly maneuverable in the subsonic region for air-to-air combat, and then perform a landing on an aircraft carrier. Other missions may call for a design that allows Vertical Takeoff and Landing (VTOL). Engine placement on a wing may greatly affect interference drag and the position of the aircraft's CG. By incorporating CFD and CAD, mission requirements can be designed into the aircraft, and expensive modifications can be avoided while allowing wind tunnel and flight testing to be more efficient (and less expensive).

Specifically, CFD uses mathematics, physics, and software to visualize how a gas or liquid affects an object as it flows around the object, allowing an engineer to monitor the changes in temperature, pressure, and density during the process. A three-dimensional model is created to help better design integral aircraft components such as airfoils and engine nacelles. An airfoil shape that may initially work well for supersonic flight may not have the structural integrity to sustain the required engine or weapon placement. As research continues into supersonic and hypersonic flight, CFD will be used extensively to understand what materials can withstand the aerodynamic heating at the stagnation points associated with these speeds. In supersonic missile development, CFD contributes to more efficient designs, allowing for greater speed and more maneuverability, as well as solving complex flow and heat transfer problems. As the proliferation of unmanned aerial vehicles (UAVs) make their way into the civilian market from the air defense sector, CFD will continue to be used to meet the regulatory challenges implemented by the Federal Aviation Administration (FAA) as well as the demands of commercial operators.

CFD has been, and still is, important in the design of commercial transport category aircraft. For example, CFD was used for the wing design of the Boeing 737 Classic in the 1980s, and instrumental in the design of the flight deck of the Boeing 757 and 767 common-type-rating. Although the aircraft may appear remarkably different from the outside, the flight decks were designed so pilots could transition from one flight deck to another with minimal difficulty. Aerodynamic drag was reduced around the flight deck of the 757, as was noise, through the use of CFD as local areas of supersonic flow were eliminated. Engine placement, engine nacelle design, and the continued growth in the composite materials industry in order to improve fuel economy all begin with CFD.

CFD continues to be used heavily in the design of the high-speed wing, specifically in the design of the nacelle and engine. As the industry continues to see ever-higher bypass ratios and fan sizes, CFD is used to mitigate wind tunnel testing as models are very accurate in predicting flow boundaries around wings and engine/nacelles during cruise flight. Components such as fairings and aft bodies are designed using variations of the Navier–Stokes equations, a long-used method of measuring the interactions of fluid velocity, pressure, temperature, and density. The use of CFD with internal flows is also becoming more prevalent, whether understanding flow boundaries during thrust reverser operation, or attempting to improve airline passenger comfort with better modeling of the Environmental Control System (ECS) in terms of noise, draft, temperature, and impacts of pollutants. Even modeling fire suppression where fuel tanks are housed along with engine nacelles and auxiliary power units (APUs) has become more efficient with improvements in CFD software. Figure 14.37 diagrams the extensive use of CFD in the development of the Boeing 787, each component or area of the aircraft listed incorporated some level of CFD design.

Low Boom Flight Development

The Low Boom Flight Demonstrator (LBFD) project involves the NASA-led construction of the X-59 Quiet Supersonic Technology (QueSST) experimental aircraft to further the development of commercially

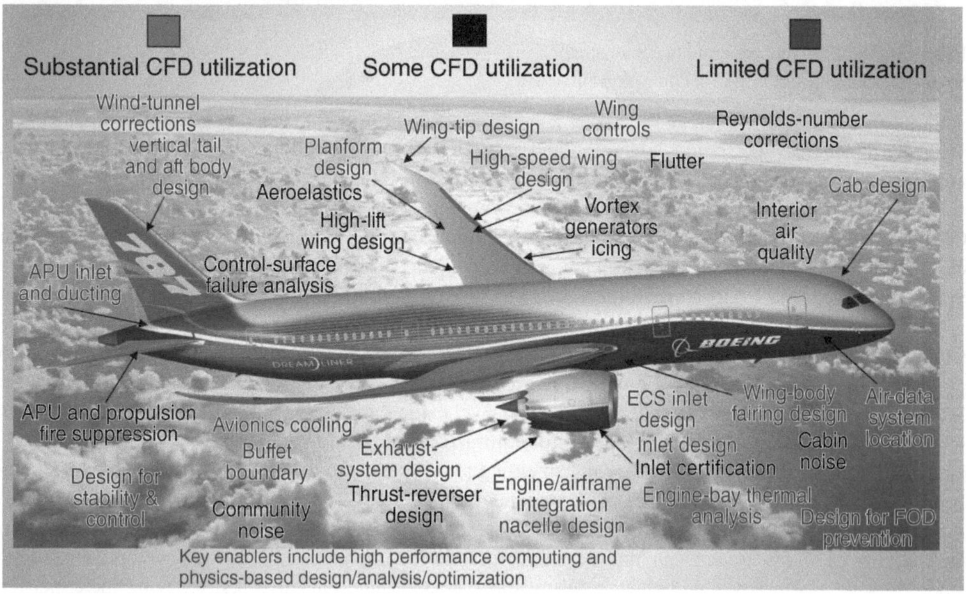

Figure 14.37. CFD use in the development of the Boeing 787.
Source: NASA/Public domain.

Figure 14.38. X-59 low boom flight demonstrator.
Source: Lockheed Martin/NASA/Public domain.

viable commercial supersonic flight over land (Figure 14.38). The primary goal of this research is to minimize the impact of the sonic boom discussed earlier in this chapter and convert what used to be a "sonic boom" into a "sonic thump." Since the retirement of the last supersonic commercial aircraft, the Concorde, manufacturers have researched improved technology to bring supersonic travel back to commercial airline travel. Some companies leading the way in this pursuit include Lockheed Martin, Exosonic, Spike Aerospace, and Boom Supersonic.

Since testing of the American variants of SST in the 1960s, noise as a result of the sonic boom over land was the primary concern during development, and along with soaring costs became a direct reason the project was eventually scrapped. Figure 14.39 provides examples of common sounds and their perceived decibel (dB) level, note the difference in the decibel level in the sonic boom of the Concorde versus the X-59 sonic thump. Features of the X-59 designed to minimize the impact of shockwave development include a long nose to shape the forward shock, wing shielding to minimize the impact of inlet spillage on the sonic boom, and a T-tail to minimize aft shock. Components and systems such as a T-38 aft canopy and ejection seat, as well as F-16 landing gear, will help minimize initial cost and certification requirements. Using high-fidelity CFD, researchers can model the structural design impact on shockwave development and can model how the shockwaves will change during speed changes and maneuvering, reducing the expensive costs of additional wind tunnel testing (Figure 14.40).

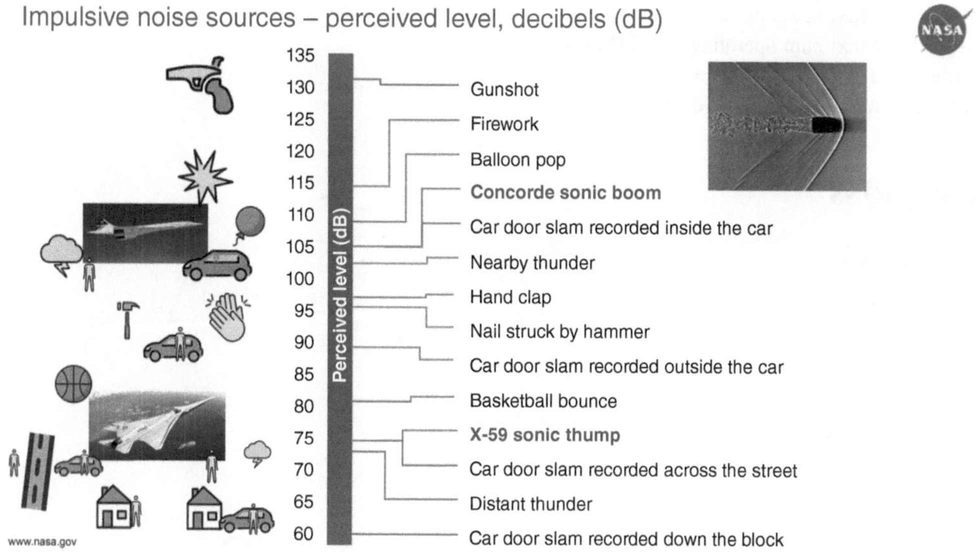

Figure 14.39. Decibel levels for common noise sources.
Source: NASA/Public domain.

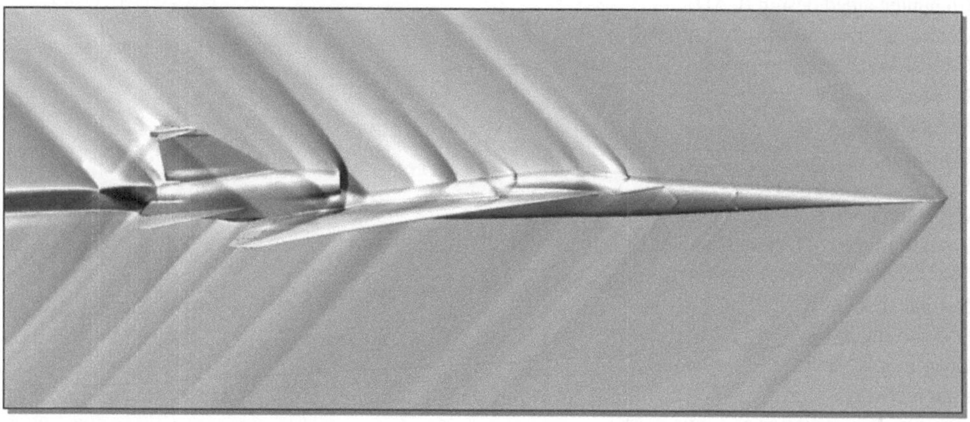

Figure 14.40. LBFD and CFD research.
Source: NASA/Public domain.

SYMBOLS

a	Speed of sound (kts.)
a_0	Speed of sound at sea level, standard day
M	Mach number (dimensionless)
M_{DD}	Drag-divergence Mach number

V_{MO}	Maximum operating speed (kts.)
M_{MO}	Maximum operating speed (Mach)
θ (theta)	Temperature ratio = T/T_0
μ (mu)	Mach wave angle (degrees)

EQUATIONS

14.1 $a = a_0\sqrt{\theta}$

14.2 $M = \frac{V}{a}$

14.3 $\sin\mu = \frac{1}{M}$

KEY TERMS

Buzz
Closed-loop system
Compressible flow
Compression wave
Computational fluid dynamics (CFD)
Computer-aided design (CAD)
Control augmentation systems (CAS)
Critical Mach number (ratio)
Divided-entrance duct
Expansion wave
Flaperons
Flutter
Hypersonic
Low Boom Flight Demonstrator (LBFD)
Mach cone
Normal shock wave
Oblique shock wave
Open-loop system
Ram recovery
Stability Augmentation Systems (SAS)
Subsonic
Supercritical airfoil
Supersonic
Transonic
Variable-geometry duct
Wave drag

PROBLEMS

Note: Answers to the problems in the first two sections are given at the end of the book.

Application of Chapter Theory

1. The speed of sound is an important factor in high-speed flight because
   a. M_{crit} occurs at $M = 1$.
   b. the pressure waves generated by the plane move at sonic speed.
   c. shock waves form when local air velocities are supersonic.
   d. Both (b) and (c).
   e. None of the above.

2. The speed of sound depends on the air density, pressure, and temperature.
   a. True
   b. False

3. Critical Mach number, M_{crit}, is the aircraft's speed when
   a. it goes supersonic.
   b. the airflow first reaches sonic speed.
   c. shock waves form.
   d. Both (b) and (c).
   e. None of the above.

4. After passing through a normal shock wave, airflow is
   a. Subsonic.
   b. not changed in direction.
   c. heated up.
   d. increased in pressure and density.
   e. All of the above.

5. Airflow passing through an oblique shock wave remains supersonic but changes in direction.
   a. True
   b. False

6. Transonic flight problems may include
   a. force divergence.
   b. increase in C_D.
   c. decrease in C_L.
   d. tuck under.
   e. buffet.
   f. control surface buzz.
   g. loss of control effectiveness.
   h. All of the above.
   i. All except (a).

7. Above about Mach 2, the straight-wing aircraft has a lower drag than a 60° swept-wing aircraft.
   a. True
   b. False

8. Mach wave angle changes with airspeed as follows:
   a. It decreases as airspeed increases.
   b. It increases as airspeed increases.
   c. It remains the same with airspeed changes.

9. All supersonic aircraft have a circular arc or biconvex supersonic airfoils.
  a. True
  b. False

10. Airflow passing through an expansion wave
  a. speeds up.
  b. increases the energy of the airstream.
  c. decreases the temperature of the air.
  d. Both (a) and (c).

11. The critical Mach number, M_{crit}, of an airplane is increased by incorporating all of the following design features except:
  a. thin airfoil sections.
  b. swept wings.
  c. winglets.
  d. supercritical airfoil design.

12. Incorporating the area rule into airplane design decreases:
  a. wave drag.
  b. form drag.
  c. induced drag.
  d. oblique shock waves.

13. When air is passing through a compression wave the static pressure will:
  a. increase.
  b. decrease.
  c. experience the same change as in an expansion wave.
  d. remain constant.

14. When an aircraft has both outboard and inboard ailerons installed, which of the following statements is true?
  a. Inboard ailerons are activated at higher speeds when the outboard ailerons may be deactivated to prevent the twisting of wing surfaces.
  b. Outboard ailerons are activated at higher speeds when the inboard ailerons may be deactivated to prevent the twisting of wing surfaces.
  c. Inboard and outboard ailerons are activated at all speeds and work in conjunction with each other to mitigate twisting forces.
  d. Inboard ailerons are deactivated at higher speeds when spoilers are installed on the aircraft.

15. The coffin corner, or "Q corner," is defined by:
  a. the apex of the triangle formed by V_{MO} and M_{MO}.
  b. the apex of the triangle formed by the stall speed and $\sim 5\% > M_{crit}$.
  c. the apex of the triangle formed by the stall speed and M_{crit}.
  d. the apex of the triangle formed by the stall speed and M_{MO}.

Calculations Using Chapter Principles

16. Calculate the speed of sound at a pressure altitude of 20 000 ft with an outside air temperature of −25 °C.

17. Using the local speed of sound found in Problem 14.16, calculate the Mach number if that same airplane has a velocity of 342 kts. TAS.

Correlation Through Scenario-Based Learning

1. Research hypersonic flight and answer the following questions:

a. Is there a definitive airspeed that marks the transition from supersonic flight to hypersonic flight?

b. What was the first manned hypersonic aircraft to have flown at hypersonic speeds? When did these flights happen?

c. Referring to the aircraft from the previous question, what was the problem with stability created by the large, ventral fin, and what was the final solution to fix this problem?

d. What is a *scramjet*?

e. What are the temperature concerns around the aircraft when traveling at hypersonic speeds?

f. What is an *ICBM*, and how does it factor into the discussion on hypersonic flight?

g. How does hypersonic flight relate to the space program and reentry vehicles?

h. What hypersonic aircraft are currently being developed?

15 Rotary-Wing Flight Theory

CHAPTER OBJECTIVES

After completing this chapter, you should be able to:

- Discuss the development of rotary-wing aircraft.
- Apply Bernoulli's principle and Newton's law of motion to how a helicopter produces lift.
- Identify the forces on a rotor system in motion.
- Compare the influence of ground effect in a helicopter with a fixed-wing airplane.
- Explain dissymmetry of lift in forward flight and summarize the methods to mitigate it.
- Describe the helicopter power-required curves and illustrate how power required varies with airspeed.
- Explain autorotation and the importance to safety of flight.
- Define dynamic rollover and provide examples of when it may occur.

Rotary-wing aircraft were envisioned by early pioneers long before one ever flew. The Wright brothers saw rotary-wing aircraft as basically toys. They are more complicated aerodynamically, but their ability to do things that other aircraft cannot do ensures they will be flying for years to come.

All wings, rotors, and propellers as subject to the same principles of the forces of flight, whether producing lift or thrust or a combination of both. This text has chosen to combine all forms of rotary-wing machines in one chapter. Each of the types of aircraft could fill an entire chapter to fully develop the somewhat unique characteristics possessed by that type.

In this text, rotary-wing aircraft are divided into four groups: Helicopters, Gyroplanes, Gyrodynes/Rotodynes, and Powered-lift (Tilt-Rotors).

The text will concentrate most of the discussion on helicopters, as they are perhaps the most ubiquitous rotary-wing aircraft at the time of this writing, used in both military and civilian operations. Also, helicopters lead by far in the category of aircraft that have received type certificates from the Federal Aviation Administration (FAA) and comparable government certification agencies in other countries. On the other

Flight Theory and Aerodynamics: A Practical Guide for Operational Safety, Fifth Edition.
Brian A. Johnson and Philip R. Fittante.

Companion website: www.wiley.com/go/flighttheoryandaerodynamics5e

hand, gyroplanes operate almost entirely in the domain of recreational/sport flying under the amateur-built category.

A helicopter (also referred to as a helo, chopper, or copter) looks strange compared to a conventional fixed-wing aircraft, but the same principles of flight apply to both. In Chapter 3, we learned that Leonardo da Vinci stated that air passing around an airfoil produces the same aerodynamic forces when the airfoil is moving through still air as it does when the air is moving past a stationary airfoil. We can carry this principle further by saying that an airfoil will produce the same aerodynamic force whether the airfoil is being moved through the air by moving the entire fixed-wing aircraft or by moving the airfoil while the rotary-wing aircraft remains stationary.

Before the aerodynamics of helicopters were completely solved, autogyros (also known as autogiros, gyrocopters, and gyroplanes) were being flown regularly. The Spaniard, Juan de la Cierva, experimented with autogyros in the 1920s and 1930s. His autogyros were later commercially built in the United States and saw relatively widespread use. Some sources credit de la Cierva as being the major contributor to solving the dissymmetry of lift problem that made rotary flight practical.

The FAA uses the term gyroplane to describe what was originally called the autogyro. Many gyroplanes today are derivatives of those developed by Igor Benson or refinements of his design made by Ken Brock. These gyroplanes are usually called gyrocopters, at least by those who fly these machines. Another famous gyroplane, with an enclosed cabin, is the "Air and Space" model. This machine was commercially built and sold as a certificated aircraft with a standard airworthiness certificate. It was later offered as a kit aircraft for homebuilders who could license the aircraft in the amateur-built experimental category.

Gyroplanes are different from helicopters in that the former does not have a powered rotor system. The rotor is in a constant state of autorotation when the gyroplane is in flight (some have a "pre-rotator" system that uses engine power to spin up the rotor for a shorter takeoff, but this system is not used for other phases of flight). The main purpose of the gyroplane engine is to drive a propeller to produce thrust for forward flight, allowing the rotor system to provide lift from the autorotating blades.

Yet another type of rotary-wing aircraft is the gyrodyne/rotodyne. The Fairey Aircraft company developed these unusual machines in the United Kingdom in the 1940s and 1950s. Very few were built. The original Fairey gyrodyne had a powered rotor and was basically a helicopter, except it had a propeller mounted on the right winglet to provide thrust in forward flight and to serve as an anti-torque device. The Fairey Jet gyrodyne was a combination of a gyroplane, helicopter, and gyrodyne with the rotor being assisted in rotation by jets of air at the rotor tips. The rotodyne was a large aircraft capable of carrying several dozen passengers. The rotodyne was similar to a gyroplane, but the rotors were augmented by jet-powered tips for vertical takeoffs and landings, making it a Vertical Takeoff and Landing (VTOL) aircraft. But they also had stubby wings to develop lift at higher airspeeds. This lift allowed for the "unloading" of the rotor system during "high speed" flight, similar to a compound helicopter. The rotor and the winglets each shared part of the load, allowing the rotor system to turn slower, thereby delaying the onset of retreating blade stall, and allowing the machine to fly faster. In the United States, a few companies experimented with compound gyroplanes, but early designs had little performance advantages over then-current helicopters.

Another type of rotary-wing aircraft is the tilt-rotor (sometimes referred to as a prop-rotor). However, the FAA classifies tilt-rotor aircraft as a separate category of aircraft called powered-lift. The tilt-rotor combines many of the flight characteristics of both airplanes and helicopters, using a relatively short-span wing for forward flight. For hovering and slow flight, the machine has large engine-powered rotors attached to the end of each wing. The rotors pivot (or tilt) about the wing tips and may be thought of as big propellers by some, hence the term prop-rotor. Somewhat similar to the tilt-rotor is the tilt-wing, which has fixed rotors attached to the ends of each wing. The entire wing pivots rather than just the engine nacelles containing the rotor system.

Incidentally, somewhat akin to the tilt-rotor were the Lockheed XFV and the Convair XFY "Pogo," experimental VTOL aircraft, which are sometimes referred to as "tail sitters." These aircraft had a large turboprop engine with contra-rotating propellers mounted atop a fighter-plane-type fuselage that sat vertically. After a vertical takeoff, and after gaining sufficient altitude, the aircraft transitioned from vertical to horizontal flight and then operated as a conventional airplane. Unfortunately, it was difficult to control in vertical flight on landing and the concept was abandoned.

MOMENTUM THEORY OF LIFT

Bernoulli's equation of lift is still a valid theory of how a helicopter produces lift, but, because the velocity of each rotor blade segment varies as the distance from the rotor hub, it is simpler to discuss the lift by using the momentum theory. The momentum theory is based on Newton's laws. The second of these states that an unbalanced force acting on a body produces an acceleration of the body in the direction of the force. The "body" in the case of the flow of air through a helicopter rotor system is the mass of air. The unbalanced force is produced by the rotor blades being driven by the engine. The result is an acceleration of the air down through the rotor disk.

Newton's third law states that for every action force, there is an equal and opposite reaction force. Thus, by causing the air to be forced downward, lift was developed upward. This is quite similar to the production of thrust from an aircraft propeller or jet engine. Figure 15.1 shows a schematic of this theory, where

1. V_o = original velocity at about two rotor diameters above the rotor
2. V_i = induced velocity at the rotor
3. V_f = final velocity below the rotor

It can be seen from the figure that there is lower pressure above the rotor disk and higher pressure below the disk, with a sudden increase at the disk. The velocity increases gradually from approximately zero at about two rotor diameters above the disk to a maximum at about the same distance below the disk.

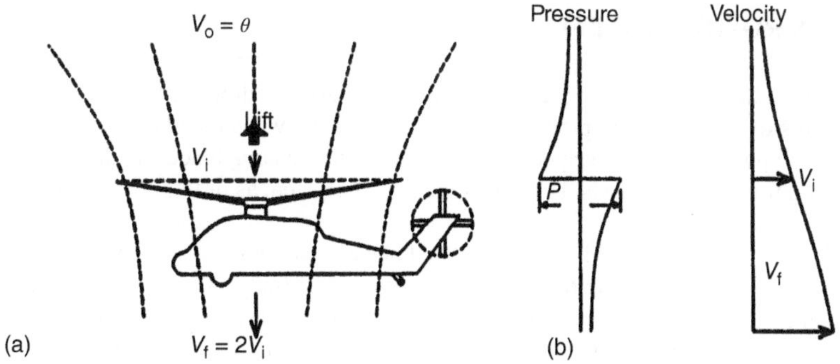

Figure 15.1. Momentum theory airflow: (a) schematic and (b) pressure and velocity distribution.

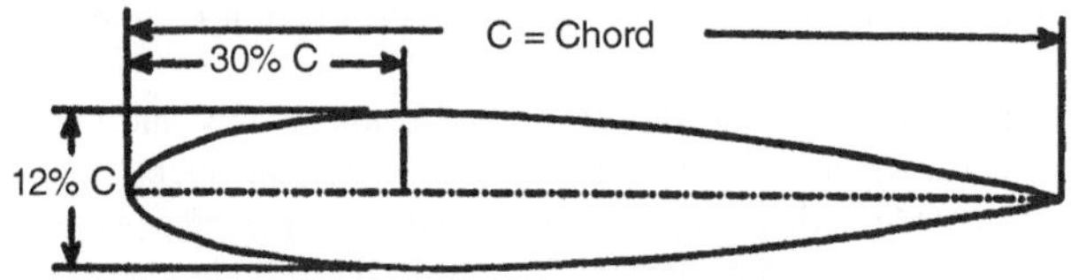

Figure 15.2. NACA 0012 airfoil.

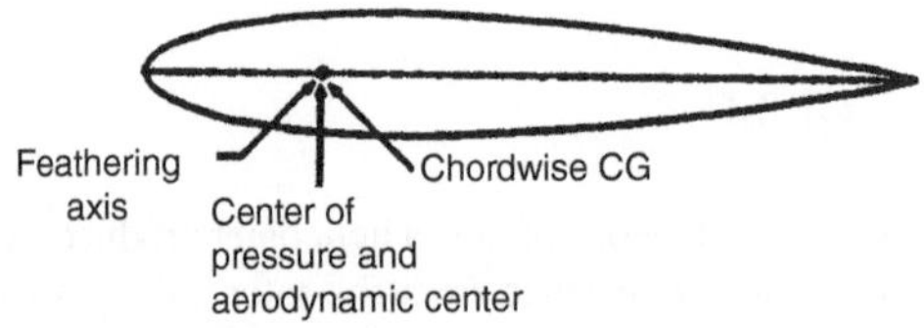

Figure 15.3. Location of critical forces on an airfoil.

AIRFOIL SELECTION

It was shown in Chapter 3 that cambered airfoils develop a nose-down pitching moment when subjected to airflow, whereas symmetrical airfoils do not (see Figures 3.30 and 3.31). These pitching moments change in value when the angle of attack changes. Later, we discuss the fact that during the rotation of the blades, the angle of attack is constantly changing. Rotor blades generally lack the stiffness to resist this moment and will twist if they are cambered. Twisting of the blades causes changes in AOA and lift, which complicates the problem, so it is avoided by using symmetrical airfoil sections. Another advantage of the symmetrical airfoil is that both the center of pressure and the aerodynamic center are at the same location and do not move with AOA changes. So, most rotor blades have symmetrical airfoil sections. Figure 15.2 shows a typical helicopter rotor airfoil.

To reduce pitching moments on rotor blades caused by such factors as the center of gravity location and axis of feathering (axis of rotation) location, it is desirable that they coincide with the center of pressure/aerodynamic center location. This is shown in Figure 15.3.

FORCES ON ROTOR SYSTEM

When the blades of a helicopter are not turning, there is a noticeable droop to them. They do not have the inherent stiffness of fixed wings to prevent this droop. However, when the blades are rotated centrifugal force (equal and opposite to the centripetal force that holds the rotating blades on the rotor shaft, and not a *real* force as introduced in Chapter 11) causes them to straighten, as in Figure 15.4. The force on only one blade is shown.

For a helicopter rotor system with fully articulated blades, devices called droop stops limit the amount of droop, preventing the blades from resting too low above the airframe. The droop stops can be observed during the spin-up and slowing down of the blades. It is common practice to have a crewmember observe the rotor system during aircraft startup and shutdown to ensure the droop stops disengage or engage as the rotor system speeds up and slows down. On shutdown, if a blade does not have the droop stop engaged, the blade may hit the fuselage.

To produce the lift required to fly the helicopter, a vertical vector must be exerted upon the rotor shaft. To produce the vertical lift vector shown in Figure 15.5, the angle of attack of the symmetrical rotor blades

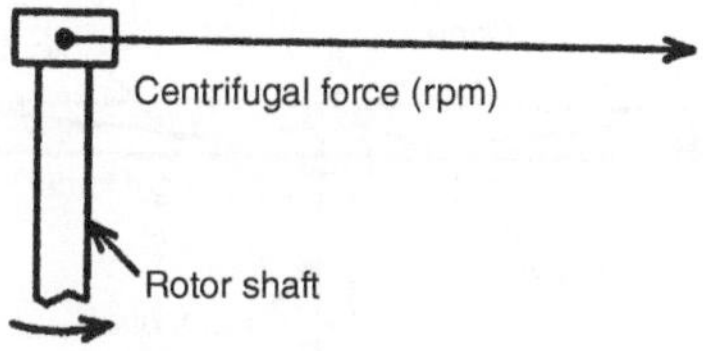

Figure 15.4. Centrifugal force straightens rotor blade.

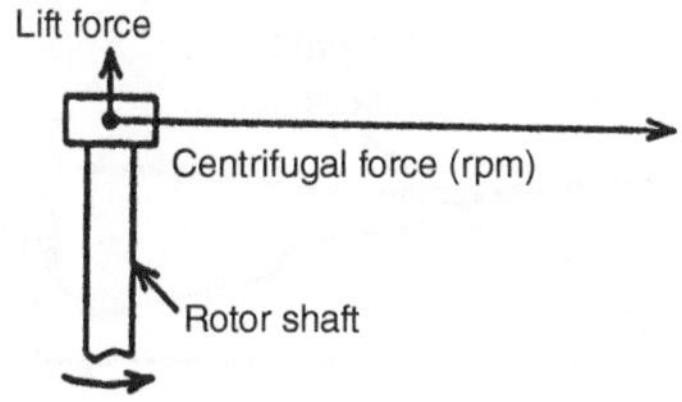

Figure 15.5. Lift force and centrifugal force.

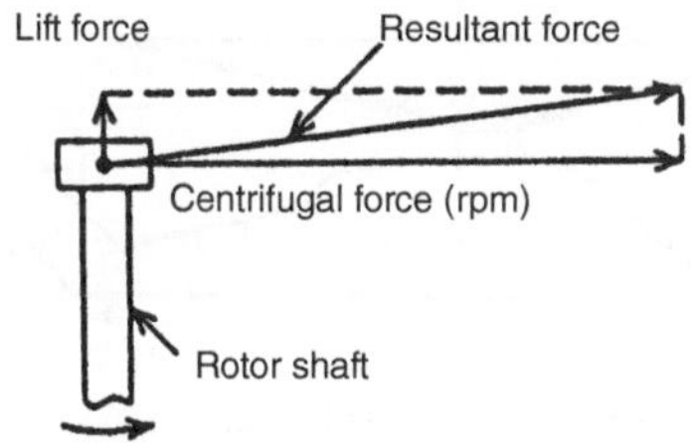

Figure 15.6. Resultant of lift and centrifugal forces.

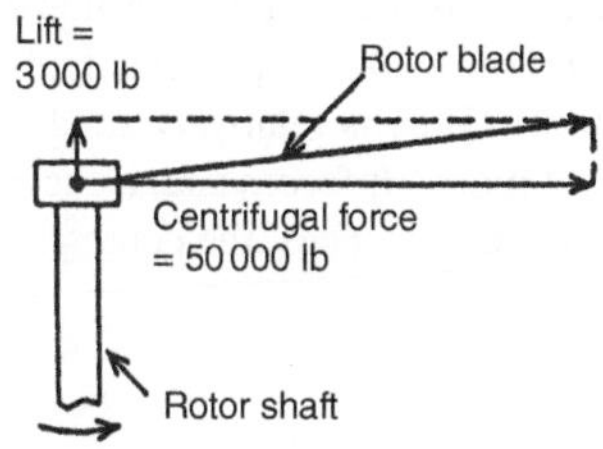

Figure 15.7. Forces acting on a lifting blade.

must be set at a positive angle of attack. The blades are now being acted upon by two force vectors: (i) the centrifugal force vector and (ii) the lift vector. The blades must assume the position of the resultant of these two vectors, as shown in Figure 15.6, where the lift vector has been exaggerated for clarity.

The centrifugal force is much larger than the lift force. In fact, the lift force is only about 6% of the centrifugal force. To illustrate this, assume that the helicopter weighs 6 000 lb and has two blades. The lift force on each blade would be 3 000 lb and the centrifugal force would be about 50 000 lb. This is shown in Figure 15.7. If the rotor blades are hinged to allow upward rotation, the lifting rotor system is shown in Figure 15.8.

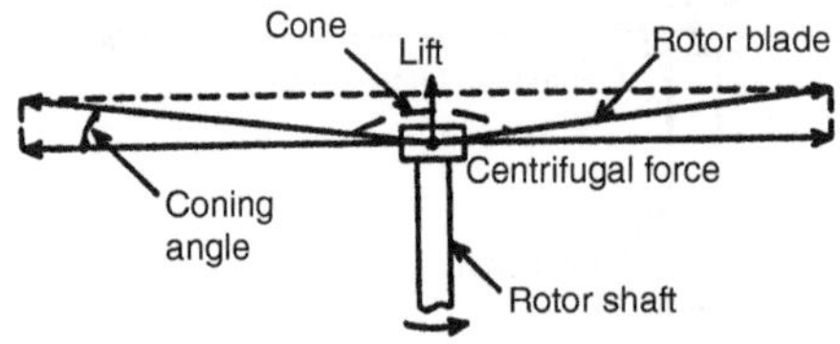

Figure 15.8. Entire lifting rotor system.

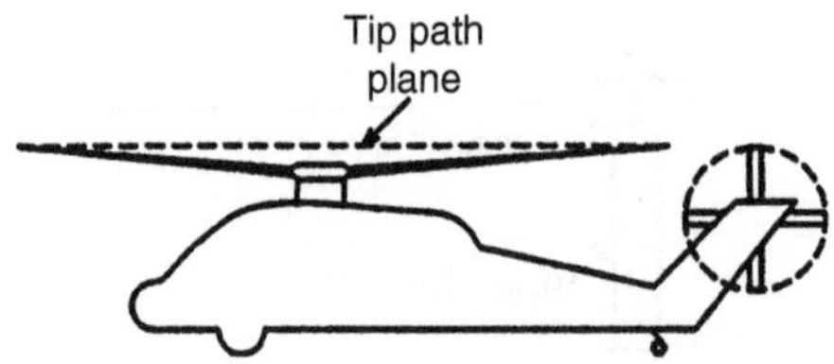

Figure 15.9. Hovering helicopter at light weight.

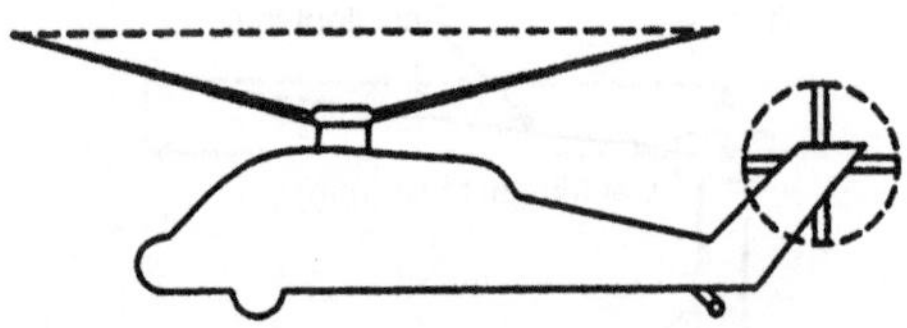

Figure 15.10. Hovering helicopter at heavyweight.

Of particular interest is the coning angle shown between the two blades and the horizontal. When viewed from the side, the rotating blades form a visual disk. A line drawn between the blade tips is called the tip-path plane. Figure 15.9 shows a hovering helicopter at a lightweight. Under lightweight conditions, the coning angle is small because the lift force is relatively small compared to the centrifugal force. In Figure 15.10, the heavy weight requires more lift forces and, if the rpm is the same, the centrifugal forces remain the same as in the lightweight condition. Thus, the coning angle is larger.

THRUST DEVELOPMENT

Up to this point, we have been discussing the forces on the rotor system while the helicopter was in a hover. To move the aircraft in any horizontal direction, we must tilt the entire rotor disk in the desired direction. We have been calling the vertical force lift. Now, we will see that in addition to the force required to overcome the weight, we also obtain a component force, which we call directional thrust. Therefore, we rename the lift force: We call it *total thrust*.

Figure 15.11 shows the total thrust vector resolved into the vertical lift component and the forward thrust component. Figure 15.12 shows that one of the first reactions that results from translating from a hover to flight in any direction is a loss of altitude. This is because the total thrust is no longer acting opposite to the weight of the helicopter as it was in the hover. Consequently, the effective lift is reduced and, if corrective action is not taken, the helicopter will hit the ground. The effective lift of 9 659 lb will

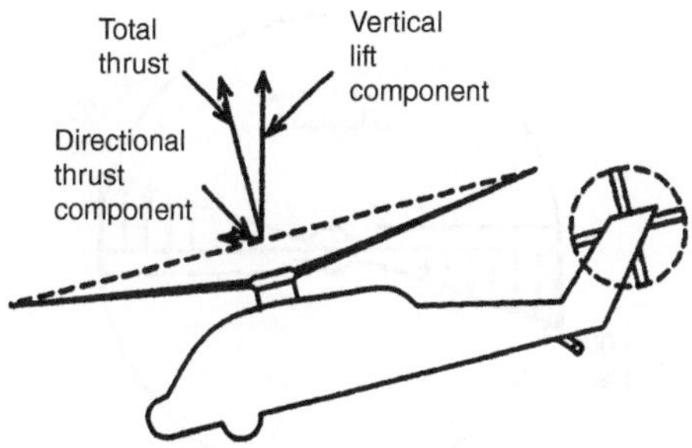

Figure 15.11. Forward flight forces.

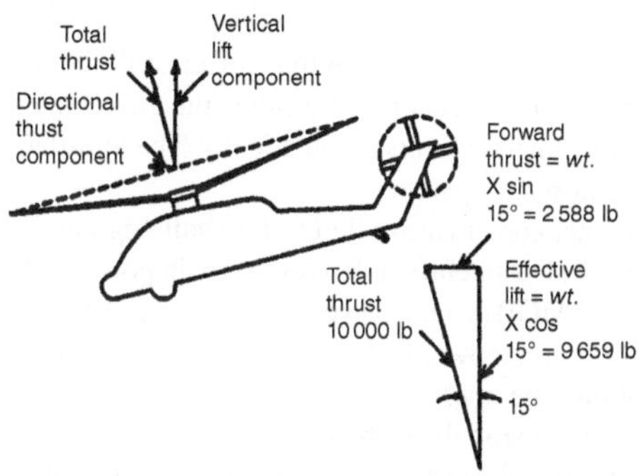

Figure 15.12. Lift component of 10 000-lb total thrust at 15°.

not support the weight of 10 000 lb, so the helicopter will lose altitude. The pilot must increase the total thrust to 10 353 lb to have the effective lift equal the weight (10 353 cos 15° = 10 000).

HOVERING FLIGHT

Hovering means maintaining a constant position, usually a few feet above the ground. To hover, the helicopter must supply lift equal to the weight of the helicopter. To accomplish this, the blades must be rotating at high rpm and the angle of attack of all the blades is increased by increasing the angle of attack with the collective stick.

The lift and weight forces are equal and opposite in a no-wind hover condition and the tip-path plane remains horizontal. If the angle of attack is increased by the pilot lifting the collective, the helicopter will climb vertically. This also requires more power or the rpm will decrease.

In some helicopters, primarily turbine-powered, there is an interlink between the collective and the throttle that automatically increases the throttle when the collective is raised. In others, generally piston-engine helicopters, the pilot must add throttle when the collective is raised. On both turbine and piston engine helicopters, the throttle may be located on the collective control. It is called a twist grip, in that it twists to regulate engine power and is similar to that found on motorcycles. Some larger turbine helicopters have throttles similar to those found in airplanes, except they are often mounted overhead in the flight deck similar to some seaplanes. The throttles may be referred to as "speed selectors."

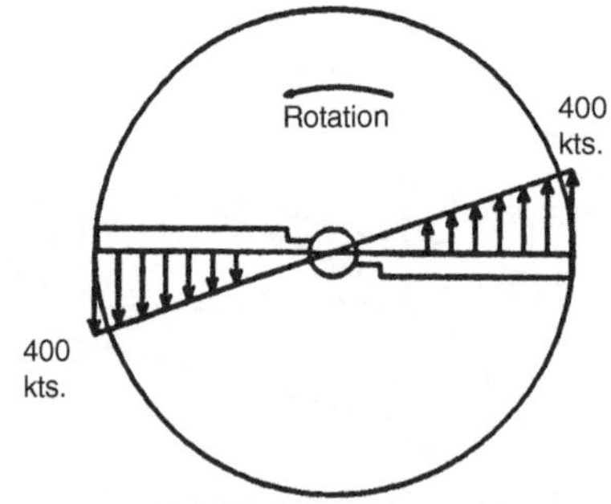

Figure 15.13. Rotor velocity distribution in hover.

The throttle(s) normally remain at a high rpm setting for most flight operations, with turbine engines operating at or near-maximum rpm, or in what may be called the governing range. In the governing range, the engine helps to keep the rotor rpm with limits automatically. The throttle(s) may be controlled manually if a malfunction occurs in the automatic system.

The turbine engines used on helicopters are called turbo-shaft engines. They have a free power turbine that, as the name implies, rotates independently or free of the main power turbine(s). The free power turbine is connected to the main gearbox (MGB) which drives the helicopter rotor system. In some helicopters, the rotor brake may be applied to the free power turbine to stop the rotation of the turbine and subsequently the rotor. The rotor brake can only be applied at low engine rpm (sometimes called "ground idle," well below the governing range) to safely stop the turbine.

In fixed-wing aircraft, the speed of the airflow over the wings is determined primarily by the airspeed of the airplane. This is not so with rotary-wing aircraft. The speed of the airflow over the rotors is determined not only by the airspeed of the helicopter but also by the speed of rotation of the rotor blades. Let us first examine the case where the helicopter is in a hover and thus the airspeed is zero.

Hovering Blade Velocity

The velocity of each blade section depends on its location with respect to the rotor shaft. The local velocity is a function of the distance from the shaft and the rpm of the shaft. Figure 15.13 shows the velocity vectors of a typical two-blade rotor system having 20-ft rotors and operating at about 322 rpm. The rotor tip speed is about 400 kts.

Blade Twist

The relative wind of each blade is equal and opposite to the blade velocity. The speed of the blade increases as its distance from the shaft and, if each section operated at the same AOA, the lift would increase as the speed squared. (Remember the lift equation?) The blades are designed with a twist to prevent this uneven lift distribution. They have higher AOA at the root, which gradually decreases toward the tip.

Figure 15.14 shows the lift distribution on an untwisted blade and on an ideally twisted blade. Note that the twisted blade develops more lift near the root and less lift at the tip than the untwisted blade. The linear increase in lift of the twisted blade is desired.

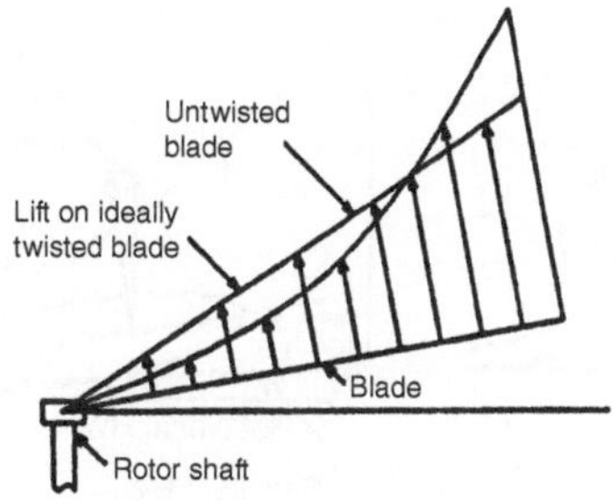

Figure 15.14. Lift distribution on a twisted/untwisted blade.

GROUND EFFECT

In the early days of helicopter development, it was found that it took much less power to hover a helicopter close to the ground than at a distance above the ground. An early theory attributed this to a build-up of a high-pressure "bubble" beneath the helicopter upon which the helicopter balanced. This also seemed to explain why the helicopter lost altitude as it moved in any direction. It "slid off the bubble." We have already disclaimed this theory as to the loss of lift (see Figure 15.12); now, we explain the reduction of power required to hover *in ground effect (IGE)*.

In Chapter 5, we discussed induced drag and ground effect for fixed-wing aircraft. In case you missed it, please review those sections of the chapter. They apply to rotary-wing aircraft just as much as they do to fixed-wing aircraft.

Figure 15.15 shows a helicopter hovering *out of ground effect (OGE)* and Figure 15.16 shows one hovering IGE. Rotor tip vortices form when the rotors are producing lift. This is quite similar to the wingtip vortices discussed in Chapter 5. The vortices cause the air to be accelerated downward causing a downwash behind the wing or rotor. In Figure 15.15, the downwash relative wind affects the rotational relative wind, and the resultant relative wind is as shown. The total thrust vector is 90° of the resultant relative wind and is tilted backward from the vertical. The effective lift is vertical and has a lower value than the total thrust. The rearward-induced drag vector is also large. To summarize, downwash results in a loss of effective lift and an increase in induced drag.

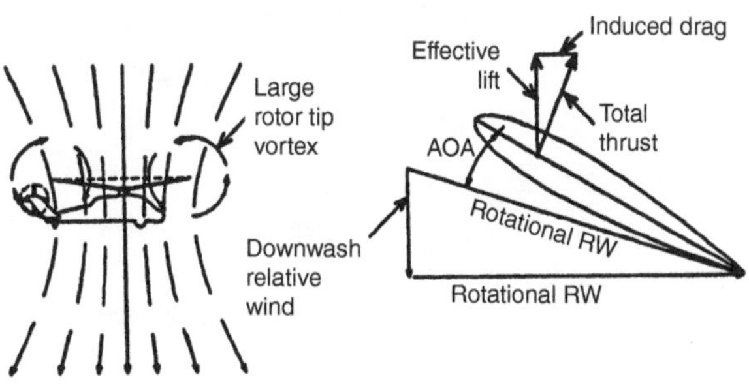

Figure 15.15. Hovering out of ground effect.

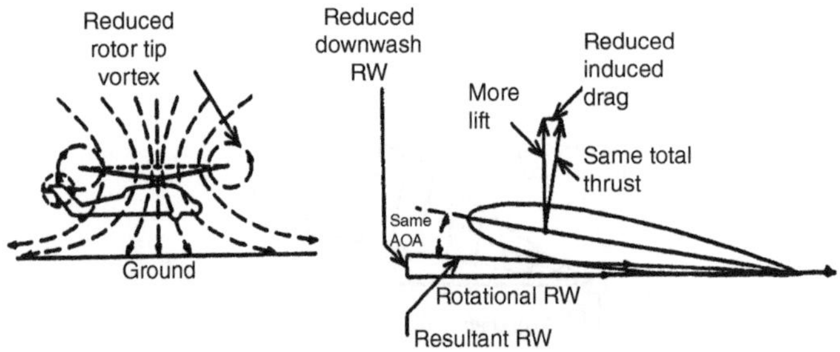

Figure 15.16. Hovering in ground effect.

If the helicopter is hovered close to the ground (or water), the downwash is physically reduced, as shown in Figure 15.16. The downwash relative wind is reduced and the angle that the resultant relative wind makes with the rotational relative wind is also reduced. Thus, the effective lift is increased and the induced drag is reduced. This reduction in drag results in a corresponding reduction in power required while IGE.

Torque

In American helicopters, the rotor blades rotate in a counterclockwise direction, as viewed from above. According to Newton's third law of action and reaction, the fuselage tends to rotate in the opposite direction. This is called torque and must be counteracted before flight is possible. There are several ways of doing this. One method is to use a dual rotor system. These systems can be either tandem or side-by-side intermeshing rotors that rotate in opposite directions.

The most common form of canceling the torque effects is the antitorque rotor shown in Figure 15.17. In this figure, the helicopter is not in equilibrium. There is a balance of moments, but not a balance of forces.

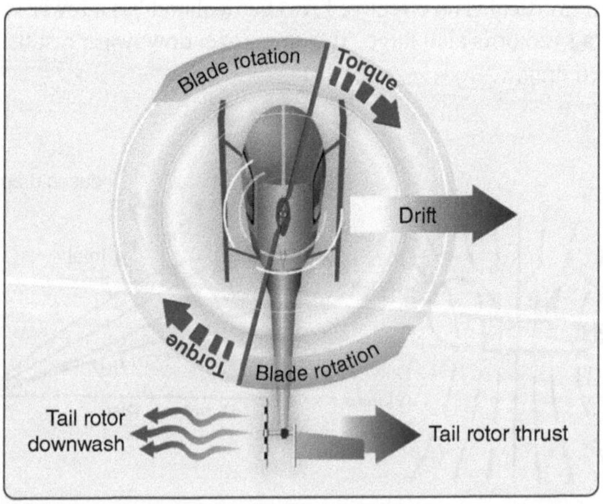

Figure 15.17. Antitorque rotor.
Source: Aviation Maintenance Technician Handbook – Airframe (2023a)/U.S. Department of Transportation Federal Aviation Administration/Public domain.

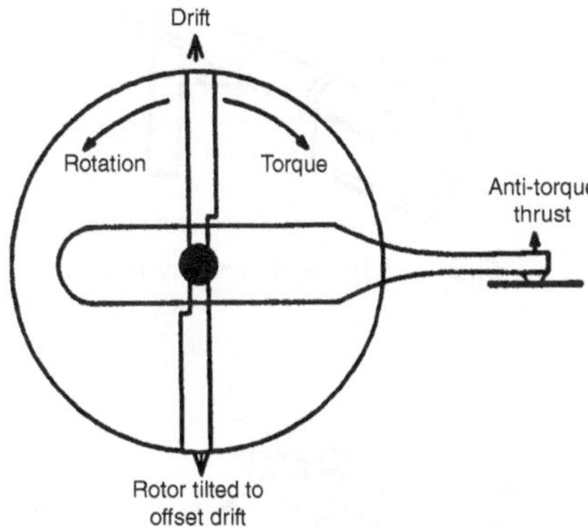

Figure 15.18. Correction for antitorque rotor drift.

The sideward force of the antitorque rotor will cause the helicopter to drift to the right unless it is balanced by a force to the left. This is called "translating tendency." Figure 15.18 shows a method of compensating for this by tilting the main rotor slightly to the left.

Basically, there are four methods to compensate for the translating tendency:

1. The pilot holds cyclic opposite to the drift.
2. The flight control system has built-in compensation that applies the correction mechanically, such as a mixing unit that blends flight control inputs.
3. An automated flight control system applies compensation through an electromechanical or other automatic system. In some helicopters, this automation may be called an automatic flight control system (AFCS) (pronounced "A-Fis"). Think of AFCS as "power steering" that not only reduces control force pressures but also assists in holding altitude, heading, etc., just like the "autopilot" on an airplane.
4. The helicopter transmission is mounted to compensate for the drift. This tilt is built into the MGB on some helicopters. For example, the MGB on one large helicopter is mounted on the airframe with a slight tilt to the left.

ROTOR SYSTEMS

There are three primary types of rotor systems in use. Each must accomplish the same task but do it in different ways.

Rigid Rotor

The simplest rotor system is the rigid rotor. It has but one degree of freedom. The angle of attack of the blades can be changed or *feathered*, but no other movement is allowed. Figure 15.19 is a schematic of the rigid rotor. If the blade is required to move up or down (flap) or to move forward or backward (hunt), the blade must be bent (flexed). The rotor itself absorbs these movements. The rigid rotor system is very strong and allowed the development of helicopters that could do aerobatic maneuvers.

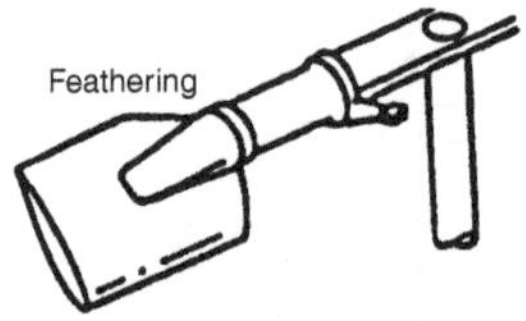

Figure 15.19. Rigid rotor system.

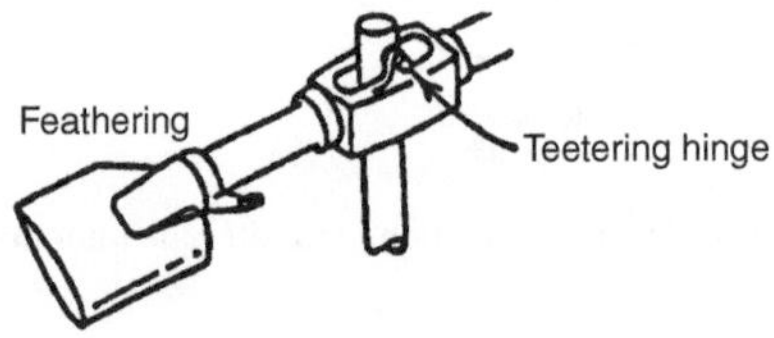

Figure 15.20. Semirigid rotor system.

The Messerschmitt-Bölkow-Blohm (MBB) Bo-5 is a widely used rigid rotor helicopter. In the 1960s and 1970s, Lockheed Aircraft developed the "Cheyenne" compound helicopter for the US Army using a rigid rotor, but the project was canceled before the aircraft entered service. A compound helicopter is one in which part of the weight of the aircraft is carried by a winglet as the machine accelerates in forward flight. This small wing, by carrying part of the weight at higher airspeeds, allows the rotor system to operate at a lower rpm, delaying the onset of retreating blade stall, and allowing the aircraft to achieve higher speeds than could be achieved by a normal helicopter. In 2015, the Sikorsky S-97 "Raider" first flew. It has coaxial, contrarotating rigid rotors and a "propeller" at the rear. Sikorsky has for many years been experimenting with ways to increase the speed, maneuverability, and other aspects of rotary-wing flight, and bills the S-97 as "game-changing." At the time of this writing, it remains to be seen if commercial and government customers purchase this helicopter in large quantities.

Semirigid Rotor (Seesaw or Teetering Hinge)

The semirigid rotor system has two degrees of freedom. In addition to the feathering axis, there is also freedom of flapping. No provision for hunting is available. Figure 15.20 shows a schematic of the semirigid system. This is the system used on many helicopters built by Bell, including the famous Bell 47 and its military version, the H-13 Sioux (most recognizable from the 1970s television show M.A.S.H.). These rotors have two blades. The teetering hinge system is also used on the Robinson series of helicopters.

Fully Articulated Rotor

Three degrees of freedom are provided by the articulated rotor system (Figure 15.21). The vertical hinges allow hunting, the horizontal hinges allow flapping, and the rotation of the blades allows them to change the angle of attack, also known as feathering. The articulated rotor is used by the Sikorsky company, among others. Famous Sikorsky models include the US Air Force HH-3E "Jolly Green Giant" and the US Army UH-60 "Black Hawk, the US Navy SH-3 Sea King, and the US Marine Corps CH-53 King Stallion, the US military's largest helicopter."

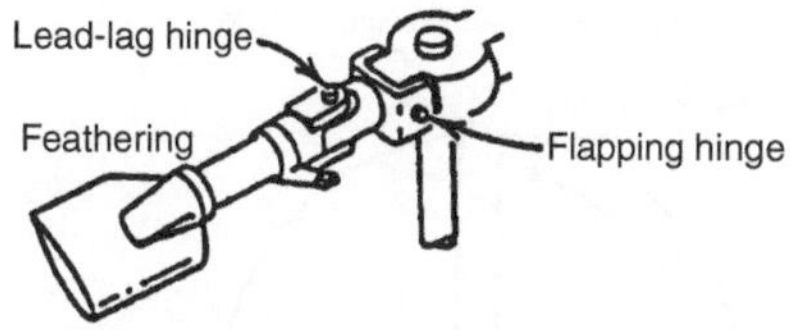

Figure 15.21. Articulated rotor system.

DISSYMMETRY OF LIFT IN FORWARD FLIGHT

Early experiments in helicopter development produced an aircraft that could hover but failed to achieve forward flight. The helicopter would roll over when it moved from a hover in any direction. To understand what caused this, we must investigate *dissymmetry of lift*.

The airflow at the rotor tips that was caused by the rotation in a hover is shown in Figure 15.22. As the helicopter starts to move in a horizontal direction, the airflow over the rotors is changed. For simplicity, consider that the helicopter is moving forward at 100 kts. The blade that is moving toward the front of the helicopter is called the *advancing blade* and the blade that is moving toward the rear is called the *retreating blade*. The advancing blade tip has an airspeed of 500 kts. (Figure 15.23), while the retreating blade-tip speed is only 300 kts. If both blades are operating at the same AOA, the advancing blade will develop more lift than the retreating blade. This condition is dissymmetry of lift and cannot be tolerated. It is quite similar to flying a fixed-wing airplane whose ailerons are jammed in the full left wing down/right wing up position. The rolling moment that is produced is shown in Figure 15.24.

Blade Flapping

The first successful solution to this problem was made by Juan de la Cierva after he crashed his first prototype in his experimental autogyros from 1923 to 1935. He adapted the principle of designing rotors so that they could pivot up or down as they rotated. This motion is called *blade flapping*. As a blade leaves the tail position and advances around the right side of the helicopter, it is influenced by the increasing airspeed and produces more lift and climbs (flaps) upward. This climbing action changes the relative wind and reduces the angle of attack and thus the lift. Similarly, the retreating blade, as it moves from the nose of the helicopter toward the tail, experiences a reduction in airspeed, loses lift, and flaps downward. This

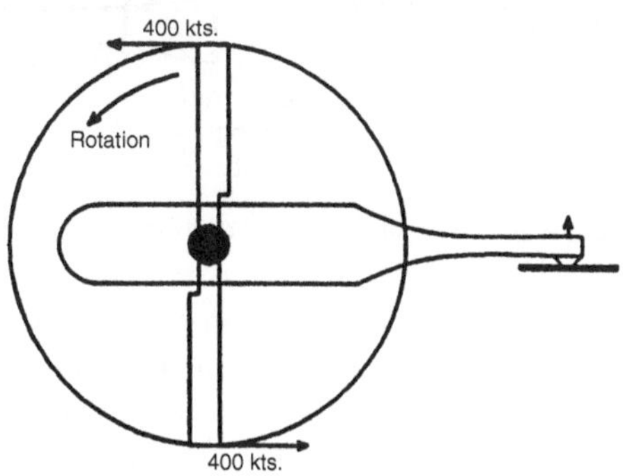

Figure 15.22. Rotor tip velocities in a hover.

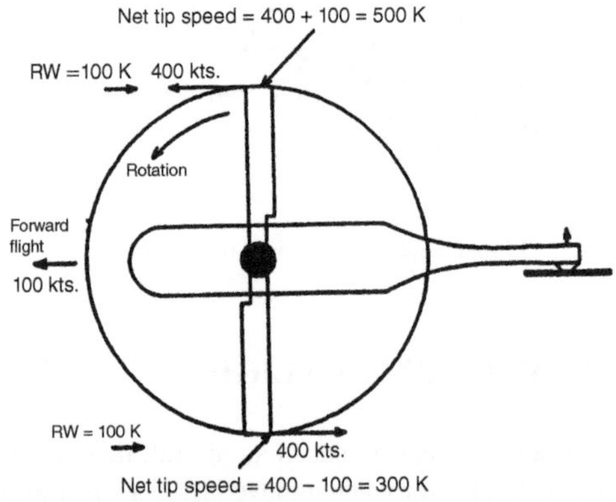

Figure 15.23. Blade-tip velocity in forward flight.

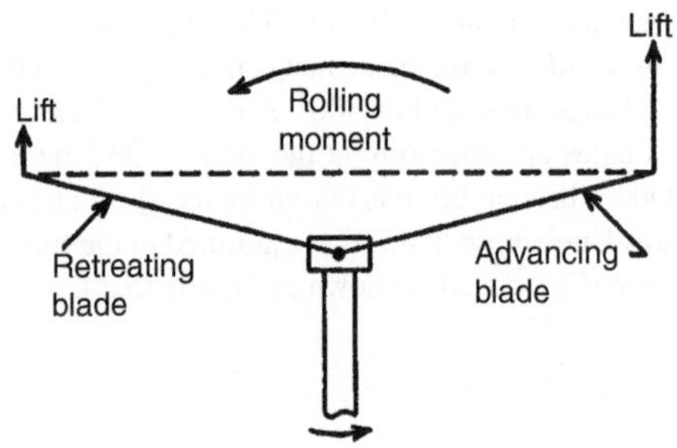

Figure 15.24. Rigid rotor rolling moment in forward flight.

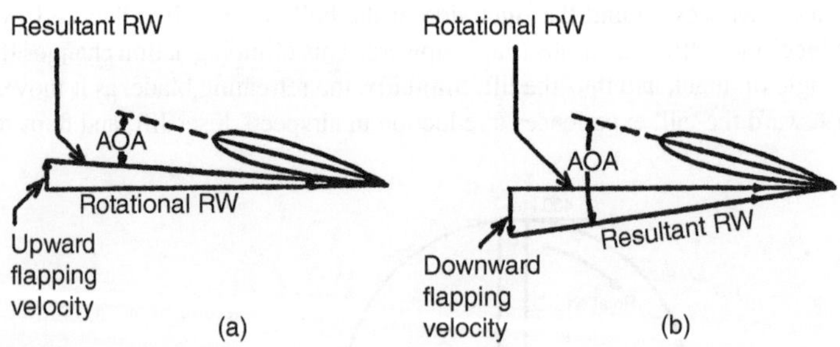

Figure 15.25. Angle of attack and flight path changes: (a) advancing blade and (b) retreating blade.

results in an increase in angle of attack and an increase in lift. The final result is that the lift is equalized, the rolling moment never materializes, and dissymmetry of lift does not exist.

Both the semirigid (seesaw) and fully articulated rotor systems have flapping hinges that automatically react to the changes in relative airspeed during blade rotation. Figure 15.25a shows the angle of attack changes on the advancing blade and Figure 15.25b shows these changes on the retreating blade.

Figure 15.25a shows that the increasing speed of the advancing blade causes the blade to flap up. Because relative wind is always equal and opposite to the flight path, this upward movement causes the

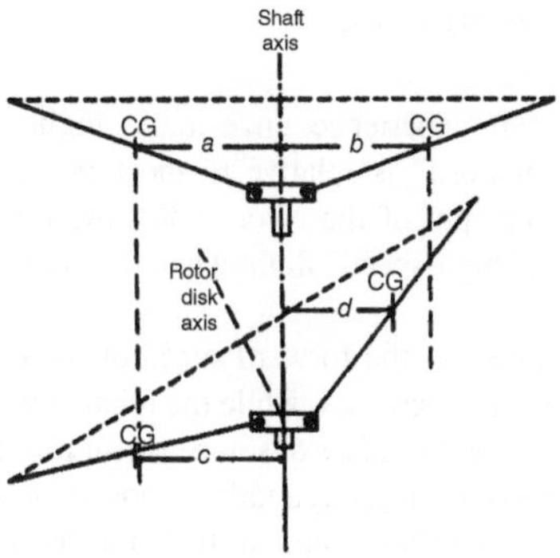

Figure 15.26. CG radius change with flapping motion.

resultant relative wind to incline downward. This effectively reduces the angle of attack of the advancing blade.

Figure 15.25b shows that the decreasing speed of the retreating blade causes a loss of lift and the blade flaps down. This downward movement causes the resultant relative wind to be inclined upward, which increases the angle of attack of the retreating blade.

Blade Lead and Lag

The motion of blade lead and lag is called *hunting*. The hinges allow this action without requiring the blades to develop large bending stresses. The rigid and semirigid systems do not have these hinges. When the helicopter is moving horizontally, the blade pitch angles are constantly changing as the blades rotate to eliminate the dissymmetry of lift. Fully articulated rotor systems have vertical hinges that allow the blades to move forward and backward. As the pitch changes, so does the drag on the blades which causes the hunting action.

Another force, called the Coriolis force, also causes the blades to lead and lag. A law of physics called the law of conservation of angular momentum states that the motion of a rotating body will continue to rotate with the same rotational velocity unless acted upon by some external force. If the mass is moved farther from the center of rotation, it will decelerate. If the mass is moved closer to the center of rotation, it will accelerate. This means that as a rotor blade flaps up, its center of gravity moves closer to the drive shaft, and it will speed up. When the blade flaps down, the opposite occurs.

Figure 15.26 shows this movement of the center of gravity (CG). The right rotor blade of the bottom sketch has flapped up and distance d is less than b. The blade will speed up (lead). The left blade has flapped down and the distance c is greater than a. The blade will slow down (lag). Figure 15.27 shows hunting.

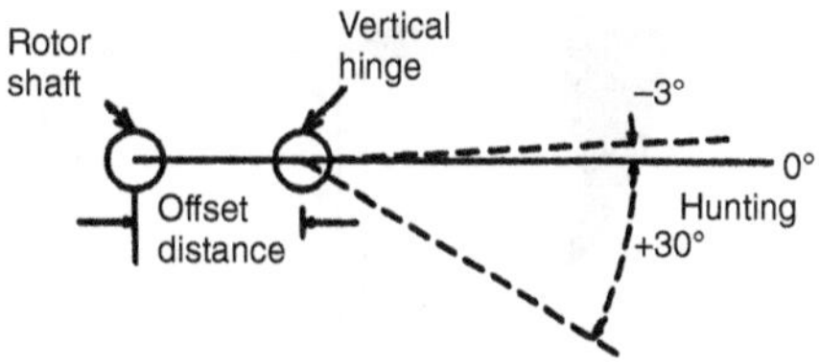

Figure 15.27. Hunting motion of a fully articulated blade.

HIGH FORWARD SPEED PROBLEMS

Both fixed- and rotary-wing aircraft experience undesirable flight characteristics in the high-speed envelope, although the term "high-speed" is relative to the type of aircraft. Rotating wings achieve "high-speed" flight due to either the rpm of the rotor and/or the forward speed of the entire aircraft, whereas fixed-wing aircraft achieve "high-speed" flight when the entire aircraft is moving through the air at a relatively rapid rate.

As far as helicopters are concerned, as the forward speed of the helicopter is increased, the relative wind speed over the advancing blade tip increases, while the relative wind speed over the retreating blade tip decreases. One of two possible problems arises at some critical forward speed: either (i) the advancing blade tip approaches supersonic speed or (ii) the retreating blade approaches stall speed.

This dilemma may be viewed as somewhat analogous to that of "coffin corner" in high-speed airplanes, where the onset of critical Mach approaches stall speed, resulting in aerodynamically induced control problems. For example, helicopters (with counterclockwise rotating blades) that experience retreating blade stall pitch up and to the left.

Advancing Blade Compressibility

As discussed previously, if the airspeed over an airfoil exceeds a certain value, called the critical Mach number ($Mach_{crit}$), shock waves form on the upper surface of the airfoil. These shock waves cause the air to separate from the airfoil and a high-speed stall results. This causes a loss of lift and an increase in drag. To help counter this problem, sweep has been added to the tips of rotor blades on some helicopters such as the H-60 Black Hawk, to reduce the onset of critical Mach, the same as airplane wings are swept to reduce the effects of $Mach_{crit}$. A more detailed discussion of high-speed problems is found in Chapter 14.

Retreating Blade Stall

As the helicopter's forward speed increases, the relative wind over the retreating blade decreases. The resulting loss in lift causes the blade to flap down further and the effective angle of attack increases. At some high angle of attack, the blades begin to become stalled. The tip stalls first and, if no corrective action is taken, the stall progresses inward. When approximately 25% of the disk is stalled, control of the helicopter is lost. Figure 15.28 shows the stall region and some typical angles of attack that the blades attain as they rotate.

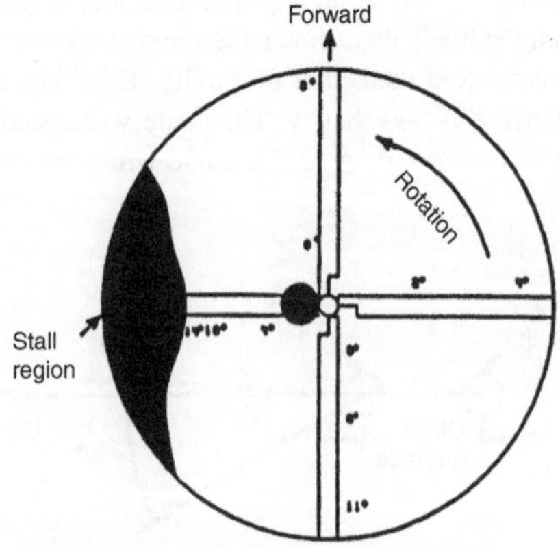

Figure 15.28. AOA distribution during a retreating blade stall.

Both advancing blade compressibility and retreating blade stall occur under similar conditions:

1. High forward speed
2. Heavy gross weight
3. Turbulent air
4. High-density altitude
5. Steep or abrupt turns

Both give warning by exhibiting abnormal vibrations under one or more of the above conditions. They differ as to rotor rpm: advancing blade compressibility occurs at high rotor rpm, whereas retreating blade stall occurs at low rotor rpm. If corrective action is not taken, then advancing blade compressibility will pitch the nose-down, and retreating blade stall will pitch the nose-up and roll the helicopter to the left. Corrective action for both advancing blade compressibility and retreating blade stall consists of

1. Checking and adjusting rotor rpm.
2. Slowing the helicopter by reducing power, not by flaring the helicopter.
3. Lowering blade AOA by using down collective.
4. Leveling the helicopter if in a turn.
5. Reducing G-loading with down collective.

The key is to unload the rotor blade.

Retreating blade stall limits the forward speed of the helicopter to about 150 kts. To counteract the effects of retreating blade stall and allow helicopters to fly faster, several companies have tried different methods. One is known as the Advancing Blade Concept (ABC). Developed by Sikorsky Aircraft, under the sponsorship of the US Army and NASA, the ABC was designed to increase the speed of helicopter to the 250 kts. range. The ABC used contrarotating rotors to, in essence, cancel out the effects of retreating blade stall. The aircraft needed no anti-torque rotor as the contrarotating rotors also canceled out the torque problem.

Gyroscopic Precession

The rigid rotor helicopter acts more like a true gyroscope than either the seesaw or fully articulated type. The flapping hinges make the difference. However, all three types exhibit some of the characteristics of a gyroscope, as shown in Figure 15.29.

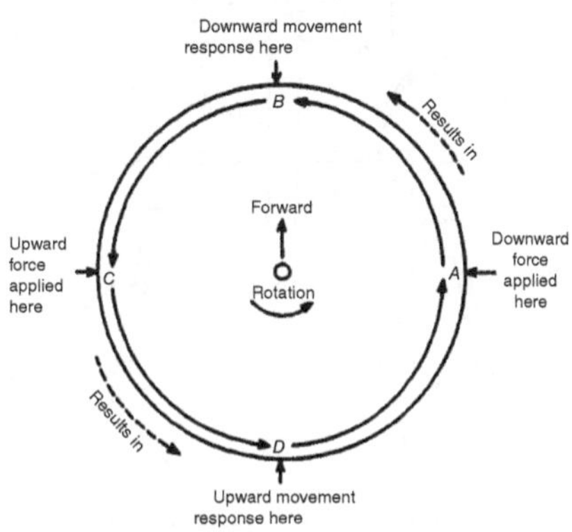

Figure 15.29. Gyroscopic precession.

A force applied to a rotating disk causes the disk to react 90° later in the direction of rotation. This explains the pitch-up and pitch-down reactions to advancing blade compressibility and retreating blade stall mentioned above. In either case, the blade stalls, and lift is lost.

It is easy to see why the retreating blade stalls at a high angle of attack, so we discuss that condition first. The blade is at the nine o'clock position when the maximum angle of attack is encountered (see Figure 15.28). Gyroscopic precession (Figure 15.29) shows that the reaction occurs 90° later at the six o'clock or tail position, which causes a download at the tail and results in pitch up. Although it may seem strange that a low angle of attack can produce a stall, that is exactly what happens when the critical Mach number is exceeded. Lift is lost at the 3 o'clock position and the reaction occurs at the 12 o'clock position, causing pitch-down.

HELICOPTER CONTROL

The control of a helicopter consists of the movement of the aircraft along the three principal axes and rotation about these axes. The rotation movements are called pitch, yaw, and roll and are discussed in some detail in Chapters 12 and 13. In the helicopter, the movement along the longitudinal axis (forward and rearward flight) and along the lateral axis (sideward flight) is controlled by the cyclic stick. Movement along the vertical axis (vertical flight) is controlled by the collective stick. Pitch and roll are also controlled by the cyclic, while yaw is controlled by the foot pedals. This book is not intended to be a flight manual, so we will describe the action of these controls, rather than how to use them.

Rotor Head Control

Rotor control can be accomplished by direct tilting of the hub or by changing the blade pitch. Blade pitch changing can be done by the swash plate, aerodynamic servo tabs, auxiliary rotors, jet flaps, and pitch links attached to a control gyro. The most common method is the swash plate (Figure 15.30).

The stationary star and the rotating star are connected together by the bearing. This allows the rotating star, pitch change arms, blades, and rotor head to be rotated by the rotor head. The stationary star, control input servos, and transmission do not rotate. Both the rotating star and stationary star can move up and down on the rotor shaft. They can also be tilted at various angles, but they always remain parallel to each other.

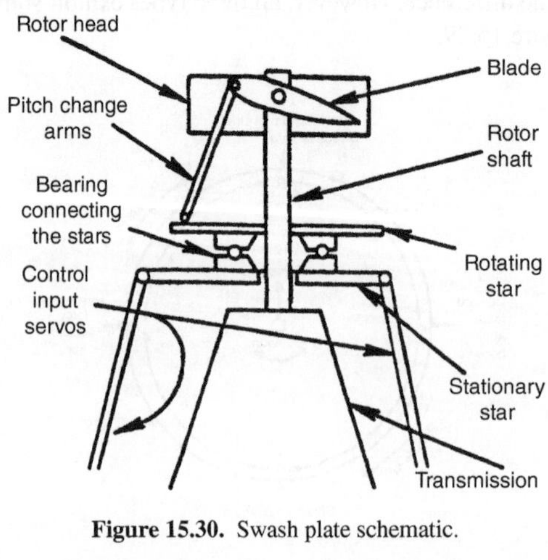

Figure 15.30. Swash plate schematic.

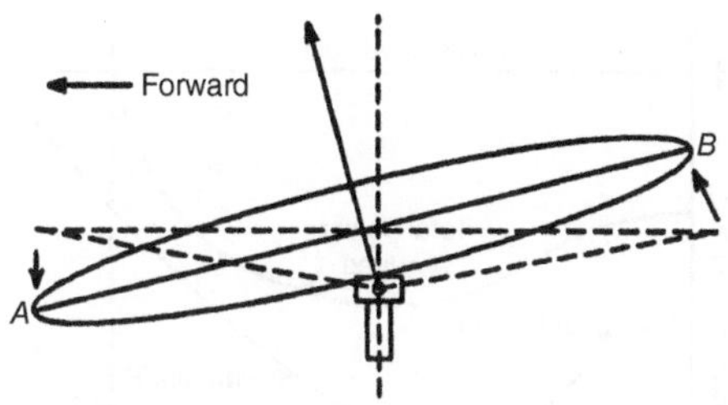

Figure 15.31. Rotor flapping caused by cyclic stick movement.

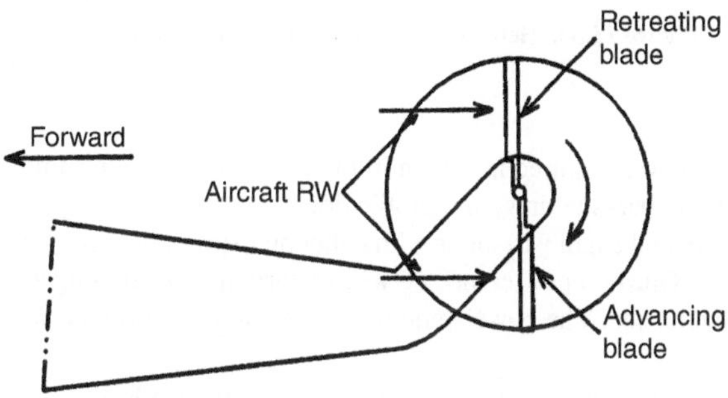

Figure 15.32. Tail rotor dissymmetry of lift.

Control of the Path

The vertical movement of the swash plate is controlled by the collective stick. If the pilot raises the collective, the pitch angle of all blades is increased and the helicopter climbs vertically. Down collective has the opposite effect. The cyclic stick tilts the stationary star through the control input servos. The rotating star remains parallel to the stationary star and changes the pitch angle of each blade by a different amount. This causes flapping of the blades and results in forces that change the tip path plane as shown in Figure 15.31. Forward cyclic movement is shown.

Both cyclic and collective inputs can be combined so that horizontal and vertical motions can be coupled. Power is controlled by a motorcycle-type twisting hand grip on the collective stick.

Foot pedals change the blade angles of the antitorque rotor, changing its trust output and thus changing the heading of the helicopter. Figure 15.32 shows that the tail rotor is subject to dissymmetry of lift, similar to the main rotor system. A seesaw rotor hub allows blade flapping to eliminate this.

HELICOPTER POWER-REQUIRED CURVES

In the case of fixed-wing aircraft, we have seen that the power required consists of the power required to overcome induced drag and the power required to overcome the parasite drag. Rotary-wing aircraft have another kind of drag: the drag caused by the rotation of the rotor system. The power required to turn the rotors is given a new name – *profile power required* (P_{ro}). It is the power required to overcome the profile drag of the rotor system. Unlike the parasite drag of the helicopter, it occurs without any forward motion

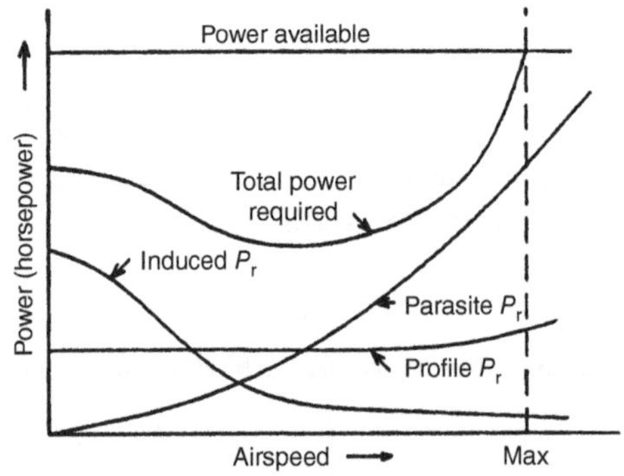

Figure 15.33. Helicopter power available and power required.

of the aircraft. The total power required is shown in Figure 15.33. It is the sum of the power required to overcome the induced drag, parasite drag, and profile drag.

Power available has several small variations with helicopter speed, but for the sake of this discussion, it is assumed to be constant. One other difference between rotary- and fixed-wing aircraft is the ability of the helicopter to fly at zero airspeed. The power-required curves do not terminate at stall speed, but continue down to zero airspeed.

Notice that more power is required to hover than in intermediate-speed forward flight. This is true in or OGE.

Translational Lift

Earlier, it was demonstrated that an IGE hover reduced the power required. The power required is also affected by forward flight. It can be seen from Figure 15.33 that there is a noticeable reduction in the total power-required curve as airspeed increases from zero (hover). This is caused by translational lift, sometimes referred to as effective translation lift. The efficiency of the hovering rotor system is improved as the helicopter moves forward. In a hover, there is much turbulence caused by the tip vortices. As the aircraft moves into a region of undisturbed air and the vortices are left behind, the airflow becomes more horizontal and the efficiency of the rotor system is improved.

The same effect can be achieved with wind. The positive effects of translational lift become effective at about 15 kts. So a wind of the same value produces the exact effect as accelerating the helicopter to 15 kts.

Stories have been told of helicopter pilots during the Vietnam War using ground effect and translational lift to carry much larger loads than they could under normal circumstances. The exigencies of combat may have required overloading the helicopter to save lives or accomplish the mission. For instance, the normal hover height for the UH-1 Iroquois (more commonly known as the "Huey" from its original designation of HU-1) was a skid height of about 4 ft above ground. With a higher-than-normal load, there may have only been power available for an IGE hover at say 1 or 2 ft.

In a zero wind condition, there may not have been enough power available to achieve translational lift with a normal takeoff. In this case, the pilot gently applies forward pressure to the cyclic control, tilting the lift vector, and causing the helicopter to move forward very slowly. All cyclic control movements (think control pressure, not movement, the same as when flying an airplane) had to be very small, as cyclic inputs generate induced drag, creating a demand for more power, which is in short supply. As the helicopter waddles forward, it slowly accelerates until translational lift is achieved and the helicopter begins a rapid

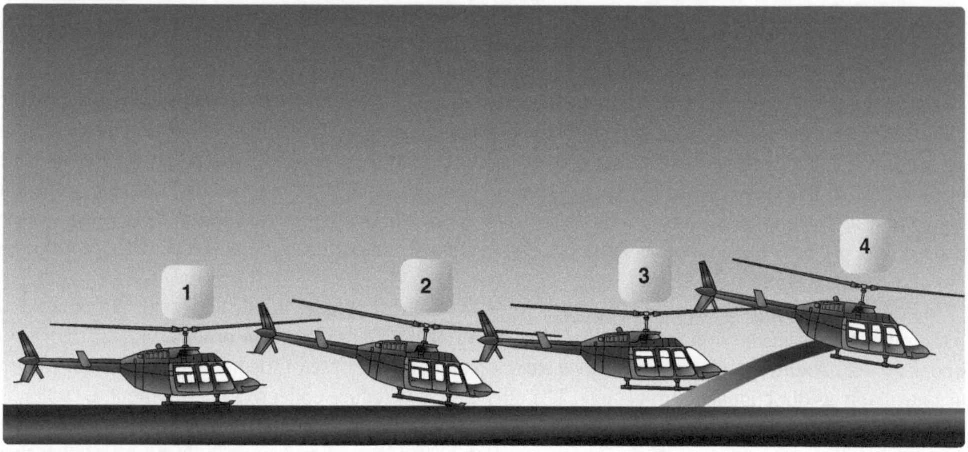

Figure 15.34. Running/rolling takeoff.
Source: Helicopter Flying Handbook (2012b)/U.S. Department of Transportation Federal Aviation Administration/ Public domain.

acceleration and a subsequent climb in the clear, undisturbed air that exists once past the transitional lift point. This is a very precise maneuver that becomes easier with practice.

The FAA *Helicopter Flying Handbook* offers a running takeoff for situations similar to the one described above. This procedure requires careful alignment of the helicopter with the flight path to prevent dynamic rollover. In Figure 15.34, the pilot first applies the collective until the helicopter is light on the skids. Second, the pilot applies a slight forward cyclic and increases collective to move the helicopter forward.

As the machine reaches translational lift, it becomes airborne (#3). Keep the helicopter IGE until it accelerates to a safe climb speed (#4).

Of course, if a helicopter has wheels rather than skids like the Huey, the translational lift can be achieved using a rolling takeoff much like an airplane. The pilot applies forward cyclic pressure to start the helicopter rolling and once through translational lift, applies slight rearward cyclic pressure and slight upward collective to take off. This procedure is particularly effective for heavily loaded helicopters in high-density altitude conditions.

POWER SETTLING, SETTLING WITH POWER, AND VORTEX RING STATE

"Power settling," "settling with power," and "vortex ring state" are terms that are confusing and often used interchangeably. There is much disagreement in the helicopter community about what the terms mean and how they are interrelated. The FAA Helicopter handbook teaches that settling with power and vortex ring state are one and the same. Other sources disagree, with the vortex ring state being very different from settling with power. And some sources use the term power settling rather than settling with power. But it is more than a problem of semantics, as helicopter pilots deal with these real issues that can lead to an accident.

This text chooses to define power settling as a situation where the power available does not equal or exceed the power required. More than one helicopter crew has planned an approach to a remote site with high-density altitude only to find that the aircraft engine(s) and rotor system "ran out of power" a few feet above the ground, and the helicopter at best landed hard because there was not enough power available to arrest the descent. In Transport Canada's Helicopter Flight Training Manual, this is called "settling with insufficient power," which may be a better way to describe what is happening.

Untrained pilots, trying to arrest this uncontrolled descent, often instinctively but incorrectly pull the collective control excessively, asking more of the engine(s) and rotor system than can be delivered. Applying too much collective results in "rotor droop" not to be confused with the droop that the blades have when at rest. In this case, the droop means the rotor rpm begins to decay or "droop" due to insufficient engine power. The engine cannot turn the blades fast enough to overcome the aerodynamic forces on the blades. Not only does the helicopter settle rapidly, but another major problem develops with rotor droop, loss of directional control.

In helicopters equipped with an anti-torque rotor, the tail rotor blades are driven by a shaft connected to the main rotor transmission (often called a gearbox). If the main rotor rpm droops, the tail rotor rpm droops as well. And when the droop exceeds a certain amount, the nose of the helicopter will start to turn uncontrollably as directional control is lost. The only cure is to reduce collective control input to "unload" the rotor system and allow the blades to accelerate. At low altitudes, there may not be enough altitude left before terrain impact, with both high downward velocity and loss of directional control, a situation that may result in an accident.

Power-required and power-available computations are extremely important for helicopter operations, especially in a remote landing zone. In order to compute power accurately for remote ops, where no precise measurements of pressure and temperature are known, it is customary for the crew to conduct what is called a "low recon." This reconnaissance takes place at a low altitude, often about 50 ft above ground level (AGL). As the crew makes the low pass (and slow, about 50 KIAS (indicated airspeed in kts.), well above translational lift), they have set 29.92″ in the altimeter to obtain the pressure altitude, and observe the outside air temperature (OAT) on the gage in the aircraft. With pressure altitude and OAT, the crew can now accurately compute density altitude for power computations, to determine if enough power is available to prevent power settling, or if not, that the approach and landing are not attempted until conditions are more favorable.

One of the most dangerous flight conditions that the helicopter pilot can encounter is known as settling with power or, technically, the vortex ring state (Figure 15.35). In this condition, the helicopter is descending into its own downwash. This can happen when the aircraft is making a vertical or near-vertical descent with low forward speed into a confined area, that is, one surrounded by obstacles such as trees or buildings. This type of operation is particularly hazardous, as it takes place almost entirely within what is known as the dead man's curve, the place in the Height/Velocity diagram where the helicopter's altitude and airspeed are too low to enter a successful autorotation (Figure 15.36). The rotor system is under some power from the engine, not in an autorotation. A helicopter pilot attempting to land at a site surrounded by high trees or other obstructions cannot make a normal approach. Forward speed cannot be maintained, so the pilot brings the helicopter to a hover at some height above the ground.

The airflow through the rotor system is shown in Figure 15.37. The maximum downward velocity is at the blade tips where the blade airspeed is highest and decreases nearer the rotor shaft. As the helicopter descends, it is acted upon by an upward relative wind that counteracts the induced flow.

With upward and downward air flows in opposite directions, there is no lift on the helicopter and it becomes a free-falling body. Preventative action calls for low descent rates when operating below translational lift to reduce the likelihood of entering the vortex ring state. Corrective action consists of increasing power (if available) to reestablish downflow. An alternative solution, and perhaps the best, is to lower collective and increase forward speed, altitude, and obstacles permitting. The pilot can also enter autorotation, again if altitude permits. The normal tendency to increase collective pitch while applying power can aggravate the power settling. If it is necessary to make a landing at a site where a forward speed of at least 15 kts. cannot be maintained (a headwind can be of great help if not blocked by obstacles), and there

Figure 15.35. Vortex ring state.
Source: Helicopter Flying Handbook (2012b)/U.S. Department of Transportation Federal Aviation Administration/Public domain.

is sufficient power available, the aircraft could be brought to a hover and a very low rate of descent used (less than 300 fpm). It is probably obvious that having OGE power makes this maneuver more feasible, but a vortex ring state is still possible with high descent rates. As in all helicopter flying, finesse with the controls is called for in this situation.

AUTOROTATION

In forward flight, the airflow through the rotor system and the force vectors are as shown in Figure 15.38. To maintain constant rotor rpm, the drag vector must be counteracted by an equal and opposite force supplied by the engine. If the engine fails or is deliberately disengaged from the rotor system, some other force must be applied to sustain the rotor rpm. This unpowered flight is similar to the glide of a fixed-wing aircraft and is called autorotation. The force that keeps the blades rotating is generated by lowering the collective pitch and thus reducing the angle of attack of all blades.

As the helicopter loses altitude, the airflow through the rotor disk changes, as shown in Figure 15.39. The airflow during descent provides the energy to overcome the blade drag and turn the rotor. The helicopter is trading the potential energy that it has due to its altitude for kinetic energy used to turn the blades. Later, we see how this kinetic energy is used to cushion the landing.

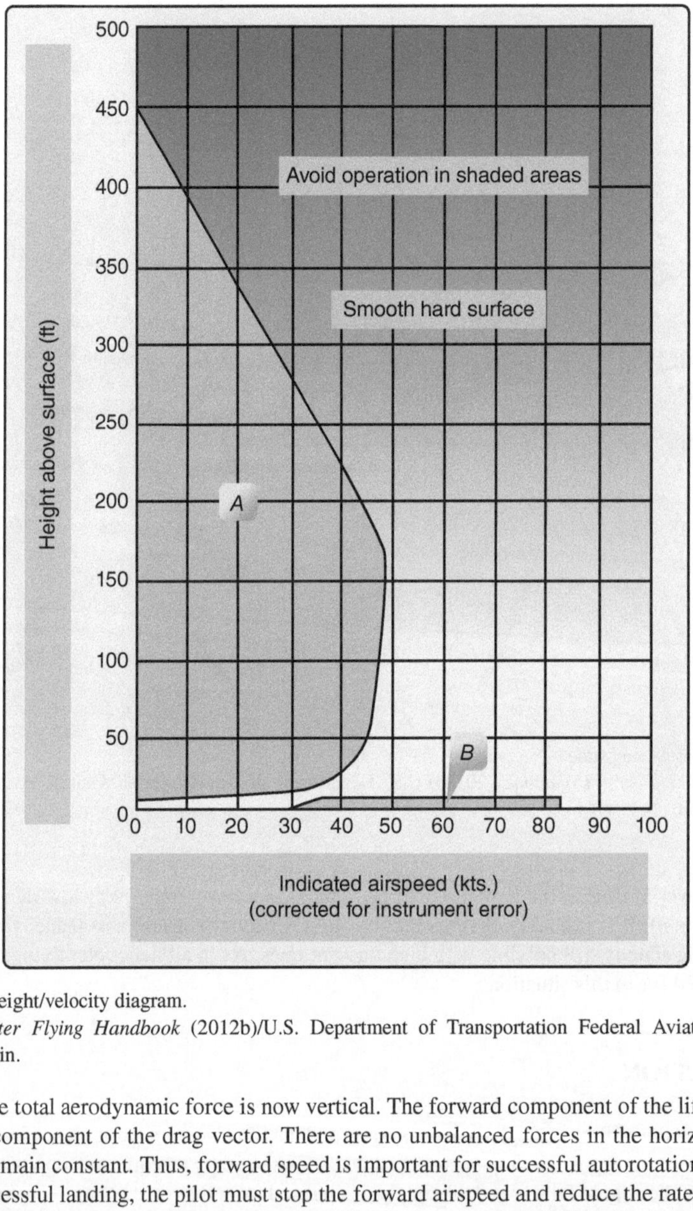

Figure 15.36. Height/velocity diagram.
Source: Helicopter Flying Handbook (2012b)/U.S. Department of Transportation Federal Aviation Administration/Public domain.

Note that the total aerodynamic force is now vertical. The forward component of the lift vector equals the backward component of the drag vector. There are no unbalanced forces in the horizontal direction and rpm will remain constant. Thus, forward speed is important for successful autorotation from altitude. To make a successful landing, the pilot must stop the forward airspeed and reduce the rate of descent just before landing. Both of these actions can be started by flaring the aircraft. This is done by moving the cyclic stick to the rear, which tilts the rotor disk to the rear and slows the forward speed. This is shown in Figure 15.40. There is also an increase in rotor rpm, which will help to cushion the landing. After the forward speed stops, the cyclic is returned to neutral and the final vertical velocity is further reduced by using up collective. The collect pitch is also used to avoid rotor overspeed.

Vertical autorotations are seldom attempted except during engine failure at hover. The pilot has much better visibility of the landing zone and better directional control if forward velocity is maintained.

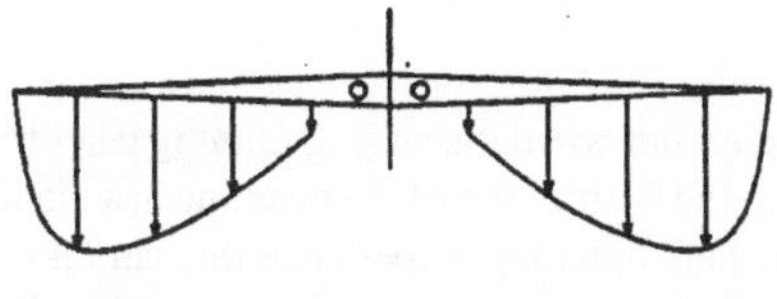

Figure 15.37. Induced flow velocity in a hover.

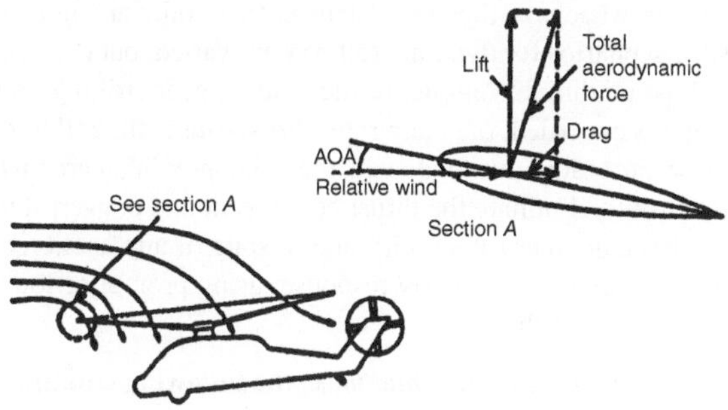

Figure 15.38. Airflow and force vectors in forward flight.

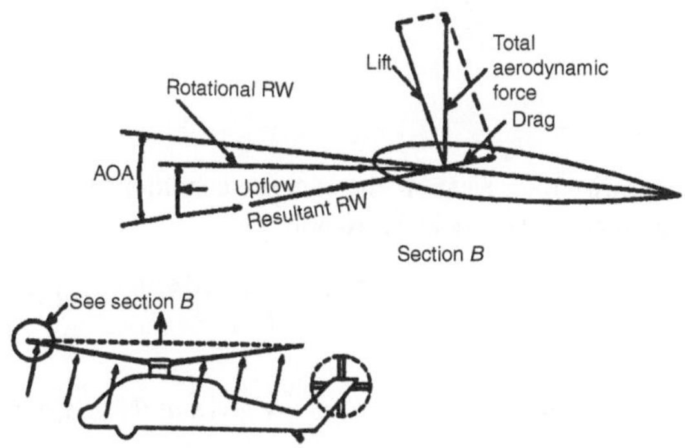

Figure 15.39. Airflow and forces in steady-state descent.

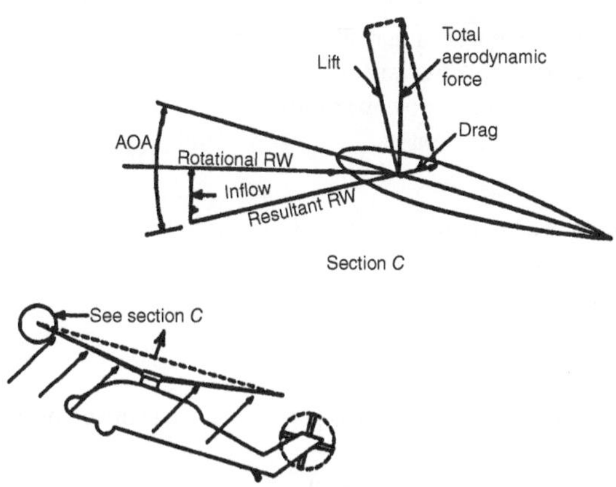

Figure 15.40. Airflow and forces during autorotative flare.

DYNAMIC ROLLOVER

Dynamic rollover was mentioned earlier when discussing running takeoffs. Dynamic rollover is a nemesis of helicopter pilots that can lead to the destruction of the machine in a matter of seconds if the pilot does not promptly correct the situation. A helicopter can experience this lateral rolling tendency, that is, dynamic rollover, whenever the helicopter is in contact with the ground. The phenomena affect helicopters with skids or wheels for landing gear. For dynamic rollover to occur, a circumstance must cause the helicopter to roll or pivot around a skid or wheel and then exceed the critical rollover angle, which varies but is often in the 5°–8° range. Possible scenarios for dynamic rollover are varied, but examples include a skid mired in the mud, or improper slope landing technique, or the landing gear striking an object during sideward flight. Once the critical angle is exceeded, the main rotor thrust causes the roll to continue. Recovery then becomes impossible because exceeding this angle puts the helicopter in an area where the cyclic does not have sufficient range of control to eliminate the thrust component and convert it to lift. If a pilot exceeds the critical rollover angle, the helicopter rolls on its side in spite of any cyclic corrections the pilot may attempt. The only cure is prevention, that is, early response, using prompt action to reduce the collective before the critical rollover angle is reached.

According to the FAA *Helicopter Flying Handbook*, the following conditions are most conducive to dynamic rollover for helicopters with counterclockwise rotor rotation:

1. Right-side skid or landing wheel down, because translating tendency adds to the rollover force.
2. Right lateral CG.
3. Crosswinds from the left.
4. Left yaw inputs.

For helicopters with clockwise rotor rotation, the opposite conditions would be true (e.g. Russian and French Helicopters, among others, have clockwise rotation).

PROBLEMS

Note: Answers to the problems in the following sections are given at the end of the book.

Application of Chapter Theory

1. Most helicopter rotors have symmetrical airfoils because
  a. they produce more lift than cambered airfoils.
  b. they can produce both upward and downward lift.
  c. they do not develop pitching moments.
  d. All of the above.

2. The lift theory most often used in helicopter aerodynamics is
  a. the gyroscopic precession theory.
  b. the momentum theory.
  c. Bernoulli's theory.
  d. the settling with power theory.

3. A helicopter can hover near the ground with less power than it can hover away from the ground because
  a. the air is denser near the ground.
  b. a high-pressure air "bubble" is produced below it.

c. the rotor blades have less induced drag.
d. None of the above.

4. The primary stress in rotor blades in flight is
 a. tension.
 b. bending.
 c. torsion.
 d. shear.

5. As a helicopter moves forward from a hover to a speed of about 15 kts., the power required is reduced because
 a. the efficiency of the rotor is improved.
 b. the helicopter is moving into a region of undisturbed air.
 c. the tip vortices are left behind.
 d. All of the above.

6. If a helicopter is flying at high speed and the engine fails, the pilot should enter an autorotation by slowing the helicopter by using back cyclic and then reducing blade angles with the collective.
 a. True
 b. False

Calculations Using Chapter Principles

7. Based on the dissymmetry of lift in forward flight discussion in this chapter, if a helicopter is moving forward in level flight at 50 kts. with rotor blades that have the same AOA and a tip speed of 500 kts., what are the respective net rotor tip speeds for the *advancing* and *retreating* blades:
 a. 550 and 450 kts.
 b. 500 and 450 kts.
 c. 450 and 500 kts.
 d. Net tip speed remains the same.

16 Unmanned Aerial Vehicle Flight Theory

CHAPTER OBJECTIVES

After completing this chapter, you should be able to:

- Explain the relationship between unmanned aerial vehicles (UAVs) and unmanned aircraft systems (UAS).
- Identify the categories of UAS as determined by the Department of Defense.
- Identify the missions best suited to different UAV fuselage and powerplant design.
- Identify the aerodynamic similarities and differences between manned rotorcraft design and UAV rotorcraft design.
- Critique the current problems with UAS and visual observation requirements and apply the advantages to BVLOS legislation and impact on mission potential.

The use of the term *Unmanned aerial vehicles* (UAV) has become ubiquitous within the aviation industry over the last 10 years and is now synonymous with *Unmanned Aircraft Systems* (UAS), drones, and Remotely Piloted Aircraft Systems (RPAS). In fact, UAV is meant to only represent the aerial vehicle itself, as opposed to the entire "system," such as the UAV, the software that lets the UAVs navigate and avoid collisions, the people operating the UAV, and the system monitoring the UAV. Because this textbook is primarily concerned about the aerodynamics of flight, detailed history and a thorough examination of UAV missions will not be presented in this textbook. The assumption is made that there is a background knowledge of the use of UAVs, thus only the aerodynamics related to UAV development and operation related to previous discussions in this book will be examined.

With the rapid expansion of "drones," within both the civilian and public/military market, regulatory guidance continues to be added and modified annually, both at the federal and state level. In August 2016, the first Part 107 rules were implemented to 14 Code of Federal Regulations (CFR), affecting those using small UAS (sUAS) for operations for-hire and operations that are not conducted purely for pleasure/recreation. Part 107 operations are limited to sUAS weighing less than 55 lb.

Flight Theory and Aerodynamics: A Practical Guide for Operational Safety, Fifth Edition.
Brian A. Johnson and Philip R. Fittante.

Companion website: www.wiley.com/go/flighttheoryandaerodynamics5e

Unmanned aircraft systems categorization chart

UA category	Maximum gross takeoff weight (lb)	Normal operating altitude (ft)	Speed (KIAS)	Representative UAS
Group 1	0–20	<1 200 AGL	100 kts	WASP III, TACMAV RQ-14A/B, Buster, Nighthawk, RQ-11B, FPASS, RQ16A, Pointer, Aqua/Terra Puma
Group 2	21–55	<3 500 AGL	<250	ScanEagle, Silver Fox, Aerosonde
Group 3	<1 320	<18 000 MSL	<250	RQ-7B Shadow, RQ-15 Neptune, XPV-1 Tern, XPV-2 Mako
Group 4	>1 320		Any airspeed	MQ-5B Hunter, MQ-8B Fire Scout, MQ-1C Gray Eagle, MQ-1A/B/C Predator
Group 5	>1 320	>18 000 MSL	Any airspeed	MQ-9 Reaper, RQ-4 Global Hawk, RQ-4N Triton

Legend

AGL	above ground level	lb	pounds
FPASS	force protection aerial surveillance system	MSL	mean sea level
ft	feet	TACMAV	tactical micro air vehicle
KIAS	knots indicated airspeed	UA	unmanned aircraft
kts.	knots	UAS	unmanned aircraft system

Figure 16.1. UAS categorization by the Department of Defense.
Source: U.S. Department of Transportation Federal Aviation Administration Public Domain.

Large public UAS may have wingspans as large as air transport category aircraft and operate at altitudes well beyond the 400 ft. restriction for civil sUAS, including well into the flight levels. Figure 16.1 shows UAS categories as set by the Department of Defense and includes examples of UAS that represent each grouping, including the RQ-4 Global Hawk (Figure 16.2) and MQ-9 Reaper.

UAV CATEGORIZATION

In this textbook, we will categorize unmanned aerial vehicles (UAVs) by their engine configuration and fuselage as opposed to size, regulation, or mission. The most common UAV recognized in the civilian

Figure 16.2. RQ-4 Global Hawk.
Source: Tom Miller/NASA/Public Domain.

market is the multi-rotor drone (quad-, hexa-, and octocopter configurations), relatively easy to control and often used by the public for photography, inspection, and recreational use. Single-rotor helicopter drones are used more commonly in the commercial market and can be used for heavier transport and agriculture research. Fixed-wing drones are often used for mapping, surveying, fire detection, security surveillance, and defense-related missions as their ability to stay aloft for longer flight times is conducive for commercial purposes. Finally, fixed-wing hybrid *Vertical Takeoff and Land* (VTOL) UAVs are used for agricultural and industrial purposes and have been the greatest segment of UAVs utilized to develop air taxis.

UAV DESIGN

The design of UAVs is based on the placement of the powerplant, whether rotors, propellers, or turbofan engines, as well as the design of the fuselage itself. The same principles introduced earlier in this textbook about lift (Chapter 4) and drag (Chapter 5) apply to UAVs in flight. Small multirotor UAVs are not designed to travel for speed or long distances, they can take off and land vertically and remain on a specific site for an extended period of time. UAVs with multi-rotor designs can be very maneuverable and are not plagued by parasite drag as they do not travel at a very high rate of speed in most cases. Fairings and fillets are not used to mitigate interference drag when the structure of the UAV incorporates 90° angles as with some fixed-wing designs presented earlier in this textbook. A large UAV platform may be needed for stability if a large amount of fuel is necessary to transport armaments or a large sensor or camera package. When inspecting infrastructure or negotiating tight spaces, a small, agile platform may be needed and range is not a consideration.

Large single-rotor helicopter UAVs, such as the Fire Scout MQ-8B/MQ-8C, operate in much the same manner as described in the prior chapter on manned rotorcraft. When in forward motion, the advancing blade in the same direction as relative motion produces greater lift, while the retreating blade produces less left, and once again rotor design must account for dissymmetry of lift. As in manned rotorcraft, the designs may incorporate a flapping hinge installed in the hub for the fully articulated rotor, or a teetering hinge in a semirigid motorhead supporting two rotor blades. To produce directional thrust, the rotor disk must be tilted in the direction of flight, and like propellers, the rotors incorporate a designed twist in the blade decreasing

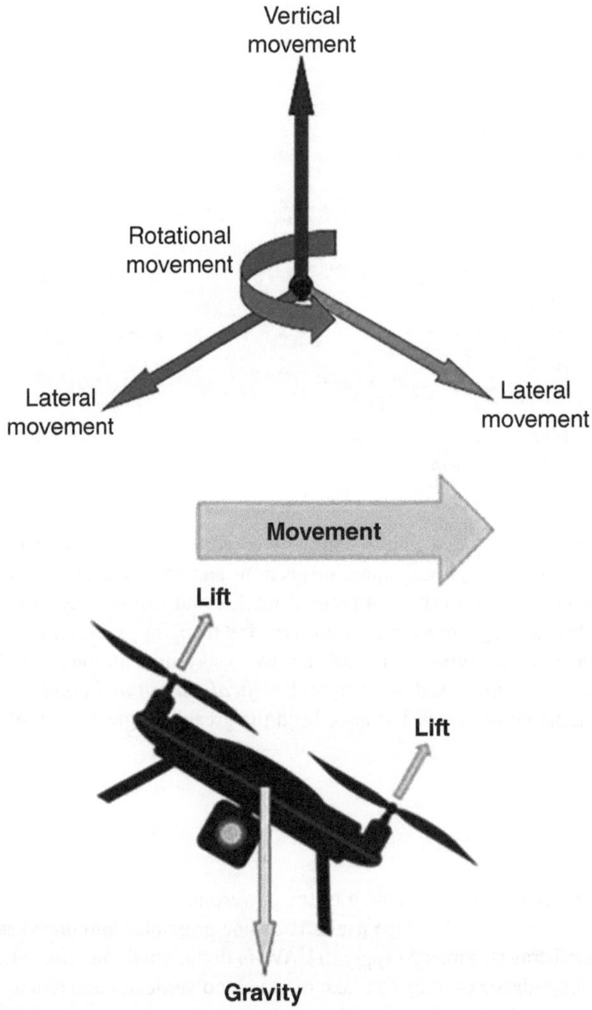

Figure 16.3. Lateral movement forces of a quadcopter.
Source: NASA Public Domain.

the blade angle from the hub to the tip. And finally, as in manned single-rotor helicopters, single-rotor UAVs incorporate a tail rotor or antitorque rotor as previously discussed in this chapter. Some UAVs incorporate contra-rotating and counterrotating designs with two main rotor systems which cancel out the torque impact of one rotor. In small, electric UAVs with fixed-pitch main rotor blades, lift is increased and decreased by varying rotor rpm. A common example of movement in quadcopters occurs when the rpms for the rotors on one side of the quadcopter are increased more than the rpms on the two rotors on the other side, the greater the combination of tilt and rotor rpm, the greater the speed of lateral movement (Figure 16.3).

Fixed-wing designs within the UAV industry vary as much as manned aircraft designs, from the flying wing to twin-boom pusher-prop designs, to V-tail and inverted T-tail designs. Because the manned element has been removed, a fuselage in shape and form as previously discussed may no longer be necessary, and a blended-wing design is built around the mission-oriented unmanned aircraft systems (UAS) platform as opposed to creature-comforts for humans and required regulatory safety enhancements. Many of the UAS platforms require the UAV to be installed with field of view (FOV) that is unobstructed, thus the UAV propulsion system that is responsible for also emitting heat, vibration, and exhaust must be mounted aft of the sensors and/or cameras. The placement of required UAS equipment and the location of the engines require the same level of understating of center of gravity (CG) placement as in manned aircraft and will be discussed further in the subsequent sections.

The flying wing design offers a more simplified manufacture and assembly compared to conventional designs. The greater aerodynamic efficiency (*L*/*D* ratio) increases range and glide capabilities. For the military, the flying wing design reduces radar cross section, minimizing the chance of detection and thus improving survivability. There is greater payload positioning flexibility though the longitudinal CG envelope may be limited. Another drawback is the inherent instability of the design necessitating advanced digital flight control systems.

Recent improvements in microprocessor technology and artificial intelligence (AI) have led to what has been termed "intelligent UAVs," meaning these UAVs could operate semi-autonomously or autonomously, making real-time decisions about mission tasks and their operating environment. By reducing the input of the human element, intelligent UAVs will be able to more efficiently complete mission requirements, to the point of swapping their own batteries (hot swapping), docking themselves to charge, or coordinating with other drones autonomously as a swarm.

AERODYNAMICS OF UAV FUSELAGE DESIGN

UAV design must still follow the same aerodynamic principles discussed in this textbook for manned aircraft, the four forces of flight are still at work, and Newton's laws still apply. UAVs operate in the same atmosphere as manned aircraft and thus are impacted the same by the viscous forces of the atmosphere. Figure 16.4 shows the upward motion of a quadcopter as explained by Newton's third law, first introduced in Chapter 1. As we have shown, the four forces of flight are interrelated, lift acts opposite weight, and thrust opposes drag (Figure 16.5). Depending on the mission and the requirements of the UAS, an increase in weight will result in reduced endurance and range, regardless of the type of propulsion system. The negative impact of induced drag and parasite drag on thrust will again influence fuselage and powerplant

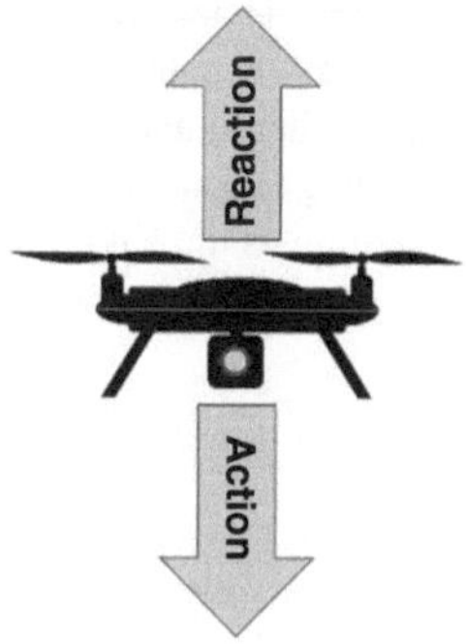

Figure 16.4. Newton's third law and a quadcopter.
Source: NASA Public Domain.

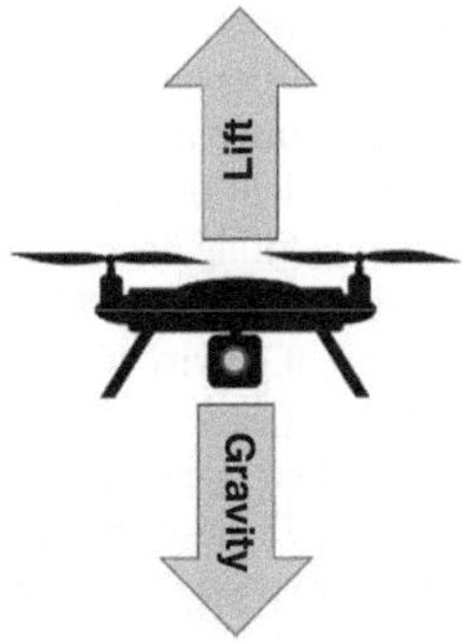

Figure 16.5. Quadcopter lift versus weight.
Source: NASA Public Domain.

design, as an increase in speed (regardless of direction), will result in an increase in parasite drag, and lifting surfaces that include fixed-wing airfoils and propellers/rotors in motion will increase induced drag. Once again aspect ratio can be analyzed to determine the amount of induced drag, with long, narrow wings producing higher aspect ratio numbers and thus reducing induced drag and extending the endurance and range of the UAV.

UAV tail designs vary more than their manned counterparts, but once again the aerodynamic principles surrounding the empennage and tail in flight remain the same. With a conventionally designed tail, the vertical and horizontal tail are airfoils as introduced in Chapter 3, providing lift and stability to the UAV. Aft-mounted engines may be installed on UAVs requiring a clear FOV, so to maintain positive longitudinal stability the tail must produce enough downward force to counteract the heavier payload toward the front of the platform. Twin-boom designs date back to the infancy of UAVs as developed by the Israelis in the 1970s. Usually powered by a pusher-propeller, with the twin booms acting as an empennage with one horizontal stabilizer joining the rudders. Many designs incorporate an adjustable set of booms so the CG can be adjusted based on the mission payload. As with the aft-mounted engine designs, twin-boom designs allow clear FOV forward for cameras or sensors. However, added weight, complexity, and parasite drag along with buffeting of the tail assembly from the pusher prop can be detractions. V-tail (Global Hawk) and inverted V-tail (Predator/Reaper) designs offer reduced drag though with the added complexity of ruddervator controls as seen on manned aircraft.

As much as we have compared the similarities in aerodynamics of UAVs to manned aircraft, there are significant differences in aerodynamic influence when an aircraft is scaled down. The wing surface area will vary inversely to the square of the scaling factor, and volume will vary inversely to the cube of the scaling factor. So, when we consider wing loading and all other factors remaining the same, weight divided by wing area, there is an increase in wing loading as the UAV is scaled down.

UAV POWERPLANT DESIGN

Most UAVs incorporate multiple propellers or rotors to develop lift and/or thrust. As we discussed in Chapter 1, the propulsive force of a propeller or rotor accelerates a mass of air, so yet again we find $F = m \times a$. Chapter 7 introduced the concept of propeller efficiency, as well as slip, the difference between geometric and effective pitch. As with manned propeller or rotor-driven aircraft, not all the shaft horsepower will be able to be converted to thrust (or lift). Propeller propulsion systems may involve fixed pitch, ground-adjustable, or constant-speed propellers. Larger propeller or rotor diameters have the potential to produce more energy but there is a trade-off with increasing tip speed when additional drag is introduced. Additional blades can be added to the propeller or rotor but then so is additional weight which may offset an increase in thrust (or lift) due to the increased mass being accelerated. The blade angle selected for a fixed-pitch propeller should be chosen based on the mission of the UAS. Lower pitch blades result in higher rpms, beneficial for lower speed operations and takeoff and landings. Higher pitch settings (lower rpm) can be more efficient for operation at higher speeds and longer cruise missions.

With the rise in multicopter UAVs, additional research has been conducted to maximize engine design and handling qualities. As the proliferation of small UAVs continues, in both the hobbyist and commercial sectors, the need for improved performance data has led to NASA testing using the Army's seven by ten-foot wind tunnel at the NASA Ames Research Center at Moffett Field in California. The goal of this research was to generate updated performance for small multicopter UAVs to improve software tools both inside NASA and outside in private industry. This research is just one example of ongoing testing within the UAV industry and is used in this textbook to show the difference in testing UAV aerodynamics versus manned aircraft from previous chapters. Table 16.1 shows the five UAVs chosen as part of the research, the parameters to be measured included the forces and moments on the multicopter rotors and airframes, as well as electrical power consumed as a function of attitude, rotor rpm, and speed. Table 16.2 lists the baseline thrust and rotor rpm of the UAVs. Figure 16.6 shows an example of a test drone mounted in the wind tunnel, as well as the direction of wind flow, directions of the forces as measured by the load cell, and moments. As was introduced in Chapters 12 and 13, the forces measured in the research used the typical aircraft axes, with F_x being positive downstream, F_z positive up, and F_y positive to the right. The value M was defined as moment, with M_x as positive roll right, M_y positive nose up, and M_z as positive nose right. The detailed results of the study can be found by accessing the Russell et al. (2016) document referenced within the bibliography and presented at the American Helicopter Society (AHS) 72nd Annual Forum in West Palm Beach, FL. Figures 16.7 and 16.8 provide a very generalized overview of the hover performance of the five test UAVs as an example of how test results were quantified. Figure 16.7 shows the test results of the five UAVs in hover with thrust as a function of rotor rpm. As noted in the study, thrust follows a quadratic trend with rotor speed. Figure 16.8 shows the thrust coefficient for each of the five test UAVs and indicates a slight increase with an increase in rpm, with a noted highest thrust coefficient by the DJI Phantom. Note that Reynolds number is included in the formula for thrust coefficient, first introduced in

Table 16.1. NASA multicopter test vehicles.

Make/model	Configuration	Rotor-to-rotor length (in.)	Rotor-to-rotor width (in.)	Rotor diameter (in.)	Nominal flight weight (lb)*
3DR SOLO	Quadcopter	11.5	11.5	10.0	3.3
DJI Phantom 3	Quadcopter	9.8	9.9	9.4	2.8
3DR Iris+	Quadcopter	10.4	16.1	9.6	2.8
Drone America x8	Octocopter	31.4	31.4	11.0	12.8
Straight Up Imaging Endurance	Quadcopter	20.1	20.1	15.0	6.0

*Without additional payload.

Source: NASA Public Domain.

Table 16.2. Test vehicle baseline thrust and rotor RPM.

Vehicle	Baseline thrust (lb)	Baseline rotor (rpm)
3DR SOLO	3.3	5 700
3DR Iris+	2.8	5 400
DJI Phantom 3 Advanced	2.8	5 300
Drone America DAx8	12.8	7 000
SUI Endurance	6.0	3 500

Source: NASA Public Domain.

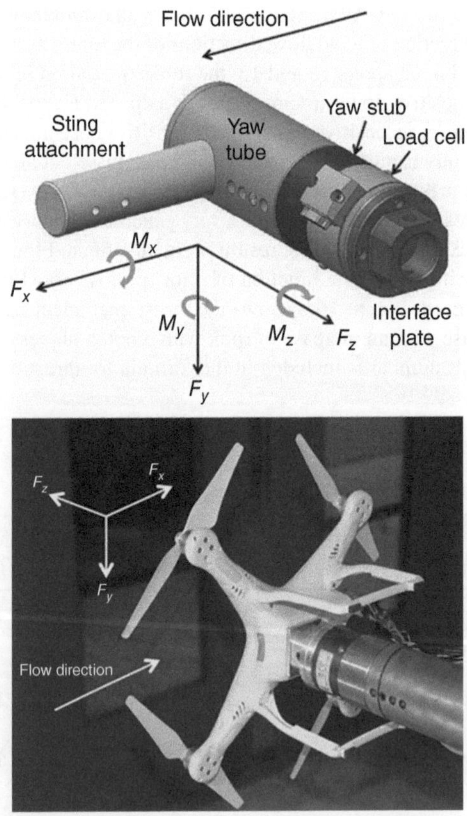

Figure 16.6. DJI Phantom 3 mounted in wind tunnel.
Source: NASA/Public Domain.

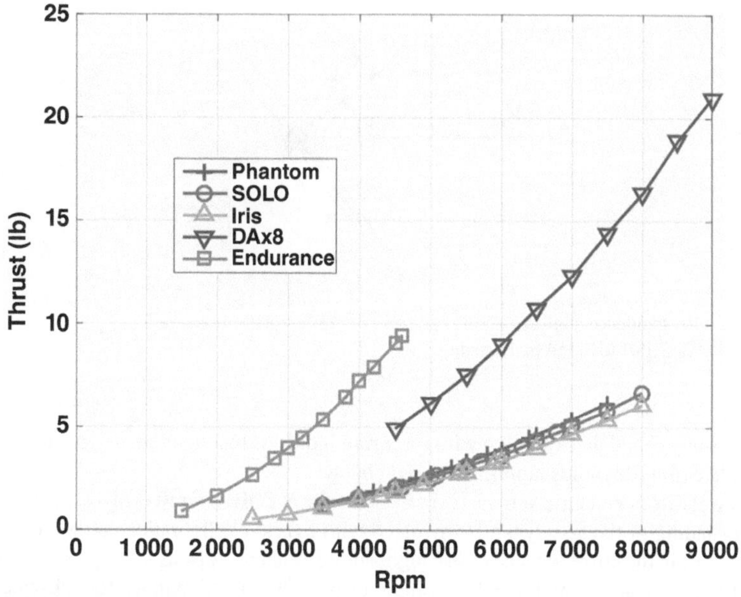

Figure 16.7. Full vehicle thrust test data.
Source: NASA Public Domain.

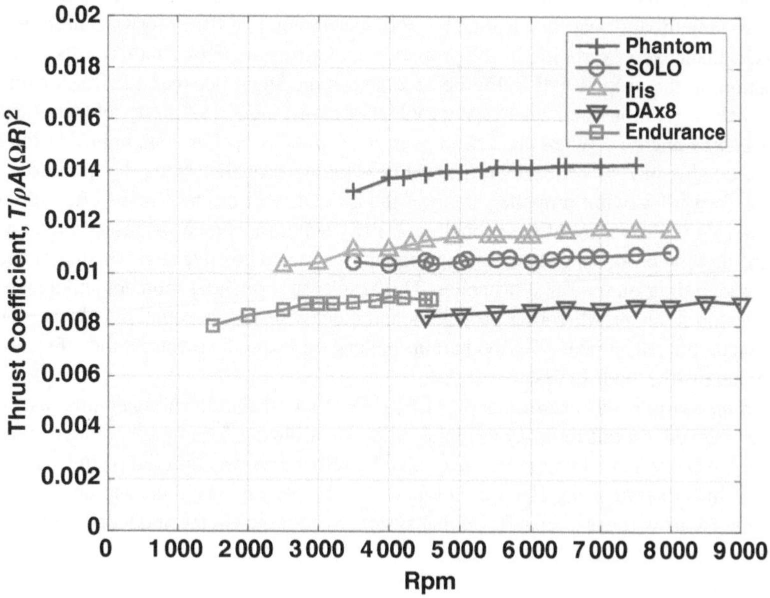

Figure 16.8. Full vehicle hover thrust coefficient.
Source: NASA Public Domain.

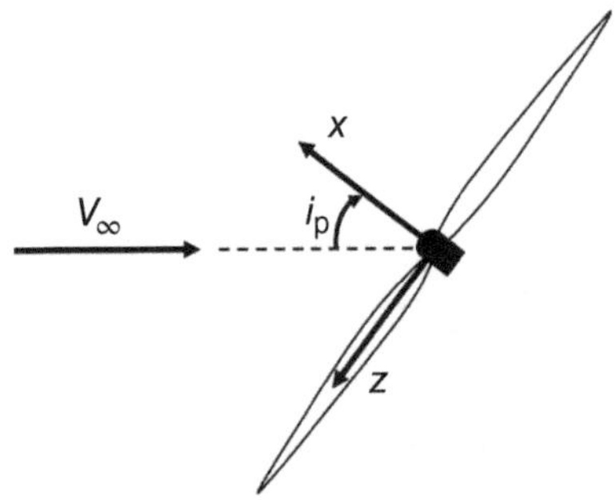

Figure 16.9. Propeller incidence angle.
Source: NASA, TM 20210014010 Public Domain.

Chapter 4. The study found that variation with rotor rpm indicates that nondimensional rotor lift increases slightly with increasing Reynolds number as speed changes.

As the future of tiltrotor electric vertical takeoff and land (eVTOL) aircraft evolves, the industry requires a better understanding of the aerodynamics involved with hybrid configurations between fixed-wing and rotary-wing aircraft in the effort to make cruise flights more efficient. A NASA study (Simmons and Hatke, 2021) aimed to investigate the high incidence angle of propeller aerodynamics for subscale eVTOL aircraft. Unlike a fixed-wing propeller in relatively constant airflow direction except for climbs, descents, and turns, a tiltrotor eVTOL will experience prolonged periods of oblique air flow, think cruise, transition, and hover, resulting in deviations of thrust and torque oriented along the propellers axes of rotation. The impact of the relative wind on the rotation of a propeller or rotor is analogous to the introduction of P-factor (asymmetrical thrust) in Chapter 13, and dissymmetry of lift in forward flight in Chapter 15. As the eVTOL maneuvers, variation in the propeller blade load angle of attack results in a nonuniform load distribution on the propeller disk. Figure 16.9 shows the angle between the freestream airflow and the propeller axis of rotation in a maneuvering eVTOL aircraft; the NASA study refers to the angle as the *propeller incidence angle* i_p. When the airflow is normal to the propeller disk, opposing the direction of axial thrust, the value of i_p is zero. Much like the descending propeller blade that produces more thrust (P-factor), the forward-moving propeller blade of the eVTOL will produce more lift, and the net thrust force is transferred to the center of the propeller. A net yawing movement is observed and acts to rotate the downward-moving propeller (when moving forward while at a positive angle of attack) and is coupled with an increase in drag on the same propeller. This results in a pitching moment and a side force as an auxiliary force and moment. Mirroring propellers and clockwise versus counterclockwise rotation along with autonomous rpm adjustment of individual propellers are methods to control the effects of imbalance with commercial off-the-shelf UAVs.

Some *medium-altitude long endurance* (MALE) UAVs that use reciprocating engines use the same turbochargers as manned aircraft to travel at higher altitudes allowing for improved range, endurance, and more efficient UAS missions. Intercoolers may also be utilized as was introduced in Chapter 7. Smaller UAV platforms may operate using electric motors or two-stroke engines; thus, engine oil is not required. Larger UAV platforms will use a separate engine system to cool the engine and lubricate parts, the engine and gearbox may utilize a single oil system or have separate dedicated oil systems. UAVs that fly at higher altitudes may need to warm the fuel before injection into the engine, and as in manned aircraft with fuel-injected engines, an engine-driven pump will be needed to inject high-pressure fuel into the combustion chamber during the compression stage to create the required mist of fuel droplets for efficient combustion. More advanced, large UAVs used by the military incorporate a *full authority digital engine control* (FADEC) control system run by multiple digital computers. FADEC systems are autonomous,

Figure 16.10. General atomics aeronautical systems MQ-9 predator B.
Source: NASA/Public Domain.

self-monitoring, redundant, and combine the throttle, propeller, and mixture controls into a single control. As the UAV changes altitude, the optimum power, prop rpm, and mixture combination is achieved.

UAVs with gas turbines are some of the most advanced, and expensive UAS on the market, and include turboprop, turboshaft, turbojet, and turbofan powerplants. These powerplants usually have the most favorable power-to-weight ratios. An introduction to these powerplants for manned aircraft can be found in Chapters 6 and 7, so a repeat of the basic operation will not be repeated here. Due to the sensitivity of the cameras and sensors on larger platform UAVs, engine reliability and vibration are paramount, so gas turbines offer the best combination of minimal vibration and excellent high-altitude performance. Turboprop-driven UAVs such as the General Atomics MQ-9 Reaper (Predator B) incorporate centrifugal compressors to achieve greater pressure stage increase than axial-flow compressors, but at lower compression efficiency, though this does allow a more compact powerplant (Figure 16.10). There are several turbofan-driven UAVs used in the military, including the General Atomics Predator C, used throughout combat zones around the world. These turbofan-driven UAVs are more fuel efficient than the previous turbojet-driven UAVs, accelerating a higher ratio mass of air for a given amount of thrust. Like the manned versions of aircraft using these powerplants, these engines are more efficient at high altitudes and are primarily utilized as a weapons platform and/or for reconnaissance.

Internal combustion engines have been widely used within the UAV market, providing high fuel energy and power density, while emitting less noise than gas turbine engines. UAVs powered by internal combustion engines are able to carry higher payloads than electric-motor-driven UAVs and they have a longer range.

As within other industries, the use of batteries and electric motors continues to proliferate within the UAV market, especially the multirotor VTOL market. A detailed review of UAV electric motor design and UAV electrical systems will not be included here, only the introduction to the use of electric motors and how this mode of propulsion differs when compared to other methods of propulsion as discussed previously. Most drone hobbyists are familiar with battery-powered, small, quadcopters. These UAVs are best utilized for shorter missions, often in the urban environment where range is not paramount, but flexibility and maneuverability are. Battery-powered drones may also be used for delivering medical supplies around a hospital campus, monitoring traffic, or with first responders during emergency responses. Electric UAVs have a reduced thermal signature compared to internal combustion engines and produce fewer pollutants.

THE FUTURE OF UAV DESIGN AND AERODYNAMICS

The future of UAV design relies heavily on the expanding mission requirements of UAS operations and the continued state and federal legislative updates to more efficiently incorporate UAVs into the National Airspace System (NAS). This large undertaking is part of the greater effort that falls under the *Unmanned Aircraft System Traffic Management* (UTM) initiative. As described by the Federal Aviation Administration (FAA), UTM is a "traffic management" ecosystem that is complementary to the FAA's Air Traffic Management (ATM) system and will ultimately identify services, roles and responsibilities, information architecture, data exchange protocols, software functions, infrastructure, and performance requirements for enabling the management of low-altitude drone operations. The expansion of UTM revolves around a partnership between the FAA, NASA, other federal agencies, and industry. A major hurdle for the continued roll-out of UTM will be the expanded ability for *beyond visual line-of-sight* (BVLOS) UAS. The design of civilian UAVs and powerplant development will depend extensively on what UAVs are needed and the level of regulation that is created to allow continued growth. Companies such as Amazon and Google are exploring the use of small UAS (sUAS) in the package delivery market, while other commercial industries such as powerline and pipeline inspection, crop management, and filming continue to expand the use of sUAS into their operations. The Concept for Operations for *Urban Air Mobility* (UAM) is a collaborative effort by government agencies and industry to focus on flight operations within urban areas and is a subset of the larger government initiative *Advanced Air Mobility* (AAM) to develop an air transportation system that moves people and cargo between local, regional, intraregional, and urban locations not previously served by aviation (Figure 16.11).

The previous NASA project, *Design Environment for Novel Vertical Lift Vehicles* (DELIVER), as well as NASA's *Revolutionary Vertical Lift Technology* (RVLT), have set a goal of improving the aerodynamic performance characteristics of multicopters. Both projects incorporate a design software tool called the

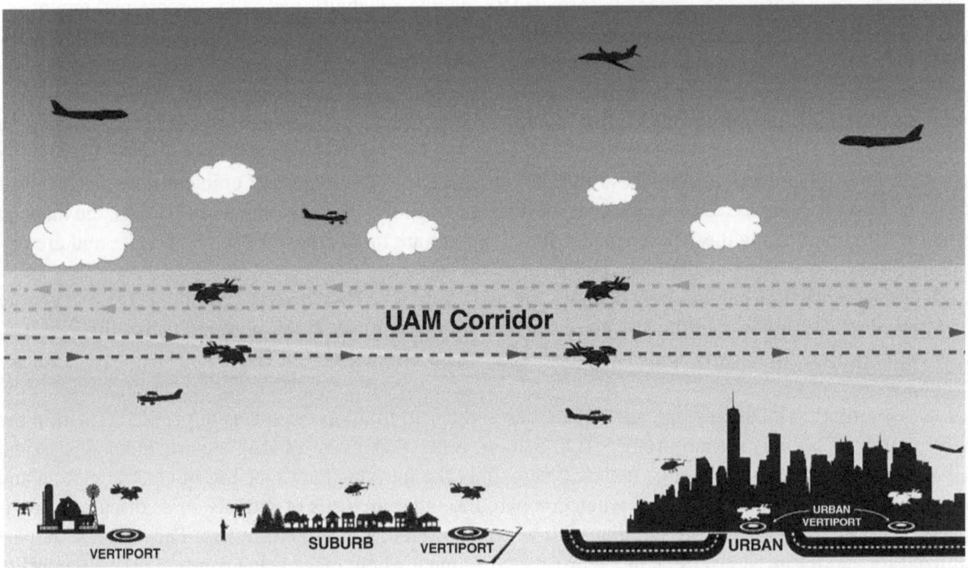

Figure 16.11. UAM corridor with passing zones.
Source: NASA/Public Domain.

Figure 16.12. AAM concept vehicle.
Source: Lillian Gipson/NASA/Public Domain.

NASA Design and Analysis of Rotorcraft (NDARC), which has been validated for large (<2 person) rotorcraft. NDARC addresses the design challenge of using current conceptual design tools for the development of large, unmanned aircraft. Mitigating the noise of large UAVs, developing autonomous operations, and incorporating hybrid-electric and all-electric options were core goals of the DELIVER project and helped set the stage for future innovation.

NASA's RVLT project focuses on large cargo-carrying vehicles and passenger-carrying taxis and is a collaborative effort by NASA, industry, and academia to develop technologies to enable revolutionary new air travel options (Figure 16.12). A critical focus of RVLT is to design eVTOL aircraft that minimize impact on the environment and reduce the noise impact when near people. Targeting propulsion system reliability and standards, RVLT focuses on enabling the use of electric propulsion systems and improving power quality standards for electric propulsion architectures. Recent improvements include a 50% reduction in the noise footprint area commonly used for commercial VTOL operations due to a high-fidelity rotor/design approach. A robust VTOL air taxi system will also require a new set of safety standards, with many regulations yet to be created by the FAA. The impact of weather, notably icing, must be addressed as the shedding of ice on rotor blades creates vibration and is a hazard to occupants. Because eVTOL aircraft are designed to carry passengers and cargo at lower altitudes, and are usually lighter weight than conventional manned aircraft, they can be more susceptible to Dutch-roll and other gust-induced oscillations. Unlike fixed-wing aircraft discussed in Chapter 13 regarding lateral/directional stability, the eVTOL aircraft do not have mechanisms to suppress oscillations such as ailerons, flaps, and rudders. NASA Ames is working on a fix that uses outboard propellers to dampen oscillations called the Active Turbulence Suppression (ATS) system. When the stable eVTOL senses a roll angle, the automated system activates the specific propeller(s) needed to adjust the speed of the propellers using a database of propeller speed profiles and roll angles. In Figure 16.13, the top image shows the activation of the right-wing outboard propellers, and the bottom image indicates the activation of the left outboard propellers. A color image depicts the change

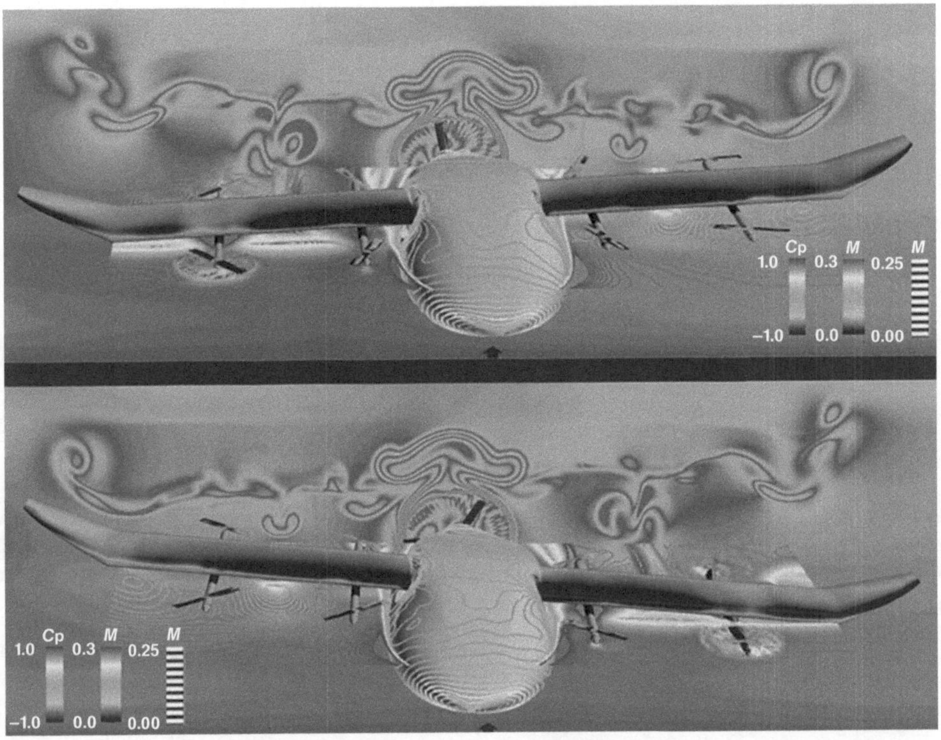

Figure 16.13. Activation of eVTOL active turbulence suppression system.
Source: Ames, Guru Guruswamy/NASA/Public Domain.

in pressure coefficient as the respective propellers are activated to maintain stable flight, while a black and white image depicts the distortion of airflow around the wing tip for the respective outboard propeller activation.

The use of composite materials in UAV production will continue to proliferate in the future. Due to their high strength-to-weight ratio, polymer composite materials such as carbon fiber reinforced polymer (CFRP) and glass fiberglass reinforced polymer (GFRP) are used in UAV wings, fuselages, and landing gear. When compared to aluminum, polymer composites can reduce the weight of UAV components by up to 45% and offer marked decreases in maintenance due to corrosion and fatigue. Another important characteristic of UAV design moving forward is the amount of stiffness allowed by the material used to produce the drone, often referred to as the stiffness-to-weight ratio or specific stiffness. For aerodynamics it is important to construct an aircraft with a higher specific stiffness, and carbon composites are much stiffer than titanium or aluminum. During the development of rotor blades or propellers, the higher the specific stiffness the better, so again the use of carbon composites allows for a more rigid design and the ability to maintain the aerodynamic shape. As the science of composite development continues to improve, more complex structures can be built, further reducing the use of other materials that require the use of fasteners or rivets. Due to the positive electromagnetic properties of composites, their use will continue to grow in both the civilian and military markets. Composites can be tuned to absorb certain electromagnetic frequencies while blocking signals from other sources.

KEY TERMS

Advanced Air Mobility (AAM)
Beyond Visual Line-of-Sight (BVLOS)
Design Environment for Novel Vertical Lift Vehicles (DELIVER)
Full Authority Digital Engine Control (FADEC)
Medium-altitude Long Endurance (MALE)
NASA Design and Analysis of Rotorcraft (NDARC)
Propeller incidence angle (i_p)
Revolutionary Vertical Lift Technology (RVLT)
Unmanned Aerial Vehicle (UAV)
Unmanned Aircraft Systems (UAS)
Unmanned Aircraft System Traffic Management (UTM)
Urban Air Mobility (UAM)
Vertical Takeoff and Land (VTOL)

Answers to Problems

CHAPTER 1

1. a; 2. b; 3. d; 4. b; 5. a; 6. d, 7. b. 8. 110 fps; 9. 119 kts.; 10. 3 546 fpm; 11. 60 sm; 12. $m = 497$ slugs; 13. $a = 12\,\text{fps}^2$; 14. Time = 16.7 seconds; 15. Takeoff roll = 1667 ft; 16. $a = 5\,\text{fps}^2$; 17. Drag = 282 lb, Lift = 102.6 lb; 18. Climb angle = 37°, Distance = 5 mi; 19. $s = 8$ ft, $F = 30$ lb; 20. $V_t = 450$ fps; 21. PE = 80×10^6 ft-lb, KE = 10×10^6 ft-lb, TE = 90×10^6 ft-lb; 22. HP = 5 000 hp; 23. $F_B = 12\,600$ lb

CHAPTER 2

1. c; 2. c; 3. a; 4. a; 5. b; 6. c; 7. d; 8. c; 9. b; 10. d; 11. b; 12. a; 13. b; 14. c; 15. $d = 0.929$, PA = 2 000 ft, $q = 1.078$, $s = 0.8618$, DA = 5 000 ft; 16. $H = 2062.8$ psf, $P_2 = 1\,931$ psf, $q_2 = 131$ psf, $A_3 = 12.5\,\text{ft}^2$, $P_3 = 2\,041$ psf, $q_3 = 21$ psf; 17. $q = 116.8$ psf; 18. EAS = 338 kts.; 19. TAS = 499 kts.; 20. Mach = 0.84

CHAPTER 3

1. b; 2. a; 3. b; 4. c; 5. b; 6. a; 7. d; 8. c; 9. d; 10. a; 11. d; 12. c; 13. e; 14. a; 15. d; 16. d; 17. b; 18. d

CHAPTER 4

1. a; 2. c; 3. b; 4. a; 5. c; 6. d; 7. d; 8. c; 9. b; 10. d; 11. a. combination, b. energy add, c. combination, d. camber chg; 12. d; 13. c; 14. 7°; 15. 1.6; 16. 5 944 lb; 17. AOA = 4°; 18. $V_s = 134$ knots; 19. $V_s = 116$ knots; 20. $R_g = 10.3 \times 10^6$; 21. 151 kts.

CHAPTER 5

1. d; 2. c; 3. d; 4. d; 5. a; 6. b; 7. d; 8. d; 9. c; 10. b; 11. a; 12. b; 13. a; 14. c; 15. AR = 7; 16. $D_p = 750$ lb; 17. $D_{min} = 1\,020$ lb; 18. $V = 172$ knots; 19. $D_i = D_p = 510$ lb;

Flight Theory and Aerodynamics: A Practical Guide for Operational Safety, Fifth Edition.
Brian A. Johnson and Philip R. Fittante.

Companion website: www.wiley.com/go/flighttheoryandaerodynamics5e

20.

V_2	(V_2/V_1)	D_p	$(V_1/V_2)^2$	D_i	D_t
125	0.530	270	1.89	963	1 233
150	0.763	389	1.31	669	1 058
172	1.0	510	1.0	510	1 020
200	1.35	690	0.74	377	1 067
300	3.04	1 562	0.33	168	1 720
400	5.42	2 766	0.184	94	2 860

CHAPTER 6

1. d; 2. d; 3. a; 4. a; 5. a; 6. a; 7. d; 8. c; 9. c; 10. b; 11. d; 12. c; 13. a; 14. b; 15. a; 16. d; 17. c; 18. b; 19. d; 20. d; 21. T = 3 000 lb, Efficiency = 77%; 22. $C1$ = 1.5, T = 4 100 lb, FF_0 = 6 150 lb/hr; 23. FF_{trop} = 1 845 lb/hr; 24. a. 500 knots, b. 240 knots, c. 19°, d. 10 940 fpm, e. 240 knots, f. 300 knots, g. 320 knots; 25. V_{12} = 325 knots, V_8 = 285 knots; 26. SR_{12} = 0.2708 nm/lb, SR_8 = 0.3455 nm/lb; 27. SR_8 = 0.3125 nm/lb, SR improvement in Problem 27 = 14.15%, SR improvement in Problem 26 = 27.6%, conclusion: reduce throttle as fuel is burned; 28. SR_0 = 0.2973, SR_{20} = 0.4108, SR improvement = 38%

CHAPTER 7

1. a; 2. d; 3. a; 4. b; 5 a; 6. c; 7. b; 8. d; 9. b; 10. b; 11. d; 12. a; 13. b; 14. c; 15. a; 16. b; 17. c; 18. c; 19. c; 20. c; 21. b; 22. b; 23. a; 24. a. 340 knots, b. 32.8°, c. 19°, d. 4 125 fpm, e. 4 950 fpm, f. 100 knots, g. 150 knots, h. 160 knots; 25. 474 hp, 488 hp, 540 hp, 657 hp, 1 588 hp, 3 520 hp; 26. V_{30} = 160 knots, V_{20} = 130 knots; 27. SR_{30} = 0.3765 nm/lb, SR_{20} = 0.5778 nm/lb; 28. SR_{20} = 0.4471 nm/lb, improvement = 18.7%, improvement in Problem 27 = 53.5%, conclusion: reduce throttle as fuel is burned; 29. SR_0 = 0.694 and SR_{20} = 0.667 nm/lb, conclusion: no change in SR with altitude (considering airframe only)

30.

V	Drag	P_r
150	2 074	957
182.7	1 923	1 081
200	1 954	1 202
300	2 948	2 721
400	4 809	5 919

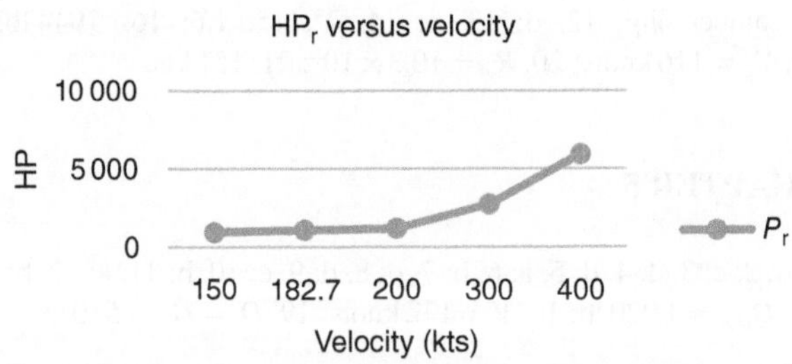

31. and 32. Not included.

CHAPTER 8

1. a 2. b; 3. c; 4. c; 5. a; 6. d; 7. b; 8. c; 9. d; 10. d; 11. a; 12. d; 13. d; 14. a; 15. a; 16. c; 17. b; 18. c; 19. s = 1 728 ft; 20. s = 2 765 ft; 21. s = 3 550 ft; 22. A = 8 fps²; 23. V = 126.5 fps (75 knots); 24. s = 2 576 ft; 25. s = 5 796 ft; 26. s = 2 087 ft; 27. s = 4 025 ft

CHAPTER 9

1. d; 2. b; 3. d; 4. a; 5. c; 6. c; 7. b; 8. b; 9. d; 10. d; 11. a; 12. a; 13. c; 14. a; 15. a; 16. c; 17. b; 18. a; 19. d; 20. a; 21. $a = -8\,\text{fps}^2$; 22. $s = 1\,774$ ft; 23. $s = 2\,883$ ft; 24. $s = 2\,147$ ft; 25. $s = 2\,218$ ft

CHAPTER 10

1. d; 2. b; 3. d; 4. b; 5. b; 6. a; 7. c; 8. b; 9. d; 10. d; 11. a; 12. a; 13. b; 14. a; 15. b; 16. a; 17. b; 18. c; 19. a; 20. c; 21. ROC = 506.5 fpm; 22. ROC = −658.5 fpm; 23. 696 fpm

CHAPTER 11

1. b; 2. c; 3. d; 4. d; 5. d; 6. a; 7. c; 8. d; 9. d; 10. b; 11. c; 12. a; 13. a; 14. b; 15. d; 16. 30.6°; 17. Verified; 18. 1.16*G*; 19. ROT = 3.23°/sec; 20. Verified

CHAPTER 12

1. b; 2. c; 3. d; 4. c; 5. a; 6. d; 7. b; 8. c; 9. a; 10. a; 11. b; 12. b; 13. c; 14. a; 15. d; 16. c; 17. 22 022 lb-in; 18. CG = 84.9 in; 19. 245 380 lb-in; 20. 83.04 in

CHAPTER 13

1. b; 2. a; 3. c; 4. a; 5. a; 6. d; 7. c; 8. a; 9. b; 10. a; 11. b; 12. a; 13. b; 14. c; 15. b; 16. c; 17. a; 18. d; 19. b; 20. b; 21. c; 22. a

CHAPTER 14

1. d; 2. b; 3. b; 4. e; 5. a; 6. h; 7. a; 8. a; 9. b; 10. d; 11. c; 12. b; 13. a; 14. a; 15. d; 16. 624 kts.; 17. $M = 0.56$

CHAPTER 15

1. c; 2. b; 3. c; 4. a; 5. d; 6. b; 7. a

Bibliography

Abbott, I.H. and VonDoenhoff, A.E. (1959). *Theory of Wing Sections*. New York: Dover.

Academic Flight (2019). Test Pilot Literature | Test Pilot School Textbooks. [online] Available at: https://academicflight.com/bibliography/test-pilot-literature/ [Accessed 1 Dec. 2024].

Airfoil Tools (2019). Airfoil Tools. [online] Airfoiltools.com. Available at: http://airfoiltools.com.

AIRPLANE FLIGHT MANUAL DA 50C Airworthiness Category: Normal Requirement: CS-23 Serial Number: ________ Registration: ________ A-2700 WIENER NEUSTADT AUSTRIA (2025). Diamond Aircraft. Available at: http://support.diamond-air.at/fileadmin/uploads/files/after_sales_support/DA50-Airplane_Flight_Manual/Basic_Manual/90101e-DA50-AFM-Issue017.pdf [Accessed 4 Jan. 2025].

Anderson, J.D. Jr. (1978). *Introduction to Flight*. New York: McGraw-Hill.

Anderson, J.D. Jr. (2007). *Fundamentals of Aerodynamics*. New York: McGraw Hill.

Anderson, J.D. Jr. (2012). *Introduction to Flight*. New York: McGraw Hill.

apps.dtic.mil (2012). USAF (United States Air Force) Stability and Control DATCOM (Data Compendium). [online] Available at: https://apps.dtic.mil/sti/citations/ADB072483.

BEA (2012). Bureau d'Enquêtes et d'Analyses pour la sécurité de l'aviation civile Final Report. [online] Available at: https://bea.aero/docspa/2009/f-cp090601.en/pdf/f-cp090601.en.pdf.

Benson, T. (2021). Boundary Layer. [online] www.grc.nasa.gov. Available at: https://www.grc.nasa.gov/www/k-12/BGP/boundlay.html.

Bent, R.D. and McKinley, J.L. (1978). *Aircraft Powerplants*, 4e, Gregg Div. New York: McGraw-Hill.

Boeing (2025). 737 MAX Software Update. [online] www.boeing.com. Available at: https://www.boeing.com/commercial/737max/737-max-update/737-max-software-updates#overview.

Carrol, R.L. (1960). *The Aerodynamics of Powered Flight*. New York: Wiley.

Coram, R. (2004) *Boyd: The Fighter Pilot Who Changed the Art of War*. Back Bay Books.

Dalton, S. (1977). *The Miracle of Flight*. New York: McGraw-Hill.

Dole, C.E. (1981). *Flight Theory and Aerodynamics*. New York: Wiley.

Dole, C.E. (1991a). *Mathematics and Physics for Aviation Personnel*. Casper, WY: IAP.

Dole, C.E. (1991b). *Fundamentals of Aircraft Material Factors*. Casper, WY: IAP.

Dole, C.E. (1993). *Flight Theory for Pilots*. Casper, WY: IAP.

Domasch, D.O., Sherby, S.S., and Connolly, T.F. (1967). *Airplane Aerodynamics*, 4e. New York: Pitman.

Dwinnell, J.H. (1949). *Principles of Aerodynamics*. New York: McGraw-Hill.

Etkin, B. (1972). *Dynamics of Atmospheric Flight*. New York: Wiley.

Flight Theory and Aerodynamics: A Practical Guide for Operational Safety, Fifth Edition.
Brian A. Johnson and Philip R. Fittante.

Companion website: www.wiley.com/go/flighttheoryandaerodynamics5e

Fielding, J.P. (1999). *Introduction to Aircraft Design*. Cambridge: Cambridge University Press.

Fujino, M., Yoshizaki, Y., and Kawamura, Y. (2003). Natural-laminar-flow airfoil development for a lightweight business jet. *Journal of Aircraft* **40** (4). https://doi.org/10.2514/1.168.

Hoerner, S.F. (1965). *Fluid-Dynamic Drag* (published by, S. F. Hoerner.).

Hoerner, S.F. and Borst, H.V. (1967). *Fluid-Dynamic Lift* (published by, Mrs. L. A. Hoerner.).

Industries, Diamond Aircraft. (n.d.). "Diamond Aircraft:: Airplane Flight Manual." Diamond Aircraft Industries, http://support.diamond-air.at/da50c_afm0+M52087573ab0.html.

klabs.org (2010). Aircraft Nomenclature. [online] Available at: https://klabs.org/DEI/References/contents_detailed/aircraft_nomenclature/index.htm [Accessed 15 Jun. 2024].

Kuethe, S. (1976). *Foundations of Aerodynamics*, 3e. New York: Wiley.

Leishman, J.G. (2023). *Introduction to Aerospace Flight Vehicles*. [online] eaglepubs.erau.edu. Embry-Riddle Aeronautical University. Available at: https://eaglepubs.erau.edu/introductiontoaerospaceflightvehicles/.

Martynov, A.K. (1965). *Practical Aerodynamics*. New York: Macmillan.

McCormick, B.W. (1979). *Aerodynamics, Astronautics, and Flight Mechanics*. New York: Wiley.

McCormick, B.W. (1995). *Aerodynamics, Astronautics, and Flight Mechanics*, 2e. New York: Wiley.

Millbrooke, A.M. (1999) *Aviation History*. Englewood, CO: Jeppesen Sanderson.

Montgomery, J.R. (1964). *Sikorsky Helicopter Flight Theory for Pilots and Mechanics*. Stratford, CT: Sikorsky Aircraft.

Nicolai, L.M. (1974). *Design of Airlift Vehicles*. Dayton, OH: University of Dayton.

Perkins, C.D. and Hage, R.E. (1949). *Airplane Performance, Stability and Control*. New York: Wiley.

Piper Aircraft Corporation (1965). *Cherokee 140 Owner's Handbook*. Vero Beach: Piper Aircraft Corporation.

Prouty, R.W. (1982). *Practical Helicopter Aerodynamics*. Peoria: PJS Publications, Inc.

R. Kurt Barnhart, Marshall, D.M., and Shappee, E. (2021). *Introduction to Unmanned Aircraft Systems*. CRC Press.

Rodrigues, C.C. and Cusick, S.K. (2012). *Commercial Aviation Safety*. New York: McGraw-Hill, Inc.

Roed, A. (1972). *Flight Safety Aerodynamics*. Kungsangen, Sweden: Aerotech.

Roland, H.E. and Detwiler, J.F. (1967). *Fundamentals of Fixed and Rotary Wing Aerodynamics*, ISSM. Los Angeles: University of Southern California.

Saravanakumar, Y.N., Sultan, M.T.H., Shahar, F.S., Giernacki, W., Łukaszewicz, A., Nowakowski, M., Holovatyy, A., and Stępień, S. (2023). Power sources for unmanned aerial vehicles: a state-of-the art. *Applied Sciences*, [online] **13** (21): 11932. https://doi.org/10.3390/app132111932.

Saunders, G.H. (1975). *Dynamics of Helicopter Flight*. New York: Wiley.

Seckel, E. (1964). *Stability and Control of Airplanes and Helicopters*. New York: Academic Press.

Shapiro, A.H. (1953). *The Dynamics and Thermodynamics of Compressible Fluid Flow*, vol. **I** and **II**. New York: Ronald Press.

Spalart, P.R. and Venkatakrishnan, V. (2016). On the role and challenges of CFD in the aerospace industry. *The Aeronautical Journal* **120** (1223): 209–232. https://doi.org/10.1017/aer.2015.10.

Smith, H.C. (1992). *The Illustrated Guide to Aerodynamics*. New York: McGraw-Hill, Inc.

Sutton, O.G. (1955). *The Science of Flight*. New York: Pelican Books (A209).

Swatton, P.J. (2011). *Principles of Flight for Pilots*. New York: Wiley.

The Boyd Institute (2023). *About John Boyd*. [online] Boydinstitute.org. Available at: https://boydinstitute.org/p/john-boyd [Accessed 29 Sep. 2024].

GOVERNMENT PUBLICATIONS

Anonymous (1965). *Aerodynamics*. Washington, DC: S. Army Material Command.

Anonymous (1966). *Interceptor*, ADCPI 62-1 through 62-16, June 1964–April. Air Defense Command, USAF.

Anonymous (1988). *Pilot Windshear Guide*, AC No. 00-54. Federal Aviation Administration.

Anonymous (1975). *Academics for Basic*, ATC P-V4A-A-AB-SW. Air Training Command, USAF.

Anonymous (1976a). *Low Level Wind Shear*, AC No. 00-50. Federal Aviation Administration.

Anonymous (1976b). *Wake Turbulence*, USAF, Pamphlet 51-12. Air Training Command.

Anonymous (1977). *Applied Aerodynamics*, ATC P-V4A-A-AA-SW. Air Training Command, USAF.

Anonymous (1979a). *Rotary Wing Flight*, Field Manual, 1-51. U.S. Army.

Anonymous (1979b). *Aerodynamics for Pilots*, ATC Manual 51-3. Air Training Command, USAF.

Bhat, B.N., Pandey, A.B., Tamirisakandala, S., Nathal, M.V., Ellis, D.L., McGill, P.B., Lee, J.A., Davis, S.E., Koo, J., Nettles, A., and Zhu, D. (2018). Chapter 2: Aerospace Materials Characteristics. *Aerospace Materials and Applications*. [online] Available at: https://ntrs.nasa.gov/citations/20180001137.

Faa.gov (2022). AC 150/5200-30D – Airport Field Condition Assessments and Winter Operations Safety – Document Information. [online] Available at: https://www.faa.gov/airports/resources/advisory_circulars/index.cfm/go/document.current/documentNumber/150_5200-30.

Faa.gov (2019). Large UAS (MGOW 55 Pounds or More). [online] Available at: https://www.faa.gov/air_traffic/publications/atpubs/aim_html/chap11_section_3.html# [Accessed 11 Dec. 2024].

FAA Aviation Safety (2022). Best Glide Speed and Distance. Available at: https://www.faa.gov/sites/faa.gov/files/2022-01/Best Glide Speed and Distance.pdf.

Federal Aviation Administration (1994). *AC 120-62 – Takeoff Training Aid*. Washington, DC: United States Government Printing Office.

Federal Aviation Administration (2004a). *Airplane Flying Handbook*. Oklahoma City: United States Department of Transportation.

Federal Aviation Administration (2004b). *Seaplane, Skiplane, and Float/Ski Equipped Helicopter Operations Handbook*. Oklahoma City: United States Department of Transportation.

Federal Aviation Administration (2004c). *Takeoff Training Safety Aid*. Oklahoma City: United States Department of Transportation.

Federal Aviation Administration (2007). *Airframe Guide for Certification of Part 23 Airplanes AC 23-19A*. Oklahoma City: United States Department of Transportation.

Federal Aviation Administration (2008a). *Pilot's Handbook of Aeronautical Knowledge*. Oklahoma City: United States Department of Transportation.

Federal Aviation Administration (2008b). *Wind Shear FAA-P-8740-40*. Oklahoma City: Federal Aviation Administration.

Federal Aviation Administration (2008c). *Balloon Flying Handbook*. Oklahoma City: United States Department of Transportation.

Federal Aviation Administration (2012a). *Aviation Maintenance Technician Handbook – Airframe*. Oklahoma City: United States Department of Transportation.

Federal Aviation Administration (2012b). *Helicopter Flying Handbook*. Oklahoma City: United States Department of Transportation.

Federal Aviation Administration (2012c). *Instrument Flying Handbook*. Oklahoma City: United States Department of Transportation.

Federal Aviation Administration (2013). *Glider Flying Handbook*. Oklahoma City: United States Department of Transportation.

Federal Aviation Administration (2014). *Aircraft Wake Turbulence AC 90-23G*. Oklahoma City: United States Department of Transportation.

Federal Aviation Administration (2015). *Pilot Guide: Flight in Icing Conditions AC 91-74B*. Oklahoma City: United States Department of Transportation.

Federal Aviation Administration (2016a). *Airplane Flying Handbook*. Oklahoma City: United States Department of Transportation.

Federal Aviation Administration (2016b). *Pilot's Handbook of Aeronautical Knowledge*. Oklahoma City: United States Department of Transportation.

Federal Aviation Administration (2016c). *Aviation Weather AC 00-6B*. Oklahoma City: United States Department of Transportation.

Federal Aviation Administration (2016d). *Weight and Balance Handbook FAA-H-8083-1B*. Oklahoma City: United States Department of Transportation.

Federal Aviation Administration (2018). *Aviation Maintenance Technician Handbook – Powerplant*. Oklahoma City: United States Department of Transportation.

Federal Aviation Administration (2019). *Helicopter Flying Handbook*. Oklahoma City: United States Department of Transportation.

Federal Aviation Administration (2020). Summary of the FAA's Review of the Boeing 737 MAX. Available at: https://www.faa.gov/sites/faa.gov/files/2022-08/737_RTS_Summary.pdf.

Federal Aviation Administration (2021). *Airplane Flying Handbook*. Oklahoma City: United States Department of Transportation.

Federal Aviation Administration (2023). Urban Air Mobility (UAM) Version 2.0 Concept of Operations. report. Available at: https://www.faa.gov/sites/faa.gov/files/Urban Air Mobility (UAM) Concept of Operations 2.0_1.pdf.

Federal Aviation Administration (2023a). *Aviation Maintenance Technician Handbook – Airframe*. Oklahoma City: United States Department of Transportation.

Federal Aviation Administration (2023b). *Aviation Maintenance Technician Handbook – General*. Oklahoma City: United States Department of Transportation.

Federal Aviation Administration (2023c). *Pilot's Handbook of Aeronautical Knowledge*. Oklahoma City: United States Department of Transportation.

Federal Aviation Administration (2023d). *Aviation Maintenance Technician Handbook – Powerplant*. Oklahoma City: United States Department of Transportation.

Federal Aviation Administration (2024). Aeronautical Information Manual – AIM. [online] Faa.gov. Available at: https://www.faa.gov/air_traffic/publications/atpubs/aim_html/.

Flechner, S.G., Jacobs, P.F., and Whitcomb, R.T. (1976). A High Subsonic Speed Wind Tunnel Investigation of Winglets on a Representative Second-Generation Jet Transport Wing. Available at: https://ntrs.nasa.gov/citations/19760019076.

Gipson, L. (2023). Revolutionary Vertical Lift Technology Project Overview. NASA, 25 July. Available at: https://www.nasa.gov/directorates/armd/aavp/armd-aavp-rvlt/revolutionary-vertical-lift-technology-project-overview/.

Gorton, S. (2015). NASA Technology for Next Generation Vertical Lift Vehicles! [online] Available at: https://ntrs.nasa.gov/api/citations/20160006951/downloads/20160006951.pdf [Accessed 28 Dec. 2024].

Harris, C. (1990). NASA. NASA Supercritical Airfoils A Matrix of Family-Related Airfoils. [online] Available at: https://ntrs.nasa.gov/api/citations/19900007394/downloads/19900007394.pdf.

Hurt, H.H. Jr. (1965). *Aerodynamics for Naval Aviators*, NAVWEPS 00-80T-80. Government Printing Office.

Johnson, W. (2022). NASA Design and Analysis of Rotorcraft Theory. [online] Available at: https://ntrs.nasa.gov/api/citations/20220000355/downloads/Johnson TP-20220000355 v1.pdf [Accessed 28 Dec. 2024].

Le Vie, L. (2014). Review of Research on Angle-of-Attack Indicator Effectiveness. [online] Available at: https://ntrs.nasa.gov/api/citations/20140011419/downloads/20140011419.pdf.

Leishman, J.G. (2023). Airfoil Geometries. eaglepubs.erau.edu, [online] 24. Available at: https://doi.org/10.15394/eaglepub.2022.1066.n21.

Malik, M. (2015). CFD Vision 2030 CFD Study a Pathway to Revolutionary Computational Aeroscienes. [online] Available at: https://ntrs.nasa.gov/api/citations/20160006942/downloads/20160006942.pdf [Accessed 29 Dec. 2024].

NASA (1970). *Evaluation of the Effect of a Yaw-Rate Damper on the Flying Qualities of a Light Twin-Engine Airplane*, NASA Technical Note TN D-5890. Washington: NASA. Available at: https://www.nasa.gov/cen-ters/dryden/pdf/87772main_H-584.pdf.

NASA (2000). FactSheet FS-2000-06-52-LaRC: Micro-Vortex Generators. Available at: https://www.nasa.gov/centers/langley/pdf/70865main_FS-2000-06-52-LaRC.pdf.

NASA (2003). NASA Facts FS-2003-08-008-DFRC: the X-29. Available at: https://www.nasa.gov/centers/dryden/pdf/120266main_FS-008-DFRC.pdf.

NASA (2004a). NASA Technology Facts – the Supercritical Airfoil. Available at: https://www.nasa.gov/pdf/89232main_TF-2004-13-DFRC.pdf.

NASA (2004b). FactSheet TF-2004-15-DFRC: Winglets. Available at: https://www.nasa.gov/centers/dryden/about/Organizations/Technology/Facts/TF-2004-15-DFRC.html.

NASA (2012). Breaking the Mishap Chain: Human Factors Lessons Learned from Aerospace Accidents and Incidents in Research, Flight Test, and Development – NASA. [online] Available at: https://www.nasa.gov/aeronautics/breaking-the-mishap-chain-human-factors-lessons-learned-from-aerospace-accidents-and-incidents-in-research-flight-test-and-development/ [Accessed 12 Oct. 2024].

NASA (2013). NASA/GA-ASI Agreement Expands Unmanned Aircraft Capabilities–NASA. [online] Available at: https://www.nasa.gov/news-release/nasa-ga-asi-agreement-expands-unmanned-aircraft-capabilities/ [Accessed 22 Dec. 2024].

NASA (2018). About Low Boom Flight Demonstrator (LBFD). [online] Available at: https://www.nasa.gov/directorates/armd/integrated-aviation-systems-program/armd-iasp-lbfd/about-low-boom-flight-demonstrator-lbfd/ [Accessed 29 Dec. 2024].

NASA (2023). C-17 Globemaster III–NASA. [online] Available at: https://www.nasa.gov/image-detail/c-17-globemaster-iii/ [Accessed 16 Nov. 2024].

NASA (2023). Ceramic Composites Revolutionize Engine Efficiency. NASA, 25 July. Available at: https://www.nasa.gov/aeronautics/ceramic-composites-revolutionize-engine-efficiency/.

NASA (2024). Aerodynamic Center. Available at: https://www1.grc.nasa.gov/beginners-guide-to-aeronautics/aerodynamic-center/#:~:text=Engineers%20call%20the%20location%20where,of%20attack%20in%20aerodynamic%20analysis.

NASA (2024). Museum in a Box Series: Principles of Flight. National Aeronautics and Space Administration. Available at: https://flight.nasa.gov/pdf/foam_wing_k-12.pdf.

NASA (2025). Active Turbulence Suppression System for Electric Vertical Take-Off and Landing (eVTOL) vehicles | T2 Portal. [online] Available at: https://technology.nasa.gov/patent/TOP2-308 [Accessed 11 Jan. 2025].

NASA | Glenn Research Center (2024). Vectored Thrust. [online] Available at: https://www1.grc.nasa.gov/beginners-guide-to-aeronautics/vectored-thrust/.

National Transportation Safety Board (NTSB) (2004). Aircraft Accident Report AAR-04/01. Available at: https://www.ntsb.gov/investigations/AccidentReports/Reports/AAR0401.pdf.

National Transportation Safety Board (2007). Runway Overrun and Collision, Aircraft Accident Report. Aircraft Accident Report NTSB/AAR-07/06. National Transportation Safety Board, p. iii. Available at: https://www.ntsb.gov/investigations/AccidentReports/Reports/AAR0706.pdf.

Richwine, D. (2020). Overview of Low-Boom Flight Demonstration Mission and X-59 QueSST Aircraft Project Manager for Technology Low Boom Flight Demonstrator Project 1. [online] Available at: https://ntrs.nasa.gov/api/citations/20205002862/downloads/LBFDaX59Overview_060920.pptx.pdf.

Russell, C., Field, M., Jung, J., Willink, G., and Glasner, B. (2016). Wind Tunnel and Hover Performance Test Results for Multicopter UAS Vehicles. [online] Available at: https://rotorcraft.arc.nasa.gov/Publications/files/72-2016-374.pdf.

Russell, C. and Sekula, M. (2017). Comprehensive Analysis Modeling of Small-Scale UAS Rotors. [online] Available at: https://rotorcraft.arc.nasa.gov/Publications/files/Russell20170011320.pdf.

Simmons, B. and Hatke, D. (2021). Investigation of High Incidence Angle Propeller Aerodynamics for Subscale eVTOL Aircraft. [online] Available at: https://ntrs.nasa.gov/api/citations/20210014010/downloads/NASA-TM-20210014010.pdf.

Somers, D.M. (2012). *Design of a slotted, Natural-Laminar-Flow Airfoil for Business-Jet Applications, NASA*. NASA/CR-2012-217559. NASA Center for AeroSpace Information. Available at: https://ntrs.nasa.gov/api/citations/20120013091/downloads/20120013091.pdf.

STEM LEARNING: Advanced Air Mobility: The Science Behind Quadcopters Reader-Student Guide. (2020). STEM LEARNING: Advanced Air Mobility: The Science Behind Quadcopters Reader – Student Guide. NASA. Available at: https://www.nasa.gov/wp-content/uploads/2020/05/aam-science-behind-quadcopters-reader-student-guide_0.pdf.

Talay, T.A. (1975). Introduction to the Aerodynamics of Flight. Available at: https://ntrs.nasa.gov/citations/19760003955.

Theodore, C.R., Aeromechanics Office, and NASA Ames Research Center (2018). A Summary of the NASA Design Environment for Novel Vertical Lift Vehicles (DELIVER) Project, AHS International Technical Conference on Aeromechanics Design for Transformative Vertical Flight. Available at: https://rotorcraft.arc.nasa.gov/Publications/files/TheodoreC_2018_TechMx.pdf.

Transport Canada (2006). *Helicopter Flight Training Manual*. Government of Canada - Transport Canada. Available at: https://www.tc.gc.ca/eng/civilaviation/publications/tp9982-menu-976.htm.

Transport Canada (2023). Aeronautical Information Manual (TC AIM). report. Available at: https://tc.canada.ca/sites/default/files/2023-03/aim-2023-1_air-e.pdf.

The National Advisory Committee for Aeronautics (NACA) - NASA (2025). Available at: https://www.nasa.gov/history/the-national-advisory-committee-for-aeronautics-naca/.

United States Air Force (1973). *Flight Manual USAF Series CH-3E/HH-3E*. United States Air Force.

Whitcomb, R.T. (1976). A Design Approach and Selected Wind Tunnel Results at High Subsonic Speeds for Wing-Tip Mounted Winglets. Available at: https://ntrs.nasa.gov/citations/19760019075.

PERIODICALS

Anonymous (1964). "Hydroplaning, Fun or Disaster," *Aerospace Safety*, USAF, AFISC.

Anonymous (1976). "Low Altitude Wind Shear," *Aerospace Safety*, USAF, AFISC.

Eggleston, B. and Jones, D.L. (1977). The design of lifting supercritical airfoils using a numerical optimization method. *Canadian Aeronautics and Space Journal* **23** (3): 172–181.

Flight Safety Foundation (2000). "Approach-and-Landing Accident Reduction Tool Kit-Wind Shear," *Flight Safety Digest*.

Freitag, W. and Schulze, E.T. (2009). Blended winglets improve performance. *Aeromagazine* 9–12.

Garrison, P. (2009). Technicalities: A Short History of Airfoils. http://www.flyingmag.com/technicalities/technicalities-short-history-airfoils.

Haines, P. and Luers, J. (1983). Aerodynamic penalties of heavy rain on landing aircraft. *Journal of Aircraft* **20** (2): 111–119.

Kocivar, B. (1976). "Super STOL," *Popular Science*.

Larson, G.C. (2001). "How Things Work: Winglets," *Air & Space Smithsonian*.

Luers, J. and Haines, P. (1983). Heavy rain influence on airplane accidents. *Journal of Aircraft* **20** (2): 187–191.

PERSONAL INTERVIEW

Saunders, S. (2016). Personal interview (26 March 2016).

Index

Note: Page references in *italics* refer to figures.

Flight Theory and Aerodynamics: A Practical Guide for Operational Safety, Fifth Edition.
Brian A. Johnson and Philip R. Fittante.

Companion website: www.wiley.com/go/flighttheoryandaerodynamics5e